UJF NATO ASI

LES HOUCHES

Session LXXIII

2000

ATOMIC CLUSTERS AND NANOPARTICLES

AGRÉGATS ATOMIQUES ET NANOPARTICULES

CONFÉRENCIERS

G.F. BERTSCH
M. BRACK
S.F. CHEKMAREV
H. FLOCARD
H. HABERLAND
P. HOBZA
M. MANNINEN
T.P. MARTIN
G.M. PASTOR
D.R. SALAHUB
A.V. SOLOV'YOV
D.J. WALES

ÉCOLE DE PHYSIQUE DES HOUCHES – UJF & INPG – GRENOBLE

a NATO Advanced Study Institute

LES HOUCHES

SESSION LXXIII

2-28 July 2000

Atomic clusters and nanoparticles

Agrégats atomiques et nanoparticules

Edited by

C. GUET, P. HOBZA, F. SPIEGELMAN and F. DAVID

EDP SCIENCES

Les Ulis, Paris, Cambridge

Springer

Berlin, Heidelberg, New York, Barcelona, Hong Kong, London Milan, Paris, Tokyo

Published in cooperation with the NATO Scientific Affair Division

ISBN 3-540-42908-5 Springer-Verlag Berlin Heidelberg New York
ISBN 2-86883-536-8 EDP Sciences Les Ulis

Printed in France

LES HOUCHES - ÉCOLE DE PHYSIQUE
ÉCOLE D'ÉTÉ DE PHYSIQUE THÉORIQUE

SERVICE INTER-UNIVERSITAIRE COMMUN
À L'UNIVERSITÉ JOSEPH FOURIER DE GRENOBLE
ET À L'INSTITUT NATIONAL POLYTECHNIQUE
DE GRENOBLE, SUBVENTIONNÉ PAR LE MINISTÈRE
DE L'ÉDUCATION NATIONALE, DE LA RECHERCHE
ET DE LA TECHNOLOGIE, LE CENTRE NATIONAL
DE LA RECHERCHE SCIENTIFIQUE ET LE COMMISSARIAT
À L'ÉNERGIE ATOMIQUE

ÉCOLE D'ÉTÉ DE PHYSIQUE THÉORIQUE
SESSION LXXIII

INSTITUT D'ÉTUDES AVANCÉES DE L'OTAN
NATO ADVANCED STUDY INSTITUTE

EURO SUMMER SCHOOL

2-28 juillet 2000

Directeurs Scientifiques de la session : Claude GUET, *CEA Ile de France, Département de Physique Théorique et Appliquée, Bruyères-le-Châtel, France ;* Pavel HOBZA, *J. Heyrovsky Institute of Physical Chemistry, Academy of Sciences, Prague, Czech Republic,* Fernand SPIEGELMAN, *Laboratoire de Physique Quantique, IRSAMC, Université Paul Sabatier, Toulouse, France ;* François DAVID, *SPhT, CEA Saclay, France*

SESSIONS PRÉCÉDENTES

1951 - 1999

I	1951	Quantum mechanics. Quantum field theory
II	1952	Quantum mechanics. Statistical mechanics. Nuclear physics
III	1953	Quantum mechanics. Solid state physics. Statistical mechanics. Elementary particle physics
IV	1954	Quantum mechanics. Collision theory. Nucleon-nucleon interaction. Quantum electrodynamics
V	1955	Quantum mechanics. Non-equilibrium phenomena. Nuclear reactions. Interaction of a nucleus with atomic and molecular fields
VI	1956	Quantum perturbation theory. Low temperature physics. Quantum theory of solids; dislocations and plastic properties. Magnesium; ferromagnetism
VII	1957	Scattering theory. Recent developments in field theory. Nuclear interaction; strong interactions. High energy electrons. Experiments in high energy nuclear physics
VIII	1958	The many body problem
IX	1959	The theory of neutral and ionized gases*
X	1960	Elementary particles and dispersion relations*
XI	1961	Low temperature physics*
XII	1962	Geophysics; the earth's environment*
XIII	1963	Relativity groups and topology*
XIV	1964	Quantum optics and electronics*
XV	1965	High energy physics*
XVI	1966	High energy astrophysics*
XVII	1967	Many body physics*
XVIII	1968	Nuclear physics*

* Sessions ayant reçu l'appui du Comité Scientifique de l'OTAN.

XIX	1969	Physical problems in biological systems*
XX	1970	Statistical mechanics and quantum field theory*
XXI	1971	Particle physics*
XXII	1972	Plasma physics*
XXIII	1972	Black holes*
XXIV	1973	Fluids dynamics*
XXV	1973	Molecular fluids*
XXVI	1974	Atomic and molecular physics and the interstellar matter*
June Inst.	1975	Structural analysis of collision amplitudes
XXVII	1975	Frontiers in laser spectroscopy*
XXVIII	1975	Methods in field theory*
XXIX	1976	Weak and electromagnetic interactions at high energy*
XXX	1977	Nuclear physics with heavy ions and mesons*
XXXI	1978	Ill-condensed matter*
XXXII	1979	Membranes and intercellular communication*
XXXIII	1979	Physical cosmology*
XXXIV	1980	Laser plasma interaction
XXXV	1980	Physics of defects*
XXXVI	1981	Chaotic behaviour of deterministic systems*
XXXVII	1981	Gauge theories in high energy physics*
XXXVIII	1982	New trends in atomic physics*
XXXIX	1982	Recent advances in field theory and statistical mechanics*
XL	1983	Relativity, groups and topology*
XLI	1983	Birth and infancy of stars*
XLII	1984	Cellular and molecular aspects of developmental biology*
XLIII	1984	Critical phenomena, random systems, gauge theories*
XLIV	1985	Architecture of fundamental interactions at short distances*
XLV	1985	Signal processing*
XLVI	1986	Chance and matter*
XLVII	1986	Astrophysical fluid dynamics*
XLVIII	1988	Liquids at interfaces*
XLIX	1988	Fields, strings and critical phenomena*
L	1988	Oceanographic and geophysical tomography*
LI	1989	Liquids, freezing and glass transition*
LII	1989	Chaos and quantum physics*
LIII	1990	Fundamental systems in quantum optics*
LIV	1990	Supernovae*

* Sessions ayant reçu l'appui du Comité Scientifique de l'OTAN.

LV	1991	Particles in the nineties*
LVI	1991	Strongly interacting fermions and high T_c superconductivity*
LVII	1992	Gravitation and quantizations*
LVIII	1992	Progress in picture processing*
LIX	1993	Computational fluid dynamics*
LX	1993	Cosmology and large scale structure*
LXI	1994	Mesoscopic quantum physics*
LXII	1994	Fluctuating geometries in statistical mechanics and quantum field theory*
LXIII	1995	Quantum fluctuations*
LXIV	1995	Quantum symmetries*
LXV	1996	From cell to brain*
LXVI	1996	Trends in nuclear physics, 100 years later
LXVII	1997	Modélisation du climat de la terre et de sa variabilité*
LXVIII	1997	Particules et interactions : le modèle standard mis à l'épreuve*
LXIX	1998	Topological aspects of low dimensional systems*
LXX	1998	Infrared space astronomy, today and tomorrow*
LXXI	1999	The primordial universe
LXXII	1999	Coherent atomic matter waves*

* Sessions ayant reçu l'appui du Comité Scientifique de l'OTAN.
Publishers: Session VIII: Dunod, Wiley, Methuen; Sessions IX & X: Herman, Wiley – Session XI: Gordon and Breach, Presses Universitaires – Sessions XII-XXV: Gordon and Breach – Sessions XXVI-LXVIII: North-Holland – Session LXIX-LXXII: EDP Sciences, Springer.

ORGANIZERS

C. GUET, CEA Ile de France, Département de Physique Théorique et Appliquée, BP. 12, 91680 Bruyères-le-Châtel, France

P. HOBZA, J. Heyrovsky Institute of Physical Chemistry, Academy of Sciences, Dolejskova 3, 18223 Prague 8, Czech Republic

F. SPIEGELMAN, Laboratoire de Physique Quantique, IRSAMC, Université Paul Sabatier, 118 route de Narbonne, 31062 Toulouse Cedex, France

F. DAVID, Service de Physique Théorique, Saclay, 91191 Gif-sur-Yvette, France

LECTURERS

G.F. BERTSCH, Institute of Nuclear Theory, University of Washington, P.O. Box 351560, Seattle, WA 98195, U.S.A.

M. BRACK, Institut für Theoretische Physik, Universität Regensburg, 93040 Regensburg, Germany

S. CHEKMAREV, Institute of Thermophysics, SB RAS, 6300090 Novosibirsk, Russia

H. FLOCARD, Division de Physique Théorique, Institut de Physique Nucléaire, 91406 Orsay Cedex, France

H. HABERLAND, Fakultät für Physik, Universität Freiburg, H. Herderstr. 3, 79104 Freiburg, Germany

P. HOBZA, J. Heyrovsky Institute of Physical Chemistry, Academy of Sciences of the Czech Republic, Dolejskova 3, 18223 Prague 8, Czech Republic

M. MANNINEN, Department of Physics, University of Jyväskylä, P.O. Box 35, 40351 Jyväskylä, Finland

T.P. MARTIN, Max-Plank-Institut für Festkörperforschung, Heisenbergstr. 1, 70569 Stuttgart, Germany

G. PASTOR, Laboratoire de Physique Quantique, IRSAMC, Université Paul Sabatier, 118 route de Narbonne, 31062 Toulouse Cedex, France

L. PIELA, Department of Chemistry, University of Warsaw, Pasteur 1, 09302 Warsaw, Poland

D. SALAHUB, Institut Steacie de Sciences Moléculaires, Conseil National des Recherches Canada, 100 Promenade Sussex, Ottawa, Ontario K1A0R6, Canada

A. SOLOVIEV, A.F. Ioffe Physical Technical Institute, Polytechnicheskaya 26, 194021 St. Petersburg, Russia

D. WALES, University Chemical Laboratories, Lensfield Road, Cambridge CB2 1EW, U.K.

SEMINAR SPEAKERS

J. BELLONI, Laboratoire de Chimie Physique, bâtiment 350, Université Paris-Sud, 91405 Orsay, France

S. BERRY, Department of Chemistry, The University of Chicago, 5735 South Ellis Av., Chicago, Illinois 60637, U.S.A.

V. BONACIC-KOUTECKY, Humbold Universität zu Berlin, Institut für Chemie, Bunsenstrasse 1, 10117 Berlin, Germany

C. BRECHIGNAC, Laboratoire Aimé Cotton, bâtiment 505, Campus d'Orsay, Université Paris-Sud, 91405 Orsay Cedex, France

F. HEKKING, CNRS-CRTBT, Université Joseph Fourier, 38042 Grenoble Cedex 9, France

J. JORTNER, Tel-Aviv University, School of Chemistry, Ramat Aviv, 69978 Tel Aviv, Israel

V. KRESIN, Department of Physics and Astronomy, University of Southern California, Los Angeles, California 90089-0484, U.S.A.

R. LAVERY, Laboratoire de Biochimie Théorique, Institut de Biologie Physico-Chimique, 113 rue Pierre et Marie Curie, 75005 Paris, France

E. SURAUD, Laboratoire de Physique Quantique, IRSAMC, 118 route de Narbonne, 31062 Toulouse Cedex, France

L. WOESTE, Institut fuer Experimentalphysik, Freie Universität Berlin, Arnimalee 14, 14195 Berlin, Germany

STUDENTS AND AUDITORS

I. ADAGIDELI, 212 Loomis Laboratory of Physics, 1110 West Green Street, Urbana, IL 61801-3080, U.S.A.

P. ANDRE, Université Pierre et Marie Curie, LSRSI, 4 place Jussieu, BP. 52, 75252 Paris Cedex 05, France

J. AUER, Institute for Tehoretical Physics, Johannes Klepler University Linz, Altenbergerstrasse 69, 4040 Linz, Austria

T. BOGDAN, The University Kiev-Mohyla Academy, vul. Skovorody 2, 04070 Kyiv, Ukraine

F. CALVO, CEA Grenoble, DRFMC/SI2A, 17 rue des Martyrs, 38054 Grenoble Cedex, France

A. CASTRO, Facultad de Ciencias, Universidad de Valladolid, 47011 Valladolid, Spain

S. CHRETIEN, Département de Chimie, Université de Montréal, CP. 6128, Succursale Centre-Ville, Montréal, Québec H3C 3J7, Canada

J. DALIGAULT, CEA/DRFMC-SI2A, 17 rue des Martyrs, 38054 Grenoble Cedex 9, France

A. DHAVALE, Department of Physics, University of Pune, Pune 411007, Maharashtra, India

J. DONGES, Fakultät für Physik, Universität Freiburg, Hermann-Heider Str. 3, 79104 Freiburg, Germany

T. DÖPNNER, University of Rostock, Department of Physics, Universitätsplatz 3, 18051 Rostock, Germany

A. EMPERADOR, Universitat de Barcelona, Facultat de Fisica, Av. Diagonal 647, 08028 Barcelona, Spain

M.Z. GEDIK, Department of Physics, Bilkent University, 06533 Bilkent, Ankara, Turkey

L. GERCHIKOV, Aviakonstruktorov Pr. 27-1-26, St. Petersburg, Russia

E. GIGLIO, 3 place Alphonse Jourdain, 31000 Toulouse, France

K. HAMADA, Konan University, Department of Physics, 8-9-1 Okamoto, Higashinada-ku, Kobe 658-8501, Japan

A. IPATOV, Experimental Physics Department, St. Petersburg State Technical University, Polytekhnicheskaja 29, 195251 St. Petersburg, Russia

A. ISACSSON, Department of Applied Physics, Chalmers University of Technology, 41296 Gothenburg, Sweden

S. KACHAN, Institute of Molecular and Atomic Physics, F. Scaryna Ave. 70, 220072 Minsk, Belarus

E.-C. KAPOSTA, Institut für Experimental Physik, Freie Universität Berlin, Arnimalle 14, 14195 Berlin, Germany

L. KARACHINSKY, St. Petersburg State Technical University, 195251 Polytechnicheskaya ul. H. 29 St. Petersburg, Russia

S.N. KLAUSEN, Riso National Laboratory, Condensed Matter Physics and Chemistry Department, Frederiksborgveij 399, P.O. Box 49, 4000 Roskilde, Denmark

M. KWATO NJOCK, P.O. Box 7978, Yaounde, Cameroun

S. LATIL, GDPC, UMR 5581, Case Courrier 26, Université Montpellier 2, Place Eugène Bataillon, 34095 Montpellier Cedex, France

C. LUPULESCU, Institut für Experimentalphysik, Freie Universität Berlin, Arnimallee 14, 14195 Berlin, Germany

A. LYALIN, Institute of Physics, St. Petersburg State University, Ulianovskaya Str. 1, St. Petersburg, Petrodvorez 198904, Russia

M. LYN, The Pennsylvania State University, 152 Davay Lab., Department of Chemistry, University Park, PA 16802, U.S.A.

D. MAKHOV, Photochemistry Center, Russian Academy of Sciences, 7A Novatorov Street, Moscow 117421, Russia

T. MERDJANOVA, Bulgarian Academy of Sciences, Central Laboratory of Solar Energy and New Energy Sources, 72 Tzarigradsko Chaussee, 1784 Sofia, Bulgaria

R. MITRIC, Humboldt Universitaet zu Berlin, Walther Nernst Institut für Physikalische und Theoretische Chemie, Bunsenstr. 1, 10117 Berlin, Germany

A. MORELLO, Kamerlingh Onnes Laboratory, Leiden Institute of Physics, Leiden University, P.O. Box 9504, 2300 RA Leiden, The Netherlands

O. MÜLKEN, Department of Physics, University of Oldenburg, 26111 Oldenburg, Germany

S. NEUKERMANS, Lab. Voor Vaste-Stoffysica en Magnetisme, K.U. Leuven, Dept. Natuurkunde, Celestijnenlaan 200D, 3001 Leuven, Belgium

J.M. NILLES, Department of Chemistry, Johns Hopkins University, Baltimore, MD 21218, U.S.A.

S. PELLEGRIN, Laboratoire Aimé Cotton, bâtiment 505, Campus d'Orsay, 91406 Orsay Cedex, France

F. RABILLOUD, Université Paul Sabatier, LCAR/IRSAMC, 118 route de Narbonne, 31062 Toulouse Cedex, France

V. RADCHENKO, Donetsk State University, Universitetskaya Street 24, 83055 Donetsk, Ukraine, and 15 Literaturnaya Str., 94075 Teplogorsk, Ukraine

J. RICHARDI, Université Pierre et Marie Curie, Laboratoire Structure et Réactivité de Systèmes Interfaciaux, 4 place Jussieu, BP. 52, 75230 Paris Cedex 05, France

R. ROBLES, Departamento Fisica Teorica, Atomica Molecular y Nuclear, Facultad de Ciencias, Universidad de Valladolid, 47011 Valladolid, Spain

T.-G. SOROP, Kamerlingh Onnes Laboratory, Universiteit Leiden, P.O. Box 9504, 2300 RA Leiden, The Netherlands

B.C. STANCA, Institut für Chemie, Humboldt Universität Berlin, Bunsen Str. 1, 10117 Berlin, Germany

A. TANGUY, Département de Physique des Matériaux, Université Lyon I, 43 boulevard du 11 novembre 1918, 69622 Villeurbanne Cedex, France

M.M. THIAM, Czech Academy of Sciences, J. Heyrovsky Institut of Physical Chemistry, Dolejskova 3, 18200 Prague 8, Czech Republic

S. TYGUBENKO, The University of Kiev-Mohyla Academy, vul. Skovorody 2, 04070 Kyiv, Ukraine

S. VONGEHR, Department of Physics and Astronomy, University of Southern California, Los Angeles, California 90089-0484, U.S.A.

T. VUKOVIC, Faculty of Physics, University of Belgrade, Studenski TRG 10-12, P.O. Box 368, Beograd 11001, Yugoslavia

W. WERNSDORFER, Laboratoire L. Néel, CNRS, 25 avenue des Martyrs, BP. 166, 38042 Grenoble Cedex 9, France

Y. YAMAGUCHI, Institut für Mikrostrukturtechnik, University of Karlsruhe, Postfach 3640, 76021 Karlsruhe, Germany

K. YOSHIZAKI, Tohoku University, Nuclear Theory Group, Department of Physics, Aramaki, Aoba-ku, Sendai 980-8578, Japan

Préface

Les agrégats atomiques sont des ensembles d'atomes ou de molécules dont le nombre de constituants est bien défini et peut varier de quelques unités à plusieurs dizaines de milliers. Les agrégats se distinguent de la matière macroscopique par le fait que leurs propriétés sont fortement affectées par la présence d'une surface impliquant une fraction importante des constituants. Ils s'en distinguent également par le spectre discret de leurs excitations électroniques. D'une part la nature finie du nombre de constituants amène des propriétés structurelles et thermodynamiques nouvelles sans équivalent dans la matière. D'autre part les agrégats constituent le pont entre la physique des atomes et celle de la matière condensée. L'étude approfondie des propriétés physiques des agrégats peut aussi nous apporter un autre éclairage sur les excitations élémentaires dans les solides and les liquides.

La physique et la chimie des agrégats atomiques et des nanoparticules couvrent un large domaine interdisciplinaire. Ce domaine est en forte croissance comme peuvent l'attester le nombre de publications, de conférences ou d'ateliers. Les axes de développement sont multiples. Certains restent purement fondamentaux, d'autres relèvent des sciences des matériaux à vocation appliquée. Avec le développement des nouvelles technologies on observe une tendance inévitable vers la miniaturisation en micro and nano électronique. On souhaite maîtriser la technologie de dispositifs de tailles si petites que leurs propriétés quantiques spécifiques deviennent pertinentes. Les propriétés magnétiques des agrégats ou des points quantiques sur des surfaces pourraient laisser entrevoir de nouveaux matériaux pour l'enregistrement à haute densité. On observe des tendances similaires dans d'autres domaines tels que la catalyse, le stockage d'énergie, ou le contrôle de la pollution de l'air.

Cette école était consacrée aux aspects fondamentaux. Elle a rassemblé des professeurs menant leurs propres travaux de recherche en physique atomique et moléculaire, en physique de la matière condensée, physique nucléaire, chimie et chimie physique, sans oublier la simulation numérique.

Le domaine s'est développé rapidement grâce à un fort couplage entre théories and expériences. Deux cours à dominante expérimentale ont permis d'apprécier ce qui est possible et ce qui deviendra possible d'étudier. Patrick Martin a présenté une revue expérimentale très large et stimulante tandis que Hellmut Haberland a construit son cours autour de récentes observations expérimentales de transitions de phase dans des agrégats métalliques.

Les cours théoriques couvrent trois principaux domaines : (i) les propriétés électroniques des agrégats métalliques et des nanostructures, (ii) les phases et les changements de phase des petits systèmes, (iii) les processus chimiques à l'échelle nanométrique.

George Bertsch a donné une série de cours très interactifs qui ont familiarisé les étudiants avec les phénomènes élémentaires dans les agrégats métalliques, les points quantiques, les fullerènes and les nanotubes. Il présente les outils théoriques capables de décrire les propriétés des agrégats et en particulier leurs excitations électroniques. Ce cours met en valeur les modèles théoriques simples ainsi que les caractères universels des systèmes finis quantiques.

La Théorie de la Fonctionnelle de la Densité est sans conteste la méthode quantique qui autorise des études réalistes des grosses molécules et des agrégats. Aujourd'hui, la combinaison de la mécanique quantique et de la mécanique moléculaire utilisant les champs de force classiques permet de comprendre en grande partie les systèmes biologiques. Denis Salahub a rendu compte de l'état actuel des réussites et progrès de cette théorie, qu'il a illustrés tout au long de son cours par des applications en chimie.

Les Agrégats métalliques et les nanosystèmes permettent de mettre en lumière les liens entre les mondes quantiques et classiques de manière spectaculaire. Matthias Brack a présenté les méthodes semiclassiques qui permettent de calculer non seulement des valeurs moyennes des observables mais aussi les effets des couches quantiques qui sont caractéristiques des systèmes fermioniques finis. Ce cours se divise en deux parties, l'une centrée sur le Modèle Thomas-Fermi Étendu, l'autre sur la Théorie des Orbites Périodiques, deux thèmes théoriques auxquels l'auteur a lui même beaucoup contribué.

Les corrélations de pairing dans les noyaux atomiques sont intimement apparentées aux phénomèmes de supraconductivité (et de superfluidité) dans les systèmes macroscopiques. Des expériences récentes sur des agrégats métalliques ont également laissé apparaître des corrélations de pairing. Hubert Flocard a consacré son cours à une étude approfondie des corrélations de pairing dans les systèmes fermioniques finis par une présentation rigoureuse des approches théoriques les plus pertinentes.

La physique des agrégats et des nanoparticules fait appel aux concepts propres à la physique de la matière condensée. Matti Manninen a présenté son cours sur la physique des nanosystèmes en adoptant un point de vue de physique des systèmes à basse dimensionnalité. Il a mis en lumière les concepts and comportements mésoscopiques qui sont communs à ces systèmes et aux systèmes finis.

Les propriétés magnétiques des petits systèmes sont sensibles à la taille, à la structure, and à la composition. Dans un cours très pédagogique Gustavo Pastor a présenté de manière détaillée les méthodes théoriques les plus efficaces pour décrire le magnétisme des agrégats, en particulier pour les métaux de transition. Le théorie des fluctuations de spin est bien adaptée au domaine des agrégats puisque les changements de spin et les fluctuations de structure y sont importants.

Les processus de diffusion sont utiles pour parvenir à comprendre les propriétés à N corps des systèmes finis, comme l'a démontré Andrey Solov'yov aux cours de ses deux exposés. Il a mis en évidence le rôle essentiel des modes de plasmon de surface et de volume dans la formation des spectres de perte d'énergie des électrons en diffusion inélastique.

Les propriétés thermodynamiques des agrégats sont d'une importance majeure. Elles exigent des approches théoriques originales des concepts de fusion, solidification, et de changement de phase dans les systèmes finis. Des variations soudaines en fonction de la taille peuvent être observées. Les notions de phase, de coexistence, d'équilibre solide-liquide, de diagrammes de phase, et leurs formulations spécifiques en termes de variables et fonctions thermodynamiques dans différents ensembles exigent le développement d'algorithmes nouveaux.

La simulation numérique de la dynamique et de la thermodynamique des agrégats est devenue un outil indispensable pour prédire et comprendre le comportement de ces derniers, à température finie. Le cours de David Wales était riche des concepts et des méthodes récentes permettant l'analyse topologique et l'échantillonnage des paysages d'énergie potentielle complexes et multi-dimensionnels.

Sergei Chekmarev a décrit une nouvelle méthode de simulation numérique. Cette méthode appliquée aux systèmes finis fournit des informations détaillées sur la surface d'énergie potentielle, les propriétés d'équilibre et la cinétique. Les biomolécules peuvent ainsi être étudiées avec une fiabilité acceptable.

Les processus chimiques entre agrégats ou nanoparticules en phase gazeuse sont l'objet d'études en forte croissance. Pour bien comprendre les propriétés des différentes familles d'agrégats atomiques, il est essentiel de bien connaître les liaisons chimiques qui sont mises en jeu. Des concepts tels que les changement de valence, la directionnalité des liens, l'hybridation, l'hypervalence, les distributions de population électronique, ou les fluctuation de charge sont essentiels. Le cours de Pavel Hobza a abordé ces concepts, méthodes et mécanismes et les a illustrés par de nombreux exemples. De son côté, Lucjan Piela a centré son cours (non publié dans ce livre) sur les effets de coopérativité, notamment dans les agrégats moléculaires.

Aux cours principaux qui sont reproduits dans ce livre, se sont ajoutés des séminaires. Ces séminaires, qui tous ont stimulé des discussions enthousiastes et argumentées ont été donnés par Jacqueline Belloni, Stephen Berry, Catherine Bréchignac, Vlasta Bonacic-Koutecky, Frank Hekking, Joshua Jortner, Vitaly Kresin, Richard Lavery, Eric Suraud, et Ludger Woeste.

Durant ces quatre semaines, étudiants et professeurs ont su nouer des relations étroites. Espérons que chaque étudiant a pu découvrir et partager l'enthousiasme que des travaux dans un autre domaine de recherches ont procuré à l'un ou l'autre des ses collègues, et qu'il ait réalisé qu'une méthode qui lui était inconnue pouvait être d'un grand intérêt pour ses propres travaux. Le format de l'École se prêtait bien à la satisfaction de ces objectifs. Le programme ne comprenait que trois cours par jour avec éventuellement un séminaire supplémentaire. Cet emploi du temps a permis l'organisation spontanée de groupes de travail, débouchant quelquefois sur un projet de recherche en collaboration.

Au nom de tous les étudiants et professeurs nous souhaitons dédier ce livre au défunt Professeur Walter D. Knight. Walter Knight et son groupe à Berkeley ont ouvert le champ des expériences avec des faisceaux d'agrégats libres de tout support. Ses travaux menés dans les années quatre-vingt ont conduit à des découvertes majeures sur les effets de taille finie et au renouveau spectaculaire de la physique des agrégats.

Cette École n'aurait pu se tenir sans :

- l'aide et le soutien du Conseil d'Administration de l'École de Physique ;
- le secrétariat de l'École de Physique : Ghyslaine d'Henry, Isabel Lelièvre et Brigitte Rousset ;
- les soutiens financiers de la Scientific and Environmental Affair Division de l'OTAN, de la Direction de la Recherche de la Commission Européenne, et de la Formation Permanente du CNRS ;
- et les soutiens à l'École de Physique par l'Université Joseph Fourier de Grenoble, le Ministère de la Recherche et de la Technologie, le CNRS et le CEA.

C. Guet
P. Hobza
F. Spiegelman
F. David

Preface

Atomic clusters are aggregates of atoms or molecules with a well-defined size varying from a few constituents to several tens of thousands. Clusters are distinguished from bulk matter in so far as their properties are strongly affected by the existence of a surface involving a large fraction of the constituents and by the discreteness of their electronic excitations. On the one hand, the finite nature of the number of constituents leads to novel structural and thermodynamic properties with no equivalent in the bulk. On the other hand, clusters bridge the physics of atoms to the physics of bulk matter. Studying thoroughly the physical properties of clusters may shed new light on elementary excitations in solids and liquids.

The physics and chemistry of atomic clusters and nanoparticles constitute a broad interdisciplinary domain. The field is growing fast, as witnessed by the number of publications and conferences or workshops. The directions of growth are numerous. Some involve pure basic science, others are motivated by more applied material science considerations. Within the fast development of new technologies one observes an unavoidable trend toward miniaturization in micro- and nano-electronics. One wishes to technologically master devices of sizes so small that their quantum specific properties are the most important. Specific magnetic properties of clusters or dots on surfaces might offer new materials for applications in high-density recording and memory devices. Similar trends are observed in other fields, such as catalysis, energy storage, and control of air pollution.

The present school was focused on basic science. It gathered lecturers with research practice in atomic and molecular physics, condensed matter physics, nuclear physics, and chemistry and physical chemistry, not to mention computational physics.

The field has developed rapidly thanks to a strong coupling between theory and experiment. Two predominantly experimental courses provided the opportunity to appreciate what is and will be experimentally possible. Patrick Martin presented a large experimental review, and Hellmut Haberland lectured on recent experimental observations of phase transitions in metal clusters.

The theoretical courses covered three main domains: (i) electronic properties of metallic clusters and nanostructures, (ii) phases and phase changes of small systems, and (iii) chemical processes in nanoscale systems. George Bertsch gave a series of quite interactive lectures that introduced the students to basic phenomena in metal clusters, quantum dots, fullerenes and nanotubes. He presented the basic theoretical tools to study cluster properties, particularly electronic excitations. He emphasized the importance of simple models that large-scale *ab initio* calculations validate, as well as the universal features of finite quantum systems.

The Density Functional Theory stands at a central position as a quantum mechanical method for practical studies of large molecules and clusters. Today, the combination of quantum mechanics and molecular mechanics using classical force fields allows us to understand biological systems to a large extent. Denis Salahub gave the state of the art of this theory, which he illustrated with chemical applications throughout his course.

Metal clusters and nanosystems are excellent physical objects for illuminating the links between the quantum and classical worlds. Matthias Brack reviewed semiclassical methods of determining both average trends and quantum shell effects in the properties of finite fermionic systems. His course was centered on two important theoretical themes: (a) the Extended Thomas–Fermi Model, and (b) the Periodic Orbit Theory (POT).

Pairing correlations in atomic nuclei are intimately related to the phenomena of superconductivity (and superfluidity) in macroscopic systems. Recent experiments on small metallic clusters also reveal pairing correlations. Hubert Flocard devoted his course to a thorough investigation of pairing correlations in finite systems within the state of the art theoretical models.

Cluster and nanoparticle physics is also part of condensed matter physics. Matti Manninen focused his course on the physics of nanosystems from the point of view of condensed matter physics, emphasizing concepts and mesoscopic features that are common between finite systems and low-dimensional systems.

Magnetic properties of small systems are very sensitive to the size, structure, and composition of the system. Within a pedagogical four-lecture course Gustavo Pastor provided a detailed account of the most powerful theoretical methods for efficiently describing the magnetic properties of clusters, particularly transition-metal clusters. The spin-fluctuation theory is quite appropriate for clusters since changes or fluctuations of structure are important. Scattering processes are also useful to gain insight on the many-body properties of finite systems, as demonstrated by Andrey Solov'yov in his two-lecture course. He emphasized the essential role of surface and volume plasmon modes in the formation of electron energy loss spectra.

The thermodynamical properties of clusters are certainly of major importance. They require theoretical approaches to the concepts of melting, freezing and phase changes in finite systems and their dependence on size. The notions of "phase-like" forms, coexistence, solid–liquid equilibrium, phase diagrams and their specific formulation in terms of the thermodynamical variables and functions in various ensembles require the development of new and sophisticated algorithms.

Computer simulations of cluster dynamics and thermodynamics have become a major tool for predicting and understanding the finite temperature behavior of clusters. David Wales' course was concerned with concepts and recent methods of achieving topological analysis and sampling of complex multi-dimensional potential energy landscapes.

Sergei Chekmarev described a novel approach to the computer simulation study of a finite many-body system that allows one to gain detailed information about this system, including its potential energy surface, equilibrium properties and kinetics. Biomolecules can now be investigated with a high degree of confidence.

The chemical processes in or with gas phase clusters and nanoscale particles are a growing field of cluster science. The key to understanding the properties of various families of atomic clusters lies in the determination and description of the chemical bond. Concepts such as valence and valence change, bond directionality, hybridization, hypervalence, electronic population distribution and charge fluctuation are essential. The course of Pavel Hobza addressed the above concepts, methods and mechanisms. Lucjan Piela focused his lecture (not published here) on cooperativity effects in quantum chemistry. This subject is indeed topical for molecular clusters.

In addition to the main courses that are the contents of this book, there were seminars not published herein. These seminars, that triggered stimulating discussions, were given by Jacqueline Belloni, Stephen Berry, Catherine Bréchignac, Vlasta Bonacic-Koutecky, Frank Hekking, Joshua Jortner, Vitaly Kresin, Richard Lavery, Eric Suraud, and Ludger Woeste.

During the four weeks the lecturers and students got to know each other pretty well. Hopefully every student discovered and shared the excitement that his (her) colleagues from other fields experienced. More practically, one would have realized that some as yet unfamiliar methods could be of great interest for one's own research. The overall organization of the school provided the best conditions to meet these goals. The daily lecture program was kept light, with no more than three main courses and sometimes an extra seminar. This schedule allowed plenty of time for discussions and organizing small working groups. We even observed that some collaborative research work had started. On behalf of all the lecturers and students we would like to dedicate this summer school to the late Professor Walter D. Knight. Walter Knight and his group at Berkeley pioneered the experimental field of cluster beams. His experiments in the 1980s led to major discoveries relating to finite size effects and to a strong revival of cluster physics.

This School would not have been possible without:

- the help and the support of the Scientific Board of the "École de Physique";
- the staff of the "École de Physique": Ghyslaine d'Henry, Isabel Lelièvre and Brigitte Rousset;

- the financial support of the Scientific and Environmental Affairs Division of NATO, the Research Directorate of the European Commission, and the Formation Permanente of CNRS; and
- the support to the "École de Physique" by the Université Joseph Fourier, the French Ministry of Research, CNRS and CEA.

C. Guet
P. Hobza
F. Spiegelman
F. David

CONTENTS

Course 5. Semiclassical Approaches to Mesoscopic Systems by M. Brack *161*

Course 6. Pairing Correlations in Finite Fermionic Systems by H. Flocard *221*

COURSE 1

EXPERIMENTAL ASPECTS OF METAL CLUSTERS

T.P. MARTIN

Max-Planck-Institut für Festkörperforschung, Heisenbergstr. 1, 70569 Stuttgart, Germany

Contents

EXPERIMENTAL ASPECTS OF METAL CLUSTERS

T.P. Martin

1 Introduction

It is not obvious that metal clusters should behave like atomic nuclei – but they do. Of course the energy and distance scales are quite different. But aside from this, the properties of these two forms of condensed matter are amazingly similar. The shell model developed by nuclear physicists describes very nicely the electronic properties of alkali metal clusters. The giant dipole resonances in the excitation spectra of nuclei have their analogue in the plasmon resonances of metal clusters. Finally, the droplet model describing the fission of unstable nuclei can be successively applied to the fragmentation of highly charged metal clusters. The similarity between clusters and nuclei is not accidental. Both systems consist of fermions moving, nearly freely, in a confined space.

Many years ago it was noticed that atomic nuclei containing either 8, 20, 50, 82 or 126 protons have very long lifetimes. It was a challenge for the nuclear physicists back in the forties to explain these so-called magic numbers. Since physicists tend to see most objects as perfectly round, it should come as no surprise that they assumed atomic nuclei are spherically symmetric. Under this assumption they had to solve only a radial Schrödinger equation.

$$\left[-\frac{\mathrm{d}^1}{\mathrm{d}r^2} + \frac{\ell(\ell+1)}{r^2} + V(r)\right] P_{n\ell}(r) = E_{n\ell} P_{n\ell}(r) \tag{1.1}$$

where ℓ is the angular momentum quantum number and $V(r)$ is the radial dependence of the potential in which the nucleons move. They assumed further that the potential could be described by a simple potential well. Some confusion can arise because nuclear physicists and atomic physicists use slightly different definitions for the principal quantum number n,

$$n \text{ (atomic)} = n \text{ (nuclear)} + \ell. \tag{1.2}$$

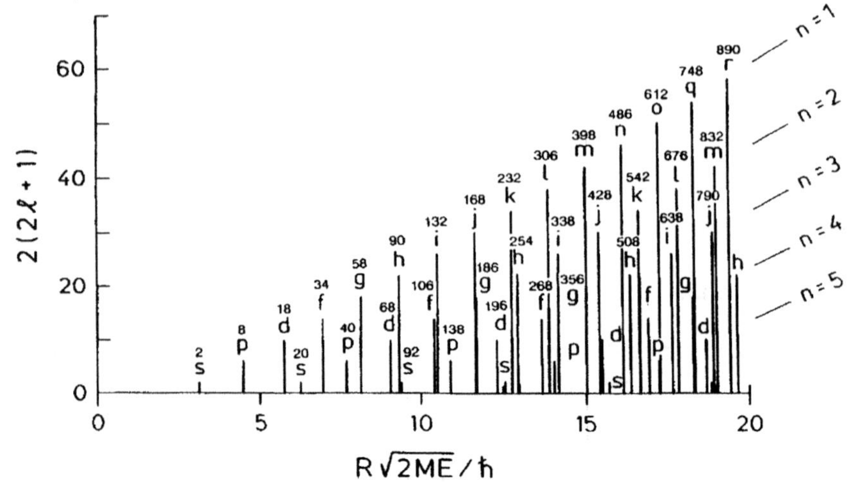

Fig. 1. The degeneracy of states of the infinitely deep spherical well on a momentum scale. The total number of fermions needed to fill all states up to and including a given subshell is indicated above each bar.

Throughout this lecture we will use the principal quantum number from nuclear physics, *i.e.*, n denotes the number of extrema in the raidal wavefunction.

Eigenstates of the radial Schrödinger equation are often called subshells. The subshells of the infinite spherical potential well are shown ordered according to momentum in Figure 1. The lowest energy state is 1s then comes 1p, 1d, 1f, 2p..., etc. This is, with 2, 8, 18, 20, 34, 40, 58, 90... nucleons, subshells are completely filled and the corresponding nuclei could be expected to be exceptionally stable. However, these are not the observed magic numbers.

In 1949 Goeppert-Mayer [1] and Haxel *et al.* [2] came up with a modified model which yielded the observed magic numbers. Their idea was that the spin-orbit interaction is unusually strong for nucleons. Subshells with high angular momentum split and the states rearrange themselves into different groups. As we shall see the original shell model, which the nuclear physicist had to discard, describes very nicely the electronic states of metal clusters [3-18].

2 Subshells, shells and supershells

If it can be assumed that the electrons in metal clusters move in a spherically symmetric potential, the problem is greatly simplified. Subshells for large values of angular momentum can contain hundreds of electrons having the same energy. The highest possible degeneracy assuming cubic symmetry is only 6. So under spherical symmetry the multitude of electronic states condenses down into a few degenerate subshells. Each subshell is characterized by a pair of quantum numbers n and ℓ. Under certain circumstances the subshells themselves condense into a smaller number of highly degenerate shells. The reason for the formation of shells out of subshells requires more explanation.

The concept of shells can be associated with a characteristic length. Every time the radius of a growing cluster increases by one unit of this characteristic length, a new shell is said to be added. The characteristic length for shells of atoms is approxiamtely equal to the interatomic distance. The characteristic length for shells of electrons is related to the wavelength of an electron in the highest occupied energy level (Fermi energy). For the alkali metals these lengths differ by a factor of about 2. This concept is useful only because the characteristic lengths are, to a first approximation, independent of cluster size.

The concept of shells can also be described in a different manner. An expansion of N, the total number of electrons, in terms of the shell index K will always have a leading term proportional to K^3. One power of K arises because we must sum over all shells up to K in order to obtain the total number of particles. One power of K arises because the number of subshells in a shell increases approximately linearly with shell index. Finally, the third power of K arises because the number of particles in the largest subshell also increases with shell index. Expressing this slightly more quantitatively, the total number of particles needed to fill all shells, k, up to and including K is

$$N_K = \sum_{k=1}^{K} \sum_{\ell=0}^{L(k)} 2(2\ell = 1) \sim K^3 \tag{2.1}$$

where $L(k)$ is the highest angular momentum subshell in shell k.

Shell structure is not necessarily an approximate and infrequent bunching of states as in the example of the spherical potential well, Figure 1. Clearly, almost none of the subshells occur exactly at the same energy for this potential. Shell structure can be the result of exactly overlapping states. Such degeneracies signal the presence of a symmetry higher than spherical symmetry. Subshells of hydrogen for which $n + \ell$ have the same value, have exactly the same energy. This additional degeneracy in the states of hydrogen is a result of the form of its potential, $1/r$, which bestows on hydrogen 0(4) symmetry. Subshells of the spherical harmonic oscillator for which $2n + \ell$ have the same value also have exactly the same energy due to

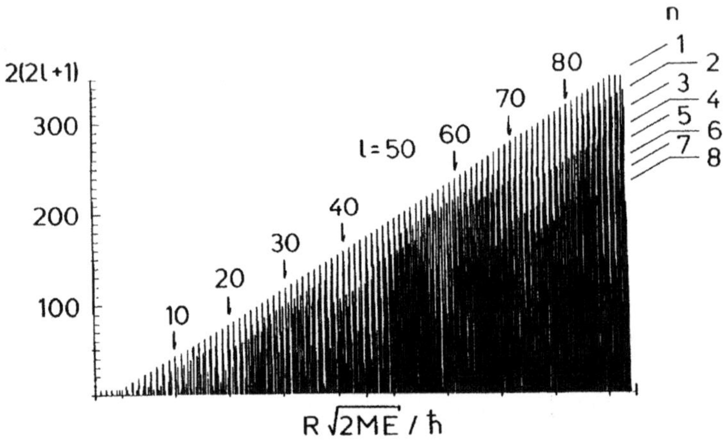

Fig. 2. The states of the infinitely deep sphereical well for very large values of ℓ. Notice the periodic bunching of states into shells. This periodic pattern is referred to as supershell structure.

the form of the potential, r^2, and the resulting symmetry, SU(3). For this reason it is said that these systems, hydrogen and oscillator, have quantum numbers $n + \ell$ and $2n + \ell$ that determine the energy. We have shown that $3n + \ell$ is an approximate energy quantum number for alkali metal clusters [16]. As the cluster increases in size, electron motion quantized in this way would finally be described as a closed triangular trajectory [19].

The grouping of large subshells into shells is illustrated in Figure 2 for the spherical potential well. Here, it can again be seen that in certain energy or momentum regions the subshells bunch together. However, the states are so densely packed in this figure that the effect is perceived as an alternating light-dark pattern. That is, for the infinite potential well, bunching of states occurs periodically on the momentum scale. The periodic appearance of shell structure is referred to as supershell structure [20,21]. Although supershell structure was predicted by nuclear physicists more than 15 years ago, it has never been observed in nuclei. The reason for this is very simple. The first supershell beat or interference occurs for a system containing 800 fermions. There exist, of course, no nuclei containing so many protons and neutrons. It is possible, however, to produce metal clusters containing such large numbers of electrons.

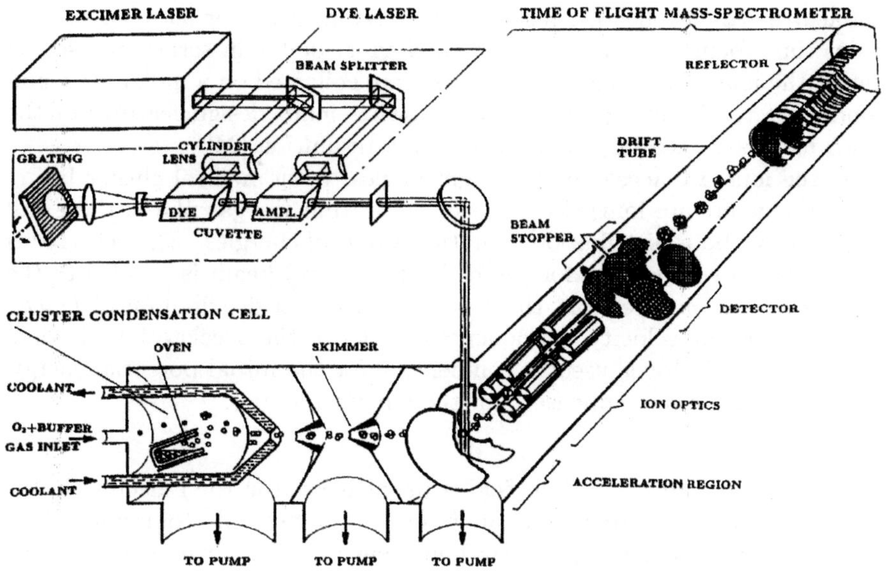

Fig. 3. Apparatus for the production, photoionization and time-of-flight mass analysis of metal clusters.

3 The experiment

The technique we have used to study shell structure in metal clusters is photoionization time-of-flight (TOF) mass spectrometry, Figure 3. The mass spectrometer has a mass range of 600 000 amu and a mass resolution of up to 20 000. The cluster source is a low-pressure, rare gas, condensation cell. Sodium vapor was quenched in cold He gas having a pressure of about 1 mbar. Clusters condensed out of the quenched vapor were transported by the gas stream through a nozzle and through two chambers of intermediate pressure into a high vacuum chamber. The size distribution of the clusters could be controlled by varying the oven-to-nozzle distance, the He gas pressure, and the oven temperature. The clusters were photoionized with a laser pulse.

Since phase space in the ion optics is anisotropically occupied at the moment of ionization, a quadrupole pair is used to focus the ions onto the detector. All ions in a volume of 1 mm^3 that have less than 500 eV kinetic energy at the moment of ionization are focused onto the detector [22].

The reflector consists of two segments with highly homogeneous electric fields, separated by wire meshes. The first segment, which is twice completely traversed by the ions, is called the retarding field, and the other segment is called reflecting field. This two-stage reflector allows a second-order time focusing of ions [23]. Two channel plates in series are used to detect the ions. The secondary electrons are collected on a metal plate and conducted to the electronics. The following main design features of the instrument are necessary to achieve such a resolution [24]:

1. the ions are accelerated at right angles to the neutral cluster beam. If clusters are ionized by a laser pulse from the gas phase, there will always be a distribution of initial *potential* energies. The reflector is used to compensate for these. If the neutral beam is parallel to the acceleration direction, there is also an initial distribution of *kinetic* energies or velocity components parallel to the acceleration diretion. If the reflector is used to compensate for the initial potential energy, it cannot also compensate for the kinetic energies;

2. a long (29 cm) retarding field segment is used in the reflector. In the vicinity of the wire meshes at the end of the two reflector segments the electric field is not perfectly homogeneous. This causes a slight deflection of ions passing through them and thus a small time error. By using along retarding field segment, the field in he vicinity of the wire meshes is lowered, and the deflection of ions passing through them is reduced.

The mass spectra which will be displayed in this paper cover a large range of masses. For this reason it will not be possible to distinguish the individual mass peaks. For example, at the top of Figure 4 we have reproduced a mass spectrum of Cs-O clusters which appears to be nothing more than a black smudge. How do we know how many oxygen atoms th clusters contain? This can be seen by graphically expanding the scale by a factor of 100, Figure 4. Because of the high resolution of our mass spectrometer, we are quite certain about the composition of the clusters examined.

4 Observation of electronic shell structure

Knight *et al.* [3] first reported electronic shell structure in sodium clusters in 1984. Electronic shell structure can be demonstrated experimentally in several ways: as an abrupt decrease in the ionization energy with increasing cluster size, as an abrupt increase or an abrupt decrease in the intensity of peaks in mass spectra. The first type of experiment can be easily understood. Electrons in newly opened shells are less tightly bound, *i.e.*, have lower ionization energies. However, considerable experimental effort is required to measure the ionization energy of even a single cluster. A complete

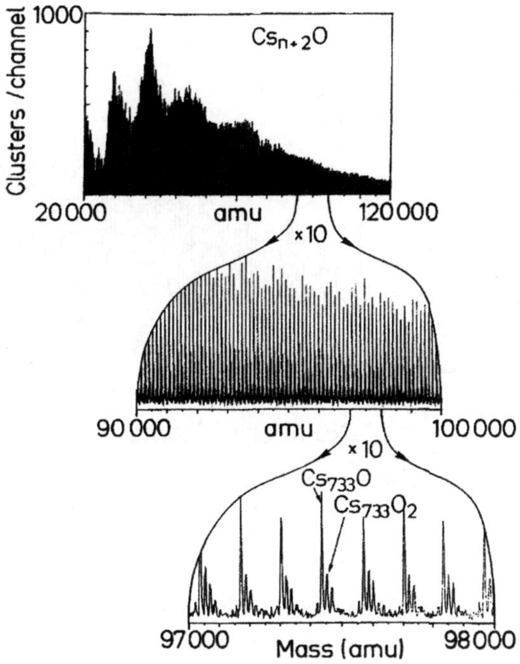

Fig. 4. Mass spectrum of Cs-O clusters.Notice that the exact composition can be determined on an expanded mass scale.

photoionization spectrum must be obtained and very often an appropriate source of tunable light is simply not available. It is much easier to observe shell closings in photoionization, TOF mass spectra. However, depending upon the intensity and wavelength of the ionizing laser pulse, the new shell is announced by either an increase or a decrease in mass peak height.

For high laser intensities, multiple-photon processes cause the mass spectra to be less wavelength sensitive and also cause considerable fragmentation of large clusters. The resulting mass spectrum reflects the stability of cluster ion fragments. Clusters with newly opened shells are less stable and are weakly represented in the mass spectra. Notice in Figure 5 that as each new shell is opened there is a sharp step downward in the mass spectrum. Remember that cluster ions containing 9, 21, 41, 59, ... sodium atoms contain the magic number (8, 20, 40, 58, ...) of electrons.

For low laser fluence and wavelengths near the ionization threshold the mass spectra have a completely different character. As each new shell is opened there is a sharp step upward in the mass spectra, Figure 6 (top). Open shell clusters have low ionization thresholds which fall below the energy of the incident photons, while closed shell clusters remain unionized.

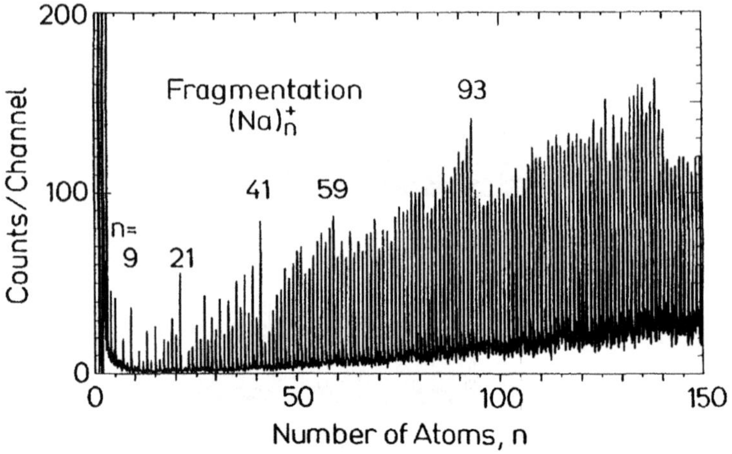

Fig. 5. Mass spectrum of $(Na)_n^+$ clusters ionized with high-intensity, 2.53 eV light. The clusters are fragmented by the ionizing laser. Fragments having closed-shell electronic configurations are particularly stable.

Finally, for low laser fluence and wavelengths well above the ionization threshold, it is possible to observe the neutral distribution of cluster sizes. If the source conditions are appropriately chosen, this distribution can peak at sizes corresponding to closed electronic shells, Figure 6 (bottom).

Cluster intensities can sometimes be increased by a factor of ten by using a seed to nucleate the cluster growth. For example, by adding less than 0.02% SO_2 to the He cooling gas, Cs_2SO_2 molecules form which apparently promote further cluster growth. Mass spectra of $Cs_{n+2}(SO_2)$ clusters obtained [15] using four different dye-laser photon energies are shown in Figure 7. Although it is not possible to distinguish the individual mass peaks in this condensed plot, it is evident that the spectra are characterized by steps. For example, a sharp increase in the mass-peak intensity occurs between $n = 92$ and 93. This can be more clearly seen if the mass scale is expanded by a factor of 50 (Fig. 8).

Notice also that the step occurs at the same value of n for clusters containing both one and two SO_2 molecules. In addition to the steps for $n = 58$ and 92 in Figure 7, there are broad minima in the 2.53 eV spectrum at about 140 and 200 Cs masses. These broad features become sharp steps if the ionizing photon energy is decreased to 2.43 eV. By successively decreasing the photon energy, steps can be observed for the magic numbers $n = 58$, 92, 138, 198 ± 2, 263 ± 5, 341 ± 5, 443 ± 5, and 557 ± 5 [15,17]. However, the steps become less well defined with increasing mass. We have studied the mass spectra of not only $Cs_{n+2}(SO_2)$ but also $Cs_{n+4}(SO_2)_2$, $Cs_{n+2}O$, and $Cs_{n+4}O_2$. They all show step-like features for the same values of n.

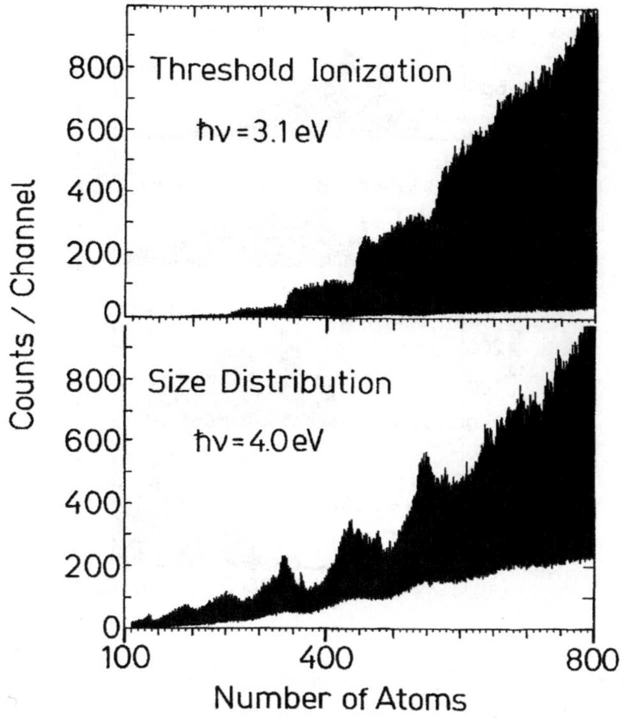

Fig. 6. Mass spectra of $(Na)_n$ clusters obtained using ionizing light near the ionization threshold (top) ($\hbar\nu = 3.1$ eV) and well above the ionization threshold (bottom) ($\hbar\nu-$ 4.0 eV). In both cases the neutral cluster beam was heated with 2.54 eV and 2.41 eV laser light.

First, we would like to offer a qualitatiave explanation for these results and then support this explanation with detailed calculation. Each cesium atom contributes one delocalized electron which can move freely within the cluster. Each oxygen atom, and each SO_2 molecule, bonds with two of these electrons. Therefore, a cluster with composition $Cs_{n+2}(SO_2)$, for example, can be said to have n delocalized electrons. The potential in which the electrons move is nearly spherically symmetric, so that the states are characterized by a well-defined angular momentum. Therefore, the delocalized electrons occupy subshells of constant angular momentum which in turn condense into shells. When one of these shells is fully populated with electrons, the ionization energy is high and the clusters will not appear in mass spectra obtained using sufficiently low ionizing photon energy.

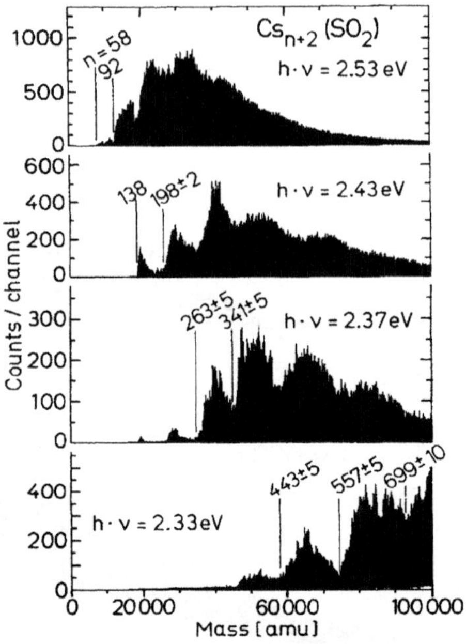

Fig. 7. Mass spectra of $Cs_{n+2}(SO_2)$ clusters with decreasing photon energy of the ionizaing laser from 2.53 eV (top) to 2.33 eV (bottom). The values of n at the steps in the mass spectra have been indicated (Ref. [15]).

In other experiments [9] the closing of small subshells of angular momentum was shown to be accompanied by a sharp step in the ionization energy for Cs-O clusters having certain sizes, namely for $Cs_{n+2}O_z$ with $n = 8$, 18, 20, 34, 58 and 92. The closing at $n = 40$ seen in all other alkali-metal clusters could not be observed, neither in the experiments nor in the calculations. The steps were observed for clusters containing from one to seven oxygen atoms.

5 Density functional calculation

Self-consistent calculations have been carried out applying the density functional approach to the spherical jellium model [10,11]. We used an exchange correlation term of the Gunnarsson-Lundqvist form and a jellium density $r_S = 5.75$ corresponding to the bulk value of cesium. This model implies two improvements over the hard sphere model discussed earlier. Firstly, electron-electron interaction is included. Secondly, the jellium is regarded to be a more realistic simplification of the positive ion background than the

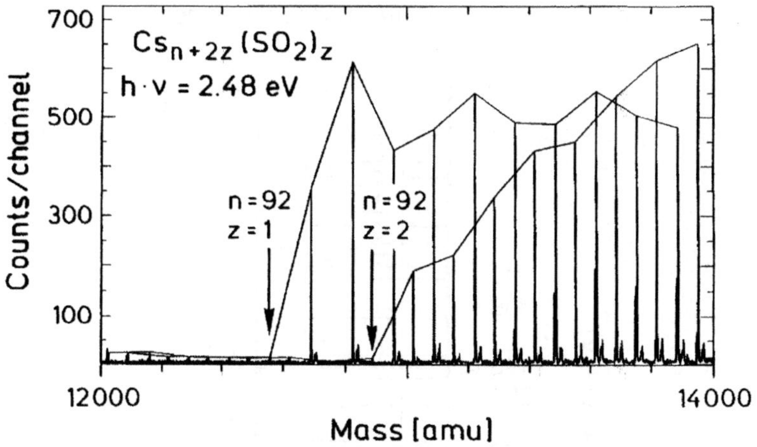

Fig. 8. Expanded mass spectra of $Cs_{n+2z}(SO_2)$ clusters for an ionizing photon energy of 2.48 eV. The lines connect mass peaks of clusters containing the same number z of SO_2 molecules. Notice that the steps for clusters containing (SO_2) and $(SO_2)_2$ are shifted by two Cs atoms (Ref. [15]).

hard sphere. The O^2 ion is taken into account only by omitting the cesium electrons presumably bound to oxygen. The calculations were performed on Cs_{600} clusters [25].

We found, that if a homogeneous jellium was used, the grouping of subshells was rather similar to the results of the infinite spherical potential well. However, a nonuniform jellium yielded a shell structure in better accordance to experimental results. We found that the subshells group fairly well into the observed shells only if the background charge distribution is slightly concentrated in the central region. This was achieved, for example, by adding a weak Gaussian (0.5% total charge density, half-width of 6 a.u.) charge distribtuion to the uniform distribution (width 48 a.u.). Figure 9 shows the ordering of subshells obtained from this potential. This leads to the rather surprising result that the Cs^+ cores seem to have higher density in the neighborhood of the center perhaps due to the existence of the O^{2-} ion. All attempts to lower the positive charge density in the central region led to an incorrect ordering of states.

The first calculation addressed the problem of the grouping of low-lying energy levels in one large Cs_{600} cluster. However, in the experiment the magic numbers were found by a rough examination of ionization potentials of the whole distribution of cluster sizes. A more direct way to explain magic numbers is to look for steps in the ionization potential curve of Cs–O

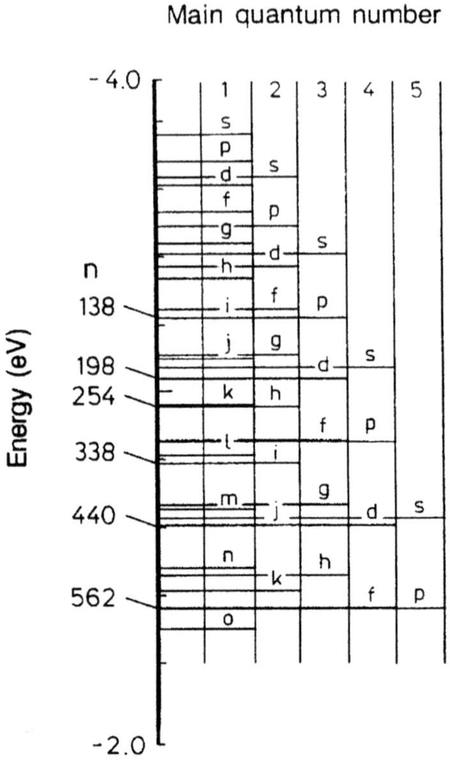

Fig. 9. The self-consistent, one-electron states of a 600 electron cesium cluster calculated using a modified spherical jellium background (Ref. [25]).

clusters. Therefore, we calculated the ionization potentials of $Cs_{n+2}O$ For $n \leq 600$ and of $(Na)_n$ for $n \leq 1100$ using the same local-density scheme described above, Figure 10. Starting from a known closed-shell configuration for $n = 18$, electrons were successively added. Three test configurations were calculated for each cluster size testing the opening of new subshells. The configuration with minimum total energy was chosen for the calculation of the ionization potential.

We found that the lower magic numbers $n = 34, 58, 92$ were well reproduced. For higher n distinct steps in the ionization potential were observed for $n = 138, 196, 268, 338, 440, 562, 704, 854$ and 1012. The absolute values of the calculated ionization potentials can be brought into better agreement with experiment by assuming that clusters have a 10–15% lower electron density than is fouond in the bulk. Magic number clusters exhibit

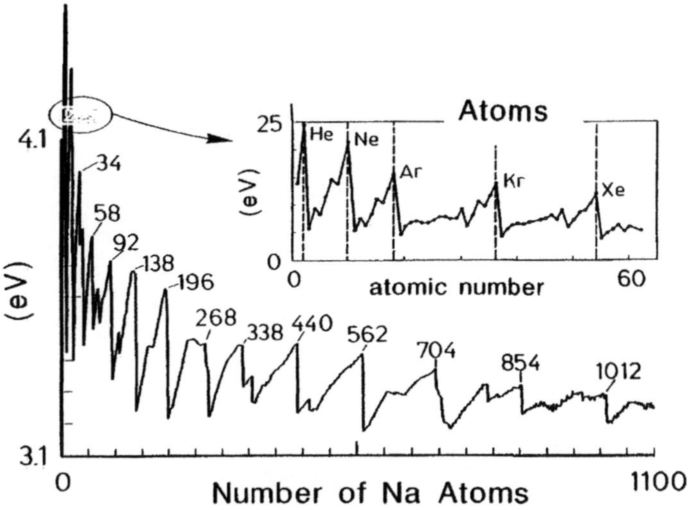

Fig. 10. Ionization potentials calculated as a function of n for $(Na)_n$ clusters. A positive background charge distribution slightly concentrated in the central region has been used. Notice the similar behavior of the ionization energies of the chemical elements (inset) (Ref. [25]).

unusually high ionization energies for the same reason rare gas atoms do: they possess a closed shell electronic configuration, Figure 10. In this sense the metallic clusters behave like giant atoms.

6 Observation of supershells

Although nuclear physicists speculated on the possible existence of supershells several decades ago, the phenomenon has never been observed in atomic nuclei for a very simple reason. No nucleus contains enough fermions to allow supershell formation. However, there is almost no limit to the number of electrons that can be contained in metal clusters.

Supershells are the periodic appearance and disappearance of shell structure in the energy density of states of a fermion system. In order to make clear the physical origin of supershells, it is necessary to go back one step to the semiclassical description of shells. Shells are associated with a characteristic length. Each time an integral number of fermi wavelengths fit into this length, a new shell has formed. The systems that we are studying are so large, that the classical picture of an electron bouncing back and forth inside a metal cluster is not completely without meaning. The characteristic length associated with a set of shells is just the length of a closed

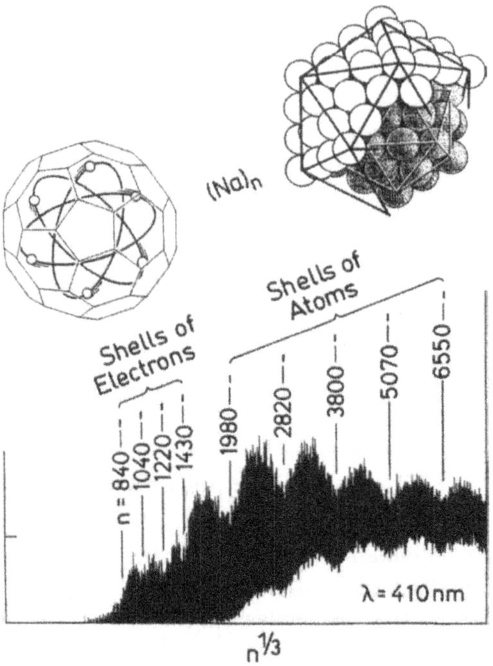

Fig. 11. Mass spectrum of $(Na)_n$ clusters photoionized with 3.02 eV photons. Two sequences of structures are observed at equally spaced intervals on the $n^{1/3}$ scale – an electronic shell sequence and a structural shell sequence.

electron trajectory within the clusters. For spheres, two closed trajectories with almost the same length turn out to be the most important – a triangualr path and a square path. This leads to two sets with nearly the same energy spacing. These two contributions interfere with one another to produce a beat pattern known as quantal supershells. The first attempts to observe supershell structure in our laboratory were hindered by the unexpected appearance of a second set of shells in clusters containing more than 1500 atoms, Figure 11.

These proved to be geometric shells of atoms that masked the weaker electronic shell structure. In the new experiments the geometric shell structure was surpressed by "melting" the clusters through heating with a continuous laser beam tuned to the plasmon frequency of the electron system.

The clusters wer warmed prior to ionization with a continuous Ar-ion laser beam running parallel to the neutral cluster beam. The laser light entred the ionization chamber through a heated window, passed through the ionization volume, through 2.2 mm ϕ and 3.0 mm ϕ skimmer aperatures,

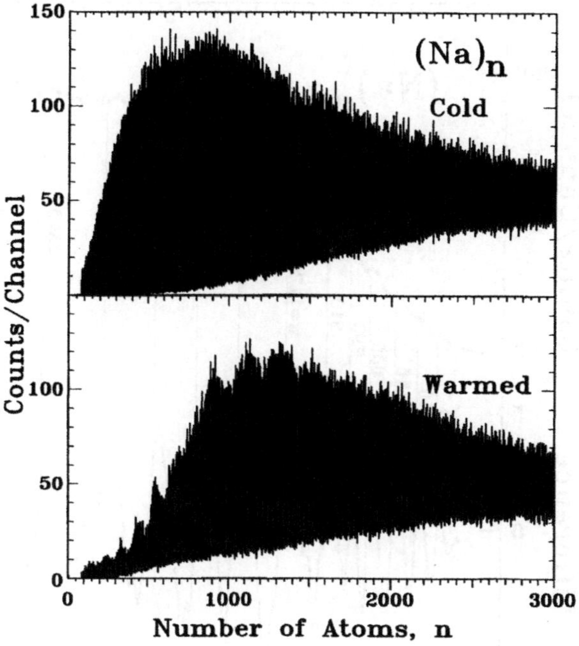

Fig. 12. Mass spectrum of $(Na)_n$ clusters using 4.0 eV ionizing light. The top spectrum shows the size distribution of cold clusters produced in the source; the bottom spectrum, after heating with 2.54 eV laser light.

through a 3.0 mm nozzle, throught he oven chamber and finally exited through a second window where the laser intensity was recorded. Short wavelength light was found to warm much more efficiently. Using the 458 nm (2.71 eV) laser line, 10 mW proved sufficient to appreciable alter the neutral size distribution.

The size distribution obtained with ionizing photons having energy well above threshold are quite different from the spectra discussed in Section 5 Without the warming laser the mass spectra are without structure, *i.e.* the size distribution of the cold clusters emerging from our source is smooth, Figure 12. If the warming laser is turned on we obtain not steps but peaks as seen in the bottom of Figure 12. We believe these peaks reflect the neutral size distribution of the laser-warmed clusters. It appears that it is usually possible to correlate a falling edge of the size disbribution with a step in the threshold ionization spectrum. Because of this correlation, we will characterize mass spectra obtained using excimer light by the number of atoms at steep negative slopes. A more extended mass spectrum of laser-warmed sodium clusters obtained with 4.0 eV ionizing photons is shown

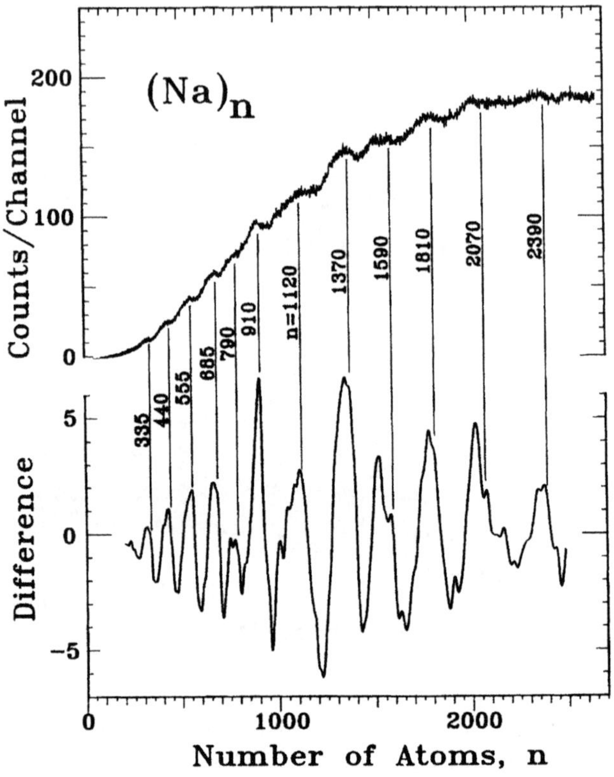

Fig. 13. Mass spectrum of $(Na)_n$ clusters using 4.0 eV ionizing light and (458 nm) 2.71 eV continuous axial warming light having an intensity of 500 mW/cm^{-2}. The spectrum has been smoothed over one-hundred 16 ns time channels (top). In order to emphasize the shell structure an envelope function (obtained by smoothing over 20 000 time channels) is subtracted from a structural mass spectrum (smoothed over 1500 time channels). The difference is shown in the bottom spectrum.

at the top of Figure 13. This spectrum has been smoothed with a spline function extending over one-hundred 16ns time channels. Notice that the structure observed does not occur at equal intrvals on a scale linear in mass. In order to present this structure in a form more convenient for analysis, the data have been procesed in the following way. First, the raw data is averaged with a spline function extending over 20 000 time channels. The result is a smooth envelope curve containing no structure. Second, the raw data is averaged with a spline over 1500 channels. Finally, the two avearges are subtracted. The result is shown in the bottom of Figure 13.

Five independent measurements were made under the same experimental conditions. The positions, relative heights and widths of features in the mass spectra were well reproducible.

The clusters in this experiment have been warmed with a continuous laser beam running parallel to the neutral cluster beam. But what is implied by "warming"? consider the fate of a typical 500-atom cluster as it moves from the nozzle to the detector.

It leaves the nozzle with the temperature of the He carrier gas (~100 K) traveling at a velocity of about 350 m/s. during its 1 ms flight to the ionization volume it undergoes no further collisions but does begint o absorb photons. We do not really know the absorption cross-section of this cluster at the warming laser wavelength (458 nm). However, 1 Å^2/atom is a typical upper limit for smaller clusters. It can be expected that the cross-section will be cluster size dependent. This size dependence will be reflected in the final mass distribution. The cluster absorbs the first 25 photons without evaporating any atoms, gaining an excess energy of about 70 eV and reaching a temperature of about 500 K. This all takes place in the first 450 μs. The temperture of the cluster remains rather constant for the last half of its journey to the ionization volume. It continues to absorb photons, of course, but after each absorption it evaporates 2 or 3 atoms returning to its original temperature before absorbing the next photon. It loses a total of 80 atoms, *i.e.* 16% of its original mass. It appears that this repeated heating and cooling through the "critical temperature for evaporation" on this time scale favors the evolution of a size distribution with relataively strong peaks near sizes corresponding to closed electronic shells.

The photon energy (4.0 eV) of the ionizing laser has been chosen so that it is well above the ionization threshold (3.0 eV) of the sodium clusters investigated. The excess energy (1 eV) insufficient to cause only one atom to evaporate. This is a neglibible loss on the mass scale we will be considering. For this reason, we believe that the magic numbers obtained reflect variations inthe size disbtribution of the neutral clusters induced by the warming laser.

The concept of shells can be associated with a characteristic length. Every time the radius of a growing cluster increases by one unit of this characteristic length, a new shell is said to be added. A good rough test of whether or not shell structure has been observed can be quickly carried out by plotting the shell index as a function of the radius or $n^{1/3}$. If the points fall on a straight line, the data is consistent with shell formation. That this is indeed the here, can be seen in Figure 14. However, an even better fit can be obtained using two straight lines with a break between shell 13 and 14. This too can be interpreted in an interesting way.

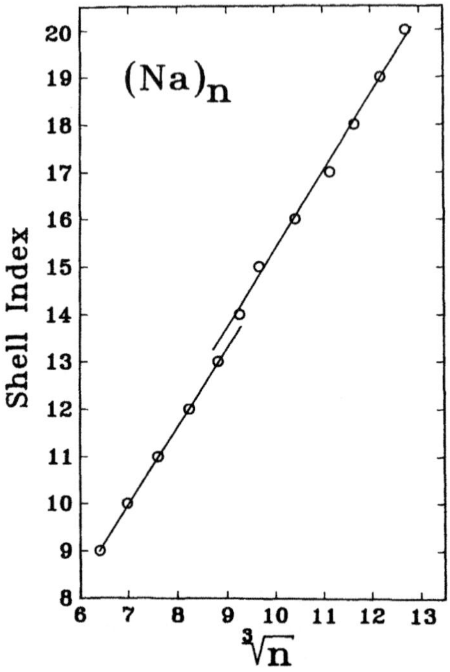

Fig. 14. The electronic shell closing falls approximately on a straight line if plotted on an $n^{1/3}$ scale. An even better fit is obtained using two straight lines with a break between shells 13 and 14. Such a break or "phase change" would be an indication of supershell structure.

It has been suggested [19-28] that shell structure might periodically appear and disappear with increasing cluster size. Such a supershell structure can be understood as a beating pattern created by the inteference of two nearly equal periodic contributions. Quantum mechanically the contributions can be described as arising from competing energy quantum numbers. Classically, the contributions can be described as arising from two closed electron trajectories within a spherical cavity. One trajectory is triangular, the other square.

7 Fission

The fission of clusters was one of the first subjects [29-43] to be investigated in the newly developing field of cluster research. It is often referred to as Coulomb explosion, since the fission is caused by the Coulomb repulsion of like charges concentrated in a cluster smaller than a critical size. The kinetic energy that the charged fragments acquire can be as high as several eV. Most of these studies have delt with the fission of doubly or triply charged clusters. Recently we have shown that it is possible to induce charges as high as +14 on large Na clusters by photoionization [44]. In this section we will discuss fission in these highly charged clusters.

The technique we have used to study fission in sodium clusters is photoionization time-of-flight (TOF) mass spectroscopy. The cluster source is a low pressure, inert gas, condensation cell. The clusters were photoionized with a 50 mJ, 15 ns, 193 nm (6.4 eV) excimer laser pulse focussed onto the neutral cluster beam with a 150 cm focal length quartz lens. The ionized clusters were heated 30 ns later with a second 5 mJ/mm^2, 470 nm (2.6 eV) laser pulse.

The energy (I) required to remove an aditional electron from a cluster that already has charge $+z$ can be written

$$I(z,R) = W + (\alpha + z)e^2/r \tag{7.1}$$

where W is the bulk work function, e is the electronic charge, and R is the radius of the cluster. Clearly, we have assumed that the cluster can be modelled as a conducting sphere. Various values of α have beenused in the literature. We will assume α is 0.5 and point out that for larg values of z the value of α used becomes unimportant. Since the radius of the cluster can be related to the number of electrons (or in our case atoms) through the Wigner–Seitz radius, $R^3 = r_s^3 n$, equation (4) can be rewritten as

$$I(z,n) = W + (\alpha + z)e^2/r_s n^{1/3}). \tag{7.2}$$

It can be seen from this expression that a larage amount of energy is required to remove electrons from small, highly charged clusters. If the amount of energy available is limited to that in one photon, then the maximum charge attainable for a cluster of a given size is

$$z_{\max} = 1 - \alpha + (h\nu - W)r_s n^{1/3}/e^2. \tag{7.3}$$

This means that a two-dimensional cluster space (n, z) can be divided by a line into clusters that can be formed with, for example, an ArF excimer laser and those that cannot, Figure 15. Also indicated in this figure is a line dividing the space into stable and unstable clusters. Notice that all of the values of n and z accessible with the ArF photons characterize stable clusters. The unstable clusters which we would like to investigate cannot be produced by direct multi-step ionization with this laser. There is, however, a way out of this dilemma.

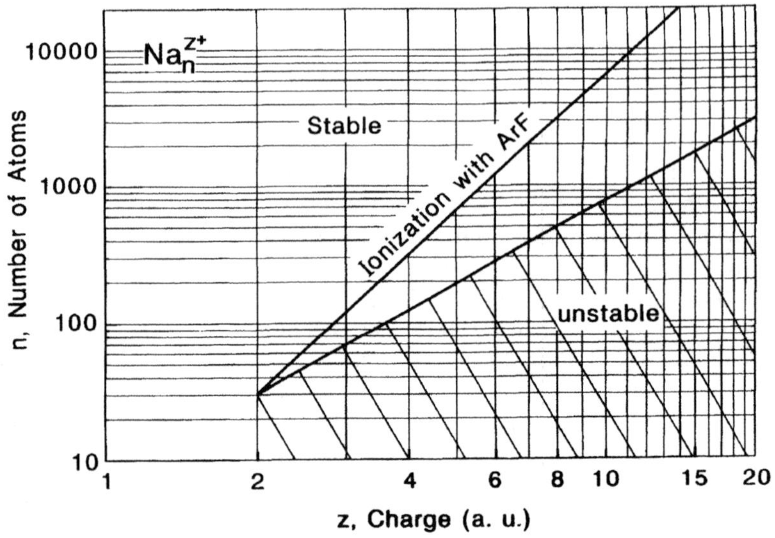

Fig. 15. The two-dimensional cluster space (n, z) can be divided by straight lines into Na_n^z clusters which can (cannot) be produced with an ArF laser and into clusters which are (are not) stable against Coulomb explosion. Notice that all clustrs (except for $z = 2$) that can be ionized with 6.4 eV photons are stable.

The ArF laser can be used to prepare a stable, highly charged, large cluster and then this large cluster can be reduced in size by heating and subsequent evaporation. A second laser pulse, containing photons with energies near the plasmon resonance of the sodium clusters, is used for heating. The clusters shrink down in size without charge until they reach a critical size at which they undergo fission. A mass spectrum, or better said, an n/z spectrum for Na_n^z clusters produced in this way is shown in Figure 16. The log scale emphasizes, perhaps even overemphasizes, the effect we wish to show.

The highest set of mass peaks belongs to singly-charged sodium clusters. The peaks which occur exactly half-way between the Na_n^+ peaks are due to Na_n^{2+} clusters. Notice that new sets of peaks appear in the spectrum at various critical values of n/z. This is perceived as a step-wise darkening of the mass spectrum. In the lower part of Figure 16 we see the threshold region for the appearance of Na_n^{5+} on an expanded scale. Another segment of the spectrum on an expanded scale is shown in Figure 17. This segment is near the threshold for the appearance of $Na^{6+}\ddot{o}n$.

In this way, by careful examination of the fine structure in the mass spectra, it is possible to determine that the critical sizes for $z = 1$, 2,

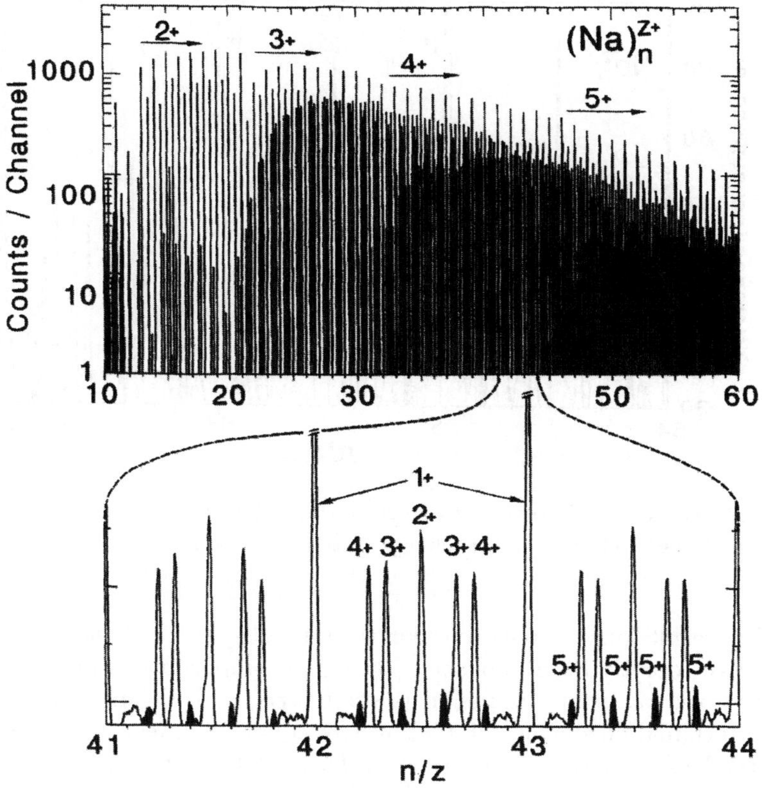

Fig. 16. An n/z spectrum of Na_n^z clusters. large clusters were first charged by multistep ionization using a high-fluence, arF laser. The clusters were then heated to reduce their size by evaporation. A portion of the spectrum is expanded to show the appearance threshold for Na_n^{5+} clusters (black filled).

3, 4, 5, 6 and 7 are 27 ± 1, 64 ± 1, 123 ± 2, 208 ± 5, 321 ± 5 and 448 ± 10 atoms, respectively. These values are plotted on a double log scale in Figure 18. They lie on a straight line with slope 2. This means that the critical condition for stability is

$$z^2/n \leq 0.125. \tag{7.4}$$

z^2/n is proportional to the so-called fissility parameter used in nuclear phyciscs as a measure of stability. It has been shown inthe past that this parameter is also useful for clusters with small total charge [40, 41]. Here we see that it continues to be applicable for values of z up to 7.

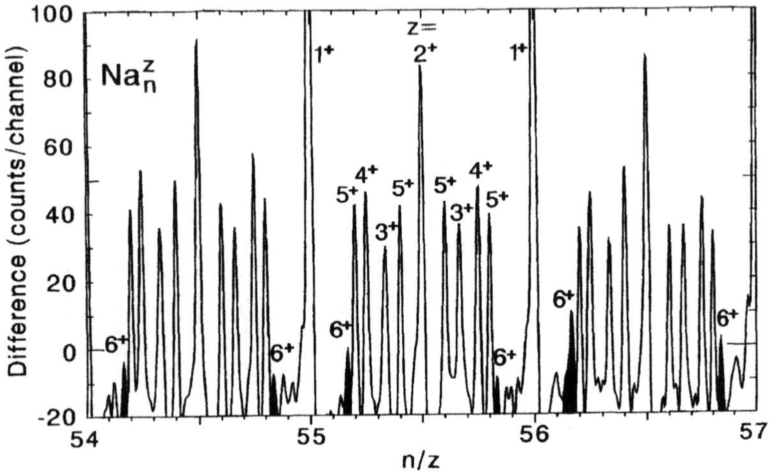

Fig. 17. An expanded portion of Figure 16 showing the first appearance on Na_n^{6+} clusters. Smaller clusters in this charge state are not stable.

The results of an extensive theoretical investigation of fission in Na clusters have recently been published [45, 46]. an important assumption made inthis work was that the fission is symmetric, *i.e.* the mass and charge of the original cluster are divided nearly equally between the fission productes. Unfortunately, we have no evidence at this time to either support or to challenge this assumption. Still, it is useful to compare the results of this calculation with our experiment, Figure 18.

Here, cluster space (n, z) has been divided into stable metastable and unstable regions according to the tunneling criteria appropriate in nuclear physics. The fission process for nuclei can be described qualitatively in terms of three energies; the initial energy (E_i) of the charged, nondeformed clusters, the final energy (E_f) which is the sum of the energies of the noninteracting fission products. The third energy necessary to characterize fission products. The third energy necessary to characterize fission is the energy (E_b) of the lowest barrier separating the initial and final states. If $E_\mathrm{b} < E_\mathrm{i}$, the nucleus is unstable. If $E_\mathrm{f} < E_\mathrm{i} < E_\mathrm{b}$, the nucleus is metastable to fission by tunneling through the barrier. Finally, if $E_\mathrm{f} > E_\mathrm{i}$, then the nucleus is stable. since tunneling for clusters has negligible probability, it is more accurate to say at zero temperature clusters are either stable or unstable, depending on whether there is a barrier or not. At finite temperature clusters can be classified as either unstable ($E_\mathrm{B} < E_\mathrm{i}$) or as metastable ($E_\mathrm{B} > E_\mathrm{i}$) to thermal hopping over the barrier. In practice it is useful to further subdivide the set of metastable clusters [36–38]. At finite

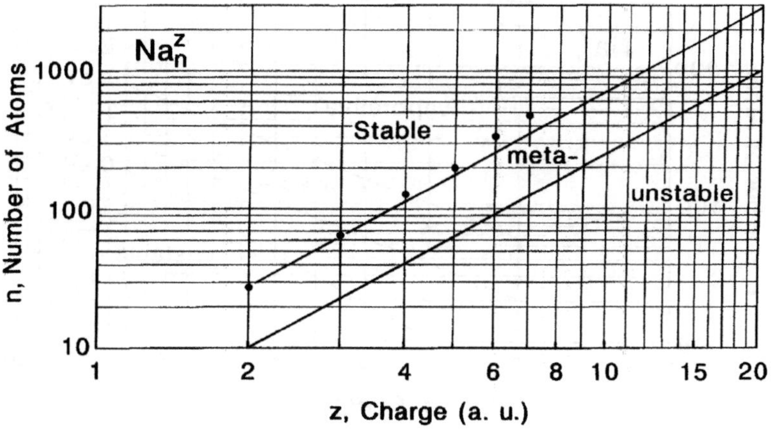

Fig. 18. The number of atoms in the smallest experimentally observed Na_n^z clusters (filled circles). The cluster space can be divided into stable, metastable and unstable regions, according to [45], using the symmetric liquid-drop model.

temperatures clusters can lose mass and thermal energy by the evaporation of neutral atoms. Evaporation will always compete with fission and will, in fact, dominate if E_B is greater than E_v, the energy needed to evaporate an atom. For this reason, we have the conditions

a) unstable to fission, $E_B < E_i$;
b) metastable to fission, $E_i < E_B < E_v$;
c) metastable to evaporation, $E_i < E_v < E_B$.

Since previous experiments [36–38] on doubly-charged Na and K clusters indicate that fission is strongly asymmetric, it would be appropriate now to consider this alternative. Even using the droplet model it is not easy to calculate the height of the barrier if the mass and charge can be distributed arbitrarily between the fission products. For this reason we will consider an energy which does not exactly characterize a real system, but it is trivial to calculate and therefore useful. It is the energy at the instant of scission, E_s. That is, starting from the final state we merely bring the fission products together until they just touch. Of course the energy increases monotonically from E_f to E_s according to Coulomb's Law. We assume that E_s is nearly equal to the barrier height for the fission process. There is no unique value of E_s for a cluster in initial state n and z. Rather, a whole set of values exist corresponding to the various ways of distributing charge and mass between the fission products. However, one value of E_s has special significance and

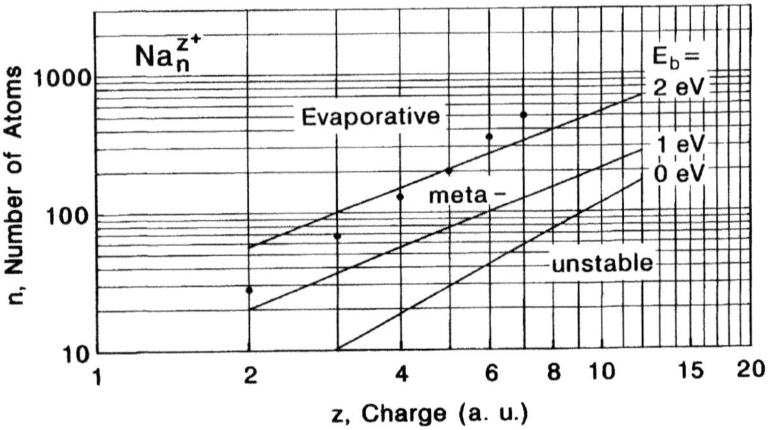

Fig. 19. Cluster space divided into evaporative and metastable und unstable regions. The barrier height has been obtained using an oversimplified model (see text). The number of atoms in the smallest experimentally observed Na_n^{z+} clusters is shown by the filled circles.

that is the minimum value. If this minimum value of $E_s < E_i$ the cluster is unstable and will spontaneously fission, even at zero temperature. If $E_s > E_i$ the cluster is stable against fission. Strictly speaking, one shuld say metastable because at finite temperatures the final state can be reached by jumping over the barrier, no matter how high. The results of these calculations are summarized in Figure 19. The initial and final energies are determined using the sphereical droplet model assuming only two fission fragments with arbitrary size and charge and assuming a surface tension parameter $\sigma = 200$ dyne/cm appropriate for sodium. One might expect that the experimental points would fall on the line corresponding to $E_B = E_v$. Clearly, this is not the case since E_v is known [36–38] to have a value of about 1 eV. That is, this rough model overestimates the barrier height by a factor of two. Various refinements are clearly needed; proper treatment of the Coulomb energy allowing for electron redistribution as the fragments move away from one another [48], a description of the asymmetric fission before scission, shell effects [49, 50] and entropy effects. Also needed are experiments demonstratinghow mass and charge are distributed between the fission fragments.

8 Concluding remarks

Clearly, cluster science has greatly benefitted from the inspired work carried out by nuclear physicists decades ago. The shell model, the liquid droplet model, and the theory of giant dipole resonances have provided a ready and appropriate framework for understanding the properties of metal clusters. Hopefully, in the future, the exchange between nuclear science and cluster science will not be so one-sided, because metal clusters offer us a unique opportunity to study well-characterized, large fermion systems.

References

[1] M. Goeppert–Mayer, *Phys. Rev.* **75** (1949) 1969L.

[2] O. Haxel, J.H.D. Jensen and H.E. Suess, **75** (1949) 1766L.

[3] W.D. Knight *et al.*, *Phys. Rev. Lett.* (1984).

[4] M.M. Kappes, R.W. Kunz and E. Schumacher, *Chem. Phys. Lett.* **91** (1982) 413.

[5] I. Katakuse *et al.*, *Int. J. Mass Spectrom. Ion Proc.* **67** (1985) 229.

[6] C. Bréchignac, Ph. Cahuzac and J.-Ph. Roux, *Chem. Phys. Lett.* **127** (1986) 445.

[7] W. Begemann *et al.*, *Z. Phys. D* **3** (1986) 183.

[8] W.A. Saunders *et al.*, *Phys. Rev. B* **32** (1986) 1366.

[9] T. Bergmann, H. Limberger and T.P. Martin, *Phys. Rev. Lett.* **60** (1988) 1767.

[10] J.L. Martins, R. Car and J. Buttet, *Surf. Sci.* **106** (1981) 265.

[11] W. Ekardt, *Ber. Bunsenges. Phys. Chem.* **88** (1984) 289.

[12] K. Clemenger, *Phys. Rev. B* **32** (1985) 1359.

[13] Y. Ishii, S. Ohnishi and S. Sugano, *Phys. Rev. B* **33** (1986) 5271.

[14] T. Bergmann and H. Limberger, *J. Chem. Phys.* **90** (1989) 2848.

[15] H. Göhlich *et al.*, *Phys. Rev. Lett.* **65** (1990) 748.

[16] T.P. Martin *et al.*, *Chem. Phys. Lett.* **72** (1991) 209.

[17] S. Bjørnholm *et al.*, *Phys. Rev. Lett.* **65** (1990) 1627.

[18] J.L. Persson *et al.*, *Chem. Phys. Lett.* **171** (1990) 147; E.C. Honea *et al.*, *Chem. Phys. Lett.* **171** (1990) 147; J. Lerme *et al.*, *Phys. Rev. Lett.* **68** (1992) 2818.

[19] R. Balian and C. Bloch, *Ann. Phys.* **69** (1971) 76.

[20] A. Bohr and B.R. Mottelson, *Nuclear Structure* (Benjamin, London, 1975).

[21] H. Nishioka, K. Hansen and B.R. Mottelson, *Phys. Rev. B* **42** (1990) 9377.

[22] T. Bergmann *et al.*, *Rev. Sci. Instrum.* **61** (1990) 2585.

[23] B.A. Mamyrin *et al.*, *Sov. Phys. JETP* **37** (1973) 45.

[24] T. Bergmann, T.P. Martin and H. Schaber, *Rev. Sci. Instrum.* **61** (1990) 2592.

[25] T. Lange *et al.*, *Z. Phys. D* **19** (1991) 113.

[26] T.P. Martin *et al.*, *Chem. Phys. Lett.* **186** (1991) 53.

[27] J. Pedersen *et al.*, *Nature* **353** (1991) 733.

[28] C. Bréchignac *et al.*, *Phys. Rev. B* **47** (1993) 2271.

[29] D. Kreisle *et al.*, *Phys. Rev. Lett.* **56** (1986) 1551.

[30] O. Echt, *Physics and Chemistry of Small Clusters*, edited by P. Jena, B.K. Rao and S.N. Khanna (Plenun Press, New York, 1987).

[31] O. Echt *et al.*, *Phys. Rev. A* **38** (1988) 3236.

[32] T.D. Märk *et al.*, *Z. Phys. D* **12** (1989) 279.

[33] N.G. Gotts, P.G. Lethbridge and A.J. Stace, *J. Chem. Phys.* **96** (1992) 408.

[34] O. Kandler *et al.*, *Z. Phys. D* **19** (1991) 151.

[35] K. Sattler *et al.*, *Phys. Rev. Lett.* **47** (1985) 160; K. Sattler, *Surf. Sci.* **156** (1985) 292.

[36] C. Bréchignac *et al.*, *Phys. Rev. Lett.* **64** (1990) 2893.

[37] C. Bréchignac *et al.*, *Z. Phys. D* **19** (1991) 1.

[38] C. Bréchignac *et al.*, *Phys. Rev. B* **44** (1991) 11386.

[39] I. Katakuse, H. Itoh and T. Ichihara, *Int. J. Mass. Spectrum. Ion Proc.* **97** (1990) 47.

[40] W.A. Saunders, *Phys. Rev. Lett.* **64** (1990) 3046.

[41] W.A. Saunders, *Z. Phys. D* **20** (1991) 111.

[42] W. Schulze, *J. Chem. Phys.* **87** (1987) 2402.

[43] I. Rabin, C. Jackschath and W. Schulze, *Z. Phys. D* **19** (1991) 153.

[44] U. Näher *et al.*, *Phys. Rev. Lett.* **68** (1992) 3416.

[45] S. Sugano, *Microcluster Physics* (Springer, Berlin, Heidelberg, 1991).

[46] M. Nakamura *et al.*, *Z. Phys. D* **19** (1991) 145.

[47] E. Lipparini and A. Vittori, *Z. Phys. D* **17** (1990) 57.

[48] F. Garcias *et al.*, *Phys. Rev. B* **43** (1991) 9459.

[49] B.K. Rao *et al.*, *Phys. Rev. Lett.* **58** (1987) 1188.

[50] R.N. Barnett, U. Landman and G. Rajagopal, *Phys. Rev. Lett.* **67** (1991) 3058.

COURSE 2

MELTING OF CLUSTERS

H. HABERLAND

Fakultät für Physik, Universität Freiburg, H.Herderstr. 3, 79104 Freiburg, Germany

Contents

MELTING OF CLUSTERS

H. Haberland

Abstract

An experiment is described which allows to measure the caloric curve of size selected sodium cluster ions. This allows to determine rather easily the melting temperatures, and latent heats in the size range between 55 and 340 atoms per cluster. A more detailed analysis is necessary to show that the cluster Na_{147}^{+} has a negative microcanonical heat capacity, and how to determine the entropy of the cluster from the data.

1 Introduction

Melting in a macroscopic object occurs at some well defined temperature. This is no longer true for a small particle or cluster. If one talks about melting one must first verify whether the concepts of *solid* or *liquid* can be transferred to small systems. The answer is yes: at low temperatures the atoms in a cluster or in a large piece of matter make only small amplitude vibrations around a fixed position. It takes a lot of energy to push an atom from its position and one has a solid. If the temperature increases, atoms in the cluster can visit neighbouring places and start a diffusive motion. There is very little resistance to external deformation forces and the cluster is liquid. This is demonstrated in Figure 1 which shows a simulation for a cold, rigid and a liquid, hot cluster. The simulation was done with the potential parameters for argon. But the main features of small harmonic oscillations at low and more diffusive motion at high temperatures are independent of the force field assumed.

Concepts like temperature, phase etc. had originally been defined only for infinitely large systems. But one has learned how to generalize these concepts to finite particles, although there are still some more semantic quarrels about nomenclature. For the solid to liquid transition of finite systems one

The work was supported by the Deutsche Forschungsgemeinschaft through SFB 276.

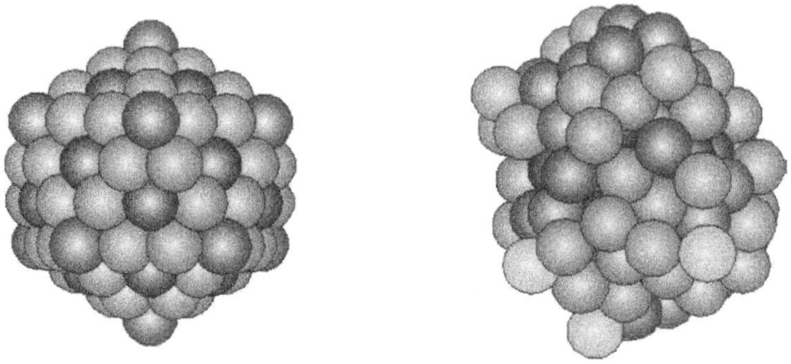

Fig. 1. A simulation of a cluster of 147 atoms at $T = 0$ K (left figure) and a high temperature. The darker the colour, the more strongly an atom is bound. The calculation was made using a Lennard–Jones potential, which is not adequate for the sodium clusters discussed in this report. So one does not know whether the low temperature structure of Na_{147}^{+} really looks like this icosahedron, whose regular, five fold symmetry can be well seen here. The general features of the solid to liquid transition however, should be independent of the force field assumed. (Calculation by Moseler, colour pictures on the WEB site of the author.)

finds four main differences with respect to the bulk counterpart, 1) the melting point is generally reduced (Sect. 4.1), 2) the latent heat is smaller (Sect. 4.2), and 3) the transition does not occur at one definite temperature but is spread out over a temperature range. The fourth difference is quite spectacular: the heat capacity can become negative, *i.e.* the temperature of the system can decrease upon energy addition (Sect. 6.1).

Experimentally, the depression of the melting point has been shown [1] as early as 1908. Many experiments have been done with particles on surfaces which contained thousands or millions of atoms [2]. Also, very many computer simulations have been performed to get a better understanding [3–6]. They have shown how the bulk concepts like temperature and melting can be generalized to finite systems. Nearly all simulations have been performed for free clusters of known size, while most earlier experiments have been performed with particles on surfaces having a rather broad size distribution.

2 Cluster calorimetry

All experiments to study the melting point of bulk material or of clusters can be divided into two classes:

1. one studies the change of some physical property across the melting point, *e.g.* change in photon or X-ray diffraction patterns, or mobility;
2. thecaloric curve $E = E(T)$ is measured, that is the clusters internal energy E as a function of its temperature T. This curve contains all the thermodynamic information, as discussed below.

The experiment discussed here, is of the second type. It measures the caloric curve $E = E(T)$ for free sodium cluster ions, where the number of atoms is exactly known.

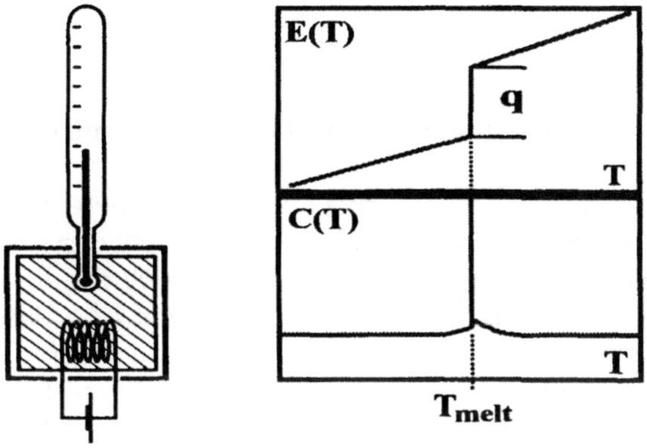

Fig. 2. Calorimeter (left), caloric curve $E = E(T)$ and heat capacity $C(T)$ for a macroscopic sample. The caloric curve increases abruptly at the sharp melting temperature ($T_{\rm melt}$). The height of the jump is given by the latent heat q of melting. The heat capacity has a delta function like peak at $T_{\rm melt}$ (see Eq. (2.1)).

2.1 The bulk limit

It is often very useful in cluster science to study the asymptotic behaviour, that is that of the atom and that of the bulk. Although there is a well understood bulk limit, the atom and the diatomic molecule do obviously not have a melting point. It seems that at least seven atoms are needed before something like a liquid behaviour is seen in numerical simulations.

It is in principle easy to measure the caloric curve for bulk materials. One puts the material into a thermally isolated container (see Fig. 2), adds energy E and measures the temperature T. After compensation for the thermal mass of the container, one can construct the caloric curve, $T = T(E)$, or $E = E(T)$.

For a large system, the caloric curve exhibits a step at the melting temperature T_{melt}. The height of the step is the latent heat q. Energy can be added to the system at T_{melt} without an increase of temperature; all the energy is used to destroy the regular solid structure, and thus increase the entropy. This fact is widely used when drinks are cooled by ice cubes. The large latent heat of ice makes it an efficient coolant. Only when the system has become completely liquid, the temperature will rise again upon addition of energy. Note, that one has phase separation for a large sample, *i.e.* water and ice coexist together at T_{melt}. As discussed below (Sect. 6.1), phase separation is not possible for a sufficiently small system. The derivative of the caloric curve is called the heat capacity $C(T)$ or also the specific heat [7]:

$$C(E) = \frac{\partial E(T)}{\partial T} \, . \tag{2.1}$$

It has a delta function like peak at the melting temperature for a bulk system.

It is possible today to construct "nano-calorimeters" using the techniques of the semiconductor chip industry, which carry the principle of Figure 2 to very small particles. In the year 2000 the lower limit of this technique is a particle size of about 1000 atoms [8]. Also electron and X-ray diffraction has been used to study clusters deposited on surfaces [8]. At present these methods are however not applicable to free, mass selected clusters in vacuum for two reasons: first no method of temperature measurement is known in this case, and also the density of mass selected clusters is so small, that it is near to impossible to collect a diffraction signal. Earlier attempts to circumvent this problem can be found in [9–13]. An experiment on electron diffraction on stored cluster ions has recently been published [32], which opens the possibility of doing a type 1 experiment on stored, thermalized ions.

2.2 Calorimetry for free clusters

The thermal experiments on free, mass selected clusters have only been performed with sodium so far. Sodium has been chosen for two reasons: 1) It was the aim of this experiment to work with mass selected clusters, *i.e.* the number of atoms in the cluster is exactly known. In this case one has to use a mass spectrometer, which forces one to work with ions. For non-metallic elements there can be quite a difference between the electronic and geometric structure of a neutral and a positively charged cluster [33].

Due to the delocalized nature of the chemical bond this is not the case in metals. 2) From all the metals available, sodium is the easiest to calculate.

The cluster calorimetry [14–19] method developed in our group does the inverse of the bulk method shown in Figures 2 and 3. Size selected clusters of known temperature are prepared and their energy (more exactly their energy distribution, averaged over an experimental broadening) is measured. The method consists essentially of two steps, as indicated in Figure 4.

Step 1 *Preparation of size selected clusters of known temperature.* Cluster ions are produced and thermalized in a heat bath. A mass spectrometer is used to select a single cluster size. This prepares clusters of known size and known temperature T.

Step 2 *Determination of their energy.* The clusters are irradiated by a laser beam and absorb several photons. This leads to the evaporation of atoms. A second mass spectrometer measures the distribution of the fragment ions produced, which has a characteristic shape as shown

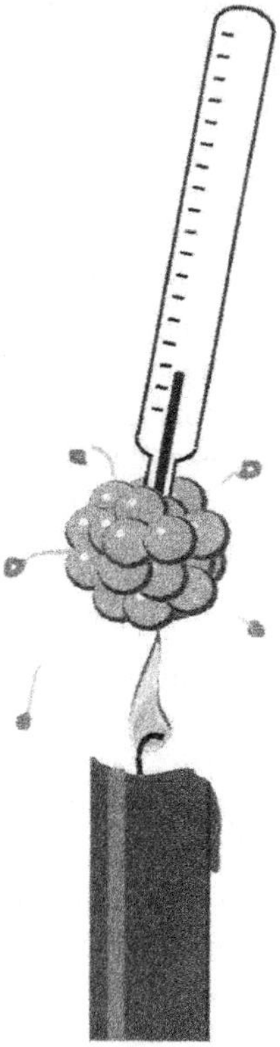

Fig. 3. One would like to make an experiment as shown here, which would be the direct generalization of the setup in Figure 2 to free clusters. Energy is supplied, here by a candle, in an experiment more probably by a laser. The temperature is measured by a thermometer, which could be a small molecule, whose states are interrogated by another laser. This method of temperature measurement has been studied but no reliable temperature scale could be established so far.

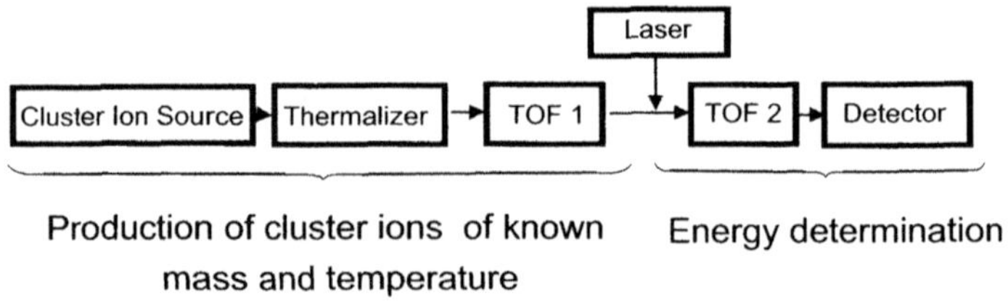

Fig. 4. Schematic of the experiment. Cluster ions are produced and thermalized in a heat bath of known temperature T. A first time of flight mass spectrometer (TOF 1) selects one cluster size. In this way, a cluster of known temperature and known mass is produced (see Step 1 above). The clusters absorb several photons from a pulsed laser, and the distribution of the charged fragments is measured by TOF 2. From these data the energy E before irradiation can be determined, as explained in Figure 5. When energy and temperature are known, the caloric curve $E = E(T)$ can be constructed (Step 2 above).

in Figure 5. Different numbers of absorbed photons lead to clearly separated groups of fragments in the mass spectrum, with the distance between two groups corresponding to exactly one photon energy. This allows one to calibrate the mass scale in terms of energy [14–17].

If the temperature of the heat bath is varied, the inner energy of the selected cluster changes and thus also the number of evaporated atoms. The fragment groups shift on the mass scale as shown in the inset of Figure 5. These twodimensional plots are the primary data of the experiment. From them the caloric curve $E = E(T)$ can be determined directly as indicated in the figure. This method of data treatment has several advantages compared to the one used earlier [14–16]; it is more robust, and avoids the averaging inherent in the use of the finite difference approximation to equation (2.1) used earlier (Eq. (4) of [14]).

3 Experiment

The schematic of the experiment is shown in Figure 4. In more detail, two Time-Of-Flight (TOF) mass spectrometers of the reflectron type are used in tandem. The detector is an Even cup [20] employing up to 30 kV post acceleration for efficient detection of the heavier clusters. The fragmentation laser is an excimer pumped dye laser. The experimental setup is similar to those used by other groups. The novel and at the same time crucial part of the experiment is the source for thermalized cluster ions.

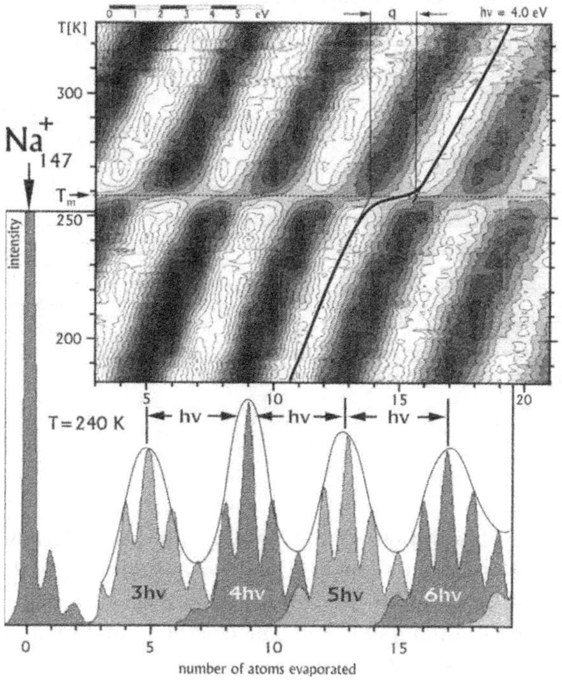

Fig. 5. The energy E of a free cluster can be obtained from its photofragmentation mass spectrum. In the experiment, a Na^+_{147} cluster is prepared at a temperature of 240 K. After absorption of several photons the cluster starts to emit neutral atoms. The intensity of the remaining charged fragments is plotted against the number of evaporated atoms (lower panel). Without laser one observes only the large peak on the left, corresponding to the intact Na^+_{147}. For four absorbed photons one has an approximately Gaussian distribution of photofragments centered around nine evaporated atoms. For five (six) absorbed photons the Gaussian is centered near thirteen (seventeen) ejected atoms. The distance between the maxima of the Gaussians corresponds to exactly one photon energy, which allows an energy calibration of the mass scale. As for all calorimeters, energy calibration is only good up to an additional constant. Inset: the mass peaks are connected by a smooth function, and plotted against the temperature of the thermalization cell of Figures 4 and 6. White implies high intensity. The caloric curve is obtained by connecting the maxima; one example is shown. The other maxima are just shifted by the photon energy. The melting temperature is indicated at the left and the latent heat q at the top. Note, that the lower panel is for 240 K, while the inset is for 180 to 330 K.

3.1 The source for thermalized cluster ions

The source is shown in Figure 6. It consist of two main parts, the cluster

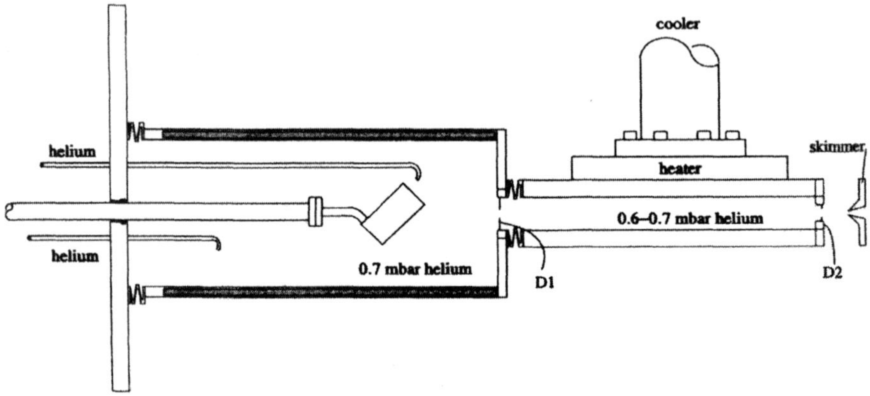

Fig. 6. The gas aggregation source used to produce thermalized cluster ions. Sodium is evaporated from a small, tilted container into a cold stream of He gas. A gas discharge (not shown) is ignited and produces charged species. Aggregation is very effective up to diaphragm D1. The cluster formation stops effectively after D1 due to the low density of sodium atoms there. The helium gas transports the clusters from the source region (left) through D1 into the thermalization region. There they make 10^5 to 10^6 collisions with He atoms, whose temperature T can be adjusted by a heater/cooler. The cluster acquire a *canonical* distribution of energies at the temperature T.

source itself (left part) and the thermalization stage. In the source, sodium is evaporated into a cold, slowly streaming helium gas of about 70 Pa. Mostly sodium atoms and a few dimers leave the boiling liquid, which aggregate in the cold He gas to clusters. The He gas can be added at two points, one at the far side of the aggregation tube, or it can be directly blown into the sodium container. One can optimize the intensity by varying the relative gas flow through the two inlets. A weak electric gas discharge (~100 V, about 1 mA) is ignited with the sodium container working as a hollow cathode. This leads to an efficient production of positively charged sodium cluster ions. Probably, charged molecular dimers, like Na_2^+ or He_2^+ act as condensation germs. Once a dimer is built, cluster formation starts rapidly. The process is similar to cloud formation in nature and will not be discussed here.

The tube containing the aggregation zone has a diameter of about 10 cm. It is cooled by liquid nitrogen, which facilitates the production of smaller clusters, and also leads to a more stable operation. The clusters together with the helium gas leave the aggregation zone through a diaphragm (D1) of variable diameter, which is an iris aperture normally used in optics. The clusters enter a copper tube, which can be cooled down to 30 K and heated up to 600 K. They leave this tube through a second variable iris (D2). Most of the helium is pumped away, and the clusters continue to fly through a skimmer into the first mass spectrometer. Both irises and the skimmer are electrically isolated and carry a small potential (less than 10 V) to compensate for surface charges and plasma effects.

The setup with the two adjustable irises is mechanically somewhat complicated, but at least the diameter of iris D2 must be adjustable during the experiment if one wants to scan over a sufficiently large temperature range. The viscosity of a gas scales as $T^{-1/2}$ and the gas flux through small holes thus diminishes with increasing temperature. The cluster intensity is a sensitive function of the gas flux, and typically every 30 K the otherwise automatic measuring process has to be interrupted in order to optimize the beam intensity by adjusting iris and gas flow.

4 Caloric curves

The primary results of the experiment are two-dimensional data fields as shown in Figure 5. One obtains the caloric curve by connecting the maxima of the fragment distributions. Figure 7 shows an example. A more detailed analysis of what is actually measured is given below (Sect. 5.3).

At low temperatures, the caloric curve increases roughly linearly, and then shows a change in slope which is due to the melting process. Near the melting temperature a relative large amount of energy is needed for a small increase of temperature. The sudden jump of bulk systems is smoothed to a finite width. The melting temperature is indicated. The latent heat q is equal to the increase in the caloric curve near T_{melt}. For higher temperatures, a nearly linear increase is again observed. The following data can be read off Figure 7: 1) the melting temperature T_{melt} as given in Figure 8, 2) the latent heat given in Figure 9, and 3) more generally, the specific heat as a function of temperature. In the temperature range studied, the heat capacity (the derivative of the caloric curve with respect to the temperature) is nearly constant below and above T_{melt}. It is generally higher above T_{melt}. For smaller cluster sizes, melting temperature and latent heat diminish. We have measured several caloric curves, (*e.g.* Na_{70}^+ and Na_{93}^+) where there seems to be only a bend (and not a smoothed out step) in the caloric curves. Data from these curves are not included below.

At a temperature above 400 K, the clusters are so hot that they can evaporate atoms without having absorbed a photon. Thus they are no longer thermalized when they arrive in the laser interaction zone. Therefore

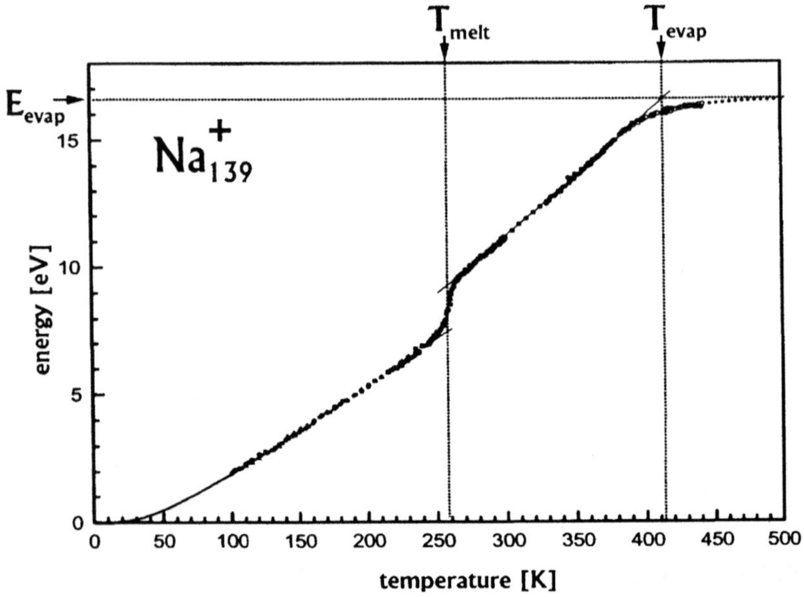

Fig. 7. Caloric curve for Na^+_{139}. Below 100 K the curve has been extrapolated using the bulk result. The soft increase near 260 K is due to the melting process. Melting temperature and latent heat can be easily read off the curve. Above 400 K the data deviate from the caloric curve as the cluster can no longer be thermalized.

the very high temperature part of the curve does not belong to the caloric curve, but describes an evaporative ensemble [21], whose temperature and energy can be read off Figure 7.

From the discussion below (Sect. 5.3), it will become apparent that Figure 7 does show for temperatures below 400 K a microcanonical caloric curve, but averaged over the unavoidable broadening due to the photofragmentation process. Therefore a possible backbending cannot be seen, and the curve looks like a canonical one. Outside $T_{\rm melt}\pm$ 20 K canonical and micro-canonical curves agree within experimental error.

4.1 Melting temperatures

The temperature corresponding to the of largest slope of the caloric curve is identified as the melting point. The data available so far are shown in Figure 8. The melting points show surprisingly large variations: one additional atom can change $T_{\rm melt}$ by up to 10 K. Two main points can be

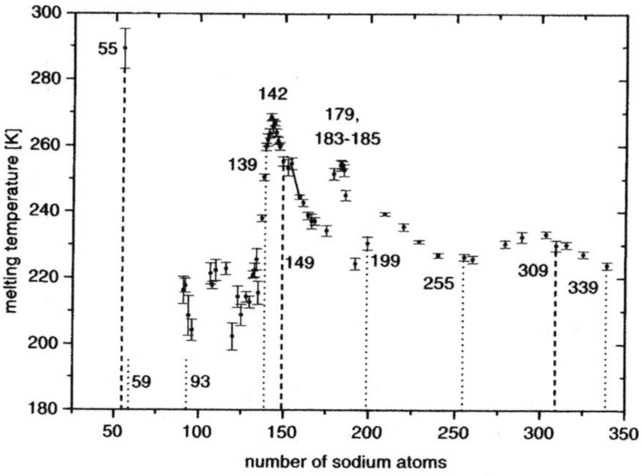

Fig. 8. The melting temperatures of Na_n^+ are plotted against the number of atoms. Large fluctuations are seen whose origin is not well understood. They do not correlate with electronic (dotted line) or geometric (dashed) shell closings, which are indicated in the figure. The bulk melting temperature is 371 K. Note, that the zero is suppressed.

observed: 1) the melting temperatures are about one third lower than in the bulk, and 2) they fluctuate by ±50 K. From less complete data it had earlier been conjectured that the melting points are high, if electronic and geometric shell closings are close to each other. This is no longer supported by the new data covering a wider mass range.

Several calculations [6,22–25] have been performed on melting of sodium clusters, but the overall structure of the data in Figure 8 has not been reproduced so far. The number of atoms for the maxima and minima in Figure 8 do not generally correspond to any known shell closings, be they electronic or atomic in origin. Also, it might be that the cluster changes its geometry near the melting temperature, as has been observed in a simulation on gold clusters [5].

Many calculated melting temperatures for a variety of cluster sizes with less than 30 atoms show a surprisingly simple scaling with the parameter $S = \overline{E}_{\text{int}}/\overline{E}_{\text{sur}}$, where $\overline{E}_{\text{int}}$ and $\overline{E}_{\text{sur}}$ are the mean potential energies of the internal atoms and the surface atoms, respectively [26]. It would be interesting to see if the larger clusters studied here behave similarly.

For two cluster sizes (70 and 93 atoms) the caloric curve does not show an increase like in Figure 7, but only a bend, which can be either interpreted as melting with a very small latent heat, or as the lower temperature of

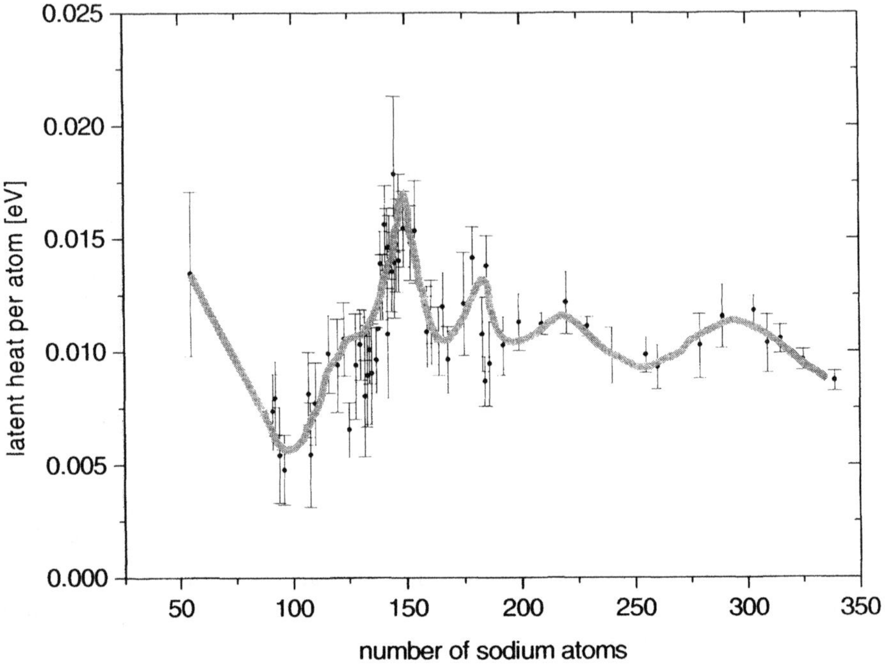

Fig. 9. The latent heat per atom is plotted against the number of atoms in the cluster. Large fluctuations are visible which correlate with those of the melting point. The line through the data is only drawn to guide the eye. The bulk value is 27.5 meV.

an extended melting range. No upper range is found in the data until the clusters evaporate at 400 and 410 K, respectively. These data are not included in Figure 8.

4.2 Latent heats

Classically, the latent heat is the energy to destroy the lattice at the melting temperature. Here it is measured as indicated in Figures 5 and 7 as the height of the increase of the caloric curve near T_{melt}. The vertical scale in Figure 7 is energy calibrated, so that the latent heats are obtained absolutely. The data are more noisy than those in Figure 8, but they seem to indicate a correlation of the extrema with those of the melting points, at least for the data above $n = 100$. This correlation is physically plausible as the higher the melting point, the more the crystal lattice can withstand thermal energy, *i.e.* the higher the latent heat.

4.3 Other experiments measuring thermal properties of free clusters

There is quite a history of earlier experiments trying to measure the melting behaviour of free cluster. The groups of Buck [9], Even/Jortner [10] looked for spectroscopic evidence. Electron diffraction from a not mass selected supersonic expansion gives Debye–Scherrer like diffraction rings, the intensity of them being a measure of cluster temperature. This was pioneered by the Farges/Torchet group [27] and later intensively studied Bartell *et al.* [12].

The Martin group [11] were the first to publish a size dependence of the melting temperatures of free clusters. They showed that the structure on mass spectra of large sodium clusters depends sensitively on the temperature, if the photon energy is chosen appropriately. The disappearance of the structure was interpreted as being due to melting.

Another method has been used to measure melting temperatures of free clusters. It has been applied to small tin clusters with the surprising result that their melting point is higher than the bulk one [28]. In this experiment, cluster ions are injected into a helium gas, and are pulled by an electric field through the gas. They make very many collisions, which produces an effective friction force. Clusters having a small collision cross section make fewer collisions, thus experience a smaller friction force and arrive first. Small Sn-clusters have a rather elongated structure which should change to near spherical upon melting. A change in collision cross section can thus be expected upon melting. No signature of melting is observed, so that the authors conclude that tin cluster ions containing 10 to 30 atoms have a melting point which is at least 50 K above the bulk one [28, 29].

The experiments discussed above all belong to class 1, as defined in the introduction to Chapter 2, *i.e.* some physical property (optical or mass spectrum, diffusion cross section...) is studied as a function of temperature. The only other experiment belonging to class 2, has been used to measure caloric curves of free tin clusters [30]. These were produced by a laser ablation source using a pulsed nozzle whose temperature is variable. The authors estimate that nozzle and cluster temperature deviate by 10 to 20 K only. Neutral clusters are studied so that no mass selector can be used. The distribution of cluster sizes is not given, but similar experiments give a $\delta m/m$ not smaller than 60%. The energy is measured very differently. The cluster impinge on a sensitive pyroelectric foil, whose temperature increase leads to a measurable voltage jump. The measured caloric curve looks very similar to the one given in Figure 7. Only the solid/liquid transition region is broader due to the cluster size distribution. The interpretation of the experiment has been questioned [31].

5 A closer look at the experiment

It was discussed above how caloric curves, melting points and latent heats can be directly obtained from the temperature dependence of the mass spectra. Here we will take a closer look at the experiment and analyse it in more detail. This is necessary, first to understand better what kind of experiment is actually performed, and second to extract more detailed information from the data.

5.1 Beam preparation

Cluster ions are produced and thermalized in a heat bath, which gives them a canonical distribution of inner energies. For the further discussion, some definitions of basic Thermodyanmics are needed, which are briefly reviewed.

5.1.1 Reminder: Canonical *versus* microcanonical ensemble

In the theory of Thermodynamics one defines several idealized situations, called ensembles, which are particular important or easy to calculate [34]. Two of them are realized in this experiment. In the *canonical* ensemble one has a collection of physical systems (atoms, molecules, clusters, spins...) which have a fixed temperature and a thermal distribution of energies. A physical realization are particles in contact with a large heat bath.

In a *microcanonical* ensemble on the other hand, the energy is fixed. A physical realization is an isolated particle in vacuum, which can neither emit nor absorb photons or atoms. Only for an infinitely large system the two ensembles give the same result, for a finite system there can be differences [4, 6, 35, 36] which will be particular important near phase transitions.

5.1.2 A canonical distribution of initial energies

In the thermalizer of Figures 4 and 6 the helium gas is in contact with the metal tube whose temperature is kept constant. Helium is a good heat conductor at the pressure of about 70 Pa. The He atoms are thermalized at the tube wall and in turn thermalize the clusters by collisions. The clusters make typically 10^5 to 10^6 collisions with the gas, more than sufficient for complete thermalization. Thus each cluster can be considered as a small system in contact with a large heat bath, and thus corresponds to a canonical ensemble. Theory tell us that in this case the small system has a *canonical* distribution of energies $P_{\mathrm{T}}(E)$, which is given by the density of states of the cluster $\Omega(E) = \exp\{S(E)/k_{\mathrm{B}}\}$ times the Boltzmann factor [34].

$$P_{\mathrm{T}}(E) \propto \Omega(E) \exp\{-E/k_{\mathrm{B}}T\} = \exp\{S(E)/k_{\mathrm{B}} - E/k_{\mathrm{B}}T\} \tag{5.1}$$

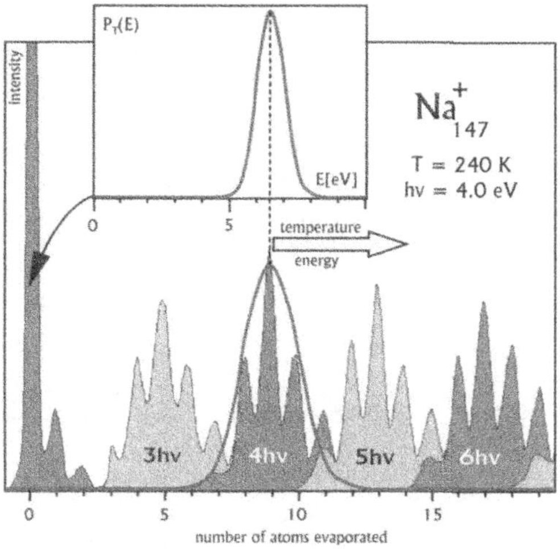

Fig. 10. Photofragmentation spectrum of Na^+_{147}. The intensity of each cluster ion is plotted against the number of atoms lost. Without laser one sees only the high peak at the left, whose distribution of inner energy $P_T(E)$ is given in the inset. The photon energy of $h\nu$ is more than four times the dissociation energy D_n so that the mass groups corresponding to a fixed number of absorbed photons are clearly separated. The mass peaks shift to the right with increasing temperature or photon energy, as indicated. The envelop of the mass peaks, indicated by the solid line, is the convolution of the initial thermal distribution $P_T(E)$ with the width $P_{evap}(E)$ due to the fragmentation process, as given by equation (5.6).

where $S(E) = k_B \ln \Omega(E)$ is the entropy, and k_B is Boltzmann's constant. The inset to Figure 10 shows an example for $P_T(E)$; far from the melting temperature it is nearly Gaussian, near a phase transition $P_T(E)$ can become bimodal as seen in Figure 11b. Equation (5.1) is quite powerful, as it gives a relation between temperature, energy and entropy. One can, for example, take the measured caloric curve $E = E(T)$, plug it into equation (5.1), and obtain entropy and $P_T(E)$.

5.1.3 Free clusters in vacuum, a microcanonical ensemble

High vacuum is needed to operate a mass spectrometer, so that the clusters have to be transferred to a low pressure for mass selection. The aim of the experiment is to measure $P_T(E)$. Thus the cluster should not make any collisions between the thermalizer and the laser interaction zone which

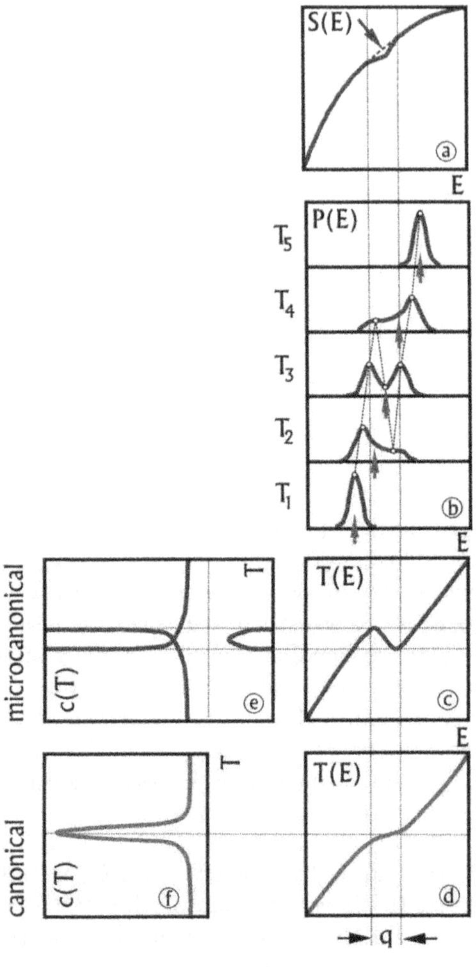

Fig. 11. a) Near a phase transition a small system can show a curious structure in the entropy, a dent where the curvature is negative. This "convex intruder" (arrow) is quite small but has far reaching consequences. b) The energy distributions $P(E)$ become bimodal near the phase transition T_3. c) For the microcanonical caloric curve one has to plot the energy of the extrema of $P(E)$ against the temperature, which results in "backbending" due to the bimodality. d) The canonical caloric curve is a plot of the mean energy [vertical arrows in b)] against the temperature, which gives always a monotonic curve. Its derivative, f) is always positive, while the derivative of the microcanonical curve [e] can become negative.

could alter this distribution. Care is taken to ensure this as far as possible, *e.g.* all nets, safe one, are removed from the ion flight path and the background pressure is kept sufficiently low. Then the energy of each *single* cluster stays effectively constant after it has left the heat bath, and one has a microcanonical ensemble. Averaging over many clusters, one regains $P_T(E)$ of equation (5.1). The energy of a single cluster is a sample of this distribution.

5.2 Analysis of the fragmentation process

The second step of the experiment (see introduction to Chap. 2 and Fig. 4) is to measure the energy E, more exactly the distribution $P_T(E)$ of equation (5.1). Photofragmentation is used for want of a better method. The processes involved will be discussed now.

5.2.1 Photo-excitation and energy relaxation

After mass selection the cluster absorbs several photons of energy $h\nu$. The excimer pumped dye laser has a pulse length of about 10 ns, which is orders of magnitude longer than electron-electron and electron-vibrational relaxation times [37]. The photon induces an excitation of the plasmon type [38–41] which decays through a chain of processes according to the following scenario: the collective plasmon excitation decays into a single electron excitation, which produces through electron-electron collisions a hot electron gas. This is turn couples to the vibrations. Only after complete relaxation, the next photon is absorbed. As the photon energy is smaller than the energy needed to eject an electron, no electron ejection is seen with a nanosecond laser under low fluence. Only if the time between the absorption of two photons becomes of the order of the electronic relaxation time, one observed electron emission. But even under these much more violent excitation, no fast emission of atoms can be observed [37].

5.2.2 Mapping of the energy on the mass scale

The original distribution $P_T(E)$ of internal energy is shifted by the absorption of n photons of energy $h\nu$ to $P_T(E - nh\nu)$. As the number of photons absorbed is not fixed, one obtains for the distribution of internal energy $f(E)$ after the laser interaction:

$$f(E) = \sum_i a_i P_T(E - ih\nu), \tag{5.2}$$

where the coefficient a_i reflect the probability of absorbing i photons. Once the energy is sufficiently high, the cluster starts to evaporate atoms [42] on the time scale of this experiment. The more energy a cluster carries before photoexcitation, the more atoms it will evaporate afterwards. Thus, the mass distribution corresponding to a fixed number of absorbed photons is a map of its internal energy distribution $U(E)$ as given by equation (5.2). The details of the mapping process are discussed in the caption to Figure 10.

5.2.3 Broadening of the mass spectra due to the statistics of evaporation

The evaporation rate $k_n(E)$ is approximately given by

$$k_n(E) = \nu g \exp\{-sD_n/U(E)\}\cdot \tag{5.3}$$

Here ν is a typical frequency (10^{12} to 10^{13} Hz), $s = 3n - 6$ is the number of vibrational degrees of freedom, g is a degeneracy factor usually taken as the number of surface atoms, and D_n is the dissociation energy of a cluster of n atoms. The time between laser excitation and ejection of the first atom is $t_n \approx k_n(E)^{-1}$, which decreases dramatically with increasing E. At each evaporation step

$$\mathrm{Na}_n^+(E) \longrightarrow \mathrm{Na} + \mathrm{Na}_{n-1}^+(E - D_n - E_{\mathrm{kin}}) + E_{\mathrm{kin}} \tag{5.4}$$

the energy decreases by essentially the dissociation energy D_n. At each step of the fragmentation chain

$$\mathrm{Na}_n^+ \xrightarrow{k_n} \mathrm{Na}_{n-1}^+ \xrightarrow{k_{n-1}} \mathrm{Na}_{n-2}^+ \xrightarrow{k_{n-2}} \mathrm{Na}_{n-3}^+ \xrightarrow{k_{n-3}} \mathrm{Na}_{n-4}^+ \cdots \tag{5.5}$$

the fragmentation rate decreases and the lifetime increases. Even if $P_{\mathrm{T}}(E)$ was a δ-function, the mass distribution after the fragmentation has a finite width $P_{\mathrm{evap}}(E)$ due to the statistical nature of the photofragmentation process. That is, more than one decay process can occur in the time window of the experiment; or in other words, the decay process described by k_{n-1} in equation (5.5) will start, before the k_n process is completed.

If the photon energy is sufficiently high (three to four times the energy needed to eject one atom from the cluster), the mass spectra $M_{\mathrm{T}}(n)$ show an oscillatory structure as function of the number n of atoms evaporated, as shown in Figures 5 and 10. The maxima and minima of the oscillations are exactly one photon energy apart, which allows an energy calibration of the mass scale. Thus one can transform $M_{\mathrm{T}}(n)$ to $M_{\mathrm{T}}(E)$. The envelope $M_{\mathrm{T}}(E)$ of the energy-calibrated mass spectrum (see Fig. 10) becomes:

$$M_{\mathrm{T}}(E) \propto P_{\mathrm{T}}(E) \otimes P_{\mathrm{evap}}(E), \tag{5.6}$$

i.e. it is given by the original distribution $P_{\mathrm{T}}(E)$ convoluted by the distribution $P_{\mathrm{evap}}(E)$ due to the evaporation process.

There are several other small contributions to the experimentally observed width: a few percent dimers are emitted at the high temperatures attained, the number of collisions with the background gas is small but not zero. All these effects can be effectively lumped together to produce an effective P_{evap}. As this distribution is quite symmetric, the maxima of $M_{\mathrm{T}}(E)$ and $P_{\mathrm{T}}(E)$ in equation (5.6) coincide for temperatures more than 10 K above or below the phase transition.

5.3 Canonical or microcanonical data evaluation

It was discussed above that the ensemble of clusters has a canonical distribution of initial energies, but that each single cluster arriving in the laser interaction zone corresponds to a microcanonical ensemble. The question thus arises, which kind of experiment is performed, a canonical or a microcanonical one? In fact neither is performed, but something more fundamental is measured: the distribution $P_{\mathrm{T}}(E)$ of which either a canonical or a microcanonical evaluation can be done.

For a plot of a caloric curve one must know E as a function of the temperature. As $P_{\mathrm{T}}(E)$ has a finite width one must have a prescription how to evaluate this energy from $P_{\mathrm{T}}(E)$. One has two different possibilities: 1) if one uses the mean energy

$$E_{\mathrm{can}} = \int E P_{\mathrm{T}}(E)\mathrm{d}E \Big/ \int P_{\mathrm{T}}(E)\mathrm{d}E \tag{5.7}$$

one has a *canonical* situation [34]. One can show analytically, that in this case the canonical caloric curve (Fig. 11d) is a monotonously increasing function [34, 43, 44] and the heat capacity (Fig. 11f) thus always positive. 2) If one uses the extrema E_{ext} of $P_{\mathrm{E}}(T)$ one has a *micro-canonical* situation. This can be seen by differentiating equation (5.1) with respect to the energy. The condition for an extremum of $P_{\mathrm{E}}(T)$,

$$T = (\partial S(E)/\partial E)^{-1} \equiv T_{\mathrm{can}}, \tag{5.8}$$

is identical to the definition of the microcanonical temperature T_{can} as the inverse of the slope of the entropy [34]. Thus, a plot of E_{ext} against the temperature T of the heat bath is identical to the microcanonical caloric curve (Fig. 11c), which is defined as a graph of E against T_{can}.

Except near the phase transition, the caloric curve is quite linear and the distributions $P_{\mathrm{T}}(E)$ are thus nearly Gaussians. Maximum and center of $P_{\mathrm{T}}(E)$ are almost at the same energy. This conclusion is not altered by the statistical evaporation process, as the distribution P_{evap} of equation (5.6) is also nearly symmetric. Thus canonical and microcanonical caloric curves obtained do not differ from each other within experimental accuracy.

The situation is more complex in the temperature range ± 10 K around the melting point. The surprising features occurring there are summarized below in Section 6.1 and in the caption to Figure 11. The distributions $P_T(E)$ are asymmetric (see Fig. 11b, temperatures T_2 to T_4), maximum and mean are no longer identical, and the two caloric curves differ from each other (see Figs. 11c and d). Additionally, neither the maximum nor the mean of $P_T(E)$ agree with the maximum of $M_T(E)$ of equation (5.6). Experimentally, both caloric curve are smeared out due to the convolution with P_{evap}. For the construction of the caloric curve in Figure 7 the maxima of the distribution $M_T(E)$ have been followed; a microcanonical curve is thus obtained, averaged by the statistical broadening.

Summarizing this part one can say that one can either make a canonical or a microcanonical evaluation of the data. All caloric curves given here and published earlier have been obtained by following the maxima in data similar to those in Figure 5. Microcanonical caloric curves are thus obtained, averaged over the distribution $P_{evap}(E)$ of equation (5.6). The averaging is only important close to the phase transition, ± 10 K away from T_{melt} canonical and microcanonical curve are the same within experimental error.

6 Results obtained from a closer look

6.1 Negative heat capacity

There exists a surprising theoretical prediction for a small system: its microcanonical heat capacity can (but must not) become negative. An increase of energy can - under certain conditions - lead to a lower temperature. Every day experience tells us the contrary, if energy is added to a system it will get warmer. But negative heat capacities have since long been known in astrophysics [43, 44], where energy can be added to a star or star cluster which then cools down. A similar effect has been calculated for melting atomic clusters [4, 45] and fragmenting nuclei [35, 46].

Only the positions of the fragment groups have been used so far for the determination of the caloric curves. As discussed elsewhere [47] a negative heat capacity cannot be observed in this case. One thus needs more information, which can be indeed obtained from the measured fragment distributions.

The entropy $S(E)$ of a small system can exhibit a curious structure near a phase transition, a dent with an inverted curvature as shown in Figure 11. This structure has been predicted by theory and has been observed in many numerical simulations [4–6, 35, 45]. The inverted curvature of the entropy has two interesting consequences 1) The microcanonical caloric curve $T_\mu(E)$ gets a negative slope (colloquially called backbending), which means that the corresponding heat capacity becomes negative. 2) The canonical energy distribution $P_{\mathrm{T}}(E)$ shows a bimodal structure [4–6, 35, 45, 48].

Since $P_{\mathrm{T}}(E)$ is mapped onto the shape of the fragment groups, these should become bimodal, too. An observation of the bimodality would therefore be a direct proof of a negative heat capacity. Fragment distributions were studied in detail for Na_{147}^{+}, a cluster for which the effect of a negative heat capacity can be expected to be particularly pronounced due to its very high latent heat [6, 15, 16]. Unfortunately, the bimodality could not be observed directly, as there are broadening mechanisms in the fragmentation process which just smear out these details. One has to apply a trick which enhances the modulation of the fragment spectra: for a certain photon energy the overlap of adjoining fragment groups produces a pattern which allows one an unambiguous decision whether or not the microcanonical caloric curve shows backbending.

The details of data evaluation [47] are a bit technical and will not be discussed here. One can indeed show that Na_{147}^{+} has an entropy with an inverted curvature and consequently a negative heat capacity in the energy range of the phase transition. A quantitative estimate for this heat capacity can be obtained from a least square fit [49] to the data: at the melting temperature an increase of the internal energy of Na_{147}^{+} by 1 eV leads to a concomitant decrease in temperature by about 10 K.

How can this negative heat capacity be interpreted? Upon melting, a large system converts added energy completely into potential energy, reducing continuously the fraction of its solid phase. The kinetic energy and thus the temperature remain constant. A small system, on the other hand, tries to avoid partly molten states and prefers to convert some of its kinetic into potential energy instead. Therefore the cluster can become colder, while its total energy increases.

Negative heat capacities have now been found for melting clusters, fragmenting nuclei, and astronomical objects. What do these widely different systems have in common? The answer is that in these systems energy is not an extensive quantity; *i.e.* if such a system is divided into arbitrary subsystems, the total energy is not simply the sum over the subsystems. The interaction between the subsystems has to be taken into account [34, 44, 50]. For example, in stars it is impossible to neglect the gravity between parts of the system [43, 44]. Similarly in clusters and nuclei the interaction between subsystems is not negligible due to their small size.

6.2 Entropy

The entropy $S(E)$ can be obtained by integrating equation (5.8):

$$S(E) = \int \mathrm{d}E/T(E). \tag{6.1}$$

Inserting the inverse of the measured caloric curve $T = T(E)$, one can calculate the entropy (up to an additional constant) from the experimental data. The result for Na_{147}^{+} looks like the one given in Figure 11a. The dent is surprisingly small, only about 3×10^{-5} eV/K at a total value of $S(n = 138, T_{\mathrm{melt}}) \approx 15$ eV/K. This method is the only experimental one, which allows one to determine entropies for clusters.

Rewriting equation (6.1) one obtains:

$$S(T) = \int \mathrm{d}T c(T)/T, \tag{6.2}$$

where c is the heat capacity defined in equation (2.1) [7]. The measured caloric curves are piecewise quite linear and the heat capacity thus constant there. This gives a $c \ln T$ dependence of the entropy and for the density of states $\Omega \propto T^c$. As the heat capacity is about proportional to the number of degrees of freedom $(3n - 6)$ one sees how dramatically the density of states increases with increasing number n of particles in a cluster.

In principle, one could also say that this experiment measures entropy directly, as the energy distribution $P_{\mathrm{T}}(E)$ produced by the source is just the exponential of the entropy times the Boltzmann factor (see Eq. (5.1)), and this $P_{\mathrm{T}}(E)$ is mapped by the fragmentation process onto the mass spectrum. Practically, the convolution integral of equation (5.6) cannot be inverted due to numerical noise problems, but outside ± 10 K of T_{melt} it can be solved numerically.

7 Unsolved problems

For several interesting problems there is no or only a partial answer today:

1. how does the caloric curve look at low and very low temperatures? A lot of interesting information is expected there. For bulk matter, the phonon heat capacity goes as T^3 while the electronic one scales as T for a metal, becoming the dominant one at sufficiently low temperature. For finite systems, one has a gap between ground and first excited state both in the electronic and vibrational density of states. Thus both heat capacities must go exponentially to zero for very low temperatures. As the vibrational gap is much smaller than the electronic one, the electronic heat capacity will be exponentially small

compared to the vibrational one at very low temperatures, even for metal clusters;

2. are there other cluster sizes with negative heat capacity near the melting transition? This is expected as many calculations have seen this effect;

3. the experiments have been carried out so far with positively charged sodium clusters. What would be the results for neutral, negatively, or doubly positively charged clusters?

4. how can the method discussed here be generalized to other elements or even molecules? This will not be a problem with the other alkalines. But for many other elements one runs into a problem: the photon energy must be at least three to four times the dissociation energy D, otherwise the modulation of the mass spectra seen in Figures 5 and 11 becomes too weak, and no caloric curve can hence be extracted from them. Dissociation energies of several volts are not uncommon, and the necessary powerful lasers are often not available at the required high photon energies. Also they will lead to ionization, further complicating the analysis;

5. can one study the caloric curve of the liquid to gas transition for a cluster? Experimentally, this is not possible. There seems to be, however, a possibility to construct this curve;

6. can other types of phase transitions be measured, *e.g.* the magnetic one, where de Heer and coworkers have published a beautiful first attempt [13]. Unluckily, the ferromagnetic materials have high dissociation energies, and one runs into the problem discussed in point 4 above.

8 Summary and outlook

The melting of a small finite system in vacuum has been discussed. The experiments have been carried out with positively charged sodium clusters, but many results should be independent of the chemical nature of the element under study and thus be quite general in nature. The main results can be summarized as:

1. the caloric curve of free, mass selected clusters in vacuum has been measured. The measuring process can be divided into two parts:

   - sodium cluster ions, Na_n^+, are thermalized, which gives them a canonical distribution of internal energies. One cluster size is selected. This prepares a cluster of known temperature T, containing an exactly known number of constituents;

- the selected cluster is irradiated with a laser, and the distribution of photofragments is measured as a function of the cluster temperature. From this the total internal energy E of the cluster can be determined.

Knowing E and T, one can construct the microcanonical caloric curve $E = E(T)$, averaged over an experimentally unavoidable broadening;

2. the melting point, and the latent heat of fusion can be directly read off the caloric curve;

3. the melting points of Na_n^+, $55 \leq n \leq 340$:

   - are about one third lower than in the bulk;
   - they fluctuate by about ±50 K;
   - the physical origin of fluctuations is not known;

4. the latent heat of fusion shows similar fluctuations as the melting point.

5. the microcanonical heat capacity is negative near T_{melt} for Na_{147}^+. Theory indicates that this should be not an uncommon behaviour for other clusters, but the broadening due to the photofragmentation has so far precluded its observation, save for this one favourable case;

6. the entropy of the size selected cluster can be determined, save near T_{melt}.

There are several other fields of science where similar phenomena are observed or applied. The atomic *nucleus* is a finite system for which phase transitions have been studied intensively. Due to its large zero point motion, the nucleus does not become solid. But it shows a liquid to gas transition, which has interesting similarities to the cluster case [46], *i.e.* it can have also a negative heat capacity. The determination of energy and temperature are more involved in the nuclear case.

Finally, are there any applications of the size dependence of the melting point of small particles? This plays an important role in the softening and melting of polymers [51]. Even a medical application has recently been proposed [52]. A drug could be encapsulated in or bound to a tiny particle, whose melting point is adjusted *via* its size to be just above the body temperature. By an external warming of a specific part of the body the particles could be forced to release the medicine exactly there. One can thus have a high drug concentration in some part and a very small one in another part of the body. As all drugs have unwanted side effects, the potential of such a technique would be great.

Evidently, the work presented results from the cooperative effort of a whole team of people. Contributed have (in alphabetical order): Jörn Donges, Thomas Hippler, Bernd von Issendorff, Werner Kronmüller, Robert Kusche, Ralph Schlipper, and Martin Schmidt. Financial support came from the Deutsche Forschungsgemeinschaft through SFB 276.

References

[1] P. Pawlow, *Z. Phys. Chem.* **65** (1909) 1.

[2] S.L. Lai, J.Y. Guo, V. Petrova, G. Ramanath and G.L. Allen, *Phys. Rev. Lett.* **77** (1996) 99.

[3] S. Berry, *Scient. Amer.* **263** (1990) 50.

[4] P. Labastie and R.L. Whetten, *Phys. Rev. Lett.* **65** (1990) 1567.

[5] C.L. Cleveland, U. Landman, T.G. Schaaf, M.N. Shafigullin, P.W. Stephens and R.L. Whetten, *Phys. Rev. Lett.* **79** (1997) 1873.

[6] F. Calvo and F. Spiegelmann, *J. Chem. Phys.* **112** (2000) 2888.

[7] It is not necessary to differentiate between heat capacity of constant volume or pressure, as these are very nearly the same for a solid or liquid.

[8] M.Yu. Efremov *et al.*, *Phys. Rev. Lett.* **85** (2000) 3560.

[9] U. Buck and I. Ettischer, *J. Chem. Phys.* **100** (1994) 6974.

[10] U. Even, N. Ben–Horin and J. Jortner, *Phys. Rev. Lett.* **62** (1989) 140.

[11] T.P. Martin, *Phys. Rep.* **273** (1996) 199.

[12] J.W. Hovick and L.S. Bartell, *J. Mol. Struct.* **413** (1997) 615.

[13] A. Hirt, D. Gerion, I.M.L. Billas, A. Châtelain and W.A. de Heer, *Z. Phys. D* **40** (1997) 160.

[14] M. Schmidt, R. Kusche, W. Kronmüller, B.V. Issendorff and H. Haberland, *Phys. Rev. Lett.* **79** (1997) 99.

[15] M. Schmidt, R. Kusche, B.V. Issendorff and H. Haberland, *Nature* **393** (1998) 238.

[16] R. Kusche, Th. Hippler, M. Schmidt, B.V. Issendorff and H. Haberland, *Eur. Phys. J. D* **9** (2000) 1.

[17] M. Schmidt *et al.*, in *The Physics and Chemistry of Clusters* Proceedings of Nobel Symposium 117 (World Scientific, Singapore, 2001), p. 326.

[18] G. Bertsch, *Science* **277** (1997) 1619.

[19] R.S. Berry, *Nature* **393** (1998) 212.

[20] See Chapter 3.2 of [53].

[21] M.F. Jarrold, Part 2.7 of [53].

[22] A. Aguado, J.M. Lopez, J.A. Alonso and M.J. Stoll, *J. Chem. Phys.* **111** (1999) 6026.

[23] A. Rytkonen, H. Hakkinen and M. Manninen, *Phys. Rev. Lett.* **80** (1998) 3940.

[24] N. Ju and A. Bulgac, *Phys. Rev. B* **48** (1993) 2721.

[25] A. Aguado, J.M. Lopez, J.A. Alonso and M.J. Stott, *J. Phys. Chem. B* **105** (2001) 2386.

[26] Y.J. Lee, E.-K. Lee, S. Kim and R.M. Nieminen, *Phys. Rev. Lett.* **86** (2001) 999.

[27] J. Farges, M.F. de Feraudy, B. Raoult and G. Torchet, *Surf. Sci.* **106** (1981) 95.

[28] A.A. Shvartsburg and M.F. Jarrold, *Phys. Rev. Lett.* **85** (2000) 2530.

[29] H. Haberland, *Physics World* (December 2000), p. 27.

[30] Th. Bachels, H.-J. Güntherodt and R. Schäfer, *Phys. Rev. Lett.* **85** (2000) 1250.

[31] R. Kofman, P. Cheyssac and F. Celestini, *Phys. Rev. Lett.* **86** (2001) 1388.

[32] S. Krückeberg, S. Schoos, M. Maier–Borst and J.H. Parks, *Phys. Rev. Lett.* **85** (2000) 4494-4497.

[33] H. Haberland, B.V. Issendorff, Th. Kolar, H. Kornmeier, Ch. Ludewigt and A. Risch, *Phys. Rev. Lett.* **67** (1991) 3290; H. Haberland, B.V. Issendorff, J. Yufeng and Th. Kolar, *Phys. Rev. Lett.* **69** (1992) 3212.

[34] Landau–Lifshitz, *Statistical Physics*, Vol. 5 of *Course of Theoretical Physics* (Pergamon Press, London, Paris, 1958).

[35] D.H.E. Gross, *Rep. Prog. Phys.* **53** (1990) 605.

[36] A. Hüller, *Z. Phys. B* **93** (1994) 401.

[37] R. Schlipper, R. Kusche, B. von Issendorff and H. Haberland, *Appl. Phys. A* **72** (2001) 255, and Ph.D. Thesis, R. Schlipper, unpublished.

[38] See the contribution of G. Bertsch in this book.

[39] M. Brack, *Rev. Mod. Phys.* **65** (1993) 677.

[40] W. de Heer, *Rev. Mod. Phys.* **65** 1993 611.

[41] H. Haberland, in *Metal Clusters*, edited by W. Ekardt (Wiley, 1999).

[42] Mainly atoms are evaporated but also about 3% of dimers. This does not change any conclusions discussed here, it only leads to an additional broadening of the mass spectra.

[43] W. Thirring, *Z. Phys.* **235** (1970) 339.

[44] D. Lynden–Bell, *Physica A* **263** (1999) 293.

[45] M. Bixon and J. Jortner, *J. Chem. Phys.* **91** (1989) 1631.

[46] M. D'Agostino *et al.*, *Phys. Lett. B* **473** (2000) 219.

[47] M. Schmidt, R. Kusche, Th. Hippler, J. Donges, W. Kronmüller, B.V. Issendorff and H. Haberland, *Phys. Rev. Lett.* (2001).

[48] A macroscopic system avoids the inverted curvature by phase separation, *i.e.* being partly liquid, partly solid, as described by the well known van Hove construction [35, 43, 44, 50]. This is not advantageous for a small system, due to the large percentage of atoms at a liquid/solid interface.

[49] M. Schmidt *et al.*, unpublished results.

[50] L. van Hove, *Physica* **15** (1949) 951.

[51] G. Strobl, *Eur. Phys. J. E* **3** (2000) 165.

[52] K. Westesen, *Colloid. Polym. Sci.* **278** (2000) 608.

[53] *Clusters of Atoms and Molecules I*, Springer Series in Chemical Physics, Vol. 52, edited by Haberland (Springer, 1994).

COURSE 3

EXCITATIONS IN CLUSTERS

G.F. BERTSCH

University of Washington, Department of Physics, FM15, Seattle, Washington 98195, U.S.A.

Contents

EXCITATIONS IN CLUSTERS

G.F. Bertsch

1 Introduction

There are two points of view on cluster physics, which are implicit when one asks the questions:

> Do the properties of clusters bridge the physics of atoms to the physics of bulk systems?
>
> Or do clusters form a world of their own, with unique properties seen neither in atoms nor in macroscopic systems?

A positive answer to the first question would give a strong motivation to study clusters in order to develop tools for studying large systems. Because clusters are finite with discrete electronic excitations, their observed properties might presumably be used more effectively to develop better theory. On the other hand, a positive answer to the second question is motivation to study clusters in their own right. In fact, depending on the properties under study, both questions have affirmative answers.

One of the unique aspects of clusters is their magic numbers, which will be discussed in detail in the lectures of Martin and Brack. A sampler of abundance spectra showing the magic number phenomena is displayed in Figure 1.

In the top panel are shown the abundances of sodium clusters in the pioneering 1984 experiment of Knight *et al.* [1]. The numbers 8, 20, 40, are clearly favored. These numbers can be associated with the group SU(3). This is a fancy way of saying that the experiment showed special stability for the closed shells in a three-dimensional harmonic oscillator potential, which happens to carry the symmetry of the group SU(3). The most important point is that spherical shell closures of delocalized electrons convey stability to the system and determines structure. In nuclear physics the discovery of shells of delocalized particles opened the door to more powerful theoretical tools to understand nuclear structure, and the same has happened in metal cluster physics.

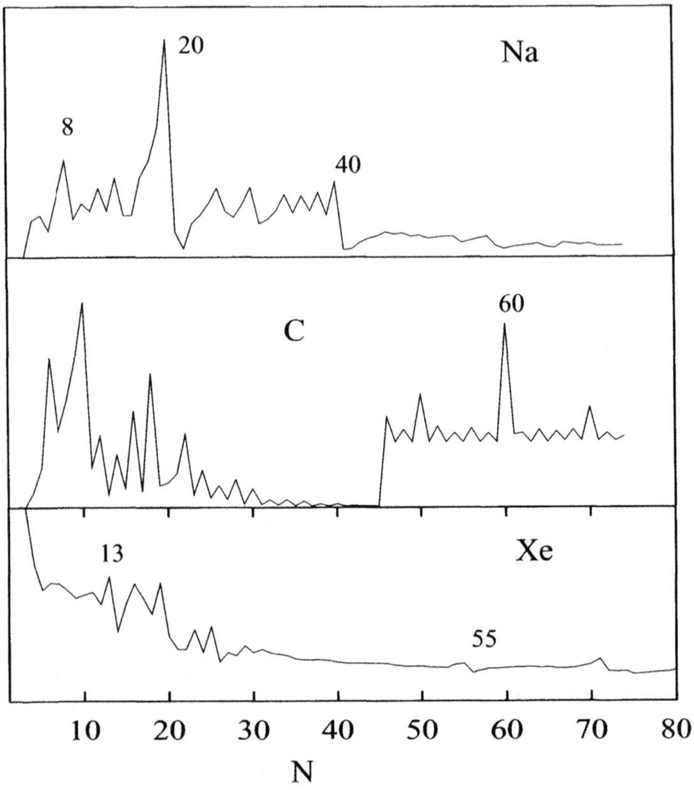

Fig. 1. Abundance spectra in cluster beams. The data are from the following references: carbon [4]; sodium [1]; xenon [3].

Magic numbers of very different origin come from the geometric arrangements of atoms in a cluster. The Mackey icosahedra [2] are the prime example of geometric packing in finite systems. Let us try to pack hard balls into a compact shape. Starting with one at the center, one finds up to twelve can be positioned around the central ball and touching it. However, the packing is imperfect in that the outer layer of balls is not close packed, and contains spaces between at least some of the balls. The space can be arranged symmetrically putting the outer balls at the vertices of an icosahedron. Each face of the icosahedron is defined by a triangle of balls. The Mackey construction adds layers by building a larger triangle on each face. These icosahedral numbers were seen first in a cluster beam experiment on Xe clusters [3], shown in the lower panel of Figure 1. The favored icosahedral numbers are 13, 55, 147, ... However, one sees in this experiment that

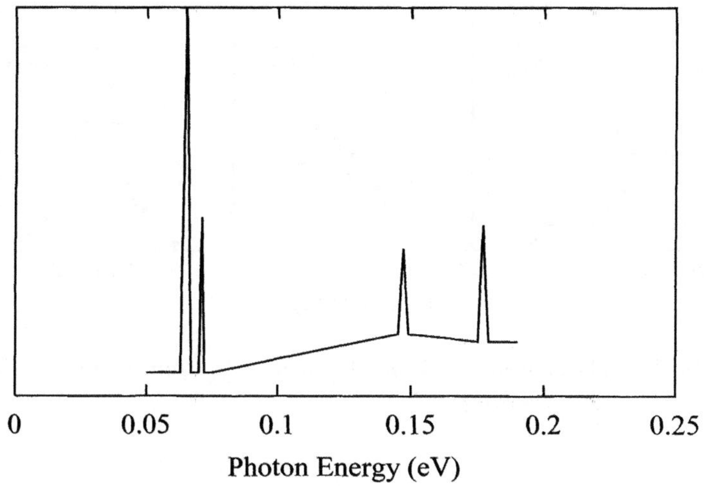

Fig. 2. Schematic infrared absorption spectrum of C_{60}, from reference [5].

the icosahedral numbers hardly stand out from others equally prominent. Martin's lectures will explain how in the basic icosahedral framework other numbers can arise as well.

The middle panel in Figure 1 shows carbon abundances with the famous peak at $N = 60$ from the fullerene molecule C_{60}. In this case, the experimentalists observed that 60 was magic, and deduced from that the structure should be icosahedral, with the carbon atoms positioned like the vertices on a soccer ball. In some respect, C_{60} is an ideal molecule to bridge the atom and the bulk: it is the largest assemblage of atoms possible in which all the atoms are equivalent. It also has the largest possible number of point group symmetries (120). The history of the synthesis of C_{60} is also interesting from another point of view. The original discoverers, Kroto, Smalley *et al.*, were chemists by background but made use of physical techniques–cluster beam apparatus and time-of-flight mass measurement–to first observe the molecule [4]. But the practical bulk synthesis was devised by physicists, Krätschmer and Huffman [5], using extraction techniques taken directly from the chemistry laboratory. Thus one sees in the study of clusters a blurring of the distinctions between physics and chemistry. Once macroscopic quantities became available, it was possible to measure properties that would be extremely difficult otherwise. An example from the original paper is the infrared absorption spectrum, shown in Figure 2. Due the high symmetry of the molecule, there are only four optical active transitions, although the number of vibrations is much larger, $3N - 6 = 174$. We will come back to this spectrum in Section 4.

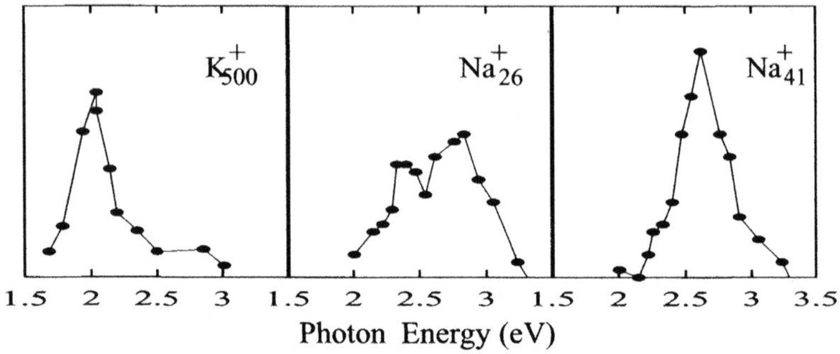

Fig. 3. Surface plasmon in alkali metals. Left, K_{500}^+ from reference [7]; middle and right $Na_{26,41}^+$ from reference [8].

There are many properties of clusters that can be measured in cluster beam experiments. The ionization potential can be measured rather directly in photoionization experiments from the threshold energy. The binding energy of an atom to the cluster is measured indirectly through the evaporation rate and its temperature dependence, using statistical theory. A recent example of the application of the theory to determine the atom separation energy in Na clusters may be found in [6].

The response of clusters to external electromagnetic fields is a large subject, and will be the main topic of my lectures. An external electric field $\mathcal{E}$ induces a dipole moment D; the linear polarizability α is the coefficient of proportionality in the expansion $D = \alpha\mathcal{E} + ...$ This is commonly measured by deflection of a beam in an inhomogeneous electric field. One can also study magnetic clusters by their deflection in an inhomogeneous magnetic field, the classic Stern–Gerlach experiment. Magnetic properties of clusters will be a topic in Pastor's lecture.

The photon absorption cross section is also measured over a frequency range starting from optical frequencies to the very far ultraviolet ($\sim$10 eV photon energy). In simple metal clusters one sees a coherence between the electrons in that there is strong peak in the response involving all of the valence electrons. This is the surface plasmon mode, and I shall have quite a bit to say about it. Figure 3 shows some typical optical absorption spectra of metal clusters. In a spherical cluster, the surface plasmon is sharply defined, as may be seen in the first panel, showing potassium spectra [7]. The plasmon is split in nonspherical clusters, as shown in the second panel. This geometric sensitivity of the plasmon may be understood at many levels, from classical to quantum mechanical. This feature of cluster behavior is also reminiscent of a similar phenomenon in nuclear physics. I will show how an analytic rough description be derived looking at the short-time behavior of the wave function.

The main aim in these lectures is to give the student some familiarity with the basic tools the theorist has at his disposal to study these cluster properties, particularly electronic excitations. In this respect one should emphasize the importance of simple models. For describing large numbers of particles, the use of models that focus on a few degrees of freedom is unavoidable. Even if one's efforts are in large-scale *ab initio* numerical calculations, the models are extremely useful to interpret the numbers and check the numerics. My own cluster research, published in the papers [9–31], has evolved from applications of simple ideas taken from nuclear physics to numerically intensive computations of *ab initio* theory. While I will show some of these *ab initio* results that are won by intensive computer computation, I think one of their main values is to validate simpler models.

Before discussing electronic excitations, I will go through statistical reaction theory, which is indispensable to interpret cluster formation and evaporation, and many electron transfer processes.

2 Statistical reaction theory

An important tool for understanding dynamic processes in clusters, as well as for extracting information about their energetics, is statistical reaction theory. Like statistical mechanics, the theory is well-known in principle. But much of the literature is confusing, and I think it is worthwhile to assemble all the relevant formulas and their derivations. The generic problem is to calculate the emission rate of an excited state of a bound system. The derivation of the statistical rate formula is based on detailed balance, which is one of the fundamental principles of statistical mechanics. If we divide the system's phase space up into microstates, the equilibrium state has all the microstates equally occupied. Looking at the detailed dynamics, the probability flux out of each microstate must equal the probability flux in. The inward flux is relatively straightforward to compute or to characterize empirically. Then making the equality gives us a way to calculate the emission rate. Note that the emission rate does not depend on the system being in equilibrium with its surroundings. It only depends on there being equilibrium among the internal microstates of the system.

We sketch a system, Figure 4, with one degree of freedom singled out, say the position of the particle that will be emitted. All the other degrees of freedom will be treated by imagining the Hamiltonian to be diagonalized, and so represent them by states. In the scattering problem these states with a fixed radial coordinate for the particle are called channels. We now put the

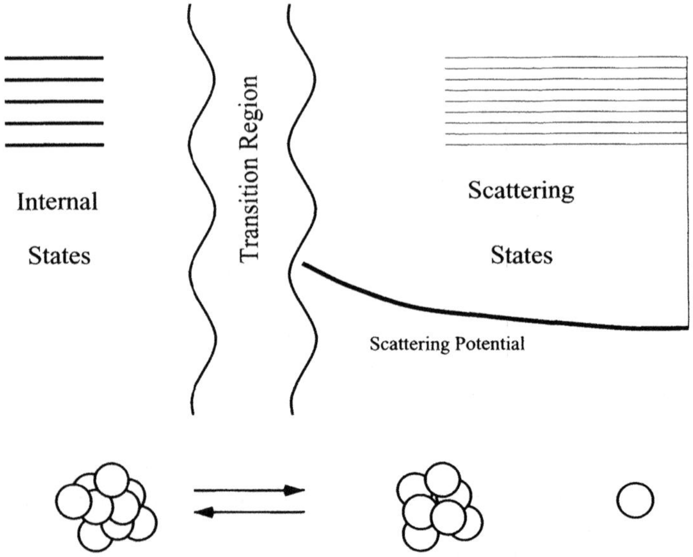

Fig. 4. Separation of internal and external states for the derivation of the statistical decay formula.

system in a box to make it easier to count states for the detailed balance argument. The thick curve in Figure 4 shows the energy of the lowest channel as a function of distance. The states of the system, separated into internal and external states, are shown with light lines. These are not the eigenstates of the system, because we have not coupled the internal to the external states. Let us now count the states and invoke detailed balance. For the external states, let us give the box a radius R. We will count the number of states in an energy interval ΔE, and we need the density of external states with respect to energy.

This is calculated as follows. For a large box, we can take the normalized asymptotic wave function to have the form[1]

$$\phi = \sqrt{\frac{2}{R}} \sin(k_n r),$$

[1] This ignores the scattering phase shift and the angular-momentum-dependent phase of the asymptotic wave function.

where n labels the different states. The density of the external levels $\mathrm{d}E/\mathrm{d}n$ is given by[2]

$$\frac{\mathrm{d}n}{\mathrm{d}E} = \frac{Rm}{\hbar^2 \pi k} .$$

Let us now find the inward current in an energy interval ΔE. The asymptotic external wave function has both incoming and outgoing components, with the incoming component given by

$$\phi = \sqrt{\frac{2}{R}} \frac{\mathrm{e}^{-ikr}}{2} .$$

The current associated with this component is

$$\langle j \rangle = -\frac{\hbar k}{4m} \frac{2}{R} .$$

When the particle reaches the cluster, it may be reflected immediately or it may be transmitted into the interior. Define the transmission coefficient T_c, with $0 \le T_\mathrm{c} \le 1$. Then the flux that goes into the target is the current of each state times the number of states times the occupation probability of each state. Calling the last quantity f_0, the inward flux is

$$f_0 T_\mathrm{c} \langle j \rangle \frac{\mathrm{d}n}{\mathrm{d}E} \Delta E = f_0 \frac{T_\mathrm{c}}{2\pi\hbar} \Delta E.$$

By detailed balance we can equate this to the outward flux, assuming that the internal states have the same occupation probability f_0. Let's define an average decay rate of an inner state, W. The total flux out is given by

$$f_0 W \rho_E \Delta E$$

where ρ_E is the density of internal states of the system. Equating the last two expressions gives the statistical formula for the decay rate W,

$$W = \frac{\sum_\mathrm{c} T_\mathrm{c}}{2\pi\hbar\rho_E} . \tag{2.1}$$

Note that the derivation does not require the system to be in equilibrium with its surrounding, only that there is an internal equilibrium among the internal states.

To use the formula, one needs to know the transmission coefficients as well as the level density of the internal states. In many applications the

[2]This may be derived from the boundary condition that the wave function vanish at the surface of the box[1], $k_n R = n\pi$, taking the asymptotic energy, $E = \hbar^2 k_n^2/2m$.

channel potential energy has a barrier, and one can make the approximation that the transmission coefficients are zero below the barrier and unity above. Then there is a correspondence between the channels and the level density of the daughter system. With these assumptions, one obtains the well-known transition state theory, called RRKM theory in chemistry. In principle the channels are discrete, and at energies close to the barrier energy the decay rate might show a step-like behavior. In fact, this was predicted in nuclear physics in the original fission theory of Bohr and Wheeler, which made use of equation (2.1). However, the discreteness of channels was never clearly seen in nuclear fission, because the quantum mechanical transmission coefficients increase smoothly as the channel opens. But in my last lecture I will show you an example from carbon structures that perfectly shows the individual channels.

The statistical decay formula also requires the density of states of the decaying system, ρ_E. Under ideal conditions when the internal states do not overlap, ($W < (\hbar\rho_E)^{-1}$), the internal states can be explicitly counted. This is the case in nuclear physics for energies close to the neutron emission barrier. For atomic clusters and molecules larger than several atoms, the vibrational density of states is huge compared to the electronic, and there is no hope to see individual internal states in the regime of electronic excitations.

2.1 Cluster evaporation rates

We now apply the formula to evaporation of atoms from a cluster. We consider a cluster having excitation energy E^*, emitting an atom of kinetic energy E_K and leaving the daughter cluster at excitation energy E_i. The energies are related by

$$E^* = E_i + E_K + D \qquad (2.2)$$

where D is the binding energy of the atom to the cluster in the ground state. In general, the atom can be emitted in many angular momentum states, and each one will have its own channel. However, in the end the incoming flux is determined by the inverse reaction cross section. To derive a simple formula, let us assume that the atoms are spinless and that the specific state of the cluster play no role in the absorption cross section. Then the transmissions coefficients depend only on the orbital angular momentum l and the energy of the atom, and the reaction cross section is given by

$$\sigma_{\rm r} = \frac{\pi}{k^2} \sum_l (2l+1) T_l(E_K)$$

where $\hbar^2 k^2/2m = E_K$. Exactly the same sum will appear when one makes an approximate sum over channels in equation (2.1). The channel are distinguished by the internal state of the daughter cluster (including its angular momentum L), the orbital angular momentum of the evaporated atom l, and the total angular momentum of the system J. For a given initial state the total angular momentum is fixed, so the channel sum will involve sums over l and L. The conditions of evaporation are generally such that the maximum l is much less than the typical J. Then the channel sum over L can be enumerating with the familiar rules of angular momentum coupling, $L = J - l, J - l + 1, ..., J + l$, giving $2l + 1$ terms altogether. If the particle has spin with g states, the spin coupling increases the number of channels to $g(2l + 1)$. Then assuming that the internal states of the daughter are independent of L, the formula (2.1) can be expressed

$$W = \frac{g \sum_i \sum_l (2l+1) T_l}{2\pi\hbar\rho_E} . \tag{2.3}$$

Here the i sum is over internal states of the daughter for some typical angular momentum L. The transmission coefficient depends on the i only because the energy of the evaporated particle depends on how much energy is left in the daughter. We now replace the sum over l by the cross section formula, to get

$$W = \frac{g \sum_i k_i^2 \sigma_{\rm r}(E_K)}{2\pi^2\hbar\rho_E} = \frac{gm \sum_i E_K \sigma_{\rm r}(E_K)}{\pi^2\hbar^3\rho_E} . \tag{2.4}$$

Here the evaporated particles momentum k and kinetic energy E_K depend on i through equation (2.2). The differential decay rate with respect to E_K is easily obtained by the replacing the sum over internal states by their level density, $\mathrm{d}\Sigma_i/\mathrm{d}E \to \rho_E$. The formula is,

$$\frac{\mathrm{d}W}{\mathrm{d}E_K} = \frac{gm\rho_{N-1}(E^* - D - E_K) E_K \sigma_{\rm r}(E_K)}{\pi^2\hbar^3\rho_N(E^*)} . \tag{2.5}$$

Here we have changed notation on the level density ρ to distinguish the parent density ρ_N and the daughter density ρ_{N-1} by a subscript. This formula was first derived by Weisskopf to describe neutron decay of excited nuclei [32].

Further reduction of this rate formula requires some specific information about the level densities. One simple limit is to assume that the excitation energy and level densities are high enough that the level density of the daughter system behaves exponentially over the range of the interest in the decay. This permits a parameterization with a temperature: $\rho_N(E^* - D - E_K) \approx \rho_N(E^* - D) \exp(-E_K/T)$, and equation (2.5) can be integrated with respect to E_K. If we further assume that the cross section is constant, the integral is elementary and the total transition rate becomes[3]

$$W = \frac{mT^2 \rho_{N-1}(E^* - D)\sigma_\mathrm{r}}{\pi^2 \hbar^3 \rho_N(E^*)} .$$

We now specialize to level densities associated with vibrations. If the vibrations are harmonic, the level density can be expressed completely analytically by Kassel's formula [33],

$$\rho_N(E) = \frac{E^{s-1}}{(s-1)! \prod_j^s (\hbar\omega_j)} . \tag{2.6}$$

Here s is the number of vibrational degrees of freedom, with $s = 3N - 6$ for triatomic and larger clusters. The sum goes over the vibrations j with their frequencies ω_j. Since the formula will be derived in Wales' lectures, I need not discuss its validity here.

Again assuming that the reaction cross section is constant, we can integrate over the density of states of the daughter system to get the following formula

$$W = \frac{\omega^3}{\pi^2}(s-1)m\sigma_\mathrm{r}\frac{(E^* - D)^{s-2}}{(E^*)^{s-1}} . \tag{2.7}$$

Here ω^3 is the ratio of the products of vibrational frequencies for parent and daughter. The decay rate in essentially this form was originally derived by Engelking [34]. Note that the quantum of action $\hbar$ has dropped out of the formula. Once one takes Kassel's density of states, the remaining physics is completely classical.

For midsize and larger clusters there are many vibrational degrees of freedom, $s \gg 1$. It also usually the case that the evaporation is observed in a regime with $E^* \gg D$. Then the conditions for an exponential approximation are satisfied, and we can write $(E^* - D)^{s-2}/(E^*)^{s-2} \approx \exp(-D/T)$ where T is defined $T = E^*/(s-2)$. Note that the relation between T and E^* is exactly what one obtains classically for the relation between temperature and energy in a system of $s - 2$ oscillators. The decay rate formula then becomes

$$W = \frac{\omega^3 m\sigma_\mathrm{r}}{\pi^2 T}\mathrm{e}^{-D/T} . \tag{2.8}$$

The most important feature of this formula is the extreme dependence of the evaporation rate on D. Let us take the example of sodium cluster of

[3] We have also assumed that the spin of the particle can be neglected ($g = 1$).

size $N \approx 100$. The binding energy of an atom is about $D \approx 1$ eV, and we will approximation ω by the Debye frequency, $\hbar\omega \approx 0.01$ eV. Let us also assume that T in the range of values near room temperature, $T = 0.025$ eV, to evaluate the prefactor in the formula. Although equation (2.8) does not have $\hbar$ in it, it is convenient to express frequencies as equivalent energies. Two more needed dimensional constants: $\hbar = 0.658$ eV-fs is needed to get a decay rate in units of s^{-1}, The mass is conveniently expressed in terms of $\hbar^2/m_{\mathrm{H}} = 4.15 \times 10^{-3}$ eV-Å^2. The inverse cross section is estimated as $\sigma_{\mathrm{r}} = \pi R^2 \approx 300$ Å, that is, assuming that all atoms that hit the cluster will stick. Putting the numbers together, we find

$$W \approx \frac{(0.01)^3(23)(300)}{\pi^2(0.00415)(0.025)} \mathrm{e}^{-D/T} \approx 10^{16} \mathrm{e}^{-D/T}.$$

Cluster beams have flight times of the order of tens to hundreds of microseconds. Thus a measurable evaporation rate requires the exponential suppression factor to be in the range $\exp(-D/T) \approx 10^{-11}$, *i.e.* $D \approx 25T$. We can see from this that a 10% change in D will change W by an order of magnitude. For $T = 0.03$ eV, the excitation energy is $E^* = 9$ eV, and the evaporation leaves the daughter clusters at the energy $E \approx 8$ eV. Thus, the chain of evaporations has very different rates for each stage, and the multiple evaporation spectrum will be very much peaked around the number that corresponds to a lifetime equal to the travel time in the cluster beam. As you will see, Haberland will use this property to accurately measure energy differences between clusters at different temperatures.

2.2 Electron emission

The statistical theory of electron emission is no different in principle, but there can be important differences in the details. Because of the Coulomb interaction, the geometric area may no longer be a good approximation to the reaction cross section.

Haberland will discuss in his lectures measurements of the electron emission in sodium clusters. He will show a very interesting result that the emission takes place when the electrons have equilibrated among themselves but before the equilibration with the vibrations has taken place.

Under these conditions, the state densities in equation (2.5) should be evaluated for the electron degrees of freedom only. The Fermion character of the electrons has two consequences. The two spin degrees of freedom gives rise to an additional factor of 2 in the formula, because there are twice as many external electron states when detailed balance is applied. Also, the level densities are quite different for Fermions. If we assume that the particles behave as a free Fermi gas, the level density has an exponential dependence on excitation energy given by Bethe's formula,

$$\rho_E \sim \exp\left(2\sqrt{\pi^2 NE/2E_\mathrm{F}}\right),$$

where E_F is the Fermi energy of the electron gas. It is interesting to note that in the Fermion case the quantum effects remain visible in the final formula as the factors of $\hbar$ no longer cancel.

A final difference is that the inverse cross section cannot be considered constant, due to the long-range Coulomb interaction between the electron and the (charged) daughter system. The semiclassical formula for the cross section when there is complete absorption at a radius R in the present of a potential field V is

$$\sigma_a b = \pi R^2 \left(1 - \frac{V(R)}{E}\right).$$

Note that this diverges at low energy when the Coulomb potential is attractive. The divergence is only apparent, because the cross section is multiplied by E in the rate formula. This theory of charged particle emission is well known in nuclear theory; it was first applied to clusters in reference [35].

2.3 Radiative cooling

The last process I will discuss is the statistical emission of a photon. To observe radiative cooling in isolated clusters, one needs longer residence times than is provided by the usual molecular beam apparatus. Observations have been reported using ion traps [36] and storage rings [37]. In these experiments, the photons were not observed directly, but rather the effect of the cooling on other processes was seen. Also, it is interesting to note that cluster radiation was proposed as a way to achieve more efficient thermal illumination [38]. We will see that in the spectrum infrared emission is suppressed compared to black body radiation. In the derivation of the photon decay rate, only the transmission factor in the dipole channel ($l = 1$) is significant, but the formula looks very similar to equation (2.3). Photons have two polarization states, making $g = 2$. Also the energy of the photon E_ph and the reduced wave number k are related by $k = E_\mathrm{ph}/\hbar c$, so the final result for The result for the radiation spectrum is

$$\frac{\mathrm{d}W}{\mathrm{d}E_\mathrm{ph}} = \frac{1}{\pi^2}\frac{E_\mathrm{ph}}{\hbar c}^2 \sigma_\mathrm{r} \frac{\rho_E(E^* - E_\mathrm{ph})}{\rho_E(E^*)}. \qquad (2.9)$$

This formula is also well-known in nuclear physics where radiative decays of hot nuclei have been studied for some time. In the cluster context, the formula was first derived in reference [39].

3 Optical properties of small particles

In the remainder of my lectures I will mainly discuss the theory of electronic excitations, but before getting into the equations to be solved, I will list for future reference a number of useful formulas for describing the electromagnetic properties of small particles. Here by small, I mean the wavelength is much larger than the size of the particle, $\lambda \gg R$. Then the electric dipole field dominates the interaction, and the optical response can be described with the dynamic polarizability $\alpha(\omega)$. For a spherical system, the polarizability is defined

$$\alpha_{zz}(\omega) = \frac{e^2}{\hbar} \sum_i |\langle 0|z|i\rangle|^2 \left(\frac{1}{-\omega - i\eta + \omega_i} + \frac{1}{\omega + i\eta + \omega_i} \right), \tag{3.1}$$

with i labeling excited states. For nonspherical systems, one can define polarizabilities for the principle axes and construct a polarizability tensor.

The photon absorption cross section is related to the polarizability by

$$\sigma_{ab} = \frac{4\pi\omega}{c} \mathrm{Im}\alpha(\omega). \tag{3.2}$$

An important property of the dipole response is the Thomas–Reiche–Kuhn sum rule. It may be derived from the operator commutator relation, $[z, [H, z]/2 = \hbar^2/2m_e$. In terms of dipole matrix elements between ground and excited states i, the sum rule reads

$$\sum_i |\langle 0|z|i\rangle|^2 \hbar\omega_i = \frac{\hbar^2}{2m_e} N, \tag{3.3}$$

where N is the number of electrons. Conventionally one defines a dimensionless oscillator strength f_i for a transition as

$$f_i = \frac{2m}{\hbar^2} |\langle 0| z |i\rangle|^2 \hbar\omega_i.$$

Then the sum rule is simply $\sum_i f_i = N$. In terms of the dynamic polarizability, an oscillator strength function can be defined

$$S_f(\omega) = \frac{2m_e\omega}{\pi\hbar e^2} \mathrm{Im}\alpha(\omega) \tag{3.4}$$

and the f-sum is

$$\int \mathrm{d}\hbar\omega S_f(\omega) = f. \tag{3.5}$$

Another handy formula is the expression for f in terms of the integral of the cross section over the photon energy,

$$f = \frac{m_e c}{2\pi^2 e^2 \hbar} \int \sigma_{ab} \mathrm{d}E. \tag{3.6}$$

A practical formula used by chemists takes for the integrand the molar extinction coefficient rather that the cross section; a derivation may be found in my book [18].

3.1 Connections to the bulk

With clusters viewed as a bridge between the atom and the bulk, it is interesting to see how the electronic response connects to the infinite medium response. There the response is characterized by the dielectric function ϵ, which gives the relation between external and internal electric fields at a perpendicular interface: $\epsilon = \mathcal{E}_{\rm int}/\mathcal{E}_{\rm ext}$. There are two connections one can make to the finite system polarizability α. The first is the dielectric function for an cubic crystal composed of polarizable particles, given by the Clausius-Mossotti relation,

$$\epsilon = \frac{1 + 8\pi\alpha n_0/3}{1 - 4\pi\alpha n_0/3} = \frac{1 + 2\alpha/r_0^3}{1 - \alpha/r_0^3} \tag{3.7}$$

where n_0 is the number density of the particles in the crystal. For convenience in the second equation the formula is expressed in terms of the Wigner-Seitz radius r_0. This is the radius of a sphere whose volume corresponds to the volume per particle in the medium, $4\pi r_0^3/3 = 1/n_0$. A similar length $r_{\rm s}$ is in common use to specify the electron density $n_{\rm e}$. This is defined $4\pi r_{\rm s}^3/3 = 1/n_{\rm e}$ and is quoted in atomic units (lengths in Bohr radii (0.529 Å); energies in Hartrees (27.2 eV).

The other connection is the polarizability of a cluster considered as a dielectric sphere. It is a textbook exercise in electrostatics to show that the polarizability of a dielectric sphere of radius R is given by

$$\alpha = R^3 \frac{\epsilon - 1}{\epsilon + 2}\cdot \tag{3.8}$$

We shall call this the Mie theory of the polarizability. One more handy formula is the dielectric function of a simple conductor. In the Drude model (independent electrons subject to a frictional force depending linearly on velocity) the dielectric function is

$$\epsilon(\omega) = 1 - \frac{\omega_{\rm p}^2}{\omega(\omega + i\gamma)} \tag{3.9}$$

where γ is the frictional rate constant and $\omega_{\rm p}$ is the plasmon frequency,

$$\omega_{\rm p}^2 = \frac{4\pi e^2 n_{\rm e}}{m_{\rm e}} . \tag{3.10}$$

In atomic units the formula is simply $\omega_{\rm p}^2 = 3/r_{\rm s}^3$.

Let us try this in equation (3.8). After some trivial algebra one finds

$$\alpha(\omega) = R^3 \frac{\omega_{\rm p}^2/3}{-\omega^2 + i\omega\gamma + \omega_{\rm p}^2/3} . \tag{3.11}$$

There are two obvious things to notice about this formula. The first is that the zero-frequency limit gives $\alpha = R^3$, which is just the formula for the polarizability of a conducting sphere. The second is that there is a resonance where the denominator vanishes, near the frequency

$$\omega_{\rm s} = \frac{\omega_{\rm p}}{\sqrt{3}} = \sqrt{\frac{e^2}{m r_{\rm s}^3}} . \tag{3.12}$$

This is the surface plasmon, and we shall call equation (3.12) the surface plasmon formula. In atomic units it is simply $\omega_{\rm s} = r_{\rm s}^{-3/2}$.

3.2 Linear response and short-time behavior

The photon absorption cross section and the polarizability are properties that only require the dynamics in weak external fields, the linear response region. This allows a variety of methods to be applied to solve the time-dependent equations, and each method has its adherents. There are two very popular methods that I will not describe in any detail in these written notes, but they need to be mentioned for completeness. In these methods one makes a matrix representation of the problem and either inverts or diagonalizes the matrix. Both methods start by Fourier transforming the Kohn-Sham equations in time, to obtain equations in the frequency domain. In the linear response region, different frequency are not coupled together except that the amplitude for a given frequency is coupled to that of its negative.

The direct solution of the time-dependent equations is possible, and I have pursued this method with my collaborator, Kazuhiro Yabana, using an algorithm originally developed for nuclear physics [40]. With the time-dependent equations, one can apply an external field of arbitrary shape and strength. If the field is weak, the response of a given frequency is obtained from the amplitude of that Fourier component in the excitation field. It is very convenient to take the perturbing field to be a δ-function in time, because it has all Fourier components with equal weight. Let us first see how this works with the one-particle Schrödinger equation,

$$H\phi \equiv -\frac{\nabla^2\phi}{2m} + U(r)\phi = i\frac{\partial\phi}{\partial t}\,.$$

Here $U(r)$ is a fixed internal potential the system. In this section I have dropped factors of $\hbar$ to simplify the formulas. I start with the ground state $\phi_0 \exp(-iE_0 t)$ and add to the Hamiltonian an impulsive external potential, $V_{\text{ext}}(r,t) = V_0(r)\delta(t)$. The differential equation then requires that there be a corresponding jump in the wave function at $t = 0_+$,

$$\phi(0_+) = (1 - iV_0)\,\phi_0.$$

The subsequent evolution can be found by solving the equation as a power series in time. The form of the power series is

$$\phi(t) = \left(1 - iV_0 - [H, V_0]t + i[H,[H,V_0]]t^2/2 + ...\right)\phi_0 e^{-iE_0 t}. \qquad (3.13)$$

The low-order terms in this expansion have direct physical interpretation in classical Newtonian mechanics. The initial perturbation of the wave function gives it the same momentum density that one would calculate from an impulsive force field $\vec{\nabla}V_{\text{ext}}$ acting on a density distribution of classical particles. This may be seen from the expansion of the expression for the momentum density to linear order in the external field,

$$\vec{p}(r) \equiv \phi^*(0_+)\left(\frac{\vec{\nabla} - \overleftarrow{\nabla}}{2i}\right)\phi(0_+) = -\vec{\nabla}V_0|\phi_0|^2. \qquad (3.14)$$

Here the right hand side is just the momentum density associated with the force field $\vec{\nabla}V_{\text{ext}}$ and the ordinary density $n_0 = |\phi_0|^2$. Next let us look at how the density varies in time. To first order in t the time-dependent density is given by

$$n(t) \equiv |\phi(t)|^2 = |\phi_0|^2 - 2\phi_0^*[H, V_0]\phi_0 t + ... = n_0 - \frac{\vec{\nabla}\cdot}{m} n_0\vec{\nabla}V_0 t + ...$$

Taking the derivative of this expression to get $\partial n/\partial t$, and comparing with equation (3.14), one can recognize the equation of continuity

$$\frac{\partial n}{\partial t} = -\vec{\nabla}\cdot\vec{p}(r)/m = -\vec{\nabla}\cdot\vec{j}$$

where $\vec{j}(r) \equiv \vec{p}(r)/m$ is the particle current. The next term in the power series will give the initial change in momentum due to the internal forces in the system. It is rather complicated to work out the double commutator for the next term in general (but see [9]), so I will specialize to the very importance case of the dipole field, $V_0 = \lambda z$. Here λ is a small parameter to remind ourselves that the response is to be calculated to linear order in the external field. The relevant commutators are

$$[H, V_0] = -\frac{\lambda}{m}\frac{\partial}{\partial z} \qquad (3.15)$$
$$[H,[H,V_0]] = \frac{\lambda}{m}\frac{\partial U}{\partial z} \cdot$$

Inserting in equation (3.13), the wave function evolves as

$$\phi(t) = \left(1 - i\lambda z + i\frac{\lambda t^2}{2m}\frac{\partial U}{\partial z}\right)\phi_0 \mathrm{e}^{-iE_0 t} + \frac{\lambda}{m}\frac{\partial \phi_0}{\partial z} t \mathrm{e}^{-iE_0 t} + \ldots$$

The generalization of this expansion to a many-particle Hamiltonian with interactions between the particles is very similar. Let us consider the Hamiltonian $H = \sum_i p^2/2m + \sum_i U(r_i) + \sum_{i<j} v(r_i - r_j)$ and call the ground state wave function Ψ_0. Then the power series expansion of the wave function is

$$\Psi(t) = \left(1 - i\lambda t^2 \sum_i z_i + i\frac{\lambda}{2m}\sum_i \frac{\partial U(r_i)}{\partial z_i}\right)\Psi_0 \mathrm{e}^{-iE_0 t}$$
$$+ \frac{\lambda}{m}\sum_i \frac{\partial \Psi_0}{z} t \mathrm{e}^{-iE_0 t} + \ldots$$

Note that the last term in parentheses only contains the one-particle potential U. The particle-particle interaction does not contribute because the relevant commutator vanishes,

$$\left[v(r_i - r_j), \frac{\partial}{\partial z_i} + \frac{\partial}{\partial z_j}\right] = 0.$$

I next write down the power series expansion of the dipole moment and the total momentum. With a little bit of algebra, these quantities can be expressed as follows

$$D(t) \equiv \langle \Psi(t)|z|\Psi(t)\rangle = \lambda\frac{N}{m}t + \ldots \qquad (3.16)$$

and

$$\langle \vec{p}\rangle \equiv \langle \Psi(t)|p_z|\Psi(t)\rangle = \frac{\lambda}{m}\left(N - \int \mathrm{d}^3 r n(r) \frac{\mathrm{d}^2 U}{\mathrm{d}z^2} t^2 + \ldots\right). \qquad (3.17)$$

It is easy to connect this time-dependent analysis with the dynamic polarizability function considered at the begin of the section. Formally, we can expand the time-dependent wave function following the impulsive dipole field in the eigenfunctions,

$$\Psi(t) = \Psi_0 - i\lambda \sum_i \Psi_i \langle 0|z|i\rangle e^{-i\omega_i t}.$$

The dipole moment and the total momentum then have the following time dependencies,

$$D(t) = \langle \Psi(t)|z|\Psi(t)\rangle = \lambda \sum_i 2\Big|\langle 0|z|i\rangle\Big|^2 \sin\omega_i t \tag{3.18}$$

$$\langle \Psi|p_z|\Psi\rangle = \lambda \sum_i 2\langle 0|p_z|i\rangle\langle i|z|0\rangle \cos\omega_i t. \tag{3.19}$$

In a power series expansion, the first term of the dipole moment is

$$\lambda \sum_f 2\omega_i \langle 0|z|i\rangle^2 t.$$

Comparing with equation (3.16), we see that they are equal if the TRK sum rule is satisfied. Thus the sum rule is just a statement about the short-time behavior of the wave function in an impulsive external field.

3.3 Collective excitations

It is possible to derive simple formulas for collective excitations just using the dynamics that was derived in the last section. From equation (3.19) we see that the power series expansion of the momentum has the first terms

$$\langle \Psi|p_z|\Psi\rangle = \lambda \left(N - t^2 \sum_i \omega_i^2 \langle 0|z|i\rangle\langle i|p_z|0\rangle + \ldots \right).$$

We can use this to estimate the frequency of the excitation, if the system is very collective. That assumption would allow us to replace the sum over states in the above equation by a single state with an excitation energy I will call ω_c. Comparing with equation (3.17), we find

$$\omega_c^2 = \frac{1}{mN} \int d^3r n(r) \frac{\partial^2 U}{\partial z^2} \cdot \tag{3.20}$$

This formula can be used to make simple estimates of collective frequencies.

Let us first consider the spherical jellium model. The background charge density is $n_0\Theta(R-r)$ and the corresponding potential is

$$U = -\frac{4\pi e^2 n_0 R^3}{3r}, \quad r > R$$
$$U = -\frac{4\pi e^2 n_0 R^2}{3}\left(\frac{3}{2} - \frac{r^2}{2R^2}\right), \quad r < R.$$

If the electrons are contained within the jellium sphere, the integral in equation (3.20) reduces to $\int_0^R \mathrm{d}^3 r n(r) = N$ and we immediately get the surface plasmon formula, equation (3.12). We thus see a condition of validity of the formula: the electrons must be contained within the jellium sphere radius. In fact the electron wave functions will spill out, decreasing the integral and thereby the frequency of the collective oscillation.

A simple extension of the model allows us to derive a formula for the splitting of the surface plasmon frequency in deformed systems. Let us suppose we have a small quadrupolar deformation with the radius vector depending on direction according to

$$R(\theta) = R(1 + \beta P_2(\cos\theta))$$

where P_2 is the Legendre function and β is a dimensionless deformation parameter. Taking β small, the charge density distortion of the jellium background is approximately

$$\delta n = \beta R n_0 P_2(\cos\theta)\delta(r-R).$$

The Coulomb potential associated with this charge density be obtained from the multipole expansion. The result for $r < R$ is

$$\delta U = -\beta\frac{4\pi e^2 n_0 r^2}{5} P_2(\cos\theta).$$

The mode in the x or y direction will now have a different frequency from the z mode because the respective second derivatives of the field $U + \delta U$ differ. Working this out, we get to leading order in β

$$\omega_\mathrm{s}^x = \omega_\mathrm{s}^y = \omega_\mathrm{s}\,(1 + 3\beta/5)$$
$$\omega_\mathrm{s}^z = \omega_\mathrm{s}\,(1 - 6\beta/5)\,.$$

4 Calculating the electron wave function

From a computational point of view, it is extremely difficult to calculate many-particle wave functions for more than a few particles, because the

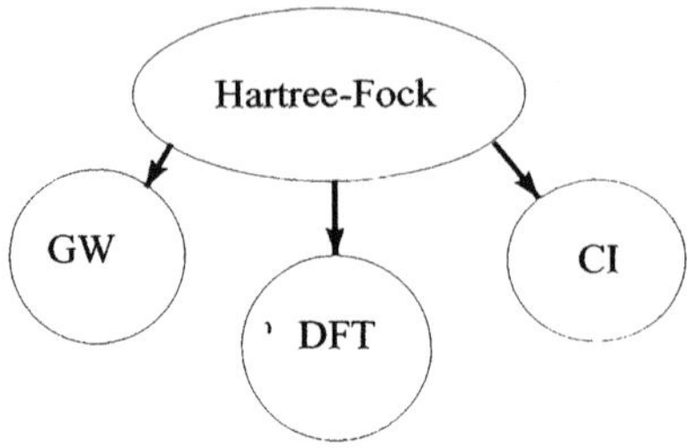

Fig. 5. Hierarchy of quantum many-body theories.

number of functional values needed in a numerical representation increases exponentially with the number of independent variables. On the other hand, the calculation of single-electron wave functions in an arbitrary three-dimensional potential is within the scope of desktop computers. Fortunately, there is a theory that is quite accurate although it only involves the solving the one-particle Schroedinger equation.

The most sophisticated version of the theory is called the time-dependent density functional theory or TDDFT for short. I will show a number of calculations done with TDDFT, but before that it is useful to review the static theory. This is the density functional theory (DFT). For systems larger than a few atoms, it has revolutionized the computation of electronic properties, as indeed was recognized in the 1998 Nobel Prize. While it has been spectacularly successful in calculating static properties, it does have well-known deficiencies in treating excitations. It is therefore important to understand the approximations and their validity.

The kinds of theory one can make for electronic structure are shown in Figure 5. In my view, Hartree–Fock theory is fundamental to all proven approaches. The theory is very simple to describe: minimize the expectation of the Hamiltonian in the space of Slater determinants. You have seen the basic objects many times before:

$$\Psi = \mathcal{A}\prod_i \phi_i(r_i), \quad \langle\phi_i|\phi_j\rangle = \delta_{ij} \tag{4.1}$$

$$\begin{aligned} E = \langle\Psi|H|\Psi\rangle &= \sum_i \int \mathrm{d}^3 r \frac{|\vec{\nabla}\phi_i|^2}{2m} + \int \mathrm{d}^3\, r n_\mathrm{e}(r) V_\mathrm{ion}(r) \\ &+ \frac{e^2}{2}\int \mathrm{d}^3 r \mathrm{d}^3\, r' \frac{n_\mathrm{e}(r) n_\mathrm{e}(r')}{|r-r'|} - \sum_{i<j}\left\langle \phi_i\phi_j \left| \frac{e^2}{|\vec{r}-\vec{r}'|}\right| \phi_j\phi_i\right\rangle \end{aligned} \tag{4.2}$$

where $n_\mathrm{e}(r) = \sum_i |\phi_i(r)|^2$ is the electron density. The variation $\delta E/\delta\phi_i^*(r) = 0$ gives the Hartree–Fock equations. The single-particle energy

Table 1. Atomization energies.

	Li_2	C_2H_2	20 simple molecules (mean absolute error)
Experimental	1.04 eV	17.6 eV	-
Theoretical errors:			
Hartree–Fock	−0.94	−4.9	3.1
LDA	−0.05	2.4	1.4
GGA	−0.2	0.4	0.35
τ	−0.05	−0.2	0.13

in the Hartree–Fock equations arises as a Lagrange multiplier to preserve the norm of the wave function. While Hartree–Fock is conceptually simple, on a practical level it is not accurate enough to be useful for chemistry or for computing cluster structures. The level of accuracy for several simple molecules is shown in Table 1, taken from Perdew [41, 42]. The mean absolute error in the atomization energies (energy difference between the molecule and the individual atoms in isolation) is 3 eV in the Hartree–Fock theory. The predicted binding of the Li_2 clusters is a factor ten too low, and another alkali metal cluster not in the table, Na_2, is incorrectly predicted to be unbound.

The three lines in Figure 5 going down from Hartree-Fock give three different approaches to improve the theory. The "CI", configuration interaction expansion, uses Hartree-Fock as a basis of many-body wave functions. This is a very systematic approach, and it gets marvelous results in small systems, as you will see in the seminar of Bonacič-Koutecky. But the number of terms needed in the CI expansion to achieve a given accuracy grows exponentially with the number of electrons. So one turns to other methods to deal with larger systems.

Many-body perturbation theory was developed in the 1950's to make possible systematic calculation of the energy of quantum systems with an infinite number of particles. One avoids the many-body wave function, but the price one pays is to deal with Green's functions that depend on a few more variables. In condensed matter physics, the most refined approximation that can be presently computed this way is called the "GW" approximation. You can find a review of it by Hedin [43]; I will not discuss it here in any detail.

The last approach to make a better theory is to keep the form of the Hartree–Fock equations, but improve the energy function that is put into the variational principle. The DFT will thus include effects of correlations by changing the Hamiltonian. In practice, there is another aspect of DFT as it is normally applied, the local density approximation (LDA). Evaluation of the exchange interaction is computationally burdensome, and approximations to make it look like an ordinary potential have been used since the time of Slater. In defining the effective exchange potential, one is guided by the energies in the uniform electron gas. The exchange energy of an electron of momentum k is [44]

$$-\sum_i \left\langle ki \left| \frac{e^2}{|r-r'|} \right| ik \right\rangle = -\frac{e^2 k_{\mathrm{F}}}{\pi} \left(1 + \frac{k_{\mathrm{F}}^2 - k^2}{2kk_{\mathrm{F}}} \log \left| \frac{k+k_{\mathrm{F}}}{k_{\mathrm{F}}-k} \right| \right), \tag{4.3}$$

where k_{F} is the Fermi momentum. The total exchange energy (per electron) E_{x} is given by

$$E_{\mathrm{x}} = -\sum_{i<j} \left\langle ij \left| \frac{e^2}{|r-r'|} \right| ji \right\rangle = \frac{-3}{4\pi} e^2 k_{\mathrm{F}} = \frac{-3}{4\pi} e^2 (3\pi^2 n_{\mathrm{e}})^{1/3}$$

where in the last equation we used the relation between electron density n_{e} and Fermi momentum, $n_{\mathrm{e}} = k_{\mathrm{F}}^3/3\pi^2$. Slater proposed making a local density approximation by using a two-body contact interaction that would have the same total energy. The one-body potential in the Schrödinger equation would then be

$$V_{\mathrm{Slater}}(r) = -\frac{3}{2\pi} e^2 (3\pi^2 n_{\mathrm{e}}(r))^{1/3}.$$

THIS IS WRONG. Going back to the variational principle, one sees that the one-body potential should be defined by the variation of E_{x},

$$V_{\mathrm{x}}(r) = \frac{\delta n E_{\mathrm{x}}}{\delta n} = -\frac{e^2}{\pi} (3\pi^2 n_{\mathrm{e}}(r))^{1/3}. \tag{4.4}$$

This is a factor 2/3 different from Slater's potential.

Kohn and Sham proposed to include the correlation energy of the electron gas in exactly the same way, determining an exchange-correlation potential V_{xc} from the exchange-correlation energy E_{xc} of the interacting Fermi gas. This is usually done through a numerical parameterization, giving us the "LDA" of density functional theory. It gives a considerable improvement over Hartree–Fock, as may be seen by the entries in Table 1. However, the accuracy is still not enough for chemical modeling. An obvious problem of the LDA is that the single-particle potential does not have the correct asymptotic behavior. The electron potential in a neutral cluster should behave as $-e^2/r$ for large separation of the electron from the cluster. But in the LDA the Coulomb potential is calculated with all the electrons and thus

Table 2. Atomic properties of the Ag atom in LDA.

	IP	First excitation
Experimental	7.75	3.74
Kohn–Sham:		
eigenvalues	$\epsilon_s = 4.6$	$\Delta\epsilon = 3.9$
total energies	8.0	4.1

vanishes outside the cluster. This makes the LDA unreliable for calculating ionization potentials from the Kohn-Sham eigenvalues.

As an example of the last problem, Table 2 shows the ionization potential (IP) of the silver atom and the excitation energy of its first excited state. According to Hartree–Fock theory, the Ag atom has a single electron in a s orbital, with an unoccupied p orbital just above and a fully occupied d orbital just below. Thus the IP should correspond to the energy of the s-orbital. For the silver atom, the Kohn–Sham eigenvalue is off by 40% from the experimental ionization potential. Nevertheless the LDA can give quite accurate ionization potentials if they are calculated a different way, namely as a difference of total energies,

$$IP = E(\mathrm{Ag}) - E(\mathrm{Ag}^+). \tag{4.5}$$

The table shows that the error is only 3% when the IP is calculated this way. For the excitation energy, which is an $s \to p$ transition, the difference of Kohn–Sham energies $\Delta\epsilon = \epsilon_p - \epsilon_s$ is fairly good. However, there are caveats that we will come back to later.

Becke [45] proposed a fix to get the $-e^2/r$ asymptotic potential by adding a term to the energy functional that depends on the gradient of the density, $\nabla n_e(r)$. His proposed form works amazingly well. This is the "generalized gradient approximation", GGA. From Table 1, we see that energies can be calculated to an accuracy of tenths of an eV. Further improvements may be possible. The Kohn–Sham energy functional depends on the nonlocal quantity

$$\tau = \sum_i |\nabla\phi_i|^2$$

in the kinetic energy term. One could think of using other functional dependencies on τ; an example is given in the last line of Table 1.

Before going on to electronic excitations, I want to show how well the LDA works for describing the infrared active vibrations in C_{60}. One first calculates the LDA potential energy surface as a function of displacement

Table 3. Excitation energies of C_{60} infrared-active vibrations (T_{1u}).

Mode	1	2	3	4
Experimental	0.065 eV	0.071	0.147	0.177
LDA error				
Ref. [22]	−2%	−2%	+2%	+10%
Ref. [46]	−3%	−7%	−8%	−7%

of the atoms from their ground state positions. Diagonalizing the Hessian matrix then gives the normal modes and their frequencies. The results of three different LDA calculations are shown in Table 3. The agreement with experiment is impressive, with mean absolute relative error on frequency only 4%. A more demanding test of the theory is the transition strength associated with the vibrations. The accuracy here is perhaps only a factor of two [22]. But that is a great improvement over previous theories that were completely unreliable.

4.1 *Time-dependent density functional theory*

Schrödinger proposed two equations in his original paper, the eigenvalue equation for static properties and the time-dependent equation for the dynamics. But the left-hand side of both equations was the same. The situation is the same for dynamic theories based on Hartree–Fock or DFT. The theories may be derived from the time-dependent variational principle,

$$\delta \int \mathrm{d}t \left\langle \Psi \middle| H - i\frac{\partial}{\partial t} \middle| \Psi \right\rangle = 0. \tag{4.6}$$

Taking Ψ to be a Slater determinant, and varying with respect to $\phi_i^*(r)$, one obtains time-dependent Hartree–Fock equations, first proposed by Dirac in 1930. The corresponding equations for DFT are the time-dependent Kohn–Sham equations,

$$-\frac{\hbar^2}{2m_\mathrm{e}}\nabla^2\phi_i(r,t) + \left(\int \mathrm{d}^3r' \frac{e^2 n(r,t)}{|r-r'|} + V_{\mathrm{xc}}[n(r,t)] + \sum_{\mathrm{ions}} V_{\mathrm{ion}}(r-R_\mathrm{I})\right)\phi_i(r,t) = i\hbar\frac{\partial}{\partial t}\phi_i(r,t). \tag{4.7}$$

The first application of TDDFT was to describe the photoionization of atoms [47]. The theory has since been widely applied to clusters, molecules,

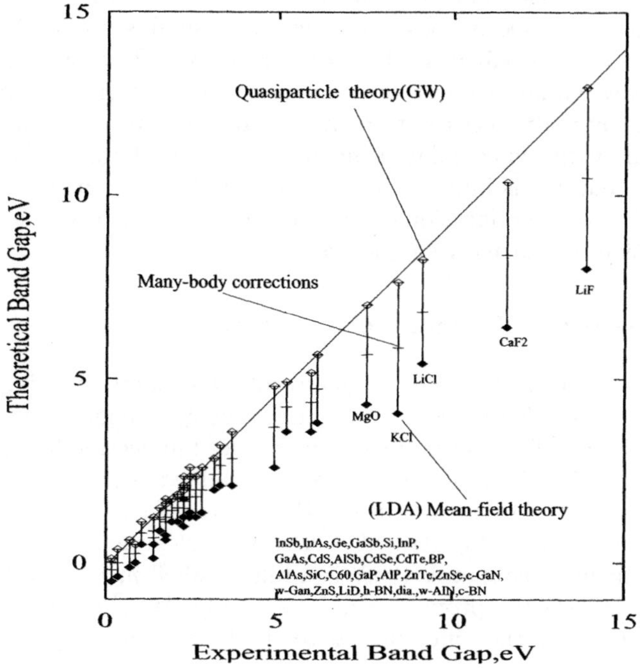

Fig. 6. Band gap in insulators and semiconductors, theoretical *versus* experimental for the GW approximation and for the DFT. From Hedin [43].

and bulk matter. There are several quite different methods to solve the equations, which I will not discuss in detail.

Before describing the applications to clusters, I want to give a cautionary remark. It is tempting to apply DFT to excitations by simply replacing the ground state single-particle wave functions by excited state orbitals in the Kohn–Sham equation. This seems to work well for small systems, as for example in the Ag atom discussed earlier, and in small carbon structures that will be discussed later. However, the theory can give poor results in large systems. This is the "band gap" problem. This may be seen in Figure 6 showing the band gaps between occupied and unoccupied bands in some insulators and semiconductors. The DFT, shown with black diamonds, is consistently low, by as much as 5 eV in the LiF crystal.

The band-gap problem originates in the LDA treatment of exchange. From equation (4.3) one sees that the exchange potential of an electron in a Fermi gas has a weak logarithmic singularity at the Fermi surface. Particle-hole excitations across the Fermi surface have a higher energy for a given momentum difference than the quadratic kinetic energy functional. Because the singularity is weak, it does not show up for small systems. Clearly, any local approximation will miss the singularity. An effective, but computationally costly method to overcome this is to calculate the electron energies from the many-body perturbation theory. In the GW approximation, the electron-self energy is calculated including exact exchange. The GW theory gives an enormous improvement to the band gap, as may be seen from Figure 6. But the computational demands of the theory has so far restricted its application to relatively simple systems.

5 Linear response of simple metal clusters

In this section I want to present some results for metal clusters. The most simple are the alkali metals, having only a single valence electron in the s-atomic orbitals. The jellium model, introduced in Section 3 and to be discussed in detail by Manninen, describes the qualitative properties of sodium clusters very well. The metals in the IB column of the periodic table (the "coinage" metals Cu, Ag, Au) also have a single valence electron in the s-orbital, but in these metals there is a filled d-shell just below the valence shell, which affects the properties quite strongly. We shall first discuss the alkalis and then see the differences with the IB metals.

5.1 Alkali metal clusters

At the crudest level, the optical response of the metals will show a surface plasmon at a frequency given by the surface plasmon formula equation (3.12). If we take the density n_e at the bulk value for sodium, the formula gives

$$\omega_s \approx 3.4 \text{ eV}.$$

In fact, the observed surface plasmon is at a lower frequency ("red-shifted"). For example, the optical absorption spectrum of the cluster Na_{41}^+ in Figure 3 shows the resonance at a frequency of 2.6 eV. One possibility to explain the red shift would be that the density of atoms in the cluster is less than in the bulk. This idea is not born out in DFT calculations of the cluster structure, so I will not discuss this possibility any further. Another mechanism to explain the red shift is the spillout of the electrons, as mentioned in Section 3. We can analyze this mechanism quantitatively by calculating the optical response of the jellium model. There is a computer code, called JellyRpa, that I wrote and distributed [11] to calculate the response of spherical jellium, making no other approximations on the dynamics. Indeed there is an effect of the spillout, as may be seen in Figure 7. The peak is shifted from

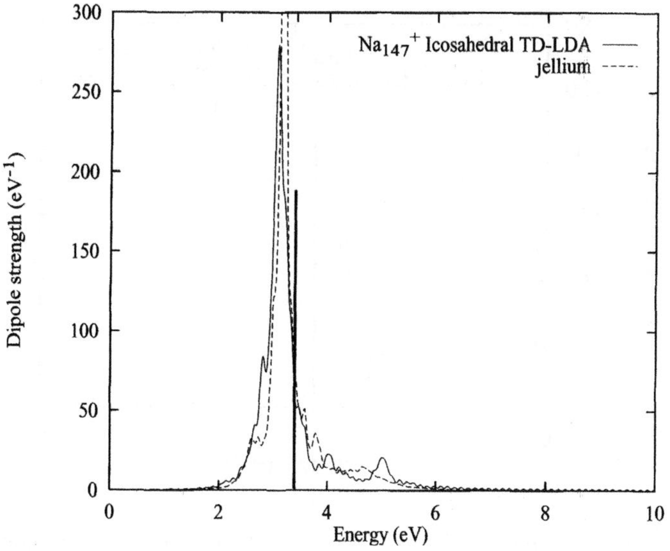

Fig. 7. Comparison of jellium model of a large sodium cluster with the surface plasmon formula and the TDDFT. The transition from the surface plasmon formula is shown with a vertical line.

3.4 to 3.0 eV, which is about half of what is needed to explain the empirical position. Another possibility is that the local density approximation might not be accurate enough. In particular, we saw in Figure 6 that the local approximation to exchange could produce serious errors in infinite systems. This question was addressed in finite clusters by Madjet *et al.* [48], who examined and compared the different treatments of exchange. They found that the LDA exchange was quite satisfactory for sodium clusters, and could not be the reason for the discrepancy.

Lastly, the jellium approximation might be inadequate; the full TDDFT of course includes a realistic treatment of the ionic potentials. Comparing the full TDDFT with the jellium, we found mixed results. Taking the ionic potential from the naive pseudopotential prescriptions in the literature, we found a surface plasmon at the same frequency as in the jellium model. This is also shown in the figure. However, there is an different prescription for the pseudopotential [49], treating the core electrons somewhat differently. This gives an additional red shift, essentially bringing the surface plasmon down to the observed position. In the end, it is rather disquieting that a seemingly small change in the ionic potential would have a very noticeable effect on the absorption spectrum.

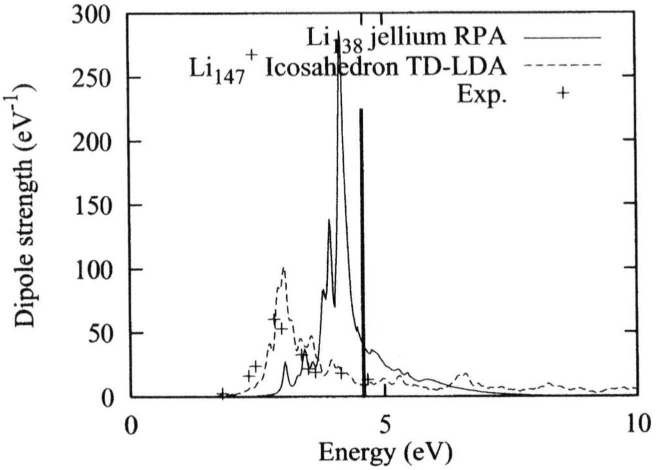

Fig. 8. Surface plasmon in Li clusters with $N \approx 150$. The prediction from equation (3.12) is shown by the vertical line.

There is an even larger red shift in lithium clusters: the formula gives 4.6 eV, while the observed peak in the absorption spectrum is at 3.0 eV [51]. Here it is much easier to understand how the shift arises from the ionic potential. The ionic potential is different for s and p waves because there is a core s orbital that is excluded from the valence wave function but no corresponding excluded p state. Thus, the ionic potential is effectively more attractive for p orbitals. This makes it easier to excite the electrons from the ground state, and lowers the excited state energies. This can be seen in Figure 8, comparing the jellium model with the full TDDFT calculation. As well as substantially shifting the peak position, the TDDFT disperses the strength somewhat on the high-frequency side. Both features are seen in the experimental data, which is rather well reproduced by the TDDFT.

The strength of the surface plasmon in sodium clusters, both theoretically and experimentally, is very close collective limit with f equal to the number of valence electrons. In lithium clusters, the effect of the ionic potential may be interpreted as a giving the electron an effective mass $m^* \approx 1.3m$, which also reduces the sum rule by the same factor (*e.g.* see [50]).

5.2 *Silver clusters*

The surface plasmon formula is also poor for silver clusters, but for a different reason. Here the formula gives 5.4 eV, but the observed surface plasmon is in the range 3.5–4.0 eV. In this case, it is not justified to neglect the polarization of other electrons, namely the filled d-shell [52]. In reference [28] we calculated spectra of silver clusters including explicitly the d-electrons in the TDDFT. The observed peaks in Ag_8, for example, are nicely reproduced by the theory. In Ag_9^+ the experimental spectrum [53] shows a single peak at 4 eV, having a width of about 1 eV. The theory gives several transitions in the range 3.6–4.4 eV that together could give a single broad peak of about the needed width. Thus the theory reproduces the energetics of the surface plasmon rather well.

I now want to show the role of the d-polarizability in the properties of the plasmon from a more analytic point of view. I start with description of the polarizability of the d-electrons in the Ag atom. This somewhat unphysical quantity was calculated in [28] by the TDDFT keeping the s-electron frozen as the ground state orbital. The polarizability of the core came out to be $\alpha_d \approx 2$ Å$^3 \approx 14$ a.u. at the surface plasmon frequency. In that work we derived the effect on the surface plasmon using the linear response technique. Here I will show another derivation using the dielectric function. We can associate a dielectric function with the d polarizability using the Clausius–Mossotti formula, equation (3.7), $\epsilon_d = (1 + 2\alpha_d/r_0^3)/(1 - \alpha_d/r_0^3)$. For silver $r_0 = r_s = 3.02$, giving $\epsilon_d \approx 4$. Thus the d-orbital polarizability screens external fields by a considerably factor.

Let us redo the dielectric theory of the response including separate contributions from the s- and the d-orbitals. To derive the dielectric function for mixed sets of charges, we go back to the definition of the dielectric function and the relation to surface charges. There will be two surface charges screening the field, σ_d and σ_s. In terms of the internal field $\mathcal{E}_{\text{int}}$ inside the conductor, the surface charges are given by

$$4\pi\sigma_{s,d} = (\epsilon_{s,d} - 1)\mathcal{E}_{\text{int}}.$$

The same equation also holds for the total charge and total dielectric function, so one can easily derive for the complete dielectric function

$$\epsilon = \epsilon_d + \epsilon_s - 1.$$

Taking ϵ_d from the Clausius–Mossotti formula and ϵ_s from the free electron model equation (3.9) the dielectric function becomes

$$\epsilon = \frac{1 + 2\alpha_d/r_0^3}{1 - \alpha_d/r_0^3} - \frac{\omega_p^2}{\omega(\omega + i\gamma)} . \tag{5.1}$$

Next we put this into the Mie formula for the polarizability. With a bit a algebra the expression becomes (dropping the $i\gamma$)

$$\alpha = R^3 \frac{1 - \epsilon}{2 + \epsilon} = R^3 \frac{\omega_s^2(1 - \alpha_d/r_0^3) - \omega^2 \alpha_d/r_0^3}{\omega_s^2(1 - \alpha_d/r_0^3) - \omega^2} . \tag{5.2}$$

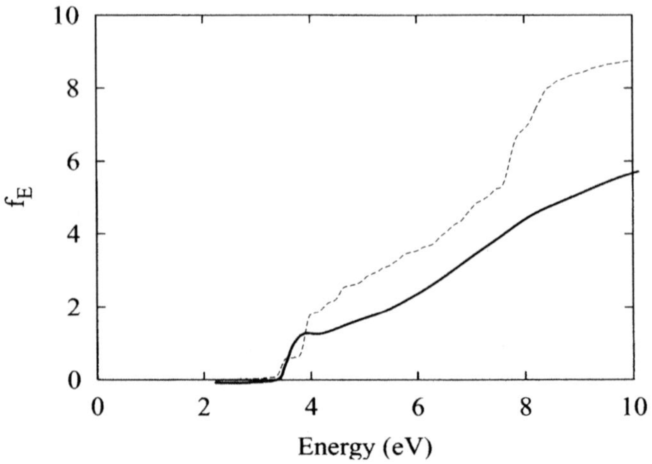

Fig. 9. Integrated strength function in Ag_8 comparing Mie theory (solid line) with TDDFT (dashed line).

From the denominator one sees immediately that the resonance energy is shifted to

$$\omega_s' = \omega_s (1 - \alpha_d / r_0^3)^{1/2}.$$

Taking the previous numerical value for ϵ_d, this gives $\omega = 3.7$ eV, which is almost exactly what is needed to describe the empirical resonance. In effect, the charges of the s-electrons are screened by a factor $\sqrt{1 - \alpha_d / r_0^3}$.

I finally want to discuss the strength of the surface plasmon. Unlike the alkali metal clusters, the strength is strongly reduced by the d-electron screening, at least in theory. Figure 9 shows the integrated strength function, $f_E = \int_0^E dE' S(E')$ for Ag_8, calculated in the TDDFT and in the Mie theory using the empirical dielectric constant for silver metal. The two curves are surprisingly similar. Thus with respect to the theoretical polarizability, a small silver cluster seems to behave as a piece of the bulk metal. In the Mie theory, the surface plasmon is seen as the step at 3.6 eV. The finite system calculation gives a split peak, with the stronger excitation slightly blue-shifted. The height of the step gives strength of the resonance; we see that it is roughly 2, which is a factor of 4 lower than the number of valence s-electrons in the clusters. This another manifestation of the screening by the d-electrons. It can also be understood rather simply by the above polarizability function. In the neighborhood of the resonance, the above polarizability can be expressed

$$\alpha \approx \frac{R^3}{2}\omega_s'(1-\alpha_d/r_0^3)\left(\frac{P}{\omega-\omega_s'}+i\pi\delta(\omega-\omega_s')\right).$$

Putting this in equation (3.4), we find for the associated strength,

$$f = N(1-\alpha_d/r_0^3)^2.$$

Since the coupling to the external field goes as e^2, each charge here is screened by a factor $1-\alpha_d/r_0^3$. Numerically, the predicted screening is a factor of 4, as found in the microscopic calculation.

The situation would be completely satisfactory except that it does not agree with experiment. The experiment of Tiggesbäumker *et al.* [53] measured not only the position of the resonance in Ag_9^+ but also its strength. The integrated strength over the resonance is about 4, more than twice the theoretical value. I am sorry that I can not give you a resolution of this disagreement. But the measurement of total strength is not easy experimentally, and so one should not give up the theory without additional confirmation.

6 Carbon structures

In this lecture I will discuss carbon structures, going from small clusters and molecules to fullerenes and nanotubes. A very simple theory, the Hückel model, turns out to be an excellent guide to the electronic structure obtained by more elaborate means such as the density functional theory. There are two requirements for the Hückel model to be applicable. The first is that each carbon and its neighbors lie in a plane. Then the symmetry with respect to plane allows the orbitals to separated into σ or π types having even or odd reflection symmetry, respectively. This classification is in fact useful even if the planar condition is not strictly met, as in the curved fullerenes and nanotubes. The other condition is that the π orbitals are at the Fermi surface. This is the case for all the systems I consider. The π orbitals on different atoms couple rather weakly, and the Hückel models treats the wave function by the amplitude of the π orbital on each carbon, constructing the eigenstates from the simple hopping Hamiltonian

$$H = -\beta\sum_s^2\sum_{j_1,j_2} a_{j_1 s}^\dagger a_{j_2 s}.$$

Here the sum j_1, j_2 runs over pairs of adjacent carbon atoms, s is a spin label, and β is a parameter giving the hopping matrix element between the atoms. The hopping parameter can depend on distance d between carbon atoms when that varies. It will turn out that a fair parameterization is given by [29]

$$\beta(d) = 2.5 \left(\frac{1.40\ \text{Å}}{d}\right)^{2.7} \text{eV}. \tag{6.1}$$

6.1 Chains

The first structures I want to consider are linear chains. If the axis of the chain is in the z-direction, both the p_x and p_y atomic orbitals will have π character. To find the spectrum within the Hückel model, we start with an infinite system. Here, by the Bloch theorem, the eigenfunctions have amplitudes on the atoms that vary from atom to atom as exponentials:

$$\phi_K(j) \sim \exp(ijK) \tag{6.2}$$

where K is a parameter of the eigenstate and j indexes the atom in the chain. Then ϕ_K is the solution of the Hamiltonian equation

$$H\phi_K = \epsilon_K \phi_K$$

with energy

$$\epsilon_K = -2\beta \cos K. \tag{6.3}$$

To find the solution for a finite chain, I use a trick. Namely, the wave function on a finite chain with N atoms at positions $j = 1, 2, ..., N$ behaves the same under the hopping Hamiltonian as a wave function on an infinite chain, provides the wave function vanishes at sites 0 and $N + 1$. There is no hopping from these sites in the finite system because there are no atoms there. The vanishing boundary conditions are satisfied by sine functions with parameters $K = \pi m/(N + 1)$, where m is an integer. The normalized wave functions are then

$$\phi_m(j) = \sqrt{\frac{2}{N+1}} \sin \frac{\pi m j}{N+1} \quad m = 1, 2, ...$$

From equation (6.2) the energies of the orbitals are

$$\epsilon_m = -2\beta \cos\left(\frac{\pi m}{N+1}\right). \tag{6.4}$$

Let's see how well this compares with the LDA energies. In Figure 10 we compare the Kohn–Sham energies of the π orbitals in the cluster C_7 with

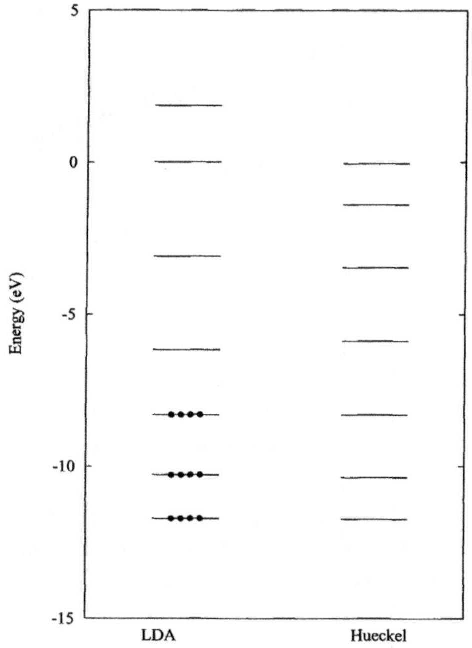

Fig. 10. Comparison of π orbital energies in the C_7 cluster.

equations (1.2–3). The hopping parameter in the Hückel model was taken from equation (6.1) and energy scale was set by aligning the highest occupied orbital, $m = 3$. The occupied orbitals are indicated by black circles in the figure. We see that the Hückel model is quite accurate in the neighborhood of the Fermi energy and below.

A very important derived property is the gap between occupied and unoccupied orbitals. For odd-numbered chains, the highest occupied orbital has $m = (N-1)/2$. Thus the gap energy is

$$\Delta\epsilon = 2\beta\left(\cos\frac{(N-1)\pi}{2(N+1)} - \cos\frac{(N+1)\pi}{2(N+1)}\right) = 2\beta\sin\frac{\pi}{N+1} \cdot \tag{6.5}$$

Asymptotically, the gap for large chains decreases as

$$\Delta\epsilon \sim \frac{1}{N} \cdot \tag{6.6}$$

The comparison of LDA [25] with the Hückel model as function of chain size is shown in Figure 11. There is one complication that shows up as an odd-even staggering in the LDA spectrum. The π orbitals in a chain have a

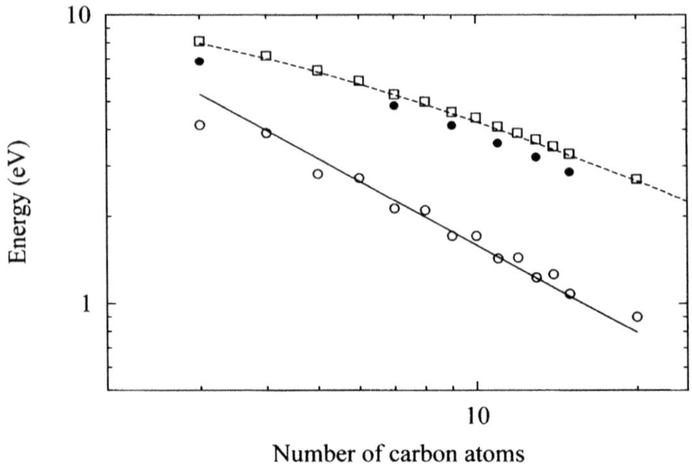

Fig. 11. Systematics of the energy gap and the collective excitation in carbon chains, compared with functional fits of the form (6.6) and (6.10). The filled circles show the experimental energies of transitions of clusters in a noble gas matrix, from references [54, 55].

two-fold spatial degeneracy (as well as the usual two-fold spin degeneracy). The highest occupied orbital is only half filled for even N, giving them a different gap.

The excitation of an electron from just below to just above the Fermi level will have a large transition strength, close to the number of electrons in the π manifold of states. Under these conditions, the frequency of the excitation will be strongly perturbed from the gap energy $\Delta\epsilon$. When the TDDFT is applied, the Coulomb interaction pushes the strength to higher energy. Of course the total strength is preserved due to the Thomas–Reiche–Kuhn sum rule.

The TDDFT calculated energies of the strong transitions are show as the boxes in Figure 11. The shift from the gap energy is factor of 2–3, making the transition very collective. The theory describes the experimental data (shown by filled circles) quite well. Note that experimental data is only available for the more stable odd-number clusters. A ring structure is probably more preferable in the midsize even-number clusters.

A qualitative description of the transition energies can be obtained using the matrix version of the TDDFT theory and considering only the single state, the excitation across the gap $\Delta\epsilon$. The matrix version of TDDFT reduces to a simple formula in that case, giving a transition energy ω

$$\omega = \sqrt{\Delta\epsilon(\Delta\epsilon + 2v)}. \tag{6.7}$$

This formula is the same as equation (4.12) in reference [18], which contains a derivation. To find the interaction v, let us assume that it depends only on the distance between carbon atoms and has the following form

$$\begin{aligned} v(j_1, j_2) &= U \quad j1 = j2 \\ &= \frac{e^2}{d|j_1 - j_2|} \quad j1 \neq j2 \end{aligned} \tag{6.8}$$

where d is the distance between carbon atoms. The matrix element for an excitation from orbital m_1 to m_2 is then

$$v = 4\sum_{j_1,j_2} \phi_{m_1}(j_1)\phi_{m_2}(j_1)\phi_{m_1}(j_2)\phi_{m_2}(j_2)v(j_1,j_2)$$

with the factor 4 coming from the degeneracy of the orbitals. Taking the two coefficients in equation (6.8) as parameters, the TDDFT calculations shown in the figure are fit with parameter values $e^2/d = 11.7$ eV and $U = 11.7$ eV. The value for e^2/d is very close to what one would expect from the distance between carbons, $d = 1.28$ Å. Note also that the on-site interaction U is not stronger than the interaction between neighboring carbons. The reason is that the exchange and correlation energies compensate the stronger onsite Coulomb interaction.

It is interesting to examine the asymptotic behavior of ω for large chains. We already found that $\Delta\epsilon$ has an $1/N$ behavior but v has a different dependence. With the form equation (6.8) for the interaction, and the sum over pairs has an asymptotic logarithmic dependence on N coming from the $1/|j_1 - j_2|$ term. The logarithmic dependence was noted in [25] and compared with an analytic formula analytic formula derived from the plasmon dispersion in a long wire,

$$\omega^2 = \frac{4\pi n e^2 q^2}{m_e} \log\frac{1}{qa} \cdot \tag{6.9}$$

Here a is the radius of the wire, and q is the wave number. For a finite chain, we may take $q \sim 1/N$, and the N-dependence of the excitation becomes

$$\omega = \frac{\sqrt{A + B\ln N}}{N} \cdot \tag{6.10}$$

One see in Figure 11 that a straight line (corresponding to $1/N$ in the logarithmic plot) gives a good fit to the N-dependence for $\Delta\epsilon$ but not for ω. The wire formula other hand describes the N-dependence of ω quite well.

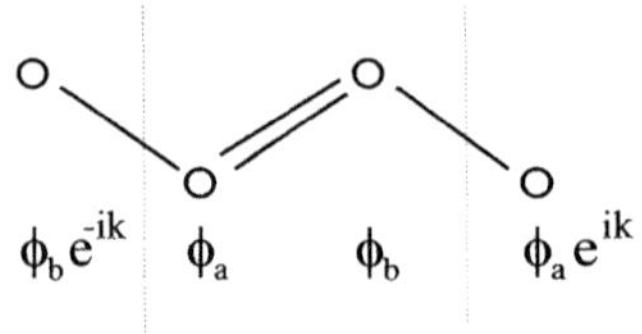

Fig. 12. Hückel wave function construction for polyacetylene.

6.2 Polyenes

The next system I will discuss are chain molecules have alternating single and double bonds between the carbons, the conjugated carbon molecules. These are the polyene molecules, beginning with the two-carbon molecule ethylene and going to the infinite chain, polyacetylene. The molecules have hydrogen atoms as well, which do not play any role in the π orbitals. As before, it is easier to solve the Hückel Hamiltonian for the infinite system. In the polyenes, the distance between the carbons alternates, depending on whether the bond is single or double. Thus there will be two hopping matrix elements, β_s and β_d. To solve the problem, we divide the chain into unit cells containing two carbon atoms attached by a double bond. The system is periodic with this unit cell so we can write the wave function for the atoms in the cell and adjacent to the cell as shown in Figure 12.

Let us apply the Hamiltonian equation $H\phi = \epsilon\phi$ at two sites in the unit cell. This gives the two equations,

$$\begin{aligned} -\beta_s \phi_b e^{-iK} - \beta_d \phi_b &= \epsilon \phi_a \\ -\beta_d \phi_a - \beta_s \phi_a e^{iK} &= \epsilon \phi_b . \end{aligned} \tag{6.11}$$

The equations are easily solved (multiplying them together) to get a formula for the energy,

$$\epsilon_k = \pm\sqrt{\beta_s^2 + \beta_d^2 + 2\beta_s\beta_d \cos K}. \tag{6.12}$$

Interestingly, it has a gap at half filling given by

$$\Delta\epsilon = 2|\beta_d - \beta_s|. \tag{6.13}$$

In this case I don't know a trick to get the finite system spectrum from this result. But the Hückel Hamiltonian is easy to diagonalize numerically, with the dimensionality of the matrix equal to the number of carbon atoms. Fitting the LDA energies of the 8-carbon polyene, one finds hopping parameters of 2.27 and 2.80 eV. These parameterization fits the LDA energy

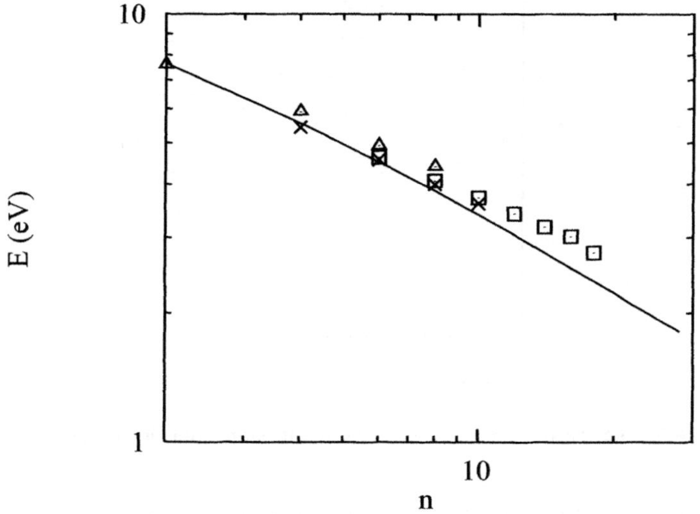

Fig. 13. Transition energy of the collective $\pi - \pi^*$ transition in polyenes, comparing theory with experiment. The TDDFT is the solid line; the points are various experimental measurements. See reference [29] for details.

gap of all of the polyenes in the range $N = 2$–40 very well. However, in polyacetylene, the predicted gap from equation (6.13) is 1.08 eV, which is rather far off the experimental value of 2.1 eV. This is undoubtedly the same gap problem of extended insulators that was mentioned in Section 2. The problem is seen in the LDA calculations of very large finite polyenes [56]. As in the case of the semiconductors, the GW approximation gives a much better gap [57].

Let us now examine the excitation energies of the strong transitions. The results of the TDDFT are compared with experimental data in Figure 13. The agreement is rather satisfactory. It would be interesting to analyze these results in the framework of the Hückel model, but I have not done so.

6.3 Benzene

We now turn to the benzene molecule, which for the Hückel molecule is a perfect hexagon of carbons (bond length 1.40 Å); the outside hydrogens hardly affect the π electrons. I will analyze the orbitals with a poor man's group theory, utilizing only the 6-fold rotation symmetry about the center of the hexagon. We can still make use the wave function from equation (6.2), but now the boundary condition is that $\phi_K(0) = \phi_K(6)$, giving $K = m\pi/3$

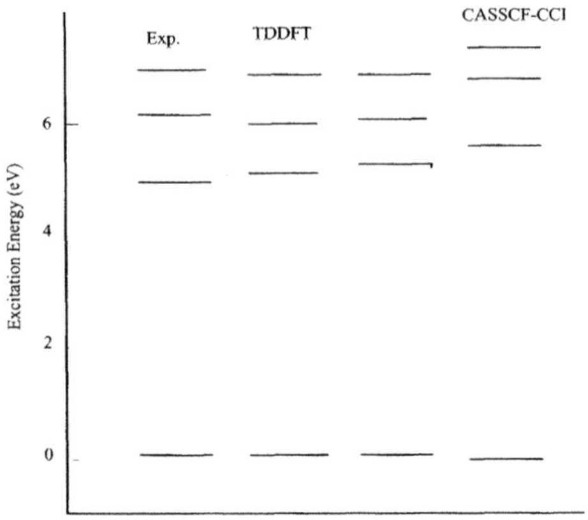

Fig. 14. Singlet $\pi - \pi^*$ vertical excitation energies of benzene.

for integers m. Here m has the same meaning as the azimuthal quantum number of the rotation group. The wave functions and orbital energies are then given by

$$\phi_m(j) = \frac{1}{\sqrt{6}} e^{imj}$$
$$\epsilon_m = -2\beta \cos \frac{m\pi}{3} \ . \tag{6.14}$$

The lowest state has $m = 0$, and then there are two degenerate states with $m = \pm 1$. The 6 π electrons in benzene fill these three orbitals (with each orbital containing two spin states). The lowest unoccupied orbital is the two-fold degenerate $m = \pm 2$ pair. The sixth and highest orbital has $m = \pm 3$. The state is the same for either sign of m. The energy gap is given simply by $\epsilon_2 - \epsilon_1 = 2\beta$.

Four distinct excitations can be made at the energy gap, due to the degeneracies of the orbitals just above and below the gap. The particle-hole states can be labeled by the change in m required to produce the state. For example, the $M = +3$ state has the electron in orbital $m = -1$ promoted to the orbital $m = +2$. Thus it appears that there should be two two-fold degenerate excitations, $M = \pm 1$ states and $M = \pm 3$ states. The former are have dipole matrix elements with the ground state and will be prominent in optical absorption spectrum. The latter two states mix together, due to the indistinctness of the ± 3 transition density. This gives an energy splitting between the states, leading to a spectrum with three excitation energies.

Figure 14 shows this energetics, and compares with the full scale time-dependent density functional theory. The uppermost state ("$^1E_{1u}$") has a large optical absorption transition strength, and can be described as a

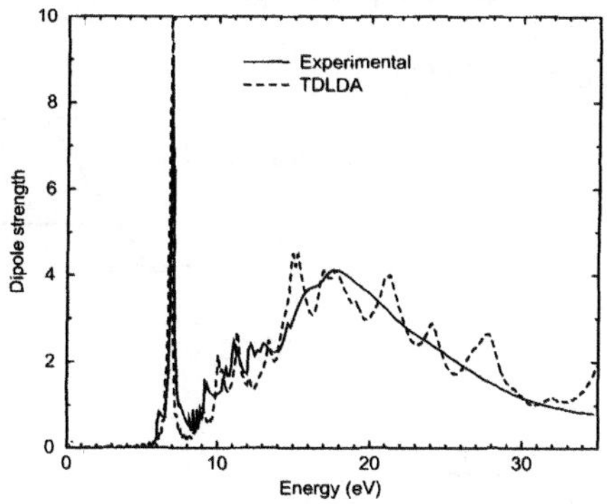

Fig. 15. Comparison of experimental and theoretical absorption spectrum in benzene, from reference [29].

$\Delta M = \pm 1$ transition. The next state down is weaker by a factor of ten, and is assigned the linear combination $|+3\rangle + |-3\rangle$ of the $\Delta M = \pm 3$ transitions. The other linear combination, $|+3\rangle - |-3\rangle$, has vanishing transition density on the carbon atoms, and is hardly shifted by the interaction from the unperturbed excitation energy $\Delta\epsilon$. The TDDFT theory gives an excellent account of these states with no free parameters. In the Hückel theory, we already get the lowest state because it is at the gap energy. It is interesting to see how well a simple Hückel model does for the energy shifts of the other states. Taking the Hückel model for the wave functions in the E_{1u} excitation, the matrix element for the $(m = 1) \rightarrow (m = 2)$ excitation is

$$v = \frac{1}{3}(v(0,0) + v(0,1) - v(0,2) - v(0,3)).$$

Taking $v(0,0)$ from Section 6.1 and the others as $v(0,j) = e^2/r_{ij}$, the matrix element is 3.6 eV and equation (6.7) gives an excitation energy of 7.6 eV, compared to 7.0 eV experimentally. So in this case the simple formula is somewhat inaccurate.

As one goes higher in energy, there are many more states that can be excited, and the TDDFT gives a prediction for the entire spectrum. The absorption spectrum is shown in Figure 15, compared with the TDDFT theory. As stated above, the sharp state at 7.0 is the collective $\pi - \pi^*$ transition. Just above, there is some tiny structure in the experimental spectrum that reminds one of tuft of grass. These are the so-called Rydberg states, having a very loosely bound electron in the Coulomb field of the ion. These states cannot be described in the LDA because of its incorrect

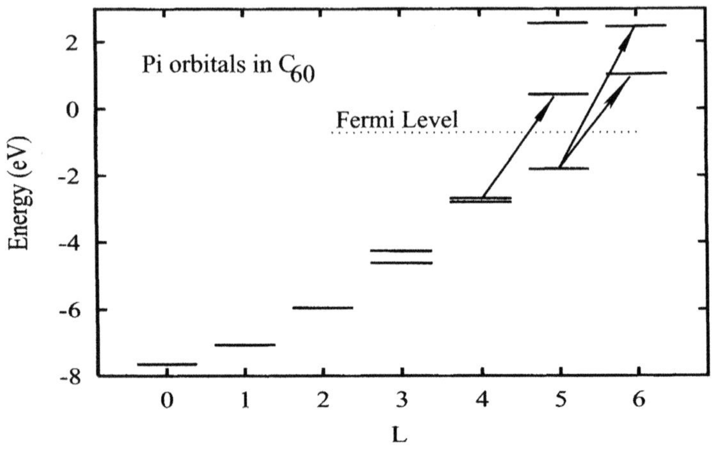

Fig. 16. Comparison of π orbital energies in the C_{60} molecule.

asymptotic potential field. The broad structure centered around 15 eV is due to the more tightly bound σ electrons. Its center position and overall width is correctly described by the theory. However, one can see that the theory has additional structure at high energy while the experimental cross section is very smooth. This is probably an artifact of the numerical method used to calculate the TDDFT, placing the system in a box.

6.4 C_{60}

The fullerenes are also carbon structures whose lowest excitations are contained in the $\pi - \pi^*$ manifold. The nearly spherical shape of C_{60} implies that angular momentum quantum numbers can be used to some extent to classify the orbitals. Thus, the lowest orbital of π character has an equal amplitude on all the carbon atoms, corresponding to $L = 0$. This classification breaks down at high L, where the discreteness of the Hückel Hamiltonian becomes visible in the spectrum. The spectrum of π orbitals is shown in Figure 16 with assigned L values. We see that the Fermi level splits the $L = 5$ orbitals. The arrows in the figure show the allowed $\pi - \pi^*$ transitions. The total transition strength occupied and empty π orbitals is about $f = 15$, which is considerably smaller than the number of π electrons (40). Experimentally, and in the TDDFT, one sees a group of transitions at about 7 eV excitation with a combined strength of about $f = 10$. Qualitatively, these transitions can be understood with the tight-binding model [12].

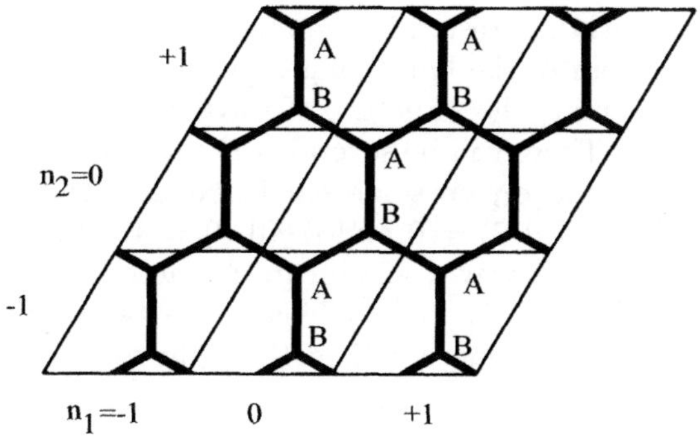

Fig. 17. Unit cells in the graphene lattice.

6.5 Carbon nanotubes

The last stop on my survey of carbon structures takes us up to extended nanostructures, the carbon nanotubes. They have very interesting electronic properties. Depending on their structure, they can be conductors or semiconductors, and one can even observe the individual conduction channels. Their properties may be understood with the Hückel model, at least on a qualitative level.

As we did before, we start with the theory for an infinite system, in this case an infinite graphite sheet. We follow the same route we used in constructing the wave functions for polyacetylene. Like polyacetylene, there are two atoms in the graphite unit cell (see Fig. 17). We label the two atoms "A" and "B" and distinguish the atomic amplitudes accordingly. Let us suppose we have an eigenstate, decomposed into A-site amplitudes $\phi_A(n_1, n_2)$ and $\phi_B(n_1, n_2)$ where (n_1, n_2) labels the cell in the two-dimensional lattice. Then the Hückel Hamiltonian only connects A and B amplitudes with each other,

$$\begin{pmatrix} 0 & H_{AB} \\ H_{AB} & 0 \end{pmatrix} \begin{pmatrix} \vec{\phi}_A \\ \vec{\phi}_B \end{pmatrix} = \epsilon \begin{pmatrix} \vec{\phi}_A \\ \vec{\phi}_B \end{pmatrix} . \tag{6.15}$$

From the structure of these equations it follows immediately that the spectrum of eigenvalues comes in pairs with equal and opposite sign. For each solution with energy ϵ we can get another state with energy $-\epsilon$ by the changing the sign of the ϕ_B's while keeping ϕ_A the same. This symmetry of the spectrum with respect to the sign of the energy was seen in benzene and polyacetylene and will be seen again in Manninen's lectures, and arises whenever the lattice can be divided into two equivalent sublattices with the Hamiltonian matrix elements zero within each sublattice.

We are particularly interested in the states at the Fermi level, which is at zero energy because of the half filling of a symmetric spectrum. So our task is now to determine the structure of wave functions of zero energy in the graphite lattice. There will be several of them, allowing arbitrary linear combinations of ϕ_A and ϕ_B on some particular unit cell. As an example we can assume that $\phi_B(0,0) = 0$ and $\phi_A(0,0) = 1$. We may also assume that the state is a Block wave, so we can characterize it with the vector $\vec{K}_F = K_1^F(\hat{x} - 1/\sqrt{3}\hat{y}) + K_2^F(2\hat{y}/\sqrt{3})$ where the two components K_1, K_2 give the phase change moving a cell to the right and to the upper right, respectively. Thus

$$\phi_{A,B}(n_1, n_2) = e^{i(K_1 n_1 + K_2 n_2)}\phi_{A,B}(0,0).$$

Then all ϕ_B amplitudes are zero, and equation (6.15) is trivially satisfied. The other condition, $H\vec{\phi}_A = 0$, reads as follows when expanded out for the B amplitude on the site (0,0):

$$-\beta\left(\phi_A(1,-1) + \phi_A(0,-1) + \phi_A(0,0)\right) = \\ -\beta\left(e^{i(K_1^F - K_2^F)} + e^{-iK_2^F} + 1\right)\phi_A(0,0) = 0.$$

The only way to satisfy this equation is to have the three terms in the middle expression be the three cube roots of one; thus

$$(K_1^F, K_2^F) = \pi(2/3, 4/3) \text{ or } \pi(-2/3, -4/3). \tag{6.16}$$

Since there is no continuous degree of freedom left in the choice of K_1^F and K_2^F, the Fermi surface is a set of isolated points. This is the well-known structure of the graphite Fermi surface: there is no energy gap between occupied and empty orbitals, but unlike a metal the Fermi surface has zero area. For other values of (K_1, K_2), the energy can be found the same way we used for polyacetylene. The result is

$$\epsilon(\vec{K}) = \pm\beta\sqrt{(1 + \cos(K_1 - K_2) + \cos K_2)^2 + (\sin(K_1 - K_2) - \sin K_2)^2}.$$

When $\vec{K} = K_1\hat{x} + K_2(\hat{x}/2 + \sqrt{3}\hat{y}/2)$is near a $\vec{K}_F$, the result can be approximated as [59]

$$\epsilon(\vec{K}) \approx \frac{\sqrt{3}}{2}\beta|\vec{K} - \vec{K}_F|. \tag{6.17}$$

We are now ready to look at nanotubes. We can construct a nanotube by rolling up a sheet of graphite, joining the sides so the hexagons align,

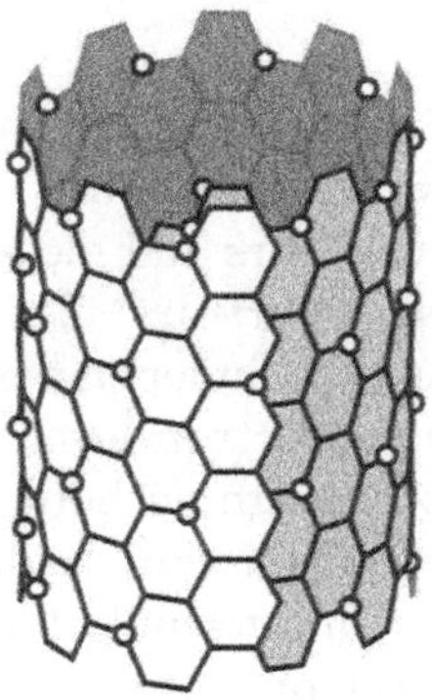

Fig. 18. Fermi surface on a carbon nanotube.

as shown in Figure 18. The various ways of aligning the hexagons can be distinguished by counting along the hexagons how far it is around the tube to come back to the starting hexagon. The counting is done along adjacent hexagons along the two directions corresponding to K_1, K_2 above.

Our $\epsilon = 0$ wave function will wrap around onto itself if the phases match on the superposed hexagon. This is the case for the tube in Figure 18, but will not be so if the hexagons are displaced one or two in any direction. The precise condition is

$$K_1 N_1 + K_2 N_2 = 2n\pi$$

which from equation (6.16) is equivalent to

$$\mathrm{mod}(N_1 - N_2, 3) = 0.$$

If it is satisfied, there is a state at the Fermi energy and thus there is a conduction channel. If the condition is not satisfied, there is a gap at the Fermi surface and the conduction channels only open when the chemical potential is changed to put the Fermi level with the allowed states of the channel.

The density of states can be measured by scanning tunneling microscopy and one can see the higher states from the peaks at the thresholds (called Van Hove singularies) [60]. In the cited reference, the authors observed a number of peaks in a nanotube that they identified as a (13,7) structure, nearly symmetrically distributed about $\epsilon = 0$. The ones closest to the middle were at $\epsilon \approx \pm 0.9$ eV. To see whether this makes sense, let's put the boundary condition for a simpler tube, the (10,10), into equation (6.17). The extra state then appears when $10\Delta K_1 + 10\Delta K_2 = 2\pi$. Taking $\Delta K_1 = \Delta K_2$, the condition becomes $\Delta K_1 = \pi/10$; and the formula gives $\beta K_1 \approx (2.5)(\pi/10) \approx 0.78$ eV, which is rather close to the measured value.

6.6 Quantized conductance

The conductance of the individual channels can be derived from the statistical reaction theory with two assumptions. Let us attach the two ends of the nanotube to (electron) reservoirs and measure the current as a function of the Fermi levels of the two reservoirs. The first assumption is that the electrons travel through the tube independently. Then we can apply the statistical theory to the single-electron level densities. The next assumption is that the electrons go through without reflection and disappear into the other reservoir, *i.e.* the transmission coefficient of the channel is unity. It is then very easy to derive the formula for the conductance considered as a statistical emission of electron from each reservoir into the other. The formula is

$$I = n_{\mathrm{c}} \frac{e^2}{2\pi\hbar} V$$

where V is the potential difference of the two reservoirs in volts and n_{c} is the number of open channels, including the two spin states of the electron as two separate channels. I will not go through the details in these notes because the algebra is just the same as in my two-page published derivation in [12]. The staircase behavior of the above expression as the channels open up is seen in many conductance measures on nanoscale structures. An example of this phenomenon seen in the carbon nanotubes is [61]. This particular experiment seems to show a case where $n_{\mathrm{c}} = 1$, which can only be if one spin state of the electron is blocked. To my knowledge, there is no explanation for a strong spin dependent transmission.

Many of the results presented here were obtained in collaboration with K. Yabana, and his contribution is gratefully acknowledged. The author also thanks D. Tomanek for discussions related to Section 6.5, C. Guet for discussions related to Section 5, and T. Döppner and Q. Huang for proofreading the manuscript. The author's research mentioned in these lectures was supported by the Department of Energy under Grant FG06-90ER-40561.

References

[1] W.D. Knight *et al.*, *Phys. Rev. Lett.* **52** (1984) 2141.
[2] A.L. Mackay, *Acta Cryst.* **15** (1962) 916.
[3] O. Echt *et al.*, *Phys. Rev. Lett.* **47** (1981) 1122, Xe clusters.
[4] H.W. Kroto *et al.*, *Nature* **318** (1985) 163.
[5] W. Krätschmer *et al.*, *Nature* **347** (1990) 354.
[6] J. Borggreen *et al.*, *Phys. Rev. A* **62** (2000) 013202.
[7] C. Brechignac *et al.*, *Phys. Rev. Lett.* **68** (1992) 3916.
[8] J. Pedersen *et al.*, *Z. Phys. D* **26** (1993) 281.
[9] G. Bertsch and W. Ekardt, *Phys. Rev. B* **32** (1985) 7659.
[10] G.F. Bertsch and D. Tomanek, *Phys. Rev. B* **40** (1989) 2749.

[11] G. Bertsch, *Comp. Phys. Comm.* **60** (1990) 247. The program JELLYRPA may be downloaded from the author's Web site, `www.phys.washington.edu/~bertsch`

[12] G.F. Bertsch, N. Oberhofer and S. Stringari, *Z. Phys. D* **20** (1991) 123.

[13] G.F. Bertsch, *J. Phys. Cond. Matt.* **3** (1991) 373.

[14] G. Bertsch, A. Bulgac, D. Tomanek and Y. Wang, *Phys. Rev. Lett.* **67** (1991) 2690.

[15] K. Yabana and G. Bertsch, *Chem. Phys. Lett.* **197** (1992) 32.

[16] Y. Wang, C. Lewenkopf, D. Tomanek and G. Bertsch, *Chem. Phys. Lett.* **205** (1993) 521.

[17] Y. Wang, G. Bertsch and K. Ieki, *Zeit. Physik D* **25** (1993) 181.

[18] G.F. Bertsch and R.A. Broglia, *Oscillations in Finite Quantum Systems* (Cambridge Univ. Press., 1994).

[19] G.F. Bertsch and K. Yabana, *Phys. Rev. A* **49** (1994) 1930.

[20] K. Yabana and G.F. Bertsch, *J. Chem. Phys.* **100** (1994) 5580.

[21] G. Bertsch, N. Onishi and K. Yabana, *Z. Phys. D* **34** (1995) 213.

[22] G.F. Bertsch, A. Smith and K. Yabana, *Phys. Rev. B* **52** (1995) 7876.

[23] K. Yabana and G.F. Bertsch, *Z. Phys. D* **32** (1995) 329.

[24] K. Yabana and G.F. Bertsch, *Phys. Rev. B* **54** (1996) 4484.

[25] K. Yabana and G.F. Bertsch, *Zeit. Phys. D* **42** (1997) 219.

[26] K. Yabana and G.F. Bertsch, *Phys. Rev. A* **58** (1998) 2604.

[27] K. Yabana and G.F. Bertsch, *Phys. Rev. A* **60** (1999) 1271.

[28] K. Yabana and G.F. Bertsch, *Phys. Rev. A* **60** (1999) 3809.

[29] K. Yabana and G.F. Bertsch, *Int. J. Quant. Chem.* **75** (1999) 55.

[30] K. Yabana, G.F. Bertsch and A. Rubio, preprint arXiv: phys/0003090.

[31] G.F. Bertsch, J.-I. Iwata, A. Rubio and K. Yabana, preprint arXiv: cond-mat/0005512.

[32] V. Weisskopf, *Phys. Rev.* **52** (1937) 295.

[33] L.S. Kassel, *J. Chem. Phys.* **32** (1928) 225; **32** (1928) 1065.

[34] P.C. Engelking, *J. Chem. Phys.* **87** (1987) 3784.

[35] J.M. Weber, K. Hansen, M. Ruf and H. Hotop, *Chem. Phys.* **239** (1998) 271.

[36] C. Walther *et al.*, *Phys. Rev. Lett.* **83** (1999) 3816.

[37] J.U. Anderson *et al.*, *Phys. Rev. Lett.* **77** (1996) 3991.

[38] B. Weber and R. Scholl, *A. Appl. Phys.* **74** (1993) 607.

[39] K. Hansen and E. Campbell, *Phys. Rev. E* **58** (1998) 5477.

[40] H. Flocard, S. Koonin and M. Weiss, *Phys. Rev. C* **17** (1978) 1682.

[41] J.P. Perdew, K. Burke and M. Ernzerhof, *Phys. Rev. Lett.* **77** (1996) 3865.

[42] J.P. Perdew, S. Kurth, A. Zupan and P. Blaha, *Phys. Rev. Lett.* **82** (1999) 2544.

[43] L. Hedin, *J. Phys. Cond. Matt.* **11** (1999) R489.

[44] A. Fetter and J.D. Walecka, *Quantum Theory of Many-Particle Systems* (McGraw-Hill, N.Y., 1971).

[45] A.D. Becke, *Phys. Rev. A* **38** (1988) 3098.

[46] J. Kohanoff *et al.*, *Phys. Rev. B* **46** (1992) 4371.

[47] A. Zangwill and P. Soven, *Phys. Rev. A* **21** (1980) 1561.

[48] M. Madjet, C. Guet and W.R. Johnson, *Phys. Rev. A* **51** (1995) 1327.

[49] S.G. Louie, S. Froyen and M.L. Cohen, *Phys. Rev. B* **26** (1982) 1738.

[50] S.A. Blunden and C. Guet, *Z. Phys. D* **33** (1995) 153.

[51] C. Brechignac *et al.*, *Phys. Rev. Lett.* **70** (1993) 2036.

[52] A. Liebsch, *Phys. Rev. Lett.* **71** (1993) 145.
[53] J. Tiggesbäumker *et al.*, *Chem. Phys. Lett.* **190** (1992) 42.
[54] D. Forney *et al.*, *J. Chem. Phys.* **104** (1996) 4954.
[55] K. Chang and W. Graham, *J. Chem. Phys.* **77** (1982) 4300.
[56] B. Champagne *et al.*, *J. Phys. Chem.* **104** (2000) 4755.
[57] M. Rohlfing and S.G. Louie, *Phys. Rev. Lett.* **82** (1999) 1959.
[58] H. Perkampus, *UV Atlas of organic compounds*, Vol. 1 (Butterworth Verlag Chemie, 1968).
[59] J.W. Mintmire and C.T. White, *Phys. Rev. Lett.* **81** (1998) 2506.
[60] P. Kim, T. Odom, J.L. Huang and C.M. Lieber, *Phys. Rev. Lett.* **82** (1999) 1225.
[61] S. Frank *et al.*, *Science* **280** (1998) 1745.

COURSE 4

DENSITY FUNCTIONAL THEORY, METHODS, TECHNIQUES, AND APPLICATIONS

D.R. SALAHUB

Steacie Institute for Molecular Sciences (SIMS), National Research Council of Canada, 100 Sussex Drive, Ottawa (Ontario) K1A 0R6, Canada, Département de Chimie, Université de Montréal, C.P. 6128, Succursale Centre-Ville, Montréal (Québec), H3C 3J7, Canada, Centre de Recherche en Calcul Appliqué (CERCA), 5160 Bd. Décarie, bureau 400, Montréal (Québec), H3X 2H9, Canada

Contents

DENSITY FUNCTIONAL THEORY, METHODS, TECHNIQUES, AND APPLICATIONS

S. Chrétien[1] and D.R. Salahub[1,2,3]

Abstract

Important concepts of Density Functional Theory (DFT) and the different types of approximation for the exchange-correlation energy functional are presented. Applications illustrate the advantages of using DFT as a computational tool and point out some limitations as well.

1 Introduction

Computer aided design of new materials and inhibitors with desired properties requires the description of chemical reactions in complex environments with a high degree of accuracy, especially if one needs to know the effect of temperature, pressure, solvation and/or pH on a reaction. At the present time, a full quantum mechanical treatment is out of the question, even with linear scaling methods [1]. To make this kind of calculation tractable on today's computers, without loss of too much accuracy, the system has to be broken into pieces and different computational techniques used for the different parts of the model. The combination of quantum mechanics and molecular mechanics (QM/MM) using classical force fields is usefull in catalysis [2] and in biological systems [3]. For example, the active site of a protein, including the inhibitor, can be computed quantum mechanically and the rest of the protein by molecular mechanics. In this model, the presence of the solvent molecules, could be added explicitly or by a continuum model. The challenge with this kind of approach is to make the techniques work together.

[1]Université de Montréal.
[2]National Research Council of Canada.
[3]CERCA.

Density Functional Theory (DFT) is often the quantum mechanical method of choice because the same accuracy as high-level traditional *ab initio* techniques may be attained but with a fraction of the computational time. Larger molecules can be studied with DFT.

In this lecture we present DFT from a chemical point of view. Chemical applications are chosen to illustrate both the successes and the limitations of DFT. For more details about electronic structure calculations and DFT the following books are recommended [4–7].

The first application we have choosen is an *ab initio* molecular dynamics (AIMD) simulation of an alanine dipeptide analogue. This calculation introduces temperature effects through molecular dynamics in which the classical force field is replaced by a quantum mechanical calculation. The study, along with others, points to the need to include polarization effects in future molecular mechanics applications. In the second example, we compare the experimental pulsed-field-ionization-zero-electron-kinetic-energy (PFI-ZEKE) spectrum of V_3 with the simulated one. This demonstrates the power of DFT to predict equilibrium structures for demanding transition metal systems. The third example is about small nickel clusters for which one sees the need for attention to technical details for these difficult systems that involve close lying states with different magnetic structure. The last example concerns preliminary results from a DFT study of the formation of benzene from acetylene catalysed by iron clusters. Although the theory, methods and applications presented in this chapter are, inevitably, far from exhaustive we think they will provide a good entry to the literature and provide an inkling of what can, and what cannot, be done with DFT as it stands today.

2 Density functional theory

For a time-independent system, the total electronic energy is obtained by the solution of the time-independent Schrödinger equation [8],

$$\hat{H}\Psi(\vec{x}_1, \vec{x}_2, \ldots, \vec{x}_N) = E\Psi(\vec{x}_1, \vec{x}_2, \ldots, \vec{x}_N). \tag{2.1}$$

$\Psi(\vec{x}_1, \vec{x}_2, \ldots, \vec{x}_N)$ is the ground-state many-electron wave function and $\vec{x}_i$ contains both the spatial, $\vec{r}_i$, and the spin, σ_i, coordinates of electron i. For a system composed of N electrons and M nuclei the non-relativistic Hamiltonian, in the Born-Oppenheimer approximation [9], is given by (in atomic units)

$$\hat{H} = \hat{T} + \hat{V}_{\text{ne}} + \hat{V}_{\text{ee}}. \tag{2.2}$$

This operator is composed of the kinetic energy operator,

$$\hat{T} = \sum_{i=1}^{N} -\frac{1}{2}\nabla_i^2, \tag{2.3}$$

the nucleus-electron attraction operator,

$$\hat{V}_{\text{ne}} = \sum_{i=1}^{N} v(\vec{r}_i) = \sum_{i=1}^{N}\sum_{\alpha=1}^{M} \frac{-Z_\alpha}{|\vec{R}_\alpha - \vec{r}_i|} \tag{2.4}$$

(where Z_α is the atomic number of atom α) and the electron-electron repulsion operator,

$$\hat{V}_{\text{ee}} = \sum_{i=1}^{N}\sum_{j>i}^{N} \frac{1}{|\vec{r}_i - \vec{r}_j|}. \tag{2.5}$$

The many-electron wave function is not an observable and is a complicated function that depends on $4N$ coordinates (x_i, y_i, z_i and σ_i for each electron) which prevents its visualization in real three-dimensional space. Exact solution of the Schrödinger equation is limited to a small number of electronic systems. On the other hand, the electronic density can be measured experimentally by X-ray diffraction and depends only on three spatial coordinates (x, y and z). Moreover, $\rho(\vec{r})$ can be obtained from the integration of the total many-electron wave function over all coordinates, except the spatial coordinate of one electron (the one we are looking at),

$$\rho(\vec{r}_1) = N \int \dots \int \mathrm{d}\sigma_1 \mathrm{d}\vec{x}_2 \mathrm{d}\vec{x}_3 \dots \mathrm{d}\vec{x}_N |\Psi(\vec{x}_1, \vec{x}_2, \dots, \vec{x}_N)|^2. \tag{2.6}$$

This function is a probability distribution that corresponds to the probability of finding simultaneously electron 1 with spin σ_1 at position $\vec{r}_1$ and electrons 2, 3, ..., N in the volume element $\mathrm{d}\vec{x}_2, \mathrm{d}\vec{x}_3, \dots, \mathrm{d}\vec{x}_N$ respectively. $\rho(\vec{r})$ is a positive function that goes to zero at infinity ($r \to \infty$) and gives the total number of electrons when integrated over all of space,

$$\int \rho(\vec{r}_1)\mathrm{d}\vec{r}_1 = N. \tag{2.7}$$

For all the properties mentioned above, $\rho(\vec{r})$ seems to be a good alternative to the complicated total many-electron wave function. But, is it theoretically justified to use the electronic density as the cornerstone of electronic structure calculations? This question was answered by the Hohenberg and Kohn (HK) theorems [10] published in 1964.

2.1 Hohenberg and Kohn theorems

The only system-dependent part in the Hamiltonian operator (2.2) is the external potential, $v(\vec{r})$, because the kinetic (2.3) and electron-electron repulsion (2.5) operators are fixed once the number of electrons is known. In traditional wave function theory, if we know $v(\vec{r})$, which fixes $\hat{H}$ and Ψ, then $\rho(\vec{r})$ is obtained from (2.6). In other words, $\rho(\vec{r})$ is a functional[4] of $v(\vec{r})$ and Ψ. Hohenberg and Kohn [10] proved the converse, $v(\vec{r})$ and Ψ are unique functionals of $\rho(\vec{r})$. The proof is very simple and Hohenberg and Kohn proceed by *reductio ad absurdum*. Suppose we have two Hamiltonian operators, $\hat{H} = \hat{T} + \hat{V}_{\rm ee} + \int v(\vec{r})\rho(\vec{r})d\vec{r}$ and $\hat{H}' = \hat{T} + \hat{V}_{\rm ee} + \int v'(\vec{r})\rho(\vec{r})d\vec{r}$, which differ only by the external potential. We also suppose that both external potentials differ by more than an additive constant but give the same ground state density, $\rho_0(\vec{r})$. Clearly, a different ground state energy, E_0 and E'_0, and ground state wave functions, Ψ_0 and Ψ'_0, is associated with each Hamiltonian operator. Using Ψ'_0 as a trial function for $\hat{H}$ we can write

$$
\begin{aligned}
E_0 < \langle\Psi'_0|\hat{H}|\Psi'_0\rangle &= \langle\Psi'_0|\hat{H}'|\Psi'_0\rangle + \langle\Psi'_0|\hat{H} - \hat{H}'|\Psi'_0\rangle \\
&= E'_0 + \int \rho(\vec{r})[v(\vec{r}) - v'(\vec{r})]d\vec{r}. \qquad (2.8)
\end{aligned}
$$

Doing the same operation for E'_0 we find

$$
\begin{aligned}
E'_0 < \langle\Psi_0|\hat{H}'|\Psi_0\rangle &= \langle\Psi_0|\hat{H}|\Psi_0\rangle + \langle\Psi_0|\hat{H}' - \hat{H}|\Psi_0\rangle \\
&= E_0 - \int \rho(\vec{r})[v(\vec{r}) - v'(\vec{r})]d\vec{r}. \qquad (2.9)
\end{aligned}
$$

Adding (2.8) and (2.9) we obtain the following contradiction

$$E_0 + E'_0 < E_0 + E'_0. \qquad (2.10)$$

This means that it is impossible that two external potentials that differ by more than an additive constant give the same ground state density. Consequently, the external potential is a unique functional of $\rho(\vec{r})$. Hence, the ground state density, $\rho_0(\vec{r})$, determines N, $v(\vec{r})$, Ψ_0, E_0 and all the ground state properties. For a given external potential, $v(\vec{r})$, the total energy expression is a functional of $\rho(\vec{r})$,

$$E_{\rm v}[\rho(\vec{r})] = \int v(\vec{r})\rho(\vec{r})d\vec{r} + F_{\rm HK}[\rho(\vec{r})] \qquad (2.11)$$

[4] A function requires a number as input and delivers a number as output. A functional requires a function as input and, as a function does, delivers a number as output.

with

$$F_{\mathrm{HK}}[\rho(\vec{r})] = \langle\Psi|\hat{T} + \hat{V}_{\mathrm{ee}}|\Psi\rangle = T[\rho(\vec{r})] + V_{\mathrm{ee}}[\rho(\vec{r})]. \tag{2.12}$$

$F_{\mathrm{HK}}[\rho(\vec{r})]$ is a universal functional of $\rho(\vec{r})$ because it is fixed when the number of electrons is known. In theory, if we were able to write down the exact mathematical form of $F_{\mathrm{HK}}[\rho(\vec{r})]$, then the ground-state energy and density would be known exactly. In practice, we do not know it, but we do know that it exists and we know some properties that it should possess.

In their 1964 paper [10], Hohenberg and Kohn presented a second theorem, a variational principle with $\rho(\vec{r})$ as the basic variable. The total energy functional (2.11) will be minimized if the trial density, $\tilde{\rho}(\vec{r})$, is the ground state density

$$E_{\mathrm{v}}[\rho_0(\vec{r})] \leq E_{\mathrm{v}}[\tilde{\rho}(\vec{r})] = \int v(\vec{r})\tilde{\rho}(\vec{r})\mathrm{d}\vec{r} + F[\tilde{\rho}(\vec{r})]. \tag{2.13}$$

The trial density is restricted to be a positive well behaved function restricted to integrate to the total number of electrons ($N[\tilde{\rho}(\vec{r})] = \int \tilde{\rho}(\vec{r})\mathrm{d}\vec{r} = N$) and must vanish at infinity.

These two very simple theorems were the first theoretical justification of using $\rho(\vec{r})$ instead of Ψ to determine all the ground state properties. But, the first theorem is limited to non-degenerate ground states and it is not guaranteed that for all "normalisable" $\rho(\vec{r})$, a physically realisable potential exists. This last uncertainty is the v-representability problem and is important because we are dealing with real systems in chemistry and this external potential determines the nuclear position according to (2.4). In the next section, we will see that Levy's constrained search [11,12] removes some of the restrictions met in the Hohenberg and Kohn theorems.

2.2 *Levy's constrained search*

We need a way to select, from all the acceptable trial densities only those that correspond to a physical external potential.

The variational principle (2.13) can be equivalently written as:

$$E_0 \leq \min_{\tilde{\rho}(\vec{r})} \left\{ \min_{\Psi\to\tilde{\rho}(\vec{r})} \langle\Psi|\hat{T} + \hat{V}_{\mathrm{ee}} + \sum_{i=1}^{N} v(\vec{r}_i)|\Psi\rangle \right\}. \tag{2.14}$$

This equation represent a two-step optimization procedure. In the first step, the inner minimization, we search all antisymmetric wave functions that yield a given trial density. In the second step we search, among all acceptable densities, the one that minimizes the total energy. We have an equality in (2.14) if the trial density is $\rho_0(\vec{r})$. Since the external potential is fixed by the density only, and will be constant for all Ψ this yields the same $\rho(\vec{r})$, we can define the universal functional

$$Q[\tilde{\rho}] = \min_{\Psi \to \tilde{\rho}(\vec{r})} \langle \Psi | \hat{T} + \hat{V}_{\text{ee}} | \Psi \rangle \cdot \quad (2.15)$$

This universal functional corresponds to F_{HK} if the trial density correspond to $\rho_0(\vec{r})$. This way of demonstrating the first Hohenberg–Kohn theorem removes the limitation to nondegenerate ground states. Moreover, using (2.15) replaces the stringent v-representability condition on $\rho(\vec{r})$ by N-representability, which is easier to satisfy. A density is N-representable if it is normalisable (2.7), acceptable and comes from an antisymmetric wave function. Using the definition for the universal functional, the variational principle (2.14) becomes:

$$E_0 \leq \min_{\tilde{\rho}(\vec{r})} \left\{ Q[\tilde{\rho}] + \int v(\vec{r})\tilde{\rho}(\vec{r})\mathrm{d}\vec{r} \right\} \cdot \quad (2.16)$$

2.3 Kohn–Sham method

The Thomas and Fermi method [13, 14] was the first density functional model. The approximations used were too crude. The simple form for the kinetic energy functional, $T_{\text{TF}}[\rho] = C_{\text{F}} \int \rho(\vec{r})^{5/3}\mathrm{d}\vec{r}$, derived from the homogeneous electron gas, is the source of part of the problem. Kohn and Sham [15] decided to compute the biggest contribution to the universal functional, $F[\rho]$ as accurately as possible. In this method, the classical coulomb repulsion energy, $J[\rho]$, and the exact kinetic energy for a system of non-interacting particles having the same density as the interacting one, $T_{\text{s}}[\rho]$, are calculated explicitly. The remaining part of $F[\rho]$ is the exchange and correlation functional, $E_{\text{xc}}[\rho]$ (2.18). This functional contains the residual kinetic energy $(T[\rho] - T_{\text{s}}[\rho])$ and non classical electron-electron repulsion energy $(V_{\text{ee}}[\rho] - J[\rho])$, *i.e.* exchange, self-interaction and correlation.

$$F[\rho] = J[\rho] + T_{\text{s}}[\rho] + E_{\text{xc}}[\rho] \quad (2.17)$$

$$E_{\text{xc}}[\rho] = T[\rho] - T_{\text{s}}[\rho] + V_{\text{ee}}[\rho] - J[\rho]. \quad (2.18)$$

To calculate the exact noninteracting kinetic energy, Kohn and Sham [15] introduced a set of one-electron orbitals, $\{\varphi_i\}$, that are eigenfunctions of the effective one-electron Hamiltonian

$$\left[-\frac{1}{2}\nabla^2 + v_{\text{eff}}(\vec{r}) \right] \varphi_i = \varepsilon_i \varphi_i. \quad (2.19)$$

In terms of these one-electron orbitals, the electronic density is:

$$\rho(\vec{r}) = \sum_{i=1}^{N} \sum_{\sigma=\uparrow,\downarrow} |\varphi_i(\vec{r},\sigma)|^2. \tag{2.20}$$

Varying $E[\rho]$ subject to the normalisation constraint (2.7), we find the effective potential,

$$v_{\text{eff}}(\vec{r}) = \frac{1}{2} \int \frac{\rho(\vec{r}')}{|\vec{r}-\vec{r}'|} \mathrm{d}\vec{r}' + v(\vec{r}) + v_{\text{xc}}(\vec{r}), \tag{2.21}$$

that minimizes the energy expression,

$$E[\rho] = T_{\text{s}}[\rho] + \frac{1}{2} \int\int \frac{\rho(\vec{r})\rho(\vec{r}')}{|\vec{r}-\vec{r}'|} \mathrm{d}\vec{r}\mathrm{d}\vec{r}' + \int v(\vec{r})\rho(\vec{r})\mathrm{d}r + E_{\text{xc}}[\rho(\vec{r})]. \tag{2.22}$$

Where the exchange-correlation potential, v_{xc}, is given by

$$v_{\text{xc}} = \frac{\delta E_{\text{xc}}[\rho]}{\delta \rho}. \tag{2.23}$$

Equations (2.19, 2.20) and (2.21) are the Kohn–Sham (KS) equations. They have to be solved self-consistently like Hartree–Fock (HF) equations. Equation (2.19) can be written in a form similar to the Hartree–Fock equation,

$$\hat{f}^{\text{KS}} \varphi_i^{\text{KS}} = \varepsilon_i^{\text{KS}} \varphi_i^{\text{KS}}. \tag{2.24}$$

Even though there is a similarity between KS and HF equations, the eigenfunctions are different. We will come back to this point in Section 5.

The Kohn–Sham equations are exact. In principle, we can get the exact ground state density. The approximations enter $E_{\text{xc}}[\rho]$ and v_{xc} for which no exact form has been found until now (we will see in Sect. 7 that they are expressed by complicated mathematical functions). However, better and better approximations exist and this explains the popularity of DFT in chemistry.

In the next five sections, we introduce concepts that should help to understand the different approaches and approximations present in different exchange-correlation functionals.

3 Density matrices and pair correlation functions

The Hamiltonian operator (2.2) contains only one- and two-electron operators which means that we can express the total energy in terms of the probability of finding one electron and pairs of electrons in space. In this section, we introduce the concept of first- and second-order spinless density matrices. For more details on density matrices, we recommend the book of Davidson [16].

Focusing on the energy, one can greatly reduce the number of variables in the Schrödinger equation by integrating the product of the many-electron wave function with its complex conjugate form over all variables, except the spatial coordinates of two electrons, and summing over spin. By doing this, we obtain the second order spinless density matrix,

$$\begin{aligned}\rho(\vec{r}_1,\vec{r}_2,\vec{r'}_1,\vec{r'}_2) &= \frac{N(N-1)}{2}\int\ldots\int\Psi(\vec{r}_1,\sigma_1,\vec{r}_2,\sigma_2,\vec{x}_3,\ldots,\vec{x}_N)\\ &\quad\Psi^*(\vec{r'}_1,\sigma_1,\vec{r'}_2,\sigma_2,\vec{x}_3,\ldots,\vec{x}_N)\mathrm{d}\sigma_1\mathrm{d}\sigma_2\mathrm{d}\vec{x}_3\ldots\mathrm{d}\vec{x}_N. \quad (3.1)\end{aligned}$$

The pair density matrix,

$$\begin{aligned}\rho_2(\vec{r}_1,\vec{r}_2) &= \rho_2(\vec{r}_1,\vec{r}_2,\vec{r}_1,\vec{r}_2) \quad (3.2)\\ &= \frac{N(N-1)}{2}\int\ldots\int|\Psi(\vec{x}_1,\vec{x}_2,\ldots,\vec{x}_N)|^2\mathrm{d}\sigma_1\mathrm{d}\sigma_2\mathrm{d}\vec{x}_3\ldots\mathrm{d}\vec{x}_N,\end{aligned}$$

is the diagonal element of the second order density matrix, $\rho_2(\vec{r}_1,\vec{r}_2,\vec{r}_1,\vec{r}_2)$, and corresponds to the probability of finding a pair of electrons, one at $\vec{r}_1$ and another at $\vec{r}_2$, when the system is in the state represented by Ψ. We introduce a short-hand notation. The subscript of ρ refers to the order of the matrix and we write $\rho_2(\vec{r}_1,\vec{r}_2)$ instead of $\rho_2(\vec{r}_1,\vec{r}_2,\vec{r}_1,\vec{r}_2)$ while keeping in mind that $\vec{r'}_1=\vec{r}_1$ and $\vec{r'}_2=\vec{r}_2$. This probability distribution contains all the information about electron correlation and gives the total number of distinct pairs of electrons when integrated over the whole space with respect to $\vec{r}_1$ and $\vec{r}_2$. Moreover, one can show that the probability of finding two electrons with the same spin at the same place in space, $\rho_2(\vec{r}_1,\vec{r}_1)$, is zero, which satisfies the Pauli principle.

The pair density matrix can be equivalently written as:

$$\rho_2(\vec{r}_1,\vec{r}_2)=\rho(\vec{r}_1)\rho(\vec{r}_2)[1+h(\vec{r}_1,\vec{r}_2)]. \quad (3.3)$$

$h(\vec{r}_1,\vec{r}_2)$ is a hole function. We will discuss its physical meaning in the next section. From (3.3), if $h(\vec{r}_1,\vec{r}_2)=0$, the pair density matrix corresponds to the limit of two uncorrelated charge distributions, $\rho(\vec{r}_1)$ and $\rho(\vec{r}_2)$, so $h(\vec{r}_1,\vec{r}_2)$ represents the departure from this uncorrelated limit. The hole function contains all the information about nonclassical effects including electron correlation ($V_{\mathrm{ee}}[\rho]-J[\rho]$) and the residual kinetic energy ($T[\rho]-T_{\mathrm{s}}[\rho]$).

The first-order spinless density matrix is obtained by integrating the product $\Psi\Psi^*$ over all variables, except the spatial coordinates of one electron, and summing over spin,

$$\begin{aligned}\rho(\vec{r}_1,\vec{r}\,'_1) &= N\int\ldots\int \Psi(\vec{r}_1,\sigma_1,\vec{x}_2,\ldots,\vec{x}_N)\\ &\quad \Psi^*(\vec{r}\,'_1,\sigma_1,\vec{x}_2,\ldots,\vec{x}_N)\mathrm{d}\sigma_1\mathrm{d}\vec{x}_2\ldots\mathrm{d}\vec{x}_N. \end{aligned} \tag{3.4}$$

The diagonal element of this matrix, $\rho_1(\vec{r}_1,\vec{r}_1)$, is the electron density (2.6),

$$\rho_1(\vec{r}_1) = \rho_1(\vec{r}_1,\vec{r}_1) = N\int\ldots\int |\Psi(\vec{r}_1,\vec{x}_2,\ldots,\vec{x}_N)|^2\mathrm{d}\sigma_1\mathrm{d}\vec{x}_2\mathrm{d}\vec{x}_3\ldots\mathrm{d}\vec{x}_N. \tag{3.5}$$

In terms of first- and second-order spinless density matrices, the total energy is given by:

$$\begin{aligned} E &= E[\rho_1(\vec{r}_1,\vec{r}\,'_1),\rho_2(\vec{r}_1,\vec{r}_2,\vec{r}\,'_1,\vec{r}\,'_2)]\\ &= \int\left[-\frac{1}{2}\nabla^2\rho_1(\vec{r}\,',\vec{r})\right]_{\vec{r}\,'=\vec{r}}\mathrm{d}\vec{r} + \int v(\vec{r})\rho_1(\vec{r})\mathrm{d}\vec{r}\\ &\quad +\frac{1}{2}\int\int\frac{\rho_2(\vec{r}_1,\vec{r}_2)}{|\vec{r}_1-\vec{r}_2|}\mathrm{d}\vec{r}_1\mathrm{d}\vec{r}_2. \end{aligned} \tag{3.6}$$

4 Adiabatic connection or coupling strength integration

The adiabatic connection method [17–20] represents one of the most important concepts in DFT. We start by introducing a coupling strength parameter, λ, that varies between 0 and 1 into the universal functional defined by the constrained search (2.15),

$$F_\lambda[\rho] = \min_{\Psi\to\rho(\vec{r})}\langle\Psi|\hat{T}+\lambda\hat{V}_{\mathrm{ee}}|\Psi\rangle = \langle\Psi^\lambda|\hat{T}+\lambda\hat{V}_{\mathrm{ee}}|\Psi^\lambda\rangle. \tag{4.1}$$

This parameter switches OFF ($\lambda = 0$) or ON ($\lambda = 1$) the electron-electron interaction. The equation above implies that for each value of λ, the many-electron wave function must integrate to the exact density. In other words, $\rho(\vec{r})$ is independant of λ. This is achieved by adjusting the external potential for each value of λ. In the two limiting cases, the universal functional is,

$$F_0[\rho] = T_\mathrm{s}[\rho] \tag{4.2}$$

$$F_1[\rho] = T[\rho] + V_{\mathrm{ee}}[\rho]. \tag{4.3}$$

With these definitions, the exchange-correlation functional (2.18) can be written as:

$$E_{\mathrm{xc}}[\rho] = F_1[\rho] - F_0[\rho] - J[\rho] = \int_0^1 \mathrm{d}\lambda \frac{\partial F_\lambda[\rho]}{\partial \lambda} - J[\rho]. \tag{4.4}$$

This formula connects the Kohn–Sham non-interacting reference system ($\lambda = 0$) with the real fully interacting system ($\lambda = 1$) by an integration over λ. The integrand for different values of λ is evaluatated at the real (fully interacting) density. This process is called the adiabatic connection because of the analogy to an adiabatic process in thermodynamics. Here, one goes from $\lambda = 0$ to $\lambda = 1$ by infinitely small variations in λ at constant density instead of constant energy.

Moreover, using the Hellmann–Feynmann theorem, one can show,

$$\frac{\partial F_\lambda[\rho]}{\partial \lambda} = \langle \Psi^\lambda | \hat{V}_{\mathrm{ee}} | \Psi^\lambda \rangle. \tag{4.5}$$

Summing over spin and integrating this definition over all spatial coordinates except two, $\vec{r}_1$ and $\vec{r}_2$, we introduce the λ-dependent second-order spinless density matrix into the definition of E_{xc},

$$E_{\mathrm{xc}}[\rho] = \int\int \frac{\mathrm{d}\vec{r}_1 \mathrm{d}\vec{r}_2}{|\vec{r}_1 - \vec{r}_2|} \int_0^1 \mathrm{d}\lambda \rho_2^\lambda(\vec{r}_1, \vec{r}_2) - J[\rho]. \tag{4.6}$$

Using the pair density matrix definition (3.3) and performing the integration over λ brings in the λ-averaged hole function, $\bar{h}(\vec{r}_1, \vec{r}_2)$,

$$\int_0^1 \mathrm{d}\lambda \rho_2^\lambda(\vec{r}_1, \vec{r}_2) = \frac{1}{2}\rho(\vec{r}_1)\rho(\vec{r}_2)[1 + \bar{h}(\vec{r}_1, \vec{r}_2)]. \tag{4.7}$$

Finally, introducing the definition of the exchange-correlation hole function,

$$\bar{\rho}_{\mathrm{xc}}(\vec{r}_1, \vec{r}_2) = \rho(\vec{r}_2)\bar{h}(\vec{r}_1, \vec{r}_2), \tag{4.8}$$

we obtain the following definition of the exchange-correlation energy functional,

$$E_{\mathrm{xc}}[\rho] = \frac{1}{2}\int\int \frac{\mathrm{d}\vec{r}_1 \mathrm{d}\vec{r}_2}{|\vec{r}_1 - \vec{r}_2|}\rho(\vec{r}_1)\bar{\rho}_{\mathrm{xc}}(\vec{r}_1, \vec{r}_2). \tag{4.9}$$

The λ-averaged exchange-correlation hole, $\bar{\rho}_{\mathrm{xc}}(\vec{r}_1, \vec{r}_2)$, is very important because it fixes the exchange-correlation energy, it contains all the information about the non-classical electron-electron interaction as well as the residual kinetic energy, and corresponds to the probability of finding an electron at $\vec{r}_2$ when the reference electron is located at $\vec{r}_1$. The hole function moves with the reference electron and we talk about a hole because it reduces the probability of a second electron coming too close to the reference one. The hole function digs a hole in the probability distribution of the second electron near the location of the reference electron.

The exact form of the exchange-correlation hole function is unknown but we do know some properties that it should posseses. For example, the hole contains exactly one electron,

$$\int \rho_{\mathrm{xc}}(\vec{r}_1, \vec{r}_2)\mathrm{d}\vec{r}_2 = \int \rho(\vec{r}_2)h(\vec{r}_1, \vec{r}_2)\mathrm{d}\vec{r}_2 = -1. \tag{4.10}$$

This property is known as the sum rule.

As we will see in Section 7, it is convenient to separate the exchange-correlation energy in terms of separate exchange and correlation functionals. This yields an exchange hole function,

$$\rho_{\mathrm{x}}(\vec{r}_1, \vec{r}_2) = -\frac{1}{2}\frac{|\rho_1(\vec{r}_1, \vec{r}_2)|^2}{\rho(\vec{r}_1)}, \tag{4.11}$$

also called the Fermi hole and a correlation hole,

$$\rho_{\mathrm{c}}(\vec{r}_1, \vec{r}_2) = \rho_{\mathrm{xc}}(\vec{r}_1, \vec{r}_2) - \rho_{\mathrm{x}}(\vec{r}_1, \vec{r}_2), \tag{4.12}$$

also called the Coulomb hole. From these definitions, the exchange and correlation holes have the following properties:

$$\rho_{\mathrm{x}}(\vec{r}_1, \vec{r}_1) = -\rho(\vec{r}_1) \tag{4.13}$$

$$\int \rho_{\mathrm{x}}(\vec{r}_1, \vec{r}_2)\mathrm{d}\vec{r}_2 = -1 \tag{4.14}$$

$$\rho_{\mathrm{x}}(\vec{r}_1, \vec{r}_2) \leq 0 \tag{4.15}$$

$$\int \rho_{\mathrm{c}}(\vec{r}_1, \vec{r}_2)\mathrm{d}\vec{r}_2 = 0. \tag{4.16}$$

Equations (4.13) and (4.14) have an important physical meaning; equation (4.13) corresponds to the Pauli exclusion principle and (4.14) is the self-interaction repulsion correction. For a graphical representation and the spin-dependent form of the exchange and correlation hole functions, see the review article of Ziegler [21].

Up to this point, we still have an exact theory. Approximations are introduced when one introduces models for the exchange-correlation hole. For the construction of a functional, we need to know the λ-average of this hole.

The adiabatic connection formula provides a means to incorporate more and more properties of the exact Kohn–Sham exchange-correlation functional into a new approximate functional. In a sense it plays an equivalent

Table 1. Comparison between KS-DFT and HF-CI.

HF-CI	KS-DFT
Reference	
HF, $\rho^{\mathrm{HF}}(\vec{r})$	KS, $\rho(\vec{r})$
Exchange	
$\langle\Phi^{\mathrm{HF}}\lvert V_{\mathrm{ee}}\rvert\Phi^{\mathrm{HF}}\rangle - J[\rho^{\mathrm{HF}}]$	$\langle\Phi^{\mathrm{KS}}\lvert V_{\mathrm{ee}}\rvert\Phi^{\mathrm{KS}}\rangle - J[\rho]$
Correlation	
$E^{\mathrm{exact}}[\rho] - E^{\mathrm{HF}}[\rho^{\mathrm{HF}}]$	$E_{\mathrm{xc}}[\rho] - E_{\mathrm{x}}[\rho]$
Correlation components	
$T_{\mathrm{corr}} = T^{\mathrm{exact}}[\rho] - T^{\mathrm{HF}}[\rho^{\mathrm{HF}}]$	$T_{\mathrm{c}}[\rho] = T^{\mathrm{exact}}[\rho] - T_{\mathrm{s}}[\rho]$
$V_{\mathrm{corr}}^{\mathrm{NE}} = \sum_{\alpha=1}^{M}\sum_{i=1}^{N}\int\frac{(\rho(\vec{r})-\rho^{\mathrm{HF}}(\vec{r}))Z_\alpha}{\lvert\vec{R}_\alpha-\vec{r}_i\rvert}$	
$V_{\mathrm{corr}}^{\mathrm{coul}} = \int\int\frac{[\rho(\vec{r}_1)\rho(\vec{r}_2)-\rho^{\mathrm{HF}}(\vec{r}_1)\rho^{\mathrm{HF}}(\vec{r}_2)]}{\lvert\vec{r}_1-\vec{r}_2\rvert}\mathrm{d}\vec{r}_1\mathrm{d}\vec{r}_2$	
Everything else	$\tilde{E}_{\mathrm{c}}[\rho(\vec{r})]$

role as adding more and more determinants in *ab initio* methods, to include more correlation.

5 Comparing and constrasting KS-DFT and HF-CI

In this section we present the formal differences between the definition of the exchange and correlation energies used in DFT and in traditional quantum chemistry or *ab initio* methods. The characteristics of both methodologies are summarized in Table 1.

The first major difference, is the reference system. In the *ab initio* world, the HF equations, and the accompanying HF density, $\rho^{\mathrm{HF}}(\vec{r})$, are taken as the reference. In KS-DFT, we use a fictitious non-interacting reference system but the reference density is the exact one.

In *ab initio* methods the exchange energy is computed exactly using the HF orbitals,

$$E_{\mathrm{x}}^{\mathrm{HF}} = -\frac{1}{2} \sum_{\sigma=\alpha,\beta} \sum_{j,k=1}^{N_\sigma} \int\int \frac{\phi_{j\sigma}^*(\vec{r}_1)\phi_{k\sigma}^*(\vec{r}_2)\phi_{k\sigma}(\vec{r}_1)\phi_{j\sigma}(\vec{r}_2)}{|\vec{r}_1 - \vec{r}_2|} \mathrm{d}\vec{r}_1 \mathrm{d}\vec{r}_2. \quad (5.1)$$

The same definition can be applied to KS-DFT using the KS orbitals, $\varphi_i^{\mathrm{KS}}(\vec{r})$. This is how one gets a part of the exchange energy in hybrid methods (see Sect. 7.5). This way of calculating the exchange energy gives rise to a nonlocal[5] potential which is in opposition to pure DFT methods where local potentials are used in the definition of $v_{\mathrm{eff}}(\vec{r})$ (2.21). However, it is possible to calculate the exact KS-DFT exchange energy through a local exchange potential using the optimized effective potential (OEP) method (see Sect. 7.6). In Table 2 we compare the exact exchange energy computed with this approach and the exact HF energy. The difference between these two quantities increases with the atomic number.

Traditionally, the correlation energy is defined as the difference between the exact many-electron wave function energy (without relativity) and the HF energy. Importantly, two densities are involved, the Hartree–Fock density and the exact density. With this definition, there are three major contributions to the correlation energy in *ab initio* methodology. The first contribution is a kinetic energy component. The exact kinetic energy involves the exact many-electron wave function, corresponding to the exact density. To calculate T_{corr}, one subtracts the Hartree–Fock kinetic energy. The last two contributions arise from the fact that we use $\rho^{\mathrm{HF}}(\vec{r})$ to compute the nuclear-electron and the electron-electron coulomb energy. These two contributions are absent in KS-DFT because we use the exact density throughout. Moreover, the kinetic energy component in the KS-DFT correlation energy does not correspond to the one previously defined. It corresponds to the difference between the exact kinetic energy of the real system and the exact kinetic energy of a fictitious non-interacting reference, both evaluated at the exact ground state density. The KS-DFT correlation energy contains also a contribution from the electron-electron interaction, $\tilde{E}_{\mathrm{xc}}$, beyond the classical coulomb interaction and the exact KS-DFT exchange definition.

In Table 3, we compare the exact correlation energy from KS-DFT and traditional quantum chemistry for atoms. One clearly sees that the following relation [26] is respected

$$|E_{\mathrm{c}}^{\mathrm{QC(exact)}}| < |E_{\mathrm{c}}^{\mathrm{DFT(exact)}}|. \quad (5.2)$$

[5] A definition of local and nonlocal terms is given in Section 7.1.

Table 2. Comparison of exact atomic HF and KS-DFT exchange energies (from Ref. [22]) with exchange energies obtained from approximate exchange functionals (from Ref. [23]) (in Hartrees). The corresponding error relative to E_x^{KS} is given in parentheses (in %).

	E_x^{HF}	E_x^{KS}	E_x^{LDA}	E_x^{PW91}
He	−1.026	−1.026	−0.884(−13.82)	−1.017(−0.88)
Be	−2.667	−2.666	−2.312(−13.26)	−2.645(−0.77)
Ne	−12.108	−12.105	−11.033(−8.85)	−12.115(0.08)
Mg	−15.994	−15.988	−14.612(−8.61)	−15.980(−0.06)
Ar	−30.185	−30.175	−27.863(−7.66)	−30.123(−0.17)
Ca	−35.211	−35.199	−32.591(−7.41)	−35.165(−0.10)
Zn	−69.641	−69.619	−65.645(−5.71)	−69.834(0.31)
Kr	−93.856	−93.833	−88.624(−5.55)	−93.831(0.00)
Sr	−101.955	−101.926	−96.362(−5.46)	−101.918(−0.01)
Cd	−148.914	−148.879	−141.543(−4.93)	−148.885(0.00)
Xe	−179.097	−179.063	−170.566(−4.75)	−178.991(−0.04)
Ba	−189.100	−189.065	−180.241(−4.67)	−189.034(−0.02)
Yb	−276.214	−276.145	−265.563(−3.83)	−276.939(0.29)
Hg	−345.304	−345.244	−332.143(−3.80)	−345.427(0.05)
Rn	−387.504	−387.450	−372.981(−3.74)	−387.417(−0.01)

Table 3. Comparison of total correlation energy (in Hartree) (from Ref. [24]) and the kinetic correlation energy (from Ref. [25]) obtained from Density Functional and traditional Quantum Chemistry.

	E_c^{DFT}	E_c^{QC}	$T_c[\rho]$
H^-	−0.041995	−0.039821	0.0280
He	−0.042107	−0.042044	0.0366
Be^{2+}	−0.044274	−0.044267	0.0415
Ne^{8+}	−0.045694	−0.045693	
Be	−0.0962	−0.0943	0.0737
Ne	−0.394	−0.390	

We compare also the kinetic energy contribution to the correlation energy. In *ab initio* methods, the virial theorem,

$$T_{coor} = -E_{coor}, \tag{5.3}$$

is satisfied. Then, the absolute values of the second column give $T_{\rm corr}^{\rm QC}$. Once again, we see that the corresponding values in KS-DFT are different.

Through Koopmans' theorem, the eigenvalues of the HF orbitals have a physical meaning for closed-shell molecule. They correspond to minus the energy required to remove an electron from that orbital. Values obtained from Koopmans' theorem are always bigger than the experimental values obtained from photoelectron spectroscopy. This is so because, in this approximation, we use the same orbitals to describe the N and the $N-1$ electron systems. Clearly, the $N-1$ electron system must be relaxed to take into account the reduction of the sceening effect created by the removal of one electron. Further discrepancy is due to the lack of correlation in the HF method. The Koopmans' theorem does not apply to DFT. However, it has been proved [27, 28] that, using the exact exchange-correlation potential, the eigenvalue of the highest occupied molecular orbital (HOMO) of the neutral molecule corresponds to minus the ionization potential (IP) and the corresponding eigenvalue in the negatively charged species gives minus the electronic affinity (EA),

$$\varepsilon_{\rm HOMO}(N) = -IP \tag{5.4}$$

$$\varepsilon_{\rm HOMO}(N+1) = -EA. \tag{5.5}$$

In Tables 4 and 5 of Section 7.3, atomic eigenvalues of the HOMO obtained from various approximations are compared with experiment.

The last difference that we want to bring to the attention of the reader is the presence of the self interaction energy in approximate exchange-correlation functionals. In the HF method the electron-electron repulsion energy is,

$$\langle \hat{V}_{\rm ee} \rangle = \frac{1}{2} \sum_{i=1}^{N} \sum_{j=1}^{N} (J_{ij} - K_{ij}), \tag{5.6}$$

where

$$J_{ij} = \int\int \phi_i^*(\vec{r}_1)\phi_j^*(\vec{r}_2) \frac{1}{|\vec{r}_1 - \vec{r}_2|} \phi_i(\vec{r}_1)\phi_j(\vec{r}_2) {\rm d}\vec{r}_1 {\rm d}\vec{r}_2 \tag{5.7}$$

$$K_{ij} = \int\int \phi_i^*(\vec{r}_1)\phi_j^*(\vec{r}_2) \frac{1}{|\vec{r}_1 - \vec{r}_2|} \phi_j(\vec{r}_1)\phi_i(\vec{r}_2) {\rm d}\vec{r}_1 {\rm d}\vec{r}_2. \tag{5.8}$$

The self coulomb energy, J_{ii}, and the self exchange energy, K_{ii}, arise because i could equal j in (5.6). These two terms cancel each other in HF theory as they should. On the other hand, the coulomb energy in DFT is computed using the density,

$$J[\rho] = \int\int \frac{\rho(\vec{r}_1)\rho(\vec{r}_2)}{|\vec{r}_1 - \vec{r}_2|} d\vec{r}_1 d\vec{r}_2, \tag{5.9}$$

and contains self interaction contributions. If one uses the exact exchange correlation energy functional, the self coulomb energy will be exactly canceled by the self exchange energy as in HF. However, because we necessarely use approximate functionals there is a contribution from the self-interaction. This is one reason that the total energy in DFT is higher than in HF.

6 Preparing new functionals

Considering the adiabatic connection procedure one can formulate a recipe to prepare new exchange-correlation functionals that can be summarized as follows:

- propose a model for $h(\vec{r}_1, \vec{r}_2)$;
- choose properties that $h(\vec{r}_1, \vec{r}_2)$ should fulfill. For example, normalisation, perfectly self-interaction free, contains opposite and parallel spin correlation, various scaling relationships. In principle, these properties can be derived from the exact wavefunction;
- determine its λ dependence and integrate over λ;
- test it on a set of reference data.

In principle, this recipe is straightforward but, in practice, the various steps present substantial difficulties.

The choice of properties for $\bar{h}(\vec{r}_1, \vec{r}_2)$ may be influenced by the nature of the applications on which we want to apply the functional. For example, the asymptotic condition is clearly important if one is interested in properties that depend on the long range behavior of v_{xc} such as the calculation of excitation energies. This property is not so crucial if one is interested in the calculation of electronic properties that do not depend on the unoccupied orbitals.

In practice, it might seem very hard to determine the λ dependence of $h(\vec{r}_1, \vec{r}_2)$. The adiabatic connection formula (4.9) *per se* does not explain how to do this. However, Levy [29] showed that by homogeneous scaling of the density using a coordinate factor γ (where γ is $1/\lambda$),

$$\rho_\gamma(\vec{r}) = \gamma^3 \rho(\gamma x, \gamma y, \gamma z), \tag{6.1}$$

we can express the adiabatic connection method as:

$$E_{\text{xc}}[\rho] = \int_0^1 E_{\text{xc}}^{\lambda}[\rho]\mathrm{d}\lambda = \int_0^1 \lambda E_{\text{xc}}^{\lambda=1}[\rho_{1/\lambda}]\mathrm{d}\lambda. \tag{6.2}$$

This formula shows that we need only know $E_{\text{xc}}^{\lambda=1}[\rho_{1/\lambda}]$ to get the value for all values of λ's. This equation implies also that knowing the value of $E_{\text{xc}}^{\lambda}[\rho]$ for some small value of λ suffices to generate all the values of $E_{\text{xc}}^{\lambda}[\rho]$. In his contribution, Levy showed that the exact exchange-correlation hole satisfies the same scaling property,

$$\rho_{\text{xc}}^{\lambda}(\gamma\vec{r}_1, \gamma\vec{r}_2) = \gamma^3 \rho_{\text{xc}}^{\lambda=1}(\gamma\vec{r}_1, \gamma\vec{r}_2). \tag{6.3}$$

In other words, once we have obtained a form for $h(\vec{r}_1, \vec{r}_2)$ that satisfies some choosen properties, one gets its λ dependence through the application of the coordinate scaling approach using (6.3).

Finally, it is very important to test the new functional on a set of reference data to assess its accuracy. For this purpose, many functionals in the literature are tested on the popular G2 [30] or, more recently, G3 [31] reference set. This set is composed of experimental values known with high accuracy such as ionization potentials and atomization energies. However, it is not *a priori* the best set for all of chemistry because, for example, there is a lack of transition-metal systems, molecules containing hydrogen bonds and other intermolecular interactions and no transition states are included in this parametrization set.

In Section 8 we will present the LAP functional which is an example of a functional developed according to this procedure.

7 Approximate exchange and correlation functionals

The following will give the flavor of the evolution of exchange-correlation functionals and present a few representative examples. A complete listing is out of the question[6]. One could refer to the work coming from the laboratories of Becke, Perdew and Scuseria, for example, as active laboratories working on the development of new exchange and correlation functionals at

[6] An extensive listing can be found at:
`http://www.centrcn.umontreal.ca/~casida/compchem/DFT.html`

the present time. We recommend some of their latest publication [32–34] and the references therein as entries into the exchange-correlation literature.

Before presenting the different approximations for E_{xc}, we should explain the difference between the terms local and nonlocal used in the literature because it could be a source of confusion. From a mathematical point of view the majority of exchange-correlation functionals, with the exception of the hybrid schemes, are local functionals because they depend on the density and its derivatives only at a given position in space. In contrast, the HF exchange energy (5.1) and the exact KS exchange energy functional are nonlocal because they also need information on other positions as well. The term nonlocal is often used to designate the Generalized Gradient Approximation (GGA) functionals. This is so, because if one thinks of a Taylor expansion of a function, then one can obtained information in the vinicity of the current position using the local information on the function and its derivatives. In the present chapter, we take the first definition and reserve the term nonlocal for mathematically nonlocal functionals.

Usually the exchange-correlation energy functional is divided into exchange and correlation parts,

$$\tilde{E}_{\mathrm{xc}} = E_{\mathrm{x}} + E_{\mathrm{c}}. \tag{7.1}$$

We will present each contribution separately. The tilde over E_{xc} indicates that from now on we are dealing with approximate E_{xc} functionals.

7.1 The Local Spin Density Approximation (LSDA)

In order to develop approximations, we need a model as a starting point. In the majority of the exchange-correlation functionals, the homogeneous electron gas is used. Particularly, in the local density approximation (LDA) the density, $\rho(\vec{r})$, is considered at each point in space and the homogeneous electron gas model is applied locally using the density only. This procedure is represented in Figure 1. In the local spin density approximation (LSDA), the electron density for individual spin components, $\rho_\alpha(\vec{r})$ and $\rho_\beta(\vec{r})$, are used.

Using the homogeneous electron gas model, the exchange energy functional is know as Dirac exchange [35] and has a very simple mathematical form,

$$E_{\mathrm{x}}^{\mathrm{LSDA}} = -\frac{3}{2}\left(\frac{3}{4\pi}\right)^{1/3} \int [\rho_\alpha(\vec{r})^{4/3} + \rho_\beta(\vec{r})^{4/3}]\mathrm{d}\vec{r}. \tag{7.2}$$

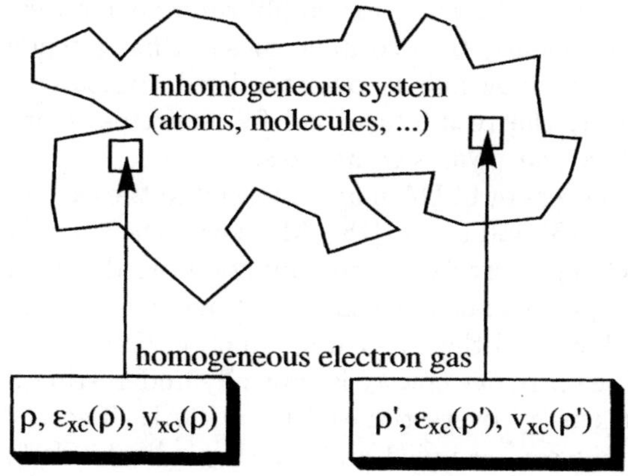

Fig. 1. Schematic representation of the local density approximation.

The general form of the local spin density approximation for the correlation energy functional has the following form,

$$E_{\rm c}^{\rm LSDA}[\rho_\alpha, \rho_\beta] = \int \rho(\vec{r})\varepsilon_{\rm c}(\rho_\alpha(\vec{r}), \rho_\beta(\vec{r}))\mathrm{d}\vec{r}, \tag{7.3}$$

where $\varepsilon_{\rm c}(\rho_\alpha(\vec{r}), \rho_\beta(\vec{r}))$ is the correlation energy per electron in a homogeneous electron gas. The exact form is unknown but many approximations to $\varepsilon_{\rm c}(\rho_\alpha(\vec{r}), \rho_\beta(\vec{r}))$ exist. The most popular is due to Vosko, Wilk and Nusair (VWN) [36] who used a Padé approximation to interpolate the results from a Monte Carlo simulation performed by Ceperley and Alder [37] on a homogeneous electron gas.

$$\begin{aligned}\varepsilon_{\rm c}(r_{\rm s}) \quad = \quad & A\left\{\ln\left(\frac{x^2}{X(x)}\right) + \frac{2b}{Q}\tan^{-1}\left(\frac{Q}{2x+b}\right)\right. \\ & - \frac{bx_0}{X(x_0)}\left[\ln\left(\frac{(x-x_0)^2}{X(x)}\right)\right. \\ & \left.\left. + \frac{2(b+2x_0)}{Q}\tan^{-1}\left(\frac{Q}{2x+b}\right)\right]\right\}. \end{aligned} \tag{7.4}$$

Where $X(x) = x^2 + bx + c$, $Q = (4c - b^2)^{1/2}$, x_0, b and c are constants. The Wigner-Seitz radius, $r_{\rm s}$, is related to the density by

$$r_{\rm s} = \left(\frac{3}{4\pi\rho}\right)^{1/3}. \tag{7.5}$$

LSDA gives fairly good results for equilibrium geometries and vibrational frequencies for covalently bonded molecules but has a tendancy to overbind atoms because the molecule is overstabilized compared to the separate atoms. Moreover, the results are bad for molecules containing hydrogen bonds [38] and van der Waals complexes.

Part of the success of LSDA may be traced to the exchange hole function which satisfies (4.13) and (4.14) [18]. Moreover, the symmetrically spherical form of the exchange-correlation hole function, needed in the calculation of E_{xc}, is a good approximation to the exact one, even though the hole itself does not reproduce well the exact hole function [39].

In LSDA, the exchange energy is typically underestimated by ~10% and the correlation energy overestimated by ~100%. The total energy is too high, the gap between occupied and unoccupied orbitals (or bands in solid state physic) is too low. Moreover, LSDA favor the $\mathrm{d}^{n+1}\mathrm{s}^1$ configuration over the $\mathrm{d}^n\mathrm{s}^2$ in the 3d transition metal atoms [40]. Gunnarsson and Jones [41, 42] showed that the major source of error in LSDA comes from the exchange energy. A source of error in the calculation of the exchange energy is the self-interaction repulsion present in LSDA. This error is related to the fact that an electron sees all other electrons including itself. The presence of such an error is easy to verify, since the exchange energy of any one-electron system should be zero. Perdew and Zunger [43] have proposed a self-interaction corrected LSDA functional that reduced the error below 3% for the exchange energy. Improvement was also obtained in the LSDA correlation energy, in the total energy, in the orbital eigenvalues, in the long range behavior of $v_{\mathrm{xc}}(\vec{r})$, in the shape of the exchange-correlation hole and consequently in the electronic density.

7.2 Gradient Expansion Approximation (GEA)

The LSDA approximation is useless for thermochemistry. This model was developed in the slowly varying density limit, which is far from the case in atoms and molecules. A logical next step is to include higher derivatives of the density to take into account the inhomogeneities. The gradient expansion approximation of the homogeneous electron gas is the following,

$$\begin{aligned} E_{\mathrm{xc}}^{\mathrm{GEA}}[\rho_\alpha;\rho_\beta] &= \int \rho(\vec{r})\varepsilon_{\mathrm{xc}}(\rho_\alpha(\vec{r}),\rho_\beta(\vec{r}))\mathrm{d}\vec{r} \\ &+ \sum_{\sigma=\alpha,\beta}\sum_{\sigma'=\alpha,\beta} C_{\mathrm{xc}}^{\sigma,\sigma'}(\rho_\alpha(\vec{r}),\rho_\beta(\vec{r}))\frac{\nabla\rho_\sigma}{\rho_\sigma^{2/3}(\vec{r})}\frac{\nabla\rho_{\sigma'}}{\rho_{\sigma'}^{2/3}(\vec{r})} \\ &+ \dots \end{aligned} \tag{7.6}$$

The second-order gradient expansion aproximation (GEA) is obtained by keeping only the first two terms of the series. The exchange hole obtained from the second-order expansion does not satisfy (4.13) and (4.14) and for that reason, little improvement over LSDA is obtained. Indeed, in practice, results are often worse than for LSDA.

7.3 Generalized Gradient Approximation (GGA)

Exchange and correlation functionals that use the second-order expansion and manage the difficulties met in the GEA, are know as generalized gradient approximations (GGA). It is convenient to express the GGA exchange energy as:

$$E_{\mathrm{x}}^{\mathrm{GGA}} = E_{\mathrm{x}}^{\mathrm{LDA}} - \sum_{\sigma=\alpha,\beta} \int F(s_\sigma)\rho_{\sigma'}^{4/3}(\vec{r})\mathrm{d}\vec{r}. \tag{7.7}$$

The argument of the function F, s_σ, is the reduced density gradient for spin σ,

$$s_\sigma(\vec{r}) = \frac{|\nabla\rho_\sigma|}{\rho_\sigma^{4/3}(\vec{r})}, \tag{7.8}$$

and corresponds to a local inhomogeneity parameter. The GGA functionals have the following general form,

$$E_{\mathrm{xc}}^{\mathrm{GGA}}[\rho] = \int f(\rho(\vec{r}), \nabla\rho(\vec{r}))\mathrm{d}\vec{r}. \tag{7.9}$$

Perdew [44] used the second-order expansion with a cut-off radius to impose conditions (4.13) and (4.14) to the exchange hole function. With this procedure, he was able to compute atomic exchange energies within 1% of the exact HF exchange energy. Further simplification of the model led to the well-know Perdew–Wang86 (PW86) [45] exchange functional,

$$\begin{aligned} F^{\mathrm{PW86}} &= \left\{1 + 1.296\left(\frac{s_\sigma}{(24\pi^2)^{1/3}}\right)^2 + 14\left(\frac{s_\sigma}{(24\pi^2)^{1/3}}\right)^4 \right. \\ &\left. + 0.2\left(\frac{s_\sigma}{(24\pi^2)^{1/3}}\right)^6\right\}^{1/15}, \end{aligned} \tag{7.10}$$

which is parameter free.

Another popular GGA exchange energy functional is due to Becke (B88) [46]. Becke imposed the correct $-1/r$ asymptotic behavior for the exchange energy density ε_x and used one parameter, β, fitted on the exact atomic

HF exchange energy of noble gas atoms.

$$F^{\mathrm{B88}} = \frac{\beta s_\sigma^2}{1 + 6\beta s_\sigma \sinh^{-1}(s_\sigma)}. \tag{7.11}$$

The GGA correlation functionals have more complicated mathematical forms. We limit ourself to mention the well-known Perdew86 (P86) [47,48] correlation functional that contains one fitted parameter and the Perdew–Wang91 (PW91) functional [49–51] that is parameter free. This last functional has an exchange counterpart.

Becke [52] showed that the inclusion of GGA exchange and correlation functionals using the combination B88 for exchange and PW91 for correlation reduced the absolute error and the overbinding tendancy of LSDA. The average absolute error in the atomization energy on a set of 56 small molecules was reduced from 35.7 kcal/mol to 5.7 kcal/mol. The improvement was not as important for the ionization energy of 46 small molecules. The average absolute error was reduced from 5.3 kcal/mol to 3.5 kcal/mol. With GGA exchange-correlation functionals we get better hydrogen bonds, good thermochemistry but still not chemical accuracy. Further improvements are needed for the van der Waals complexes and activation energies.

The LDA exchange-correlation potential does not have the right asymptotic behavior at long distance [53,54]. For that reason, equations (5.4) and (5.5) are not satisfied as can be seen from Tables 4 and 5. From Table 5 it is demonstrated that the HOMO has a positive eigenvalue (unbound) for anions. This is the case also for all negatively charged molecules. Adding the gradient correction does not lead to significant improvement since the bad asymptotic behavior of LDA is not corrected by the addition of gradient corrections [53, 54]. For example, the exchange functional of Becke, B88, falls off too fast asymptotically ($-1/r^2$ instead of $-1/r$) [22]. Moreover, Engel *et al.* [22] and van Leeuwen and Baerends [55] showed that it is impossible for a GGA functional defined as (7.7), *i.e.* containing only $s_\sigma(\vec{r})$, to satisfy simultaneoulsy the asymptotic behavior of the exchange-correlation energy functional and the exchange-correlation potential. These conditions can be fulfilled by the addition of the Laplacian of the density into the function F. Using an asymptotically corrected exchange-correlation potential like LB94 [55] improves the results (see also Ref. [56]). Asymptotically corrected potentials are also essential for a proper description of electronic excitation spectra [57]. In Table 4 we present the value of $-\epsilon_{\mathrm{HOMO}}^{\mathrm{OEP}}$ obtained with the exact KS-DFT exchange potential (without correlation) which has the right asymptotic behavior. In DFT, in contrast to HF, no correction due to electronic relaxation has to be added. Discrepancies between $-\epsilon_{\mathrm{HOMO}}^{\mathrm{OEP}}$ and experimental IPs are due to the lack of correlation in the calculations.

Table 4. Comparison of the ionization energies (in a.u.) obtained from the highest occupied molecular orbital energies (HOMO) with the experimental ionization potential (IP). (HF, experimental, LDA, B88P86 and LB94 results are taken from Ref. [55] and OEP from Ref. [58])

Atom	$-\epsilon_{\mathrm{HOMO}}$HF	IP	$-\epsilon^{\mathrm{OEP}}_{\mathrm{HOMO}}$	$-\epsilon^{\mathrm{LDA}}_{\mathrm{HOMO}}$	$-\epsilon^{\mathrm{B88P86}}_{\mathrm{HOMO}}$	$-\epsilon^{\mathrm{LB94}}_{\mathrm{HOMO}}$
H	0.500	0.500		0.234	0.280	0.440
He	0.918	0.903	0.918	0.571	0.585	0.851
Be	0.309	0.343	0.309	0.206	0.209	0.321
Ne	0.850	0.792	0.851	0.490	0.496	0.788
Ar	0.591	0.579	0.591	0.381	0.380	0.577
Kr	0.524	0.517	0.523	0.346	0.344	0.529
Xe	0.457	0.446	0.456	0.310	0.308	0.474

Table 5. Comparison of the electron affinity (EA) (in a.u.) obtained from the highest occupied molecular orbital energies (HOMO) with the experimental EAs. (From Ref. [55])

Anion	EA	$-\epsilon^{\mathrm{LDA}}_{\mathrm{HOMO}}$	$-\epsilon^{\mathrm{B88P86}}_{\mathrm{HOMO}}$	$-\epsilon^{\mathrm{LB94}}_{\mathrm{HOMO}}$
F^-	0.125	−0.097	−0.099	0.128
Cl^-	0.133	−0.022	−0.023	0.140
Br^-	0.124	−0.008	−0.009	0.140
I^-	0.112	+0.005	+0.004	0.139

7.4 meta-Generalized Gradient Approximation (meta-GGA)

We can go beyond the GGA approximation by adding an explicit dependence on the Laplacian of the density, $\nabla^2\rho(\vec{r})$, and the kinetic energy density[7],

$$\tau_\sigma(\vec{r}) = \sum_{i=1}^{occ} |\nabla\varphi_i^\sigma(\vec{r})|^2, \tag{7.12}$$

for the calculation of exchange-correlation energy. These two quantities exhibit atomic shell structure and may be viewed as inhomogeneity

[7] We should mention that different definitions exist in the literature for the kinetic energy density, here we take the definition of Becke [59].

parameters used to get nonlocal information about the exchange-correlation hole function.

Exchange-correlation energy functionals that could be written in the following general form are commonly called the meta-generalized gradient approximation (meta-GGA),

$$E_{\mathrm{xc}}^{\mathrm{meta-GGA}}[\rho] = \int f(\rho(\vec{r}), \nabla\rho(\vec{r}), \nabla^2\rho(\vec{r}), \tau(\vec{r}))\mathrm{d}\vec{r}. \qquad (7.13)$$

Adding the kinetic energy density into the exchange-correlation energy functional brings extra computational work in the KS calculation. In the SCF procedure, the exchange-correlation potential, $v_{\mathrm{xc}}(\vec{r})$, has to be calculated. The problem arises because τ is not an explicit functional of the density. Thus, approximations have to be programmed.

Among this kind of functional we mention the exchange energy functional of Becke and Roussel [60], the exchange-correlation energy functionals of Perdew *et al.* [33], Van Hoorhis and Scuseria (VSXC) [61] (which contains 21 parameters...), and the correlation energy functionals of Proynov *et al.* (LAP, Bmτ1) [62–67] and of Lee, Yang and Parr (LYP) [68]. This family of functionals solve the majority of the problems met with GGA. Part of the success lies in the orbital-dependent exchange-correlation energy. meta-GGA exchange-correlation energy functionals seem to be a good alternative to the GGA and hybrid schemes which are the subject of the next section.

7.5 Hybrid functionals

Another kind of approximation to go beyond the GGA was proposed by Becke [69]. He suggested to mix some of the exact KS exchange energy with a traditional exchange-correlation energy functional. His argumentation was based on the adiabatic connection formula. He showed [70] that the local electron gas model is inapropriate near $\lambda = 0$ and used the H_2 molecular system to illustrate this. At $\lambda = 0$, the noninteracting limit, the exchange-correlation energy is the exact KS exchange energy without correlation. For H_2, the exact exchange hole function, $\rho_{\mathrm{x}}(\vec{r}_1, \vec{r}_2)$, is the negative of the σ_{g} orbital density and, consequently, is symmetrically delocalized over both atoms. On the other hand, the exchange hole function of the local electron gas model is more or less localized by its nature. This hole is a bad approximation for the exact hole near the $\lambda = 0$ limit. As a remedy, Becke suggested to use the exact KS exchange energy at $\lambda = 0$ and LSDA [69] or GGA [70] at $\lambda = 1$. As an improvement of his first half-and-half model [69] he proposed [70],

$$E_{\mathrm{xc}}^{\mathrm{B3PW91}} = E_{\mathrm{xc}}^{\mathrm{LSDA}} + a_0(E_{\mathrm{x}}^{\mathrm{Exact}} - E_{\mathrm{x}}^{\mathrm{LSDA}}) + a_{\mathrm{x}}\Delta E_{\mathrm{x}}^{\mathrm{B88}} + a_{\mathrm{c}}\Delta E_{\mathrm{c}}^{\mathrm{PW91}}, \quad (7.14)$$

where $a_0 = 0.20$, $a_x = 0.72$ and $a_c = 0.81$. These parameters were determined by a linear least-squares fit on an experimental reference data set [30,71] rather than from the adiabatic connection formula. The parameter a_0 represents the amount of exact exchange needed for the correction of LSDA or GGA near the $\lambda = 0$ limit. A value of 20% is typical for all hybrid schemes as shown by the recent development of nonempirical (no fitting procedure on a set of reference data) hybrid functionals [72,73]. B3LYP [74] is the most popular hybrid scheme. The difference with the above hybrid scheme is the replacement of the PW91 correlation energy functional by the LYP correlation functional. The values of the parameters are however the same. B3LYP gives better results than B3PW91 and other GGA functionals for organic molecules. Results for transition metal compounds and for hydrogen bonds, for example, do not appear to represent any improvement over GGA functionals.

7.6 The Optimized Effective Potential method (OEP)

The exchange energy is the largest component of exchange-correlation energy. In addition to the exact calculation of T_s proposed by Kohn and Sham [15], one may think of computing exactly the KS exchange energy using,

$$E_x^{\text{exact}}[\rho] = -\frac{1}{2} \sum_{\sigma=\alpha,\beta} \sum_{j,k=1}^{N_\sigma} \int\int \frac{\varphi_{j\sigma}^*(\vec{r}_1)\varphi_{k\sigma}^*(\vec{r}_2)\varphi_{k\sigma}(\vec{r}_1)\varphi_{j\sigma}(\vec{r}_2)}{|\vec{r}_1 - \vec{r}_2|} \mathrm{d}\vec{r}_1 \mathrm{d}\vec{r}_2. \tag{7.15}$$

Then the only approximation in a KS calculation comes from the approximate form for the correlation energy functional. One of the major problems with this kind of approach, is the exchange-correlation potential needed in a KS calculation. It implies the solution of complicated equations. The exact derivation of $v_{xc}(\vec{r})$ leads to the optimized effective potential (OEP) [75,76] method. It should be emphasized that, in the OEP method, the energy expression, E_x, involves the full non-local exchange but the potential, v_{xc} is local. We recommend the paper of Grabo *et al.* [77] as a good introduction to the OEP formalism.

The OEP equation must be solved numerically because no simple analytical form exists. This explains why its application has been limited to atoms. A good approximation is due to Krieger *et al.* [78] known as the KLI approximation. This method suffers from the computational effort which is very high compared to GGA and meta-GGA functionals. The application of the KLI method is limited to small molecules. Finally, we should mention the problem related to finding a suitable correlation functional because the cancellation of errors between exchange and correlation contributions which make the success of LDA is absent. However, better total energy, orbital energies (see Tabs. 4 and 5), excitation spectra and all the properties that depend on the KS orbitals could be obtained with better accuracy than with any other type of functional presented here. In contrast to the hybrid scheme, the OEP-based exchange potentials are local, which respects the pure Kohn–Sham DFT philosophy.

7.7 Comparison between various approximate functionals

In this section, we illustrate the performance of different KS-DFT schemes for thermochemistry. The listing is not exhaustive because there are many comparisons in the literature and we just want to give an idea of the accuracy of the approximations previously presented.

In Table 6, we compare different exchange-correlation approximations for thermochemistry. We should mention that these functionals do not contain any fitted parameters and the experimental data are corrected for the zero point energy. The lack of correlation makes HF useless. LSDA performs better than HF without increasing the computational time. One can see the overbinding tendency of this approximation. The GGA functional improves quantitatively the accuracy but it is still far from chemical accuracy ($\pm$ 1 kcal/mol). The inclusion of the Laplacian of the density seems to be necessary for a quantitative description. Further improvement is needed. We mention that the popular empirical hybrid scheme B3LYP gives a mean absolute error (MAE) of 2 kcal/mol on this set of molecules [61]. But remember that this functional contains parameters fitted to reproduce atomization energies of a set of molecules, including those in Table 6.

8 LAP correlation functional

In this section we present the LAP correlation functional family [62–67] which is a meta-GGA correlation energy functional developed according to the adiabatic connection method, following the recipe given in Section 6.

First we reformulate the adiabatic connection method as,

$$E_c[\rho] = \int_0^1 E_c^\lambda[\rho]\mathrm{d}\lambda, \tag{8.1}$$

to justify its application to the correlation energy alone. The exchange energy functional is taken out of the integral because the exact KS exchange energy is defined as the $\lambda = 0$ limit of the adiabatic connection method

Table 6. Atomization energies (in kcal/mol) of small molecules computed with various exchange-correlation functionals. (UHF, LSDA and PW91 results are taken from Ref. [79]). Experimental and meta-GGA from Ref. [33]).

Molecule	ΔE^{UHF}	ΔE^{LSDA}	ΔE^{PW91}	$\Delta E^{\text{meta-GGA}}$	ΔE^{expt}
H_2	84	113	105	115	109.5
LiH	33	60	53	58	57.8
CH_4	328	462	421	421	419.3
NH_3	201	337	303	299	297.4
OH	68	124	110	108	106.4
H_2O	155	267	235	230	232.2
HF	97	162	143	139	140.8
Li_2	3	23	20	23	24.4
LiF	89	153	137	128	138.9
Be_2	−7	13	10	5	3.0
C_2H_2	294	460	415	401	405.4
C_2H_4	428	633	573	562	562.6
HCN	199	361	326	312	311.9
CO	174	299	269	256	259.3
N_2	115	267	242	229	228.5
NO	53	199	171	159	152.9
O_2	33	175	143	131	120.5
F_2	−37	78	54	43	38.5
P_2	36	142	120	118	117.3
Cl_2	17	81	64	59	58.0
MAE	71.2	31.4	8.0	3.1	

and therefore, is independent of λ. The spin-dependent correlation energy functional can be divided as,

$$E_c[\rho] = E_c^{\uparrow\uparrow}[\rho] + 2E_c^{\uparrow\downarrow}[\rho] + E_c^{\downarrow\downarrow}[\rho], \tag{8.2}$$

where $E_c^{\uparrow\uparrow}$ and $E_c^{\downarrow\downarrow}$ represent the correlation energy of electrons with parallel spin beyond the exchange energy and $E_c^{\uparrow\downarrow}$ represents the correlation energy for electrons with opposite spin. This last contribution is bigger than the other two and was the only correlation component included in LAP1 [62,63]. The parallel spin correlation is included in LAP3 [66] and $\tau 1$ [67] which are the recent improvements of the LAP1 functional.

For each different type of contribution to the correlation energy, a spin-dependent pair correlation function of Colle and Salvetti [80] is defined and the λ-dependence is introduced through the coordinate scaling approach [29]. The inhomogeneities are taken into account by local thermodynamic analogy [5] using the Laplacian of the density and the kinetic energy density defined as,

$$\tau_\sigma = \frac{1}{8}\sum_i^{occ}(\nabla\varphi_{i\sigma})^2 - \frac{1}{8}\nabla^2\rho_\sigma. \tag{8.3}$$

There are (at least...) two points of view on the construction of a functional. A functional free of parameters that is based on physical arguments should give better results in general but can be less accurate for a specific set of reference data when compared to a functional fitted to reproduce the results of this reference set. The LAP family of functionals is a good compromise between these two approaches. The equations are derived based on physical arguments and then a relatively small set of molecules (N_2, F_2, O_2, HF, CN, H_2O, NH_3, CH_4, C_2H_4, C_2H_6, C_6H_6, $(H_2O)_2$) has been used to parametrize the correlation energy functional and to synchronize the LAP correlation energy functional with the exchange functional of Becke (B88) [46] and Perdew–Wang86 (PW86) [45]. Such synchronization is essential when exchange and correlation energy functionals are developed separately.

The LAP correlation functionals are then tested on a set of 40 small and medium sized molecules which represent difficult cases at the LSDA and GGA level and for which very accurate experimental atomization energies can be extrapolated to 0 K. For difficult cases like hydrogen bonds, activation barriers and some transition metal compounds the results are comparable to CCSD(T) [81–84]. We should mention that the results for transition metal compounds are good even though no transition metals were included in the parametrization step, an indication of the good physics incorporated in the functional.

9 Solving the Kohn–Sham equations

Choosing an exchange-correlation energy functional based on the properties of interest is not the only decision to make in a typical KS-DFT calculation. In this section, we give an overview of the possible choices based on our own KS-DFT program, deMon-KS[8] [85–87]. These choices, together with representative references, are given in Figure 2. Sources of error related to these choices are compiled in Table 7. These choices influence directly the

[8] The program is available at: `http://www.sims.nrc.ca/sims/deMon/`

computational time. Informed decisions balancing accuracy and speed have to be made for the majority of every-day applications.

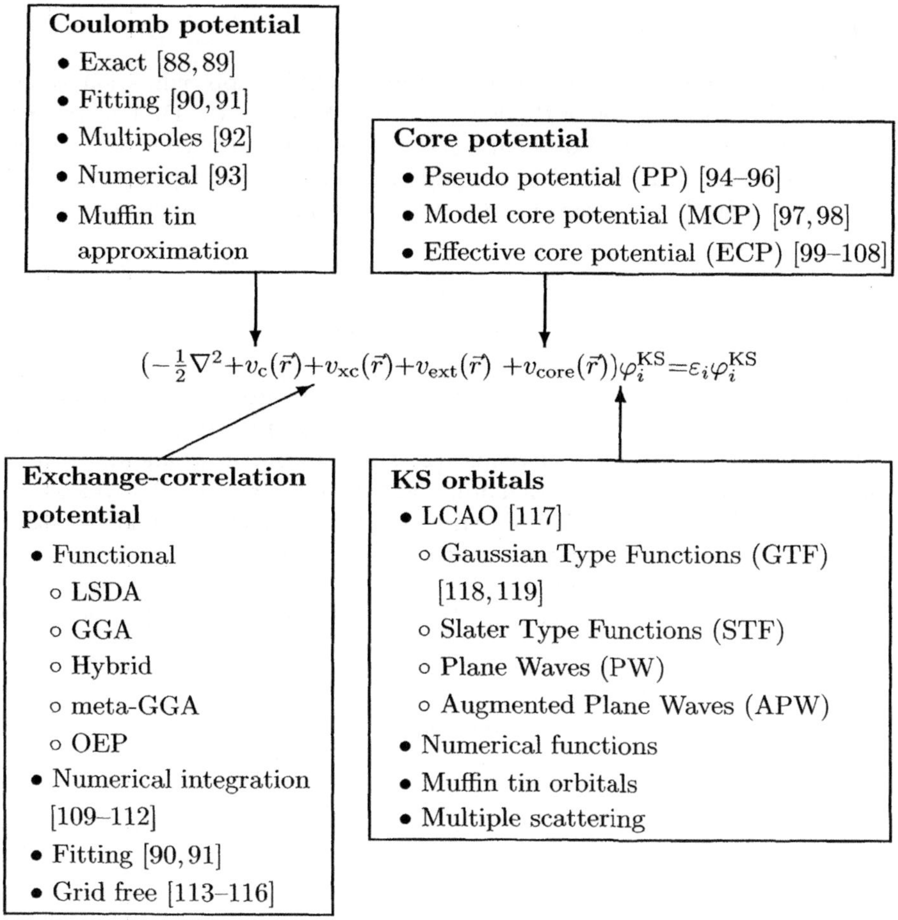

Fig. 2. Schematical representation of the KS choices.

First of all, we should mention that it is possible to solve the KS equation fully by numerical integration. This is how KS-DFT calculation are performed within the NUMOL program [120] of Becke. This kind of calculation does not suffer from basis set incompleteness but is subject to numerical noise coming from the numerical integration procedure. However, this is the method of choice for the determination of the performance of a new exchange-correlation functional because it excludes all the errors coming from using a basis set. This way of solving the KS equation is very demanding on computational ressources which limits its application to small molecules.

9.1 The Kohn–Sham orbitals

The linear combination of atomic orbitals (LCAO) approximation is the most used representation of the KS orbitals, $\varphi_i^{\mathrm{KS}}(\vec{r})$, in KS-DFT programs. This approach is due to Roothaan [117] and was first applied to traditional *ab initio* methods. A good detailed description is given in [4]. In the LCAO approach each $\varphi_i^{\mathrm{KS}}(\vec{r})$ is expanded in a linear combination of atom-centered basis function $\chi_\mu(\vec{r})$,

$$\varphi_i(\vec{r},\sigma) = \sum_{\mu}^{N_\mu} C_{\mu i}^{\sigma} \chi_\mu(\vec{r}), \tag{9.1}$$

where $C_{\mu i}^{\sigma}$ are coefficients determined by the SCF procedure. Each atomic basis function is also a linear combination of contracted primitives. The most popular choices are Gaussians [121], because integrals and derivatives are analytical and easy to calculate, and Slater functions [122] which correspond to the exact solution of hydrogen-like atoms and then are a better representation of the radial part of atomic functions. Gaussians[9] are generally prefered over Slater functions because their manipulation is easier even though more functions are needed. Using cartesian Gaussian functions,

$$g(\vec{r},\vec{l},\alpha) = N_{nl} x^{l_x} y^{l_y} z^{l_z} \mathrm{e}^{-\alpha r^2} \tag{9.2}$$

with the normalization factor given by,

$$N_{nl} = \left(\frac{2\alpha}{\pi}\right)^{3/4} (4\alpha)^{(l_x+l_y+l_z)/2} [(2l_x-1)!!(2l_y-1)!!(2l_z-1)!!]^{-1/2}, \tag{9.3}$$

an atomic basis function can be written as

$$\chi_\mu(\vec{r}) = \sum_{p}^{N_{\mathrm{p}}} \mathrm{d}_{\mu\mathrm{p}} g_{\mathrm{p}}(\vec{r},\vec{l},\alpha_{\mathrm{p}}). \tag{9.4}$$

The contraction coefficients, $d_{\mu\mathrm{p}}$, and the exponents, α_{p}, are optimized separately only once and are kept fixed for all successive calculations. The optimization procedure of atomic basis sets using Gaussian functions is described in [123]. The linear combinations in equations (9.1) and (9.4) are not infinite which introduces an error known as the basis set incompleteness.

---

[9] Different basis sets using Gausssian functions are available at `http://www.emsl.pnl.gov:2080/forms/basisform.html`

Table 7. Sources of error in KS-DFT calculations.

Sources of error
Exchange-correlation functional
Orbital basis set incompleteness
Auxiliary basis set incompleteness
Basis set superposition error
Fitting procedure
Numerical integration
Molecular grid
Relativity
SCF convergence criteria
Geometry convergence criteria
Solvation
Finite system

Introducing the LCAO approximation (9.1) into the KS equations (2.24), multiplying to the left by an arbitrary basis function, $\chi_\nu(\vec{r})$, and integrating over the whole space, we obtain a matrix equation,

$$\mathbf{F}^{\mathbf{KS}}\mathbf{C} = \mathbf{SC}\varepsilon, \tag{9.5}$$

where the matrix elements are given by

$$F^{\mathrm{KS}}_{\nu\mu} = \int \chi_\nu(\vec{r}) f^{\mathrm{KS}} \chi_\mu(\vec{r}) \mathrm{d}\vec{r} \tag{9.6}$$

$$S_{\nu\mu} = \int \chi_\nu(\vec{r}) \chi_\mu(\vec{r}) \mathrm{d}\vec{r}, \tag{9.7}$$

which constitute a linear system of equations to be solved by standard techniques. Efficient algorithm design for the integration of (9.6) and (9.7) is presented in [88,89]. The time required to solve (9.5) is directly related to the total number of atomic functions[10]. This number can be reduced simply by using an atomic basis set optimized for DFT calculations (such as the one presented in [118, 119]) instead of a big basis set needed for post-HF methods, where higher flexibility is needed to get as much as possible of the

[10]The diagonalization of the KS matrix $F^{\mathrm{KS}}_{\nu\mu}$ scales as $\mathrm{O}(N^3)$, where N is the number of atomic basis function. This step of the SCF procedure can be reduced to $\mathrm{O}(N)$ for large system using special numerical techniques because in this case, the KS matrix is sparse [124].

total electronic correlation. Moreover, using an atomic basis set for DFT reduces the basis set superposition error (BSSE). The choice of a basis set is as important as the choice of an exchange-correlation functional.

9.2 Coulomb potential

Using the following definition for the electronic density in the LCAO approach,

$$\rho(\vec{r}) = \sum_{i=1}^{N} \sum_{\mu}^{N_\mu} \sum_{\nu}^{N_\nu} (n_i^\alpha C_{\mu i}^\alpha C_{\nu i}^\alpha + n_i^\beta C_{\mu i}^\beta C_{\nu i}^\beta) \chi_\mu(\vec{r}) \chi_\nu(\vec{r}), \tag{9.8}$$

the coulomb potential can be evaluated exactly [88,89] using

$$V_c(\vec{r}) = \frac{1}{2} \int \frac{\rho(\vec{r}_2)}{|\vec{r}_1 - \vec{r}_2|} d\vec{r}_2 \tag{9.9}$$

$$= \sum_{i=1}^{N} \sum_{\mu}^{N_\mu} \sum_{\nu}^{N_\nu} (n_i^\alpha C_{\mu i}^\alpha C_{\nu i}^\alpha + n_i^\beta C_{\mu i}^\beta C_{\nu i}^\beta) \int \frac{\chi_\mu(\vec{r}_2) \chi_\nu(\vec{r}_2)}{|\vec{r}_1 - \vec{r}_2|} d\vec{r}_2. \tag{9.10}$$

Consequently, the evaluation of the coulomb potential depends on the square of the total number of basis functions (N^2) and its contribution to the evaluation of $\mathbf{F}^{\mathrm{KS}}$ grows as N^4. To reduce the computational time, one may approximate the electronic density by a fit on an auxiliary atomic centered basis set [90,91],

$$\rho(\vec{r}) \approx \tilde{\rho}(\vec{r}) = \sum_{k}^{N_k} a_k f_k(\vec{r}), \tag{9.11}$$

where the a_k are fitting coefficients and $f_k(\vec{r})$ are fitting functions (Gaussian, Slater, ...). The tilde indicates a fitted function. An optimization procedure similar to the orbital atomic basis set is employed to get the exponent of the primitive functions, $f_k(\vec{r})$. The fitted density is equal to the exact density in the limit of a complete auxiliary basis set which is not the case in practice. The coulomb potential with the fitted density becomes

$$V_c(\vec{r}_1) \approx \tilde{V}_c(\vec{r}_1) = \int \frac{\tilde{\rho}(\vec{r}_2)}{r_{12}} d\vec{r}_2 = \sum_{k}^{N_k} a_k \int \frac{f_k(\vec{r}_2)}{r_{12}} d\vec{r}_2. \tag{9.12}$$

With the fitted density, the cost of the evaluation the coulomb energy needed in $\mathbf{F}^{\mathrm{KS}}$ is reduced to $N^2 N_k$. Since the number of fitting functions is smaller than the number of basis functions, this procedure saves a lot of CPU time.

One has to use the fitted density to get the N^3 scaling attributed to DFT in the literature; otherwise it is formally N^4 as for HF.

One can go one step further, knowing that electron-electron coulomb repulsion is a short range interaction, by using the fast multipole method [92]. This approach is used in almost all order-N methods.

9.3 Exchange-correlation potential

The exchange-correlation potential needed for the solution of the KS equation (2.24) is a much more complex mathematical function than the definition of the exchange-correlation energy functional presented in Section 7. For that reason, the exchange-correlation potential and energy are calculated numerically on a grid[11] and constitute the bottleneck in the solution of the KS equations. As already mentioned, the results can be influenced by numerical noise if a grid with not enough points is used. To reduce the computational time during the SCF procedure, one can calculate the exchange-correlation potential on a reduced grid and then use a fitting procedure [90],

$$v_{\mathrm{xc}}^{\sigma}(\vec{r}) \approx \tilde{v}_{\mathrm{xc}}^{\sigma}(\vec{r}) = \sum_{k}^{N_k} b_k^{\sigma} g_k(\vec{r}), \tag{9.13}$$

where b_k are fitting coefficients determined by least squares fit and $g_k(\vec{r})$ are fitting functions. Once again, error is introduced because the new auxiliary basis set in not complete. This procedure is accurate only if one or several extra iterations are performed on an augmented grid when the SCF calculation is converged. (This is how the exchange energy contribution is calculated in deMon-KS). This procedure is still time consuming even though the fitting procedure requires a smaller grid than a full numerical intregration. To further decrease the computational time, one can avoid the numerical evaluation of the exchange-correlation potential and energy on a grid using a grid-free technique [113–116].

9.4 Core potential

Core electrons play a limited role in the chemistry of molecular systems because they are tightly bound to the nucleus. For that reason, it is attractive to separate an atom into core and valence regions. Then, the effects of the core electrons are simulated by an analytical one-electron potential,

[11] Explanation of the details of the three dimensional numerical integration in molecular systems is beyond the scope of this chapter. The reader is refered to recent contributions [109–112] for technical details.

$V_{core}(\vec{r})$, and a truncated basis set is used to treat the valence electrons explicitly. The smaller size of the basis set compared to an all-electron calculation leads to an appreciable computational time saving. Moreover, they can also include relativistic effects through an approximate treatment of mass-velocity and Darwin terms during the optimization procedure. This effect is very important for a good description of compounds containing second and third row transition metals. The different methodologies can be divided into three categories, pseudo-potential (PP) [94–96], model core potential (MCP) [98] and effective core potential (ECP) [101–104,106–108]. This last type of core potential is usually optimized for *ab initio* techniques. We recommend the review articles of Frenking *et al.* [125] and Cundari *et al.* [126] for more details.

9.5 Other choices and sources of error

As mentioned in Section 2.3, KS equations have to be solved self-consistently. The iterative process is stopped when the change in the total energy and in the density is less than a threshold supplied by the user. Typical values are 1×10^{-8} and 1×10^{-7} a.u. for the energy and the density respectively. There is also a value to give for the determination of the convergence during a geometry optimization. Depending on the program, there is between two to four quantities to check for convergence. For example, the maximum atomic contribution to the gradient, the root mean square (RMS) of the gradient, the maximum displacement, and the RMS of the displacement. In general, all the convergence criteria mentioned so far possess default values in DFT programs. These values are not necessarily adapted for your specific problem and must be checked.

Beside technical errors, errors can arise from the choice of the model. For example, using an over simplified model for a biological system can result in a bad description of the cooperative effects [127]. Solvation is also critical in this kind of system. Finite size systems can also be a source of error in particular situations. For example, a cluster model is not necessarily the best model to study the chemistry that occurs on a surface since its behaviour is not the same as the bulk when small sized clusters are considered. Attention has to be paid when selecting a model.

9.6 Functionality

Finally, we conclude this section by giving a list of functionalities available in different DFT programs with some representative references:

- geometry optimization (minima, transition states and saddle points) [128];

- *ab initio* molecular dynamics simulation [129,130];
- molecular vibrations, Infrared and Raman intensities [131];
- charge analysis [132,133];
- dipole moments, polarisability and hyperpolarisability [134–136];
- molecular electrostatic potential and its critical points [137];
- electronic excitations [136,138];
- NMR and EPR spectra [139–143];
- ionization potential, electronic affinity, atomization and binding energy;
- reaction path following (IRC) [144];
- X-ray absorption [145] and emission [146] spectra;
- vibrational Circular Dichroism [74,147];
- solvent effects [142,148].

10 Applications

Here we present just a few examples to indicate some areas where DFT has been used to solve real chemical problems and to provide valid insight into properties and processes. The literature is rich, and the reader should have no difficulty finding other exciting paths to follow.

10.1 Ab initio molecular dynamics for an alanine dipeptide model

DFT and *ab initio* calculations are usually performed in a static mode. That is, they yield information on extrema, like minima and transition states, but not on properties that depend on the dynamics of the system. To get such information, a common approach is to use classical molecular dynamics (CMD) simulations by solving Newton's equations of motion,

$$m_i \frac{\mathrm{d}^2 x_i}{\mathrm{d}t} = -\frac{\partial E}{\partial x_i}, \tag{10.1}$$

using a classical force field to get the forces ($-\partial E/\partial x_i$) felt by an atom (where m_i is the atomic mass of atom i and x_i is one of its 3 cartesian coordinates). The ability of this kind of calculation to reproduce all the interactions present in the system under consideration, is related to the quality of a given force field, which depends on the parametrization and the nature of the terms included in it. Recent work in several laboratories is revealing the importance of polarization terms.

In an AIMD simulation, a DFT or an *ab initio* electronic calculation is performed at each time step and the forces between atoms are obtained from the first derivatives of the total energy with respect to the nuclear positions. This approach has been introduced by Car and Parrinello [129]. With the use of DFT it is possible to compute all the interactions from first principles. The description of hydrogen bonds is now under reasonable control with the new generation of functionals [83] but further improvements are needed for van der Waals interactions which are largely dominated by dispersion.

Recently, Wei *et al.* [149] have performed an AIMD simulation on (S)-α-(formylamino)propamide, an alanine dipeptide analogue, using DFT. A scan of the potential energy surface reveals two low energy conformers, $C7_{eq}$ and C5, (represented in Fig. 3) characterized by two dihedral angles, ϕ and ψ, defined by C_3-C_2-N_1-C_1 and N_7-C_3-C_2-N_1 respectively. The value of the angles (ϕ, ψ) is $(-80, 80)$ and $(-160, 160)$ for $C7_{eq}$ and C5 respectively.

Using the combination B88 [46] for exchange and P86 [47, 48] for correlation, the $C7_{eq}$ conformer is lower in energy by 1.7 kcal/mol. The result of the AIMD simulation is presented in Figure 4.

A hydrogen bond is formed each time the distance between $O_1 \cdots H_7$ (solid line) or $O_2 \cdots H_2$ (dotted line) is around 2 Å. From Figure 4, we can see that both conformer regions are frequently visited at room temperature and the transformation takes about 3 ps to be completed. In contrast, no transition from $C7_{eq}$ to C5 conformer is observed for a CMD simulation lasting many nanoseconds, using popular classical force fields, at 298.15 K. But, at 375 K a CMD simulation using CHARMM [150], showed a conformational transformation between $C7_{eq}$ and C5. This difference is not coming from a bigger activation barrier since the value obtained with

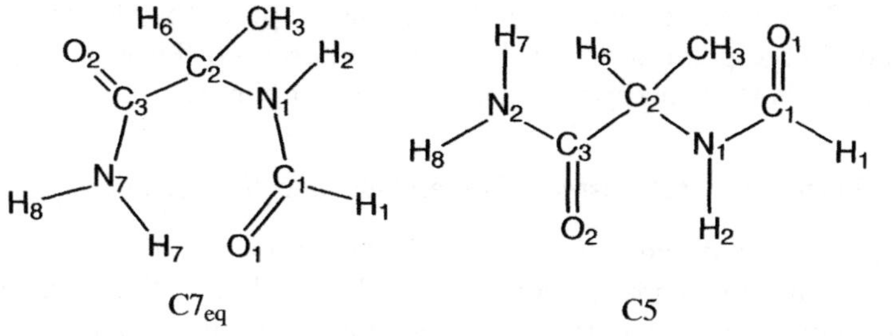

Fig. 3. 2D schematical representation of $C7_{eq}$ and C5 isomers of (S)-α-(formylamino)propanamide.

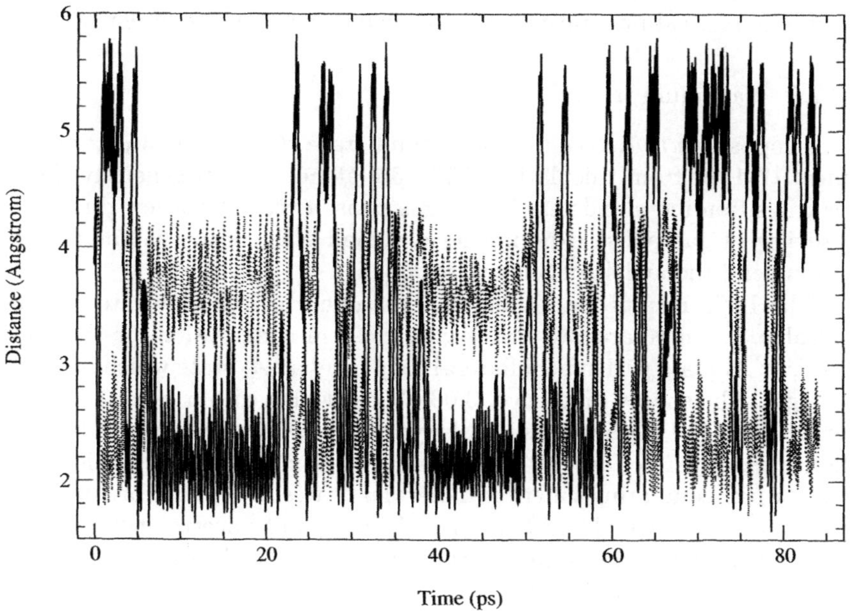

Fig. 4. Time evolution of the $O_1 \cdots H_7$ (solid line) and $O_2 \cdots H_2$ (dotted line) hydrogen bonds at 298.15 K.

CHARMM force field, 2.4 kcal/mol, is similar to the value of 2.5 kcal/mol obtained with DFT. Then, as propose by Wei *et al.* [149], force fields are probably not appropriate for gas phase transformations. This is not surprising because usually, force fields are parametrized from experiment, which include solvent effect and as a consequence, there is less flexibility in the whole molecule coming from constraints of the dihedral angles. This example showed the importance of treating properly the electronic correlation and charge polarization in biological molecules. There is a great need for force fields that incorporate this added complexity.

10.2 Transition metal clusters: The ecstasy, and the agony...

The next two examples concern small transition metal clusters. Clusters with less than 20 atoms do not behave like the bulk. They are characterized by shorter bond length and higher magnetic moment per atom when compared to the bulk values and there are many low-lying excited states. For these reasons, the accuracy of the results may be highly influenced by the technical details. For example, many tests have to be performed in order to determine a good basis set and a grid that contains enough points for the numerical integration of the exchange-correlation potential.

10.2.1 Vanadium trimer

It is impossible to determine the ground state structure of V_3 from experiment [151] or from calculation [152–156] alone. This is generally the case for transition metal clusters in the gas phase except for a few dimers. For a review of gas phase experiments on small clusters in the gas phase we recommend the article of Simard *et al.* [157].

The DFT results for V_3 and V_3^+ obtained in Ref. [156] are presented in Table 8. There are two states very close in competition for the ground state, a $^2A_1'$ state with an equilateral shape and a $^4A_2$ state which is an acute triangle. The energy difference, 0.03 eV, is clearly within typical error bars for DFT (GGA) applications to this type of system (errors around 0.5 eV for dissociation energies are not uncommon and, while some cancellation should occur when comparing different molecular states, it is probably prudent to keep 0.5 eV ($\simeq$10 kcal/mol) as a rough margin of uncertainty). The bond energy computed from both states are very close to the experimental value of 1.42 $\pm$ 0.10 eV [158]. Moreover, the authors point out the importance of the basis set to get the correct energy ordering of the two states of V_3. They showed that a basis set optimized for LSDA calculations [118] was not good enough, preparing a GGA-optimized one proved to be worthwhile.

The situation is also ambiguous for V_3^+. A $^3A_2'$ state could be chosen as the ground state based on the energy difference but a $^1A_1'$ state based on the bond energy. The experimental bond energy of V_3^+ is 2.27 $\pm$ 0.09 eV.

Recently, the comparison between the experimental PFI-ZEKE spectra and the simulated ones from KS-DFT calculations, has been successfully applied to determine the ground state structure of Nb_3O, Nb_3O^+, Nb_3C_2 and $Nb_3C_2^+$ [159, 160]. Calaminici et al. [156] have applied the same procedure to V_3 and V_3^+. The experimental [151] and the simulated spectra for

Table 8. KS-DFT results for V_3 and V_3^+ from [156].

	V_3		V_3^+	
	$^2A_1'$	$^4A_2$	$^3A_2'$	$^1A_1'$
Relative energy	0.00	0.03	5.61	5.79
Bond energy	1.59	1.60	2.10	2.27
$d(V_1 - V_2)$	2.169	2.335	2.180	2.211
$\theta(V_1V_2V_3)$	60.0	49.7	60.0	60.0

V_3 ($^2A_1'$) $\rightarrow$ V_3^+ ($^3A_2'$) and V_3 ($^4A_2$) $\rightarrow$ V_3^+ ($^3A_2'$) are presented in Figure 5. The very good agreement between the experimental and the first simulated spectra allowed Calaminici *et al.* to determine the ground state of V_3 and V_3^+ unambiguously as $^2A_1'$ and $^3A_2'$ respectively. Clearly, modern DFT can be a powerful partner for modern experiment.

10.2.2 Nickel clusters

A systematic study on transition metal clusters implies the calculation of many structures without symmetry constraints, in order to allow the cluster to relax, for example, according to the Jahn–Teller theorem. Each of the starting geometries must be fully optimized for many multiplicities and all the different optimized structures must be characterized by vibrational analysis to determine the nature of the stationary point found. These are difficult systems, requiring close attention to technical details. The lack of detailed computational information or protocol [161] followed makes it hard to assign the quality of some of the published computational studies, for example, on Ni_4 and Ni_5.

The lowest energy structure of Ni_4 obtained with at least four different DFT programs, various exchange-correlation functionals, basis sets and grids, are presented in Table 9. This Table illustrates the need to carefully choose the computational details for the characterization of transition metal clusters.

The ground state structure of Ni_4 is unknown experimentally. A distorted tetrahedron (D_{2d}) is obtained as the lowest energy structure using different computational choices [162–167]. Reddy *et al.* [168] obtained a perfect tetrahedral structure (T_d) probably because they used the fractional occupation of molecular orbitals to help the SCF convergence. The use of this procedure tends to favor highly symmetric structures when it is not removed completely at the end of the calculation. The absence of vibrational frequencies make this result questionable. Other authors founded an

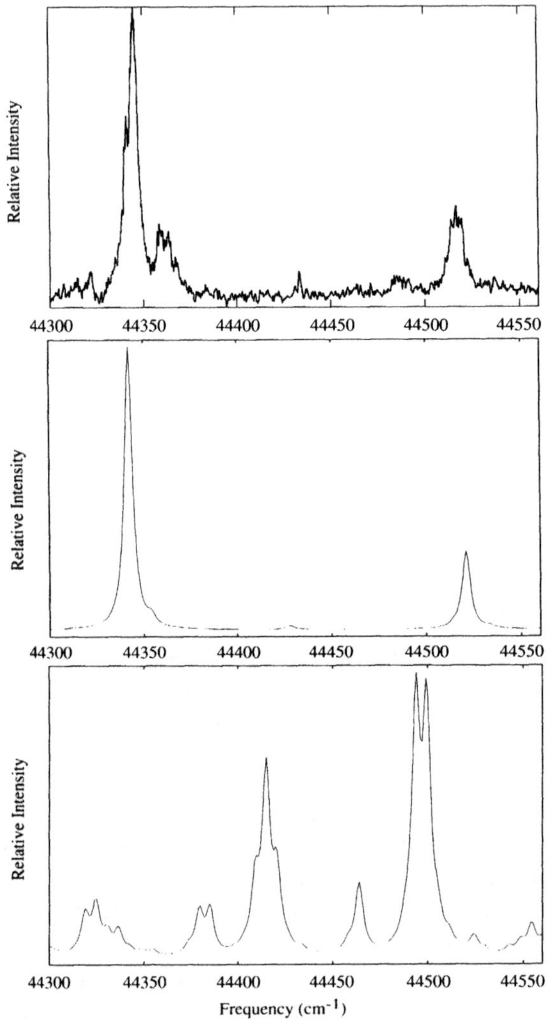

Fig. 5. Experimental (top panel) and simulated PFI-ZEKE spectra of V_3 at 298.15 K. Middle panel: $V_3(^2A_1') \to V_3^+(^3A_2')$, lower panel: $V_3(^4A_1') \to V_3^+(^3A_2')$.

imaginary frequency for that structure [164, 165] which indicates a Jahn–Teller active structure. From [164] the relaxed structure, a distorted tetrahedron, is lower in energy by only 0.025 eV at the GGA level.

It is hard to comment on the details of the geometry of Table 9 because different functionals, basis sets and grids have been used. Comparing the LSDA results obtained with deMon-KS and DGAUSS using the same basis set, one may think that the difference is coming from the small grid used by Castro *et al.* [164], 832 points per atom compare to at least 7760 points by Cisneros *et al.* [166]. We have repeated the calculation of Castro *et al.* with the same KS-DFT program, basis set and functional but we used more stringent convergence criteria and we increased the total number of points per atom to 20 952. The results are the same at the LSDA level but with the GGA PW86P86 functional we get slightly different results; $a = 2.23$ Å and $b = 2.39$ Å closer to the DGAUSS results. The remaining differences could be due to the functional used or to the small differences in the computational details. At the GGA level, calculations using an ECP agree well with each other but they seem to favor a larger value of the parameter "a" when compared to all electron calculations.

Concerning the electronic state, all the computational studies obtained a cluster with 4 unpaired electrons (multiplicity of 5) [163–168] except Reuse and Khanna [162] who obtained 6. In [163, 168] it is mentioned that Reuse and Khanna have done again the calculations using non relativistic all electron basis set and their results are in agreement with all other studies. At the present time it is not clear if the discrepancy is coming from the presence of relativistic effects or from the lack of core-valence polarization in the pseudo-potential. In principle, the issue could be resolved by a converged all-electron basis set calculations, using a relativistic treatment but that is beyond the state of the art for systems like Ni_4. The ground state of Ni_4 cannot be assigned with certainty as a distorted tetrahedron with 4 unpaired electrons because there are a lot of low-lying states. For example, Michelini, Diez and Jubert found three isomers within 0.1 eV above the lowest state. As already mentioned for V_3, an energy difference less than 0.5 eV leaves uncertainty in the identity of the ground state.

All DFT studies on Ni_5 [162–164, 168] agree on a trigonal bipyramid as the lowest energy structure (see Tab. 10). In three of the studies [163, 164, 168], the lowest energy structure possesses 4 unpaired electrons giving rise to a multiplicity of 5. As already mentioned just above, Reuse and Khanna have re-investigated their calculations performed in [162] and they have found also 4 unpaired electrons instead of the 8 obtained previously. Recently, Aspel *et al.* [173] have measured the magnetic moment per atom in nickel clusters containing 5 to 740 atoms. In the case of Ni_5 they obtained 1.6 μ_B per atom corresponding to 8 unpaired electrons. Many potential explanations have been proposed beyond experimental error and errors coming from the calculations. It is possible that the experiment does not detect the ground state but, rather, a higher energy structure. This is possible because we do not know the temperature exactly in this kind of experiment. During the school, Gustavo Pastor pointed out the possibility of a significant contribution to the magnetism coming from the orbital

Table 9. Comparison of the lowest energy structure of Ni_4 obtained from different programs, basis sets and functionals.

Program	XC func.	Sym.	R_e(Å)	N_s	Ref.
	LSDA[a] [37]	D_{2d}	$a = 2.12$; $b = 2.74$	6	[162]
ADF [169]	LSDA[b] [36]	D_{2d}	$a = 2.19$; $b = 2.31$	4	[165]
deMon-KS [85–87]	LSDA[c] [36]	D_{2d}	$a = 2.15$; $b = 2.28$	4	[164]
DGAUSS [170]	LSDA[c] [36]	D_{2d}	$a = 2.18$; $b = 2.26$	4	[166]
deMon-KS [85–87]	PW86P86[c]	D_{2d}	$a = 2.21$; $b = 2.41$	4	[164]
DGAUSS [170]	B88P86[c]	D_{2d}	$a = 2.24$; $b = 2.35$	4	[166]
G94 [171]	B88PW91[d]	D_{2d}	$a = 2.28$; $b = 2.38$	4	[163]
	BLYP[e]	D_{2d}	$a = 2.26$; $b = 2.40$	4	[167]
DMOL [172]	BLYP[f]	T_d	2.24	4	[168]

[a]Norm-conserving pseudo-potential from [94].
[b]Frozen core approximation used up to 3p.
[c]All electron basis set from [118].
[d]Effective core potential.
[e]Effective core potential from [100].
[f]Numerical basis set.

Table 10. Comparison of the lowest energy structure of Ni_5 obtained from different programs, basis sets and functionals.

Program	XC func.	Sym.	R_e(Å)	N_s	Ref.
	LSDA[a] [37]	D_{3h}	$a = 2.23$; $b = 2.29$	8	[162]
deMon-KS [85–87]	PW86P86[b]	D_{3h}	$a = 2.31$; $b = 2.36$	4	[164]
G94 [171]	B88PW91[c]	D_{3h}	$a = 2.30$; $b = 2.38$	4	[163]
DMOL [172]	BLYP[d]	D_{3h}	$a = 2.24$; $b = 2.29$	4	[168]

[a]Norm-conserving pseudo-potential from [94].
[b]All electron basis set from [118].
[c]Effective core potential from [100].
[d]Numerical basis set.

angular momemtum in free clusters. At the present time this hypothesis needs further investigation and we refer the reader to the chapter of Pastor, in the present book.

Clearly, a systematic study is needed to investigate the effect of ECP *vs.* all electron basis set, the effect of various gradient correction schemes, the effect of a given grid type and size, convergence criteria, etc. on the geometrical description of Ni_4 and Ni_5. The various DFT programs have to be compared and the "numerics" must eventually converge to the same results. This is essential to assess the accuracy of DFT for the description of the difficult case of transition metal clusters and to propose a computational protocol.

10.3 The conversion of acetylene to benzene on Fe clusters

In the last example, KS-DFT is used to study a chemical reaction involving a transition metal cluster in the gas phase. The absence of solvent in a gas phase experiment allows a better comparison with theoretical studies which are usually performed in a reference system corresponding to the vacuum at 0 K.

The formation of benzene on naked transition metal clusters in the gas phase has been observed from acetylene on Fe^+ [174] and from ethylene on Fe_4^+ [175–178], W^+ [179] and U^+ [180]. The reaction mechanism is unknown but reaction intermediates have been observed. The formation of benzene from ethylene implies the dehydrogenation of ethylene to form benzene. We think that a detailed study of the formation of benzene from acetylene on iron clusters, Fe_n^+ ($n = 1 - 4$), can served as a good model for ethylene and other transition metal clusters in the gas phase. To do so, we have studied the reaction mechanism presented in Figure 6. Results on the cyclodimerization of two acetylene units (step D in Fig. 6) will be discussed in more detail.

All-electron calculations were performed with deMon-KS [85–87] which uses the Linear Combination of Gaussian Type Orbitals-Kohn–Sham-Density Functional method (LCGTO-KS-DFT). We used Perdew and Wang for exchange [45], and Perdew for correlation [47, 48], denoted PW86P86. Moreover, we have also included the results obtained with the LAP3 correlation functional [66] synchronized with the popular GGA exchange functionals of Becke [46] (BLAP3) and Perdew and Wang for the exchange [45] (PLAP3). The orbital basis sets were (63321/5211*/311+), (7111/411/1*) and (41/1*) respectively for iron, carbon and hydrogen atoms. The basis sets were optimized as described in [118]. The density and the exchange-correlation potential were fitted using the auxiliary basis set; (5,5; 5,5) for iron, (4,3; 4,3) for carbon, and (3,1; 3,1) for hydrogen. Following the notation $(k_1, k_2;\ l_1, l_2)$, k_1 (l_1) is the number of s-type Gaussians in the density (exchange-correlation) basis and k_2 (l_2) is the number of s-, p-, and d-type Gaussians constrained to have the same exponent in the density (exchange-correlation) basis. During the SCF calculation, the density

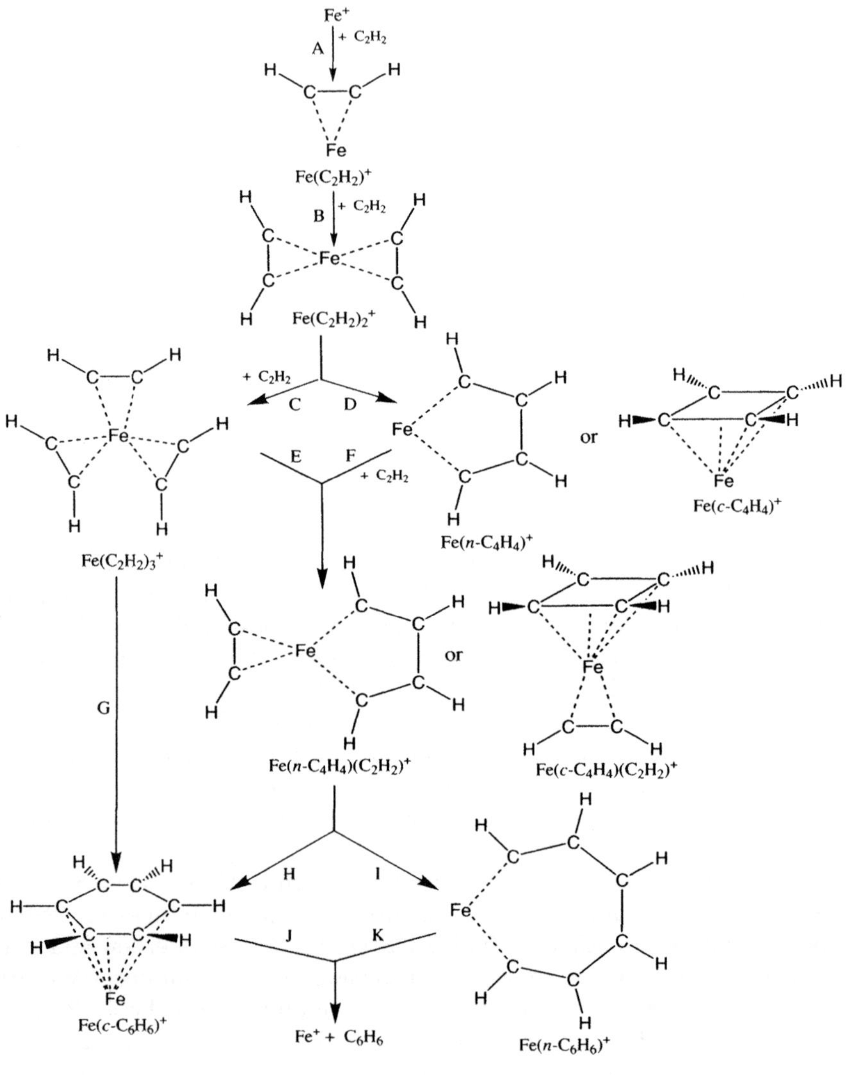

Fig. 6. Schematic representation of the mechanism studied for the formation of benzene from acetylene on Fe^+.

was fitted analytically while the exchange-correlation potential was fitted numerically on a grid containing at least 1066 points per atom. At the SCF convergence, the exchange-correlation potential is evaluated on a grid containing more than 4000 points per atom. This is done to evaluate the final exchange-correlation energy and the contribution of the exchange-correlation energy to the gradient accurately. In the angular grid each shell of angular points is rotated randomly. This procedure destroys slightly the symmetry of the molecule. In our cases, the difference is a few thousandths of an Ångstrom for interatomic distances and 1 degree for the angle, which are acceptable values. We choose this option because, sometimes, it facilitates the SCF convergence.

In order to obtain some minima on the potential energy surface, each geometry of Figure 6 (and many others) has been fully optimized, without symmetry and geometric constraints, for different multiplicities (different number of unpaired electrons) using the Broyden–Fletcher–Goldfarb–Shanno algorithm [128]. During the optimization, the norm of the gradient and the maximum atomic contribution to the norm of the gradient was minimized below 10^{-4}. The convergence criteria for the density was fixed to 10^{-6} while it was 10^{-7} for the energy. Vibrational analysis was performed, in the harmonic approximation, for all the optimized structures. If one or several imaginary frequencies were detected, the structure was perturbed along each mode and re-optimized. The second derivatives were calculated by finite differences of the analytical gradient using a displacement of 0.03 a.u. Transition states have been located with GAUSSIAN94 [171] using the combination of B88 [46] for exchange and P86 for correlation [47,48] (B88P86). We should mention that the density and the exchange-correlation potential are not fitted in GAUSSIAN94. This could introduce very small difference between the results obtained by both programs.

Reaction profiles for the cyclodimerization of two acetylenes on an iron cation computed with the B88P86 functional are presented in Figure 7. Corresponding values for PW86P86, PLAP3 and BLAP3 are collected in Table 11. N_s corresponds to the number of unpaired electrons. The lowest energy structure for the adsorption of a single acetylene possesses three unpaired electrons. The addition of a second acetylene will bring us on the path represented by a dotted line ($N_s = 3$). The energy gain coming from the addition of a second acetylene is larger than any of the activation energies involved in this step of the reaction. Consequently, $Fe(C_2H_2)_2{}^+$ possesses enough internal energy to cross the barriers. Supposing that we will stay on this path until the addition of a third acetylene (step C and F in Fig. 6), one can clearly see from Figure 7 why the cyclodimerization of two acetylenes does not seem to be important experimentally [174], the energy increases when going from $Fe(C_2H_2)_2{}^+$ to any of the cyclic forms, $Fe(n\text{-}C_4H_4)^+$ or $Fe(c\text{-}C_4H_4)^+$. It is possible to go on the second path ($N_s = 1$), and observe cyclodimerization of acetylenes, only if spin-orbit coupling is important or if the first acetylene molecule binds to a highly excited iron cation. Since the formation of a cycle, $Fe(n\text{-}C_4H_4)^+$ or $Fe(c\text{-}C_4H_4)^+$, was not observed quantitatively in the experiment, we think that both situations are not

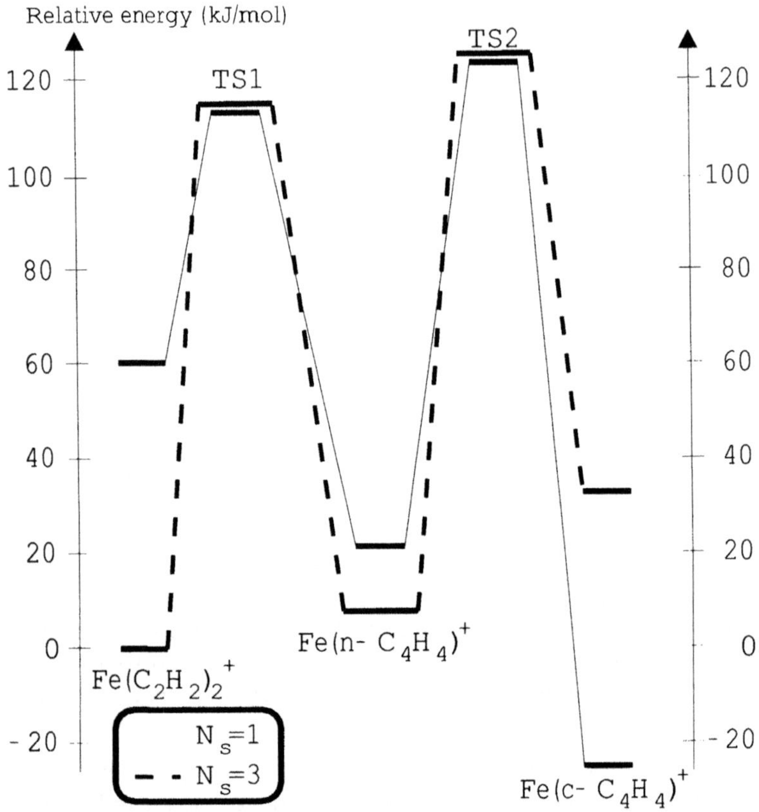

Fig. 7. Relative energy (in kJ/mol) for the cyclodimerization of acetylene on Fe^+ obtained with B88P86 exchange-correlation functional. The solid line and the dot line correspond to $N_s = 1$ and $N_s = 3$ respectively.

frequently met. All the attempts to find a transition state to connect directly the $Fe(C_2H_2)_2^+$ and $Fe(c\text{-}C_4H_4)^+$ complexes have failed so far or end on TS1 or TS2. For the moment, we do not rule out this path but we get the same problem when we tried to find a transition state between the $Fe(C_2H_2)_3^+$ and $Fe(C_2H_2)(c\text{-}C_4H_4)^+$ complexes.

Comparing different exchange-correlation functionals (Tab. 11), the cyclodimerization of acetylene seems to be possible with PLAP3 and BLAP3. In contrast to PW86P86, BLAP3 and B88P86 where $Fe(c\text{-}C_4H_4)^+$ with $N_s = 1$ is the most stable structure by many kJ/mol, PLAP3 gives four nearly degenerate structures. We notice a variation of 50 kJ/mol at most in the relative energies for a given complex and a given value of N_s which

Table 11. Relative and activation energies (in kJ/mol) for the cyclodimerization of acetylene on Fe^+ computed with various functionals.

	N_s	PW86P86	PLAP3	BLAP3	B88P86
	1	46.9	65.1	73.1	59.9
$Fe(C_2H_2)_2{}^+$	3	0.0	0.0	0.0	0.0
	5	168.0		164.9	169.7
TS1	1	113.2[a]	146.3[a]	128.0[a]	114.5
	3	132.0	163.1	140.1	114.8
	1	0.2	−5.9	−7.9	21.4
$Fe(n\text{-}C_4H_4)^+$	3	2.7	−3.6	−5.6	8.0
	5	73.8	65.4	59.4	75.9
TS2	1	125.5	165.4	152.6	126.1
	3	130.6[a]	148.5[a]	141.9[a]	127.4
	1	−37.1	−4.1	−28.3	−26.0
$Fe(c\text{-}C_4H_4)^+$	3	37.6	59.2	35.7	33.4
	5	109.5	109.3	93.2	137.1

[a]Energy calculated at the B88P86 optimized structure.

corresponds to the 0.5 eV for the margin of uncertainty introduced in the previous section for GGA functionals. The correct energetic description of these complexes may require multiconfigurational treatment which is included formally in KS-DFT method through the exact density but is not the case for approximate exchange-correlation functionals. We are currently performing calculations using the Coupled Cluster Singles, Doubles and Triples[12] (CCSD-T) method on that step of the reaction to better assess the energetics of the reaction. We should mention that some of the relative energies of the transition states were obtained by an evalutation of total energy at the B88P86 optimized structure. The difference between these values and the ones obtained from a complete optimization was only a few kJ/mol for the other ones where we have optimized the structure, thus this difference is negligible compared to the 50 kJ/mol of uncertainty.

[12]The triple excitations are included non-iteratively.

Table 12. Optimized structure (distances in Ångstroms and angles in degrees) and frequencies (in cm^{-1}) of the TS corresponding to the cyclodimerization of acetylene on Fe^+, TS1, computed with various functionals ($N_s = 3$).

	PW86P86	PLAP3	BLAP3	B88P86
$r(FeC_1)$	1.925	1.928	1.924	1.925
$r(C_1C_2)$	1.303	1.294	1.288	1.303
$r(C_2C_3)$	1.901	1.854	1.854	1.822
$r(C_1H_1)$	1.090	1.073	1.072	1.089
$r(C_2H_2)$	1.090	1.074	1.073	1.092
$\Theta(FeC_1C_2)$	76.0	77.4	77.2	77.0
$\Theta(C_2FeC_3)$	55.1	52.8	53.6	52.6
$\Theta(H_1C_1C_2)$	140.4	139.6	139.9	139.0
$\Theta(C_1C_2C_3C_4)$	−75.0	−67.7	−70.0	−66.1
$\bar{\nu}$	−365	−383	−344	−324

The differences between the optimized structure for the minima computed with various functionals are less than 0.01 Ångstrom for bond lengths and 1 degree for angles. These values should be taken as error bars in our calculations. The optimized structure of TS1 ($N_s = 3$) is presented in Table 12. The difference in the bond lengths calculated with various functionals is less than 0.01 Ångstrom except for the bond that is formed during the cyclodimerization process (C_2-C_3). The dihedral angle corresponding to the departure from planarity of the carbon skeleton is also very sensitive to the functional. There is a difference of 60 cm^{-1} in the frequency corresponding to the transition state mode. This value is bigger than the differences observed for the minima and for TS2 ($N_s = 1$). When comparing PW86P86 with B88P86 and PLAP3 with BLAP3 (*i.e.* changing the exchange functional) we see that the exchange energy functional is responsible for 2/3 of the difference and the remaining difference is due to the correlation functional. Clearly, further work is needed before DFT can claim universal quantitative accuracy for transition-metal reactions.

11 Conclusions

The creation of very accurate exchange-correlation energy functionals and efficient implementations makes DFT a very useful computational tool for chemical applications. But, there is still room for improvement. Chemical accuracy, errors less than 1 kcal/mol, is still a challenge and there is an increasing need for correlation energy functionals suitable to use in combination with exact KS exchange. The meta-GGA seems to be a promising alternative to the hybrid schemes that require more computational effort. The example of V_3 showed that a proper combination of experiment and theory can somethimes be the key to success. Nickel clusters were used to illustrate the need to carefully choose the computational details. KS-DFT can be used for the determination of reaction mechanisms but more work is needed for the determination of relative energies.

We thank Dongqing Wei for help with the data of the AIMD simulation and Patrizia Calaminici for the PFI-ZEKE simulation data. We thank NSERC, FCAR, NRC/NSERC, CFI, CIPI (NCE), Boehringer-Ingelheim and Astra-Zeneca for support of some of the work reported or reviewed in this paper.

References

[1] S. Goedecker, *Rev. Mod. Phys.* **71** (1999) 1085.

[2] T.K. Woo, P.M. Margl, L. Deng, L. Cavallo and T. Ziegler, *Catalysis Today* **50** (1999) 479.

[3] C. Alhambra, L. Wu, Z.-Y. Zhang and J. Gao, *J. Am. Chem. Soc.* **120** (1998) 3858.

[4] A. Szabo and N.S. Ostlund, in *Modern Quantum Chemistry: Introduction to Advanced Electronic Structure Theory* (McGraw-Hill, New York, 1989).

[5] R.G. Parr and W. Yang, in *Density-Functional Theory of Atoms and Molecules* (Oxford University Press, Oxford, 1989).

[6] R.M. Dreizler and E.K.U. Gross, in *Density Functional Theory: An Approach to the Quantum Many-Body Problem* (Springer-Verlag, Berlin, 1990).

[7] W. Koch and M.C. Holthausen, in *A Chemist's Guide to Density Functional Theory* (Wiley-VCH, New York, 2000).

[8] E. Schrödinger, *Ann. Phys.* **79** (1926) 361.

[9] M. Born and R. Oppenheimer, *Ann. Phys.* **4** (1927) 457.

[10] P. Hohenberg and W. Kohn, *Phys. Rev.* **136** (1964) B864.

[11] M. Levy, *Proc. Natl. Acad. Sci. USA* **76** (1979) 6062.

[12] M. Levy, *Phys. Rev. A* **26** (1982) 1200.

[13] L.H. Thomas, *Proc. Camb. Phil. Soc.* **23** (1926) 542.

[14] E. Fermi, *Zeits. für Physik* **48** (1928) 73.

[15] W. Kohn and L.J. Sham, *Phys. Rev.* **140** (1965) A1133.

[16] E.R. Davidson, in *Reduced Density Matrices in Quantum Chemistry* (Academic Press, New York, 1976).

[17] J. Harris and R.O. Jones, *J. Phys. F: Metal Phys.* **4** (1974) 1170.

[18] O. Gunnarsson and B.I. Lundqvist, *Phys. Rev. B* **13** (1976) 4274.

[19] D.C. Langreth and J.P. Perdew, *Phys. Rev. B* **15** (1977) 2884.

[20] J. Harris, *Phys. Rev. A* **29** (1984) 1648.

[21] T. Ziegler, *Chem. Rev.* **91** (1991) 651.

[22] E. Engel, J.A. Chevary, L.D. Macdonald and S.H. Vosko, *Z. Phys. D* **23** (1992) 7.

[23] E. Engel and S.H. Vosko, *Phys. Rev. B* **47** (1993) 13164.

[24] E.K.U. Gross, M. Petersilka and T. Grabo, in *Chemical Applications of Density-Functional Theory, ACS Symposium Series 629*, edited by B.B. Laird, R.B. Ross and T. Ziegler (American Chemical Society, Washington, DC, 1996), Chap. 3, p. 42.

[25] Q. Zhao and R.G. Parr, *Phys. Rev. A* **46** (1992) 2337.

[26] V. Sahni and M. Levy, *Phys. Rev. B* **33** (1986) 3869.

[27] M. Levy, J.P. Perdew and V. Sahni, *Phys. Rev. A* **30** (1984) 2745.

[28] C.-O. Almbladh and U. von Barth, *Phys. Rev. B* **31** (1985) 3231.

[29] M. Levy, *Phys. Rev. A* **43** (1991) 4637.

[30] L.A. Curtiss, K. Raghavachari, G.W. Trucks and J.A. Pople, *J. Chem. Phys.* **94** (1991) 7221.

[31] L.A. Curtiss, K. Raghavachari, P.C. Redfern, V. Rassolov and J.A. Pople, *J. Chem. Phys.* **109** (1998) 7764.

[32] A.D. Becke, *J. Chem. Phys.* **112** (2000) 4020.

[33] J.P. Perdew, S. Kurth, A. Zupan and P. Blaha, *Phys. Rev. Lett.* **82** (1999) 2544.

[34] C. Adamo, M. Ernzerhof and G.E. Scuseria, *J. Chem. Phys.* **112** (2000) 2643.

[35] P. Dirac, *Proc. Camb. Philos. Soc.* **26** (1930) 376.

[36] S.H. Vosko, L. Wilk and M. Nusair, *Can. J. Phys.* **58** (1980) 1200.

[37] D.M. Ceperley and B.J. Alder, *Phys. Rev. Lett.* **45** (1980) 566.

[38] F. Sim, A. St-Amant, I. Papai and D.R. Salahub, *J. Am. Chem. Soc.* **114** (1992) 4391.

[39] O. Gunnarsson, M. Jonson and B.I. Lundqvist, *Phys. Rev. B* **20** (1979) 3136.

[40] J. Harris and R.O. Jones, *J. Chem. Phys.* **68** (1978) 3316.

[41] O. Gunnarsson and R.O. Jones, *Phys. Rev. B* **31** (1985) 7588.

[42] R.O. Jones and O. Gunnarsson, *Phys. Rev. Lett.* **55** (1985) 107.

[43] J.P. Perdew and A. Zunger, *Phys. Rev. B* **23** (1981) 5048.

[44] J.P. Perdew, *Phys. Rev. Lett.* **55** (1985) 1665.

[45] J.P. Perdew and W. Yue, *Phys. Rev. B* **33** (1986) 8800.

[46] A.D. Becke, *Phys. Rev. A* **38** (1988) 3098.

[47] J.P. Perdew, *Phys. Rev. B* **33** (1986) 8822.

[48] J.P. Perdew, *Phys. Rev. B* **34** (1986) 7406E.

[49] J.P. Perdew *et al.*, *Phys. Rev. B* **46** (1992) 6671.

[50] J.P. Perdew *et al.*, *Phys. Rev. B* **48** (1993) 4978E.

[51] J.P. Perdew, K. Burke and Y. Wang, *Phys. Rev. B* **54** (1996) 16533.

[52] A.D. Becke, *J. Chem. Phys.* **97** (1992) 9173.

[53] C.J. Umrigar and X. Gonze, *Phys. Rev. A* **50** (1994) 3827.

[54] C.J. Umrigar and X. Gonze, in *High Performance Computing and its Applications in the Physical Sciences: Proceedings of the Mardi Gras 1993 Conference*, edited by D.A. Browne *et al.* (World Scientific, Singapore, 1994), p. 43.

[55] R. van Leeuwen and E.J. Baerends, *Phys. Rev. A* **49** (1994) 2421.

[56] A. Lembarki, F. Rogemond and H. Chermette, *Phys. Rev. A* **52** (1995) 3704.

[57] M.E. Casida, C. Jamorski, K.C. Casida and D.R. Salahub, *J. Chem. Phys.* **108** (1998) 4439.

[58] E. Engel and S.H. Vosko, *Phys. Rev. A* **47** (1993) 2800.

[59] A.D. Becke, *J. Chem. Phys.* **104** (1996) 1040.

[60] A.D. Becke and M.R. Roussel, *Phys. Rev. A* **39** (1989) 3761.

[61] T. Van Voorhis and G.E. Scuseria, *J. Chem. Phys.* **109** (1998) 400.

[62] E.I. Proynov, A. Vela and D.R. Salahub, *Chem. Phys. Lett.* **230** (1994) 419.
[63] E.I. Proynov, A. Vela and D.R. Salahub, *Chem. Phys. Lett.* **234** (1995) 462E.
[64] E.I. Proynov and D.R. Salahub, *Phys. Rev. B* **49** (1994) 7874.
[65] E.I. Proynov and D.R. Salahub, *Phys. Rev. B* **57** (1998) 12616E.
[66] E.I. Proynov, S. Sirois and D.R. Salahub, *Int. J. Quant. Chem.* **64** (1997) 427.
[67] E.I. Proynov, H. Chermette and D.R. Salahub, *J. Chem. Phys.* **113** (2000) 10013.
[68] C. Lee, W. Yang and R.G. Parr, *Phys. Rev. B* **37** (1988) 785.
[69] A.D. Becke, *J. Chem. Phys.* **98** (1993) 1372.
[70] A.D. Becke, *J. Chem. Phys.* **98** (1993) 5648.
[71] P.M.W. Gill, B.G. Johnson, J.A. Pople and M.J. Frisch, *Int. J. Quant. Chem. Quant. Chem. Symp.* **26** (1992) 319.
[72] J.P. Perdew, M. Ernzerhof and K. Burke, *J. Chem. Phys.* **105** (1996) 9982.
[73] M. Ernzerhof, J.P. Perdew and K. Burke, *Int. J. Quant. Chem.* **64** (1997) 285.
[74] P.J. Stephens, F.J. Devlin, C.F. Chabalowski and M.J. Frisch, *J. Phys. Chem.* **98** (1994) 11623.
[75] R.T. Sharp and G.K. Horton, *Phys. Rev.* **90** (1953) 317.
[76] J.D. Talman and W.F. Shadwick, *Phys. Rev. A* **14** (1976) 36.
[77] T. Grabo, T. Kreibich, S. Kurth and E.K.U. Gross, in *Strong Coulomb Correlations in Electronic Structure: Beyond the Local Density Approximation*, edited by V.I. Anisimov (Gordon & Breach, Tokyo, 1998).
[78] J.B. Krieger, Yan Li and G.J. Iafrate, *Phys. Rev. A* **46** (1992) 5453.
[79] J.P. Perdew, K. Burke and M. Ernzerhof, *Phys. Rev. Lett.* **77** (1996) 3865.
[80] R. Colle and O. Salvetti, *Theoret. Chim. Acta* **37** (1975) 329.
[81] H. Guo, S. Sirois, E.I. Proynov and D.R. Salahub, in *Theoretical Treatments of Hydrogen Bonding*, edited by D. Hadzi (Wiley, New York, 1996).
[82] S. Sirois, E.I. Proynov, D.T. Nguyen and D.R. Salahub, *J. Chem. Phys.* **107** (1997) 6770.
[83] D.R. Salahub, S. Chrétien, A. Milet and E.I. Proynov, in *Transition State Modeling for Catalysis, ACS Symposium Series 721*, edited by Donald G. Truhlar and Keiji Morokuma (American Chemical Society, Washington, DC, 1999), Chap. 2, p. 20.
[84] D. Wei, E.I. Proynov, A. Milet and D.R. Salahub, *J. Phys. Chem. A* **104** (2000) 2384.
[85] A. St-Amant and D.R. Salahub, *Chem. Phys. Lett.* **169** (1990) 387.
[86] Alain St-Amant, Ph.D. Thesis (Université de Montréal, Montréal, Québec, Canada, 1992).
[87] M.E. Casida, C.D. Daul, A. Goursot, A.M. Köster, L.G.M. Pettersson, E.I. Proynov, A. St-Amant, D.R. Salahub, H.A. Duarte, N. Godbout, J. Guan, C. Jamorski, M. Leboeuf, V.G. Malkin, O.L. Malkina, F. Sim and A. Vela, *deMon Software-deMon-KS3 Module* (Université de Montréal, Montréal, Québec, Canada, 1996).
[88] A.M. Köster, *J. Chem. Phys.* **104** (1996) 4114.
[89] S. Obara and A. Saika, *J. Chem. Phys.* **84** (1986) 3963.
[90] B.I. Dunlap, J.W.D. Connolly and J.R. Sabin, *J. Chem. Phys.* **71** (1979) 3396.
[91] H. Sambe and R.H. Felton, *J. Chem. Phys.* **62** (1975) 1122.
[92] J.M. Pérez-Jordá and W. Yang, *J. Chem. Phys.* **107** (1997) 1218.
[93] A.D. Becke, *J. Chem. Phys.* **88** (1988) 2547.
[94] G.B. Bachelet, D.R. Hamann and M. Schlüter, *Phys. Rev. B* **26** (1982) 4199.

[95] D. Vanderbilt, *Phys. Rev. B* **41** (1990) 7892.

[96] A. Höck and E. Engel, *Phys. Rev. A* **58** (1998) 3578.

[97] V. Bonifacic and S. Huzinaga, *J. Chem. Phys.* **60** (1974) 2779.

[98] J. Andzelm, E. Radzio and D.R. Salahub, *J. Chem. Phys.* **83** (1985) 4573.

[99] W.R. Wadt and P.J. Hay, *J. Chem. Phys.* **82** (1985) 284.

[100] P.J. Hay and W.R. Wadt, *J. Chem. Phys.* **82** (1985) 299.

[101] L.F. Pacios and P.A. Christiansen, *J. Chem. Phys.* **82** (1985) 2664.

[102] M.M. Hurley, L.F. Pacios, P.A. Christiansen, R.B. Ross and W.C. Ermler, *J. Chem. Phys.* **84** (1986) 6840.

[103] L.A. LaJohn, P.A. Christiansen, R.B. Ross, T. Atashroo and W. C. Ermler, *J. Chem. Phys.* **87** (1987) 2812.

[104] R.B. Ross *et al.*, *J. Chem. Phys.* **93** (1990) 6654.

[105] W.J. Stevens, M. Krauss, H. Basch and P.G. Jasien, *Can. J. Chem.* **70** (1992) 612.

[106] N.S. Mosyagin, A.V. Titov and A.V. Tulub, *Phys. Rev. A* **50** (1994) 2239.

[107] I.I. Tupitsyn, N.S. Mosyagin and A.V. Titov, *J. Chem. Phys.* **103** (1995) 6548.

[108] N.S. Mosyagin, A.V. Titov and Z. Latajka, *Int. J. Quant. Chem.* **63** (1997) 1107.

[109] B. Delley, *J. Chem. Phys.* **92** (1990) 508.

[110] O. Treutler and R. Ahlrichs, *J. Chem. Phys.* **102** (1998) 346.

[111] J.M. Pérez-Jordá, A.D. Becke and E. San-Fabián, *J. Chem. Phys.* **100** (1994) 6520.

[112] M. Krack and A.M. Köster, *J. Chem. Phys.* **108** (1998) 3226.

[113] Y.C. Zheng and J.E. Almlöf, *J. Mol. Struc.* **388** (1996) 277.

[114] K.S. Werpetinski and M. Cook, *J. Chem. Phys.* **106** (1997) 7124.

[115] K.R. Glaesemann and M.S. Gordon, *J. Chem. Phys.* **112** (2000) 10738.

[116] G. Berghold, J. Hutter and M. Parrinello, *Theor. Chem. Acc.* **99** (1998) 344.

[117] C.C.J. Roothaan, *Rev. Mod. Phys.* **23** (1951) 69.

[118] N. Godbout, D.R. Salahub, J. Andzelm and E. Wimmer, *Can. J. Chem.* **70** (1992) 560.

[119] J. Andzelm, E. Radzio and D.R. Salahub, *J. Comp. Chem.* **6** (1985) 520.

[120] A.D. Becke, *Int. J. Quant. Chem. Quant. Chem. Symp.* **23** (1989) 599.

[121] S.F. Boys, *Proc. R. Soc. London Ser. A* **200** (1950) 542.

[122] J.C. Slater, *Phys. Rev.* **36** (1930) 57.

[123] S. Huzinaga *et al.*, in *Gaussian basis sets for molecular calculations*, edited by S. Huzinaga (Elsevier, Amsterdam, 1984), Vol. 16, p. 1.

[124] J.M. Millan and G.E. Scuseria, *J. Chem. Phys.* **106** (1997) 5569.

[125] G. Frenking *et al.*, in *Reviews in Computational Chemistry*, edited by K.B. Lipkowitz and D.B. Boyd (VCH Publishers, New York, 1993), Vol. 8, Chap. 2, p. 63.

[126] T.R. Cundari, M.T. Benson, M.L. Lutz and S.O. Sommerer, in *Reviews in Computational Chemistry*, edited by K.B. Lipkowitz and D.B. Boyd (VCH Publishers, New York, 1993), Vol. 8, Chap. 3, p. 145.

[127] H. Guo, N. Gresh, B.P. Roques and D.R. Salahub, *J. Phys. Chem. B* **104** (2000) 9746.

[128] H.B. Schlegel, in *Modern Electronic Structure Theory, Advanced Series in Physical Chemistry*, edited by D.R. Yarkony (World Scientific, River Edge, NJ, 1995), Chap. 8, p. 459.

[129] R. Car and M. Parrinello, *Phys. Rev. Lett.* **55** (1985) 2471.

[130] P.E. Blöchl, *Phys. Rev. B* **50** (1994) 17953.

[131] E.B. Wilson Jr., J.C. Decius and P.C. Cross, in *Molecular Vibrations: The Theory of Infrared and Raman Vibrational Spectra* (Mc Graw-Hill, Toronto, 1955).

[132] P.O. Löwdin, *J. Chem. Phys.* **18** (1950) 365.

[133] R.S. Mulliken, *J. Chem. Phys.* **23** (1955) 1833.

[134] A.M. Lee and S.M. Colwell, *J. Chem. Phys.* **101** (1994) 9704.

[135] J. Guan, M.E. Casida, A.M. Köster and D.R. Salahub, *Phys. Rev. B* **52** (1995) 2184.

[136] C. Jamorski, M.E. Casida and D.R. Salahub, *J. Chem. Phys.* **104** (1996) 5134.

[137] M. Leboeuf, A.M. Köster, K. Jug and D.R. Salahub, *J. Chem. Phys.* **111** (1999) 4893.

[138] M.E. Casida, in *Recent Advances in Density Functional Methods*, Vol. 1 of *Advanced series in physical chemistry*, edited by D.P. Chong (World Scientific, Singapore, 1995), p. 155.

[139] V.G. Malkin, O.L. Malkina, M.E. Casida and D.R. Salahub, *J. Am. Chem. Soc.* **116** (1994) 5898.

[140] G. Schreckenbach and T. Ziegler, *J. Phys. Chem.* **99** (1995) 606.

[141] G. Rauhut, S. Puyear, K. Wolinski and P. Pulay, *J. Phys. Chem.* **100** (1996) 6130.

[142] J.B. Foresman, T.A. Keith, K.B. Wiberg, J. Snoonian and M.J. Frisch, *J. Phys. Chem.* **100** (1996) 16098.

[143] T. Helgaker, M. Jaszuński and K. Ruud, *Chem. Rev.* **99** (1999) 293.

[144] L. Deng and T. Ziegler, *Int. J. Quant. Chem.* **52** (1994) 731.

[145] D. Fuhrmann *et al.*, *J. Chem. Phys.* **108** (1998) 2651.

[146] A. Föhlisch *et al.*, *Phys. Rev. B* **61** (2000) 16229.

[147] F.J. Devlin, P.J. Stephens, J.R. Cheeseman and M. J. Frisch, *J. Phys. Chem. A* **101** (1997) 9912.

[148] T. Mineva, N. Russo and M. Toscano, *Int. J. Quant. Chem.* **56** (1995) 663.

[149] D. Wei, H. Guo and D.R. Salahub, *Phys. Rev. E* (submitted) (2001).

[150] A.D. MacKerell Jr. *et al.*, *J. Phys. Chem. B* **102** (1998) 3586.

[151] D.-S. Yang, A.M. James, D.M. Rayner and P.A. Hackett, *Chem. Phys. Lett.* **231** (1994) 177.

[152] H. Grönbeck and A. Rosén, *J. Chem. Phys.* **107** (1997) 10620.

[153] S.E. Weber *et al.*, *J. Phys.: Condens. Matter* **9** (1997) 10739.

[154] S.E. Weber *et al.*, *J. Phys.: Condens. Matter* **10** (1998) 9591E.

[155] X. Wu and A.K. Ray, *J. Chem. Phys.* **110** (1999) 2437.

[156] P. Calaminici *et al.*, *J. Chem. Phys.* **114** (2001) 4036.

[157] B. Simard, S.A. Mitchell, D.M. Rayner and D.-S. Yang, in *Metal-Ligand Interactions in Chemistry, Physics and Biology*, Vol. 546 of *Nato Science Series C: Mathematical and Physical Sciences*, edited by N. Russo and D.R. Salahub (Kluwer Academic Publishers, The Netherlands, 2000), p. 239.

[158] C.-X. Su, D.A. Hales and P.B Armentrout, *J. Chem. Phys.* **99** (1993) 6613.

[159] D.-S. Yang *et al.*, *J. Chem. Phys.* **103** (1995) 5335.

[160] D.-S. Yang *et al.*, *J. Chem. Phys.* **105** (1996) 10663.

[161] J.E. Boggs, *Pure and Appl. Chemistry* **70** (1998) 1015.

[162] F.A. Reuse and S.N. Khanna, *Chem. Phys. Lett.* **234** (1995) 77.

[163] S.E. Weber and P. Jena, *Chem. Phys. Lett.* **281** (1997) 401.

[164] M. Castro, C. Jamorski and D.R. Salahub, *Chem. Phys. Lett.* **271** (1997) 133.

[165] M.C. Michelini, R.P. Diez and A.H. Jubert, *Int. J. Quant. Chem.* **70** (1998) 693.

[166] G.A. Cisneros, M. Castro and D.R. Salahub, *Int. J. Quant. Chem.* **75** (1999) 847.

[167] B. Chen, A.W. Castleman Jr. and S.N. Khanna, *Chem. Phys. Lett.* **304** (1999) 423.

[168] B.V. Reddy, S.K. Nayak, S.N. Khanna, B.K. Rao and P. Jena, *J. Phys. Chem. A* **102** (1998) 1748.

[169] *ADF 2.3.0*, Theoretical Chemistry, Vrije Universiteit, Amsterdam, 1995.

[170] J.W. Andzelm, in *Density Functional Methods in Chemistry*, edited by J.K. Labanowski and J.W. Andzelm (Springler-Verlag, New York, 1991).

[171] M.J. Frisch, G.W. Trucks, H.B. Schlegel, P.M.W. Gill, B.G. Johnson, M.A. Robb, J.R. Cheeseman, T. Keith, G.A. Petersson, J.A. Montgomery, K. Raghavachari, M.A. Al-Laham, V.G. Zakrzewski, J.V. Ortiz, J.B. Foresman, J. Cioslowski, B.B. Stefanov, A. Nanayakkara, M. Challacombe, C.Y. Peng, P.Y. Ayala, W. Chen, M.W. Wong, J.L. Andres, E.S. Replogle, R. Gomperts, R.L. Martin, D.J. Fox, J.S. Binkley, D.J. Defrees, J. Baker, J.P. Stewart, M. Head-Gordon, C. Gonzalez and J.A. Pople, *Gaussian 94*, revision e.2 ed., Gaussian, Inc., Pittburgh PA, 1995.

[172] *DMOL code*, Biosym Technologies, Inc., San Diego.

[173] S.E. Aspel, J.W. Emmert, J. Deng and L.A. Bloomfield, *Phys. Rev. Lett.* **76** (1996) 1441.

[174] D. Schröder, D. Sülze, J. Hrušák, D. K. Böhme and H. Schwarz, *Int. J. Mass Spectrom. Ion Processes* **110** (1991) 145.

[175] P. Schnabel, M.P. Irion and K.G. Weil, *J. Phys. Chem.* **95** (1991) 9688.

[176] P. Schnabel and M.P. Irion, *Ber. Bunsenges. Phys. Chem.* **96** (1992) 1101.

[177] P. Schnabel, M.P. Irion and K.G. Weil, *Chem. Phys. Lett.* **190** (1992) 255.

[178] O. Gehret and M.P. Irion, *Chem. Phys. Lett.* **254** (1996) 379.

[179] C. Berg *et al.*, *Chem. Phys. Lett.* **231** (1994) 139.

[180] C. Heinemann, H.H. Cornehl and H. Schwarz, *J. Organomet. Chem.* **501** (1995) 201.

COURSE 5

SEMICLASSICAL APPROACHES TO MESOSCOPIC SYSTEMS

M. BRACK

Institute for Theoretical Physics,
University of Regensburg,
9304 Regensburg,
Germany

Contents

SEMICLASSICAL APPROACHES TO MESOSCOPIC SYSTEMS

M. Brack

Abstract

We review semiclassical methods of determining both average trends and quantum shell effects in the properties of finite fermion systems. *I. Extended Thomas–Fermi model (ETF):* the average, selfconsistent mean field can be determined by density variational calculations using the semiclassical gradient-expanded ETF density functional for the kinetic energy. From this, average ground-state properties such as binding energies, densities, separation energies, etc. can be derived. *II. Periodic orbit theory (POT):* quantum oscillations in a mean-field system can be obtained from the semiclassical trace formula that expresses the quantum-mechanical density of states in terms of the periodic orbits of the corresponding classical system. Only the shortest periodic orbits with highest degeneracy are important for the coarse-grained level density, *i.e.*, for the gross shell effects. Particular uniform approximations are required to treat systems with mixed classical dynamics due to the effects of symmetry breaking and orbit bifurcations. *III. Local-current approximation (LCA):* the collective dynamics of the fermions can be described in linear-response theory, approximating the particle-hole excitation operators semiclassically by local current distributions. The method is suitable in combination with both the ETF density variational approach or with the Kohn–Sham density functional approach for the ground state, and allows one to describe optic response properties such as static polarizabilities and plasmon resonances. Applications of all methods to metal clusters and various mesoscopic nanostructures are given.

1 Introduction

We want to describe properties of finite fermion systems semiclassically. We start from a mean-field approach where the particles move independently in a selfconsistent potential. There are essentially two standard methods to derive such a selfconsistent potential (see, *e.g.* [1]): *i*) the Hartree–Fock (HF) theory in which the exchange is treated exactly, and *ii*) the Kohn–Sham (KS) density functional theory (DFT) in which the exchange often is treated only in the local density approximation (LDA), but other correlations can also be included. In the following, we write the mean-field Hamiltonian as the sum of the kinetic energy operator and the selfconsistent mean field V (that includes possible external fields)

$$\hat{H} = \hat{T} + V. \tag{1.1}$$

The single-particle energies ε_i and wavefunctions φ_i are given by the (HF or KS) Schrödinger equation

$$\hat{H}\varphi_i(\mathbf{r}) = \varepsilon_i \varphi_i(\mathbf{r})\,. \tag{1.2}$$

In HF theory, V contains also the non-local Fock exchange term, whereas in KS-DFT theory $V = V(\mathbf{r})$ is a *local* potential (the total KS potential, which includes also an exchange-correlation part). For simplicity we assume that the spectrum ε_i is discrete.

The fundamental quantity that contains all the quantum information of the energy spectrum is the single-particle *level density*

$$g(E) = \mathrm{tr}\,\delta(E - \hat{H}) = \sum_i \left\langle i|\delta(E - \hat{H})|i\right\rangle = \sum_i \delta(E - \varepsilon_i)\,, \tag{1.3}$$

where $\delta(x)$ is the Dirac delta function. The summation over i here goes over the complete spectrum including all degeneracies.

Following very general arguments of Strutinsky [2], the level density may be split into an average part $\widetilde{g}(E)$ and an oscillating part $\delta g(E)$:

$$g(E) = \widetilde{g}(E) + \delta g(E). \tag{1.4}$$

Although the two parts arise naturally within the periodic orbit theory (POT), as will be discussed in Section 3, it is customary to consider them separately. The average part $\widetilde{g}$ may be calculated numerically from the ε_i by the Strutinsky averaging method [2]. It is equivalent [3,4] to its semiclassical approximation obtained in the extended Thomas–Fermi (ETF) model which will be discussed in Section 2. The oscillating part δg, which contains all the quantum shell effects coming from the non-uniform distribution of the single-particle levels ε_i (including their degeneracies), is obtained semiclassically in the POT discussed explicitly and illustrated in Section 3. A semiclassical approach to the *collective* excitation spectrum, using a local-current approximation (LCA), is finally presented in Section 4.

2 Extended Thomas–Fermi model for average properties

The ETF model allows one to obtain average properties such as binding and deformation energies, separation energies, etc., that do not include quantum oscillations. The main idea of the TF model is to take an average over the classical phase space with a unit occupation probability per unit cell h^3, where h is Planck's constant (we assume in the following a 3-dimensional system), multiplied by a factor 2 to take the spin degeneracy into account. Including finite $\hbar$-correction terms yields the extended TF model that contains average quantum corrections without, however, generally allowing for quantum shell effects.

2.1 Thomas–Fermi approximation

The TF approximation for the level density is obtained by taking the trace in (1.3) in the phase space:

$$g_{\mathrm{TF}}(E) = \frac{2}{h^3} \int \delta[E - H_{\mathrm{cl}}(\mathbf{r}, \mathbf{p})] \, \mathrm{d}^3r \, \mathrm{d}^3p \,, \tag{2.1}$$

where we have replaced $\hat{H}$ by the classical Hamiltonian H_{cl}

$$H_{\mathrm{cl}}(\mathbf{r}, \mathbf{p}) = \frac{1}{2m}\mathbf{p}^2 + V(\mathbf{r}). \tag{2.2}$$

If the mean field in (1.1) contains non-local terms (*e.g.*, the Fock term in the case of HF theory), then the potential will also be momentum dependent, $V = V(\mathbf{r}, \mathbf{p})$, and can be obtained from (1.1) by a Wigner transform [1]. In the following we assume $V(\mathbf{r})$ to be local and analytical. In the ground state of a mean-field system, all levels up to the Fermi energy E_{F} are occupied. The total number of particles N is then given by the integral

$$N = \int_0^{E_{\mathrm{F}}} g(E) \, \mathrm{d}E = \frac{2}{h^3} \int \Theta \left[E_{\mathrm{F}} - H_{\mathrm{cl}}(\mathbf{r}, \mathbf{p})\right] \, \mathrm{d}^3r \, \mathrm{d}^3p, \tag{2.3}$$

where $\Theta(x)$ is the Heavyside step function. The momentum integration can be done immediately with (2.2), using its spherical symmetry in momentum space, and yields

$$N = \frac{1}{3\pi^2} \left(\frac{2m}{\hbar^2}\right)^{3/2} \int \left[E_{\mathrm{F}} - V(\mathbf{r})\right]^{3/2} \mathrm{d}^3r. \tag{2.4}$$

Omitting the spatial part of the integral in (2.4) yields the TF approximation of the spatial density:

$$\rho_{\mathrm{TF}}(\mathbf{r}) = \frac{1}{3\pi^2}\left(\frac{2m}{\hbar^2}\right)^{3/2} [E_{\mathrm{F}} - V(\mathbf{r})]^{3/2}\, \Theta\left[E_{\mathrm{F}} - V(\mathbf{r})\right]. \tag{2.5}$$

Note the step function which implies that the TF density is zero outside the classical turning points $\mathbf{r}_i$ defined by $E_{\mathrm{F}} = V(\mathbf{r}_i)$. In the same way we can also obtain the total kinetic energy E_{kin} of the system by the integral

$$E_{\mathrm{kin}} = \frac{2}{h^3}\int \frac{p^2}{2m}\,\Theta\left[E_{\mathrm{F}} - H_{\mathrm{cl}}(\mathbf{r},\mathbf{p})\right]\, \mathrm{d}^3r\, \mathrm{d}^3p \tag{2.6}$$

which after momentum integration yields

$$E_{\mathrm{kin}} = \frac{1}{5\pi^2}\left(\frac{2m}{\hbar^2}\right)^{3/2} \int [E_{\mathrm{F}} - V(\mathbf{r})]^{5/2}\, \mathrm{d}^3r\,, \tag{2.7}$$

so that the kinetic energy density becomes

$$\tau_{\mathrm{TF}}(\mathbf{r}) = \frac{1}{5\pi^2}\left(\frac{2m}{\hbar^2}\right)^{3/2} [E_{\mathrm{F}} - V(\mathbf{r})]^{5/2}\, \Theta\left[E_{\mathrm{F}} - V(\mathbf{r})\right]. \tag{2.8}$$

From equations (2.5, 2.8) we can eliminate the Fermi energy and the potential to obtain the TF functional for the kinetic energy density

$$\tau_{\mathrm{TF}}[\rho] = \frac{\hbar^2}{2m}\,\kappa\,\rho^{5/3}\,, \qquad \text{with} \quad \kappa = \frac{3}{5}\,(3\pi^2)^{2/3}. \tag{2.9}$$

Equations (2.5) and (2.8, 2.9) are the expressions which were used by Thomas and Fermi [5] to describe the electronic density and energy of the atom semiclassically in a selfconsistent way (see, *e.g.*, the book by March [6] for a very nice presentation). Although the TF model gives a nice qualitative description of the average properties of an atom, it is not very good. (It has been shown, though, that the total TF energy for the neutral atom becomes *exact* in the mathematical limit of infinite nuclear charge $Z \to \infty$; see [7] for a review). It is possible, however, to improve the TF model by taking into account a series of quantum corrections, leading to the so-called "extended Thomas–Fermi" (ETF) model which we will summarize in the following subsections.

2.2 Wigner–Kirkwood expansion

There are various ways to derive the quantum corrections to the TF model; we use the method developed by Wigner and Kirkwood [8] which starts from expansion in powers of $\hbar$ of the single-particle Bloch density matrix around its classical TF value. We refer to a recent monograph [4] for a detailed presentation of this method and summarize just the general idea and results. The basic idea is to relate the quantities of interest to the Bloch density matrix which is defined as the statistical operator $\mathrm{e}^{-\beta\hat{H}}$ in coordinate-space representation:

$$C(\mathbf{r},\mathbf{r}';\beta) = \sum_i \varphi_i^*(\mathbf{r}')\,\varphi_i(\mathbf{r})\,\mathrm{e}^{-\beta\varepsilon_i} = \left\langle \mathbf{r}|\mathrm{e}^{-\beta\hat{H}}|\mathbf{r}'\right\rangle, \tag{2.10}$$

where the sum goes again over the complete spectrum. Here β is a complex mathematical variable. The inverse Laplace transform (see, *e.g.* [9, 10]) of $C(\mathbf{r},\mathbf{r}';\beta)$ yields the single-particle density matrix

$$\begin{aligned}\rho(\mathbf{r},\mathbf{r}') &= \sum_{\varepsilon_i\le E_\mathrm{F}} \varphi_i^*(\mathbf{r}')\,\varphi_i(\mathbf{r}) = \mathcal{L}_{E_\mathrm{F}}^{-1}\left[\frac{1}{\beta}C(\mathbf{r},\mathbf{r}';\beta)\right] \\ &= \frac{1}{2\pi i}\int_{c-i\infty}^{c+i\infty}\frac{1}{\beta}\,C(\mathbf{r},\mathbf{r}';\beta)\,\mathrm{e}^{\beta E_\mathrm{F}}\mathrm{d}\beta \qquad (c>0).\end{aligned} \tag{2.11}$$

From this one obtains easily the local density

$$\rho(\mathbf{r}) = \sum_{\varepsilon_i\le E_\mathrm{F}} |\varphi_i(\mathbf{r})|^2 = \rho(\mathbf{r},\mathbf{r}) \tag{2.12}$$

and the kinetic energy density

$$\tau(\mathbf{r}) = -\frac{\hbar^2}{2m}\sum_{\varepsilon_i\le E_\mathrm{F}} \varphi_i^*(\mathbf{r})\,\Delta\varphi_i(\mathbf{r}) = -\frac{\hbar^2}{2m}\,\Delta_\mathbf{r}\rho(\mathbf{r},\mathbf{r}')|_{\mathbf{r}'=\mathbf{r}}, \tag{2.13}$$

where the subscript $\mathbf{r}$ here and in the following indicates that an operator acts only on the variable $\mathbf{r}$ on its right. Note that these are so far the exact quantum-mechanical expressions. The TF approximation and its extensions may now be obtained in the following way. Take the matrix element in (2.10) in a plane-wave representation (including the spin factor 2):

$$C(\mathbf{r},\mathbf{r}';\beta) = \frac{2}{h^3}\int \mathrm{d}^3p\,\mathrm{e}^{-i\mathbf{p}\cdot\mathbf{r}'/\hbar}\,\mathrm{e}^{-\beta\hat{H}_\mathbf{r}}\,\mathrm{e}^{i\mathbf{p}\cdot\mathbf{r}/\hbar}. \tag{2.14}$$

Due to the potential $V(\mathbf{r})$ in (1.1), the plane waves are no eigenfunctions of $\hat{H}$. If we replace $\hat{H}_\mathbf{r}$ in the exponent above by the classical Hamilton function $H_\mathrm{cl}(\mathbf{r},\mathbf{p})$ and perform the $\mathbf{p}$ integration, we obtain the classical Bloch density matrix

$$C_\mathrm{cl}(\mathbf{r},\mathbf{r}';\beta) = 2\left(\frac{m}{2\pi\hbar^2\beta}\right)^{3/2}\exp\left[-\beta V(\mathbf{r}) - \frac{m}{2\hbar^2\beta}(\mathbf{r}-\mathbf{r}')^2\right]. \tag{2.15}$$

Inserting this into equation (2.11) yields the TF densities. To obtain quantum corrections to the TF results, one starts from the ansatz

$$e^{-\beta\hat{H}}\, e^{i\mathbf{p}\cdot\mathbf{r}/\hbar} = e^{-\beta H_{\rm cl}(\mathbf{r},\mathbf{p})}\, e^{i\mathbf{p}\cdot\mathbf{r}/\hbar}\, w(\mathbf{r},\mathbf{p};\beta), \tag{2.16}$$

and expands $w(\mathbf{r},\mathbf{p};\beta)$ into a series in powers of $\hbar$:

$$w = 1 + \hbar w_1 + \hbar^2 w_2 + \ldots \tag{2.17}$$

Hereby the w_n are functions of $\mathbf{r}$, $\mathbf{p}$, and β, containing up to n gradients acting on the potential $V(\mathbf{r})$. The series (2.17) may, *e.g.*, be obtained by expanding the left side of equation (2.16) in terms of repeated commutators of the kinetic energy operator with $V(\mathbf{r})$ (see Ref. [1]). Another way, solving iteratively a differential equation for $w(\mathbf{r},\mathbf{p};\beta)$ (the so-called Bloch equation), is described in [4]. Only the even terms of the series (2.17) with $n = 2m$, $m = 1, 2, 3, \ldots$ survive the $\mathbf{p}$ integration for the local density and kinetic energy density, and give the ETF corrections to (2.5) and (2.8). These correction terms diverge at the classical turning points, but they can be integrated to yield finite corrections to the particle number, the total energies and other integrated quantities. The densities obtained in this way by truncation of the series (2.17) at some order are usually called the ETF densities $\rho_{\rm ETF}(\mathbf{r})$ and $\tau_{\rm ETF}(\mathbf{r})$ (see, *e.g.* [4] for their explicit forms).

2.3 Gradient expansion of density functionals

Eliminating the potential and its gradients from the ETF densities, which takes some algebra (*cf.* [11] where this has been done systematically also for momentum and spin dependent potentials), one arrives at the kinetic energy-density functional $\tau_{\rm ETF}[\rho]$ including gradient correction terms. We quote here only its integrated form as it contributes to the total kinetic energy functional $T_{\rm ETF}[\rho]$. Keeping terms up to fourth order in $\hbar$ one obtains after some partial integrations [12, 13]

$$\begin{aligned} E_{\rm kin} &= T_{\rm ETF}[\rho] = \frac{\hbar^2}{2m}\int {\rm d}^3r \left\{ \kappa\rho^{5/3} + \frac{1}{36}\frac{(\nabla\rho)^2}{\rho} \right. \\ &\left. + \frac{1}{6480}(3\pi^2)^{-2/3}\rho^{1/3}\left[8\left(\frac{\nabla\rho}{\rho}\right)^4 - 27\left(\frac{\nabla\rho}{\rho}\right)^2\frac{\Delta\rho}{\rho} + 24\left(\frac{\Delta\rho}{\rho}\right)^2 \right] \right\}. \end{aligned} \tag{2.18}$$

In the first term of (2.18) we recognize the TF functional (2.9). The second term, coming from the $\hbar^2$ corrections, has been derived for the first time by von Weizsäcker [14], albeit with a different coefficient. Equation (2.18) has the correct coefficient in the limit of slowly varying densities, whereas the original Weizsäcker coefficient (1/4 instead of 1/36) is correct in the limit of rapid density oscillations with small amplitude (see [15] for a detailed discussion). The terms shown on the second line in (2.18) come from the $\hbar^4$ terms of the semiclassical expansion (2.17). Although the derivation of the functional (2.18) sketched above is valid only within the classically allowed region, it can be shown from its extension to finite temperature [16, 17], where the ETF densities stay finite and are defined in the whole space, that the zero-temperature limit leads to equation (2.18) also in the classically forbidden region (see also [18, 19]).

The functional (2.18) is only expected to be correct for average densities without quantum shell oscillations, since it has been derived in a semiclassical framework that has no room for shell effects. Nevertheless, it has been used in connection with atomic HF densities [20] and found to reproduce the total kinetic energies quite well. A careful study of (2.18) using numerically Strutinsky-averaged quantum-mechanical densities $\widetilde{\rho}(\mathbf{r})$ for various forms of the potential $V(\mathbf{r})$ was done in [21]. The Strutinsky-averaged kinetic energies $\widetilde{T}$ were found to be well reproduced when all terms up to fourth order in (2.18) were taken into account, as is demonstrated in Figure 1. The dashed curve in the right panel of the figure shows the exact quantum-mechanical kinetic energy which exhibits pronounced shell effects; the ETF functional (2.18) clearly serves to reproduce only the *average* part of the kinetic energy.

Similarly, one obtains also gradient corrections to the exchange-correlation energy functional occurring in DFT [1], or corrections arising in connection with nonlocal potentials (*e.g.*, spin-orbit and effective mass terms; see [19] for the case of effective Skyrme interactions in nuclear physics). Note that nonlocal potentials give also corrections to the kinetic energy functional (2.18) [11, 19]. Relativistic functionals $T_{\mathrm{ETF}}[\rho]$ have also been derived [1, 22, 23].

2.4 Density variational method

Within the framework of the density functional theory (DFT), the Hohenberg–Kohn theorem [24] tells us that the total energy of a fermion system is uniquely given as a functional of the local density $\rho(\mathbf{r})$. We write it here as the sum of kinetic and potential parts:

$$E_{\mathrm{tot}} = T_{\mathrm{s}}[\rho] + E_{\mathrm{pot}}[\rho] = \int \{\tau(\mathbf{r}) + \mathcal{E}_{\mathrm{pot}}[\rho]\}\ \mathrm{d}^3r, \tag{2.19}$$

where $T_{\mathrm{s}}[\rho]$ is the kinetic energy of a non-interacting system with density $\rho(\mathbf{r})$ and $E_{\mathrm{pot}}[\rho]$ the sum of all potential energy terms including the contribution from any external one-body potential $V_{\mathrm{ext}}(\mathbf{r})$, which just gives a term $V_{\mathrm{ext}}(\mathbf{r})\rho(\mathbf{r})$ under the integral above, and the exchange-correlation

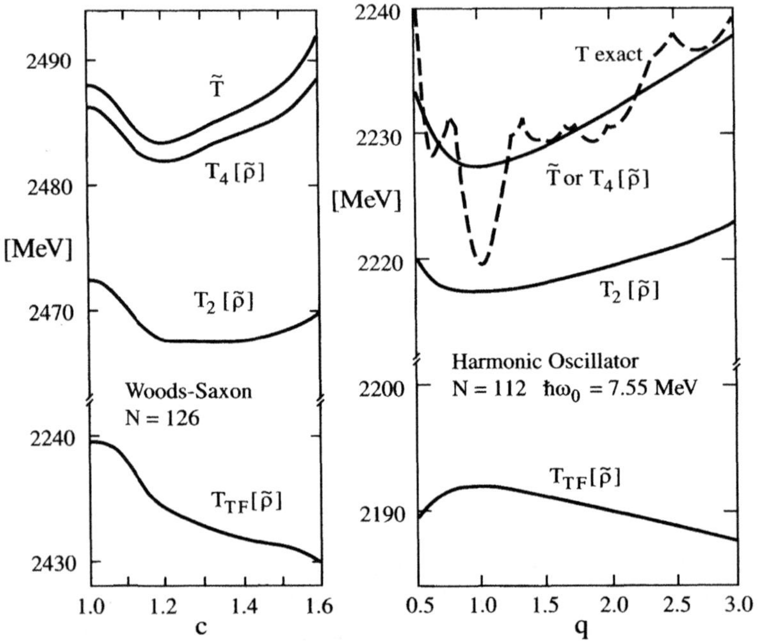

Fig. 1. Tests of the kinetic energy functional $\tau_{\mathrm{ETF}}[\rho]$ (2.18). *Left:* deformed Woods–Saxon potential with $N = 126$ nucleons (after [13]; see the definition there of the deformation parameter c). *Right:* axially symmetric harmonic-oscillator potential with frequency ratio q with $N = 112$ particles (after [21]). The index in $T_{2m}[\widetilde{\rho}]$ gives the maximum order of $\hbar$ corrections included ($m = 0$ is TF). $\widetilde{T}$ is the Strutinsky-averaged kinetic energy (not distinguishable from $T_4[\widetilde{\rho}]$ on the r.h. side). Note the shell effects in the exact kinetic energy T shown by a dashed line.

energy (that also takes into account the interacting part of the kinetic energy, see [1]). The exact functional $T_s[\rho]$ is not explicitly known; therefore one has to express it *via* $\tau(\mathbf{r})$ in equation (2.13) in terms of single-particle wavefunctions $\varphi_i(\mathbf{r})$ and vary the total energy with respect to these, which leads to the Kohn–Sham (KS) equations.

If we are not interested in the quantum shell effects, we can replace $T_s[\rho]$ by the semiclassical ETF functional (2.18), so that the energy (2.19) becomes an explicit functional of $\rho(\mathbf{r})$. We can then directly apply the variational principle by making $E[\rho]$ stationary with respect to an arbitrary local variation $\delta\rho(\mathbf{r})$ of the (average) density. To keep the number of particles N constant we include a Lagrange multiplier μ which is just the Fermi energy of the average system. This yields the Euler–Lagrange variational equation

$$\frac{\delta}{\delta\rho(\mathbf{r})}\left[E[\rho(\mathbf{r})] - \mu\int\rho(\mathbf{r})\,\mathrm{d}^3r\right] = 0. \tag{2.20}$$

Its solution gives the *average* ground-state energy and density of the system within the ETF approximation. To keep the notation simple, we still write $\rho(\mathbf{r})$ but keep in mind that $\rho(\mathbf{r})$ and $E[\rho]$ only represent the average quantities. The variation of the potential energy gives, quite generally, the local mean-field potential $V(\mathbf{r})$ which depends in a selfconsistent way on the density itself

$$\frac{\delta}{\delta\rho(\mathbf{r})}E_{\mathrm{pot}}[\rho(\mathbf{r})] = V[\rho(\mathbf{r})]\,. \tag{2.21}$$

$V[\rho]$ is formally identical to the total KS potential (except that it is evaluated here for the average density); it depends, of course, on the particular system under consideration and on the two-body interaction. The variation of the kinetic energy part, however, is universal and can be calculated directly from the explicit form of the ETF functional given above. To avoid lengthy formulae, we give here explicitly only the result obtained up to the Weizsäcker term. Assuming the density variation $\delta\rho(\mathbf{r})$ to vanish on the boundary (which usually is at infinity) and using the standard techniques of variational calculus, one gets the following differential equation

$$\frac{\hbar^2}{2m}\left\{\frac{5}{3}\kappa\rho^{2/3} + \frac{1}{36}\left[\frac{(\nabla\rho)^2}{\rho^2} - 2\frac{\nabla^2\rho}{\rho}\right]\right\} + D_4[\rho] + V[\rho] = \mu\,, \tag{2.22}$$

where D_4 is the contribution from the fourth-order gradient term, $D_4[\rho] = \delta T_4[\rho]/\delta\rho$, whose explicit form (containing seven terms with up to fourth-order gradients of ρ) may be found in [19]. Equation (2.22) is a partial differential equation which in general can only be solved numerically. To lowest TF order, neglecting all gradient terms, the solution of (2.22) for the density $\rho(\mathbf{r})$ is algebraic and gives, as it should, exactly the TF density $\rho_{\mathrm{TF}}(\mathbf{r})$ given in (2.5).

If the gradient terms are kept in equation (2.22), the nature of the solutions depends on the behaviour of the potential $V(\mathbf{r})$. For the following discussions and applications, we shall assume that $V(\mathbf{r})$ is finite in the interior of the system and goes to zero at large distances. Then the asymptotic fall-off of the density is always governed by the highest gradient term included in the variational equation (2.22). (See [1] for the problems that arise in atomic systems from the $1/r$ divergence at the center due to the external nuclear potential). Keeping just the Weizsäcker term, the solutions for the density are found to decay exponentially. This is because the constant Fermi energy μ can only be cancelled by the highest gradient term in equation (2.22). The rate of the exponential fall-off turns, however, out to be too fast in comparison to that of exact quantum-mechanical densities. One pragmatic way out of this problem is to artificially increase the coefficient of the Weizsäcker term, adjusting it by optimizing the results. The dilemma hereby is that the same constant cannot be used for densities and for integrated energies. Nevertheless, this procedure has been widely used in physics and physical chemistry (see [1] for details and applications). It violates, however, the aspect of universality which is one of the nice features of the ETF model, and can easily be avoided by including the higher-order gradient terms in equation (2.22).

Including the fourth-order contribution $D_4[\rho]$, the density of a spherical system falls off only with the sixth inverse power of r, *i.e.*, $\rho(r) \sim r^{-6}$. More generally, the highest derivative term in $\tau_{2m}[\rho]$ goes radially like [21]

$$\tau_{2m}[\rho(r)] \sim [\rho(r)]^{(5-2m)/3} \frac{1}{\rho(r)} \frac{\mathrm{d}^{2m}\rho(r)}{\mathrm{d}r^{2m}}, \tag{2.23}$$

and the solution of equation (2.20) is found to fall off like

$$\rho(r) \sim r^{-3m/(m-1)} \quad \text{for} \quad r \longrightarrow \infty \qquad (m \geq 2). \tag{2.24}$$

Inserting this into (2.23), one finds that the selfconsistent kinetic energy density $\tau(r)$ goes to zero with the same power law as $\rho(r)$ in (2.24). It can therefore always be integrated to give a finite kinetic energy. (The fact that the sixth- and higher-order gradient corrections to $\tau_{\mathrm{ETF}}[\rho]$ yield diverging integrals for densities that decay exponentially like $\rho(r) \sim \mathrm{e}^{-ar}$ has repeatedly been noted in the literature and interpreted as a break down of the ETF gradient expansion. This is an erroneous conclusion, since the selfconsistent solutions of (2.22) for $m > 2$ do *not* fall off exponentially.)

The inverse power law (2.24) found for $m \geq 2$ is, of course, not realistic for most systems, as the quantum-mechanical density usually falls of exponentially. It has turned out, however, that the asymptotic solution (2.24) is reached only mathematically very far away from the actual surface of a realistic system, and that the inclusion of the fourth-order term definitely improves the variational results. This has been shown in numerical solutions of the ETF variational equation (2.22) up to fourth order, both for atomic nuclei [18,25] and for metallic clusters within the jellium model [26]. Spherical symmetry was assumed in both cases so that the differential equation was only one-dimensional. The applications to metal clusters will be discussed explicitly in the remainder of this section.

One basic problem of the ETF variational method is that there is effectively no lower bound to the total energy, different from the Hartree–Fock method where this is granted by the Ritz principle. *A priori*, the Hohenberg–Kohn theorem [24] tells us that the total energy (2.19) has its variational minimum at the value of the exact energy of the system. But this holds only true if the *exact* density functional $E[\rho]$ is used which, however, is unknown. Due to the approximations necessary to put the density functional method into use – and here in particular due to the ETF approximation – this lower bound is no longer given. As a consequence, the variational ETF results are typically overbound, *i.e.*, the (absolute) binding energies are too large [19, 25]. The amount of overbinding in nuclei is, however, much smaller than one percent of the total energy.

We should like to emphasize again that due to the very nature of the ETF model, the Euler equation (2.22) cannot reproduce any of the quantum shell oscillations typical of finite fermion systems. The ETF variational method is thus limited to yield average properties of finite fermion systems. A perturbative way to include the shell effects, without going through the fully self-consistent microscopic HF or KS scheme, makes use of Strutinsky's shell-correction method (see Ref. [4]).

The extension of the ETF model to systems at finite temperature that can be described by a grand-canonical ensemble is straightforward, though algebraically quite tedious. We refer to the literature [4, 17, 18] for its presentation and, in particular, to [18] where the functional $T_{\mathrm{ETF}}[\rho]$ with its explicitly temperature-dependent coefficients can be found, as well as applications to nuclear physics. An example of the results obtained there is given in Figure 5 below.

2.5 Applications to metal clusters

2.5.1 Restricted spherical density variation

The numerical solution of the nonlinear fourth-order differential equation (2.22) becomes very difficult, if not practically impossible, for deformed systems without symmetry. For that case one may perform the variation in (2.20) not exactly in $\mathbf{r}$ space, but in a restricted space of parameterized variational densities. The choice of the variational density is then a matter of physical intuition and, of course, not free of a certain bias. A spherical density profile that has proved useful for both nuclei [19] and metal clusters [27, 28] is a generalized Fermi function:

$$\rho(r) = \frac{\rho_0}{\left[1+\exp\left(\frac{r-R}{a}\right)\right]^{\gamma}}. \tag{2.25}$$

Of the four parameters above, only three are independent due to the constraint of the particle number conservation. The independent parameters are then determined variationally by minimizing the total energy of the system under consideration, and no adjustable parameter is used anywhere. With this restricted variational procedure up to fourth order in the ETF

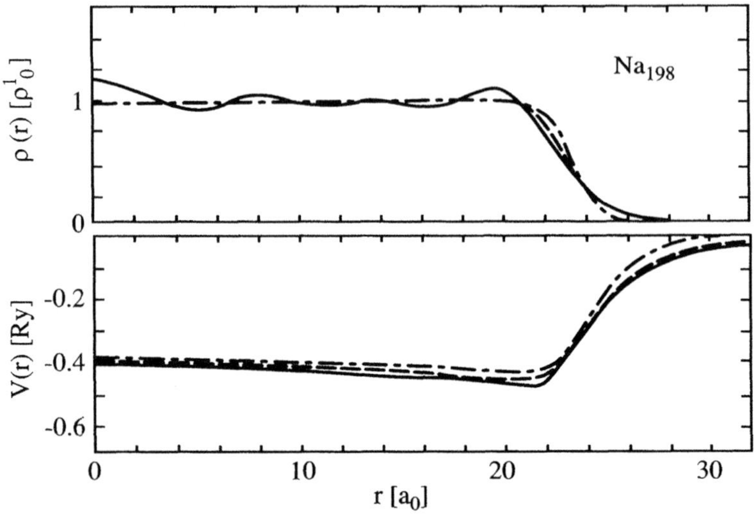

Fig. 2. Comparison of selfconsistent densities and potentials for the spherical Na_{198} cluster obtained in the jellium model. $V(r)$ is in Rydberg units, $\rho(r)$ in units of the ionic bulk density ρ_0^{I}. *Solid lines:* quantum-mechanical KS results [29]. *Dashed lines:* semiclassical ETF results including up to 4th-order gradient terms, and *dashed-dotted lines:* only up to 2nd-order gradient terms in the functional (2.18). (From [27]).

functionals, density profiles and energies were obtained that agree very closely with the fully variational ETF results [25, 26] on one hand (see the following subsection), and with the averaged results of fully microscopic KS calculations for metal clusters [29,30] and HF calculations for nuclei [19,25] on the other hand. In particular, the profiles of the variationally optimized densities (2.25) match the selfconsistent quantum-mechanical ones over the entire region of the surface, demonstrating that the mathematically inadequate asymptotic inverse-power decays (2.24) of the exact solutions of (2.22) have no practical significance.

As an example we show in Figure 2 the densities and potentials obtained in a large sodium cluster within the spherical jellium model (see, *e.g.* [31] for a detailed discussion of this model), both by solving the fully quantum-mechanical KS equations [29] and by the ETF density variational method using the profile (2.25) [27]. When including the 4th-order gradient terms in (2.18), the ETF solution agrees perfectly with the KS solution in the outer surface region, whereas the 2nd-order solution falls off too rapidly (as mentioned above). Note the shell oscillations in the quantum results.

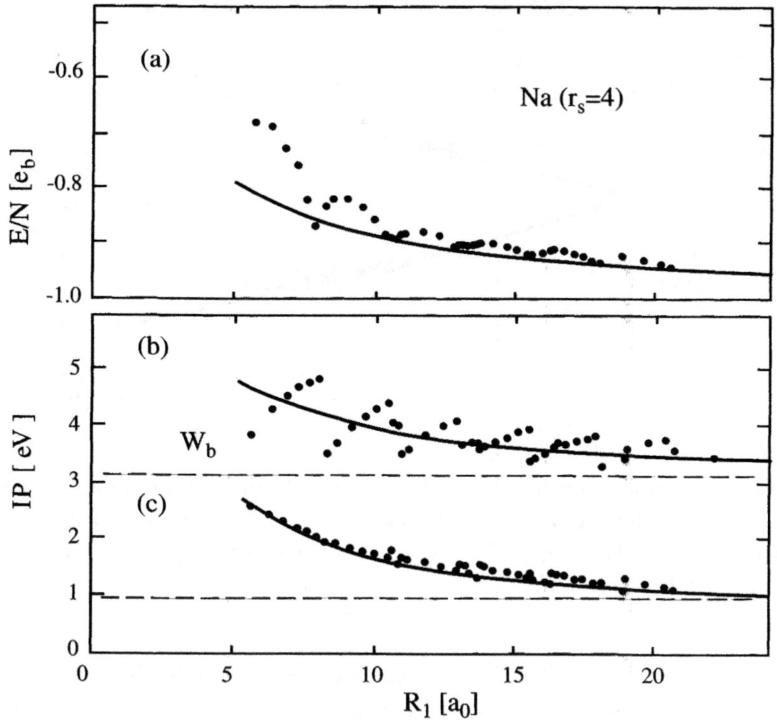

Fig. 3. *Upper part* (a): energies per atom E/N in units of the bulk value e_b; *lower part* (b): ionization potentials IP (in eV) for spherical jellium clusters with $r_s = 4$ a.u., shown *versus* cluster radius R_I. W_b is the bulk work function. Part (c) gives the electrostatic parts of the ionization potentials. Dots and small circles are the quantum-mechanical KS results [29]; the solid lines are the 4th-order ETF results [27]. (After [27]).

Figure 3 shows a similar comparison for (a) the total energy per particle E/N and (b) the ionization potential IP for neutral spherical sodium clusters with N atoms obtained in the jellium model, plotted *versus* the cluster radius $R_I = r_s N^{1/3}$ (with a Wigner–Seitz radius of $r_s = 4$ a.u.). Again we see that the ETF results are smooth curves that miss the periodic shell effects but represents the correct average behaviour. The asymptotic limit of the IP for large clusters is the bulk work function W_b. Note that the electrostatic part of the IP shown at the bottom (c) oscillates much less than the other energies. The fact that the semiclassical total energy seems to lie lower than the average of the quantum results is due to the imposed spherical symmetry. Indeed, the minima of the quantum results correspond

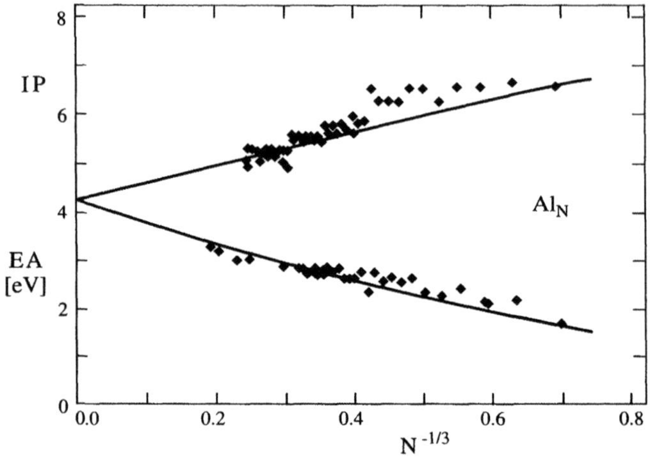

Fig. 4. Ionization potentials IP and electron affinities EA of aluminum clusters with N atoms, plotted *versus* the inverse cluster radius. *Black diamonds:* experimental results, *solid lines:* calculated ETF values in the spherical jellium model with $r_s = 2.99$. (From [32]).

the spherically "magic" clusters that have filled shells; in the regions in between, the clusters have only partially filled spherical shells and want to be deformed due to the Jahn–Teller effect. If a spontaneous symmetry breaking of the mean field is allowed in the KS calculation, their energies are systematically lowered, so that the ETF results come to lie much closer to the averaged KS results (see [31] and Sect. 3.5.2 for a discussion of the deformation effects).

A comparison of experimental ionisation potentials IP and electron affinities EA (see their definitions in Eq. (2.27) below) of aluminum clusters with the variational ETF results [32], obtained in the spherical jellium model, is shown in Figure 4 *versus* the inverse average cluster radius (which is proportional to $N^{1/3}$). The experimental results exhibit, of course, oscillations that come both from electronic shell effects and from the ionic structure that cannot be described by the jellium model. It is quite amazing that a simple semiclassical ETF calculation can capture the average experimental trends even down to the smallest molecules with $N = 3$, corresponding to the rightmost experimental point. (Note, however, that the calculated values have been vertically adjusted to fit the experimental bulk work function W_b at the intersection with the vertical axis, since the jellium model yields a wrong value of W_b. In this way, the average effects of the ionic structure has been taken into account implicitly (see [32] for details). See also [33] for improved results in the "stabilized" jellium model).

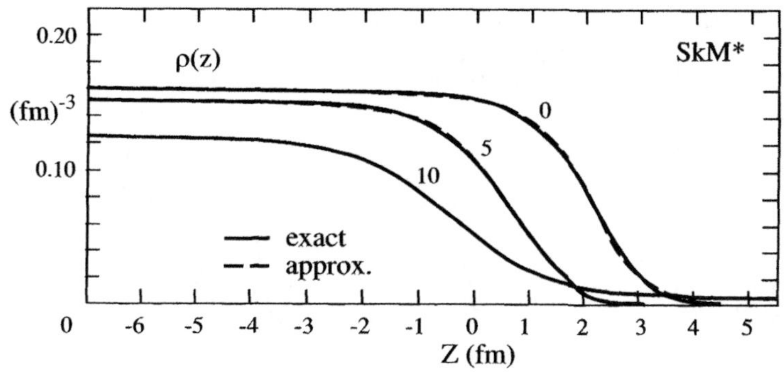

Fig. 5. Density profiles $\rho(z)$ for an interface of nuclear matter and a gas of nucleons in thermal equilibrium, calculated at the temperatures $T = 0$, 5, and 10 MeV. it Solid lines: exact solutions of the Euler–Lagrange equation (2.22) up to fourth order; *dashed lines:* results using a restricted parametrization of $\rho(z)$ corresponding to equation (2.25). (From [18]).

2.5.2 Unrestricted spherical density variation

A test of the parametrization (2.25) of the spherical density profiles has been provided by unrestricted density variational calculations of Engel and Perdew [26] in the framework of the spherical jellium model. These authors solved numerically the Euler–Lagrange equation (2.22) including the fourth-order gradient terms. Their density profiles $\rho(r)$, and quantities derived from them, are extremely well reproduced by the results of the restricted variation using the profiles (2.25). A similar test was done in nuclear physics [18] using the finite-temperature version of the ETF model. In Figure 5 we show the density profiles $\rho(z)$ across the plane interface (positioned at $z = 0$) of nuclear matter in thermal equilibrium with a gas of nucleons, calculated for three different temperatures. The solid lines give the results obtained from the exact numerical solutions of the Euler–Lagrange variational equations (in one-dimensional planar geometry); the dashed lines are obtained using a restricted variation with the one-dimensional analogue of the parametrization (2.25).

When the ionic structure is included in the theoretical description of metal clusters, the electronic density profiles can no longer be given by simple parametrizations, and unrestricted density variations become a necessity. This has been done by various groups [34–37] who performed variational (E)TF calculations for the electrons, in connection with molecular dynamics for the ions in the Born–Oppenheimer approach. As an example of the results of such calculations, we show in Figure 6 the density profile of the highly deformed charged cluster Na_{18}^{++}, which has been called a

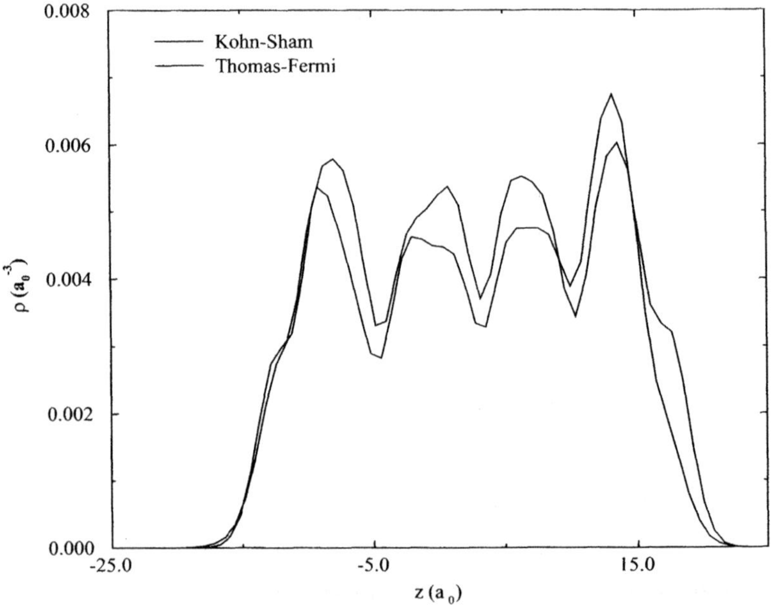

Fig. 6. Ground-state electron density $\rho(z)$ of the fission isomer of Na_{18}^{++}, computed quantum-mechanically in Kohn–Sham DFT (heavy line) and semiclassically in TF theory (thin line). (From [35]).

"fission isomer" [35]. It has an approximate axial symmetry, and the density is shown here along the direction of the symmetry axis. The heavy line was obtained from solving the quantum-mechanical KS equations, whereas the thin line is obtained in the simple TF approximation.

2.5.3 Liquid drop model for charged spherical metal clusters

We finally mention a useful application of ETF variational calculations to derive liquid-drop model (LDM) expansions of the total energy or other observables of a fermion system. When its density is of the form seen in Figures 5 or 2 with a constant saturated value ρ_0 in the interior and a well-defined surface region a around the radius R where it falls off most steeply, one can perform analytically a so-called "leptodermous" expansion [19, 38] of all integrated quantities in powers of the small parameter a/R. For metal clusters this has been done systematically by Seidl [28, 39] within the spherical jellium model. The total energy of a cluster with N atoms and Z

valence electrons is found to be:

$$
\begin{aligned}
E(N,Z) &= e_{\mathrm{b}}N + a_{\mathrm{s}}N^{2/3} + a_{\mathrm{c}}N^{1/3} + a_0 - W_{\mathrm{b}}(Z-N) \\
&+ \left\{ a_{-1} + \frac{e^2}{r_{\mathrm{s}}}\left[C(Z-N) + \frac{1}{2}(Z-N)^2 \right] \right\} N^{-1/3} \\
&+ O\left(N^{-2/3}\right). \qquad (2.26)
\end{aligned}
$$

For the ionization potential IP and the electron affinity EA, defined by

$$
\mathrm{IP} = E(N,N-1) - E(N,N)\,, \quad \mathrm{EA} = E(N,N) - E(N,N+1)\,, \qquad (2.27)
$$

one finds the following expansions up to order $1/R = O(N^{-1/3})$:

$$
\begin{aligned}
\mathrm{IP}\,(R) &= W_{\mathrm{b}} + \left(\frac{1}{2} - C\right)\frac{e^2}{R} + O\left(R^{-2}\right), \\
\mathrm{EA}\,(R) &= W_{\mathrm{b}} - \left(\frac{1}{2} + C\right)\frac{e^2}{R} + O\left(R^{-2}\right). \qquad (2.28)
\end{aligned}
$$

These expressions have been found to fit rather well the average experimental data for large clusters; the terms of order $O(R^{-2})$ become, however, particularly important for the electron affinities of small clusters [40] (see also Fig. 4 above).

All the parameters appearing in equations (2.26, 2.28) can be related to bulk properties (energy per particle e_{b}, Wigner–Seitz radius r_{s}) or properties of the infinite planar metal surface such as the work function W_{b}, the surface tension σ in $a_{\mathrm{s}} = 4\pi r_{\mathrm{s}}^2\sigma$, and other integrals [19,38] over the infinite surface density profile $\rho_\infty(x)$ (x being the distance from the plane with $x = 0$ at the jellium edge). In particular, the parameter C is given by

$$
C = \frac{2}{3}\frac{a_{\mathrm{s}}}{e^2/r_{\mathrm{s}}} - 4\pi\int_0^\infty \mathrm{d}x\,\left[x^2\rho_\infty(x) + xf(x)\right], \qquad (2.29)
$$

where the function $f(x)$ appears in the asymptotic expansion of the density profile $\rho_R(r-R)$ of a large spherical cluster with radius R [26,33,41]:

$$
\rho_R(x) = \rho_\infty(x) + \frac{f(x)}{R} + O(R^{-2}). \qquad (2.30)
$$

Note that C is *not* equal to 1/8, as was derived from a wrong classical image-potential argument which unfortunately still floats around in the literature and in the heads of some researchers. (See [42], where the flaw in this argument has been uncovered and corrected). In fact, the quantity C, which in (2.28) above gives small deviations from the (correctly calculated) classical charging energy $e^2/2R$ of a sphere, is of purely quantum-mechanical origin with contributions from the kinetic and exchange-correlation energies of the electrons and the non-classical tail of their density.

The LDM expansion has also been discussed in the framework of the so-called "stabilized jellium model" [33] which yields improved bulk work functions and surface energies, and therefore provides a better starting point than the simple jellium model.

3 Periodic orbit theory for quantum shell effects

In this section, we shall describe the semiclassical approach to quantum shell effects. Historically, the "old" quantum theory by Bohr [43] started from a semiclassical approximation to the energy spectrum of the atom, requiring the action integral of the electron on its supposedly classical orbit around the nucleus to be an integer multiple of Planck's constant h (see [44] for a remarkable historical account). This semiclassical quantization scheme was later developed into the WKB method (named after Wenzel *et al.* [45]) for one-dimensional systems and, more than 30 years later, into the EBK quantization or "torus quantization" scheme for more-dimensional integrable systems. This method is named after Einstein *et al.* [46], who revived the idea of semiclassical quantization using the very fundamental observation by Einstein [47] that the classical motion of an integrable system is bound to a torus in phase space. (See, *e.g.* [4] for a presentation of the basic ideas of semiclassical quantization).

The EBK method does not apply if a system with N degrees of freedom is not classically integrable, *i.e.*, if it does not possess the same number N of constants of the motion which are independent and which Poisson-commute pairwise. The question if a non-integrable system can be quantized semiclassically remained therefore unanswered for some time. A break-through was accomplished by Gutzwiller [48], who derived a "trace formula" that connects the oscillating part of the quantum-mechanical level density to a sum over all periodic orbits of a classically non-integrable system. Independently, similar trace formulae were derived later by Balian and Bloch for particles in a cavity with reflecting walls (*i.e.*, for three-dimensional billiards) [49], and by Berry and Tabor for integrable systems. Starting from EBK quantization, these authors first showed that the sum over the EBK spectrum can be converted quite generally in a sum over periodic orbits [50], and later derived the same result starting from Gutzwiller's semiclassical Green function [51] (see Sect. 3.1 below). Since the classical periodic orbits play a fundamental role in this semiclassical approach, it has been called "periodic orbit theory" (POT). Including a sufficient number of orbits in the trace formula, it can in principle yield the quantized energies of the system in the form of singularities (the peaks of the sum of delta functions). This works, however, only in a limited number of cases as the trace formulae represent asymptotic series which usually do not converge [52].

The applications of the POT which we want to discuss here do not aim at the quantization, but at a semiclassical description of *gross-shell* features in mean-field systems, which can be obtained by convoluting the level density with a smooth function over a finite energy range γ. Then, the periodic orbit sums converge [53] and only the orbits with the shortest periods and largest degeneracies become important.

We shall not derive here the trace formulae but refer to the original literature. For the approach of Gutzwiller, the reader may consult a monograph by the same author [52], and the very readable series of lectures which he presented a an earlier Les Houches Summer School [54]. The original trace formula of Gutzwiller [48] applies only to systems in which all orbits are isolated; this is the case for a classically chaotic system. A series of extensions to Gutzwiller's theory to take into account degenerate orbits, which appear in systems that possess continuous symmetries, have been developed over the past 30 years [55, 56]. Their basic ideas are discussed in [4].

In Section 3.1 we shall present the semiclassical Green function developed by Gutzwiller, which is the basic ingredient to all further developments in the POT. In Section 3.2 we present the trace formulae for the level density, particle number and total energy of a mean-field system, and in Section 3.3 we briefly discuss practical methods for finding periodic orbits and calculating their stabilities. Section 3.4 contains a summary of the problems connected with symmetry breaking transitions and orbit bifurcations, and indicates that these can be overcome by uniform approximations. The remaining parts are reserved for applications to metal clusters and mesoscopic semiconductor devices.

3.1 Semiclassical expansion of the Green function

The quantum-mechanical level density $g(E)$ given in equation (1.3) can be related to the imaginary part of the trace of the energy-dependent single-particle Green function $G(\mathbf{r}, \mathbf{r}'; E)$ for a system in D dimensions:

$$g(E) = -\frac{1}{\pi} \Im \int G(\mathbf{r}, \mathbf{r}; E + i\epsilon)\, \mathrm{d}^D r \qquad (\epsilon > 0). \tag{3.1}$$

The Green function itself is defined by

$$G(\mathbf{r}, \mathbf{r}'; E) = \sum_i \varphi_i^*(\mathbf{r}')\, \varphi_i(\mathbf{r})\, \frac{1}{(E - \varepsilon_i)} \tag{3.2}$$

and can be related to the single-particle propagator $K(\mathbf{r},\mathbf{r}';t)$ by a (half-sided) Fourier transform:

$$G(\mathbf{r},\mathbf{r}';E) = -\frac{i}{\hbar}\lim_{\epsilon\to 0}\int_0^\infty K(\mathbf{r},\mathbf{r}';t)\,\exp\left[\frac{i}{\hbar}(E+i\epsilon)t\right]\mathrm{d}t. \tag{3.3}$$

Semiclassical approximations to the propagator go back to Van Vleck in 1928 [57]. Gutzwiller rederived Van Vleck's approximation from Feynman's path integral approach to quantum mechanics and improved it by the incorporation of some important phases. The basic idea of his derivation is the stationary phase approximation, applied to integrals containing a rapidly oscillating phase. Doing also the Fourier transform in (3.3) by the stationary phase method, he arrived at the following semiclassical approximation to the Green function [48]:

$$G_{\mathrm{scl}}(\mathbf{r},\mathbf{r}';E) \simeq \frac{2\pi}{(2\pi i\hbar)^{\frac{D+1}{2}}}\sum_\alpha \sqrt{|\mathcal{D}_\alpha|}\,\exp\left[\frac{i}{\hbar}S_\alpha(\mathbf{r},\mathbf{r}',E) - i\mu_\alpha\frac{\pi}{2}\right]\cdot \tag{3.4}$$

The sum goes over all classical trajectories α that lead from the point $\mathbf{r}'$ to the point $\mathbf{r}$ at fixed energy E; the functions $S_\alpha(\mathbf{r},\mathbf{r}',E)$ are the action integrals along these trajectories:

$$S_\alpha(\mathbf{r},\mathbf{r}';E) = \int_{\mathbf{r}'}^{\mathbf{r}}\mathbf{p}''\cdot\mathrm{d}\mathbf{r}'' = \sum_{i=1}^{D}\int_{r_i'}^{r_i} p_i(\mathbf{r}'',E)\,\mathrm{d}r_i'', \tag{3.5}$$

where r_i, p_i are the components of phase-space vectors $\mathbf{r}$ and $\mathbf{p}$, respectively. In equation (3.4), $\mathcal{D}_\alpha$ is the determinant of the $(D{+}1)$-dimensional matrix of second partial derivatives of the action integral with respect to all its arguments

$$\mathcal{D}_\alpha(\mathbf{r},\mathbf{r}';E) = \begin{vmatrix} \dfrac{\partial^2 S_\alpha}{\partial\mathbf{r}'\partial\mathbf{r}} & \dfrac{\partial^2 S_\alpha}{\partial\mathbf{r}'\partial E} \\[2ex] \dfrac{\partial^2 S_\alpha}{\partial E\partial\mathbf{r}} & \dfrac{\partial^2 S_\alpha}{\partial E^2}\end{vmatrix}. \tag{3.6}$$

The quantities μ_α in (3.4) count the number of caustic (or "conjugate") points along the trajectory α; they are connected to the Maslov indices appearing in the EBK quantization of integrable systems [46].

3.2 *Trace formulae for level density and total energy*

It remains only to insert the semiclassical Green function (3.4) into the relation (3.1) and to perform the trace integral. Taking $\mathbf{r}' = \mathbf{r}$ under the integral automatically selects only *closed* orbits from the sum over trajectories α. If these closed orbits are isolated in phase space, then all spatial integrals transverse to the orbits can again be done using the stationary phase approximation; only the integration along the orbit is done exactly. The stationary condition for the exponent in (3.4) leads, using equation (3.5), to

$$\left[\frac{\partial S}{\partial \mathbf{r}}\right]_{\mathbf{r}_0} = \left[\frac{\partial S(\mathbf{r},\mathbf{r}',E)}{\partial \mathbf{r}} + \frac{\partial S(\mathbf{r},\mathbf{r}',E)}{\partial \mathbf{r}'}\right]_{\mathbf{r}'=\mathbf{r}=\mathbf{r}_0} = \mathbf{p} - \mathbf{p}' = 0, \qquad (3.7)$$

where $\mathbf{r}_0$ is the point where S becomes stationary. Equation (3.7) implies that the orbits are not only closed but *periodic*, *i.e.*, they have to start and end with the same momentum: $\mathbf{p}' = \mathbf{p}$. The transformation to a coordinate system co-moving with each periodic orbit, the evaluation of the saddle-point integrations, and expressing the result in a canonically invariant form – all this requires quite some tedious algebra and finally leads to Gutzwiller's famous trace formula for isolated orbits [48]:

$$\delta g_{\text{scl}}(E) \simeq \frac{1}{\hbar\pi}\sum_{po}\frac{T_{ppo}}{\sqrt{|\det(\widetilde{\mathbf{M}}_{po} - \mathbf{1})|}}\cos\left(\frac{1}{\hbar}S_{po} - \sigma_{po}\frac{\pi}{2}\right). \qquad (3.8)$$

The sum here goes over all periodic orbits *po*, including their repeated cycles. $\widetilde{\mathbf{M}}_{po}$ is the stability matrix discussed in Section 3.3 below and $\mathbf{1}$ the unit matrix with the same dimension, S_{po} is the action integral along the (repeated) periodic orbit

$$S_{po}(E) = \oint_{po} \mathbf{p}\cdot d\mathbf{r}, \qquad (3.9)$$

$T_{ppo} = \partial S_{ppo}/\partial E$ is the period of the *primitive* periodic orbit (*i.e.*, its first cycle), and σ_{po} is the so-called Maslov index which arises as a sum of the phase index μ_{po} occurring in the semiclassical Green function (3.4) plus an index ν_{po} (being 1 or 0 for a two-dimensional system) that stems from the stationary-phase evaluation of the trace integral. (See also [58], where it has been shown that the total Maslov index $\sigma_{po} = \mu_{po} + \nu_{po}$ is a canonical and topological invariant, whereas μ_{po} and ν_{po} alone may depend on the starting point along the orbit or on the choice of the coordinate system). The index σ_{po} may also be obtained as a winding number [58, 59]. All quantities appearing on the r.h.s. of the trace formula (3.8) depend smoothly on the energy E. (The Maslov indices are integer constants in finite energy intervals but usually change abruptly by one or more units at bifurcations; see Sect. 3.4 below).

For systems that contain continuous symmetries, most (or all) periodic orbits are not isolated in phase space but appear in degenerate families.

The single orbits within these families can be smoothly transformed into each other by a symmetry operation (translation, rotation, etc.) that does not affect their properties (actions, stabilities, Maslov indices). Consequently, the stationary-phase approximation to some of the trace integrations perpendicular to the orbits breaks down (namely the directions in which the orbits can be displaced by these symmetry operations and in which the actions remain constant). As a result, the determinant in the denominator of (3.8) becomes zero and the trace formula diverges. Similar problems arise also in isolated situations (*e.g.*, at specific energies or values of some other continuous system parameter) in connection with orbit bifurcations and symmetry-breaking (or symmetry-restoring) transitions. For these latter situations special uniform approximations can be developed, which will be briefly discussed in Section 3.4.

The problem of continuous orbit degeneracies can be solved by performing *exactly* the trace integrations in those directions where the action remains constant [55,56]. (See [4] for a detailed discussion of different techniques used hereby). The results of these extensions of the Gutzwiller theory are trace formulae of the general form, which also applies to those derived earlier for billiards [49] and integrable systems [50,51]:

$$\delta g_{\rm scl}(E) \simeq \sum_{po} \mathcal{A}_{po}(E) \cos\left[\frac{1}{\hbar} S_{po}(E) + \sigma_{po}\frac{\pi}{2}\right] . \qquad (3.10)$$

Here the sum goes over all distinct periodic orbit families (plus isolated orbits if they exist). The overall Maslov indices σ_{po} may also become half-integer. The precise form of the amplitudes $\mathcal{A}_{po}$ depends on the particular symmetry of the system. The inverse power of $\hbar$ contained in these amplitudes equals $1+f/2$ where f is the degree of degeneracy of a given orbit family, *i.e.*, the number of independent continuous parameters describing its degeneracy (or, equivalently, the number of different symmetry operations possible, or the dimension of the group creating these symmetry operations). Note that families with different degrees of degeneracy and isolated orbits with $f = 0$ may coexist in one and the same system. For example, in a three-dimensional system with spherical symmetry, the most degenerate orbit families occur with a three-fold continuous degeneracy ($f = 3$), because their possible orientations are described by the three Euler angles; the group creating the corresponding rotations is the SO(3) group; the amplitudes $\mathcal{A}$ of these orbits are proportional to $\hbar^{-5/2}$. In a spherical billiard, there is also the linear orbit bouncing along a diameter; its degeneracy is only $f = 2$ since rotation about its own direction does not create a new orbit. If the potential is a smooth and attractive central field $V(r)$, there exists also an isolated orbit with $f = 0$, namely the circular orbit with a radius corresponding to the minimum of the effective potential $V(r) + L^2/2mr^2$ for a given non-zero angular momentum L.

When applying the POT to a fermionic mean-field system, one is also interested in integrated quantities obtained from the level density. Its first integral gives the number of particles, see equation (2.3). Performing the energy integration in the saddle-point approximation yields the trace formula for the oscillating part of the function $N(E_{\mathrm{F}})$:

$$\delta N_{\mathrm{scl}}(E_{\mathrm{F}}) \simeq \sum_{po} \left(\frac{\hbar}{T_{po}}\right) \mathcal{A}_{po}(E_{\mathrm{F}}) \sin\left[\frac{1}{\hbar} S_{po}(E_{\mathrm{F}}) + \sigma_{po}\frac{\pi}{2}\right], \qquad (3.11)$$

where T_{po} is the period of the (repeated) periodic orbit (family) at the Fermi energy E_{F}. The oscillating part δE of the total binding energy of an interacting system in the mean-field approximation can be shown [2,55] to be given, to leading order in the oscillating part of the level density δg, by

$$\begin{aligned} \delta E_{\mathrm{scl}} &\simeq \int_0^{E_{\mathrm{F}}} (E - E_{\mathrm{F}})\, \delta g(E)\, \mathrm{d}E \\ &\simeq \sum_{po} \left(\frac{\hbar}{T_{po}}\right)^2 \mathcal{A}_{po}(E_{\mathrm{F}}) \cos\left[\frac{1}{\hbar} S_{po}(E_{\mathrm{F}}) + \sigma_{po}\frac{\pi}{2}\right]. \end{aligned} \qquad (3.12)$$

(See also [4] for a discussion of the "Strutinsky theorem" from which the first equation above can be derived). Note that in the above two trace formulae, only the orbit properties at the Fermi energy E_{F} are required.

In most realistic physical systems, the periodic orbits and their properties can only be calculated numerically (see Sect. 3.3) – which, as already mentioned, can be a very difficult task. For systems which are only weakly perturbed away from an analytically known integrable limit, a perturbative trace formula has been developed by Creagh [60], for which the perturbed orbits need not be determined. Assume a Hamiltonian of the form

$$\hat{H} = \hat{H}_0 + \epsilon \hat{H}_1, \qquad (3.13)$$

where $\hat{H}_0$ is integrable and ϵ is a small dimensionless number. The idea now is that to lowest order in ϵ, the amplitudes and Maslov indices in the trace formula will not be affected by the perturbation, and that only the lowest-order change δS_{po} induced by $\epsilon \hat{H}_1$ has to be retained in the phase. The perturbed trace formula is then of the form

$$\delta g_{\mathrm{scl}}(E) \simeq \Re\left\{ \sum_{po} \mathcal{A}_{po}^{(0)}(E)\, \mathcal{M}_{po}(x) \mathrm{e}^{i\left[\frac{1}{\hbar} S_{po}^{(0)}(E) + \sigma_{po}^{(0)} \frac{\pi}{2}\right]} \right\}, \qquad (3.14)$$

where the quantities with superscript "(0)" are those appearing in the trace formula for the unperturbed system $\hat{H}_0$, and $\mathcal{M}_{po}(x)$ is a – generally complex – modulation factor defined as the average of the phase change due to δS_{po} over the unperturbed orbit family:

$$\mathcal{M}_{po}(x) = \left\langle \mathrm{e}^{i\delta S_{po}/\hbar} \right\rangle_{po} . \tag{3.15}$$

The argument x of the modulation factor is proportional to the lowest power of ϵ for which the action change δS_{po} is non-zero, inverse proportional to $\hbar$, and further depends on the energy E. The perturbative trace formula for the total energy is correspondingly given by

$$\delta E_{\mathrm{scl}} \simeq \Re \left\{ \sum_{po} \mathcal{A}_{po}^{(0)}(E_{\mathrm{F}})\, \mathcal{M}_{po}(x) \left(\frac{\hbar}{T_{po}} \right)^2 \mathrm{e}^{i\left[\frac{1}{\hbar} S_{po}^{(0)}(E_{\mathrm{F}}) + \sigma_{po}^{(0)} \frac{\pi}{2}\right]} \right\} . \tag{3.16}$$

By construction, the unperturbed trace formulae are recovered in the limit $\epsilon \to 0$. For large values of ϵ, this approach will eventually break down and uniform approximations must be used to recover the correct amplitudes of the trace formula applying to the strongly perturbed system (see Sect. 3.4).

All the above trace formulae are remarkable in the sense that they relate the quantum-mechanical oscillations contained in their left-hand side with a sum over periodic orbits on their right-hand side that includes only properties of the classical system. They represent asymptotic series which can be expected to become exact in the limit $S_{po} \gg \hbar$. As already mentioned in the introduction to this section, these series usually do not converge, and in many systems the problem of enumerating all periodic orbits and calculating their properties is practically not solvable. There exist a few soluble (integrable and non-integrable) systems for which the trace formula can be shown to converge to a sum of delta functions if the smooth part $\widetilde{g}(E)$ of the level density is added to it – and for a few of those, the result is even identical to the exact quantum-mechanical level density [4, 52]. But these are rare exceptions.

For the applications discussed here, both sides of the trace formulae are convoluted with a normalized Gaussian function $\exp\left[-(E/\gamma)^2\right]/\sqrt{\pi}\gamma$. Using saddle-point integration for the convolution of the periodic orbit sum, one obtains an extra factor

$$\exp\left[-\left(\frac{\gamma}{2} \frac{T_{po}}{\hbar} \right)^2 \right] \tag{3.17}$$

under the summations in all trace formulae above. This factor suppresses the orbits with longer periods and usually leads to the convergence of the orbit sums (*cf.* Ref. [53]). In particular, it shows that the *gross-shell* effects are governed by the shortest orbits. Exact quantization by summing over all orbits can still be attempted by taking the limit $\gamma \to 0$.

Concerning the smooth part $\widetilde{g}(E)$ of the level density that was the subject of Section 2, it has been shown [49, 61] to arise within the periodic orbit theory from the closed orbits of *zero length* (and hence zero action) that mathematically also contribute to the sum of trajectories in the semiclassical Green function (3.4) when taking the trace integral in (3.1). Since their actions (and also their Maslov indices) are zero, they do not contribute in an oscillatory fashion to the level density but just constitute its phase-space average which yields the (extended) TF approximation as discussed in Section 2. This becomes particularly transparent in the derivation of the trace formula for integrable systems by Berry and Tabor [50] (see [4] for some exactly soluble models).

3.3 Calculation of periodic orbits and their stability

We briefly present here the linear stability analysis of periodic orbits and how the stability matrix can be used to find periodic orbits. We start from a holonomic Hamiltonian with a local potential $V(\mathbf{r})$ in D dimensions:

$$H(\mathbf{r}, \mathbf{p}) = \frac{1}{2}\mathbf{p}^2 + V(\mathbf{r})\,. \tag{3.18}$$

(We put the mass equal to unity and assume the kinetic energy to be quadratic in $\mathbf{p}$ just for simplicity of the presentation here; any general dependence of $\mathbf{p}$ and $\mathbf{r}$ is allowed in principle). Hamilton's equations of motion

$$\dot{\mathbf{p}} = -\frac{\partial H}{\partial \mathbf{r}}\,, \qquad \dot{\mathbf{r}} = \frac{\partial H}{\partial \mathbf{p}} \tag{3.19}$$

are in general non-linear. Assume now that we have found a periodic orbit, *i.e.*, a T-periodic solution

$$\mathbf{r}_0(t) = \mathbf{r}_0(t+T), \qquad \mathbf{p}_0(t) = \mathbf{p}_0(t+T), \tag{3.20}$$

and that $V(\mathbf{r})$ in equation (3.18) is holomorphic around $V(\mathbf{r}_0)$. (We shall turn further below to the question how to find such a periodic orbit practically). We now want to investigate the stability of this solution with respect to some small perturbations $\delta\mathbf{r}$, $\delta\mathbf{p}$:

$$\mathbf{r} = \mathbf{r}_0 + \delta\mathbf{r}, \qquad \mathbf{p} = \mathbf{p}_0 + \delta\mathbf{p}. \tag{3.21}$$

Substituting (3.21) into (3.19) and expanding up to second orde with respect to the perturbations, we find the $2D$ *linearized* differential equations for $\delta\mathbf{r}$

and $\delta\mathbf{p}$, also called the Poincaré variational equations:

$$\frac{\mathrm{d}}{\mathrm{d}t}\begin{pmatrix}\delta\mathbf{r}\\ \delta\mathbf{p}\end{pmatrix} = \begin{pmatrix}\mathbf{0} & \mathbf{1}\\ -\mathbf{V}''_0(t) & \mathbf{0}\end{pmatrix}\begin{pmatrix}\delta\mathbf{r}\\ \delta\mathbf{p}\end{pmatrix}, \tag{3.22}$$

where $\mathbf{0}$ and $\mathbf{1}$ are the D-dimensional zero and unit matrices, respectively, and $\mathbf{V}''_0(t)$ is the matrix of second derivatives of the potential $V(\mathbf{r})$, taken at $\mathbf{r}_0(t)$:

$$(\mathbf{V}''_0)_{ij}(t) = \left.\frac{\partial^2 V(\mathbf{r})}{\partial r_i \partial r_j}\right|_{\mathbf{r}=\mathbf{r}_0(t)} \qquad (i,j = 1,2,\ldots,D). \tag{3.23}$$

Equation (3.22) is a set of linear differential equations with T-periodic coefficients. From their general theory (see, *e.g.* [62]) one learns that the evolution of some initial perturbation $\delta\mathbf{r}$, $\delta\mathbf{p}$ at time $t=0$ up to the time t is given by

$$\begin{pmatrix}\delta\mathbf{r}(t)\\ \delta\mathbf{p}(t)\end{pmatrix} = \mathbf{X}(t)\begin{pmatrix}\delta\mathbf{r}(0)\\ \delta\mathbf{p}(0)\end{pmatrix}, \tag{3.24}$$

where $\mathbf{X}(t)$ is the so-called "matrizant" of dimension $2D$. From equation (3.22) it is seen that the matrizant fulfills the differential equation

$$\frac{\mathrm{d}}{\mathrm{d}t}\mathbf{X}(t) = \begin{pmatrix}\mathbf{0} & \mathbf{1}\\ -\mathbf{V}''_0(t) & \mathbf{0}\end{pmatrix}\mathbf{X}(t) \tag{3.25}$$

with the initial conditions

$$(\mathbf{X})_{mn}(0) = \delta_{mn} \qquad (m,n = 1,2,\ldots,2D). \tag{3.26}$$

The value of the matrix $\mathbf{X}(t)$ at the time of the period T is called the "monodromy matrix" $\mathbf{M}$:

$$\mathbf{M} = \mathbf{X}(T). \tag{3.27}$$

$\mathbf{M}$ is a symplectic $(2D\times 2D)$ matrix. According to the Lyapounov–Poincaré theorem, its eigenvalues λ_m $(m = 1,2,\ldots,2D)$ are pairwise inverse. Furthermore, two of the eigenvalues are always equal to unity, corresponding to small perturbations that either occur along the orbit or amount to a rescaling of the energy. Thus, $\mathbf{M}$ can always be brought into the form

$$\mathbf{M} = \begin{pmatrix}\widetilde{\mathbf{M}} & \cdots\\ \mathbf{0} & \begin{pmatrix}1 & \cdots\\ 0 & 1\end{pmatrix}\end{pmatrix}, \tag{3.28}$$

where the $2D-2$ dimensional submatrix $\widetilde{\mathbf{M}}$ is the stability matrix appearing in the denominator of Gutzwiller's trace formula (3.8). The stability matrix thus describes the propagation of a small perturbation $(\mathbf{r}_\perp, \mathbf{p}_\perp)$, transverse (in phase space) to the given periodic orbit, over one period T:

$$\begin{pmatrix} \delta\mathbf{r}_\perp(T) \\ \delta\mathbf{p}_\perp(T) \end{pmatrix} = \widetilde{\mathbf{M}} \begin{pmatrix} \delta\mathbf{r}_\perp(0) \\ \delta\mathbf{p}_\perp(0) \end{pmatrix}. \qquad (3.29)$$

Its eigenvalues give us the information on the stability of the periodic orbit. Note that $\det \widetilde{\mathbf{M}} = 1$, which expresses the fact that the Poincaré mapping defined by equation (3.29) is area preserving.

For two-dimensional systems ($D = 2$), the stability matrix is also two-dimensional and the stability analysis becomes particularly simple. Its two eigenvalues are just λ and $1/\lambda$, so that

$$\mathrm{tr}\, \widetilde{\mathbf{M}} = \lambda + 1/\lambda \qquad (D = 2). \qquad (3.30)$$

For *stable* orbits, the eigenvalues lie on the complex unit circle, so that $\lambda = \mathrm{e}^{\pm i\kappa}$ and a small perturbation just keeps oscillating around the periodic orbit. The real number κ is called the stability angle. For *unstable* orbits, the eigenvalues are of the form $\lambda = \pm \mathrm{e}^{\pm\chi}$, where the real $\chi > 1$ is the Lyapounov exponent which measures the rate at which a small initial perturbation is growing exponentially in magnitude. Actually, one does not need to compute the eigenvalues of $\widetilde{\mathbf{M}}$ to find out if an orbit is stable or not: since, as is easily seen from the above, $|\mathrm{tr}\, \widetilde{\mathbf{M}}| > 2$ for unstable and $|\mathrm{tr}\, \widetilde{\mathbf{M}}| < 2$ for stable orbits, it is enough to know the trace of $\widetilde{\mathbf{M}}$. When $|\mathrm{tr}\, \widetilde{\mathbf{M}}| = 2$, an orbit is called *neutrally stable* or *marginally stable*; this is the case for all families of degenerate orbits occurring in (integrable) systems with continuous symmetries. It is easily seen that the quantity under the root in the denominator of the trace formula (3.8) for $D = 2$ becomes $|\det(\widetilde{\mathbf{M}}_{po} - \mathbf{1})| = |2 - \mathrm{tr}\, \widetilde{\mathbf{M}}|$, so that the formula diverges for all contributions from marginally stable orbit families. The other situation where the same divergence occurs is, as already mentioned in Section 3.2, when $\mathrm{tr}\, \widetilde{\mathbf{M}} = +2$ for an isolated orbit. This happens in connection with orbit bifurcations which we will discuss briefly in Section 3.4.

Using the stability matrix we can now give a practical recipe for finding periodic orbits of a Hamiltonian system. The idea is to exploit equation (3.29) for a Newton–Raphson iteration procedure (see, *e.g.* [59]). For this, we have to calculate the stability matrix during our search for periodic orbits. This is done by solving simultaneously the Hamilton equations of motion (3.19) *and* the equation of motion for the matrizant $\mathbf{X}(t)$ given in equation (3.25). (Actually, to get rid of the two irrelevant degrees of freedom along the orbit, one may transform immediately to the coordinates $\mathbf{r}_\perp$, $\mathbf{p}_\perp$ orthogonal to the orbit, to obtain the reduced part $\widetilde{\mathbf{X}}(t)$ which after one period T gives $\widetilde{\mathbf{M}} = \widetilde{\mathbf{X}}(T)$; see [59] for the equations of motion for $\widetilde{\mathbf{X}}(t)$).

Let us introduce a short notation $\mathbf{z} = (\mathbf{r}_\perp, \mathbf{p}_\perp)$ for the orthogonal phase-space vector. Usually, the search of periodic orbits is done on a Poincaré surface of section, on which small displacements are automatically transverse to the searched orbit. We now assume that we have found an approximate periodic solution with the approximate period T. If we are close enough to the fixed point $\mathbf{z}^*$ corresponding to the exact periodic solution, the remaining error of our approximate first starting point $\mathbf{z}_1$ will propagate according to equation (3.29). After one period, the starting point $\mathbf{z}_1$ is mapped onto $\mathbf{z}'_1$, so that we have moved over an interval $\Delta\mathbf{z}_1 = \mathbf{z}'_1 - \mathbf{z}_1$. The distances of the two points from the unknown fixed point $\mathbf{z}^*$ are $\delta\mathbf{z}_1 = \mathbf{z}_1 - \mathbf{z}^*$ and $\delta\mathbf{z}'_1 = \mathbf{z}'_1 - \mathbf{z}^*$. If $\delta\mathbf{z}_1$ and $\delta\mathbf{z}'_1$ are sufficiently small, they obey the linear differential equation (3.29), so that

$$\delta\mathbf{z}'_1 = \widetilde{\mathbf{M}}\,\delta\mathbf{z}_1, \tag{3.31}$$

from which we get, using $\Delta\mathbf{z}_1 = \delta\mathbf{z}'_1 - \delta\mathbf{z}_1$,

$$\Delta\mathbf{z}_1 = (\widetilde{\mathbf{M}} - \mathbf{1})\,\delta\mathbf{z}_1. \tag{3.32}$$

This equation can be inverted to obtain the error $\delta\mathbf{z}_1$ of our starting point

$$\delta\mathbf{z}_1 = (\widetilde{\mathbf{M}} - \mathbf{1})^{-1}\Delta\mathbf{z}_1, \tag{3.33}$$

which now can be used to correct the starting point to $\mathbf{z}_2 = \mathbf{z}_1 - \delta\mathbf{z}_1$. Iterating in this way, the fixed point is usually found within very few steps. The only problem with this procedure is that for an orbit with two eigenvalues $\lambda = 1$, the inversion of $(\widetilde{\mathbf{M}} - \mathbf{1})$ is not possible, giving exactly the same divergence that appears in the trace formula (3.8). In practice, this is not a big handicap; when working with a sufficient numerical accuracy, one can come sufficiently close to the point where $\mathrm{tr}\,\widetilde{\mathbf{M}} = 2$.

The Newton-Raphson iteration converges very fast, once one comes close enough to a periodic orbit (*i.e.*, a fixed point). It is therefore possible to search periodic orbits just by scanning a given Poincaré surface of section, either systematically in small steps, or just randomly. The practical experience shows that, different from what one might expect at first thought, stable and unstable fixed points are found with equal probability.

3.4 Uniform approximations

A classical system that is neither integrable nor fully chaotic is usually called a "mixed system", or a system with mixed classical dynamics. It is characteristic of such systems that periodic orbits change their stability upon variation of a smooth parameter. Such a parameter may be the energy, any parameter appearing in the potential $V(\mathbf{r})$ (*e.g.*, a deformation parameter), or the strength of an externally applied electric or magnetic field. Let us call this parameter a. Whenever a pair of eigenvalues of the stability matrix for a periodic orbit has the value $\lambda = 1$ (*i.e.*, $\mathrm{tr}\,\widetilde{\mathbf{M}} = 2$ for a two-dimensional system) at $a = a_0$, then the orbit undergoes a bifurcation. What typically happens at a bifurcation is that an orbit changes from stable to unstable (or *vice versa*), and hereby one or several new orbits are "born". (Orbits may also just "touch" the value $\mathrm{tr}\,\widetilde{\mathbf{M}} = 2$ at $a = a_0$ without changing their stability, or a pair of stable and unstable orbits may emerge for $a \geq a_0$ and not exist at all for $a < a_0$). Infinitesimally close to the bifurcation point a_0, the new "child" orbits have the same shape and action as the "parent" orbit (but usually a different Maslov index). As one changes a away from a_0, the new orbits develop independently, and eventually change their stability again at a new bifurcation point a_1. Often, periodic orbits proliferate this way in chains of repeated bifurcations, leading to an increased degree of chaoticity (*cf.* the so-called Feigenbaum scenario [63, 64] and [65]).

This behaviour of periodic orbits in mixed systems makes the semiclassical POT difficult, since the trace formulae diverge at each bifurcation point. Mathematically speaking, the divergence comes from the fact that one of the stationary phase integrations used in their derivation (see Sect. 3.2) breaks down. The remedy is, in principle, to expand the action integral in the exponent of the semiclassical Green function (3.4) up to higher order and then to do the corresponding integration. For instance, when integrating along a direction x, one may have to expand the exponent up to third order in $(x - x_0)$, where x_0 is the stationary point; the integral over x then would lead to an Airy function. More generally, one expands the action integral around the stationary point $\mathbf{r}_0$, $\mathbf{p}_0$ in phase space (corresponding to the bifurcation point a_0) into so-called normal forms [66] and then performs the trace integrations (if possible, analytically). This locally removes the singularity at $a = a_0$, but it does not yet guarantee that the result can be used further away: when a is sufficiently different from a_0 one wants to recover the original Gutzwiller amplitudes $\mathcal{A}_{po}$. To guarantee this, more refined techniques are required, yielding so-called "uniform approximations". Such approximations have been developed systematically for all generic types of orbit bifurcations by Sieber and Schomerus [67], and for pairs of interfering bifurcations by Schomerus [68]. We refer to their articles for further details; the resulting trace formulae have finite amplitudes at the bifurcation points and go over into the standard Gutzwiller trace formula away from the bifurcations.

Uniform approximations are also required when the change of a smooth parameter, or its being switched from zero to non-zero (or *vice versa*) leads to the breaking (or restoring) of a continuous symmetry. As we have discussed in Section 3.2 after equation (3.10), the power of $\hbar$ contained in the

Gutzwiller amplitude $\mathcal{A}_{po}$ depends on the number f of continuous symmetry parameters describing the degeneracy of orbit families. When f is changed by one or more units upon variation of a parameter, the original form of the amplitude becomes invalid and therefore diverges. Start, *e.g.*, from a two-dimensional billiard with oval boundary, which is non-integrable (with $f = 0$) so that the Gutzwiller trace formula (3.8) applies. Now, when letting its shape go to a circle, $\mathrm{tr}\,\widetilde{\mathbf{M}}$ of all the isolated periodic orbits will approach the value +2. Thus the trace formula diverges in the limit where the circular U(1) symmetry is restored and the billiard becomes integrable (with $f = 1$). (The amplitudes there are all proportional to $\hbar^{-3/2}$).

Such a symmetry breaking arises generically if we perturb an integrable Hamiltonian like in equation (3.13) by switching on the parameter ϵ. For small values of ϵ, the perturbative trace formula (3.14) solves this problem but, as already mentioned, for large values ϵ its amplitudes do not go over to the correct Gutzwiller amplitudes of the perturbed system. The techniques used to achieve this are similar to those used to derive the uniform approximations for bifurcations. For the breaking of U(1) symmetry, a general uniform approximation has been derived by Tomsovic *et al.* [69], and uniform approximations for SU(2) and SO(3) breaking have been derived for some specific systems in [70].

Finally, we mention that in [71] an analytical trace formula has been given for the two-dimensional elliptic billiard in which – although it is integrable! – an infinity of bifurcations occur that cumulate in the limit of zero eccentricity (*i.e.*, the circular billiard), so that both bifurcations and symmetry breaking had to be overcome.

3.5 Applications to metal clusters

The examples given here for metal clusters will be short, since they have already been discussed extensively at various occasions [4, 31, 72].

3.5.1 Supershell structure of spherical alkali clusters

Based upon the semiclassical trace formula derived by Balian and Bloch [49] for the oscillating level density of a spherical billiard, and extending the semiclassical POT analysis to Woods–Saxon type shell-model potentials for large clusters, Nishioka *et al.* [73] predicted that supershell structure should be experimentally observable in large metal clusters. This was, indeed, verified in the experiments of the Copenhagen-Stuttgart collaboration [74]. Figure 7 shows the oscillating part of the cluster abundance in an adiabatic expansion source. Shown are the statistically averaged fist differences of the logarithmic yields I_N, multiplied by a factor that accounts for the temperature suppression effects (see [30, 75] for a theoretical investigation

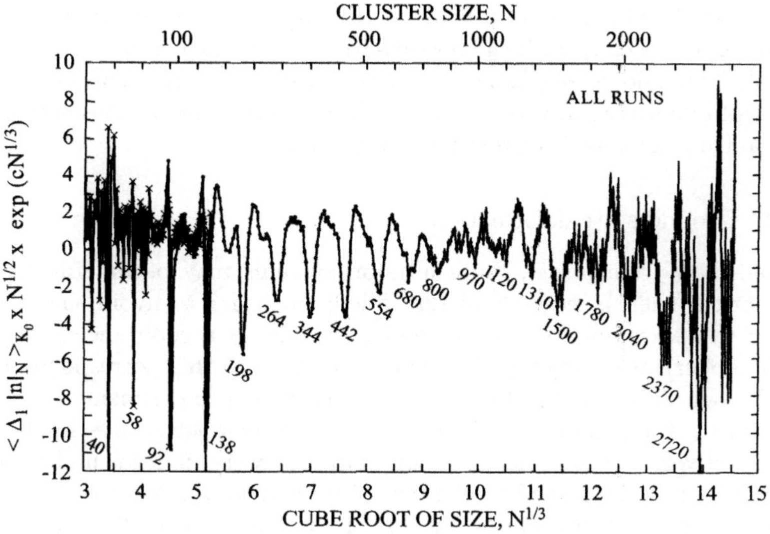

Fig. 7. Experimental observation of the supershells. Shown are the logarithmically derived relative yields of sodium clusters from an adiabatic expansion source, scaled by an extra factor $N^{1/2}\exp(cN^{1/3})$ that takes account of the finite temperature effects. (After [74]).

of these effects in finite-temperature Kohn–Sham calculations for canonical ensembles of spherical jellium clusters).

The negative peaks in Figure 7 correspond to the most stable clusters which are understood as those with filled spherical *electronic* shells with the "magic numbers" indicated. Note that they appear at equal distances on the scale $N^{1/3}$ which is proportional to the cluster radius. The constant spacing $s = \Delta N^{1/3}$ is experimentally found to be $s = 0.61 \pm 0.01$, also for other metal clusters (Li, Ga). The beating amplitude of the shell oscillations is understood in the POT from the interference of the shortest periodic orbits with largest amplitude, which here are the triangular and square orbits. Their *average length* determines the shell spacing s; from the trace formula of a spherical billiard [49] one finds the value $s = 0.603$ which is independent of the Wigner–Seitz parameter r_s of the metal. The calculations of Nishioka *et al.* yield $s = 0.607$; the same value is found in the jellium-model DFT calculations for spherical Na clusters [30]. (Some finer details in connection with the Ga experiments are discussed in [76]). The beat length, *i.e.*, the period of the supershells, is given essentially by the *difference* between the lengths of the leading orbits. It does depend on the metal type (see a detailed investigation in [77]).

The effect of a magnetic field on the supershell structure was investigated semiclassically with the perturbative trace formula of equation (3.14) in [78]. The field strengths necessary to detect interesting effects are, however, not experimentally available. (This is different for the mesoscopic sizes of quantum dots, as discussed further below).

3.5.2 Ground-state deformations

As we know from nuclei, finite fermion systems may become deformed if a spherical shell is not completely filled: in order to avoid the degeneracy of the wavefunction, the system responds by a spontaneous breaking of the spherical symmetry. In molecular physics, this corresponds to the well-known Jahn–Teller effect. In the pioneering experiments on sodium clusters by the group of Knight [79, 80], the finer structures in the mass yields between the spherically magic numbers were, indeed, understood in terms of a deformed shell model (see [40] for details). Most of the smaller minima appearing in Figure 7 for cluster sizes up to $N \sim 800$ could, in fact, also be attributed to deformed cluster shapes [81].

The POT analysis for deformed potentials becomes more complicated, as these usually are not integrable. Simple estimates of cluster deformations are, however, possible in terms of integrable deformed model potentials. This has been done successfully in nuclear physics by Strutinsky *et al.* [55]. Assume that only one periodic orbit, or one family of periodic orbits, is dominating the shell-correction energy δE given by (3.12). The condition for finding its extrema then reduces to

$$\delta S_{po} = 0. \tag{3.34}$$

This idea has been used for metal clusters in [82]. Figure 8 compares semiclassical estimates of the ground-state quadrupole deformations δ with results of quantum-mechanical calculations. For small clusters up to $N \sim 40$, the selfconsistent Kohn–Sham field is well approximated by an axially deformed harmonic oscillator whose exact trace formula is known analytically [83]. In the limit of large clusters, the mean field has very steep walls, as demonstrated by the success of the spherical billiard model for the supershells. For large deformed clusters it may therefore be approximated by a spheroidal billiard, whose approximate trace formula is also known analytically [84]. The heavy lines in Figure 8 give the semiclassical estimates of δ using these exact trace formulae and equation (3.34). The thin lines are the quantum-mechanical results from KS calculations in the deformed jellium model [85] for $N \leq 40$, and for $N > 40$ from calculations using Strutinsky's shell-correction method [2] in a phenomenological modified Nilsson model for large clusters that yields the spherical KS spectra [81].

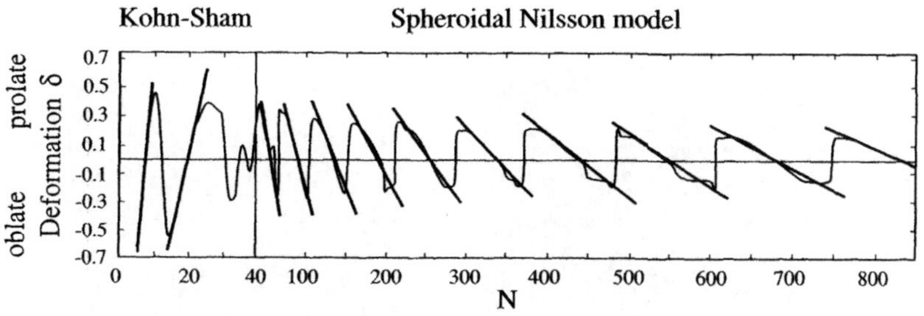

Fig. 8. Ground-state quadrupole deformations δ of sodium clusters *versus* cluster size N. The solid thin lines for $N \leq 40$ are the results from KS calculations using the spheroidal jellium model [85], and for $N > 40$ those from microscopic Strutinsky calculations using a modified Nilsson model that is adjusted to the spherical jellium-KS results [81]. The heavy lines give the semiclassical estimates based on equation (3.34) using the trace formula of an axial harmonic oscillator for $N < 26$ and that of a spheroidal billiard for $N > 40$. (From [82]).

More realistic deformed potentials have mixed classical dynamics, which leads to the problems discussed in Section 3.4 above. The POT analysis then depends on finding the most important short orbits and treating their bifurcations correctly. However, for the ground states of larger systems the deformations are usually not very large – as can be recognized, *e.g.*, from the decreasing amplitudes of δ seen in Figure 8. It is therefore possible to treat the deformed part of the potential in perturbation theory, using the trace formula (3.16) discussed above. For a slightly quadrupole-deformed billiard the modulation factor (3.15) can be calculated analytically [72, 86]; hereby the unperturbed system is the spherical billiard. In Figure 9 we show by contour plots the shell-correction energy δE as a function of particle number N and quadrupole deformation ε_2. The agreement of the quantum-mechanical (left) and semiclassical results (right) demonstrates the quantitative predictive power of the semiclassical theory.

3.6 Applications to two-dimensional electronic systems

The recent developments in the fabrication of semiconductor devices on micro- and nanometer scales have opened up interesting new studies of the interplay of classical and quantum mechanics. Quantum dots and other nanostructures with any wanted geometry can be produced at scales where on one hand the coherence length of the electrons is larger than their confinement size, so that quantum interference takes place, but on the other hand their Fermi wave length is small enough so that high quantum numbers are involved and the classical limit is within reach. In addition, these structures

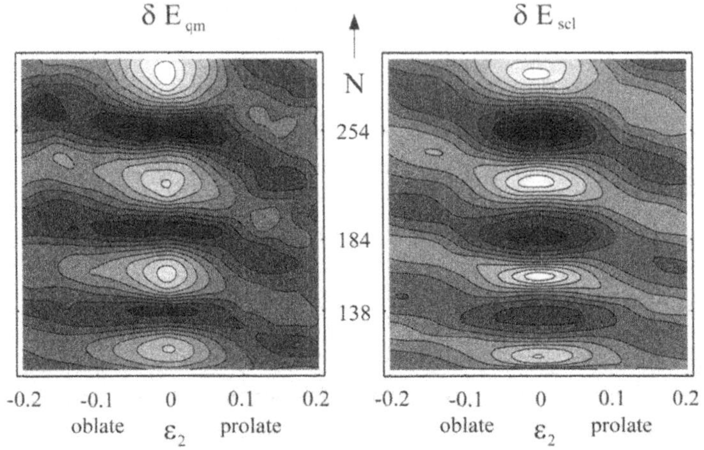

Fig. 9. Contour plots of the shell-correction energy δE for an axially quadrupole-deformed cavity *versus* particle number N and quadrupole deformation ε_2. *Left:* quantum-mechanical result, *right:* semiclassical result using the perturbative trace formula (3.16); the unperturbed system is the spherical billiard [49]. (From [86]).

are of the order of, or bigger than, the magnetic length scales that can be reached experimentally. Hence, the strength of an applied external magnetic field represents a new degree of freedom for the observation of interesting quantum effects, such as the quantum Hall effect (QHE) and its various manifestations in connection with a laterally confined two-dimensional electron gas (2DEG).

In the following we shall give three applications of the POT to such mesoscopic systems, where the magnetic field plays an important role. In the first one, a circular quantum dot, the observed conductance oscillations could be directly related to the density of states of the confined electrons and hence be interpreted in terms of the trace formula (3.10). The other two examples concern open systems where the conductance is described within the Kubo formalism. The semiclassical description of such systems has been developed in [87, 88]. In the quantum-mechanical Kubo formula based on the linear response theory, the Green function is replaced by its semiclassical approximation (3.4). The oscillating parts of the longitudinal and transverse conductivity $\delta\sigma_{xx}$ and $\delta\sigma_{xy}$, respectively, with respect to an applied external voltage in the x direction ($z = 0$ is the plane of the 2DEG) can then be expressed in terms of periodic orbits through the following "semiclassical Kubo formulae" (including a spin degeneracy factor 2):

$$\delta\sigma_{xx} = \frac{4e^2}{h}\frac{1}{A}\sum_{po} \mathcal{C}_{xx} \frac{R_{po}(\tau_\beta)\,F_{po}(\tau_s)}{|\det(\widetilde{\mathbf{M}}_{po}-\mathbf{1})|^{1/2}} \cos\left(\frac{S_{po}}{\hbar}-\sigma_{po}\frac{\pi}{2}\right), \tag{3.35}$$

$$\delta\sigma_{xy} = \frac{4e^2}{h}\frac{1}{A}\sum_{po}\left(\frac{1}{e}\frac{\partial S_{po}}{\partial B}+\mathcal{C}_{xy}\right)\frac{R_{po}(\tau_\beta)\,F_{po}(\tau_s)}{|\det(\widetilde{\mathbf{M}}_{po}-\mathbf{1})|^{1/2}} \cos\left(\frac{S_{po}}{\hbar}-\sigma_{po}\frac{\pi}{2}\right).$$

Here S_{po}, σ_{po}, and $\widetilde{\mathbf{M}}_{po}$ have the same meaning as in the trace formulae presented in Section 3.2 above; all are evaluated at the Fermi energy E_F. A finite temperature T is included in the factor $R_{po}(\tau_\beta) = (T_{po}/\tau_\beta)/\sinh(T_{po}/\tau_\beta)$ with the period $T_{po} = \partial S_{po}/\partial E$ and $\tau_\beta = \hbar/(\pi k T)$. Damping due to a finite mean-free path is included by the factor $F_{po}(\tau_s) = \exp[-T_{po}/(2\tau_s)]$, where $\tau_s = m^*\mu/e$ is the scattering time extracted from the experimental mobility μ, and m^* is the effective mass of the electrons. A is the effective area of the investigated device, and $\mathcal{C}_{ij}$ are the velocity-velocity correlation functions of the periodic orbit, defined by

$$\mathcal{C}_{ij} = \int_0^\infty \mathrm{d}t\, \mathrm{e}^{-t/\tau_s} \int_0^{T_{po}} \mathrm{d}\tau\, v_i(\tau)\, v_j(t+\tau) \qquad (i,j = x,y) \tag{3.36}$$

in terms of the velocities $v_i(t) = p_i(t)/m^*$ of the periodic orbits.

3.6.1 Conductance oscillations in a circular quantum dot

In a circular quantum dot with approximately 1200 – 1500 confined electrons, the conductance was measured [89, 90] as a function of an applied external (perpendicular) magnetic field B and of the applied gate voltage V_{g1} that defined the effective radius of the confined 2DEG. Regular oscillations of the conductance as a function of both V_{g1} and weak field strength B in the milli-Tesla domain were observed, as shown in the left-hand part of Figure 10. A semiclassical reproduction of these oscillations in terms of the trace formula of a circular billiard was given in [91] and is shown in the right-hand part of Figure 10. Both the experiment and the semiclassical analysis are well described in the quoted literature; we just summarize the essential conclusions here.

Like in the large metal clusters, the selfconsistent mean field of this quantum dot is expected to have steep walls due to the strong mutual Coulomb repulsion of the trapped electrons. Hence, a two-dimensional circular billiard is a good first-order model. The period of the oscillation in the vertical direction, *i.e.*, as a function of the dot radius R, is given exactly like for the spherical metal clusters by the average length of the leading periodic orbits. In contrast to the three-dimensional spherical billiard, all orbits in a two-dimensional circular billiard have the same one-dimensional degeneracy,

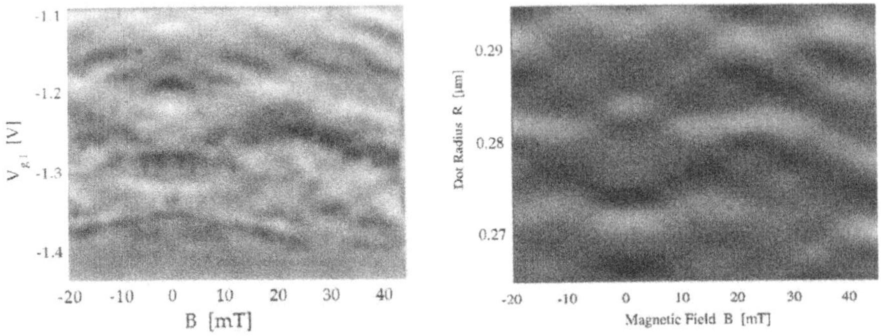

Fig. 10. Gray-scale plots of conductance of a circular quantum dot, plotted *versus* gate voltage (vertical axis) and magnetic field (horizontal axis). *Left:* experimental result, *right:* semiclassical result using the circular billiard model including the magnetic field in first-order perturbation. (From [91]).

and the leading orbit here is the diameter orbit. Using the average length of the diameter and the triangular orbits leads, indeed, to the correct period of the vertical oscillations seen in Figure 10. (Note, however, that the number of electrons and the radii R were only known approximately here). The oscillation as a function of a weak magnetic field B is well understood in perturbation theory where only the first-order corrections to the actions S_{po} are taken into account. The modulation factor (3.15) then becomes [91]

$$\mathcal{M}_{po} = \cos\left(\frac{e}{\hbar c} F_{po} B\right) , \tag{3.37}$$

where F_{po} is the *area* included by the orbit *po*. The origin of this modulation factor is just the Aharonov–Bohm (AB) phase of a charged particle surrounding a magnetic field, *i.e.*, the magnetic flux included by the electron's orbit. Here the phase adds to the classical action, the sign of the flux depending on the orientation of the orbit with respect to the B field. The summation over both directions yields the cosine factor in (3.37). The period of these oscillations for weak B – often called "AB oscillations" – are here essentially given by the shortest orbit with a non-zero area, *i.e.*, by the triangular orbit. This example shows how through the introduction of a weak magnetic field, one is able to "measure" not only the average length, but independently also the average area of the leading periodic orbit(s).

Similar AB oscillations have also been observed in a quantum dot with an equilateral triangular shape [92,93]. In the steep-wall limit, one may use a triangular billiard whose exact trace formula is known analytically [4]; the period of the measured AB oscillations could, indeed, be understood by the area of the shortest periodic orbit of this system. A challenging speculation

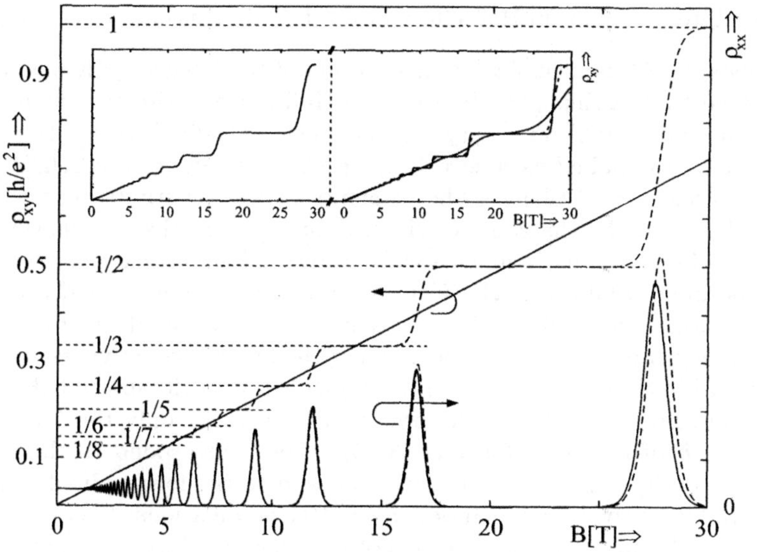

Fig. 11. Semiclassical calculation of the transverse Hall resistivity ρ_{xy} (left axis) and the longitudinal resistivity ρ_{xx} (right axis) of the 2DEG in a transverse magnetic field B. ρ_{xx} shows the typical $1/B$-periodic Shubnikov–de Haas oscillations. ρ_{xy} exhibits the plateaux corresponding to the integer QHE. Solid lines are obtained including only the lowest-order terms in the semiclassical Kubo formula (3.35). Dashed lines include the only $\hbar$-correction to the first term of $\delta\sigma_{xy}$ in (3.35). The inserts illustrate small variations of the mobility of the electrons (left) and of the temperature (right). (From [99]).

was proposed for the study of a triangular quantum dot with open contacts and a small number of electrons [92,94]. Neglecting the Coulomb repulsion, the potential of the electrons then is given to lowest order by the famous Hénon–Heiles potential [95] which has mixed classical dynamics. The semiclassical description of its level density using the Gutzwiller theory has been very successful (see [70] and earlier papers quoted therein). It would be interesting to observe the transition from regular to chaotic motion by tuning the Fermi energy of the electrons. This has, however, not been realized experimentally so far.

Before turning to open nanostructures, we conclude this section on quantum dots by mentioning that for the disk billiard in arbitrarily strong homogeneous magnetic fields, an analytical trace formula has been developed in [96]. Magnetic properties of quantum dots and dot arrays have also been investigated within the POT [97,98].

3.6.2 Integer quantum Hall effect in the two-dimensional electron gas

An unexpected semiclassical interpretation of the integer QHE was recently proposed by Blaschke [99]. In a typical Hall bar experiment, one measures the transverse (Hall) resistivity ρ_{xy} and the longitudinal resistivity ρ_{xx} with respect to an applied electric voltage in the x direction; the 2DEG is confined to the $z = 0$ plane. The elements of the resistivity tensor ρ_{ij} are given by inverting the conductivity tensor σ_{ij}, and their oscillating parts can thus be calculated semiclassically from the Kubo trace formulae (3.35). The periodic orbits of a 2DEG in a transverse magnetic field are simply the cyclotron orbits whose properties are easily calculated; their continuous degeneracy cancels the factor A. Adding the average resistivities, which are given within the classical Drude model, yields the results shown in Figure 11. The solid lines give the results obtained by keeping only the lowest-order terms in $\hbar$ of the trace formula (3.35). They correspond to the classical Hall resistivity ρ_{xy} which is linear in B, and to the longitudinal resistivity ρ_{xx} exhibiting the typical Shubnikov–de Haas oscillations. (These are easily understood semiclassically in terms of the magnetic flux surrounded by the cyclotron orbits which is proportional to $1/B$; see, *e.g.* [4]). To lowest order in $\hbar$, the POT is not able to reproduce the plateaux in the Hall resistivity. However, when adding the (only! – see [99]) $\hbar$ correction coming into the term proportional to $\partial S_{po}/\partial B$ in the trace formula (3.35) for $\delta\sigma_{xy}$ (dashed lines), one obtains the plateaux in ρ_{xy} characteristic of the integer QHE; the influence of this correction on ρ_{xx} is very small.

When the 2DEG is modulated laterally by an "antidot superlattice", interesting commensurability peaks can be observed in the longitudinal resistivity, which correspond to electrons trapped on specific cyclotron orbits that fit around one or more antidots [100]. At low temperatures, small AB oscillations in ρ_{xx} become observable and can be qualitatively reproduced [101] by the interference of the most important periodic orbits *via* the trace formulae equation (3.35). (See [101] for a mini-review of the so-called "Weiss oscillations".)

3.6.3 Conductance oscillations in a channel with antidots

Our last example is a mesoscopic arrangement in which the electrons of a 2DEG are confined laterally to a channel of width $\sim$1.0 μm. Two antidots represent obstacles to the electric current through the channel; the effective radius of these antidots can be regulated by an applied gate voltage V_g. Figure 12 shows an SEM photograph of the experimental gate structure [102]. The longitudinal conductance G_{xx} along the channel has been measured for various strengths of a perpendicular magnetic field B and gate voltages V_g [102, 103]. Like in the case of antidot supperlatices [100], a

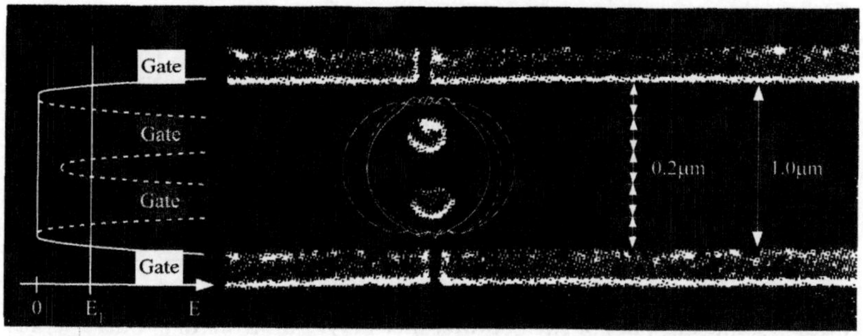

Fig. 12. SEM photograph of the gate structure of a mesoscopic channel with two antidots [102] (without contacts). *Left:* model potential used for the calculations. *Center:* typical periodic orbits encircling the antidots. (From [104]).

commensurability minimum in the average conductance has been observed near those values of B for which a cyclotron orbit fits around the antidots.

Small observed AB oscillations around the average part of G_{xx} can again be interpreted semiclassically [99, 104] as interferences between the leading periodic orbits (a few of them are shown in Fig. 12 by solid and dashed white lines) through the trace formulae (3.35). The result of the semiclassical calculation of δG_{xx} is shown in Figure 13; hereby the parameters of the model potential used to describe the lateral confinement (including the antidots) have been optimized [99]. Like in the case of the Weiss oscillations [101], the amplitude had to be adjusted since the semiclassical Kubo formula is not able to reproduce quantitatively the amplitude of the oscillations.

An interesting phenomenon is observed when varying both the magnetic field B and the gate voltage V_g, and plotting the loci of the oscillation maxima in δG_{xx}. These arrange themselves, as seen in Figure 14a, along smooth lines whose slopes are well understood in terms of the B and V_g dependence of the actions S_{po} of the leading periodic orbits. However, some characteristic dislocations occur at apparently random places in the (B, V_g) plane, as emphasized by the boxes. In the semiclassical analysis, they originate from successive bifurcations of periodic orbits: the different orbit generations lead to different slopes in Figure 14b, and these do not match near the loci in the (B, V_g) plane (shown for some leading orbits by gray-shaded thick lines) along which the bifurcations occur. Although the theory does not fit the experiment globally (at least 10 different orbit families contribute), the local agreement near the dislocations is excellent; see the box in Figure 14c. Thus, orbit bifurcations *can* be "seen" in the experiment!

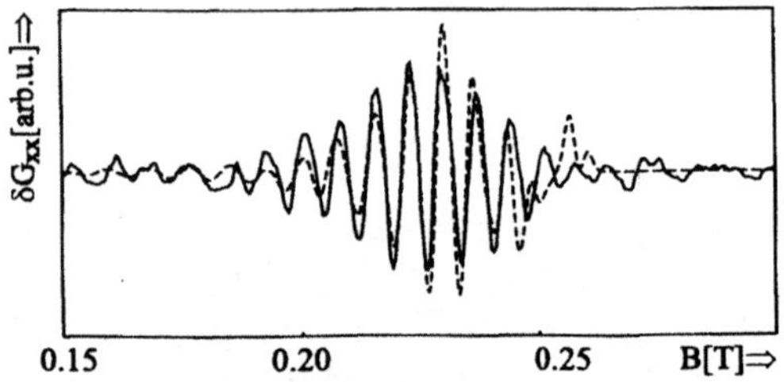

Fig. 13. Oscillating part of the conductance δG_{xx} (in arbitrary units) *versus* magnetic field strength B (in Tesla). *Solid line:* experiment for $V_g = -1.50$ Volts [103]; *dashed line:* semiclassical result with optimized potential parameters [99].

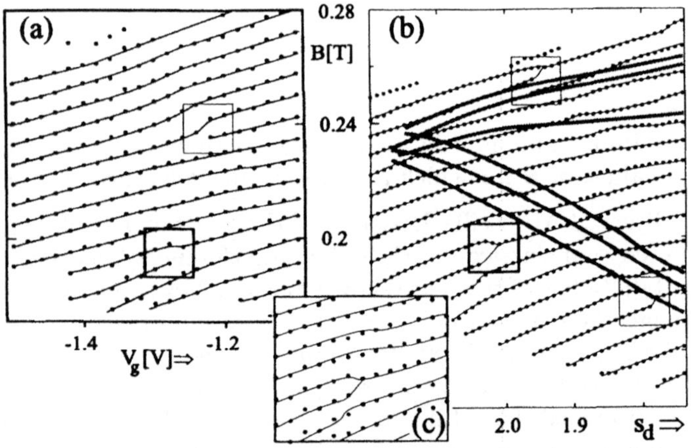

Fig. 14. Maximum positions of δG_{xx} *versus* B (vertical axes) and V_g (horizontal axes). (a) Experimental values [103]. (b) Semiclassical results [104]; s_d is the antidot radius regulated by V_g; approximately we have $s_d \propto V_g$; the gray-shaded lines correspond to the loci of bifurcations of some leading orbit families. (c) Behaviour near a dislocation (dots: experiment; lines: semiclassical results). (From [104]).

4 Local-current approximation for linear response

We shall in this final section present a semiclassical approach to the collective excitation spectrum of a finite interacting fermion system. The semiclassical aspect here pertains to the calculation of the excitation energies, starting from a ground state that has been obtained either quantum-mechanically or semiclassically in the mean-field approximation. This approach is based on the use of generalized sum rules [105] and is particularly well suited for the description of collective excitations such as giant resonances in nuclei (see [106] for a review) or plasmon resonances in metal clusters (see Sect. IV.C of the review article [31] on metal clusters). For the latter application, this approach has been developed first within the ETF framework [27] and later in connection with the Kohn–Sham formalism, where it was shown [107] to be an approximation to the random phase approximation (RPA). Since it makes use of localized collective current distributions, it was termed "local RPA" (LRPA, see also [108, 109]). The method can, however, be derived [110] from the most general quantum-mechanical equations of motion that yield the exact excitation spectrum; the basic semiclassical assumption is, as previously, that the collective currents are given by local functions of $\mathbf{r}$. We therefore propose [111] to term this approach the "local-current approximation" (LCA) which does not restrict its name to further approximations inherent in the RPA.

We first state in Section 4.1 the quantum-mechanical equations of motion and relate them to a variational principle, and in Section 4.2 we sketch briefly the variational equations that are obtained if the basic excitation operators are assumed to be local functions of $\mathbf{r}$. In Section 4.3 we give a practical way to solve these equations approximately in finite basis sets and give a physical interpretation of the LCA, and in the remaining parts we shall present some results of applications to metal clusters.

4.1 Quantum-mechanical equations of motion

In this and the next subsection we follow closely [110]. We consider a many-particle system defined by the Hamiltonian $H = T + V$, where V includes external one-body potentials and the two-body interaction. (For simplicity, we omit in the following the "hat" symbol "$\hat{\ }$" to indicate operators.) The exact ground state $|0\rangle$ and excited states $|\nu\rangle$ are given by the stationary Schrödinger equation

$$H|\nu\rangle = E_\nu|\nu\rangle = (E_0 + \hbar\omega_\nu)|\nu\rangle\,. \qquad \nu = 0, 1, 2, \dots \tag{4.1}$$

It is a matter of standard many-body theory (see, *e.g.* [112–114]) to rewrite (4.1) in the form of the following equations of motion

$$\langle 0|\mathcal{O}_\nu[H, \mathcal{O}_\nu^\dagger]|0\rangle \;=\; \hbar\omega_\nu\langle 0|\mathcal{O}_\nu\mathcal{O}_\nu^\dagger|0\rangle\,, \tag{4.2}$$

$$\langle 0|\mathcal{O}_\nu[H, \mathcal{O}_\nu]|0\rangle \;=\; \hbar\omega_\nu\langle 0|\mathcal{O}_\nu\mathcal{O}_\nu|0\rangle = 0\,, \tag{4.3}$$

where the operators $\mathcal{O}_\nu$ and $\mathcal{O}_\nu^\dagger$ are defined as the creation and annihilation operators of the excited states:

$$\mathcal{O}_\nu^\dagger|0\rangle = |\nu\rangle, \qquad \mathcal{O}_\nu|\nu\rangle = |0\rangle, \qquad \text{and} \qquad \mathcal{O}_\nu|0\rangle = 0. \tag{4.4}$$

The exact solutions of the equations (4.2, 4.3) are unknown for most systems, as are those of the Schrödinger equation (4.1). Several approximative ways to find the excited energies $\hbar\omega_\nu$ and/or their eigenstates have been developed; they differ in the approximations used for the ground state $|0\rangle$ and the explicit form of the excitation operators $\mathcal{O}_\nu^\dagger$. In the Tam–Dancoff scheme one starts from the HF approximation, where the ground state $|0\rangle = |\mathrm{HF}\rangle$ is a Slater determinant, and the excitation operators are taken to be simple one-particle-one-hole ($1p$-$1h$) operators. In the small-amplitude limit of the time-dependent HF theory, the so-called random phase approximation (RPA) is obtained by admitting $2p$-$2h$ excitations in the ground state

$$|0\rangle = |\mathrm{RPA}\rangle = \left(1 + \sum_{pp'hh'} \gamma^{pp'hh'} a_p^\dagger a_{p'}^\dagger a_h a_{h'}\right) |\mathrm{HF}\rangle\,, \tag{4.5}$$

and taking the excitation operators to be linear combinations of $1p$-$1h$ excitations of the form:

$$\mathcal{O}_\nu^{\dagger\,(\mathrm{RPA})} = \sum_{ph} \left(x_\nu{}^{ph} a_p^\dagger a_h - y_\nu{}^{ph} a_h^\dagger a_p \right). \tag{4.6}$$

In the above definitions, $a^\dagger$ and a are the single-particle creation and annihilation operators for particle states (with $\varepsilon_p, \varepsilon_{p'} > E_\mathrm{F}$) and hole states (with $\varepsilon_h, \varepsilon_{h'} \le E_\mathrm{F}$), respectively, with respect to the HF ground state. We refer again to the literature for the details of these approximations [114].

For the following it is now essential that a variational principle can be formulated [110] which is exactly equivalent to solving the general equations of motion (4.2, 4.3). For that we introduce a Hermitian operator Q that can be interpreted as a generalized coordinate. To find the lowest excited state, one has to solve the variational equation

$$\frac{\delta E_3[Q]}{\delta Q} = 0, \tag{4.7}$$

where $E_3[Q]$ is defined by

$$E_3[Q] = \sqrt{\frac{m_3[Q]}{m_1[Q]}}, \tag{4.8}$$

and the "moments" m_1 and m_3 (see the discussion of their physical significance in Sect. 4.3 below) are defined as expectation values of multiple commutators

$$m_1[Q] = \frac{1}{2} \langle 0| \left[Q, [H, Q]\right] |0\rangle, \tag{4.9}$$

$$m_3[Q] = \frac{1}{2} \langle 0| \left[[H, Q]\,, [[H, Q]\,, H]\right] |0\rangle\cdot \tag{4.10}$$

The minimum value of E_3 after variation gives the first excitation energy $\hbar\omega_1$. The corresponding operator Q_1 is written as the linear combination

$$Q_1 \propto \mathcal{O}_1^\dagger + \mathcal{O}_1 \tag{4.11}$$

of the creation and annihilation operators for the first excited state. The second excitation with energy $\hbar\omega_2$ can then be obtained from variation of Q in an operator space that has been orthogonalized to Q_1, and so on. In this way the whole excitation spectrum $\hbar\omega_\nu$ can be successively constructed (see [110, 111] for the details).

In principle, the *exact* excited states and excitation energies of the system are given by the solutions of the variational equation (4.7). For practical calculations, however, one is forced to make some ansatz for the generally non-local operator Q. If it is taken from the space of particle-hole operators of the form (4.6), then equation (4.7) leads precisely to the RPA equations [107]. (Note that the RPA, in a consistent use within the DFT, includes the exchange-correlations effects. In the condensed matter literature, however, the corresponding approach is more often referred to as the "time-dependent LDA" or TDLDA approach, whereas the name RPA is reserved to an approximation which ignores the exchange-correlations effects. Nuclear physicists usually include the exact exchange in their RPA treatment, starting from the HF ground state as implied in equation (4.5) above). Even in this approximation, the equations of motion are very difficult to solve for clusters without any spatial symmetry, *e.g.*, including the ionic structure. In the following we shall discuss a semiclassical approach which makes use of a local approximation to the operator Q and has been successfully used for collective excitations both in nuclear [106] and cluster physics.

4.2 Variational equation for the local current density

Starting from the exact equations of motion defined by the variational principle equations (4.7–4.10), we now make the following two approximations. First, we write the ground-state energy as usual in DFT as a functional of the local density, $E = E[\rho(\mathbf{r})]$ as in (2.19), using a standard approximation for the exchange-correlation functional (*e.g.*, the LDA). Second, we assume Q to be a *local function of* $\mathbf{r}$:

$$Q = Q(\mathbf{r}). \tag{4.12}$$

Then, for an electronic system where the external potentials are local and the two-body interaction is just the bare Coulomb interaction, the commutator $[H, Q]$ is only given by its kinetic part (*i.e.*, the potential part in H

commutes with Q):

$$[H,Q] = [T,Q] = \frac{1}{2}(\nabla \cdot \mathbf{u}) + \mathbf{u} \cdot \nabla, \tag{4.13}$$

where

$$\mathbf{u}(\mathbf{r}) = -\frac{\hbar^2}{m}\nabla Q(\mathbf{r}) \tag{4.14}$$

is a local velocity field, as will be discussed in Section 4.3 below. (Note that only the gradient of the velocity potential $Q(\mathbf{r})$ enters the following development.) The moments (4.9, 4.10) are then functionals of the local density $\rho(\mathbf{r})$ and the velocity field $\mathbf{u}(\mathbf{r})$. The moment m_1 is easily found to be

$$m_1[Q] = m_1[\mathbf{u}] = \frac{m}{2\hbar^2}\int \mathbf{u}(\mathbf{r}) \cdot \mathbf{u}(\mathbf{r})\,\rho(\mathbf{r})\,\mathrm{d}^3r. \tag{4.15}$$

The moment $m_3[\mathbf{u}]$ becomes more complicated but can be readily calculated from $\rho(\mathbf{r})$, $\mathbf{u}(\mathbf{r})$, their derivatives, and from the KS wavefunctions, using the current-density functional theory [115]. (See [110] for the most general expressions for $m_3[\mathbf{u}]$; special cases within the ETF model have been given in [27].) Performing the variation (4.7) then leads to the following eigenvalue equations for the excitation energy $\hbar\omega$:

$$\frac{\delta m_3[\mathbf{u}]}{\delta u_j(\mathbf{r})} = (\hbar\omega)^2\frac{m}{\hbar^2}\,\rho(\mathbf{r})\,u_j(\mathbf{r}) = m\omega^2\,\rho(\mathbf{r})\,u_j(\mathbf{r}) \qquad (j = x, y, z) \tag{4.16}$$

(see [110,111] for the explicit form of the left-hand side above). The solution of these equations yields the lowest excitation energy $\hbar\omega = \hbar\omega_1$ and the corresponding velocity field $\mathbf{u}_1(\mathbf{r})$. As sketched above, the second excitation energy $\hbar\omega_2$ is then found by restricting the solution $\mathbf{u}(\mathbf{r})$ of (4.16) to be orthogonal to $\mathbf{u}_1(\mathbf{r})$, and so on.

Due to the complexity of $m_3[\mathbf{u}]$, equations (4.16) represent a set of coupled nonlinear fourth-order partial differential equations for the three spatial components of $\mathbf{u}(\mathbf{r})$, which are close in spirit to the variational equations [116] derived from the so-called fluid dynamical approach [117, 118] developed in nuclear physics. Different from the approximations used there, however, equations (4.16) can be solved using the exact quantum-mechanical KS wavefunctions of the ground state. The semiclassical approximation implied by (4.12) then affects only the collective excitation mechanism.

Equations (4.16) are still quite formidable to solve numerically without further restrictions on the form of the velocity fields $\mathbf{u}(\mathbf{r})$. (For some first attempts, see [110] and [111]). An appreciable simplification arises when $\mathbf{u}(\mathbf{r})$ or, equivalently, the operator $Q(\mathbf{r})$, is expanded on a finite basis of local functions, as discussed in the following section.

4.3 Secular equation using a finite basis

We now expand the local excitation operator $Q(\mathbf{r})$ in a finite set of basis functions $Q_n(\mathbf{r})$:

$$Q(\mathbf{r}) \in \{Q_n(\mathbf{r})\} \qquad n = 1, 2, ..., M. \tag{4.17}$$

The variational equation (4.7) then becomes simply a secular equation for coupled harmonic vibrations generated by these operators:

$$\det|\mathcal{K}_{mn} - \omega_\nu^2 \mathcal{B}_{mn}| = 0, \qquad m, n = 1, 2, ..., M \tag{4.18}$$

yielding the collective excitation energies $\hbar\omega_\nu$ that represent the finite-basis LCA to the exact spectrum. Here the mass tensor $\mathcal{B}_{mn}$ and the restoring force tensor $\mathcal{K}_{mn}$ are given by

$$\mathcal{B}_{mn} = \hbar^2 \langle 0| [Q_m, [H, Q_n]] |0\rangle , \tag{4.19}$$

$$\mathcal{K}_{mn} = \langle 0| [[H, Q_m], [[H, Q_n], H]] |0\rangle , \tag{4.20}$$

which, apart from some reshuffling of constants, are the nondiagonal generalizations of the moments m_1 and m_3 defined in equations (4.9, 4.10). After solving equation (4.18) one finds the intrinsic local excitation operators $Q_\nu(\mathbf{r})$, which create the states $|\nu\rangle$, as linear combinations of the basis operators $Q_n(\mathbf{r})$:

$$Q_\nu(\mathbf{r}) = \sum_{n=1}^{M} c_n^\nu Q_n(\mathbf{r}) . \tag{4.21}$$

The physical meaning (see also [27, 106, 107]) of the operators $Q_\nu(\mathbf{r})$ shall in the following be illustrated in the so-called "scaling approach" [105, 106]. This is the simple case where a collective excitation is created by one single operator $Q(\mathbf{r})$, *i.e.*, using $M = 1$ in (4.17) and (4.18). The collective flow of the particles can be described by a velocity field $\mathbf{v}_\alpha(\mathbf{r}, t)$ which is proportional to $\mathbf{u}(\mathbf{r})$

$$\mathbf{v}_\alpha(\mathbf{r}, t) = \dot{\alpha}(t)\, \mathbf{u}(\mathbf{r}) , \tag{4.22}$$

so that $Q(\mathbf{r})$ (apart from a constant factor) plays the role of a velocity potential The collective coordinate $\alpha(t)$, which undergoes harmonic oscillations

$$\alpha(t) = \alpha_0 \sin(\omega t), \tag{4.23}$$

defines the time dependence of the single-particle wavefunctions by

$$\varphi_i(\mathbf{r}, t) = \varphi_i(\mathbf{r}, \alpha(t)) = \mathrm{e}^{-\alpha(t) S} \varphi_i(\mathbf{r}), \tag{4.24}$$

where the "scaling operator" S is defined by

$$S = [T, Q] = \frac{1}{2}(\nabla \cdot \mathbf{u}) + \mathbf{u} \cdot \nabla. \tag{4.25}$$

Taking the time derivative of (4.24), the time dependent density

$$\rho_\alpha(\mathbf{r}, t) = \rho(\mathbf{r}, \alpha(t)) = \sum_{\varepsilon_i \leq E_\mathrm{F}} |\varphi_i(\mathbf{r}, \alpha(t))|^2 \tag{4.26}$$

is immediately seen to fulfill the continuity equation

$$\frac{\partial}{\partial t}\rho_\alpha(\mathbf{r}, t) + \nabla \cdot \mathbf{j}_\alpha(\mathbf{r}, t) = 0, \tag{4.27}$$

where

$$\mathbf{j}_\alpha(\mathbf{r}, t) = \rho_\alpha(\mathbf{r}, t)\, \mathbf{v}_\alpha(\mathbf{r}, t) \tag{4.28}$$

is the *local current distribution* describing the collective flow of the excited system. This is the reason why we call the present scheme the "local-current approximation" (LCA) – in contrast to the RPA scheme where the currents are implicitly given by nonlocal particle-hole operators. The mass parameter $\mathcal{B}$ is now given by (*cf.* (4.15))

$$\mathcal{B} = m \int \mathbf{u}(\mathbf{r}) \cdot \mathbf{u}(\mathbf{r})\, \rho(\mathbf{r})\, \mathrm{d}^3 r, \tag{4.29}$$

and the restoring force parameter $\mathcal{K}$ is obtained from the "scaled" ground-state energy $\langle \alpha | H | \alpha \rangle = E[\rho(\mathbf{r}, \alpha)]$ according to

$$\begin{aligned} \mathcal{K} &= \langle 0 | \, [S, [S, H]] \, | 0 \rangle = \left[\frac{\mathrm{d}^2}{\mathrm{d}\alpha^2} \langle 0 | \, \mathrm{e}^{\alpha S} H \mathrm{e}^{-\alpha S} \, | 0 \rangle \right]_{\alpha=0} \\ &= \left[\frac{\mathrm{d}^2}{\mathrm{d}\alpha^2} \langle \alpha | H | \alpha \rangle \right]_{\alpha=0} = \left[\frac{\mathrm{d}^2}{\mathrm{d}\alpha^2} E[\rho(\mathbf{r}, \alpha)] \right]_{\alpha=0} . \end{aligned} \tag{4.30}$$

Since $E[\rho(\mathbf{r}, \alpha)]$ on the right-hand side can be considered as the potential energy $V(\alpha)$ of a collective Hamiltonian

$$H_\mathrm{coll}(\alpha, \dot{\alpha}) = \frac{1}{2}\mathcal{B}\dot{\alpha}^2 + V(\alpha), \tag{4.31}$$

we see that the solution of equatoin (4.18) in this one-mode picture is just the first excitation energy of (4.31) in the harmonic approximation: $\hbar\omega = E_3(Q) = \sqrt{\mathcal{K}/\mathcal{B}}$. If one identifies the intrinsic excitation operator $Q(\mathbf{r})$ with the external operator Q_ext by which the system is probed, this corresponds exactly to the simple sum rule picture [105] (or, in solid state physics, the so-called plasmon pole approximation).

The extension to the LCA scheme with several basis operators (4.17) is now evident: one couples several harmonic oscillators, each corresponding to one individual collective degree of freedom $\alpha_n(t)$, described by a velocity potential $Q_n(\mathbf{r})$, and diagonalizes this system *via* equation (4.18). This corresponds to a collective Hamiltonian with M coupled degrees of freedom, which in the harmonic approximation becomes

$$H_{\text{coll}} = \frac{1}{2} \sum_{m,n=1}^{M} \left(\mathcal{B}_{mn}\, \dot{\alpha}_m \dot{\alpha}_n + \mathcal{K}_{mn}\, \alpha_m \alpha_n \right). \tag{4.32}$$

We now use this method to calculate the response of the system to an external field, characterized by an operator Q_{ext}. The strength function $S_{Q_{\text{ext}}}(E)$ is defined by

$$S_{Q_{ext}}(E) = \sum_{\nu \neq 0} |\langle \nu | Q_{ext} | 0 \rangle|^2 \, \delta(E - \hbar\omega_\nu). \tag{4.33}$$

Hereby the spectrum $\{|\nu\rangle, \hbar\omega_\nu\}$ may be the exact one given by equations (4.2–4.4), or that obtained in any approximation. From the strength function one defines the k-th energy-weighted moments $m_k(Q_{\text{ext}})$:

$$m_k(Q_{\text{ext}}) = \int_0^\infty E^k \, S_{Q_{\text{ext}}}(E) \, \mathrm{d}E = \sum_{\nu \neq 0} (\hbar\omega_\nu)^k \, |\langle \nu | Q_{\text{ext}} | 0 \rangle|^2 , \tag{4.34}$$

which can be related to experimental observables. For instance, the photoabsorption cross section $\sigma(\omega)$ is in the long-wavelength limit given by

$$\sigma(\omega) = \frac{4\pi\omega}{3c} \, S_D(E = \hbar\omega), \tag{4.35}$$

where

$$D = Q_{\text{ext}} = e\, z \tag{4.36}$$

is the electric dipole operator and ω the frequency of the external electric field (assumed to be polarized in the z-direction). The above moments are then simply given as the following integrals over the cross section:

$$m_1(D) = \int \sigma(\omega) \, \mathrm{d}\omega \,, \qquad m_3(D) = \int \omega^2 \sigma(\omega) \, \mathrm{d}\omega. \tag{4.37}$$

The moment $m_1(D)$ can be calculated immediately and yields the model-independent Thomas–Reiche–Kuhn sum rule (in atomic physics called the "f-sum rule"):

$$m_1(D) = \int \sigma(\omega) \, \mathrm{d}\omega = 2\pi^2 \, \frac{e^2}{mc} \, Z, \tag{4.38}$$

where Z is the number of electrons taking part in the collective excitation. The experimental verification of the sum rule (4.38) thus helps to identify the collective nature of a resonance. The observed plasmon resonances in alkali clusters typically account for $\sim 60 - 80\%$ of the total dipole strength (see, *e.g.*, Sect. VIII of [40]).

The moment $m_{-1}(Q_{\text{ext}})$ can be shown [119, 120] to be proportional to the static polarizability α_{pol} of the ground state with respect to the external field Q_{ext}

$$m_{-1}(Q_{\text{ext}}) = \frac{1}{2}\,\alpha_{\text{pol}}(Q_{\text{ext}}). \tag{4.39}$$

This gives a convenient way to calculate static polarizabilities once the collective spectrum of a system is known.

By construction, the spectrum $\{|\nu\rangle, \hbar\omega_\nu\}$ obtained by solving the LCA secular equations (4.18) fulfills the sum rules $m_1(Q_{\text{ext}})$ and $m_3(Q_{\text{ext}})$ exactly, if the operator Q_{ext} is contained in the basis $\{Q_n(\mathbf{r})\}$. (This is, of course, trivial in the one-mode picture discussed above, where the only local operator $Q(\mathbf{r})$ is taken to be the external excitation operator.) It is now a matter of physical intuition to guess the form of the intrinsic excitation operators of a given system by making a suitable choice of the local basis $\{Q_n(\mathbf{r})\}$. In connection with the optic response, it is obvious that the dipole operator D (4.36) should be contained in this basis. Suitable basis sets for applications to metal clusters will be given in the following subsection.

4.4 Applications to metal clusters

For applications of the finite-basis LCA to metal clusters using the jellium model, the following set of local operators in polar coordinates (r, θ, ϕ) has proven to be very efficient and simple to use [27, 107]:

$$Q^p_{lm}(\mathbf{r}) = -e\,r^p\,\frac{1}{2}\left[Y_{lm}\,(\vartheta,\varphi) + Y^*_{lm}\,(\vartheta,\varphi)\right], \tag{4.40}$$

where $Y_{lm}(\theta, \phi)$ are the spherical harmonics in polar coordinates (r, θ, ϕ); $m = 0$ and $m = \pm 1$ correspond to the different polarizations of the external dipole field. For spherical clusters, where the angular momentum l is a conserved quantum number, a set of four to eight basis operators with $l = 1$ and $p = 1, 3, 5, \ldots$ or $p = 1, 4, 7, \ldots$ are often sufficient to get a good estimate of the lowest states of the collective dipole spectrum, in particular in the jellium model. For deformed clusters, several values of l must, of course, be included in the basis (see, *e.g.* [121] for calculations in the deformed jellium model). In calculations where the ionic structure is taken into account [122, 123], the following set of operators has been added to the basis set (4.40):

$$Q^k_{lm}(\mathbf{r}) = -e\, j_l(kr)\, \frac{1}{2} \left[Y_{lm}\,(\vartheta,\varphi) + Y^*_{lm}\,(\vartheta,\varphi) \right] , \tag{4.41}$$

where $j_l(x)$ are the spherical Bessel functions. We shall in the following present some results obtained with these basis sets.

4.4.1 Optic response in the jellium model

As a first test of the LCA, we show in Figure 15 the collective spectrum obtained in [107] for the spherical neutral clusters Na_8 and Na_{20}, calculated in the spherical jellium model by solving the KS equation in LDA. The upper parts give the results of a fully microscopical RPA calculation [124]; the lower parts give the results of the LCA using the basis (4.40). The main peak is split in the RPA result for Na_{20}; this is understood in terms of an interference of the collective oscillation with a specific $1p1h$ excitation [124] (so-called "Landau fragmentation"). It remains to be seen if a solution of the unrestricted LCA variational equations (4.16) is able to reproduce this splitting [111]. Apart from this, the agreement between RPA and LCA is excellent; in particular, the m_3 and m_1 dipole sum rules exhausted by the main peak(s) below 3.2 eV agree within less than one percent.

Figure 16 shows static electric dipole polarizabilities of sodium clusters, plotted *versus* the cluster radii R_I. Black dots with error bars are the experimental values; the other symbols are results of spherical jellium-model calculations. The square boxes were obtained in the TDLDA [29, 125, 126]. They are perfectly well reproduced by the LCA calculations of [107] *via* the m_{-1} sum rule relation (4.39), using the operator basis (4.40) and the full KS approach for the ground state. When the latter is replaced by a variational ETF calculation, the average results shown by the dashed line are obtained [27].

The main discrepancy between theory and experiment is here due to the lack of ionic structure in the jellium model. Inclusion of ionic structure gives a substantial improvement of the theoretical results [127, 128], in particular if the finite temperature is also taken into account [129, 130].

4.4.2 Optic response with ionic structure

We finally give some examples of LCA calculations that include the ionic structure. In the "cylindrically averaged pseudopotential scheme" (CAPS) [131], the ions are treated three-dimensionally by Monte-Carlo simulated annealing using a local pseudopotential, but only the axially averaged total ionic potential is included into the DFT treatment of the electrons. This has the advantage that the KS equations only have to be solved on

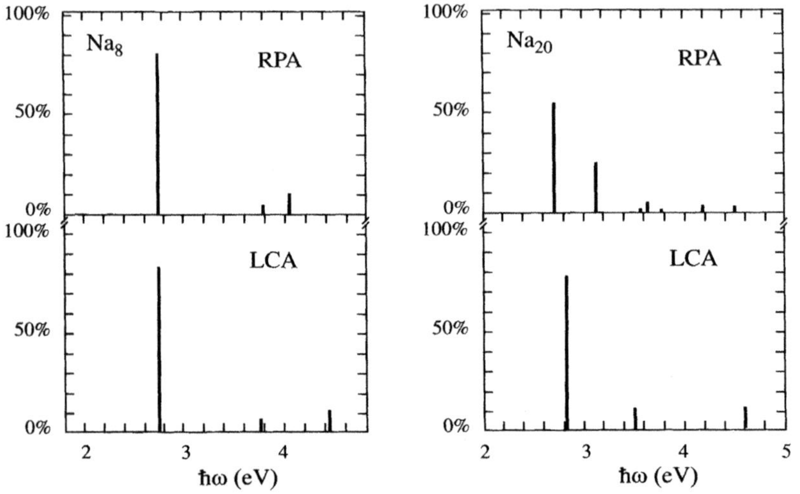

Fig. 15. Collective spectra of the spherical clusters Na_8 and Na_{20} obtained in the jellium model. The positions of the vertical lines give the eigenenergies found by solving the secular equation (4.18), their heights give their percentage of the dipole m_1 sum rule. The ground state was obtained by solving the KS equations. *Upper panels:* RPA spectrum [124], *lower panels:* LCA spectrum [107].

a two-dimensional grid (with cylindrical symmetry), which allows to treat larger systems with an affordable numerical effort. Kümmel [110] has constructed a local "soft" pseudopotential that reproduces the correct bulk properties of sodium and at the same time the 3s energy level of the valence electron in the sodium atom (see also [123, 132]). It should therefore be suitable for the use in clusters that interpolate from the atom to the bulk. Indeed, the ground-state structures obtained with this pseudopotential in CAPS calculations reproduce the experimental bond lengths in the smallest sodium clusters as well as all-electron *ab initio* calculations [110, 127].

In Figure 17 (top part) we show the photoabsorption spectra obtained for Na_2 and Na_8 in CAPS calculations using the LCA for the collective excitation spectrum (note that the widths are phenomenological like in most TDLDA or RPA calculations). The positions of the main peaks reproduce the experimental peaks (bottom part) very well. Also shown (middle part) is the result of a typical TDLDA calculation [133] using the Troullier–Martins pseudopotential, which yields a slight blue shift of the peaks due to a systematic underestimation of the bond lengths.

Figure 18 shows the structures and collective dipole spectra obtained for the charged clusters Na_{27}^+ and Na_{41}^+ using the same scheme. Again, the LCA calculations [122] yield an excellent reproduction of the experimental peaks

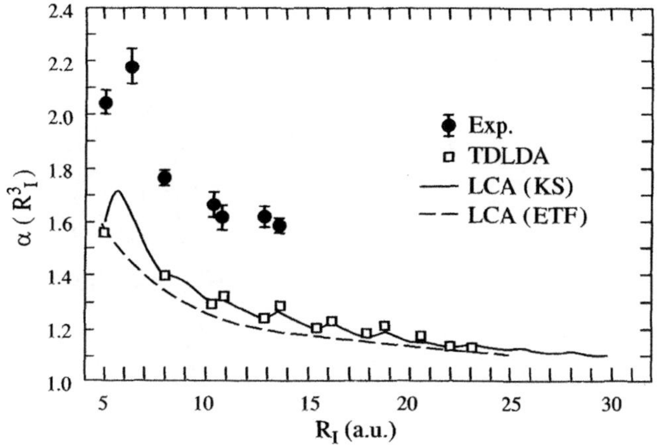

Fig. 16. Static electric dipole polarizabilities α of sodium clusters. *Dots with error bars:* experimental results [80]. *Open squares:* TDLDA results [29,125,126]. *Solid line:* LCA result using the m_{-1} sum rule from the KS ground state [107]. *Dashed line:* LCA result using the m_{-1} sum rule from the ETF ground state [27].

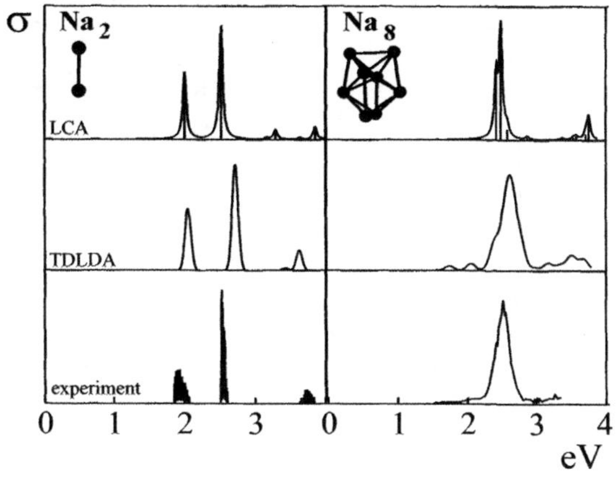

Fig. 17. Photoabsorption cross section σ (in arbitrary units) *versus* energy (in eV) for the neutral clusters Na_2 (left) and Na_8 (right). *Top panels:* LCA results [110,127]; *center panels:* TDLDA results [133]; *bottom panels:* experimental results [134]. (From [110]).

in the photoabsorption cross section, including the high-energy shoulder seen in Na_{41}^+.

We conclude by noting that an approximate icosahedral structure was found in the same set of CAPS calculations for the ground state of the Na_{55}^+

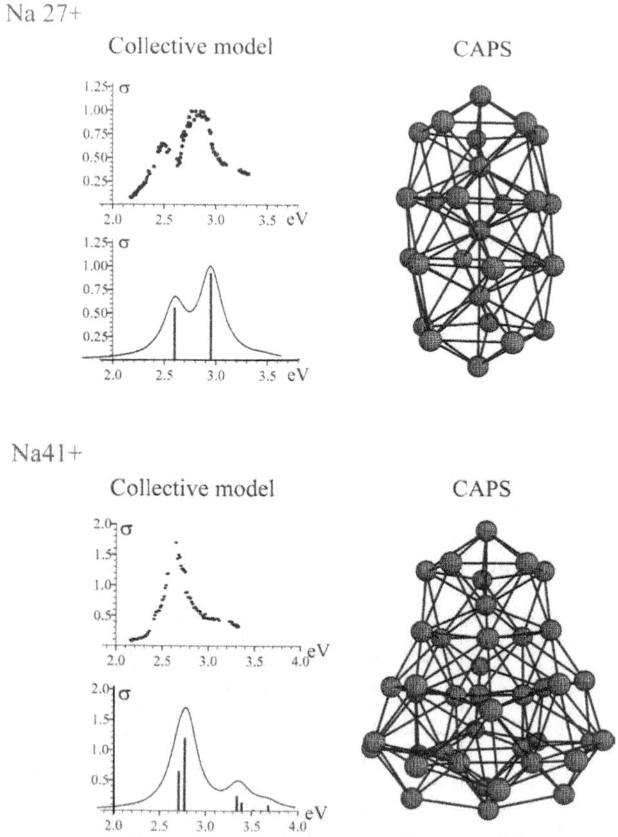

Fig. 18. *Right:* ionic structures of Na_{27}^+ and Na_{41}^+ obtained in the CAPS [122]. *Left:* photoabsorption cross sections σ in arbitrary units, plotted *versus* energy in eV. *Upper parts, dots:* experimental results [136]; *lower parts:* theoretical results obtained in the LCA [122].

cluster [123, 132]; it has been confirmed (with only minor non-axial deviations) in DFT-LDA molecular dynamics calculations without symmetry restrictions [135].

I have profited substantially from my collaborators and students. Without their contributions and encouragement much of the material included in these lectures would not exist. I am particularly indebted to Joachim Blaschke and Stephan Kümmel for allowing me to include some partially unpublished results of their Ph.D. work. Several parts of the work presented here were supported by the Deutsche Forschungsgemeinschaft.

References

[1] R.M. Dreizler and E.K.U. Gross, *Density Functional Theory* (Springer, Berlin, 1990).

[2] V.M. Strutinsky, *Nucl. Phys. A* **122** (1968) 1, and earlier references quoted therein.

[3] R.K. Bhaduri and C.K. Ross, *Phys. Rev. Lett.* **27** (1971) 606.

[4] M. Brack and R.K. Bhaduri, *Semiclassical Physics*, Frontiers in Physics, Vol. 96 (Addison-Wesley, Reading, USA, 1997).

[5] L.H. Thomas, *Proc. Camb. Phil. Soc.* **23** (1926) 195 ; E. Fermi, *Z. Phys.* **48** (1928) 73.

[6] N.H. March, *Self-consistent Fields in Atoms* (Pergamon Press, Oxford, 1975).

[7] E.H. Lieb, *Rev. Mod. Phys.* **53** (1981) 603.

[8] E. Wigner, *Phys. Rev.* **40** (1932) 749; J.G. Kirkwood, *Phys. Rev.* **44** (1933) 31.

[9] M. Abramowitz and I.A. Stegun, *Handbook of Mathematical Functions* (Dover Publications, New York, 9th printing, 1970).

[10] B. Van der Pohl and H. Bremmer, *Operational Calculus* (Cambridge University Press, Cambridge, 1955).

[11] B. Grammaticos and A. Voros, *Ann. Phys. (N. Y.)* **123** (1979) 359; **129** (1980) 153.

[12] C.H. Hodges, *Can. J. Phys.* **51** (1973) 1428.

[13] M. Brack, B.K. Jennings and Y.H. Chu, *Phys. Lett. B* **65** (1976) 1.

[14] C.F.v. Weizsäcker, *Z. Phys.* **96** (1935) 431.

[15] R.O. Jones and O. Gunnarsson, *Rev. Mod. Phys.* **61** (1989) 689.

[16] F. Perrot, *Phys. Rev. A* **20** (1979) 586.

[17] M. Brack, *Phys. Rev. Lett.* **53** (1984) 119; *ibid.* (Erratum) **54** (1985) 851.

[18] J. Bartel, M. Brack and M. Durand, *Nucl. Phys. A* **445** (1985) 263.

[19] M. Brack, C. Guet and H.-B. Håkansson, *Phys. Rep.* **123** (1985) 275.

[20] D. R. Murphy and W. P. Wang, *J. Chem. Phys.* **72** (1980) 429.

[21] C. Guet and M. Brack, *Z. Phys. A* **297** (1980) 247.

[22] M. Centelles, X. Viñas, M. Barranco and P. Schuck, *Ann. Phys. (N. Y.)* **221** (1993) 165.

[23] J. Caro, E. Ruiz Arriola and L.L. Salcedo, *J. Phys. G* **22** (1996) 981.

[24] P. Hohenberg and W. Kohn, *Phys. Rev.* **136** (1964) B864.

[25] M. Centelles, M. Pi, X. Viñas, F. Garcias and M. Barranco, *Nucl. Phys. A* **510** (1990) 397.

[26] E. Engel and J.P. Perdew, *Phys. Rev. B* **43** (1991) 1331.

[27] M. Brack, *Phys. Rev. B* **39** (1989) 3533.

[28] M. Seidl and M. Brack, *Ann. Phys. (N. Y.)* **245** (1996) 275.

[29] W. Ekardt, *Phys. Rev. B* **29** (1984) 1558.

[30] O. Genzken and M. Brack, *Phys. Rev. Lett.* **67** (1991) 3286.

[31] M. Brack, *Rev. Mod. Phys.* **65** (1993) 677.

[32] M. Seidl, K.-H. Meiwes-Broer and M. Brack, *J. Chem. Phys.* **95** (1991) 1295.

[33] M. Seidl, J.P. Perdew, M. Brajczewska and C. Fiolhais, *Phys. Rev. B* **55** (1997) 13288; *J. Chem. Phys.* **108** (1998) 8182.

[34] P. Blaise, S.A. Blundell and C. Guet, *Phys. Rev. B* **55** (1997) 15856.

[35] A. Domps, E. Suraud and P.-G. Reinhard, *Eur. Phys. J. D* **2** (1998) 191; A. Domps, P.-G. Reinhard and E. Suraud, *Phys. Rev. Lett.* **80** (1998) 5520.

[36] A. Aguado, J.M. López, J.A. Alonso and M.J. Stott, *J. Chem. Phys.* **111** (1999) 6026.

[37] S. A. Blundell, C. Guet and R. Zope, *Phys. Rev. Lett.* **84** (2000) 4826.

[38] W.D. Myers and W.J. Swiatecki, *Ann. Phys. (N. Y.)* **55** (1969) 395.

[39] M. Seidl, Ph.D. Thesis (University of Regensburg, 1994, unpublished).

[40] W.A. de Heer, *Rev. Mod. Phys.* **65** (1993) 611.

[41] C. Fiolhais and J.P. Perdew, *Phys. Rev. B* **45** (1992)6207.

[42] M. Seidl and J.P. Perdew, *Phys. Rev. B* **50** 5744 (1994).

[43] N. Bohr, *Phil. Mag. (Series 6)* **26** (1913) 857.

[44] A. Pais, *Niels Bohr's Times* (Clarendon Press, Oxford, 1991).

[45] G. Wentzel, *Z. Phys.* **38** (1926) 518; H. Kramers, *Z. Phys.* **39** (1926) 828; L. Brillouin, *Compt. Rend.* **183** (1926) 24.

[46] J.B. Keller, *Ann. Phys. (N. Y.)* **4** (1958) 180.

[47] A. Einstein, *Verh. Dtsch. Phys. Ges.* **19** (1917) 82.

[48] M.C. Gutzwiller, *J. Math. Phys.* **12** (1971) 343; and earlier references quoted therein.

[49] R. Balian and C. Bloch, *Ann. Phys. (N. Y.)* **69** (1972) 76.

[50] M.V. Berry and M. Tabor, *Proc. R. Soc. Lond. A* **349** (1976)101 .

[51] M.V. Berry and M. Tabor, *Proc. R. Soc. Lond. A* **356** (1977) 375.

[52] M.C. Gutzwiller, *Chaos in classical and quantum mechanics* (Springer, New York, 1990).

[53] M. Sieber and F. Steiner, *Phys. Rev. Lett.* **67** (1991) 1941; M. Sieber, *Chaos* **2** (1992) 35.

[54] M.C. Gutzwiller, in *Chaos and quantum physics*, Les Houches Session LII (1989), edited by M.-J. Giannoni *et al.* (North-Holland, Amsterdam, 1991) 201.

[55] V.M. Strutinsky, *Nukleonika (Poland)* **20** (1975) 679; V.M. Strutinsky and A.G. Magner, *Sov. J. Part. Nucl.* **7** (1976) 138; V.M. Strutinsky, A.G. Magner, S.R. Ofengenden and T. Døssing, *Z. Phys. A* **283** (1977) 269.

[56] S.C. Creagh and R.G. Littlejohn, *Phys. Rev. A* **44** (1991) 836; *J. Phys. A* **25** (1992) 1643.

[57] J.H. Van Vleck, *Proc. Natl. Acad. Sci. USA* **14** (1928) 178.

[58] S.C. Creagh, J.M. Robbins and R.G. Littlejohn, *Phys. Rev. A* **42** (1990) 1907.

[59] H. Friedrich and D. Wintgen, *Phys. Rep.* **183** (1989) 39.

[60] S.C. Creagh, *Ann. Phys. (N. Y.)* **248** (1996) 60.

[61] M.V. Berry and K.E. Mount, *Rep. Prog. Phys.* **35** (1972) 315.

[62] V.A. Yakubovich and V.M. Starzhinskii, *Linear differential equations with periodic coefficients*, Vols. 1 and 2 (J. Wiley & Sons, New York, 1975).

[63] M.J. Feigenbaum, *J. Stat. Phys.* **19** (1978) 25; *Physica D* **7** (1983) 16.

[64] T.C. Bountis, *Physica D* **3** (1981) 577; J.M. Greene, R.S. McKay, F. Vivaldi and M.J. Feigenbaum, *Physica D* **3** (1981) 468.

[65] M. Brack, *Foundations of Physics* **31** (2001) 209.

[66] A.M. Ozorio de Almeida and J.H. Hannay, *J. Phys. A* **20** (1987) 5873; see also A.M. Ozorio de Almeida, *Hamiltonian Systems: Chaos and Quantization* (Cambridge University Press, Cambridge, 1988).

[67] M. Sieber, *J. Phys. A* **29** (1996) 4715; H. Schomerus and M. Sieber, *J. Phys. A* **30** (1997) 4537; M. Sieber and H. Schomerus, *J. Phys. A* **31** (1998) 165; M. Sieber, *J. Phys. A* **30** (1997) 4563.

[68] H. Schomerus, *Europhys. Lett.* **38** (1997) 423; *J. Phys. A* **31** (1998) 4167.

[69] S. Tomsovic, M. Grinberg and D. Ullmo, *Phys. Rev. Lett.* **75** (1995) 4346; see also D. Ullmo, M. Grinberg and S. Tomsovic, *Phys. Rev. E* **54** (1996) 135.

[70] M. Brack, P. Meier and K. Tanaka, *J. Phys. A* **32** (1999) 331.

[71] A. Magner, S.N. Fedotkin, K. Arita, T. Misu, K. Matsuyanagi, T. Schachner and M. Brack, *Prog. Theor. Phys. (Japan)* **102** (1999) 551.

[72] M. Brack, S.C. Creagh, P. Meier, S.M. Reimann and M. Seidl, in *Large Clusters of Atoms and Molecules*, edited by T.P. Martin (Kluwer, Dordrecht, 1996) 1.

[73] H. Nishioka, K. Hansen and B.R. Mottelson, *Phys. Rev. B* **42** (1990) 9377.

[74] J. Pedersen, S. Bjørnholm, J. Borggreen, K. Hansen, T.P. Martin and H.D. Rasmussen, *Nature* **353** (1991) 733.

[75] M. Brack, O. Genzken and K. Hansen, *Z. Phys. D* **21** (1991) 65.

[76] M. Pellarin, B. Baguenard, C. Bordas, M. Broyer, J. Lermé and J.L. Vialle, *Phys. Rev. B* **48** (1993) 17645.

[77] E. Koch and O. Gunnarsson, *Phys. Rev. B* **54**, 5168 (1996).

[78] K. Tanaka, S.C. Creagh and M. Brack, *Phys. Rev. B* **53** (1996) 16050.

[79] W.D. Knight, K. Clemenger, W.A. de Heer, W.A. Saunders, M.Y. Chou and M.L. Cohen, *Phys. Rev. Lett.* **52** (1984) 2141.

[80] W.D. Knight, K. Clemenger, W.A. de Heer and W.A. Saunders, *Phys. Rev. B* **31** (1985) 2539.

[81] S.M. Reimann, M. Brack and K. Hansen, *Z. Phys. D* **28** (1993) 235.

[82] S.M. Reimann and M. Brack, *Comp. Mat. Sci.* **2** (1994) 433.

[83] M. Brack and S.R. Jain, *Phys. Rev. A* **51** (1995) 3462.

[84] A.G. Magner, S.N. Fedotkin, F.A. Ivanyuk, P. Meier, M. Brack, S.M. Reimann and H. Koizumi, *Ann. Physik (Leipzig)* **6** (1997) 555; A.G. Magner, S.N. Fedotkin, F.A. Ivanyuk, P. Meier and M. Brack, Czech. *J. Phys.* **48** (1998) 845.

[85] T. Hirschmann, M. Brack and J. Meyer, *Ann. Phys. (Leipzig)* **3** (1994) 336.

[86] P. Meier, M. Brack and S.C. Creagh, *Z. Phys. D* **41** (1997) 281.

[87] K. Richter, *Europhys. Lett.* **29** (1995) 7.

[88] G. Hackenbroich and F. von Oppen, *Europhys. Lett.* **29** (1995) 151.

[89] M. Persson, J. Pettersson, B. von Sydow, P.E. Lindelof, A. Kristensen and K. Berggreen, *Phys. Rev. B* **52** (1995) 8921; and earlier references quoted therein.

[90] P.E. Lindelof, P. Hullmann, P. Bøggild, M. Persson and S.M. Reimann, in *Large Clusters of Atoms and Molecules*, edited by T.P. Martin (Kluwer, Dordrecht, 1996) 89.

[91] S.M. Reimann, M. Persson, P.E. Lindelof and M. Brack, *Z. Phys. B* **101** (1996) 377.

[92] P. Bøggild, A. Kristensen, P.E. Lindelof, S.M. Reimann and C.B. Sørensen, in *8th International Conference on Physics of Semiconductors (ICPS)*, Proceedings (Berlin, 1997) 1533.

[93] H. Linke, L. Christensson, P. Omling and P.E. Lindelof, *Phys. Rev. B* **56** (1997) 1440.

[94] M. Brack, J. Blaschke, S.C. Creagh, A.G. Magner, P. Meier and S.M. Reimann, *Z. Phys. D* **40** (1997) 276.

[95] M. Hénon and C. Heiles, *Astr. J.* **69** (1964) 73.

[96] J. Blaschke and M. Brack, *Phys. Rev. A* **56** (1997) 182; *Physica E* **1** (1997) 288.

[97] K. Richter, D. Ullmo and R.A. Jalabert, *Phys. Rep.* **276** (1996) 1.

[98] K. Tanaka, *Ann. Phys. (N. Y.)* **268** (1998) 31.

[99] J. Blaschke, Ph.D. Thesis (Regensburg University, 1999), available at ⟨`http://www.joachim-blaschke.de`⟩.

[100] D. Weiss, M.L. Roukes, A. Menschig, P. Grambow, K. von Klitzing and G. Weimann, *Phys. Rev. Lett.* **66** (1991) 2790; D. Weiss, K. Richter, A. Menschig, R. Bergmann, H. Schweizer, K. von Klitzing and G. Weimann, *Phys. Rev. Lett.* **70** (1993) 4118.

[101] D. Weiss and K. Richter, *Physica D* **83** (1995) 290.

[102] C. Gould *et al.*, *Phys. Rev. B* **51** (1995) 11213.

[103] G. Kirczenov *et al.*, *Phys. Rev. B* **56** (1997) 7503.

[104] J. Blaschke and M. Brack, *Europhys. Lett.* **50** (2000) 294.

[105] O. Bohigas, A.M. Lane and J. Martorell, *Phys. Rep.* **51** (1979) 267.

[106] P. Gleissl, M. Brack, J. Meyer and P. Quentin, *Ann. Phys. (N. Y.)* **197** (1990) 205.

[107] P.-G. Reinhard, M. Brack and O. Genzken, *Phys. Rev. A* **41** (1990) 5568.

[108] P.-G. Reinhard and Y. Gambhir, *Ann. Phys. (Leipzig)* **1** (1992) 598.

[109] P.-G. Reinhard, O. Genzken and M. Brack, *Ann. Phys. (Leipzig)* **5** (1996) 576.

[110] S. Kümmel, Ph.D. Thesis, Regensburg University, *Structural and Optical Properties of Sodium Clusters Studied in Density Functional Theory* (Logos, Berlin, 2000).

[111] S. Kümmel and M. Brack, *Phys. Rev. A* **64** (2001) 022506.

[112] A. L. Fetter and J. D. Walecka *Quantum theory of many particle systems* (McGraw-Hill, New York, 1971).

[113] A.M. Lane, *Nuclear Theory* (Benjamin, New York, 1964).

[114] D.J. Rowe, *Nuclear collective motion* (Methuen, London, 1970).

[115] G. Vignale and W. Kohn, *Phys. Rev. Lett.* **77** (1996) 2037; G. Vignale, C.A. Ullrich and S. Conti, *Phys. Rev. Lett.* **79** (1997) 4878.

[116] J.P. da Providência and G. Holzwarth, *Nucl. Phys. A* **398** (1983) 59; **439** (1985) 477.

[117] G.F. Bertsch, *Nucl. Phys. A* **249** (1975) 253.

[118] H. Sagawa and G. Holzwarth, *Prog. Theor. Phys.* **59** (1978) 1213.

[119] D.J. Thouless, *Nucl. Phys.* **22** (1961) 78.

[120] E.R. Marshalek and J.P. da Providência, *Phys. Rev. C* **7** (1973) 2281.

[121] T. Hirschmann, M. Brack and P.-G. Reinhard, *Z. Phys. D* **40** (1997) 254.

[122] S. Kümmel, M. Brack and P.-G. Reinhard, *Phys. Rev. B* **58** (1998) R1774.

[123] S. Kümmel, M. Brack and P.-G. Reinhard, *Phys. Rev. B* **62** (2000) 7602.

[124] C. Yannouleas, M. Brack, R.A. Broglia and P.F. Bortignon, *Phys. Rev. Lett.* **63** (1989) 255.

[125] D.E. Beck, *Sol. State Commun.* **49** (1984) 381; *Phys. Rev. B* **30** (1984) 6935.

[126] M. Manninen, R.M. Nieminen and M.J. Puska, *Phys. Rev. B* **33** (1986) 4297.

[127] S. Kümmel, T. Berkus, P.-G. Reinhard and M. Brack, *Eur. Phys. J. D* **11** (2000) 239.

[128] D. Rayane *et al.*, *Eur. Phys. J. D* **9** (1999) 243.

[129] S. Kümmel, J. Akola and M. Manninen, *Phys. Rev. Lett.* **84** (2000) 3827.

[130] S. Blundell, C. Guet and R.R. Zope, *Phys. Rev. Lett.* **84** (2000) 4826.

[131] B. Montag and P.-G. Reinhard, *Z. Phys. D* **33** (1995) 265.

[132] S. Kümmel, P.-G. Reinhard and M. Brack, *Eur. Phys. J. D* **9** (1999) 149.

[133] I. Vasiliev, S. Öğüt and J.R. Chelikowsky, *Phys. Rev. Lett.* **82** (1999) 1919.

[134] W.R. Frederickson and W.W. Watson, *Phys. Rev.* **30**, (1927) 429; C.R.C. Wang, S. Pollack, D. Cameron and M.M. Kappes, *Chem. Phys. Lett.* **166** (1990) 26.

[135] J. Akola, A. Rytkönen, H. Häkkinen and M. Manninen, *Eur. Phys. J. D* **8** (2000) 93.

[136] M. Schmidt and H. Haberland, *Eur. Phys. J. D* **6** (1999) 109.

COURSE 6

PAIRING CORRELATIONS IN FINITE FERMIONIC SYSTEMS

H. FLOCARD

Groupe de Physique Théorique,
Institut de Physique Nucléaire,
91406 Orsay Cedex, France

Contents

PAIRING CORRELATIONS IN FINITE FERMIONIC SYSTEMS

H. Flocard

1 Introduction

Superconductivity has been discovered almost a century ago. The understanding of the underlying mechanisms took approximately fifty years of efforts and required the input of several outstanding physicists such as London, Landau, Bogoliubov, Abrikosov, Bardeen, Cooper and Shrieffer to name only a few. In the mean time, the existence of the phenomenon has been established in several areas of physics [1, 2]. Most often, the size of the system is much larger than that of the microscopic building block of superconductivity: the Cooper pair. For instance, this is the case for the aluminium, copper or nobium wires at very small temperatures which are widely used in magnets when high magnetic fields are needed. In the second half of this century, the long searched superconductivity of liquid He_3 involving pairs with new spin properties was established for its two predicted phases. More recently, and somewhat unexpectedly, new families of materials have been discovered which display superconducting properties at temperature sometimes exceeding hundred Kelvin. Superconductivity is also conjectured to exist in neutron stars. In particular, it has been invoked to explain the glitches detected in the long term evolution of the frequency of some pulsars.

At the other end of the length scale, many nuclear features are influenced by pairing correlations [3]. Those show up prominently when properties of odd and even nuclei are compared. Among them, one can mention level densities near the ground state, binding energies, charge radii, moments of inertia, spontaneous alpha decay lifetimes and fission cross sections. As a matter of fact, the last property can be said to have led to the first application of (admitedly nature controled, rather than man made) superconductivity with a significant impact on the life of the average citizen. Indeed, fissile nuclei in the thermal nuclear reactors (producing today approximately 15% of the world electricity) are always neutron-odd

(^{233}U, ^{235}U, ^{239}Pu and to a lesser degree ^{241}Pu) while fertile nuclei (^{230}Th, ^{238}U) are neutron-even. Although the even-odd proton character plays a smaller role in neutron induced fission, the dynamics which leads to the splitting of an heavy nucleus into two smaller fragments is favored for even Z elements such as Th, U, Pu and Cm as compared to Pa, Np and Am. More recently, with the advent of large size arrays of gamma detectors, the investigation of high spin properties has established the strong similarity between the reaction of a nucleus to a rotation field and that of a metallic superconductor to a magnetic field. In particular, the nuclear moments of inertia are very sensitive to growing angular momentum [4]. Although it was much more poorly documented as it is today, the relevance of the superconductivity concepts to the understanding of nuclear structure was evident to many discerning physicists already fifty years ago. As a result, within a year of the publication of the BCS founding paper, Bohr, Mottelson and Pines [5–7] had given a description of the microscopic structure of nuclear Cooper pairs which is still valid today.

On the other hand, nuclear dynamics is concerned with rather different conditions than those prevalent in studies of metallic superconductors. It takes place at zero or very small (compared to the pairing gap) excitation energy and more importantly in systems where a canonical description (as compared to grand canonical) is very much needed. For instance, it is clear that a meaningful discussion of odd *versus* even effects requires that experiment and/or formalism keeps a good (albeit not necessary absolute) control of the number of particles in the system under consideration. In that respect, recent experimental breakthroughs in the production and control of superconducting microscopic metallic islands and aluminium grains at the nanometer scale have established a link between the nuclear and condensed matter perspectives as regards superconductivity [8–11]. For instance, the spectroscopy of tiny grains subjected to a magnetic field display features reminiscent of that of a rotating nuclei. These systems are also of great help to understand why and how odd-even staggering effects loose importance as the system size and the temperature grow. We shall see that a characteristic of such small systems is that their dimensions become comparable and even smaller than that of a Cooper pair. This makes impossible an observation of the effects connected with that spatial dependence of the phase of pairs which is known to be essential in the most spectacular manifestations of superconductivity in macroscopic devices. With nuclei and nanoscopic grains, we are considering a domain for which Anderson once coined the appelation "dirty supraconductors" [12]. In relation with them, he suggested that an interesting problem was to establish how small a system could be and still remain supraconducting. This question can be also reformulated in a more theoretical langage by asking the relevance and the limitations for finite systems of the symmetry breaking concept which is basic to our present understanding of superconductivity.

At the outset, I have to say that this lecture does not provide the answers to these important questions. Rather, in qualitative terms, it outlines the ingredients of the problem and use them to review a number of the approaches which can be used to study superconductivity in small systems. These notes also contain a summary of a recent discussion on the connection between the odd-even staggerings of binding energies of nuclei and of atomic clusters, and the existence and the magnitude of pairing correlations. Finally, some lately published theoretical works for systems at zero and finite temperature in relation with Anderson's question will also be briefly described. My hope is that despite its limited scope, this text will give the student some insight into the range of methods which have been devised to investigate superconductivity concepts for finite systems and help him find which one is the most appropriate for a description of the pair correlated object he might be interested to analyze.

A realistic description of physical systems often requires complex effective Hamiltonians and much more elaborate methods than those which are presented here. For instance, rather sophisticated Hamiltonians are essential to obtain an accurate description of nuclear properties. As a matter of fact, the analysis of the influence of pairing correlations on the high angular momentum behavior of nuclei is one of the most useful tool for the determination of the spin, isospin and spatial dependence of the nuclear two-body effective interaction [13]. Nevertheless, for pedagogical reasons, these notes use a very schematic dynamics which mimics only those basic features of nuclear or condensed matter physics which are essential for my purpose. For instance, the mean-field part of the Hamiltonian is simplified to a diagonal one-body operator. In addition, the lectures describe the effect of pairing correlations by means of the simplest two-body potential which retains the features necessary to start an investigation of superconductivity. For the same reasons, the formalism is kept at the simplest possible level so that formal calculations involve only two by two matrices and no refined mathematics. By doing so, I have kept the size of the presented material within limits compatible with the teaching schedule and, may be, have set up a workframe such that students can easily perform for themselves the calculations required to cover the intermediate steps skipped in the notes.

2 Basic mechanism: Cooper pair and condensation

For the sake of simplicity, in these notes, we consider that the one particle subspace has a finite dimension Ω. Nevertheless, when it is convenient, the sums are replaced by integrals and vice versa without worry on problems which one might incur when a continuum instead of a discrete formulation is used. Sections 2 and 3 give a short and schematic presentation of the BCS ideas and formalism and establish our notations.

2.1 Condensed matter perspective: Electron pairs

Let us start with a reminder of the mechanism which leads to the formation of Cooper pairs [1, 14]. At zero temperature T, we consider two electrons in orbitals above the spherical fermi sea of a system with N_0 fermions. We take the kinetic energy as zero order hamiltonian $\hat{H}_0$ and we assume that the individual orbitals are described by the tensor product of a plane wave and a spin vector $|\sigma = \pm 1/2\rangle$. The energy of a single particle state with momentum $\vec{p} = \hbar\vec{k}$ is

$$\epsilon_k = \hbar^2\vec{k}^2/2m, \tag{2.1.1}$$

where m is the mass of the particle.

Following Cooper's argument, we want to prove that *even the weakest attractive interaction between the two fermions will bind them into a pair.* This is a remarkable feature since in free space two particles only form a pair when the attraction between them is strong enough to overcome the quantum fluctuations associated with the kinetic energy. An order of magnitude of the energy associated with such fluctuations is $\hbar^2/(2m\,d^2)$ where d is the typical distance at which attraction is maximal. In particular, it is very large for short range interactions.

From the wave-functions of the two particles, one can build a pair wave-function $\Psi_{1,2}$ which is translation invariant. Here, we will impose the total momentum to be zero since we are looking for the state of lowest energy. Then, the general form for $\Psi_{1,2}$ is

$$\Psi_{1,2} \propto \sum_{|\vec{k}|>k_F} f_{\vec{k}}\, e^{i\vec{k}.(\vec{r}_1 - \vec{r}_2)}\, \Xi(S), \tag{2.1.2}$$

where the $f_{\vec{k}}$'s are expansion coefficients still to be determined. The Pauli principle imposes that only unoccupied states (above the Fermi surface) are involved in the expansion. The symbol $\Xi(S)$ stands for the total spin function of spin S. When $S = 0$ the state is spin antisymetric while it is spin symmetric when $S = 1$. Since $\Psi_{1,2}$ must be antisymmetric with respect to the exchange of both space and spin coordinates, when $S = 0$ (2.1.2) becomes

$$\Psi_{1,2} \propto \sum_{|\vec{k}|>k_F} f_{\vec{k}} \cos[\vec{k}.(\vec{r}_1 - \vec{r}_2)]\, \Xi(S = 0). \tag{2.1.3}$$

For $S = 1$, the cosine is replaced by a sine function which vanishes when $\vec{r}_1 = \vec{r}_2$. Because it is short range, the effective attractive force between the electrons favors the cosine spatial dependence and therefore the total spin $S = 0$. Thus, the optimal $\Psi_{1,2}$ should be searched as a mixture of two-body states involving electrons of opposite momentum and spin.

For a given potential $V(|\vec{r}_1 - \vec{r}_2|)$, the Schroedinger equation which determines the energy $E_{(2)}$ of the hypothesized pair is

$$(E_{(2)} - 2\epsilon_k)\, f_{\vec{k}} = \sum_{|\vec{k}'|>k_{\rm F}} V_{\vec{k},\vec{k}'} f_{\vec{k}'}, \tag{2.1.4}$$

where $V_{\vec{k},\vec{k}'}$ is the Fourier transform of the interaction. For the sake of simplicity let us take the extreme case of a zero range attractive force $V \propto -G\delta(\vec{r}_1 - \vec{r}_2)$. Then, $V_{\vec{k},\vec{k}'} = -G$ for any pair $\vec{k}$, $\vec{k}'$ and equation (2.1.4) can be solved as

$$f_{\vec{k}} = G\, \frac{\sum f_{\vec{k}'}}{2\epsilon_k - E_{(2)}}. \tag{2.1.5}$$

In equation (2.1.5), the sum over $\vec{k}'$ can be replaced by an integral over the kinetic energy $\int_{\epsilon_{\rm F}}^{\epsilon_{\rm F}+\Lambda} w\, {\rm d}\epsilon$ where w denotes the density of (doubly degenerated because $\sigma = \pm 1/2$) levels *at the Fermi surface.* Note that to account for the schematic character of the two-body potential, one has introduced an energy cutoff Λ. In the limit when Λ is much larger than the expected binding energy of the pair

$$\Lambda \gg 2\epsilon_{\rm F} - E_{(2)}, \tag{2.1.6}$$

the solution of equation (2.1.5) is

$$E_{(2)} - 2\epsilon_{\rm F} = -2\Lambda {\rm e}^{-\dfrac{2}{wG}}. \tag{2.1.7}$$

It is obviously negative for any (positive) value of G and w.

The expression (2.1.7) shows that that the Cooper mechanism depends on the existence of a fermi sea. Indeed, $E_{(2)}$ vanishes as w goes to zero. For a set of N_0 spin 1/2 fermions homogenously filling a volume $\mathcal{V}$, the density at $T = 0$ is given by

$$\frac{N_0}{\mathcal{V}} = \frac{k_{\rm F}^3}{3\pi^2} = \frac{(2m\epsilon_{\rm F})^{3/2}}{3\pi^2\hbar^3}, \tag{2.1.8}$$

where $k_{\rm F}$ and $\epsilon_{\rm F}$ are the Fermi momentum and energy, respectively. As a consequence, the level density $w = 1/2\,(\partial N_0/\partial\epsilon_{\rm F})$ is

$$w = \frac{3N_0}{4\epsilon_{\rm F}} = \mathcal{V}\frac{m\, k_{\rm F}}{2\hbar^2\pi^2}. \tag{2.1.9}$$

Thus w (and $E_{(2)} - 2\epsilon_{\mathrm{F}}$) tends to zero when N_0, or ϵ_{F}, or $\mathcal{V}$ vanish.

Using equation (2.1.5), a calculation of the expansion coefficients $f_{\vec{k}}$ can also be made. It shows that these coefficients are the largest for the orbitals at the Fermi surface and they vanish rapidly as ϵ_k increases. Such a concentration of the probabilities on orbitals near the Fermi surface is also an essential feature of the BCS approximation and all the other which are described below.

2.2 *Nuclear physics perspective: Two nucleons in a shell*

In most nuclei, the number of neutrons N is larger than the proton number Z and the two Fermi surfaces occur for shells with different quantum numbers. Then, from overlap wave-function considerations, pairing between identical particles is favored over neutron-proton pairing which is only expected to play a role in $N = Z$ nuclei. In these notes, we will thus only consider pairs made of identical particles.

Nuclei are almost spherical objects. Therefore a natural description of individual neutron or proton states involves spherical shells which are labeled by their orbital quantum numbers l, their spin j and a radial quantum number n. The degeneracy of a spherical shell is $2j+1$. For two identical nucleons (protons or neutrons) in a such a shell, the number of configurations compatible with the Pauli principle is $j(2j+1)$. Let us consider a zero range attractive interaction $-V\,\delta(\vec{r}_1 - \vec{r}_2)$ between these two particle [15]. Since it only acts in states of relative angular momentum $l_{\mathrm{r}} = 0$, the total spin of the pair must be $S = 0$ for the same reason as above. Then the total angular and orbital momentums are equal ($J = L$). The paired states can be organized in multiplets associated with even values of J ($J = 0, 2, 4, \ldots, J_M = (2j-1)$). A lengthy, not particularly instructive in itself, and straightforward manipulation of the angular momentum algebra [16] provides the energy of the $2J+1$ degenerate paired states associated with the total angular momentum J

$$E_J = -G\,\frac{(2j+1)^2}{2}\begin{pmatrix} j & j & J \\ 0 & 0 & 0 \end{pmatrix}^2 , \quad G \equiv V \int_0^\infty \mathrm{d}r\, u_{nlj}(r)^4. \qquad (2.2.1)$$

In this expression, there appears a Racah $3j$ symbol and an integral involving the fourth power of the radial wave function $u_{nlj}(r)$ of orbitals in the shell.

For heavy nuclei, we can consider the limit $j \gg 1/2$. Then, by approximating the $3j$ symbol by its asymptotic expression one obtains

$$E_0 = -G\left(j + \frac{1}{2}\right), \quad E_2 \simeq -\frac{1}{4}G\,j\,, \quad E_J \simeq -G\frac{\sqrt{J_M^2 - J^2}}{2\pi J}. \qquad (2.2.2)$$

The non degenerate $J = 0$ pair is thus approximately four times more bound than any other state. Moreover, the energy of the states vanishes as their angular momentum J grows. This analysis indicates that the privileged pair structure taking full advantage of the short range attractive nature of the effective nucleon nucleon interaction corresponds to the wave-function

$$\Psi_{1,2} = u_{nlj}(r_1)u_{nlj}(r_2)\,\Xi(S=0) \sum_{m=-l}^{l} Y^*_{l,m}(\vartheta_1,\varphi_1)\,Y_{l,-m}(\vartheta_2,\varphi_2), \quad (2.2.3)$$

where (r_i, θ_i, ϕ_i) denotes the spherical coordinates of the particle i and $Y_{l,m}$ are the spherical harmonics. This two-body wave function corresponds to a fully symmetric combination of pairs involving $\{(m,\sigma),(-m,-\sigma)\}$ pairs just as superconductivity in condensed matter involves the quantum numbers $\vec{k},\sigma$ and $-\vec{k},-\sigma$.

2.3 Condensation of Cooper's pairs

From now on, we shall work with the formalism of the second quantification and consider states as vectors in the Fock space. This space is built from a vacuum of particles $|0\rangle$ by means of the repeated action of a set of particle creation operators. These operators denoted $a^\dagger_p$ and $a^\dagger_{\bar{p}}$ are associated with some reference single particle basis. In practice, it is convenient to select the basis which diagonalizes the one-body operator corresponding to our reference Hamiltonian

$$\hat{H}_0 = \sum_p \epsilon_p \left(a^\dagger_p a_p + a^\dagger_{\bar{p}} a_{\bar{p}}\right) . \quad (2.3.1)$$

In equation (2.3.1), for the sake of simplicity, the energies of both states $a^\dagger_p|0\rangle$ and $a^\dagger_{\bar{p}}|0\rangle$ are taken equal to ϵ_p. The correspondance with Sections 2.1 and 2.2 is thus

$$\{p,\bar{p}\} \leftrightarrow \left\{(\vec{k},\sigma),\,(-\vec{k},-\sigma)\right\} \leftrightarrow \{(n,l,m,\sigma),\,(n,l,-m,-\sigma)\}\cdot \quad (2.3.2)$$

The degeneracy assumed in equation (2.3.1) holds in condensed matter (ϵ_k only depends on $|\vec{k}|$) and in spherical nuclei. As a matter of fact, the levels in the mean-field of deformed (axial or triaxial) nuclei calculated for instance with the Hartree-Fock method also exhibits a twofold Kramers degeneracy. In experiments performed on small aluminium grains, $\hat{H}_0$ does not correspond to the plane wave Hamiltonian (2.1.1). The shape of the well in which electrons move is randomly determined by the process of congealation of the grain. Still each state should be spin degenerate. The degeneracy is only lifted by time-odd external operators such as a magnetic or a rotation field.

In such cases, a description of pairing requires a formalism which is only slightly more complicated than what is described below.

From the above discussion, we are led to select a two body state built from elementary pairs $\{a_p^\dagger a_{\bar{p}}^\dagger\}$ and to define

$$|\Psi_{1,2}\rangle = \hat{S}^\dagger|0\rangle, \tag{2.3.3}$$

with

$$\hat{S}^\dagger = \sum_p \frac{v_p}{u_p} a_p^\dagger a_{\bar{p}}^\dagger. \tag{2.3.4}$$

Our notation for the coefficients entering the definition of $\hat{S}^\dagger$ anticipates that of the BCS formalism and introduces coefficients u_p and v_p which satisfy the relation

$$u_p^2 + v_p^2 = 1. \tag{2.3.5}$$

Based on the intuition that a fermion pair should in first approximation behave as a single boson, people have tried to approach the exact N_0-body wave-function by a condensate of such pairs, that is $|\Psi_{N_0}\rangle \propto (\hat{S}^\dagger)^{N_0/2}|0\rangle$. In practice, because calculations with $|\Psi_{N_0}\rangle$ are difficult to handle, Bardeen, Cooper and Schrieffer introduced an approximation which retains the underlying physics and, at the same time, allows the use of the well established and efficient set of mean-field methods [17]. The BCS wave-function that they proposed in order to describe a superconducting system at $T = 0$, is

$$|\Psi_{[\mathrm{BCS}]}\rangle \propto \exp(\hat{S}^\dagger)|0\rangle \propto \prod_p \left(u_p + v_p\, a_p^\dagger a_{\bar{p}}^\dagger\right)|0\rangle\cdot \tag{2.3.6}$$

In contrast with $(\hat{S}^\dagger)^{N_0/2}|0\rangle$, the BCS approximation violates the particle number symmetry and $|\Psi_{[\mathrm{BCS}]}\rangle$ does not have a well defined particle number N_0. One must therefore constrain the coefficients u_p and v_p to satisfy the equation

$$\left\langle \Psi_{[\mathrm{BCS}]}|\hat{N}|\Psi_{[\mathrm{BCS}]}\right\rangle = N_0, \tag{2.3.7}$$

where $\hat{N}$ is the particle number operator

$$\hat{N} = \sum_p \left(a_p^\dagger a_p + a_{\bar{p}}^\dagger a_{\bar{p}}\right). \tag{2.3.8}$$

3 Mean-field approach at finite temperature

Much of the information on small superconducting systems has been obtained at very small temperatures well below the critical temperature. Strictly speaking, on the earth (as opposed to stars), nuclei are observed at zero temperature. The description of their excitation properties by means of a formalism involving the concept of temperature, is always an approximation whose validity depends very much on the experimental conditions. Nevertheless, temperature is a useful tool whenever experiments only access properties averaged over a fraction or all of the excitation spectrum. Moreover, understanding temperature effects in superconducting finite systems has a relevance in connection with Anderson's question and is certainly useful for condensed matter physics. Therefore, I have chosen to work within a finite temperature formalism. This section begins with a discussion of the BCS theory and introduces notations and techniques employed in the methods which are proposed later to improve the description of pairing in finite systems.

3.1 *Family of basic operators*

At finite temperature, physical systems are described by their density operator $\hat{D}$ which contains all the information on their dynamics and accounts for thermal statistics averaging effects. The BCS theory seeks to determine an approximation of $\hat{D}$ which gives a good prediction of the most important properties associated with superconductivity.

3.1.1 Duplicated representation

Because BCS breaks the particle number symmetry, it is convenient to consider a set of $\Omega/2$ subspaces of dimension four, which take into account both the creation and the destruction of particles associated with pairs $(p, \bar{p})$. Thus, we introduce the following column and line vectors

$$\alpha^{(p)} = \begin{pmatrix} a_p \\ a_{\bar{p}} \\ a_p^\dagger \\ a_{\bar{p}}^\dagger \end{pmatrix}, \qquad \alpha^{(p)\dagger} = \begin{pmatrix} a_p^\dagger & a_{\bar{p}}^\dagger & a_p & a_{\bar{p}} \end{pmatrix}, \tag{3.1.1}$$

whose components satisfy generalized fermion anticommutation relations

$$\left\{\alpha^{(p)}{}_\lambda, \alpha^{(q)\dagger}{}_\mu\right\} = \delta_{pq}\delta_{\lambda\mu}\,, \qquad \left\{\alpha^{(p)}{}_\lambda, \alpha^{(q)}{}_\mu\right\} = \delta_{pq}\sigma_{\lambda\mu}\,, \tag{3.1.2}$$

with

$$\sigma = \begin{pmatrix} 0 & 0 & 1 & 0 \\ 0 & 0 & 0 & 1 \\ 1 & 0 & 0 & 0 \\ 0 & 1 & 0 & 0 \end{pmatrix}. \tag{3.1.3}$$

The matrix σ (not to be confused with the spin magnetic component used above) satisfies $\sigma^2 = 1$ and $\sigma^T = \sigma$ where the letter T stands for matrix transposition.

3.1.2 Basic operators

Let us now consider a family of operators acting in the Fock space which we will use to construct various approximations of the density operator. The operator $\hat{\mathcal{T}}^{[d]}$ which is the representative of this family is defined in terms of a c-number ${l_0}^{[d]}$ and a set of p 4×4 matrices ${\mathcal{L}_p}^{[d]}$ as

$$\hat{\mathcal{T}}^{[d]} = \mathrm{e}^{-{l_0}^{[d]} - \frac{1}{2}\sum_p \alpha^{(p)\dagger} {\mathcal{L}_p}^{[d]} \alpha^{(p)}} . \tag{3.1.4}$$

The anticommutation properties (3.1.2) impose that ${\mathcal{L}_p}^{[d]}$ satisfies $\sigma {\mathcal{L}_p}^{[d]} \sigma = -{\mathcal{L}_p}^{[d]T}$. In addition, one often requests that $\hat{\mathcal{T}}^{[d]}$ and therefore ${\mathcal{L}_p}^{[d]}$ be hermitian. These conditions restrict the number of degrees of degrees of freedom in ${\mathcal{L}_p}^{[d]}$ to six real parameters.

For the simplest form of the BCS theory, one can work with a parametrization which relies on only two real parameters $({l_p}^{[d]}, \vartheta_p^{[d]})$ and take ${\mathcal{L}_p}^{[d]}$ equal to

$${\mathcal{L}_p}^{[d]} = {l_p}^{[d]} \begin{pmatrix} {C_p}^{[d]} & 0 & 0 & {S_p}^{[d]} \\ 0 & {C_p}^{[d]} & -{S_p}^{[d]} & 0 \\ 0 & -{S_p}^{[d]} & -{C_p}^{[d]} & 0 \\ {S_p}^{[d]} & 0 & 0 & -{C_p}^{[d]} \end{pmatrix} , \tag{3.1.5}$$

where the coefficients ${C_p}^{[d]}$ and ${S_p}^{[d]}$ are related to $\vartheta_p^{[d]}$ by

$${C_p}^{[d]} = \cos(\vartheta_p^{[d]}) , \quad {S_p}^{[d]} = \sin(\vartheta_p^{[d]}) . \tag{3.1.6}$$

Below, we will also need a simple extension of this parametrization involving another angle φ_p with the replacement

$${S_p}^{[d]} \quad \rightarrow \quad {S_p}^{[d]} \exp(i\varphi_p^{[d]}) \tag{3.1.7}$$

in the upper right 2×2 matrix in equation (3.1.5). The coefficients in the lower left submatrix are then adjusted accordingly when the hermiticity of ${\mathcal{L}_p}^{[d]}$ must be preserved.

The other degrees of freedom, which we will not consider in this text, are:

- when the p, $\bar{p}$ degeneracy is lifted (a situation evoked in Sect. 2.3) it is necessary to introduce two distinct coefficients ${l_p}^{[d]}$ and $l_{\bar{p}}^{[d]}$;
- the last two degrees of freedom describe the mixing within the $p, \bar{p}$ subspaces. They do not bring any new physical information at the level of this simple version of the BCS approximation (they correspond to a rotation in the two dimensional degenerate subspace and therefore a redefinition of states p and $\bar{p}$). On the other hand, they become interesting when extensions of BCS are considered (*i.e.* Hartree–Fock–Bogoliubov) [18].

The form (3.1.5) of ${\mathcal{L}_p}^{[d]}$ introduces a decoupling which allows one to only work with 2×2 matrices:

$$-{l_0}^{[d]} - \frac{1}{2}\sum_p \alpha^{(p)\dagger}{\mathcal{L}_p}^{[d]}\alpha^{(p)} = -{l_0}^{[d]} - \sum_p {l_p}^{[d]} \begin{pmatrix} a_p^\dagger & a_{\bar{p}} \end{pmatrix} {\mathcal{U}_p}^{[d]} \begin{pmatrix} a_p \\ a_{\bar{p}}^\dagger \end{pmatrix}, \tag{3.1.8}$$

with

$${\mathcal{U}_p}^{[d]} = \begin{pmatrix} {C_p}^{[d]} & {S_p}^{[d]} \\ {S_p}^{[d]} & -{C_p}^{[d]} \end{pmatrix}. \tag{3.1.9}$$

Note that the matrix ${\mathcal{U}_p}^{[d]}$ is unitary: ${\mathcal{U}_p}^{[d]}{{\mathcal{U}_p}^{[d]}}^\dagger = 1$.

3.1.3 BCS coefficients; quasi-particles

The wave-function (2.3.6) is written in terms of standard BCS coefficients which are connected to ${C_p}^{[d]}$ and ${S_p}^{[d]}$ by the relations

$${C_p}^{[d]} \equiv {u_p^{[d]}}^2 - {v_p^{[d]}}^2, \quad {S_p}^{[d]} \equiv 2u_p^{[d]}v_p^{[d]}, \tag{3.1.10}$$

so that equation (2.3.5) is verified. From u_p and v_p, we define a pair of quasiparticle creation operators ${b_p^{[d]}}^\dagger$ and ${b_{\bar{p}}^{[d]}}^\dagger$

$$\begin{aligned} b_p^{[d]} &= u_p^{[d]}a_p + v_p^{[d]}a_{\bar{p}}^\dagger \\ b_{\bar{p}}^{[d]} &= u_p^{[d]}a_{\bar{p}} - v_p^{[d]}a_p^\dagger \end{aligned}, \qquad \begin{aligned} a_p &= u_p^{[d]}b_p^{[d]} - v_p^{[d]}{b_{\bar{p}}^{[d]}}^\dagger \\ a_{\bar{p}} &= u_p^{[d]}b_{\bar{p}}^{[d]} + v_p^{[d]}{b_p^{[d]}}^\dagger \end{aligned}, \tag{3.1.11}$$

which satisfy fermion anticommutation properties.

With the choice (3.1.11), the exponent of $\hat{T}^{[d]}$ takes a diagonal form

$$\hat{T}^{[d]} = \mathrm{e}^{[-{l_0}^{[d]} + \sum_p {l_p}^{[d]}]}\,, \mathrm{e}^{-\sum_p {l_p}^{[d]}({b_p^{[d]}}^\dagger b_p^{[d]} + {b_{\bar{p}}^{[d]}}^\dagger b_{\bar{p}}^{[d]})}. \tag{3.1.12}$$

3.2 Wick theorem

One of the interesting features of the operators $\hat{T}^{[d]}$ is that calculations of expectation values of one, two, many body observables can be effected by means of the Wick theorem. The form (3.1.4) is one of the simplest having this property. Other operators allowing the usage of the Wick theorem are exponentials of more general quadratic forms of creation and destruction operators. A possible starting point to demonstrate the Wick theorem is the relation

$$\begin{pmatrix} a_p \\ a_{\bar{p}}^\dagger \end{pmatrix} \hat{T}^{[d]} = \mathcal{T}_p{}^{[d]} \hat{T}^{[d]} \begin{pmatrix} a_p \\ a_{\bar{p}}^\dagger \end{pmatrix}, \tag{3.2.1}$$

where the 2×2 matrix $\mathcal{T}_p{}^{[d]}$ is defined as

$$\mathcal{T}_p{}^{[d]} = \exp\left(-l_p{}^{[d]} \mathcal{U}_p{}^{[d]}\right) = \cosh l_p{}^{[d]} - \sinh l_p{}^{[d]} \mathcal{U}_p{}^{[d]}. \tag{3.2.2}$$

Hint: to prove equation (3.2.1), consider $\hat{T}^{[d]-1} a_p \hat{T}^{[d]}$ and $\hat{T}^{[d]-1} a_{\bar{p}}^\dagger \hat{T}^{[d]}$ as functions of $l_p{}^{[d]}$, derive a differential equation and solve it.

Any thermodynamic property can be obtained from the partition function Tr $\hat{D}$ where the notation Tr indicates a trace taken over the Fock space. The evaluation of the quantity Tr $\hat{T}^{[d]}$ is thus a prerequisite to any thermal analysis. Using the quasi-particle representation (3.1.12) one finds

$$\mathrm{Tr}\ \hat{T}^{[d]} = \exp(-l_0{}^{[d]}) \prod_p \det\left(1 + \mathcal{T}_p{}^{[d]}\right). \tag{3.2.3}$$

The Wick theorem states that the expectation values of an observable $\hat{Q}$

$$\langle \hat{Q} \rangle^{[d]} \equiv \mathrm{Tr}\ \left(\hat{T}^{[d]} \hat{Q}\right) / \mathrm{Tr}\ \hat{T}^{[d]}, \tag{3.2.4}$$

can be expressed as a polynomial of the matrix elements of the 2×2 matrices

$$\mathcal{R}_p{}^{[d]} \equiv \left\langle \begin{pmatrix} a_p^\dagger a_p & a_{\bar{p}} a_p \\ a_p^\dagger a_{\bar{p}}^\dagger & a_{\bar{p}} a_{\bar{p}}^\dagger \end{pmatrix} \right\rangle^{[d]} = \frac{1}{\mathcal{T}_p{}^{[d]^{-1}} + 1}. \tag{3.2.5}$$

Note that the last equality in (3.2.5) generalizes the usual form of the occupation factor in Fermi statistics. Hint: use equation (3.2.1).

For instance, for the particle number operator (2.3.8), one obtains

$$\langle \hat{N} \rangle^{[d]} = \sum_p \left[\langle a_p^\dagger a_p \rangle^{[d]} + \left(1 - \langle a_{\bar{p}} a_{\bar{p}}^\dagger \rangle^{[d]}\right) \right] \equiv \mathcal{N}\left\{\mathcal{R}^{[d]}\right\}. \tag{3.2.6}$$

For the energy of the superconducting system, we consider the simplest possible type of pairing Hamiltonian

$$\hat{H} = \hat{H}_0 - \sum_{pq} G_{pq} a_p^\dagger a_{\bar{p}}^\dagger a_{\bar{q}} a_q \ , \tag{3.2.7}$$

where $\hat{H}_0$ is defined in equation (2.3.1). The matrix elements of the pairing interaction satisfy $G_{pq} = G_{qp}$ and (only justified by my desire to shorten the formulas in these lectures) $G_{pp} = 0$. Such an interaction scatters the pair from any fully occupied level q to any empty p level.

In fact, instead of $\hat{H}$, we will rather work with the shifted operator

$$\hat{K} = \hat{H} - \mu\hat{N}, \tag{3.2.8}$$

which allows one to treat with similar notations the canonical ($\mu = 0$) or grand canonical ($\mu \neq 0$) situations. By means of the Wick theorem, we readily obtain

$$\begin{aligned}
\langle \hat{K} \rangle^{[d]} &= \sum_p (\epsilon_p - \mu) \left[\langle a_p^\dagger a_p \rangle^{[d]} + 1 - \langle a_{\bar{p}} a_{\bar{p}}^\dagger \rangle^{[d]} \right] \\
&\quad - \sum_{pq} G_{pq} \langle a_p^\dagger a_{\bar{p}}^\dagger \rangle^{[d]} \langle a_{\bar{q}} a_q \rangle^{[d]} \\
&\equiv \mathcal{E}\{\mathcal{R}^{[d]}\}.
\end{aligned} \tag{3.2.9}$$

Below, we will have to consider first order variations of quantities such as $\mathcal{E}\{\mathcal{R}^{[x]}\}$ with respect to arbitrary variations of the matrices $\delta\mathcal{R}_p{}^{[x]}$. From such a variation, the matrices

$$\mathcal{H}\{\mathcal{R}_p^{[d]}\} = \begin{pmatrix} \epsilon_p - \mu & \Delta_p^{[d]} \\ \tilde{\Delta}_p^{[d]} & -\epsilon_p + \mu \end{pmatrix}, \tag{3.2.10}$$

naturally appear. In equation (3.2.10) the quantity $\Delta_p^{[d]}$ is the *pairing gap* associated with $\hat{T}^{[d]}$ and $\hat{H}$. The quantities $\Delta_p^{[d]}$ and $\tilde{\Delta}_p^{[d]}$ are related to $\mathcal{R}_p{}^{[d]}$ by

$$\Delta_p^{[d]} = -\sum_p G_{pq} \langle a_{\bar{q}} a_q \rangle^{[d]}, \quad \tilde{\Delta}_p^{[d]} = -\sum_p G_{pq} \langle a_q^\dagger a_{\bar{q}}^\dagger \rangle^{[d]}. \tag{3.2.11}$$

When $\hat{T}^{[d]}$ is hermitian $\tilde{\Delta}_p^{[d]}$ and $\Delta_p^{[d]*}$ are equal. This is the case when $\mathcal{L}_p{}^{[d]}$ is given by (3.1.5) (in addition $\Delta_p^{[d]}$ is real).

With these definitions, the variation of the energy (3.2.9) takes a compact form

$$\delta\mathcal{E}\{\mathcal{R}^{[d]}\} = \sum_p \mathrm{tr}\ \mathcal{H}\{\mathcal{R}_p^{[d]}\} \delta\mathcal{R}_p{}^{[d]} \tag{3.2.12}$$

where tr stands for the trace of the 2×2 matrice.

3.3 BCS finite temperature equations

The grand canonical BCS theory relies on the simplest possible approximation for the statistical density operator:

$$\hat{\mathcal{D}}^{[\mathrm{BCS}]} = \hat{\mathcal{T}}^{[d]}, \tag{3.3.1}$$

where $\hat{\mathcal{T}}^{[d]}$ is optimal according to a standard variational principle of the thermodynamics. Within BCS, the parameters $\{l_p{}^{[d]}, \theta_p{}^{[d]}\}$ are those which minimize the approximate Gibbs free energy F_{G} calculated with $\hat{\mathcal{T}}^{[d]}$, where F_{G} is

$$F_{\mathrm{G}} \equiv \langle \hat{K} \rangle - \frac{1}{\beta} S \equiv -\frac{1}{\beta} \ln Z_{\mathrm{G}} \ , \tag{3.3.2}$$

whith the entropy defined by

$$S \equiv -\mathrm{Tr}\ (\hat{\mathcal{D}} \ln \hat{\mathcal{D}})/\mathrm{Tr}\ \hat{\mathcal{D}} + \ln \mathrm{Tr}\ (\hat{\mathcal{D}}). \tag{3.3.3}$$

(Note: Eq. (3.3.3) takes into account that $\hat{\mathcal{D}}$ is not necessarily normalized.)

3.3.1 Density operator, entropy, average particle number

From (3.2.3), one obtains the expression of the partition function

$$\mathrm{Tr}\ \hat{\mathcal{D}}^{[\mathrm{BCS}]} = \exp\left(-l_0{}^{[d]}\right) \prod_p \left(2 \cosh \frac{l_p{}^{[d]}}{2}\right)^2 . \tag{3.3.4}$$

The density $\mathcal{R}_p{}^{[d]}$ (3.2.5) is calculated explicitly as

$$\mathcal{R}_p{}^{[d]} = \frac{1}{2}\left(1 - t_p{}^{[d]} \mathcal{U}_p{}^{[d]}\right), \quad t_p{}^{[d]} = \tanh \frac{l_p{}^{[d]}}{2} . \tag{3.3.5}$$

Then, one calculates the BCS entropy $S^{[\mathrm{BCS}]}$ and the necessary expectation values $\langle \hat{Q} \rangle^{[\mathrm{BCS}]} = \langle \hat{Q} \rangle^{[d]}$. The results are

$$S^{[\mathrm{BCS}]} = \sum_p \left(2 \ln\left(2 \cosh \frac{l_p{}^{[d]}}{2}\right) - l_p{}^{[d]} t_p{}^{[d]}\right), \tag{3.3.6}$$

for the entropy,

$$\langle \hat{N} \rangle^{[\mathrm{BCS}]} = \sum_p \left(1 - t_p{}^{[d]} C_p{}^{[d]}\right), \tag{3.3.7}$$

for the particle number and

$$\langle \hat{K} \rangle^{[\mathrm{BCS}]} = \sum (\epsilon_p - \mu)\left(1 - t_p{}^{[d]} C_p{}^{[d]}\right) - \frac{1}{4} \sum_{pq} G_{pq} t_p{}^{[d]} S_p{}^{[d]} t_q{}^{[d]} S_q{}^{[d]} \ , \tag{3.3.8}$$

for the (shifted) energy.

3.3.2 BCS equations

We can now write ${F_{\rm G}}^{[{\rm BCS}]}$ from its definition (3.3.2). and express its stationarity with respect to the parameters which define $\hat{T}^{[d]}$. We note that, in the BCS formalism, the normalization factor ${l_0}^{[d]}$ does not play any role.

Variation with respect to ${\theta_p}^{[d]}$
It yields the equation

$$0 = (\epsilon_p - \mu){S_p}^{[d]} - \Delta_p^{[{\rm BCS}]}{C_p}^{[d]} , \qquad (3.3.9)$$

where the gap defined by equation (3.2.11) is

$$\Delta_p^{[{\rm BCS}]} \equiv \frac{1}{2}\sum_q G_{pq}{t_q}^{[d]}{S_q}^{[d]} . \qquad (3.3.10)$$

After introducing the quasi particle energies

$$e_p^{[{\rm BCS}]} \equiv \sqrt{(\epsilon_p - \mu)^2 + (\Delta_p^{[{\rm BCS}]})^2} , \qquad (3.3.11)$$

one obtains an implicit solution of equation (3.3.9) for the quantities ${C_p}^{[d]}$ and ${S_p}^{[d]}$

$${S_p}^{[d]} = \frac{\Delta_p^{[{\rm BCS}]}}{e_p^{[{\rm BCS}]}} , \quad {C_p}^{[d]} = \frac{\epsilon_p - \mu}{e_p^{[{\rm BCS}]}}. \qquad (3.3.12)$$

Equation (3.3.10) can now be rewritten in the standard form of the finite temperature *gap equation*

$$\Delta_p^{[{\rm BCS}]} = \frac{1}{2}\sum_q G_{pq}\frac{{\Delta_q}^{[{\rm BCS}]}}{{e_q}^{[{\rm BCS}]}}{t_q}^{[d]}. \qquad (3.3.13)$$

Variation with respect to ${l_p}^{[d]}$
It gives a second equation

$$\frac{1}{\beta}{l_p}^{[d]} = (\epsilon_p - \mu){C_p}^{[d]} + \Delta_p^{[{\rm BCS}]}{S_p}^{[d]} , \qquad (3.3.14)$$

which by means of equation (3.3.12) shows that ${l_p}^{[d]}$ is equal to $e_p^{[{\rm BCS}]}$ up to a temperature factor

$${l_p}^{[d]} = \beta e_p^{[{\rm BCS}]}. \qquad (3.3.15)$$

These results can be reformulated by means of the definition (3.2.10) and the equations (3.1.9) and (3.3.12). The matrix $\mathcal{H}\{\mathcal{R}_p^{[{\rm BCS}]}\}$ satisfies

$$\mathcal{H}\{\mathcal{R}_p^{[{\rm BCS}]}\} = e_p^{[{\rm BCS}]}{\mathcal{U}_p}^{[d]} . \qquad (3.3.16)$$

An equivalent form is obtained by using equations (3.3.15, 3.2.2) and (3.2.5). The BCS density matrix ${\mathcal{R}_p}^{[\mathrm{BCS}]}$ is solution of the self-consistent equation

$$ {\mathcal{R}_p}^{[\mathrm{BCS}]} = \frac{1}{\mathrm{e}^{\beta \mathcal{H}\{\mathcal{R}_p^{[\mathrm{BCS}]}\}} + 1}. \tag{3.3.17} $$

3.3.3 Discussion; problems for finite systems

Many textbooks present and analyze the rich physical content of the BCS formalism in the context of condensed matter physics (see for instance [1,2]). Here, I only discuss some of its limitations when it is used for finite systems. These are related to the breaking of particle number symmetry which, for instance, makes an interpretation of the quasi particle energies $e_p^{[\mathrm{BCS}]}$ somewhat ambiguous.

For a discussion of this point, consider Figure 1. On the left side, the quasi particle spectrum $e_p^{[\mathrm{BCS}]}$ is plotted (with a 90 degree tilt) *versus* the single particle energies ϵ_p. Both spectra are discrete so that only the intersections of the solid curve with the short horizontal segments have to be considered. To investigate the meaning of $e_p^{[d]}$, we first consider a situation with the pairing interaction turned off. The solid line is then replaced by the broken dotted line which is the union of the bisectors of two quadrants. On the right part of the figure we show that the upper dotted line contains information on the excitation energy of the state with $N+1$ particles *versus* the state with N particles. Such excitations, can for instance, be measured as the Q factors of a pick-up reaction transfering one particle from a projectile into the N-particle target. On the other hand, the lower part of the dotted line tells about the excitation energies of the system with $N-1$ particle (as obtained for instance by means of a stripping reaction) with respect to the N-particle system. No part of the curve has any relation with the excitation spectrum in the N-particle system (rightmost spectrum) which involves particle-hole excitations. Returning now from the dotted curve to the solid line, one sees that its physical meaning evolves continuously from top to bottom. When $\epsilon_p \gg \mu$, the curve tells about the system with $N+1$ particles. On the other hand, its lower part is more related to the $N-1$ system. Although it remains an interesting outcome of the theory to discuss properties of odd systems, the interpretation of the curve for $\epsilon \approx \mu$ is not without ambiguity. This is in contrast with infinite systems, where its shape can be tested by means of an analysis of the density of conduction states [1].

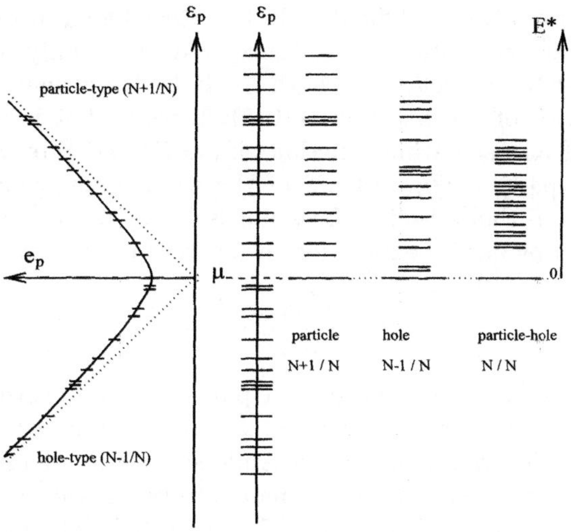

Fig. 1. Interpretation of the BCS quasiparticle energy spectrum for finite systems. From left to right are plotted the quasiparticle energy spectrum e_p (horizontal axis) *versus* the discretized single particle energy ϵ_p (horizontal segments crossing the vertical axis). The three rightmost spectra correspond to the excitation energy of the $N+1$, $N-1$ and N systems in the absence of two-body (pairing) interaction. In the three cases, the reference energy is the ground state energy of the system with N particles.

3.3.4 Discussion; size of a Cooper pair

There is another notion introduced in the discussion of an infinite superconducting medium whose importance requires some reevaluation in the context of small systems: the correlation length, sometimes also called the size of the Cooper pair. This quantity ξ is the typical dimension of a zone such that two particles in the quantum state $\Psi_{1,2}$ can be simultaneously extracted from (destroyed in) the many-body state. The quantity which describes this process is

$$K(\vec{r},\vec{r'}) = \langle \bar{\Phi}(\vec{r'})\Phi(\vec{r})\rangle^{[\mathrm{BCS}]} \ , \tag{3.3.18}$$

where $\Phi(\vec{r})$ (resp. $\bar{\Phi}(\vec{r'})$) is the field destroying one particle of p type (resp. $\bar{p}$) at point $\vec{r}$ (resp. $\vec{r'}$). The expression of these fields is

$$\Phi(\vec{r}) = \sum_p a_p \phi_p(\vec{r}) \ , \quad \bar{\Phi}(\vec{r}) = \sum_p a_{\bar{p}} \phi_{\bar{p}}(\vec{r}) \tag{3.3.19}$$

where $\phi_p(\vec{r})$ (resp. $\phi_{\bar{p}}(\vec{r})$ are the single particle functions associated with p (resp. $\bar{p}$) state. In finite systems, K depends on the two coordinates $\vec{r}$ and

$\vec{r'}$ and its interpretation is difficult. On the other hand, in infinite systems, where because of translational invariance K depends only on $|\vec{r} - \vec{r'}|$, it is easier to define the correlation length ξ. Indeed, an analysis of the long distance behavior of the function $K(0, \vec{r})$ shows that it is proportional to the product of an oscillating function $\sin(k_F|\vec{r}|)/|\vec{r}|^{3/2}$ by an exponential function $\exp(-|\vec{r}|/\xi)$. The range ξ of the exponential which governs the dominant term in the fall-off of K is precisely the correlation length. The same analysis provides its expression in terms of dynamical parameters

$$\xi = \frac{\hbar^2 k_F}{m\Delta} = \frac{\hbar v_F}{\Delta} \tag{3.3.20}$$

where $v_F = \hbar k_F/m$ is the speed of a particle at the Fermi surface. As a matter of fact, the formula (3.3.20) can also be found by means of the following heuristic argument due to Anderson. The energy variation δE associated with the destruction of a pair into two asymptotically free particles is $\Delta = \delta E = \hbar^2 k_F \delta k/m = v_F \delta|\vec{p}| = v_F \hbar/\delta|\vec{r}|$. The quantity $\delta|\vec{r}|$ can be assimilated to the size ξ of the Cooper pair $\Psi_{1,2}$. Note that ξ increases as Δ decreases.

Let us now apply the formula (3.3.20) to finite superconducting systems.

- In nuclei, the gap Δ is approximately equal to 1 MeV. The Fermi level being typically 40 to 50 MeV excited with respect to the bottom of the nuclear well, the Fermi velocity v_F is about 0.3 the light velocity. We thus find that an order of magnitude for ξ is 50 fm; it is 5 times larger than the radius of the heaviest nucleus;
- For a small spherical metallic grain, by means of equation (2.1.9), one can write the dimensionless ratio of the correlation length to the radius a of the grain as $\xi/a \simeq 2.8\,10^4\, w/(\Delta\, a^4)$ where w^{-1} and Δ are given in meV and a in nanometers. Typical values realized in experiments [10] are $\Delta = 0.2$ meV, $w\Delta = 2$ and $a = 10$ nm. In such a case, the ratio ξ/a exceeds hundred. The situation is even more extreme than in nuclear physics.

As a conclusion of this section, one can say that in small systems such as nuclei and nanoscopic grains, the notion of correlation length looses relevance. It is the physical dimension of the system which sets the microscopic scale of the superconductivity phenomenon.

3.4 *Discussion; low temperature BCS properties*

In order to prepare comparisons with the methods exposed later, let us analyze the limiting behavior of the BCS theory when $T = 1/\beta \to 0$. Since the quasi particle energies are positive by definition, the limit of the thermal factors ${t_p}^{[d]}$ is $1 - 2\exp(-\beta e_p^{[d]})$. In nuclei or very small aluminium grains, the temperature T can become smaller than any other energy scale. Then, an expansion of the gap equation (3.3.13) gives

$$\Delta_p^{[\mathrm{BCS}]} = \frac{1}{2}\sum_q G_{pq}\frac{{\Delta_q}^{[\mathrm{BCS}]}}{{e_q}^{[\mathrm{BCS}]}}\left(1 - 2\mathrm{e}^{-\beta {e_q}^{[\mathrm{BCS}]}}\right). \tag{3.4.1}$$

Similarly, the entropy (3.3.6) becomes

$$S^{[\mathrm{BCS}]} = 2\beta\sum_p e_p^{[\mathrm{BCS}]}\mathrm{e}^{-\beta e_p^{[\mathrm{BCS}]}}. \tag{3.4.2}$$

For larger systems such as microscopic metallic islands, one may instead use the continuous limit of condensed matter physics. In this limit, T is smaller than all energy scales except the single particle energy spacing $1/w$ (see Eq. (2.1.9)). In general, as in the discussion of Section 2.1 one also takes all matrix elements G_{pq} equal to the same value G over an energy range $[-\Lambda/2, \Lambda/2]$ centered at the Fermi surface. We will use the BCS gap Δ at $T = 0$ as our reference pairing energy. This gap is the solution of the equation

$$\frac{2}{G} = \int_{-\Lambda/2}^{\Lambda/2}\frac{w\,\mathrm{d}\epsilon}{\sqrt{\epsilon^2+\Delta^2}}. \tag{3.4.3}$$

The condensed matter low temperature limit corresponds to the conditions

$$\frac{1}{\Lambda} \ll \frac{1}{\Delta} \ll \beta \ll w. \tag{3.4.4}$$

In such a situation, a useful expansion formula is provided by the approximation

$$\mathrm{e}^{-\beta\sqrt{\epsilon^2+\Delta^2}} \simeq \sqrt{\frac{2\pi\Delta}{\beta}}\,\mathrm{e}^{-\beta\Delta}\left[\left(1+\frac{3}{8\beta\Delta}\right)\delta(\epsilon) + \frac{\Delta}{2\beta}\delta''(\epsilon)\right], \tag{3.4.5}$$

where $\delta(\epsilon)$ and $\delta''(\epsilon)$ are the dirac distribution and its second derivative, respectively. By means of equation (3.4.5), the equations (3.4.1) and (3.4.2) are transformed into

$$\Delta^{[\mathrm{BCS}]}(\beta) = \Delta\left[1 - \sqrt{\frac{2\pi}{\beta\Delta}}\,\mathrm{e}^{-\beta\Delta}\right] \tag{3.4.6}$$

and

$$S^{[\mathrm{BCS}]}(\beta) = 2\,w\,\Delta\sqrt{2\pi\beta\Delta}\,\mathrm{e}^{-\beta\Delta}. \tag{3.4.7}$$

These formulas as well as equations (3.4.1) and (3.4.2) display an unsatisfactory feature of the BCS theory which restricts its application for small superconducting systems: the temperature dependence is governed by the factor $e^{-\beta\Delta}$. As in the discussion on the interpretation of the quasi particle energies, one finds that the BCS low temperature expansion formulas are appropriate neither for an even system for which the argument of the exponential should be $-2\beta\Delta$, nor for an odd system for which the argument is expected to be close to $-\beta/(2w^2\Delta)$.

4 First attempt at particle number restoration

4.1 Particle number projection

Working in canonical formalism corrects the problems which in BCS are caused by the breaking of particle number symmetry. To this end one introduces the projection on the subspace with N_0 particles

$$\hat{P}_{N_0} = \frac{1}{2\pi}\int_0^{2\pi} d\varphi\, e^{i\varphi(\hat{N} - N_0)} \,. \tag{4.1.1}$$

This operator is hermitian ($\hat{P}_{N_0} = \hat{P}^{\dagger}_{N_0}$) and satisfies $\hat{P}^2_{N_0} = \hat{P}_{N_0}$. From the expression (2.3.8) of $\hat{N}$, one sees that the exponent within the integral can be written in a form similar to (3.1.8)

$$i\varphi\hat{N} = i\varphi\Omega + \sum_p i\varphi \begin{pmatrix} a_p^\dagger & a_{\bar{p}} \end{pmatrix} \begin{pmatrix} 1 & 0 \\ 0 & -1 \end{pmatrix} \begin{pmatrix} a_p \\ a_{\bar{p}}^\dagger \end{pmatrix} . \tag{4.1.2}$$

A comparison with (3.1.6, 3.1.8) and (3.2.2) leads directly to the following correspondance table

$$\begin{aligned} l_0{}^{[d]} &\to -i\varphi\Omega \\ l_p{}^{[d]} &\to -i\varphi \\ \vartheta_p^{[d]} &\to 0. \end{aligned} \tag{4.1.3}$$

If one introduces the following notation for the non hermitian operator

$$\hat{T}^{[\varphi]} = e^{i\varphi\hat{N}} \,, \tag{4.1.4}$$

which satisfies the relation $\hat{T}^{[\varphi]\dagger} = \hat{T}^{[-\varphi]}$, the projection can therefore be written

$$\hat{P}_{N_0} = \frac{1}{2\pi}\int_0^{2\pi} d\varphi\, e^{-i\varphi N_0}\, \hat{T}^{[\varphi]} \,. \tag{4.1.5}$$

The matrices $\mathcal{T}_p$ associated with $\hat{T}^{[\varphi]}$ are diagonal and p independent

$$\mathcal{T}_p{}^{[\varphi]} = \begin{pmatrix} e^{i\varphi} & 0 \\ 0 & e^{-i\varphi} \end{pmatrix} . \tag{4.1.6}$$

4.2 Projected density operator

Let us define a projected mean-field density operator by

$$\begin{aligned}\hat{\mathcal{D}}^{[N_0]} &= \hat{P}_{N_0}\hat{\mathcal{T}}^{[d]}\hat{P}^{\dagger}_{N_0}\\ &= \frac{1}{4\pi^2}\int_0^{2\pi}\int_0^{2\pi} \mathrm{d}\varphi''\,\mathrm{d}\varphi'\,\mathrm{e}^{-i(\varphi''-\varphi')N_0}\,\hat{\mathcal{T}}^{[\varphi'']}\,\hat{\mathcal{T}}^{[d]}\,\hat{\mathcal{T}}^{[-\varphi']}.\end{aligned} \tag{4.2.1}$$

One notes that both sides of $\hat{\mathcal{T}}^{[d]}$ are multiplied by the projection because $\hat{\mathcal{T}}^{[d]}$ does not commute with $\hat{N}$ as soon as one of the $\vartheta_p^{[d]}$'s is non zero.

Using the group structure of the set of $\hat{\mathcal{T}}^{[x]}$ operators, one defines

$$\hat{\mathcal{T}}^{[\varphi'',d,-\varphi']} \equiv \hat{\mathcal{T}}^{[\varphi'']}\hat{\mathcal{T}}^{[d]}\hat{\mathcal{T}}^{[-\varphi']}. \tag{4.2.2}$$

The associated $\mathcal{T}_p$ matrices are also obtained by a multiplication law

$$\mathcal{T}_p{}^{[\varphi'',d,-\varphi']} = \mathcal{T}_p{}^{[\varphi'']}\mathcal{T}_p{}^{[d]}\mathcal{T}_p{}^{[-\varphi']}. \tag{4.2.3}$$

With the notation $\varphi = \varphi'' - \varphi'$, according to equation (3.2.3), the partition function can be written as

$$\begin{aligned}\mathrm{Tr}\,\hat{\mathcal{D}}^{[N_0]} &= \frac{1}{4\pi^2}\int_0^{2\pi}\int_0^{2\pi} \mathrm{d}\varphi''\,\mathrm{d}\varphi'\,\mathrm{e}^{[-l_0{}^{[d]} + i\varphi(\Omega - N_0)]}\\ &\quad\times \textstyle\prod_p \det\left(1 + \mathcal{T}_p{}^{[\varphi'',d,-\varphi']}\right).\end{aligned} \tag{4.2.4}$$

Switching from the variables (φ'', φ') to (φ, φ'), one obtains a more compact expression

$$\mathrm{Tr}\,\hat{\mathcal{D}}^{[N_0]} = \frac{1}{2\pi}\int_0^{2\pi} \mathrm{d}\varphi\,\mathrm{e}^{[-l_0{}^{[d]} + i\varphi(\Omega - N_0)]}\prod_p D_p^{(d,\varphi)}, \tag{4.2.5}$$

with

$$D_p^{(d,\varphi)} = 2\left(1 + \cosh l_p{}^{[d]}\,\cos\varphi - i\sinh l_p{}^{[d]}\,C_p{}^{[d]}\,\sin\varphi\right). \tag{4.2.6}$$

From equation (3.2.5) we obtain the generalized density matrices

$$\mathcal{R}_p{}^{[\varphi'',d,-\varphi']} = \left\langle\begin{pmatrix} a_p^{\dagger}a_p & a_{\bar{p}}a_p \\ a_p^{\dagger}a_{\bar{p}}^{\dagger} & a_{\bar{p}}a_{\bar{p}}^{\dagger}\end{pmatrix}\right\rangle^{[\varphi'',d,-\varphi']} = \frac{1}{\mathcal{T}_p{}^{[\varphi'',d,-\varphi']^{-1}} + 1}. \tag{4.2.7}$$

whose explicit expression is

$$\mathcal{R}_p{}^{[\varphi'',d,-\varphi']} = \frac{1}{D_p^{(d,\varphi)}} \times \begin{pmatrix} 1+\left(\cosh l_p{}^{[d]} - \sinh l_p{}^{[d]}\, C_p{}^{[d]}\right) e^{i\varphi} & -\sinh l_p{}^{[d]}\, S_p{}^{[d]}\, e^{i\varphi}\, e^{2i\varphi'} \\ -\sinh l_p{}^{[d]}\, S_p{}^{[d]}\, e^{-i\varphi}\, e^{-2i\varphi'} & 1+\left(\cosh l_p{}^{[d]} + \sinh l_p{}^{[d]}\, C_p{}^{[d]}\right) e^{-i\varphi} \end{pmatrix}. \tag{4.2.8}$$

4.3 Expectation values

For any function $f(\varphi)$ of the gauge angle φ, let us introduce the more compact notation

$$\langle\langle f(\varphi)\rangle\rangle^{[d]} \equiv \frac{1}{2\pi}\int_0^{2\pi} d\varphi\, e^{[-l_0{}^{[d]} + i\varphi(\Omega - N_0)]} \prod_q D_q^{(d,\varphi)}\, f(\varphi)\,, \tag{4.3.1}$$

so that, for instance, one has $\mathrm{Tr}\ \hat{\mathcal{D}}^{[N_0]} = \langle\langle 1\rangle\rangle^{[d]}$. With this notation, the expectation value of an operator $\hat{Q}$ for the projected density operator (4.2.1) is

$$\langle\hat{Q}\rangle^{[N_0]} \equiv \frac{\mathrm{Tr}\ \hat{\mathcal{D}}^{[N_0]}\hat{Q}}{\mathrm{Tr}\ \hat{\mathcal{D}}^{[N_0]}} = \frac{1}{\langle\langle 1\rangle\rangle^{[d]}} \left\langle\left\langle \frac{1}{2\pi}\int_0^{2\pi} d\varphi'\, \langle\hat{Q}\rangle^{[\varphi+\varphi',d,-\varphi']} \right\rangle\right\rangle^{[d]}. \tag{4.3.2}$$

The quantity $\langle\hat{Q}\rangle^{[\varphi+\varphi',d,-\varphi']}$ being defined as in equation (3.2.4) can be calculated by means of the Wick theorem with the density matrix (4.2.8). In most cases, it turns out that the integral on φ' in (4.3.2) can be performed trivially.

For the calculation of the expectation value of the particle number $\langle\hat{N}\rangle^{[\varphi+\varphi',d,-\varphi']}$, we use (3.2.6) and obtain

$$N_0 = \langle\hat{N}\rangle^{[N_0]} = \frac{1}{\langle\langle 1\rangle\rangle^{[d]}} \left\langle\left\langle \sum_p 2\frac{1+(\cosh l_p{}^{[d]} - \sinh l_p{}^{[d]} C_p{}^{[d]})e^{i\varphi}}{D_p^{(d,\varphi)}} \right\rangle\right\rangle^{[d]}. \tag{4.3.3}$$

It is important to note that (4.3.3) *is neither an equation nor a constraint.* It is *an identity* valid for any set $\{l_p{}^{[d]},\, C_p{}^{[d]}\}$ as long as $\hat{T}^{[d]}$ has a component with N_0 particles ($\hat{\mathcal{D}}^{[N_0]} \neq 0$).

For the average value $\langle \hat{H} \rangle^{[\varphi+\varphi',d,-\varphi']}$ of the energy, we use (3.2.9). The projected energy is then obtained

$$\langle \hat{H} \rangle^{[N_0]} = \frac{1}{\langle\langle 1 \rangle\rangle^{[d]}} \left\langle\left\langle \sum_p 2\epsilon_p \frac{1 + (\cosh {l_p}^{[d]} - \sinh {l_p}^{[d]} {C_p}^{[d]}) \mathrm{e}^{i\varphi}}{D_p^{(d,\varphi)}} \right.\right. \\ \left.\left. - \sum_{pq} G_{pq} \frac{\sinh {l_p}^{[d]} {S_p}^{[d]} \sinh {l_q}^{[d]} {S_q}^{[d]}}{D_p^{(d,\varphi)} D_q^{(d,\varphi)}} \right\rangle\right\rangle^{[d]} . \tag{4.3.4}$$

4.4 *Projected BCS at $T = 0$, expectation values*

It would seem that we are in position to obtain all the contributions to the Gibbs free energy (3.3.2). This is not the case. The direct approach based on the choice (4.2.1) for the projected density operator turns out to be a dead end. Indeed, one does not know how to evaluate the entropy (3.3.3) because it involves the calculation of $\langle \ln(\hat{P}_{N_0} \hat{T}^{[d]} \hat{P}^\dagger_{N_0}) \rangle^{[N_0]}$. Thus, the solution of the finite temperature projected BCS problem will have to rely on other methods such as those which are described below. Nevertheless, the results obtained in Sections 4.2, 4.3 are sufficient to treat the problem at $T = 0$. Indeed, the determination of the ground state only requires a minimization of the projected energy (4.3.4).

We still have to transform the results of the preceding section and obtain their limit at zero temperature. According to equation (3.3.15) this is obtained by letting all the quantities ${l_p}^{[d]}$ go to $+\infty$ in proportion to the variable $\beta = 1/T$. Keeping the dominant terms in all the expressions, we find that it is convenient to define the quantity

$$D_{0\,p}^{(d,\varphi)} \equiv \cos\varphi - i{C_p}^{[d]} \sin\varphi, \tag{4.4.1}$$

and for any function $f(\varphi)$ to introduce the notation

$$\langle\langle f(\varphi) \rangle\rangle_0^{[d]} \equiv \frac{1}{2\pi} \int_0^{2\pi} \mathrm{d}\varphi \; \mathrm{e}^{i\varphi(\Omega - N_0)} \prod_q D_{0\,q}^{(d,\varphi)} f(\varphi). \tag{4.4.2}$$

Then, the partition function becomes

$$\mathrm{Tr}\; \hat{\mathcal{D}}^{[N_0]} \propto \mathrm{e}^{[-{l_0}^{[d]} + \sum_p {l_p}^{[d]}]} \langle\langle 1 \rangle\rangle_0^{[d]} . \tag{4.4.3}$$

The diverging prefactor $\exp[-{l_0}^{[d]} + \sum_p {l_p}^{[d]}]$ is an integral part of the limiting process. However, because it appears in the numerator and in the denominator of all the calculated expressions, it disappears from the physical quantities.

Then, the limit of (4.2.8) gives the zero temperature density matrices

$$\mathcal{R}_p{}^{[\varphi'',d,-\varphi']} = \frac{1}{2D^{(d,\varphi)}_{0\,p}} \begin{pmatrix} (1 - C_p{}^{[d]})\mathrm{e}^{i\varphi} & -S_p{}^{[d]}\,\mathrm{e}^{i\varphi}\,\mathrm{e}^{2i\varphi'} \\ -S_p{}^{[d]}\,\mathrm{e}^{-i\varphi}\,\mathrm{e}^{-2i\varphi'} & (1 + C_p{}^{[d]})\mathrm{e}^{-i\varphi} \end{pmatrix}, \tag{4.4.4}$$

from which one derives the expression of the expectation value of $\hat{N}$ using (3.2.6)

$$N_0 = \sum_p (1 - C_p{}^{[d]}) \frac{1}{\langle\langle 1 \rangle\rangle_0^{[d]}} \left\langle\left\langle \frac{\mathrm{e}^{i\varphi}}{D^{(d,\varphi)}_{0\,p}} \right\rangle\right\rangle_0^{[d]} . \tag{4.4.5}$$

As (4.3.3), this is not an equation but an identity valid for any set of coefficients $\{C_p{}^{[d]}\}$.

For the projected energy we use (3.2.9) and obtain

$$\begin{aligned} \langle \hat{H} \rangle^{[N_0]} &= \sum_p \epsilon_p (1 - C_p{}^{[d]}) \frac{1}{\langle\langle 1 \rangle\rangle_0^{[d]}} \left\langle\left\langle \frac{\mathrm{e}^{i\varphi}}{D^{(d,\varphi)}_{0\,p}} \right\rangle\right\rangle_0^{[d]} \\ &\quad - \frac{1}{4} \sum_{pq} G_{pq} S_p{}^{[d]} S_q{}^{[d]} \frac{1}{\langle\langle 1 \rangle\rangle_0^{[d]}} \left\langle\left\langle \frac{1}{D^{(d,\varphi)}_{0\,p}\, D^{(d,\varphi)}_{0\,q}} \right\rangle\right\rangle_0^{[d]} . \end{aligned} \tag{4.4.6}$$

4.5 *Projected BCS at $T = 0$, equations*

The variation of (4.4.5) with respect to $\vartheta_p^{[d]}$ provides the following set of Ω *equalities*

$$\begin{aligned} &N_0 \frac{\partial}{\partial \vartheta_p^{[d]}} \ln \langle\langle 1 \rangle\rangle_0^{[d]} = \\ &S_p{}^{[d]} \frac{1}{\langle\langle 1 \rangle\rangle_0^{[d]}} \left[\left\langle\left\langle \frac{\mathrm{e}^{i\varphi}}{D^{(d,\varphi)}_{0\,p}} \right\rangle\right\rangle_0^{[d]} + \sum_{q \neq p} (1 - C_q{}^{[d]}) \left\langle\left\langle \frac{i \sin\varphi\, \mathrm{e}^{i\varphi}}{D^{(d,\varphi)}_{0\,p} D^{(d,\varphi)}_{0\,q}} \right\rangle\right\rangle_0^{[d]} \right] . \end{aligned} \tag{4.5.1}$$

On the other hand, the minimization of the energy (4.4.6) gives a set of Ω *equations*

$$
\begin{aligned}
0 = &-\langle \hat{H} \rangle^{[N_0]} \frac{\partial}{\partial \vartheta_p^{[d]}} \ln \langle\langle 1 \rangle\rangle_0^{[d]} \\
&- \frac{C_p{}^{[d]}}{\langle\langle 1 \rangle\rangle_0^{[d]}} \frac{1}{2} \sum_q G_{pq}\, S_q{}^{[d]} \left\langle\left\langle \frac{1}{D_{0\,p}^{(d,\varphi)}\, D_{0\,q}^{(d,\varphi)}} \right\rangle\right\rangle_0^{[d]} \\
&+ \frac{S_p{}^{[d]}}{\langle\langle 1 \rangle\rangle_0^{[d]}} \left\langle\left\langle \frac{1}{D_{0\,p}^{(d,\varphi)}} \left(\epsilon_p e^{i\varphi} + i \sin\varphi \left[\sum_{q \neq p} \frac{\epsilon_q (1 - C_q{}^{[d]}) e^{i\varphi}}{D_{0\,q}^{(d,\varphi)}} \right.\right.\right.\right. \\
&\left.\left.\left.\left. - \frac{1}{4} \sum_{q,r \neq p} G_{qr} \frac{S_q{}^{[d]}\, S_r^{[d]}}{D_{0\,q}^{(d,\varphi)}\, D_{0\,r}^{(d,\varphi)}} \right] \right) \right\rangle\right\rangle_0^{[d]} .
\end{aligned}
\tag{4.5.2}
$$

4.6 Projected BCS at $T = 0$, generalized gaps and single particle shifts

The elimination of $\partial/\partial\vartheta_p^{[d]}(\ln \langle\langle 1 \rangle\rangle_0^{[d]})$ in equation (4.5.2) yields the variational equations which have to be solved in terms of $C_p{}^{[d]}$ and $S_p{}^{[d]}$:

$$
0 = S_p{}^{[d]} (\epsilon_p - \mu_p^{[N_0]}) - C_p{}^{[d]} \Delta_p^{[N_0]} . \tag{4.6.1}
$$

These equations are formaly similar to the BCS ones (3.3.9). However, they contain a state dependent energy shift $\mu_p^{[N_0]}$

$$
\begin{aligned}
\mu_p^{[N_0]} = &\frac{\langle \hat{H} \rangle^{[N_0]}}{N_0} - \left\langle\left\langle \frac{e^{i\varphi}}{D_{0\,p}^{(d,\varphi)}} \right\rangle\right\rangle^{[d]\,-1} \\
&\left\langle\left\langle \sum_{q \neq p} \left(\epsilon_q - \frac{\langle \hat{H} \rangle^{[N_0]}}{N_0} \right) \left(1 - C_q{}^{[d]} \right) \frac{i \sin\varphi\, e^{i\varphi}}{D_{0\,p}^{(d,\varphi)} D_{0\,q}^{(d,\varphi)}} \right.\right. \\
&\left.\left. - \frac{1}{4} \sum_{q,r \neq p} G_{qr}\, S_q{}^{[d]}\, S_r^{[d]} \frac{1}{D_{0\,p}^{(d,\varphi)} D_{0\,q}^{(d,\varphi)}\, D_{0\,r}^{(d,\varphi)}} \right\rangle\right\rangle_0^{[d]} .
\end{aligned}
\tag{4.6.2}
$$

One should note that $\mu_p^{[N_0]}$ is not a Lagrange parameter as μ is in the standard BCS formalism. Indeed, there is no constraint on the average particle number to fix the $\mu_p^{[N_0]}$'s since the $T=0$ limit of the density operator $\hat{\mathcal{D}}^{[N_0]}$ corresponds to a state with exactly N_0 particles. The quantities $\mu_p^{[N_0]}$ are *self consistently calculated* from the ${C_p}^{[d]}$ and ${S_p}^{[d]}$.

The gap is also state dependent and equal to

$$\Delta_p^{[N_0]} = \frac{1}{2}\sum_q G_{pq}\, {S_q}^{[d]} \left\langle\!\left\langle \frac{e^{i\varphi}}{D_{0\,p}^{(d,\varphi)}} \right\rangle\!\right\rangle^{[d]^{-1}} \left\langle\!\left\langle \frac{1}{D_{0\,p}^{(d,\varphi)}\, D_{0\,q}^{(d,\varphi)}} \right\rangle\!\right\rangle^{[d]} . \tag{4.6.3}$$

The projected BCS solution is therefore formally identical to BCS (3.3.12)

$${S_p}^{[d]} = \frac{\Delta_p^{[N_0]}}{e_p^{[N_0]}} , \quad {C_p}^{[d]} = \frac{\epsilon_p - \mu_p^{[N_0]}}{e_p^{[N_0]}} \tag{4.6.4}$$

once quantities $e_p^{[N_0]}$ similar to BCS quasiparticle energies have been introduced

$$e_p^{[N_0]} \equiv \sqrt{\left(\epsilon_p - \mu_p^{[N_0]}\right)^2 + \left(\Delta_p^{[N_0]}\right)^2} . \tag{4.6.5}$$

One can also write a generalized $T=0$ gap equation

$$\Delta_p^{[N_0]} = \frac{1}{2}\sum_q G_{pq} \frac{{\Delta_q}^{[N_0]}}{{e_q}^{[N_0]}} \left\langle\!\left\langle \frac{e^{i\varphi}}{D_{0\,p}^{(d,\varphi)}} \right\rangle\!\right\rangle^{[d]^{-1}} \left\langle\!\left\langle \frac{1}{D_{0\,q}^{(d,\varphi)}\, D_{0\,r}^{[d,\varphi]}} \right\rangle\!\right\rangle^{[d]} . \tag{4.6.6}$$

Finally, one notes that the average value of the particle number calculated with the unprojected density operator $\hat{T}^{[d]}$ (*i.e.* $\langle\hat{N}\rangle^{[d]} = \sum_p(1-{C_p}^{[d]})$) is not equal to N_0. The equality is that given by equation (4.4.5).

The method described in this section is only one among the several which have been devised to study projected BCS (or Hartree–Fock–Bogoliubov) equations at $T=0$ [18,19,21]. Other methods rely on the Thouless theorem and the steepest descent technique. On the other hand, in most calculations, additional constraints on $\langle\hat{N}\rangle^{[d]}$ are introduced in order to stabilize the evolution of the numerics.

In some cases (strong pairing), the integrals over φ can be approximated with good accuracy by a gaussian approximation about $\varphi=0$. In other limits (weak pairing), it may be sufficient to replace the integrals by sums over a limited number (4 to 8) of angles φ.

5 Stationary variational principle for thermodynamics

5.1 General method for constructing stationary principles

In Section 4.4, we have seen that, in the canonical formalism, the straightforward minimization of the Gibbs Free Energy is facing difficulties linked to our unability to evaluate the entropy (3.3.3) for density operators such as (4.2.1). In this section, we describe an alternative variational principle which does not require the direct calculation of the entropy. It involves the search for a saddle point rather than a minimum solution.

We first sketch a general procedure for constructing variational principles which has been analyzed by Balian and Vénéroni over a series of articles [22, 23]. It concerns the following class of problems: given a quantity of interest $f(x^i)$, function of some variables $\{x^i\}$, one wants to determine the value of f for the solutions $\{x^i_0\}$ of a set of equations $\{g_j(x^i) = 0\}$. In many interesting cases, solving exactly this problem is impossible because, either one cannot solve the equations $\{g_j(x^i) = 0\}$, one cannot calculate the value of f for the solution x^i_0, or both reasons.

In the method discussed by Balian and Vénéroni, an additional variable ϖ^j is associated with each equation $g_j(x^i)$. The action $\mathcal{J}$ is then defined as

$$\mathcal{J}\{x^i, \varpi^j\} \equiv f(x^i) - \sum_j \varpi^j\, g_j(x^i). \tag{5.1.1}$$

It is a function of both sets of variables $\{x^i\}$ and $\{\varpi^j\}$. Formally, the latter play the rôle of Lagrange parameters. When a full variation of $\mathcal{J}$ with respect to x^i and ϖ^j is done, the equations $g_j = 0$ are recovered. Then, since the stationary values of the variables $x^i_{\rm s}$ and $\varpi^j_{\rm s}$ are such that $x^i_{\rm s} = x^i_0$, one has $\mathcal{J}\{x^i_{\rm s}, \varpi^j_{\rm s}\} = f(x^i_0)$.

In general, a full variation is not possible. On the other hand, it can sometimes be performed for restricted trial spaces of the x^i's and the ϖ^j's. Then, depending on the adequacy of these trial spaces, the approximate stationary solution may be sufficiently close to the exact saddle point of the functional $\mathcal{J}$ to yield a good approximation of the quantity $f(x^i_0)$. Figure 2 illustrates the spirit of the method.

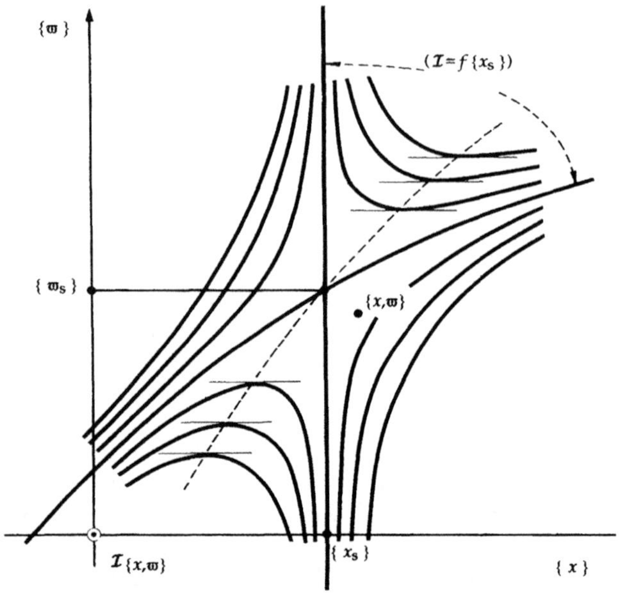

Fig. 2. Contour lines of the functional (5.1.1). The axes w and ϖ represent the spaces associated with the variables x^i and ϖ^j, respectively. The surface $\mathcal{J}$ exhibits a saddle point behaviour. The height of the saddle point is the quantity of interest $f(x_s^i)$. The point marked with $\{x, \varpi\}$ stands for an approximate solution which lies in the vicinity of the saddle point. The value of the action $\mathcal{J}$ evaluated at $\{x, \varpi\}$ is equal to the seeked value $f(x_s)$ up to second order terms in $x - x_s$ and $\varpi - \varpi_s$.

5.2 Stationary action

We now apply this general procedure to the evaluation of the thermodynamics properties of a system [24].

5.2.1 Characteristic function

At equilibrium, the density operator of a system is given by the Boltzmann Gibbs distribution calculated with the Hamiltonian $\hat{K}$ (3.2.8)

$$\hat{D} = \exp(-\beta \hat{K}), \tag{5.2.1}$$

where $\hat{K} = \hat{H}$ (resp. $\hat{K} = \hat{H} - \mu \hat{N}$) in the canonical (resp. grand canonical) formalism. The thermal averages of any set of relevant operators $\{\hat{Q}_\gamma\}$ can be derived from the characteristic function

$$\exp(\varphi(\xi)) \equiv \mathrm{Tr}\ \hat{A}(\xi)\hat{D}, \tag{5.2.2}$$

in which appears the generating operator

$$\hat{A}(\xi) \equiv \mathrm{e}^{-\sum_{\gamma} \xi_\gamma \hat{Q}_\gamma} . \tag{5.2.3}$$

Indeed, when all quantities ξ_γ vanish, the characteristic function $\varphi(0)$ gives the partition function Tr $\hat{D}$. Furthermore, the first derivative of φ with respect to ξ_γ provides the expectation value of the operator $\hat{Q}_\gamma$

$$-\frac{\mathrm{d}\varphi(\xi)}{\mathrm{d}\xi_\gamma}|_{\xi=0} \equiv \langle \hat{Q}_\gamma \rangle, \tag{5.2.4}$$

while second derivatives yield correlations and fluctuations of the $\{\hat{Q}_\gamma\}$ operators. When feasible, higher derivatives give access to all other cumulants.

5.2.2 Transposition of the general procedure

For the problem that we are considering, the variables x^i introduced in Section 5.1 are the matrix elements of density operators in an appropriate subspace of the Fock space. For instance, for the canonical case, this subspace is defined by the projector $\hat{P}_{N_0}$ if N_0 is the number of particles. The values $\{x_0^i\}$ are the matrix elements of the Boltzmann Gibbs operator $\hat{D}$ (5.2.1) in this subspace.

On account of the discussion in the previous section, we select the characteristic function (5.2.2) as quantity of interest playing the role of the function f in Section 5.1.

For the constraints $g_j\{x^i\}$ we take the Bloch equation

$$\frac{\mathrm{d}\hat{\mathcal{D}}(\tau)}{\mathrm{d}\tau} + \frac{1}{2}\left[\hat{K}\hat{\mathcal{D}}(\tau) + \hat{\mathcal{D}}(\tau)\hat{K}\right] = 0, \tag{5.2.5}$$

whose solution over the interval $[0, \beta]$ determines the matrix elements of the Boltzmann Gibbs operator. This equation is solved with the boundary condition

$$\hat{\mathcal{D}}(0) = \hat{\mathbf{1}}, \tag{5.2.6}$$

where $\hat{\mathbf{1}}$ is the unity operator in the Fock subspace associated with all the *exactly* preserved observables. For instance, in the canonical case, one has $\hat{\mathbf{1}} = \hat{P}_{N_0}$. The number of equations $g_j = 0$ associated with the Bloch equation is therefore the number of matrix elements of the density operator $\hat{\mathcal{D}}$ times the "number" of pseudo times values τ. Thus, as Lagrange parameters ϖ^j, we can take the matrix elements of a τ-dependent operator $\hat{\mathcal{A}}(\tau)$.

According to the prescription (5.1.1), the functional $\mathcal{J}$ is then

$$\mathcal{J}\{\hat{\mathcal{D}}(\tau), \hat{\mathcal{A}}(\tau)\} \equiv \mathrm{Tr}\ \hat{\mathcal{D}}(\beta)\hat{\mathcal{A}}(\xi) - \int_0^\beta \mathrm{d}\tau\, \mathrm{Tr}\ \hat{\mathcal{A}}(\tau) \times\left(\frac{\mathrm{d}\hat{\mathcal{D}}(\tau)}{\mathrm{d}\tau} + \frac{1}{2}[\hat{K}\hat{\mathcal{D}}(\tau) + \hat{\mathcal{D}}(\tau)\hat{K}]\right), \tag{5.2.7}$$

where Tr denotes now a trace in the relevant Fock subspace. In $\mathcal{J}$, the variables are $\hat{\mathcal{D}}(\tau)$ $(0 < \tau \leq \beta)$ and $\hat{\mathcal{A}}(\tau)$ $(0 \leq \tau \leq \beta)$.

The variation of this action with respect to $\hat{\mathcal{A}}(\tau)$ gives the equation

$$\mathrm{Tr}\ \delta\hat{\mathcal{A}}(\tau)\left(\frac{\mathrm{d}\hat{\mathcal{D}}(\tau)}{\mathrm{d}\tau} + \frac{1}{2}\left[\hat{K}\hat{\mathcal{D}}(\tau) + \hat{\mathcal{D}}(\tau)\hat{K}\right]\right) = 0. \tag{5.2.8}$$

As expected from Section 5.1, the most general variation with respect to $\hat{\mathcal{A}}(\tau)$ gives the equation (5.2.5). The variation of the action with respect to $\hat{\mathcal{D}}(\tau)$ yields

$$\mathrm{Tr}\ \delta\hat{\mathcal{D}}(\tau)\left(\frac{\mathrm{d}\hat{\mathcal{A}}(\tau)}{\mathrm{d}\tau} - \frac{1}{2}\left[\hat{K}\hat{\mathcal{A}}(\tau) + \hat{\mathcal{A}}(\tau)\hat{K}\right]\right) = 0. \tag{5.2.9}$$

(Hint: first, transform the action (5.2.7) by means of a part integration over τ, then vary) while the variation with respect to $\hat{\mathcal{D}}(\beta)$ gives

$$\mathrm{Tr}\ \delta\hat{\mathcal{D}}(\beta)(\hat{\mathcal{A}}(\xi) - \hat{\mathcal{A}}(\beta)) = 0. \tag{5.2.10}$$

5.2.3 General properties

In this section, we discuss the properties of the formalism when some general conditions (satisfied by most trial subspaces) are imposed on the variational spaces of $\hat{\mathcal{D}}$ and $\hat{\mathcal{A}}$.

We consider trial spaces for $\hat{\mathcal{D}}$ and $\hat{\mathcal{A}}$ such that

$$\delta\hat{\mathcal{A}} \propto \hat{\mathcal{A}}\ , \quad \delta\hat{\mathcal{D}} \propto \hat{\mathcal{D}}\ , \tag{5.2.11}$$

are acceptable variations (*i.e.* multiplication by constant is allowed) and such that $\hat{\mathbf{1}}$ belongs to space of operators $\hat{\mathcal{A}}$. Then, one derives immediately the following equation

$$\mathrm{e}^{\varphi(\xi)} = \mathrm{Tr}\ \hat{\mathcal{A}}(\xi)\hat{\mathcal{D}}(\beta) = \mathrm{Tr}\ \hat{\mathcal{A}}(0)\ , \tag{5.2.12}$$

which for $\xi = 0$ determines the partition function in terms of $\hat{\mathcal{D}}(\beta)$

$$\mathrm{e}^{\varphi(0)} = \mathrm{Tr}\ \hat{\mathcal{D}}(\beta)_{\xi=0}\ . \tag{5.2.13}$$

Other properties concern the evaluation of the expectation values of the Hamiltonian $\hat{K}$, of the number of particle $\hat{N}$ and of the operators $\hat{Q}_\gamma$. These are a direct consequence of the stationarity of the action within the trial subspaces. Indeed, the *total* derivative of the *stationary* value of the action $\mathcal{J}(x^i, \varpi^j)$ with respect to any parameter (here the parameters are ξ_γ, β or μ) is equal to the *partial* derivative of $\mathcal{J}$ with respect to this parameter evaluated at the stationary values of the variables $\{x^i_{\rm s}, \varpi^j_{\rm s}\}$. Using this property, one derives the following equalities:

$$-\frac{\mathrm{d}\varphi}{\mathrm{d}\xi_\gamma}|_{\xi=0} \equiv \langle \hat{Q}_\gamma \rangle = \frac{\mathrm{Tr}\ \hat{\mathcal{D}}(\beta)\hat{Q}_\gamma}{\mathrm{Tr}\ \hat{\mathcal{D}}(\beta)}|_{\xi=0}, \tag{5.2.14}$$

and

$$-\frac{\mathrm{d}\varphi(0)}{\mathrm{d}\beta} = -\frac{\mathrm{d}}{\mathrm{d}\beta}\ln \mathrm{Tr}\ \hat{\mathcal{D}}(\beta)|_{\xi=0} = \frac{\mathrm{Tr}\ \hat{\mathcal{D}}(\beta)\hat{K}}{\mathrm{Tr}\ \hat{\mathcal{D}}(\beta)}|_{\xi=0} = \langle \hat{K} \rangle \equiv E. \tag{5.2.15}$$

We can now *define* the free energy and the entropy as

$$F_{\rm G} \equiv -\frac{1}{\beta}\varphi(0), \tag{5.2.16}$$

and

$$S \equiv \beta(E - F_{\rm G}) = \varphi(0) - \beta\frac{\mathrm{d}\varphi(0)}{\mathrm{d}\beta}. \tag{5.2.17}$$

Thus, in this method, the entropy is a *secondary* quantity obtained only after the calculation is completed. In addition, it does not have to be related to $\hat{\mathcal{D}}(\beta)$ by equation (3.3.3). Nevertheless, our definition guarantees that for any trial space compatible with the above conditions, the entropy (5.2.17) satisfies the thermodynamic consistency relation

$$S = -\frac{\partial F_{\rm G}}{\partial T}. \tag{5.2.18}$$

6 Variational principle applied to extended BCS

Here the adjective "extended" indicates that we are concerned by more than the partition function since the equations are designed for an evaluation of the characteristic function $\varphi(\xi)$ for any value of the parameters $\{\xi\}$. Thus, in addition to the partition function, the action (5.2.7) gives access to expectation values, correlations and fluctuations.

6.1 Variational spaces and group properties

The choice of the variational spaces for $\hat{\mathcal{D}}$ and $\hat{\mathcal{A}}$ is inspired from Section 3.3

$$\begin{aligned} \hat{\mathcal{A}}(\tau) &= \hat{\mathcal{T}}^{[a]}(\tau) \leftrightarrow \left\{ l_0{}^{[a]},\, l_p{}^{[a]},\, \theta_p{}^{[a]} \right\} \\ \hat{\mathcal{D}}(\tau) &= \hat{\mathcal{T}}^{[d]}(\tau) \leftrightarrow \left\{ l_0{}^{[d]},\, l_p{}^{[d]},\, \theta_p{}^{[d]} \right\}. \end{aligned} \tag{6.1.1}$$

We now introduce notations which take into account the group property of the $\hat{\mathcal{T}}^{[d]}$-like operators:

$$\hat{\mathcal{T}}^{[ad]} \equiv \hat{\mathcal{T}}^{[a]}\,\hat{\mathcal{T}}^{[d]}\,, \qquad \hat{\mathcal{T}}^{[da]} \equiv \hat{\mathcal{T}}^{[d]}\,\hat{\mathcal{T}}^{[a]}. \tag{6.1.2}$$

The group is non commutative since in general $\hat{\mathcal{T}}^{[da]} \neq \hat{\mathcal{T}}^{[ad]}$. We have therefore

$$\mathcal{T}_p{}^{[ad]} = \mathcal{T}_p{}^{[a]}\,\mathcal{T}_p{}^{[d]}\,, \qquad \mathcal{T}_p{}^{[da]} = \mathcal{T}_p{}^{[d]}\,\mathcal{T}_p{}^{[a]}. \tag{6.1.3}$$

In order to manipulate the algebra more conveniently, it is preferable to work with the following expression of $\mathcal{T}_p{}^{[d]}$

$$\mathcal{T}_p{}^{[d]} = \cosh l_p{}^{[d]} - \sinh l_p{}^{[d]} \hat{u}_p^{[d]}.\vec{\sigma}, \tag{6.1.4}$$

which is equivalent to equation (3.2.2). It involves a vector of length unity $\hat{u}_p^{[d]}$ determined by $\theta_p{}^{[d]}$. In particular, this form of $\mathcal{T}_p{}^{[d]}$ takes naturally into account an extension to complex values for the quantities $S_p{}^{[d]}$ such as mentioned in (3.1.7). Using the property of Pauli matrices $(\sigma_x, \sigma_y, \sigma_z)$

$$(\vec{b}.\vec{\sigma})(\vec{c}.\vec{\sigma}) = (\vec{b}.\vec{c}) + i(\vec{b} \times \vec{c}).\vec{\sigma}, \tag{6.1.5}$$

which is valid for any pair of vectors $\vec{b}$ and $\vec{c}$ as well as some of its consequences such as

$$\det(\gamma + \vec{c}.\vec{\sigma}) = \gamma^2 - \vec{c}^2, \quad (\gamma + \vec{c}.\vec{\sigma})^{-1} = \frac{1}{\gamma^2 - \vec{c}^2}(\gamma - \vec{c}.\vec{\sigma}), \tag{6.1.6}$$

(here γ is a c-number) one obtains an explicit expression of $\mathcal{T}_p{}^{[ad]}$

$$\begin{aligned} \mathcal{T}_p{}^{[ad]} = {} & \cosh l_p{}^{[d]} \cosh l_p{}^{[a]} + \sinh l_p{}^{[d]} \sinh l_p{}^{[a]} \hat{u}_p^{[a]}.\hat{u}_p^{[d]} \\ & - \Big[\cosh l_p{}^{[d]} \sinh l_p{}^{[a]} \hat{u}_p^{[a]} + \sinh l_p{}^{[d]} \cosh l_p{}^{[a]} \hat{u}_p^{[d]} \\ & - i \sinh l_p{}^{[d]} \sinh l_p{}^{[a]} (\hat{u}_p^{[a]} \times \hat{u}_p^{[d]}) \Big] .\vec{\sigma}. \end{aligned} \tag{6.1.7}$$

The matrix $\mathcal{T}_p{}^{[ad]}$ being the product of two unitary matrices is also unitary. Thus, when the bracket is written as $\sinh l_p{}^{[ad]} \hat{u}_p^{[ad]}$, the first term is equal to

$\cosh l_p{}^{[ad]}$ and the form (6.1.4) is retrieved. The noncommutativity property $\mathcal{T}_p{}^{[ad]} \neq \mathcal{T}_p{}^{[da]}$ manifests itself in the last term of the bracket.

According to equation (3.2.5), the associated density matrices are

$$\mathcal{R}_p{}^{[ad]} = \left(\mathcal{T}_p{}^{[ad]^{-1}} + 1\right)^{-1}, \quad \mathcal{R}_p{}^{[da]} = \left(\mathcal{T}_p{}^{[ad]^{-1}} + 1\right)^{-1}, \tag{6.1.8}$$

and the required evaluations of traces are obtained from equation (3.2.3)

$$Z = \mathrm{Tr}\ \hat{\mathcal{A}}\hat{\mathcal{D}} = \mathrm{e}^{-l_0{}^{[a]} - l_0{}^{[d]}} \prod_p \det\left(1 + \mathcal{T}_p{}^{[ad]}\right). \tag{6.1.9}$$

6.2 *Extended BCS functional*

We now have to specify the generating operator $\hat{A}(\xi)$ (5.2.3). We assume that it also belongs to space of $\hat{T}^{[x]}$ operators as all the $\hat{\mathcal{A}}(\tau)$'s. This means that the operators $\hat{Q}_\gamma$ are one-body operators:

$$\hat{Q}_\gamma = \sum_p q_p^{(\gamma)} \left(a_p^\dagger a_p + a_{\bar{p}}^\dagger a_{\bar{p}}\right), \tag{6.2.1}$$

where $\{q_p^{(\gamma)}\}$ is a collection of Ω c-numbers. From equation (6.1.9) one calculates $Z^{[A]} = \mathrm{Tr}\ \hat{A}(\xi)\hat{\mathcal{D}}$ by means of an appropriate introduction of quantities such as $l_0{}^{[A]}$, $l_p{}^{[A]}$, $(\theta_p{}^{[A]} = 0)$and $\mathcal{T}_p{}^{[A]}$.

The first term in the integral of the action is obtained by deriving $l_0{}^{[d]}$ and $\mathcal{T}_p{}^{[d]}$ with respect to τ in the expression (6.1.9). After some work, one gets

$$\mathrm{Tr}\ \hat{\mathcal{A}}\frac{\mathrm{d}\hat{\mathcal{D}}}{\mathrm{d}\tau} = Z\left(-\frac{\mathrm{d}l_0{}^{[d]}}{\mathrm{d}\tau} + \sum_p \mathrm{tr}\ \mathcal{T}_p{}^{[d]^{-1}} \mathcal{R}_p{}^{[da]} \frac{\mathrm{d}\mathcal{T}_p{}^{[d]}}{\mathrm{d}\tau}\right). \tag{6.2.2}$$

(Hint: use the following properties (I) Z depends on $l_0{}^{[d]}$ only through the multiplicative factor $\exp(-l_0{}^{[d]})$; (II) $\det A = \exp \mathrm{tr} \ln A$ for any matrix A; (III) if δ stands for any kind of variation then $\delta \mathrm{tr} \ln A = \mathrm{tr}\ A^{-1}\delta A$; (IV) $\delta\mathcal{R}_p{}^{[ad]} = \mathcal{T}_p{}^{[d]^{-1}}\mathcal{R}_p{}^{[da]}\mathcal{T}_p{}^{[d]}$).

As the Wick theorem is valid for any operator of the $\hat{T}^{[x]}$ type, from equation (3.2.9) one obtains the pseudo energy

$$\frac{1}{2}\mathrm{Tr}\ \hat{\mathcal{A}}\left[\hat{K}\hat{\mathcal{D}} + \hat{\mathcal{D}}\hat{K}\right] = \frac{1}{2}\mathrm{Tr}\ \hat{K}\left[\hat{\mathcal{D}}\hat{\mathcal{A}} + \hat{\mathcal{A}}\hat{\mathcal{D}}\right] = \frac{1}{2}\left(\mathcal{E}\{\mathcal{R}^{[da]}\} + \mathcal{E}\{\mathcal{R}^{[ad]}\}\right), \tag{6.2.3}$$

and the extended BCS functional $\mathcal{I}$ (5.2.7)

$$\mathcal{J}\{\hat{\mathcal{D}},\hat{\mathcal{A}}\} = Z^{[A]} - \int_0^\beta \mathrm{d}\tau\, Z \left(-\frac{\mathrm{d}l_0{}^{[d]}}{\mathrm{d}\tau} + \sum_p \mathrm{tr}\ \mathcal{T}_p{}^{[d]}{}^{-1}\mathcal{R}_p{}^{[da]}\frac{\mathrm{d}\mathcal{T}_p{}^{[d]}}{\mathrm{d}\tau} + \frac{1}{2}\left(\mathcal{E}\{\mathcal{R}^{[da]}\} + \mathcal{E}\{\mathcal{R}^{[ad]}\}\right)\right). \tag{6.2.4}$$

By means of a part integration with respect to the variable τ one obtains the equivalent expression

$$\mathcal{J}\{\hat{\mathcal{D}},\hat{\mathcal{A}}\} = Z^{[A]} - Z(\beta) + Z(0) + \int_0^\beta \mathrm{d}\tau\, Z \left(-\frac{\mathrm{d}l_0{}^{[a]}}{\mathrm{d}\tau} + \sum_p \mathrm{tr}\ \mathcal{T}_p{}^{[a]}{}^{-1}\mathcal{R}_p{}^{[ad]}\frac{\mathrm{d}\mathcal{T}_p{}^{[a]}}{\mathrm{d}\tau} - \frac{1}{2}\left(\mathcal{E}\{\mathcal{R}^{[da]}\} + \mathcal{E}\{\mathcal{R}^{[ad]}\}\right)\right) \tag{6.2.5}$$

which is more convenient when variations with respect to $\mathcal{T}_p{}^{[d]}$ have to be performed.

6.3 Extended BCS equations

Let us use equation (6.2.4) and express the stationary conditions of $\mathcal{J}$ with respect to $\hat{\mathcal{A}}(\tau)$. The variation with respect $l_0{}^{[a]}(\tau)$ is easily performed since this quantity only appears in $Z(\tau)$.

$$\frac{\mathrm{d}l_0{}^{[d]}}{\mathrm{d}\tau} = \sum_p \mathrm{tr}\ \mathcal{T}_p{}^{[d]}{}^{-1}\mathcal{R}_p{}^{[da]}\frac{\mathrm{d}\mathcal{T}_p{}^{[d]}}{\mathrm{d}\tau} + \frac{1}{2}\left(\mathcal{E}\{\mathcal{R}^{[da]}\} + \mathcal{E}\{\mathcal{R}^{[ad]}\}\right). \tag{6.3.1}$$

To vary $\mathcal{J}$ with respect to $\mathcal{T}_p{}^{[a]}(\tau)$, one can use the properties $\delta\mathcal{E}\{\mathcal{R}^{[x]}\} = \sum_p \mathrm{tr}\ \mathcal{H}\{\mathcal{R}_p^{[x]}\}\delta\mathcal{R}_p{}^{[x]}$ and $\delta\mathcal{R} = (1-\mathcal{R})\delta\mathcal{T}(1-\mathcal{R})$ to obtain

$$\frac{\mathrm{d}\mathcal{T}_p{}^{[d]}}{\mathrm{d}\tau} = -\frac{1}{2}\left[\mathcal{H}\{\mathcal{R}_p^{[da]}\}\mathcal{T}_p{}^{[d]} + \mathcal{T}_p{}^{[d]}\mathcal{H}\{\mathcal{R}_p^{[ad]}\}\right]. \tag{6.3.2}$$

The corresponding equations deduced from a variation of the action with respect to $\hat{\mathcal{D}}(\tau)$ are obtained in similar way from the form (6.2.5) of $\mathcal{J}$. They are

$$\frac{\mathrm{d}l_0{}^{[a]}}{\mathrm{d}\tau} = \sum_p \mathrm{tr}\ \mathcal{T}_p{}^{[a]}{}^{-1}\mathcal{R}_p{}^{[ad]}\frac{\mathrm{d}\mathcal{T}_p{}^{[a]}}{\mathrm{d}\tau} - \frac{1}{2}\left(\mathcal{E}\{\mathcal{R}^{[da]}\} + \mathcal{E}\{\mathcal{R}^{[ad]}\}\right), \tag{6.3.3}$$

and

$$\frac{\mathrm{d}\mathcal{T}_p{}^{[a]}}{d\tau} = \frac{1}{2}\left[\mathcal{H}\{\mathcal{R}_p^{[ad]}\}\mathcal{T}_p{}^{[a]} + \mathcal{T}_p{}^{[a]}\mathcal{H}\{\mathcal{R}_p^{[da]}\}\right]. \tag{6.3.4}$$

It remains to specify the form of the boundary conditions associated with equations (5.2.6) and (5.2.10). For the extended BCS problem as it is considered here, they are

$$\mathcal{l}_0{}^{[d]}(0) = 0\,, \quad \mathcal{T}_p{}^{[d]}(0) = 1, \tag{6.3.5}$$

and

$$\mathcal{l}_0{}^{[a]}(\beta) = \mathcal{l}_0{}^{[A]}\,, \quad \mathcal{T}_p{}^{[a]}(\beta) = \mathcal{T}_p{}^{[A]}. \tag{6.3.6}$$

6.4 Properties of the extended BCS equations

The equations (6.3.2) and (6.3.4) can be solved first independently from equations (6.3.1) and (6.3.3). Then, $\mathcal{l}_0{}^{[d]}$ and $\mathcal{l}_0{}^{[a]}$ are obtained by means of an integration as

$$\begin{aligned}\mathcal{l}_0{}^{[d]}(\tau) = \frac{1}{2}\int_0^\tau \mathrm{d}\tau' \Big(& \mathcal{E}\{\mathcal{R}^{[ad]}\} + \mathcal{E}\{\mathcal{R}^{[da]}\} \\ & - \sum_p \mathrm{tr}\left(\mathcal{H}\{\mathcal{R}_p^{[da]}\}\mathcal{R}_p{}^{[da]} + \mathcal{H}\{\mathcal{R}_p^{[ad]}\}\mathcal{R}_p{}^{[ad]}\right)\Big)\end{aligned} \tag{6.4.1}$$

and

$$\mathcal{l}_0{}^{[a]}(\tau) = \mathcal{l}_0{}^{[A]} + \mathcal{l}_0{}^{[d]}(\beta) - \mathcal{l}_0{}^{[d]}(\tau). \tag{6.4.2}$$

From an analysis of the equations one checks the following two conservation laws

$$\frac{\mathrm{d}Z}{\mathrm{d}\tau} = 0, \qquad \frac{\mathrm{d}}{\mathrm{d}\tau}\left(\mathcal{E}\{\mathcal{R}^{[ad]}\} + \mathcal{E}\{\mathcal{R}^{[da]}\}\right) = 0, \tag{6.4.3}$$

which can be used to constrain the numerical algorithms. Combining equations (6.3.2) and (6.3.4) one finds that the derivatives of $\mathcal{T}_p{}^{[ad]}$ and $\mathcal{T}_p{}^{[da]}$ can be expressed in a *commutator* form:

$$\frac{\mathrm{d}\mathcal{T}_p{}^{[ad]}}{\mathrm{d}\tau} = \frac{1}{2}\left[\mathcal{H}\{\mathcal{R}_p^{[ad]}\},\ \mathcal{T}_p{}^{[ad]}\right], \tag{6.4.4}$$

and a similar equation (with a minus sign in front of the commutator) for $\mathcal{T}_p{}^{[da]}$.

For the BCS case considered here, alternative formulations are possible. Indeed, from the expression (3.2.10) of $\mathcal{H}\{\mathcal{R}_p^{[x]}\}$, this matrix can be written in terms of an appropriately defined vector $\vec{h}_p^{[x]}$ as

$$\vec{h}_p^{[x]}.\vec{\sigma} \equiv \mathcal{H}\{\mathcal{R}_p^{[x]}\}. \tag{6.4.5}$$

Then, by means of the representation (6.1.4) of the matrices $\mathcal{T}_p{}^{[d]}$ and $\mathcal{T}_p{}^{[a]}$, equations (6.3.2) and (6.3.4) can be expressed as a set of 2Ω coupled ordinary differential equations plus torque equations acting on the normed vectors $\hat{u}_p^{[d]}$ and $\hat{u}_p^{[a]}$. For instance equation (6.3.2) becomes

$$\frac{\mathrm{d}l_p{}^{[d]}}{\mathrm{d}\tau} = \frac{1}{2}\left(\vec{h}_p^{[da]} + \vec{h}_p^{[ad]}\right).\hat{u}_p^{[d]}$$

$$\frac{\mathrm{d}\hat{u}_p^{[d]}}{\mathrm{d}\tau} = \frac{1}{2}\left(\coth l_p{}^{[d]}\left[\hat{u}_p^{[d]} \times \left(\vec{h}_p^{[da]} + \vec{h}_p^{[ad]}\right)\right] + i\left(\vec{h}_p^{[da]} - \vec{h}_p^{[ad]}\right)\right) \times \hat{u}_p^{[d]}, \tag{6.4.6}$$

while equation (6.4.4) becomes

$$\begin{aligned} \frac{\mathrm{d}l_p{}^{[ad]}}{\mathrm{d}\tau} &= 0 \\ \frac{\mathrm{d}\hat{u}_p^{[ad]}}{\mathrm{d}\tau} &= -i\vec{h}_p^{[ad]} \times \hat{u}_p^{[ad]}. \end{aligned} \tag{6.4.7}$$

6.5 *Recovering the BCS solution*

The BCS solution is obtained from the above equations when the problem is reduced to an evaluation of the partition function. Thus, all the parameters ξ_γ are taken equal to zero and

$$\hat{A}(0) = \hat{\mathbf{1}} \ . \tag{6.5.1}$$

Then, one checks that if one introduces the operator

$$\hat{W} = h_0 + \sum_p \left(a_p^\dagger a_{\bar{p}}\right) \mathcal{H}\{\mathcal{R}_p^{[\mathrm{BCS}]}\} \begin{pmatrix} a_p \\ a_{\bar{p}}^\dagger \end{pmatrix}, \tag{6.5.2}$$

where the matrix $\mathcal{H}\{\mathcal{R}_p^{[\mathrm{BCS}]}\}$ has been defined by equation (3.3.16) and h_0 is a c-number to be determined below, the operators

$$\hat{\mathcal{D}}(\tau) = \mathrm{e}^{-\tau\hat{W}} , \tag{6.5.3}$$

and

$$\hat{\mathcal{A}}(\tau) = \mathrm{e}^{-(\beta-\tau)\hat{W}} , \tag{6.5.4}$$

are solutions of the equations of Section 6.3 and are compatible with the boundary conditions.

Indeed, from equations (6.5.3) and (6.5.4) one has $\hat{\mathcal{T}}^{[ad]} = \hat{\mathcal{T}}^{[da]} = \exp(-\beta\hat{W})$ and thus

$$\mathcal{T}_p{}^{[ad]} = \mathcal{T}_p{}^{[da]} = \mathrm{e}^{-\beta\mathcal{H}\{\mathcal{R}_p^{[\mathrm{BCS}]}\}}. \tag{6.5.5}$$

As a consequence ${\mathcal{R}_p}^{[ad]}$ and ${\mathcal{R}_p}^{[da]}$ defined according to equation (3.2.5) are τ-independent and equal to ${\mathcal{R}_p}^{[\mathrm{BCS}]}$ which satisfies the selfconsistent equation (3.3.17). Thus one has

$$\mathcal{H}\{\mathcal{R}_p^{[ad]}\} = \mathcal{H}\{\mathcal{R}_p^{[da]}\} = \mathcal{H}\{\mathcal{R}_p^{[\mathrm{BCS}]}\} \tag{6.5.6}$$

so that the matrix ${\mathcal{T}_p}^{[d]} = \exp(-\tau\mathcal{H}\{\mathcal{R}_p^{[\mathrm{BCS}]}\})$ is a solution of equation (6.3.2).

According to equation (6.4.1) the value of the c-number h_0 is determined by the variational equations (in contrast with the situation in Section 3.3 for the BCS theory) and is given by

$$h_0 = \mathcal{E}\{\mathcal{R}^{[\mathrm{BCS}]}\} - \sum_p \mathrm{tr}\ \mathcal{H}\{\mathcal{R}_p^{[\mathrm{BCS}]}\}{\mathcal{R}_p}^{[\mathrm{BCS}]}. \tag{6.5.7}$$

From the general equation (5.2.4), the expectation value of any operator of the type (6.2.1) is given by

$$\frac{\mathrm{Tr}\ \hat{Q}_\gamma \hat{\mathcal{D}}(\beta)}{\mathrm{Tr}\ \hat{\mathcal{D}}(\beta)} = \sum_p q_p^{(\gamma)} \left(1 - {t_p}^{[d]}{C_p}^{[d]}\right) \tag{6.5.8}$$

where ${C_p}^{[d]}$ is given by equation (3.3.12) This expression is the same as given by standard BCS formalism.

6.6 Beyond the BCS solution

The first difference between the predictions of the extended and standard BCS formalisms appears when correlations and fluctuations are evaluated. Two situations are possible: the operators $\hat{Q}_\gamma$ commute with the Hamiltonian or they do not commute.

In the first case, it is possible to reformulate the evaluation of $\exp(\varphi(\xi))$ as the calculation of the partition function $\exp(\tilde{\varphi}(0))$ associated with the shifted Hamiltonian

$$\hat{\tilde{K}} = \hat{K} + \sum_\gamma \frac{\xi_\gamma}{\beta}\hat{Q}_\gamma\ . \tag{6.6.1}$$

The quantities $-\xi_\gamma/\beta$ play thus the rôle of μ for the operator $\hat{N}$. One can then use directly the results of the previous section For instance, the fluctuation $\Delta\hat{N}^2$ is given by the formula

$$\Delta\hat{N}^2 = \frac{1}{\beta^2}\frac{\partial^2\varphi}{\partial\mu^2} = \frac{1}{\beta}\frac{\partial}{\partial\mu}\langle\hat{N}\rangle. \tag{6.6.2}$$

Thus, the grand canonical *approximate* fluctuation is proportional to the derivative of the *approximate* expectation value just as it is required by the exact thermodynamical relation. The result (6.6.2) differs from the value $\langle(\hat{N}-\langle\hat{N}\rangle^{[\mathrm{BCS}]})^2\rangle^{[\mathrm{BCS}]}$ which would be obtained from a naive usage of the density operator $\hat{\mathcal{D}}(\beta)=\hat{\mathcal{T}}^{[\mathrm{BCS}]}$. For instance, at $T=0$, this formula gives a non zero fluctuation, while according to equation (6.6.2) $\Delta\hat{N}^2$ vanishes in agreement with thermodynamics.

When the operators $\hat{Q}_\gamma$ do not commute with $\hat{K}$, the evaluation of the fluctuation is more complicated. I refer the interested reader to Sections 4.2–4.4 of [24] for a proof. I will only state the result. With the notations of the Liouville space, the fluctuation of an operator $\hat{Q}$ is given by

$$\Delta\hat{Q}^2=\frac{1}{2}\mathcal{Q}:\left(\mathbf{K}\coth\frac{1}{2}\beta\mathbf{K}\right)\mathbf{F}^{-1}:\mathcal{Q}, \tag{6.6.3}$$

where $\mathbf{K}$ is the RPA matrix, $\mathbf{F}$ the stability matrix and $\mathcal{Q}$ the matrix associated with the operator $\hat{Q}$.

When $\hat{Q}$ commutes with $\hat{K}$, it corresponds to a zero energy mode of the RPA: $\mathcal{Q}:\mathbf{K}=0$. Then, the parenthesis in equation (6.6.3) reduces to $2/\beta$ and one retrieves the usual expression

$$\Delta\hat{Q}^2=\frac{1}{\beta}\mathcal{Q}:\mathbf{F}^{-1}:\mathcal{Q}\,. \tag{6.6.4}$$

Therefore, the mean field formalism of generalized BCS takes also into account the two body correlations at the level of the RPA.

The evaluation of the fluctuations by means of equations (6.6.3) and (6.6.4) can be difficult since it involves working in the Liouville space whose dimension is Ω^2. In practice, it may be easier to return to equations (6.3.4) and (6.4.4), to solve them for several (at least three) values of ξ, to evaluate the function $\varphi(\xi)$ and to compute the fluctuation by means of a numerical differentiation.

7 Particle number projection at finite temperature

7.1 Particle number projected action

The generalized BCS method discussed in Section 5 can be extended to the canonical framework, by the following choice of trial spaces for $\hat{\mathcal{D}}$ and $\hat{\mathcal{A}}$:

$$\begin{aligned}\hat{\mathcal{A}}(\tau)&=\hat{\mathcal{T}}^{[a]}(\tau)\leftrightarrow\left\{l_0{}^{[a]},\,l_p{}^{[a]},\,\theta_p{}^{[a]}\right\}\\ \hat{\mathcal{D}}(\tau)&=\hat{P}_{N_0}\hat{\mathcal{T}}^{[d]}(\tau)\hat{P}_{N_0}\leftrightarrow\left\{l_0{}^{[d]},\,l_p{}^{[d]},\,\theta_p{}^{[d]}\right\}.\end{aligned} \tag{7.1.1}$$

It involves the same number of variational parameters as (6.1.1) With notations similar to those of Section 4.2 the explicit form of $\hat{\mathcal{D}}(\tau)$ is

$$\hat{\mathcal{D}}(\tau)=\frac{1}{4\pi^2}\int_0^{2\pi}\int_0^{2\pi}\mathrm{d}\varphi\,\mathrm{d}\varphi'\,\mathrm{e}^{-i(\varphi+\varphi')N_0}\,\hat{\mathcal{T}}^{[\varphi]}\,\hat{\mathcal{T}}^{[d]}(\tau)\,\hat{\mathcal{T}}^{[\varphi']}. \tag{7.1.2}$$

We again introduce the more compact notations

$$\hat{\mathcal{T}}^{[a,\varphi,d,\varphi']} \equiv \hat{\mathcal{T}}^{[a]}\hat{\mathcal{T}}^{[\varphi]}\,\hat{\mathcal{T}}^{[d]}\,\hat{\mathcal{T}}^{[\varphi']}, \quad \mathcal{T}_p{}^{[a,\varphi,d,\varphi']} \equiv \mathcal{T}_p{}^{[a]}\mathcal{T}_p{}^{[\varphi]}\,\mathcal{T}_p{}^{[d]}\,\mathcal{T}_p{}^{[\varphi']}, \quad (7.1.3)$$

which express the group properties. In equation (7.1.3), $\mathcal{T}_p{}^{[a]}$ and $\mathcal{T}_p{}^{[d]}$ are defined as in equation (6.1.4) and $\mathcal{T}_p{}^{[\varphi]}$ is given by equation (4.1.5).

Then, a straightforward generalization of equation (6.1.9) yields

$$Z = \mathrm{Tr}\,\hat{\mathcal{A}}\hat{\mathcal{D}} = \frac{1}{4\pi^2}\int_0^{2\pi}\int_0^{2\pi} \mathrm{d}\varphi\,\mathrm{d}\varphi' \mathrm{e}^{-i(\varphi+\varphi')N_0}\, Z^{[a,\varphi,d,\varphi']}, \quad (7.1.4)$$

with

$$Z^{[a,\varphi,d,\varphi']} = \mathrm{e}^{i(\varphi+\varphi')\Omega - l_0{}^{[a]} - l_0{}^{[d]}} \prod_p \det\left(1 + \mathcal{T}_p{}^{[a,\varphi,d,\varphi']}\right). \quad (7.1.5)$$

In order to write the action explicitly, we notice that it would have been equivalent to take $\hat{\mathcal{A}} = \hat{P}_{N_0}\hat{\mathcal{T}}^{[a]}\hat{P}_{N_0}$ and $\hat{\mathcal{D}} = \hat{\mathcal{T}}^{[d]}$. Therefore, from a direct transposition of (6.2.4) we obtain

$$\begin{aligned}\mathcal{J}\{\hat{\mathcal{D}}, \hat{\mathcal{A}}\} = Z^A - \frac{1}{4\pi^2}\int_0^{2\pi}\int_0^{2\pi} \mathrm{d}\varphi\,\mathrm{d}\varphi' \mathrm{e}^{-i(\varphi+\varphi')N_0} \int_0^{\beta} \mathrm{d}\tau\, Z^{[a,\varphi,d,\varphi']} \\ \times\Bigg[-\frac{\mathrm{d}l_0{}^{[d]}}{\mathrm{d}\tau} + \sum_p \mathrm{tr}\;\mathcal{T}_p{}^{[d]^{-1}}\mathcal{R}_p{}^{[d,\varphi',a,\varphi]}\frac{\mathrm{d}\mathcal{T}_p{}^{[d]}}{\mathrm{d}\tau} \\ +\frac{1}{2}\left(\mathcal{E}\{\mathcal{R}^{[d,\varphi',a,\varphi]}\} + \mathcal{E}\{\mathcal{R}^{[a,\varphi,d,\varphi']}\}\right)\Bigg].\end{aligned} \quad (7.1.6)$$

In this equation, Z^A is obtained by replacing $\hat{\mathcal{T}}^{[a]}$ by $\hat{\mathcal{A}}$ and $\tau = \beta$ in equations (7.1.4) and (7.1.5). In addition, we have used the property

$$\mathcal{E}\{\mathcal{R}^{[\varphi',a,\varphi,d]}\} = \mathcal{E}\{\mathcal{R}^{[a,\varphi,d,\varphi']}\} \quad (7.1.7)$$

which results from the commutation of $\hat{K}$ and $\hat{N}$.

7.2 *Number projected stationary equations: sketch of the method*

The variation of the action (7.1.6) with respect to $l_0{}^{[a]}$ is easily performed since this quantity only appears as a multiplicative factor $\exp(-l_0{}^{[a]})$ in

$Z^{[a,\varphi,d,\varphi']}$. One obtains the equation

$$Z\,\frac{\mathrm{d}l_0{}^{[d]}}{\mathrm{d}\tau} = \frac{1}{Z}\frac{1}{4\pi^2}\int_0^{2\pi}\int_0^{2\pi} \mathrm{d}\varphi\,\mathrm{d}\varphi' \mathrm{e}^{-i\,(\varphi+\varphi')\,N_0}\, Z^{[a,\varphi,d,\varphi']}$$

$$\left[\sum_p \mathrm{tr}\; \mathcal{T}_p{}^{[d]^{-1}} \mathcal{R}_p{}^{[d,\varphi',a,\varphi]} \frac{\mathrm{d}\mathcal{T}_p{}^{[d]}}{\mathrm{d}\tau} + \frac{1}{2}\left(\mathcal{E}\{\mathcal{R}^{[d,\varphi',a,\varphi]}\} + \mathcal{E}\{\mathcal{R}^{[a,\varphi,d,\varphi']}\}\right)\right]. \tag{7.2.1}$$

It is a generalization of equation (6.3.1) and, as this equation, it implies that $l_0{}^{[d]}(\tau)$ is obtained by an integration once $\mathcal{T}_p{}^{[a]}(\tau)$ and $\mathcal{T}_p{}^{[d]}(\tau)$ are known.

The variation of $\mathcal{J}$ with respect to $\mathcal{T}_p{}^{[a]}$ gives a matrix equation of the type (6.3.2) or (6.4.6). It is however much more complicated since it involves angle averaged operators of the $\mathcal{H}$ type. For a detailed derivation we refer the reader to the Sections 5.2 and 5.3 of [24]. An analytic solution of these equations appears difficult and a numerical solution is not yet available.

8 Number parity projected BCS at finite temperature

In the introduction, we emphasized that in finite systems, many signatures of pairing correlations were detected in and measured from the staggering of properties with respect to the parity of the particle number. For binding energies, a critical evaluation of this point, will be done in the Section 9.2. At any rate, it appears that an interesting improvement over BCS of the physics of pairing can already be obtained by enforcing exactly the correct number parity while keeping a grand canonical formalism to constrain the the average particle number to have the correct value N_0.

8.1 Projection and action

The projection which extracts the correct number parity component out of a state is

$$\hat{P}_\eta = \frac{1}{2}\left(\hat{\mathbf{1}} + \eta\,\mathrm{e}^{\mathrm{i}\pi\hat{N}}\right), \tag{8.1.1}$$

where η is equal to $+1$ for even systems and -1 for odd ones.

In this section, we specialize the action (5.2.7) for the calculation of the thermodynamic function and the expectation value of one-body observables. Then, according to the discussion in Section 5.2, the generating operator is equal to the unit operator: $\hat{A} = \hat{\mathbf{1}}$.

The other boundary condition concerns the operator $\hat{\mathcal{D}}_\eta$ for $\tau = 0$. It must be equal to the unity operator in the Fock subspace with the correct number parity. This implies that $\hat{\mathcal{D}}_\eta(0) = \hat{P}_\eta$.

Because the projection (8.1.1) commutes with operators of the $\hat{T}^{[d]}$ type, one can use the action (5.2.7) with an Ansatz very similar to the choice made in Section 6.5. As was done in this section, we introduce an operator

$$\hat{W}_\eta = h_0{}^{[\eta]} + \sum_p e_p^{[\eta]} \left(a_p^\dagger a_{\bar{p}}\right) \mathcal{U}_p{}^{[\eta]} \begin{pmatrix} a_p \\ a_{\bar{p}}^\dagger \end{pmatrix} \tag{8.1.2}$$

where $h_0{}^{[\eta]}$, $\vartheta_p^{[\eta]}$ (implicit in $\mathcal{U}_p{}^{[\eta]}$) and $e_p^{[\eta]}$ are the quantities which must be determined from the stationarity condition of the action. We look for solutions $\hat{\mathcal{D}}_\eta(\tau)$ and $\hat{\mathcal{A}}_\eta(\tau)$ of the form

$$\hat{\mathcal{D}}_\eta(\tau) = \hat{P}_\eta \mathrm{e}^{-\tau \hat{W}_\eta}\ , \quad \hat{\mathcal{A}}_\eta(\tau) = \mathrm{e}^{-(\beta - \tau)\hat{W}_\eta}\ . \tag{8.1.3}$$

One notes that, because the projection commutes with $\hat{W}_\eta$ and satisfies $\hat{P}_\eta^2 = \hat{P}_\eta$, it is not necessary to have a projection on both sides of $\exp(-\tau\hat{W}_\eta)$ as in equation (7.1.1). The operators $\hat{\mathcal{D}}_\eta\hat{\mathcal{A}}_\eta$ and $\hat{\mathcal{A}}_\eta\hat{\mathcal{D}}_\eta$ are equal and do not depend on τ

$$\hat{\mathcal{D}}_\eta(\tau)\hat{\mathcal{A}}(\tau) = \hat{\mathcal{A}}_\eta(\tau)\hat{\mathcal{D}}_\eta(\tau) = \hat{P}_\eta \hat{T}^{[\eta]}\ , \tag{8.1.4}$$

where $\hat{T}^{[\eta]}$ is defined as

$$\hat{T}^{[\eta]} \equiv \mathrm{e}^{-\beta\hat{W}_\eta}\ . \tag{8.1.5}$$

To calculate the action $\mathcal{J}$, one remarks that

$$\frac{\hat{\mathcal{D}}_\eta}{\mathrm{d}\tau} = -\hat{W}_\eta\hat{\mathcal{D}}_\eta\ , \tag{8.1.6}$$

so that the integrand in definition (5.2.7) is constant. The integral is performed trivially and the action can be expressed as

$$\mathcal{J}_\eta = \mathrm{Tr}\ \hat{P}_\eta \hat{T}^{[\eta]} \left[1 + \beta(\hat{W}_\eta - \hat{K})\right]. \tag{8.1.7}$$

The density operator $\hat{P}_\eta\hat{T}^{[\eta]}$ has two components for the values 0 and π of the gauge angle in the operator $\exp(i\varphi\hat{N})$ introduced in Section 4.1. With the conventions of this section, we denote these two components $\hat{T}^{[\eta,0]}$ and $\hat{T}^{[\eta,\pi]}$ respectively. They are defined in terms of $\hat{T}^{[\eta]}$ by

$$\hat{T}^{[\eta,0]} \equiv \hat{T}^{[\eta]}\ , \quad \hat{T}^{[\eta,\pi]} \equiv \mathrm{e}^{i\pi\hat{N}}\hat{T}^{[\eta]}\ . \tag{8.1.8}$$

From equation (3.2.3, 3.3.4) and (3.3.5), for $\varphi = 0$, one obtains an explicit formula for the trace

$$Z^{[\eta,0]} \equiv \mathrm{Tr}\ \hat{T}^{[\eta,0]} = \mathrm{e}^{-\beta h_0{}^{[\eta]}} \prod_p \left(2\cosh\beta\frac{e_p^{[\eta]}}{2}\right)^2, \tag{8.1.9}$$

and for the density matrix

$$\mathcal{R}_p{}^{[\eta,0]} = \frac{1}{2}\left(1 - t_p{}^{[\eta]}\mathcal{U}_p{}^{[\eta]}\right), \quad t_p{}^{[\eta]} = \tanh\beta\frac{e_p^{[\eta]}}{2}. \tag{8.1.10}$$

The definition (3.2.9) provides the expression of $\mathcal{E}\{\mathcal{R}^{[\eta,0]}\}$ and gives

$$\begin{aligned}\mathrm{Tr}\ \hat{T}^{[\eta,0]}\left[1+\beta(\hat{W}_\eta - \hat{K})\right] &= \\ Z^{[\eta,0]}\left[1+\beta\left(h_0{}^{[\eta]} - \textstyle\sum_p e_p^{[\eta]} t_p{}^{[\eta]} - \mathcal{E}\{\mathcal{R}^{[\eta,0]}\}\right)\right]&.\end{aligned} \tag{8.1.11}$$

In order to evaluate the second part the action, one notes that according to equation (4.1.5) the matrix $\mathcal{T}_p{}^{[\pi]}$ is equal to -1, Then, one obtains the following equalities

$$\mathcal{T}_p{}^{[\eta,\pi]} = -\mathcal{T}_p{}^{[\eta,0]}, \tag{8.1.12}$$

and

$$\mathcal{R}_p{}^{[\eta,\pi]} = \frac{\mathcal{R}_p{}^{[\eta,0]}}{2\mathcal{R}_p{}^{[\eta,0]} - 1} = \frac{1}{2}\left(1 - t_p{}^{[\eta]-1}\mathcal{U}_p{}^{[\eta]}\right). \tag{8.1.13}$$

Using equations (3.2.3) and (8.1.12) we calculate the second trace

$$Z^{[\eta,\pi]} \equiv \mathrm{Tr}\ \hat{T}^{[\eta,\pi]} = \mathrm{e}^{-\beta h_0{}^{[\eta]}} \prod_p \left(2\sinh\beta\frac{e_p^{[\eta]}}{2}\right)^2 = Z^{[\eta,0]}\, r_\eta\,, \tag{8.1.14}$$

where the c-number r_η is the product of all the thermal factors

$$r_\eta = \prod_p t_p{}^{[\eta]2}. \tag{8.1.15}$$

The calculation of the action can now be completed. One obtains

$$\mathcal{J}_\eta = \bar{Z}_\eta\left[1+\beta\left(h_0{}^{[\eta]} - \sum_p e_p^{[\eta]}\frac{t_p{}^{[\eta]} + \eta\, r_\eta\, t_p^{[\eta]-1}}{1+\eta\, r_\eta} + \bar{\mathcal{E}}_\eta\right)\right], \tag{8.1.16}$$

where we have introduced the following number parity averaged quantities

$$\bar{Z}_\eta = Z^{[\eta,0]}\frac{1+\eta\, r_\eta}{2}, \quad \bar{\mathcal{E}}_\eta = \frac{\mathcal{E}\{\mathcal{R}^{[\eta,0]}\} + \eta\, r_\eta\, \mathcal{E}\{\mathcal{R}^{[\eta,\pi]}\}}{1+\eta\, r_\eta}. \tag{8.1.17}$$

8.2 *Variational equations*

The discussion of the variational principle in Section 5.2 and in particular equation (5.2.16) has established that the opposite of the logarithm of the stationary value of the action $\mathcal{J}_{\eta\,\mathrm{st}}$, for a given set of trial spaces, should be equal to the optimal value of the product of the Gibbs Free energy F_{G} by the temperature. $\mathcal{J}_{\eta\,\mathrm{st}}$ is determined by the stationarity condition of the action (8.1.16) with respect to the parameters which appear in $\hat{W}_{\eta}$;

a) Stationarity with respect to $h_0{}^{[\eta]}$

As $h_0{}^{[\eta]}$ appears explictly in the bracket and implicitly in $\bar{Z}_{\eta}$ through a multiplicative exponential factor, the variation is easy to perform. It yields the equation

$$h_0{}^{[\eta]} = \sum_p e_p^{[\eta]} \frac{t_p{}^{[\eta]} + \eta\, r_\eta\, t_p^{[\eta]-1}}{1+\eta\, r_\eta} + \bar{\mathcal{E}}_\eta. \tag{8.2.1}$$

Reporting this equation into equation (8.1.16) one ontains the expected result:

$$\mathcal{J}_{\eta\,\mathrm{st}} = \bar{Z}_\eta. \tag{8.2.2}$$

b) Stationarity with respect to $\vartheta_p[\eta]$

This angle only appears in $\mathcal{U}_p{}^{[\eta]}$. According to equation (3.2.11) one has to introduce the two quantities

$$\Delta_p^{[\eta,0]} = \frac{1}{2}\sum_q G_{pq} S_q{}^{[\eta]} t_q^{[\eta]}\,, \quad \Delta_p^{[\eta,\pi]} = \frac{1}{2}\sum_q G_{pq} S_q{}^{[\eta]} t_q^{[\eta]-1}. \tag{8.2.3}$$

Then, one checks that the following definition of a number parity projected gap

$$\Delta_{\eta\,p} \equiv \frac{\Delta_p^{[\eta,0]} + \eta r_\eta t_p^{[\eta]-2}\,\Delta_p^{[\eta,\pi]}}{1+\eta r_\eta t_p^{[\eta]-2}}, \tag{8.2.4}$$

allows one to write the stationarity equation in a form similar to equations (3.3.9) and (4.6.1):

$$0 = (\epsilon_p - \mu) S_p{}^{[\eta]} - \Delta_{\eta\,p} C_p{}^{[\eta]}. \tag{8.2.5}$$

A formal solution of equation (8.2.5) is

$$S_p{}^{[\eta]} = \frac{\Delta_{\eta\,p}}{\tilde{e}_{\eta\,p}}, \quad C_p{}^{[\eta]} = \frac{\epsilon_p - \mu}{\tilde{e}_{\eta\,p}} \tag{8.2.6}$$

where we have introduced the BCS-like energies

$$\tilde{e}_{\eta\,p} \equiv \sqrt{(\epsilon_p - \mu)^2 + \Delta_{\eta\,p}{}^2}. \tag{8.2.7}$$

c) Stationarity with respect to $e_p^{[\eta]}$

This variation is more toilsome because $e_p^{[\eta]}$ is present in almost all the terms in the action. Nevertheless, it is technically straightforward and it leads to equation analog to equations (3.3.14)and (3.3.15):

$$e_p^{[\eta]} + \frac{2\eta\, r_\eta\, k_\eta}{t_p{}^{[\eta]} - \eta\, r_\eta t_p^{[\eta]-1}} = (\epsilon_p - \mu) C_p{}^{[\eta]} + \frac{\Delta_p^{[\eta,0]} - \eta\, r_\eta\, \Delta_p^{[\eta,\pi]}}{1 - \eta\, r_\eta t_p^{[\eta]-2}} S_p{}^{[\eta]}\,, \quad (8.2.8)$$

where the c-number k_η is given by

$$k_\eta = \sum_q \frac{t_q^{[\eta]-1} - t_q^{[\eta]}}{1 + \eta\, r_\eta} \left[\left[e_q{}^{[\eta]} - (\epsilon_q - \mu) C_q{}^{[\eta]} \right] - \frac{1}{2} S_q{}^{[\eta]} \left(\Delta_q{}^{[\eta,0]} + \Delta_q{}^{[\eta,\pi]} \right) \right]. \quad (8.2.9)$$

One notes that no equation equivalent to (3.3.14) and (8.2.8) appears in the zero temperature formalisms of Sections 4.5 and 4.6.

The set of equations is now complete. As with BCS, the numerical solution of the problem is more easily performed when equations (8.2.5) and (8.2.9) have been given another more standard form *via* the elimination of $C_p{}^{[\eta]}$ and $S_p{}^{[\eta]}$. From the definitions (8.2.3) and (8.2.4) and equation (8.2.6), we obtain

$$\Delta_{\eta\, p} = \frac{1}{2} \sum_q G_{pq} \frac{\Delta_{\eta\, q}}{\tilde{e}_{\eta\, q}} t_q^{[\eta]} \frac{1 + \eta\, r_\eta\, t_p^{[\eta]-2}\, t_q^{[\eta]-2}}{1 + \eta\, r_\eta\, t_p^{[\eta]-2}}\,, \quad (8.2.10)$$

which is the analog of the BCS gap equation (3.3.3) from which it differs only through the last ratio in the R.H.S.. Another equation

$$\Delta_p^{[\eta,\pi]} - \Delta_p^{[\eta,0]} = \frac{1}{2} \sum_q G_{pq} \frac{\Delta_{\eta\, q}}{\tilde{e}_{\eta\, q}} \left(t_q^{[\eta]-1} - t_q^{[\eta]} \right), \quad (8.2.11)$$

has no BCS counterpart. Finally, it is convenient to rewrite the quantity k_η as

$$k_\eta = \left(\sum_p \frac{t_p^{[\eta]-1} - t_p{}^{[\eta]}}{2(1 + \eta r_\eta)} \frac{1 + \eta r_\eta t_p^{[\eta]-2}}{1 - \eta r_\eta t_p^{[\eta]-2}} \frac{\Delta_{\eta\, p} \left(\Delta_p^{[\eta,0]} - \Delta_p^{[\eta,\pi]} \right)}{\tilde{e}_{\eta\, p}} \right) \times \left(1 + \frac{2\eta r_\eta}{1 + \eta r_\eta} \sum_p \frac{t_p^{[\eta]-2} - 1}{1 - \eta r_\eta t_p^{[\eta]-2}} \right)^{-1} \quad (8.2.12)$$

and to express the quasi-particle energies as

$$e_p^{[\eta]} = \tilde{e}_{\eta\, p} - \frac{2\eta\, r_\eta}{t_p{}^{[\eta]} - \eta\, r_\eta t_p^{[\eta]-1}} \left[k + \frac{\Delta_{\eta\, p} (\Delta_p^{[\eta,\pi]} - \Delta_p^{[\eta,0]})}{\left(t_p{}^{[\eta]} + \eta\, r_\eta t_p^{[\eta]-1} \right) \tilde{e}_{\eta\, p}} \right]. \quad (8.2.13)$$

In this expression of $e_p^{[\eta]}$, the second term in the bracket can be evaluated by means of equation (8.2.11).

One notes that this set of equations reduces to standard BCS when one sets $\eta = 0$. In the gap equation (8.2.10) the modification introduced by the number parity projection concerns only the last ratio in the R.H.S.. Its dependence on $t_q^{[\eta]-2}$ shows that contributions to the sum only come from terms such that $\beta e_q^{[\eta]}$ is small. As expected, only single particle states near the Fermi surface play a significant rôle. The dependence on η is such that the fraction is larger (resp. smaller) than one for $\eta = +$ (resp. $\eta = -$). The gap is thus increased (resp. decreased) with respect to BCS.

8.3 Average values and thermodynamic potentials

Using the expressions for ${C_p}^{[\eta]}$ and ${S_p}^{[\eta]}$ resulting from the solution of equation (8.2.6), one can write expectation values of all operators. For the energy, equation (8.1.7) yields

$$\langle \hat{K} \rangle^{[\eta]} = \bar{\mathcal{E}}_\eta = \sum_p \left[(\epsilon_p - \mu) - \frac{(\epsilon_p - \mu)^2 + \frac{1}{2}{\Delta_{\eta\, p}}^2}{\tilde{e}_{\eta\, p}} \frac{{t_p}^{[\eta]} + \eta\, r_\eta\, t_p^{[\eta]-1}}{1 + \eta\, r_\eta} \right], \tag{8.3.1}$$

while the particle number average value is

$$\langle \hat{N} \rangle^{[\eta]} = \frac{\mathrm{Tr}\ \hat{\mathcal{D}}_\eta \hat{N}}{\mathrm{Tr}\ \hat{\mathcal{D}}_\eta} = \mathcal{N}\{\bar{\mathcal{R}}_\eta\} = \sum_p \left[1 - \frac{\epsilon_p - \mu}{\tilde{e}_{\eta\, p}} \frac{{t_p}^{[\eta]} + \eta\, r_\eta\, t_p^{[\eta]-1}}{1 + \eta\, r_\eta} \right]. \tag{8.3.2}$$

In the latter equation, the second equality results from equation (3.2.6) used with the parity projected density matrices

$$\bar{\mathcal{R}}_{\eta\, p} = \frac{{\mathcal{R}_p}^{[\eta,0]} + \eta\, r_\eta\, {\mathcal{R}_p}^{[\eta,\pi]}}{1 + \eta\, r_\eta}. \tag{8.3.3}$$

Since, although $\hat{\mathcal{D}}^{[\eta]}(\beta)$ has the correct number parity, one is still working in the grand canonical formalism, the equation (8.3.2) must be used to determine μ through the constraint

$$\langle \hat{N} \rangle^{[\eta]} = N_0\ . \tag{8.3.4}$$

From equations (5.2.13, 5.2.16) and (8.2.2) one obtains

$$F_{\mathrm{G}}{}^{[\eta]} = -\frac{1}{\beta} \ln \bar{Z}_\eta, \tag{8.3.5}$$

which according to equations (8.1.9, 8.1.17, 8.2.1, 8.2.7) and (8.2.7) satisfies the thermodynamics relation

$$\frac{\partial {F_{\mathrm{G}}}^{[\eta]}}{\partial \mu} = -\langle \hat{N} \rangle^{[\eta]} = -N_0. \tag{8.3.6}$$

Finally, equation (5.2.17) can be used to get an explicit form for the mean-field parity number projected entropy

$$\begin{aligned} S^{[\eta]} &= \beta(\bar{\mathcal{E}}_\eta - F_{\mathrm{G}\,\eta}) \\ &= \sum_p \left(2\ln\left(2\cosh\beta\frac{e_p^{[\eta]}}{2}\right) - \beta e_p^{[\eta]} \frac{{t_p}^{[\eta]} + \eta\, r_\eta\, t_p^{[\eta]-1}}{1+\eta\, r_\eta} \right) \\ &\quad + \ln\left[\frac{1}{2}\left(1+\eta\, r_\eta\right)\right] \end{aligned} \tag{8.3.7}$$

to be compared with the BCS expression (3.3.6). In contrast with the derivation of BCS equations in Section 3.3 which were based on the *prior* evaluation of an expression of the entropy associated with the operator $\hat{T}^{[d]}$, here, the entropy is only obtained at the *end* of the calculation as pointed out at the end of Section 5.2.

Similar, although not quite equivalent, equations have been given and discussed in [25–30].

8.4 *Small temperatures*

Since, up to now, small superconducting systems are experimentally investigated mostly at zero or small (compared to the gap) temperatures, it is worthwhile to analyze the large β behavior of the equations derived in Section 8.4

8.4.1 Even number systems

For even systems, it turns out that first order terms in the expansions vanish. These should therefore be carried out to second order in order to extract the dominant temperature behavior. From the expansion of the thermal factors

$${t_p}^{[+]} \simeq 1 - 2\mathrm{e}^{-\beta e_p^{[+]}} + 2\mathrm{e}^{-2\beta e_p^{[+]}} + o\left(2\mathrm{e}^{-2\beta e_p^{[+]}}\right), \tag{8.4.1}$$

one deduces that

$$\ln r_+ \simeq -4 \sum_p \mathrm{e}^{-\beta e_p^{[+]}} + o\left(2\mathrm{e}^{-2\beta e_0^{[+]}}\right), \tag{8.4.2}$$

where $e_0^{[+]}$ is the smallest of the quasiparticle energies $e_p^{[+]}$.

The extreme low (see meaning of this adjective in Section 3.4) temperature form of the projected gap equation (8.2.10) is

$$\Delta_{+p} = \frac{1}{2} \sum_q G_{pq} \frac{\Delta_{+q}}{\tilde{e}_{+p}} \left(1 - 2\mathrm{e}^{-2\beta e_q{}^{[+]}} - 4 \sum_{r \neq p,q} \mathrm{e}^{-\beta(e_q{}^{[+]} + e_r^{[+]})}\right). \tag{8.4.3}$$

For $T = 0$, one recovers the BCS equation which yields the solution (2.3.6) which is obviously a superposition of states with an even number of particles. On the other hand, in contrast with equation (3.4.1) the temperature behavior is governed by terms involving the sum of two quasiparticle energies. These correspond to the simplest excitations in an even system (see discussion of Fig. 1). As for BCS, an increase of the temperature induces a decrease of the gaps.

At small T, the entropy (8.3.7)

$$S^{[+]} = 2\beta \sum_p e_p^{[+]} \left[\mathrm{e}^{-2\beta e_p^{[+]}} + 2 \sum_{q \neq p} \mathrm{e}^{-\beta(e_p^{[+]} + e_q{}^{[+]})}\right], \tag{8.4.4}$$

also has a correct temperature dependence in contrast with the BCS expression (3.4.2).

The condensed matter low temperature limit (defined by the inequalities (3.4.4)) of the even parity number gap is

$$\Delta_+(\beta) = \Delta\left(1 - 4\pi \frac{w}{\beta} \mathrm{e}^{-2\beta\Delta}\right), \tag{8.4.5}$$

where Δ is the $T = 0$ BCS gap defined by equation (3.4.3).

In the same limit, the even number parity entropy is

$$S^{[+]} = 8\pi (w\Delta)^2 \mathrm{e}^{-2\beta\Delta}, \tag{8.4.6}$$

which can be compared to equation (3.4.7).

8.4.2 Odd number systems

When $\eta = -$, the analysis of the last ratio in the R.H.S of equation (8.2.10) requires some care. Indeed, since it takes the undetermined form 0/0, one must first extract from the numerator and the denominator the dominant terms which behave as $\exp(-\beta e_0^{[-]})$ where $e_0^{[-]}$ is the lowest quasi particle energy among the $e_p^{[-]}$'s.

Then the extreme low temperature limit of the equation (8.2.10) for $p \neq 0$ is

$$\Delta_{-p} = \frac{1}{2}\sum_{q\neq 0} G_{pq}\frac{\Delta_{-q}}{\tilde{e}_{-p}} + \left(G_{p0}\frac{\Delta_{-0}}{\tilde{e}_{-0}} - G_{pq}\frac{\Delta_{-q}}{\tilde{e}_{-q}}\right) \mathrm{e}^{-\beta(e_q{}^{[-]} - e_0^{[-]})}. \tag{8.4.7}$$

At $T = 0$, one does not recover the usual BCS gap equation. Instead, one obtains a "blocked" equation in which the single particle levels 0 and $\bar{0}$ are eliminated from the sum. Thus, our formalism recovers automatically the variational equation which determines the coefficients u_p and v_p ($p \neq 0$) of the state

$$|\Psi_{[\text{Blocked BCS}]}\rangle = a_0^\dagger \prod_{p\neq 0}(u_p + v_p a_p^\dagger a_{\bar{p}}^\dagger)|0\rangle \tag{8.4.8}$$

which replaces $|\Psi_{[\text{BCS}]}\rangle$ given by equation (2.3.6) for a BCS description of odd systems. The gap equation (8.2.10) not only contains a blocked equation as a limit, but it also selects the single particle state which must be blocked. The blocking of a state near the Fermi surface leads to a reduction of the gaps Δ_{-p} at $T = 0$ as compared to $\Delta_{+p} = \Delta$.

In equation (8.4.7), the temperature dependence is governed by the quasi particle energy differences $e_p^{[-]} - e_0^{[-]}$. In a perturbative scheme, this corresponds to a promotion of the unpaired particle from the state 0 to the state p.

In the same low T limit the entropy (8.3.7) becomes

$$S^{[-]} = \ln 2 + \beta \sum_p \left(e_p^{[-]} - e_0^{[-]}\right) \mathrm{e}^{-\beta(e_p^{[-]} - e_0^{[-]})}, \tag{8.4.9}$$

where the term $\ln 2$ reflects the degeneracy of the ground state.

In order to extract the condensed matter limit expressions of the gap equation and of the entropy, one must start from the original equations (8.2.10) and (8.3.7). Indeed, the condition (3.4.4) which insures that the temperature is smaller than $e_p^{[+]} + e_p^{[q]}$ does not imply that it is less than $e_p^{[-]} - e_0^{[-]}$ as it has been assumed to obtain equations (8.4.7) and (8.4.9). Using the formula (3.4.5). the quantity r_- can be approximated as

$$\ln r_- = -4\,w\sqrt{\frac{2\pi\Delta_-}{\beta}}\,\mathrm{e}^{-\beta\Delta_-}\left(1 + \frac{3}{8\beta\Delta_-}\right). \tag{8.4.10}$$

Then, after a somewhat lengthy calculation, one obtains

$$\Delta_-(\beta) = \Delta - \frac{1}{2w} + \frac{1}{4\beta\, w\, \Delta}\cdot \tag{8.4.11}$$

One checks that at $T = 0$, the gap Δ_- is smaller than Δ. This is a consequence of the blocking effect which eliminates one level at the Fermi surface. As a matter of fact, $\Delta_-(0)$ is solution of the equation

$$\frac{2}{G} = \int_{-\Lambda/2}^{-1/2w} \frac{w\,\mathrm{d}\epsilon}{\sqrt{\epsilon^2 + \Delta_-(0)^2}} + \int_{1/2w}^{\Lambda/2} \frac{w\,\mathrm{d}\epsilon}{\sqrt{\epsilon^2 + \Delta_-(0)^2}}\,, \tag{8.4.12}$$

which only differs from equation (3.4.3) by the suppression of an interval of width $1/w$ at $\epsilon = 0$.

An analysis of the entropy performed along the same lines gives

$$S^{[-]} = \ln 2 + \frac{1}{2} + \frac{1}{2} \ln\left[\frac{2\pi\, w^2 \Delta}{\beta}\right]\cdot \tag{8.4.13}$$

Apart from the contribution $\ln 2$ associated with the ground state degeneracy, one finds an expression like the Sackur Tetrode entropy for a classical gas of *one* particle moving on a line. Thus, the limit at low temperature of the heat capacity $C = T\mathrm{d}S^{[-]}/\mathrm{d}T$ is $1/2$. Of course, this result is only valid as long as the condition (3.4.4) is satisfied. At temperatures such that $wT \ll 1$, the correct formula for the entropy is (8.4.9).

8.5 Numerical illustration

A solution of equations (8.2.4, 8.2.7, 8.2.10–8.2.13) can be easily performed by standard iteration methods for any type of single particle spectrum. To pursue the analysis, we consider a simple model such that the energies ϵ_p the twofold degenerate levels are regularly distributed with a density w. For the numerical application presented here, we have adjusted the interaction strength so that the quantity $w\Delta$ is equal to 10. The active space is defined by a cutoff energy Λ such that $\Lambda/\Delta = 20$ (*i.e.* 201 $p, \bar{p}$ levels belong to the active space). This is realized for a pairing strength $G_{pq} = \tilde{G}(1 - \delta_{pq})$ with $w\tilde{G} = 0.34$. This corresponds to a situation intermediate between nuclei ($w\Delta \approx 4$) and superconducting metallic islands $w\Delta > 50$. When temperatures are measured in units of the BCS gap at $T = 0$, the BCS critical temperature T_c corresponds to the value 0.56.

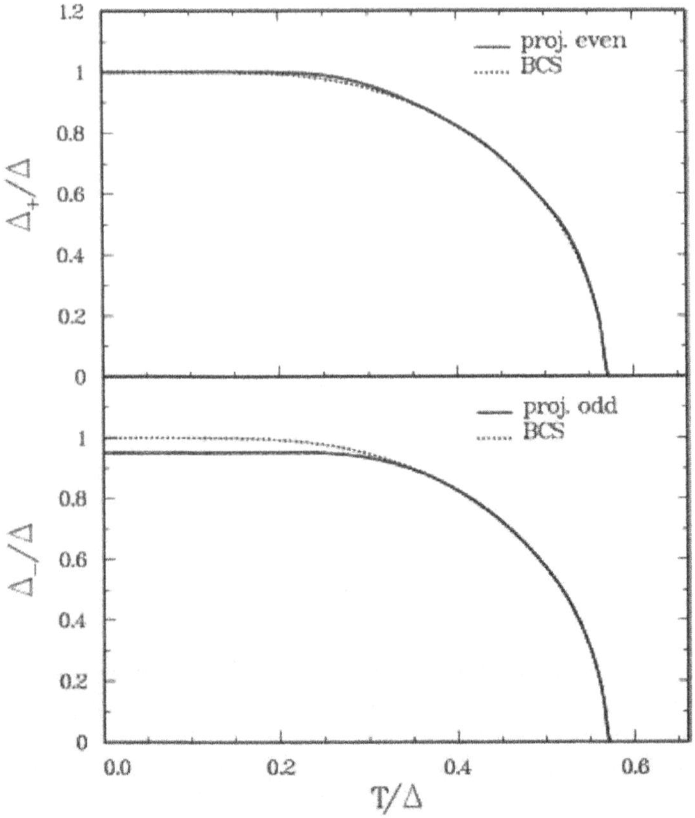

Fig. 3. Comparison of BCS and number-parity projected gaps at the Fermi surface *versus* temperature for $w\Delta = 10$ and $w\Lambda = 200$. See text.

The upper part of Figure 3 shows the gap for an even system with two hundred particles. Thus, half of the levels in the pairing active space are filled in the absence of pairing interaction. According to equation (8.2.10) the values of the gap depend on p. On the other hand, direct computation shows that the variation *versus* p never exceeds a few % and can therefore be neglected as will be done in the following analysis. We plot our results for the level 0 corresponding to the smallest quasi particle energy. The even number parity projected gap is larger than the BCS value as is expected from the discussion at the end of Section 8.2. However, the effect is very small.

The lower part of the same figure shows the results obtained when $\eta = -$. At $T = 0$, due to blocking, the gap Δ_- is less than Δ. Then, as the temperature increases, so does the gap $\Delta_-(\beta)$ until the curve merges with that of the BCS approximation at a temperature $T/\Delta \simeq 0.35$.

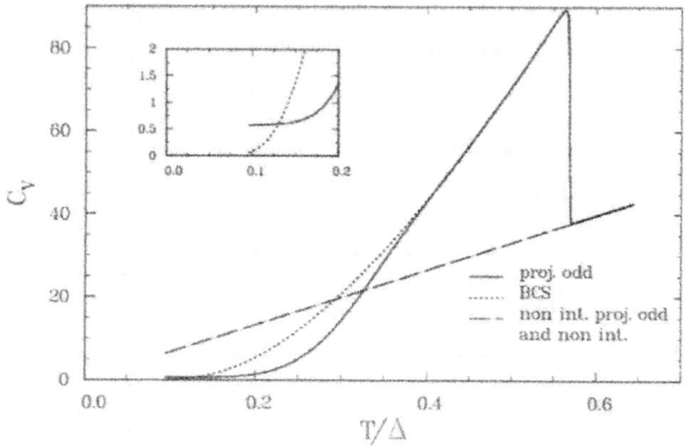

Fig. 4. Specific heat of an odd-number system for $w\Delta = 10$. See text.

Figure 4 illustrates our discussion of the entropy in odd systems of Section 8.4. It displays the evolution of the specific heat C_V calculated as $T(\mathrm{d}S^{[-]}/\mathrm{d}T)$ *versus* temperature for the same system. From the transition temperature downward, the odd number projected specific heat first becomes smaller than the BCS value. On the other hand, the inset shows that at small T, it crosses the BCS curves and approaches the limit 1/2 as predicted by equation (8.4.13). As compared with the specific heat of a non interacting fermion system ($S = (2\pi^2 wT)/3$) ("non int." curve) the BCS and odd number parity projected curves exhibit the well known discontinuity at the transition temperature. We remind the reader, that, according to equation (8.4.9), at $T = 0$, the odd number parity entropy reaches the value ln2 associated with the degeneracy of the ground state and that C_V vanishes.

9 Odd–even effects

9.1 Number parity projected free energy differences

Up to now we have been working in the grand canonical formalism and have been concerned mostly with the Gibbs free energies ${F_G}^{[\eta]}$. For systems with a definite number of particles N_0, one must instead consider the free energy $F^{[\eta]}$ which is related to ${F_G}^{[\eta]}$ by a Legendre transform

$$F^{[\eta]}(N_0) = {F_G}^{[\eta]}(\mu) + \mu N_0, \tag{9.1.1}$$

where the chemical potential μ is adjusted so that equation (8.3.6) is satisfied.

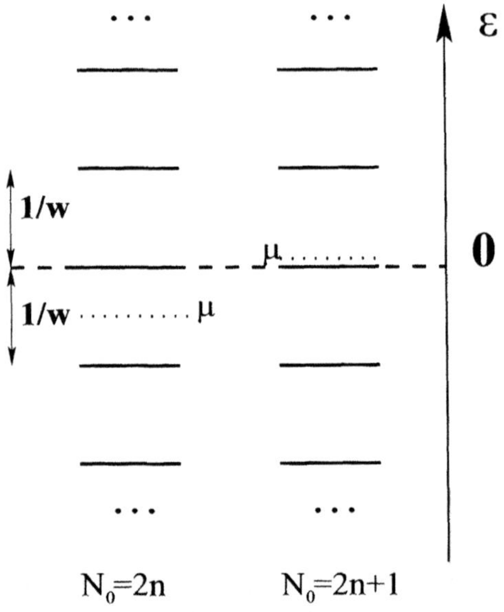

Fig. 5. Evolution of the Fermi level μ from an even ($N_0 = 2n$) to an odd ($N_0 = 2n+1$) value of $\langle \hat{N} \rangle^{[\eta]}$ for $\eta = +, -$ or 0.

Although $F^{[+]}$, ${F_G}^{[+]}$ only have a physical meaning for even systems and $F^{[-]}$, ${F_G}^{[-]}$ for odd systems, they can as well be evaluated for any (even non integer) value of the expectation value $\langle \hat{N} \rangle^{[\eta]}$ by means of a continuous modification of μ. We have done it for the physical system analyzed in Section 8.5. For both $\eta = \pm$ (and $\eta = 0$ in the BCS case) we have let $\langle \hat{N} \rangle^{[\eta]}$ vary from $2n = 200$ to 201. In the three cases, we find, as intuition suggests, that the chemical potential μ increases continuously in the interval $[-1/2w, 0]$ as sketched in Figure 5. An accurate approximation of the relation between μ and the average number of particle $\langle \hat{N} \rangle^{[\eta]}$ is provided by the relation

$$\mu(2n + x) = -\frac{1}{2w}(1 - x). \tag{9.1.2}$$

At finite temperature, the relevant difference between odd and even neighbour systems is

$$\delta F = F^{[-]}(2n + 1) - F^{[+]}(2n), \tag{9.1.3}$$

with the free energies $F^{[\eta]}$ defined by equation (9.1.1). For the system considered in Section 8.5, it can be computed directly from the numerical solutions of the equations derived in Section 8.2. The result is plotted as the solid curve in Figure 6. Two regimes are clearly visible. At small temperatures, the difference (9.1.3) decreases almost linearly from a value very close to Δ at $T = 0$, until it vanishes at a crossover temperature close to 0.35Δ. Above this temperature the difference remains constant and equal to a small negative value $\simeq -0.05\Delta$. Although superconductivity still exists up to the transition temperature $T_c = 0.56\Delta$ (see Fig. 4), δF does not exhibit any modification when the pairing disappears. In other words, above the crossover temperature and below T_c, the odd-even difference of the free energies does not reflect the still existing pairing correlations.

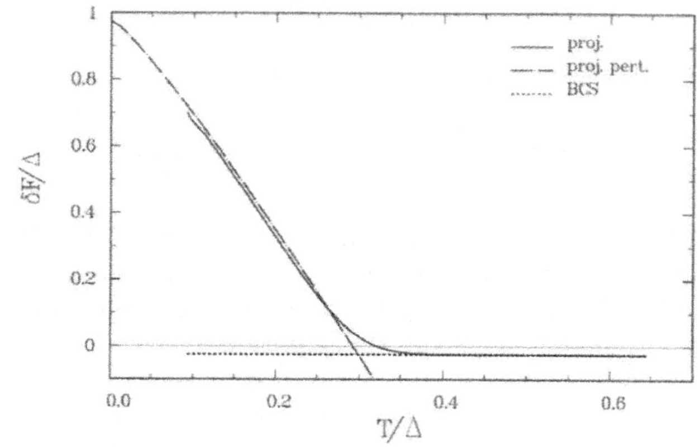

Fig. 6. Difference between the free energies of neighbouring odd ($N_0 = 201$) and even ($N_0 = 200$) systems as a function of temperature for $w\Delta = 10$ and $w\Lambda = 200$.

This phenomenon has been mapped out carefully in the experiments reported in [9].

We will now attempt a semiquantitave analysis of this result. Based on the discussion at the beginning of this section, we consider a path which connects the systems $N_0 = 2n$ and $N_0 = 2n + 1$. The difference δF is written as a sum of two contributions

$$\delta F = \delta_1 F + \delta_2 F$$

$$\delta_1 F = F^{[-]}(2n+1) - F^{[+]}(2n+1) \ , \quad \delta_2 F = F^{[+]}(2n+1) - F^{[+]}(2n) \ . \tag{9.1.4}$$

Since for both values of η the chemical potential $\mu(\langle \hat{N} \rangle^{[\eta]} = 2n+1)$ vanishes, one obtains

$$\delta_1 F = F_G{}^{[-]}(\mu = 0) - F_G{}^{[+]}(\mu = 0) = -\frac{1}{\beta} \ln \frac{\bar{Z}_-}{\bar{Z}_+} \simeq -\frac{1}{\beta} \ln \frac{1 - r_-}{1 + r_+} \ , \tag{9.1.5}$$

by means of equations (8.3.5) and (8.1.7) and with the assumption that $\ln Z^{[+,0]} \simeq \ln Z^{[-,0]}$. As a matter of fact, from the numerical solution one can check the validity of this assumption. For small temperatures, it is justified to use the approximation (8.4.10) of $\ln r_-$. We then obtain (in [24], we assumed $\Delta_- = \Delta$ instead of equation (8.4.11))

$$\delta_1 F \simeq \Delta - \frac{1}{2w} - \frac{T}{2}\ln\left(\frac{8(\pi w \Delta)^2}{\beta\Delta}\right). \tag{9.1.6}$$

The contribution $\delta_2 F$ can be evaluated from the thermodynamic relation valid for the free energy

$$\frac{\partial F^{[\eta]}}{\partial N_0} = \mu. \tag{9.1.7}$$

Using the approximation (9.1.2), we calculate

$$\delta_2 F = \int_{2n}^{2n+1} \mu(N_0)\mathrm{d}N_0 = -\frac{1}{4w}. \tag{9.1.8}$$

In Figure 6, the dotted curve corresponds to $\delta_2 F$ (it is labelled BCS since the value $-1/4w$ is the result obtained for the BCS theory at any temperature) while the dash-dot curve is $\delta_1 F$. The quality of the agreement vindicates our analysis, although it relies on the definition of a path such that, except for the end points, there exists no corresponding physical system.

Below the crossover temperature, pairing effects are detectable in the difference δF. Above this temperature, there only remain effects associated with the discreteness of the spectrum measured by level density w. The range of temperature over which pairing correlations affect δF is determined by the slope of the curve $\delta_1 F$. This slope increases as the size of the system grows (w decreases). For sufficiently large systems, the temperature range over which odd-even effects are visible will be out of reach for present experimental cooling techniques. In such a situation, number-parity projected BCS and standard BCS make identical predictions.

9.2 Nuclear odd–even energy differences

Although the last section was concerned with the temperature dependence of the odd-even free energy difference, Figure 6 shows that, at $T = 0$, δF which becomes the difference between the binding energies $\delta\langle\hat{H}\rangle$, is equal to $\Delta - 1/2w$. This quantity involves contributions from both pairing and single particle spectrum origin. In this section, we report on a detailed analysis of this point in the context of nuclear physics.

Until recently, within the nuclear physics community there existed a consensus as to how the observed odd-even staggering of the binding energies should be interpreted and used to extract information on pairing properties [3, 31]. This state of affair was somewhat disturbed by a work which, among several other points, analyzed the well established odd-even staggering of clusters binding energies [32] and explained it solely in terms of shape deformation [33]. Its authors asserted that connecting the observed staggering to pairing correlations was not justified for clusters and therefore probably also not for nuclei. Since the reasons to believe in the importance of pairing concepts for the dynamics of nuclei extends well beyond the odd-even mass staggering effects (see Introduction), the authors of [34, 35] took up the challenge and decided to review the analysis of nuclear data which had been performed in the old days. It is their work that this section summarizes.

Rather than the simple difference discussed in Section 9.1, for systems at zero temperature it is common to measure the amplitude of the odd-even staggering by means of "filters" on the binding energy. The simplest of these filters is an average of differences involving the binding energies of three neighbouring systems

$$\Delta_\nu^{(3)}(N) = \frac{(-1)^N}{2}\left[E(N-1) - 2E(N) + E(N+1)\right]. \tag{9.2.1}$$

For nuclei, the index ν specifies the type of particle (neutron or proton) whose number N is involved in the differences.

Even in the absence of pairing, the twofold Kramers degeneracy of single particle levels naturally leads to odd-even oscillations of $\Delta_\nu^{(3)}$. For instance, for the one-body Hamiltonian $\hat{H}_0$ (2.3.1) (we assume here that the energies ϵ_p are increasing with the index p), one has

$$\begin{aligned} \Delta_\nu^{(3)}(2n+1) &= 0 \\ \Delta_\nu^{(3)}(2n) &= \frac{1}{2}(\epsilon_{n+1} - \epsilon_n). \end{aligned} \tag{9.2.2}$$

Hence, for a non interacting Hamiltonian, the filter (9.2.1) vanishes for odd particle numbers and gives half the single-particle spacings at even particle numbers. A first conclusion is that the non zero value of the filter does not establish the existence of pairing correlations.

A straightforward generalization of the relation (9.2.2) to the full Hamiltonian (3.2.7), leads to the prescription (9.2.3) which is pictured in Figure 7: the filters centered at odd values determine pairing gaps while the difference between consecutive filters gives access to the single particle spacings

$$\begin{aligned} \Delta &= \Delta_\nu^{(3)}(2n+1) \\ \epsilon_{n+1} - \epsilon_n &= 2\left[\Delta_\nu^{(3)}(2n) - \Delta_\nu^{(3)}(2n+1)\right]. \end{aligned} \tag{9.2.3}$$

As shown by [34], it is a remarkable and somewhat surprising fact that that this simple prescription remains the correct one for strongly correlated

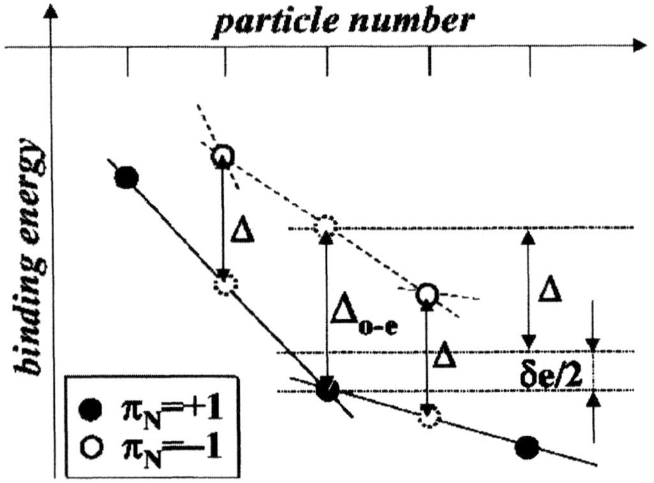

Fig. 7. From [34] Principle of the analysis of the odd-even staggering of the nuclear binding energies according to equation (9.2.3). The particle number corresponds either to neutron or proton.

systems such as the nuclei. On the other hand, it cannot be expected to be valid for any system and any Hamiltonian. Indeed, in general, because particles are interacting between themselves or with partners (proton and neutrons in nuclei) or with the background (interaction with surface for the clusters), one should expect a deviation from the relation (9.2.3) even when pairing forces are absent. An example is provided by the metallic clusters where there are good reasons to believe that the pairing mechanism is either not present or of little importance and for which the odd and even centered filters have approximately the same absolute value (Ref. [33]).

Reference [34] analyzes separately the contributions to the filters of the several components of the total energy of a nucleus. One part of this work relies on the decomposition of the Hartree–Fock energy proposed by Strutinsky [36]. According to this author, the total energy can first be written as the sum of the pairing energy and a mean-field energy. The latter is again split into three contributions associated respectively with i) a one-body Hamiltonian similar to $\hat{H}_0$ (2.3.1) (although it is in principle self consistently defined), ii) a single-particle smooth background and iii) a liquid drop bulk energy. Then, from the well established knowledge of the liquid drop and the single particle densities, the authors of [34] show that the contributions to the filters of the last two terms cancel each other almost exactly. They conclude that, effectively, the only non vanishing contributions to the filter (9.2.1) come from the one-body and the pairing contribution.

For nuclei, it is therefore justified to analyze data by means of the simple prescription (9.2.3).

Entering the details of this analysis goes beyond the scope of these notes. Moreover, Ref. [34] gives another more direct justification of the prescription (9.2.3) based on a numerical instead of a theoretical analysis. Figure 8 displays the filter $\Delta_\nu^{(3)}$ calculated from numerical Hartree-Fock (HF) energies computed with the effective interaction SIII [37]. This force has been tested with success on the nuclear energies over the entire chart of isotopes (that is for binding energies varying over almost two orders of magnitude) as well as on many other nuclear properties including multipole deformations. Such HF energies take into account the contributions of the one-body and two-body parts of the Hamiltonian but not those due to the pairing interaction. Indeed, the HF formalism (in contrast with Hartree Fock Bogoliubov) automatically eliminates the pairing terms from the mean-field description. The figure shows that the odd filters vanish and thus confirms the analysis performed on the smooth Strutinsky contributions. In addition, the authors of Ref. [34] show that the even filters give a good description of the Hartree-Fock single particle spacing.

Now that the prescription (9.2.4) has been established on firm ground for nuclei, it can be used to analyze the experimental binding energies [38]. This gives an access both to the pairing gap and to the single-particle spectrum.

In Figure 9, the black dots associated with odd centers filters $\Delta_\nu^{(3)}(2n+1)$ provide an experimental estimate of the neutron pairing gap. One expects the curve to display minima at magic numbers since the reduced level density decreases the effectiveness of the pairing Hamiltonian. This is verified for the neutron numbers $N = 28$ and 50 and (not visible on this figure) the numbers 82 and 126. On the other hand, there is no indication of a minimum at $N = 20$ while there is a deep minimum near $N = 16$. This result may be related with the experimentally established weakness of the magic number 20 which has been particularly investigated because of its possible implications on the astrophysical nucleosynthesis processes [39].

The vertical difference between the white dot curve and the black dot curve yields information on the neutron single particle level spacing at the fermi surface. One expects this difference to be large near magic nuclei. This is indeed the case for $N = 28$, 50, 82 and 126. One observes a structure at $N = 20$ which provides an evidence for an increased level spacing although still smaller for instance than at $N = 16$.

Each point in Figure 9 is an average over several isotopes with even values of the proton number Z. This averaging procedure may affect the conclusion when the nuclear structure is locally dependent on the number of protons. Nevertheless, the picture which results from the present analysis

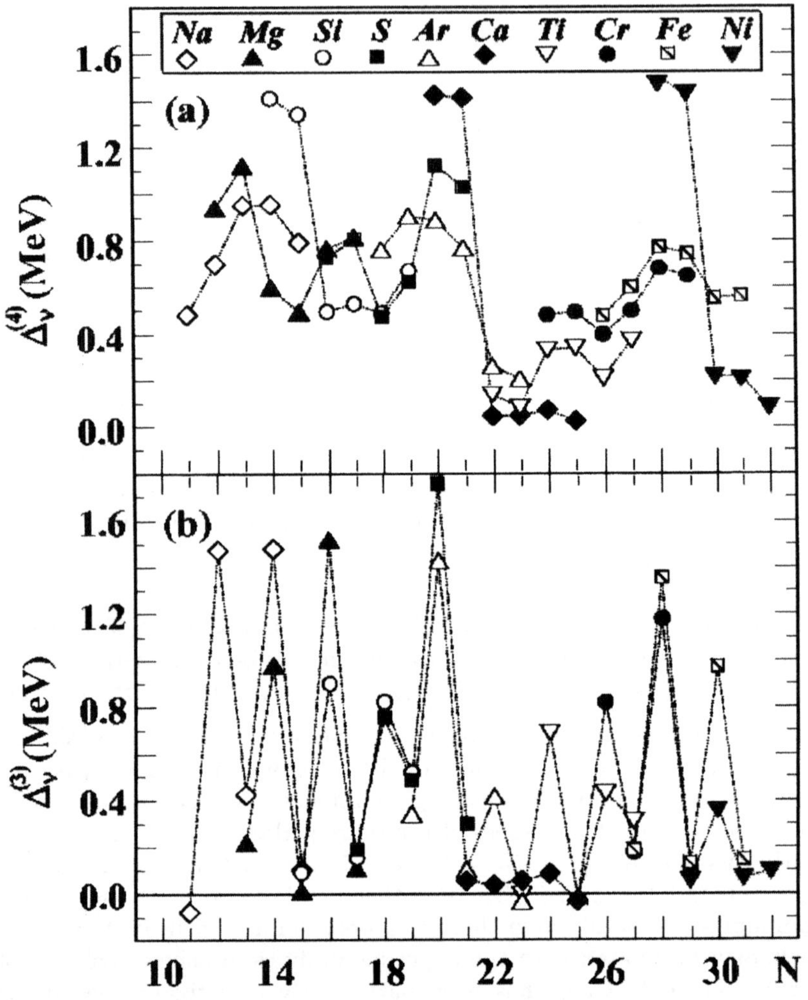

Fig. 8. From [34]. Odd and even centered Filters $\Delta_\nu^{(3)}$ evaluated from Hartree–Fock binding energies with the effective interaction SIII.

is much more convincing than that provided by the fourth order filter

$$\Delta_\nu^{(4)}(N) = \frac{1}{2}\left[\Delta_\nu^{(3)}(N) + \Delta_\nu^{(3)}(N+1)\right], \tag{9.2.4}$$

which, until now, was considered to yield "cleaner" values of the gap because it corresponds to a higher order Taylor expansion [3, 31]. The above analysis shows that $\Delta_\nu^{(4)}$ is a somewhat unsatisfactory quantity which mixes

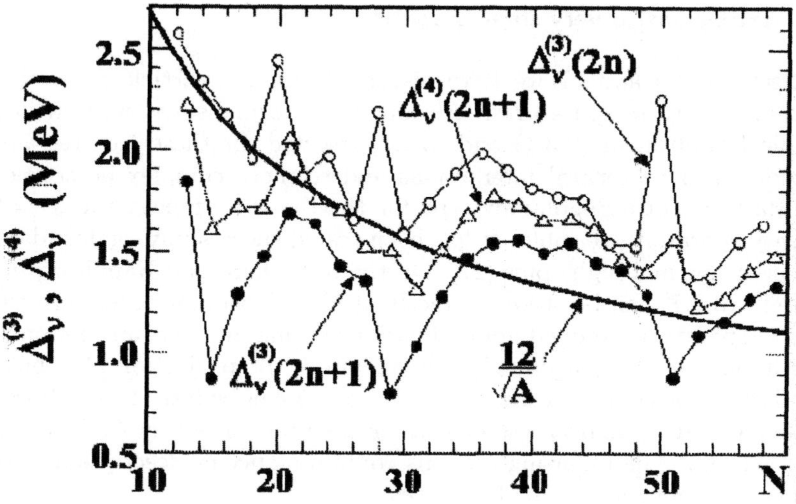

Fig. 9. From [34]. Neutron odd and even centered filters $\Delta_\nu^{(3)}$ and odd centered filter $\Delta_\nu^{(4)}$ evaluated from experimental nuclear binding energies.

contributions with distinct physical contents. Moreover, if one keeps in mind that the addition of only a few nucleons sometimes leads to a significant reorganization of the mean field potential and of the single particle spectra, one has other reasons to remain cautious when using filters which involve a large number of nuclei.

In the past, results based on the filter $\Delta_\nu^{(4)}$ have been used to establish a reference function giving the average pairing gap over the chart of isotopes: $\Delta \simeq 12/\sqrt{A}$ $(A = N + Z)$ [3]. In the light of the analysis of [34], a slower decrease *versus* A reproduces more accurately the average behaviour of the data. It turns out that this is consistent with theoretical findings whether they are estimated by means of a zero range force or obtained with the most refined Hartree–Fock Bogoliubov calculations [40].

To conclude this section, one may say that the nuclear physics community can be grateful to the cluster physics community; not so much because of the somewhat controversial conclusions on nuclear pairing that some members of the latter community infered from the findings in their field, but because the wave they generated has led nuclear physicists to reassess what seemed to be a firmly established corpus of knowledge. Scientific progress travels on unexpected routes.

10 Extensions to very small systems

In Sections 4.4 and 7 we have discussed number particle projection at zero and finite temperatures. Although the formalism is written for the simple Hamiltonian (2.3.1) and (3.2.7) selected for these lectures, an extension to more general interactions and to more complex description of the Hartree–Fock–Bogoliubov type for the the wave-function at $T = 0$ or density operator is feasible. It has been tested successfully against data for nuclei at low and high spins [21]. On the other hand, an Hamiltonian very close to (3.2.7) can be studied exactly at $T = 0$. Moreover, for zero and finite temperatures, refined approximations taking into account not only the fluctuations of the gauge angle φ by means of the projection $\hat{P}_{N_0}$, but also the fluctuations of the magnitude of the gap Δ can be introduced. Recently, these methods have been used to attack the problem set by Anderson question [44–50]. The following sections contain a succint description of those works and comment some of their results.

10.1 Zero temperature

From now on, we consider the schematic Hamiltonian

$$\hat{H} = \hat{H}_0 - G\hat{P}^\dagger\hat{P} \ , \quad \hat{P}^\dagger = \sum_p a_p^\dagger a_{\bar{p}}^\dagger, \tag{10.1.1}$$

where $\hat{H}_0$ is defined as in (2.3.1) and G is a positive constant. (The main difference between (10.1.1) and (3.2.7), is that in the latter we chose the matrix elements G_{pp} to be zero. This ensures that there is no contribution of the two-body interaction to the average particle potential (no state dependent shift of the single particle energies ϵ_p) and it leads to shorter formulas).

In the Hamiltonian (10.1.1), the two-body potential scatters pairs from any fully occupied level p to any fully empty level q, while it leaves half filled levels unchanged. In order to expand exact solutions with a particle number equal to N_0, it is therefore convenient to label the basis states by the number of unpaired particles N_u and the list of levels $[n_j]$ which are partially filled:

$$|[n_j]_{N_u}[m_i]_{N_P}\rangle = \prod_{j=1}^{N_u} a_{\tilde{n}_j}^\dagger \prod_{i=1}^{N_P} a_{m_i}^\dagger a_{\bar{m}_i}^\dagger |0\rangle. \tag{10.1.2}$$

Here $\tilde{n}_j$ specifies which of the two states n_j or $\bar{n}_j$ is occupied. The indices $[m_j]$ characterize the fully occupied single particle states. The collections of indices $[n_j]$ and $[m_i]$ must not overlap. The quantities N_u and N_P are connected to the particle number by the relation

$$N_\mathrm{u} + 2N_\mathrm{P} = N_0. \qquad (10.1.3)$$

Since the subspaces of the Hilbert associated with distinct collections of indices $\tilde{n}_j$ are unconnected by the Hamiltonian, N_u (or N_P) is an invariant which can be used to sort the eigenstates into families of solutions. For a given set of indices $\tilde{n}_j$ the solution of the problem amounts to the diagonalization of a matrix with dimension $\binom{\Omega-N_\mathrm{u}}{N_\mathrm{P}}$. This matrix is sparse since, apart from the diagonal, only the matrix elements between states differing by at most one index in their collections of indices m_j are connected by the two-body force. On a given line of the matrix, the number of non zero matrix elements (all equal to $-G$) is $N_\mathrm{p} \times (\Omega - N_\mathrm{u} - N_\mathrm{p})$. This number is thus much smaller than the dimension of the matrix.

At $T = 0$, for negative values of G the ground state of an even system will always be in the subspace of dimension $\binom{\Omega}{N_0/2}$ corresponding to states $|[m_j]_{N_0/2}\rangle$. For an odd system, several (at most Ω) calculations may be necessary in subspaces of dimension $\binom{\Omega-1}{(N_0-1)/2}$ after selection of an unpaired level occupied by only one particle. In practice, fewer calculations suffice, since the optimal state to be blocked must be in the vicinity of the Fermi level. For studies at finite temperature, in principle, one has to diagonalize the Hamiltonian in all the subspaces compatible with the relation (10.1.3). The canonical thermal averages of operators are then obtained by sums with Boltzmann weight factors. At low temperature, it is not necessary to diagonalize $\hat{H}$ in the Hilbert subspaces such that the energy of the lowest eigenstate is much larger than the temperature.

Although for large values of Ω and N_0 the problem remains of considerable size, it is within reach of today's computational means for dimensions compatible with systems of interest (from one digit to three-digit numbers of interacting particles and levels). Note that, in shell model calculations of nuclear structure, one now routinely solves diagonalization problems involving much more complex Hamiltonians than (10.1.1) and matrices with dimensions exceeding 10^{15} [51]. As a matter of fact, for the simple Hamiltonian (10.1.1), a remarkable analysis of the problem by Richardson, based on a quasi boson representation, has shown that the determination of the ground state energy and wave-function in every subspace can be effected by means of a coupled set of $\Omega - N_\mathrm{u}$ non linear equations [52]. Using one of the aforementioned numerical methods, in recent years, the $T = 0$ problem has been thorougly discussed by several groups when $\hat{H}_0$ corresponds to a regularly spaced single particle spectrum (picket fence) [44–48]. Some results extracted from [47], in relation with the Anderson's question mentioned in the Introduction, are presented in Figure 10.

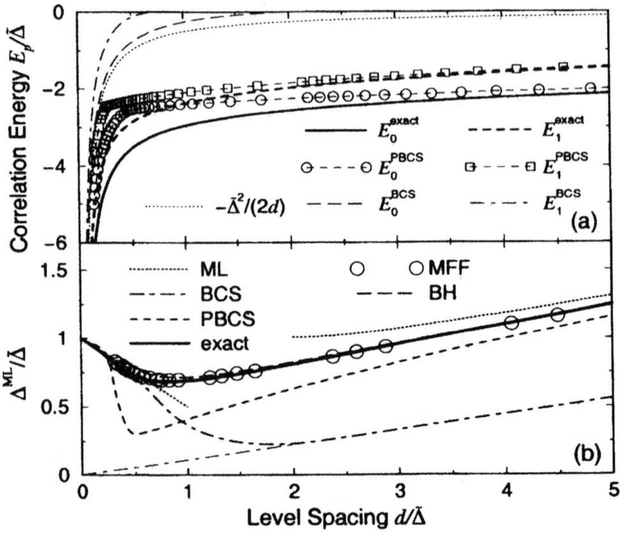

Fig. 10. From [47]. Comparison between exact and several approximate results at $T = 0$ for the condensation energy and the odd-even binding energy staggering. The Hamiltonian is given in equation (10.1.1) (see text).

Since there are several ways to define the limit of a small superconducting system, we indicate here the choice made in [47]. We have seen in Section 2.1 that, in first approximation, the volume of the system is inversely proportional to the distance $d = 1/w$ between single particle levels. According to Anderson's suggestion [12], one would like to investigate the region where d becomes equal or larger than some quantity measuring the strength of pairing correlations. Reference [47] defines this quantity as the continuous spectrum pairing gap solution of equation (3.4.3) $\Delta = \Lambda/[2\sinh(1/wG)]$ (in the figure, the notation is $\tilde{\Delta}$). Δ is kept constant as $d = 1/w$ varies by keeping both Λ and wG unchanged ($wG = .224$). Thus, as the abscissa $x \equiv d/\Delta$ increases, so does G. In contrast, the density w, the volume and the number of particles $N = w\Lambda$ decrease while the density of fermions *involved in the pairing phenomenon* remains constant. The last equation implies that the Fermi level does not move within the single particle spectrum. In [47] it is assumed to remain approximately at mid spectrum.

The range of abscissae investigated correspond to a rather stretched scale. For macroscopic and even microscopic superconducting systems the value of x is almost zero. The case investigated numerically in Section 8.5 corresponds to $x = 1/(w\Delta) = 0.1$ and, for nuclei, the value of x varies between 0.2 to 0.4 depending on the difference between the values of N or

Z and a magic number. It is not yet clear which values of x are reached in nanoscopic grains.

The upper part of the figure displays the difference between the ground state energy and the first order perturbation corrected energy (*i.e.* the expectation value of the full Hamiltonian (10.1.1) in the ground state of $\hat{H}_0$). One sees that the odd and even exact curves are well reproduced by all pairing approximations when x is less than one. For values of the abscissa larger than two, the projected approximations are of good quality while standard BCS underestimates the fluctuation energy. For $x \approx 1$, none of the approximate methods can be said to work very well.

In the lower part of the figure, the filter $\Delta_\nu^{(3)}(2n+1)$. (In Fig. 10, $\Delta_\nu^{(3)}(2n+1)$ is denoted Δ^{ML}) is plotted as a function of x. For $x < 1$, the dot curve corresponds to the approximation $\Delta_\nu^{(3)}/\Delta \approx 1 - 1/(2x) = \Delta_-(0)/\Delta$ given by equation (8.4.11). For these small values of d/Δ, BCS (blocked BCS is used for odd systems) and all other approximations reproduce well the exact results. When x is larger than one, the BCS result approaches the value $G/2$ which correspond to the filter $\Delta_\nu^{(3)}$ calculated from ground state energies evaluated by means of a first order perturbation method. This corresponds to a situation where pairing correlations have vanished. The deviation from the exact result is significant. For large values of x, the particle number projected BCS method provides a reasonably good approximation. On the other hand, although this method predicts correctly the zone of the transition between the two regimes, the quality of the quantitative estimate of the filter is modest for $x \approx 1$. The results obtained by solving the Richardson equations (labeled "exact") are reproduced accurately by the numerical diagonalization (open circles) as well as by the density matrix renormalization [46] to the point that the curves almost undistinguishable.

From Figure 10, one can draw the conclusion that under the conditions selected for this calculation there exist two distinct dynamical regimes. On the other hand, as d/Δ increases from 0.5 to 2, the transition between them is very smooth.

As mentioned above, it is still not clear whether there exist a set of systems which would allow us to probe the evolution displayed in the lower part of Figure 10. Note also that because the model requires that the strength G grows as the inverse of the single particle energy density, no part of the curve can be said to correspond to a perturbation regime.

The results we have just discussed have been obtained for a picket fence single particle spectrum. Recently, additional information has been collected which, in metallic grains, suggests a more complex structure with some evidences for spin orbit structure as in atoms and nuclei [53, 54]. In addition, the production technique of the grains does not warrant a regular shape for the grains. For this reason, several groups have investigated the transition problem assuming randomly distributed single particle spacings [47,55,56].

Another approach to the same problem is the Generator Coordinate Method (GCM) [18,57,58]. The approximate wave function is constructed as a linear superposition

$$|\Psi\rangle = \int_0^\infty d\tilde{G} f(\tilde{G}) \hat{P}_{N_0} |\mathrm{BCS}(\tilde{G})\rangle, \tag{10.1.4}$$

where the state $\hat{P}_{N_0}|\mathrm{BCS}(\tilde{G})\rangle$ is the projected BCS solution calculated with the auxiliary Hamiltonian

$$\hat{\tilde{H}} = \hat{H}_0 - \tilde{G}\hat{P}^\dagger \hat{P}\ , \quad (0 < \tilde{G} < \infty). \tag{10.1.5}$$

The unknown function $f(\tilde{G})$ which minimizes the energy E calculated with the Hamiltonian (10.1.1)

$$\frac{\delta E}{\delta f} = 0, \quad E = \frac{\langle\Psi|\hat{H}|\Psi\rangle}{\langle\Psi|\Psi\rangle}, \tag{10.1.6}$$

is solution of the Hill Wheeler integral equation. Thanks to a generalized Wick theorem [59], the calculation of the kernels entering the integral equation can be evaluated for Hamiltonians much more sophisticated than (10.1.1) [60]. For $\hat{H}$ given by (10.1.1), a comparison with exact calculations [61] shows that the GCM method provides a very accurate description of the exact ground state and first excited states. Indeed, it can be shown that the overlaps of the exact and the GCM many-body wave functions $|\Psi\rangle$ differ from unity by very small amounts (typically less than 10^{-3}).

10.2 Finite temperatures

One of the result of [47] is that, at $T = 0$, as the size of the system decreases, there is a smooth transition between two regimes. A similar prediction for the temperature dependence of pairing effects in small system had been made as early as 1972 [62]. In this section, we describe a recent work, which reaches the same conclusion by means of a fully microscopic analysis [49,50].

First we note that by introducing the quantities

$$v_1 = v_2 = \frac{G}{2}$$

$$\tilde{\epsilon}_p = \epsilon_p - \frac{1}{2}G\ , \quad \hat{\tilde{H}}_0 = \sum_p \tilde{\epsilon}_p (a_p^\dagger a_p + a_{\bar{p}}^\dagger a_{\bar{p}}) \tag{10.2.1}$$

$$\hat{Q}_1 = \hat{P} + \hat{P}^\dagger\ , \quad \hat{Q}_2 = i(\hat{P} - \hat{P}^\dagger),$$

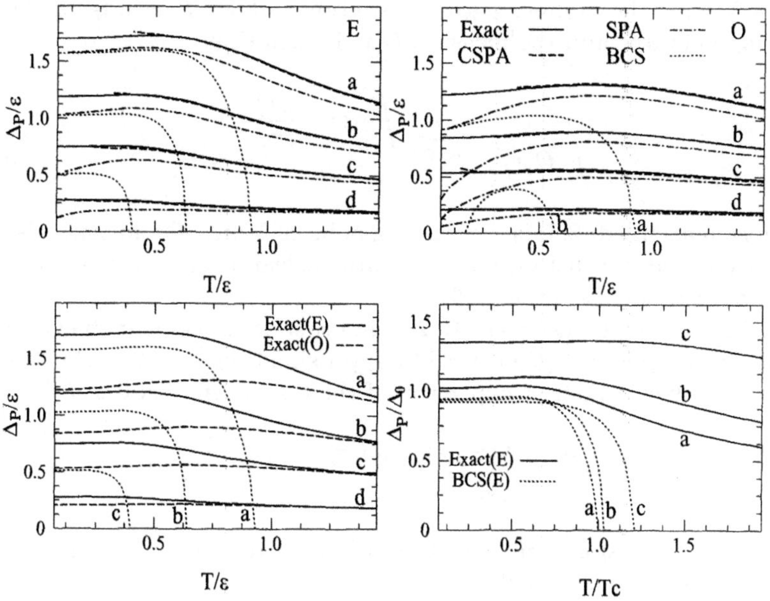

Fig. 11. From [50] Temperature evolution of the pairing gap equivalent quantity (10.2.9) for an exact calculation and several approximations for an odd system. The letters a to d refer to decreasing pairing intensities (see text). For the intensities c and d, the (number parity projected) BCS gap vanish identically.

up to a c-number, the Hamiltonian 10.1.1 can be put into the following form

$$\hat{H} = \hat{\tilde{H}}_0 - \frac{1}{2} \sum_{\nu=1,2} v_\nu \hat{Q}_\nu^2 . \tag{10.2.2}$$

We now introduce the τ-dependent ($0 \le \tau \le \beta$) auxiliary fields $\{x_\nu\}$ and the family of one-body Hamiltonians

$$\hat{H}(\{x(\tau)\}) = \hat{\tilde{H}}_0 + \sum_{\nu=1,2} \left[\frac{x_\nu(\tau)^2}{2v_\nu} - x_\nu \hat{Q}_\nu \right] . \tag{10.2.3}$$

Then, by means of Hubbard–Stratonovitch transformation [63], the exact number parity projected partition function Z_η can be expressed as a functional integral over the fields $\{x_\nu\}$

$$Z_\eta = \mathrm{Tr}\ \hat{P}_\eta \, \mathrm{e}^{-\beta \hat{H}} = \int D[x]\ \mathrm{Tr} \hat{P}_\eta \hat{\mathcal{T}} \mathrm{e}^{-\int_0^\beta \mathrm{d}\tau \, \hat{H}(\{x(\tau)\})} , \tag{10.2.4}$$

in which $\hat{P}_\eta$ is the number parity projection (8.1.1), $\hat{\mathcal{T}}$ stands for the time ordering operator and the measure $D[x]$ is such that

$$\int D[x]\,\mathrm{e}^{-\sum_{\nu=1,2}\int_0^\beta \mathrm{d}\tau\,\frac{x_\nu(\tau)^2}{2v_\nu}} = 1\,. \tag{10.2.5}$$

The formulation (10.2.4) is the starting point of Monte-Carlo methods which have been used, for instance, to calculate nuclear level densities relevant for astrophysical nucleosynthesis [64].

Reference [50] transforms (10.2.4) in order to implement the Correlated Static Path Approximation (CSPA) approximation [65–72]. This transformation involves a Fourier expansion of the fluctuating fields $x_\nu(\tau)$

$$x_\nu(\tau) = x_\nu^{(0)} + \sum_{n\neq 0} x_\nu^{(n)}\mathrm{e}^{-i\omega_n\tau}\,, \quad \omega_n = \frac{2\pi n}{\beta}. \tag{10.2.6}$$

From this expansion, an expression of the partition function equivalent to (10.2.4) can be obtained

$$Z_\eta = \int \prod_{\nu=1,2}\mathrm{d}[x_\nu^{(0)}] \prod_{\nu=1,2,\,n\neq 0}\mathrm{d}[x_\nu^{(n)}]\;\mathrm{Tr}\hat{P}_\eta\hat{\mathcal{T}}\mathrm{e}^{-\int_0^\beta \mathrm{d}\tau\left[\hat{H}\left(\{x_\nu^{(0)}\}\right)+\sum_{\nu=1,2,\,n\neq 0}\left(\frac{x_\nu^{(-n)}x_\nu^{(n)}}{2v_\nu} - x_\nu^{(n)}\mathrm{e}^{-i\omega_n\tau}\hat{Q}_\nu\right)\right]} \tag{10.2.7}$$

with

$$\mathrm{d}[x_\nu^{(0)}] = \left(\frac{\beta}{2\pi v_\nu}\right)^{1/2}\mathrm{d}x_\nu^{(0)},\quad \mathrm{d}[x_\nu^{(n)}] = \left(\frac{\beta}{2\pi v_\nu\, i}\right)^{1/2}\mathrm{d}x_\nu^{(n)},\ (n\neq 0). \tag{10.2.8}$$

Within the CSPA, the fluctuations associated with the static fields $x_\nu^{(0)}$ are exactly taken into account by integration, while for the other fields $x_\nu^{(n)}$ the integral is performed in the Gaussian approximation. The CSPA is therefore essentially exact at high temperature (small β values). On the other hand, because it also accounts for the quadratic fluctuations of the fields compnents $x_\nu^{(n)}$ over the interval $0 \leq \tau \leq \beta$, its domain of validity can extend to rather small temperatures.

Among several other results, reference [50] presents a numerical comparison of CSPA and exact results for a picket fence model involving ten single particle levels with a spacing $\epsilon = 1/w$ and ten or eleven particles. The results displayed in Figures 11 and 12 have been calculated for several values

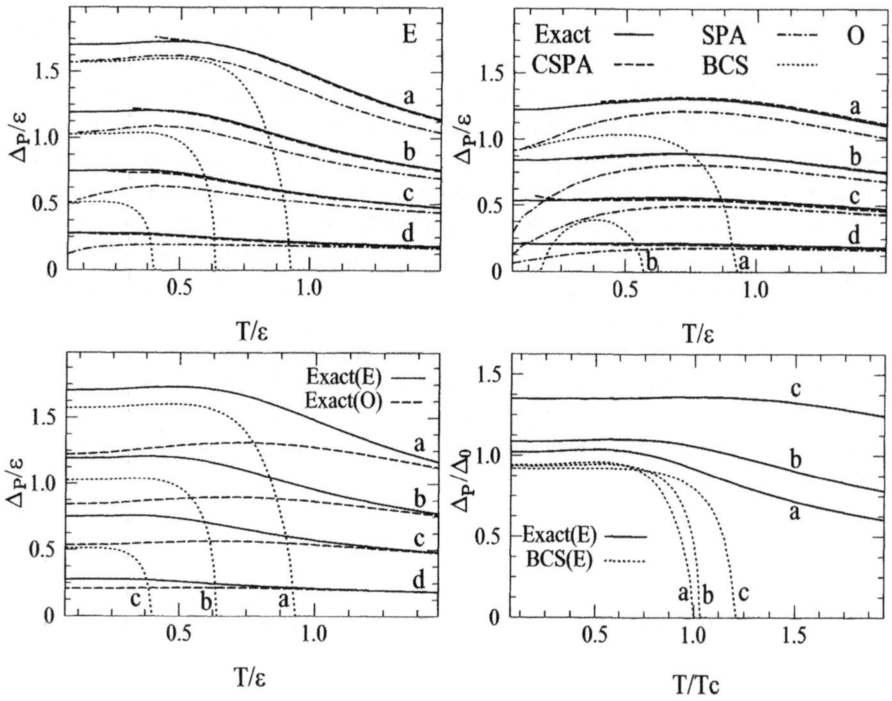

Fig. 12. From [50] Temperature evolution of the pairing gap equivalent quantity (10.2.9) for an exact calculation and several approximations for an even system. The symbols for the curves are the same as in Figure 11. The letters a to d refer to decreasing pairing intensities (see text). For the intensity d, the (number parity projected) BCS gaps vanishes identically.

of the strength G selected to emulate small paired systems. In the figures, they correspond to the labels a ($wG = 0.55$), b ($wG = 0.45$), c ($wG = 0.35$) and d ($wG = 0.2$), respectively. Since the pairing gap Δ is defined only within the mean-field methods, the authors introduce the quantity

$$\Delta_P = G\sqrt{\langle \hat{P}^\dagger \hat{P}\rangle_{(G)} - \langle \hat{P}^\dagger \hat{P}\rangle_{(G=0)}} \tag{10.2.9}$$

which can be evaluated for any method. It is a measure of the intensity of pairing correlations which can be shown to reduce to the BCS gap Δ in the BCS approximation.

The figures show an excellent agreement between exact and CSPA calculations down to small temperatures. The curves do not display any sign of a phase transition. Both in the odd and even cases and for the four strengthes, one only observes a gentle decrease of $\Delta_{\mathrm{P}}(T)$ as the temperature exceeds the BCS critical temperature ($\approx 0.56\Delta_{\mathrm{P}}(0)$).

Two other sets of curves have been plotted. One of them corresponds to the Static Path Approximation (SPA). In this approximation, the contribution of the Fourier components $x_\nu^{(n)}$ $n \neq 0$ of the fields are neglected. As expected, SPA gives a reasonable agreement at high temperature and displays the same smooth evolution as CSPA. At zero temperature, for even systems (and not too small values of G), it joins the number parity projected BCS curve (BCS). This latter curve is systematically below the exact (and CSPA) results. It also displays the transition typical of the mean field methods which turns out to be unphysical in such very small systems, *i.e.* for values of $w\Delta$ of the order or smaller than unity. It will be interesting to apply SPA and CSPA to larger systems in order to determine the limit of validity of the number-parity projected (or particle-number projected) BCS method.

11 Conclusions and perspectives

From this short review, I hope to have convinced you that, if not tomorrow, at least very soon, we will have at our disposal a set of methods appropriate for handling pairing Hamiltonians for condensed matter systems of any size from the microscopic to the nanoscopic domain and for nuclei as well.

In this lecture, I have also tried to stress that, when finite systems are considered, it is important to restore the symmetry which is broken at the level of the mean field approximation (BCS). The order parameter Δ is characterized both by a phase φ and an amplitude. Restoration of particle number symmetry is achieved by means of an integration over the phase to ensure that the particle number is exactly (and not on the average) equal to N_0. A useful first approximation to this projection for systems of microscopic size (and also for nuclei) consists in the projection on the particle number parity.

On the other hand, for very small system ($w\Delta \leq 1$) the fluctuations of the amplitude $|\Delta|$ of the gap have also to be taken into account. For Hamiltonians whose complexity precludes an exact solution (in contrast with $\hat{H}$ given by Eq. (10.1.1)), the Generator Coordinate Method (at $T = 0$) or path integral methods such as the CSPA appear to have an interesting potential.

Therefore, the unsolved problems ahead of us do not appear to belong to the realm of methods and solution techniques but rather to that of the pairing dynamics. This is a question that this lecture has carefully avoided to address. The excuse I gave you was that, beyond the schematic Hamiltonian used in these notes, any attempt to introduce a more realistic description of the pairing interaction would be specific of either the nuclear physics or the condensed matter physics domain and therefore not fit the goals of this school (especially if one takes into account that my knowledge of the question is restricted to nuclear aspects). It is my conviction that it is indeed in the case. I do not see why a phonon mediated attraction between screened electrons relates to the multipole structure of an interaction between nucleons acting in different isospin channels. The most important question is how each field can effectively acquire a better knowledge of the relevant pairing Hamiltonian. Traditionally, two lines of attack can be considered.

The first is of a phenomenological nature. It relies on the confrontation of theoretical results with a sufficiently large body of data. In nuclear physics, the availability of segmented gamma ray detectors has already allowed a very precise determination of many rotational bands. Several groups in Europe and in North America are competing to extract and gather these data. Thus, significant improvement of our knowledge of rotational structures can be expected in a near future. From it, and in particular from the dependence of the moments of inertia on both angular momentum and the single particle structure, one can hope to obtain enough information to constrain and characterize the pairing Hamiltonian. In condensed matter physics, the corresponding approach would rely on an analysis of magnetic properties od small superconducting grains. However, although the feasibility of such studies has already been demonstrated, the situation does appear as mature for small superconducting grains as it is for nuclei. The competence in making these grains and studying them is still very much confined. In addition, while a nucleus is a reproducible object, no two metallic grains are identical. Sorting out pairing contributions from dynamical information associated with a mean field which varies randomly from one experiment to the next may be a difficult task despite the ingenuity displayed by both the experimentalists and the theoreticians of the field.

In another approach, one could try to determine the Hamiltonian from first (or at least prior) principles and a better understanding of the microscopic mechanisms which generate pair attraction. My opinion is that the possibility for this to happen in nuclear physics, is small at least in a near future. Indeed, there is still a large theory gap between Quantum Chromodynamics or even Boson Exchange Theories and the effective nucleon interaction which determines the structure of nuclei. In addition, the size of the community working on the subject is rather limited. Although (or perhaps because) it is not my domain, I think that the situation is more favorable in condensed matter physics. Indeed, one can hope that the important theoretical effort presently devoted to explaining the mechanisms underlying high T_c superconductivity in compound crystals will yield some quantitative improvements of the overall understanding of the pairing mechanisms which may turn out to be relevant for small metallic systems as well.

To conclude I would like to return to the question asked by Anderson more than forty years ago [12]. In his article in *J. Phys. Chem. Solids*, one could read "An interesting question is [at] what size of particles and at what degree of scattering will superconductivity actually cease". I have the feeling that over forty years most the tools necessary to solve problems related with the "size" aspect have been developped. On the other hand, there is still some important work ahead of us to understand that part of the question which is related to the "degree of scattering".

I am indebted to R. Balian and M. Vénéroni for an enjoyable collaboration of many years and for their steady support. I also want to express my thanks to J. Dobaczewski, J. Dukelski, P.H. Heenen, P. Ring, R. Rossignoli and W. Satula for useful discussions and for providing some of the figures which illustrate this lecture.

References

[1] M. Tinkham, *Introduction to Superconductivity* (McGraw Hill, 1975).

[2] D.R. Tilley and J. Tilley, *Superfluidity and Supraconductivity* (IOP publishing Ltd., 1990).

[3] A. Bohr and B.R. Mottelson, *Nuclear Structure*, Vol. II (W.A. Benjamin, 1975).

[4] P. Janssens and T.L. Khoo, *Ann. Rev. Nucl. Part. Sci.* **41** (1991) 321; C. Backtash, B. Haas and W. Nazarewicz, *Ann. Rev. Part. Sci.* **45** (1995) 485.

[5] A. Bohr, B.R. Mottelson and D. Pines, *Phys. Rev.* **110** (1958) 936.

[6] B.R. Mottelson, *The Many Body Problem*, Les Houches, Methuen (Wiley and Dunod, 1958), 283.

[7] S.T. Beliaev, *The Many Body Problem*, Les Houches Methuen (Wiley and Dunod, 1958), 377.

[8] M.T. Tuominen, J.M. Hergenrother, T.S. Tighe, and M. Tinkham, *Phys. Rev. Lett.* **69** (1992) 1997.

[9] P. Lafarge, P. Joyez, D. Esteve, C. Urbina and M.H. Devoret, *Phys. Rev. Lett.* **70** (1993) 994.

[10] D.C. Ralph, C.T. Black and M. Tinkham, *Phys. Rev. Lett.* **74** (1995) 3241; **76** (1996) 688; **78** (1997) 4089.

[11] M. Devoret, D. Esteve and C. Urbina, *Mesoscopic Quantum Physics*, Les Houches Lectures Session LXI (Elsevier, 1994), 435.

[12] P.W. Anderson, *Phys. Rev.* **112** (1958) 1900; *J. Phys. Chem. Solids* **11** (1959) 18.

[13] See for instance; T. Duguet, P. Bonche and P.H. Heenen, nucl-th/0005040, accepted for publication in *Nucl. Phys. A*.

[14] L.N. Cooper, *Phys. Rev.* **104** (1956) 1189.

[15] M.G. Mayer, *Phys. Rev.* **78** (1950) 22.

[16] A.R. Edmonds, *Angular Momentum in Quantum Mechanics* (Princeton University Press, Princeton, N.J. 1957); D.M. Brink and G.R. Satchler, *Angular Momentum* (Oxford University Press, Third Ed., 1971).

[17] J. Bardeen, L.N. Cooper and J.R. Schrieffer, *Phys. Rev.* **106** (1957) 162; **108** (1957) 1175.

[18] P. Ring and P. Schuck, *The Nuclear Many-Body Problem* (Springer-Verlag, New-York/Berlin, 1980).

[19] K. Dietrich, H.J. Mang and J. Pradal, *Phys. Rev.* **135** (1964) 22.

[20] H.J. Mang, *Phys. Rep.* **18** (1975) 325.

[21] J.L. Egido and P. Ring, *Nucl. Phys. A* **383** (1982) 189.

[22] E. Gerjuoy, A.R.P. Rau and L. Spruch, *Rev. Mod. Phys.* **55** (1983) 725; R. Balian and M. Vénéroni, *Ann. Phys. (N.Y.)* **187** (1988) 29.

[23] R. Balian and M. Vénéroni, *Nucl. Phys. B* **408** (1993) 445.

[24] R. Balian, H. Flocard and M. Vénéroni, *Phys. Rep.* **317** (1999) 251; cond-mat/9802006.

[25] K. Sugawara–Tanabe, K. Tanabe and H.J. Mang, *Nucl. Phys. A* **357** (1981) 20.

[26] C. Esebbag and J.L. Egido, *Nucl. Phys. A* **552** (1993) 205.

[27] D.V. Averin and Yu.V. Nazarov, *Phys. Rev. Lett.* **69** (1992) 1993; *Physica B* **203** (1994) 310.

[28] B. Jankó, A. Smith and V. Ambegaokar, *Phys. Rev. B* **50** (1994) 1152.

[29] D.S. Golubev and A.D. Zaikin, *Phys. Lett. A* **195** (1994) 380.

[30] J. von Delft, A.D. Zaikin, D.S. Golubev and W. Tichy, *Phys. Rev. Lett.* **77** (1996) 3189.

[31] S.G. Nilsson and O. Prior, *Mat. Fys. Dan. Vid. Selsk.* **32** (1961) 16.

[32] W.A. de Heer, *Rev. Mod. Phys.* **65** (1993) 611.

[33] H. Häkkinen, J. Kolehmainen, M. Koskinen, P.O. Linas and M. Manninen, *Phys. Rev. Lett.* **78** (1997) 1034.

[34] W. Satula, J. Dobaczewski and W. Nazarewicz, *Phys. Rev. Lett.* **81** (1998) 3599.

[35] J. Dobaczewski, P. Magierski, W. Nazarewicz, W. Satula and Z. Szymanski, nucl-th/0003019.

[36] V.M. Strutinsky, *Nucl. Phys. A* **95** (1967) 420; *Nucl. Phys. A* **122** (1968) 1.

[37] M. Beiner, H. Flocard, Nguyen Van Giai and Ph. Quentin, *Nucl. Phys. A* **238** (1975) 29.

[38] G. Audi and A.H. Wapstra, *Nucl. Phys. A* **595** (1995) 409.

[39] D. Guillemaud-Mueller *et al.*, *Nucl. Phys. A* **426** (1984) 37; T. Motobayashi *et al.* *Phys. Lett. B* **346** (1995) 9.

[40] J. Decharge and D. Cogny, *Phys. Rev. C* **21** (1980) 1568; J.F. Berger, M. Girod and D. Cogny, *Nucl. Phys. A* **428** (1984) 23c; J.F. Berger, M. Girod, S. Hilaire, W. Satula and P. Schuck to be published.

[41] F. Braun, J. von Delft, D.C. Ralph and M. Tinkham, *Phys. Rev. Lett.* **79** (1997) 921.

[42] K.A. Matveev and A.I. Larkin, *Phys. Rev. Lett.* **78** (1997) 3749.

[43] S.D. Berger and B.I. Halperin, *Phys. Rev. B* **58** (1998) 5213.

[44] A. Mastelllone, G. Falci and R. Fazio, *Phys. Rev. Lett.* **80** (1998) 4542.

[45] F. Braun and J. von Delft, *Phys. Rev. Lett.* **81** (1998) 4712; *Phys. Rev. B* **59** (1999) 9527; cond-mat/9907402, cond-mat/9911058.

[46] J. Dukelsky and G. Sierra, *Phys. Rev. Lett.* **83** (1999) 172; *Phys. Rev. B* **61** (2000) 12302.

[47] G. Sierra, J. Dukelsky, G.G. Dussel, J. von Delft and F. Braun, *Phys. Rev. B* **61** (2000) 11890.

[48] H. Boyaci, Z. Gedik, I.O. Kerlik, cond-mat/9909386; I.O. Kulik, H. Boyaci and Z. Gedik, cond-mat/0006003.

[49] R. Rossignoli, N. Canosa and P. Ring, *Phys. Rev.* **80** (1998) 1853.

[50] R. Rossignoli, N. Canosa and P. Ring, *Ann. Phys. (N.Y.)* **275** (1999) 1.

[51] E. Caurier *et al.*, *Acta Physica Polonica* **30** (1999) 705; *Phys. Rev. C* **59** (1999) 2033.

[52] R. Richardson, *Phys. Rev.* **141** (1966) 949; *Phys. Rev.* **144** (1966) 874; *Phys. Rev.* **159** (1966) 792; *J. Math. Phys.* **18** (1977) 1802.

[53] D. Davidovic and M. Tinkham, cond-mat/9905043; cond-mat/9910396.

[54] D.G. Salinas, S. Guéron, D.C. Ralph, C.T. Black and M. Tinkham, cond-mat 9905158.

[55] R.A. Smith and D.V. Ambegaokar, *Phys. Rev. Lett.* **77** (1996) 4962.

[56] K.B. Efetov, *Adv. Phys.* **32** (1983) 53.

[57] D.L. Hill and J.A. Wheeler, *Phys. Rev.* **89** (1953) 1102.

[58] J.J. Griffin and J.A. Wheeler, *Phys. Rev.* **108** (1957) 311.

[59] R. Balian and E. Brézin, *Nuovo Cimento B* **64** (1969) 37.

[60] A. Valor, H. Flocard and P.H. Heenen (to be published).

[61] H. Flocard and N. Onishi (to be published).

[62] B. Mühlschlegel, D.J. Scalapino and R. Denton, *Phys. Rev. B* **6** (1972) 1767.

[63] J. Hubbard, *Phys. Rev. Lett.* **3** (1959) 77; R.L. Stratonovitch, *Sov. Phys. Dokl.* **2** (1958) 458.

[64] see for instance, S.E. Koonin, D.J. Dean and K. Langanke, *Phys. Rep.* **278** (1997) 1.

[65] Y. Alhassid and J. Zingman, *Phys. Rev. C* **39** (1984) 684.

[66] P. Arve, G. Bertsch, B. Lauritzen and G. Puddu, *Ann. Phys. (N.Y.)* **183** (1988) 309.

[67] B. Lauritzen, P. Arve and G. Bertsch, *Phys. Rev. Lett.* **61** (1988) 2835.

[68] B. Lauritzen and G. Bertsch, *Phys. Rev. C* **39** (1989) 2412.

[69] R. Rossignoli and P. Ring, *Ann. Phys. (N.Y.)* **235** (1994) 350.

[70] B.K. Agrawal and P.K. Sahu, *Phys. Lett. B* **351** (1995) 1.

[71] R. Rossignoli, N. Canosa and J.L. Egido, *Nucl. Phys. A* **605** (1996) 1.

[72] H. Attias and Y. Alhassid, *Nucl. Phys. A* **625** (1997) 565.

COURSE 7

MODELS OF METAL CLUSTERS AND QUANTUM DOTS

M. MANNINEN

Department of Physics, University of Jyväskylä, P.O. Box 35, 40351 Jyväskylä, Finland

Contents

MODELS OF METAL CLUSTERS AND QUANTUM DOTS

M. Manninen

Abstract

The electronic structure of simple metal clusters and quantum dots is studied on the basis of the density functional theory and simple models. It is demonstrated that single-particle models explain well the gross features of deformation and magnetism in small clusters, nuclei and quantum dots and that the local density approximation can give valuable information of the internal structure of the many-body state.

1 Introduction

The observation of magic numbers in alkali metal clusters [1] started an extensive study of their electronic properties both by using simple models, like the jellium model [2], and by performing accurate ab inition electronic structure calculations [3]. The same happened earlier in the case of atomic nuclei and more recently in the case of semiconductor quantum dots. Many physicists have worked on all these finite fermion systems.

The purpose of these lectures is to show that many key properties can be most easily undestood by using simple models. *Ab initio* many-body calculations are definitely necessary but the best understanding is often obtained by comparing their results with those obtained by simple models.

The density functional Kohn–Sham method [4] gives a formal way to transform the many-body problem into a single-particle problem. Most of the models in these lectures are based on the Kohn–Sham formalism. Special emphasis is put on the symmetry broken solutions and their relevance in giving information of the internal structure of the many-body wave function.

2 Jellium model and the density functional theory

The first model to explain the electronic shell structure in metal clusters was the jellium model. Before going to the applications of this model to clusters it is interesting to review the use of this model in studying surfaces and defects in bulk metals. The background of the jellium model is the fact that many properties of simple metals can be understood by studying a homogeneous electron gas, or nearly free electrons [5]. The band structure can be described with weak pseudopotentials: the electron energy levels form free electron energy bands, which are only slightly deformed close to the Brillouin zone boundaries. At the boundary an energy gap is opened. The size of this gap is $2|V_{\mathbf{G}}|$, *i.e.* twice the value of the pseudopotential component corresponding to the reciprocal lattice vector $\mathbf{G}$. In alkali metals, with only one valence electron per atom, the Fermi surface fits well inside the first Brillouin zone and is very accurately spherical, as has been observed using several different methods.

The cohesive energy of simple metals can be approximated using model pseudopotentials and assuming the electron density to be homogeneous. In calculating the total energy, however, it is important to evaluate the energy of the interacting electron gas. The use of interacting electron gas is essential also for understanding collective excitations. The jellium model means a homogeneous electron gas with a compensating homogeneous background charge. The physical approximation is that the charge of the positive ions is replaced with a homogeneous positive charge.

A metal surface in the jellium model is descibed by cutting off the background charge from the vacuum site of the surface. Usually this is done with a discontinuous step function

$$n_+(z) = \theta(z)n_0, \tag{2.1}$$

where n_+ is the positive background charge, θ a step function and n_0 the density of the homogeneous electron gas. The density is described by a density parameter r_s ($n_0 = 3/4\pi r_s^3 a_0^3$, a_0 being the Bohr radius), which varies between 2 and 6 for normal metals. In a similar fashion a vacancy is described by a spherical hole in the background density and a metal cluster by a homogeneous sphere of the background charge.

In order to describe adequately the interacting electron system the density functional Kohn–Sham method [4] can be used. We do not go in the details of the density functional method but only mention here the key equations needed (for details see [6]). The total electron density is described as a sum over the occupied single-electron states, separately for each spin component

$$n(\mathbf{r}) = n_\uparrow(\mathbf{r}) + n_\downarrow(\mathbf{r}), \tag{2.2}$$

where

$$n_\sigma(\mathbf{r}) = \sum_i^{\text{occ}} |\psi_{i\sigma}(\mathbf{r})|^2. \tag{2.3}$$

The single particle states are self-consistent solutions of the Schrödinger equation

$$-\frac{\hbar}{2m}\nabla^2\psi_{i\sigma}(\mathbf{r}) + V^{\sigma}_{\mathrm{eff}}(\mathbf{r})\psi_{i\sigma}(\mathbf{r}) = \epsilon_{i\sigma}\psi_{i\sigma}(\mathbf{r}), \tag{2.4}$$

where V_{eff} is an effective potential consisting of the Coulomb potential and of the so-called exchange-correlation potential. In the local spin density approximation (LSDA) the exchange-correlation potential depends locally on the electron spin density and can be calculated using existing interpolation formulas [6–8]. In cases where the system does not have a spin-polarization the local-density approximation (LDA) can be used. In this case the role of the spin is simply to add a degeneracy two to each single particle state.

The surface of simple metals in the jellium model was studied first by Lang and Kohn [9, 10]. They solved the self-consistent electron density, surface energy and work function as a function of the electron density of the metal. While the mere jellium model gave good results for alkali metals it failed in high-density metals, *e.g.* by resulting to a negative surface energy. This drawback could be corrected by introducing ion pseudopotentials as a perturbation [10] or by taking them into account from the very beginning in the variational method [11]. It was observed that the average of the pseudopotential in the bulk was the key incredient to correct surface energies [12]. The deficiency of the jellium model in describing energetics was caused by the fact that in the bulk the rigid positive background stabilized the electron gas, but at the surface it could relax to a density closer to the minimun energy density of the homogeneous gas (coresponding to $r_s \approx 4.2$). Later, so-called stabilized jellium model [13, 14] was introduced to account the problem of the average pseudopotential (it was later applied also for clusters [15]).

The jellium model has also been used to study spherical defects inside the metal. A vacancy or a spherical void is then described as a spherical hole in the background density [12, 16]. The Kohn–Sham equations have been solved self-consistently and the results can be related to those of plane surfaces. However, the void formation energy could not be well described with pseudopotentials. The reason is that in large spherical voids the void surface (cut in a regular lattice) becomes very rough when the void size increases, causing the Coulomb energy eventually to diverge [12]. Similar problems are expected in the case when the jellium model is applied to planar surfaces with large Miller indexes (say [710]-surface). It is interesting to note that the same problem was met in the first application of the jellium model to metal clusters [17].

The electron density close to a planar surface has slowly decaying Friedel oscillations

$$n(z) - n_0 \propto \frac{\cos(2k_F z)}{k_F z}, \tag{2.5}$$

where z is the distance from the surface and k_F the Fermi wave vector. Similar oscillations will result around point defects in metals. In the linear screening theory they can be understood as arising from the logarithmic divergence of the derivative of the dielectric response function. In the Kohn-Sham theory they are a result of the phase shifts of the single particle wave functions: Far away from a defect, say vacancy, the radial wave functions are

$$R_{k\ell} \propto \frac{\cos(kr + \delta_\ell(k))}{r^3}, \tag{2.6}$$

where the scattering phase shifts obey the Friedel sum rule

$$\sum_\ell \delta_\ell(k_F) = \int d^3r(n(\mathbf{r}) - n_0) = Z, \tag{2.7}$$

where n is the electron density around the vacancy and n_0 the bulk density, and Z the total charge removed with the background density at the vacancy. Related Friedel oscillations will allways exist at cluster surfaces.

The applications of the jellium model to surfaces and defects are inherently scattering problems. There is a continuous spectrum of energy levels from zero to the Fermi level. Consequently, the electronic properties of spherical voids in bulk (jellium) are smooth functions of the size. There are no magic sizes of vacancy clusters caused by the electronic structure. The vacancy clusters have scattering resonances but since in a large void the repulsive effective potential is higher than the Fermi level, sharp resonances can only occur above the Fermi level and they do not directly affect on the stability of the ground state. In fact similar scattering resonances will accur also in spherical jellium clusters and have to be taken into account in calculating for example the photoionization cross section [18]. Note that the Friedel sum rule also applies for the scattering phase shifts in the case of a jellium cluster.

3 Spherical jellium clusters

A spherical jellium cluster consists of N interacting electrons moving in an external potential of a homogeneously charged sphere (positive background charge). The electrostatic potential energy of an electron in this background charge is

$$V(r) = \begin{cases} \dfrac{3e^2Z}{8\pi\epsilon_0 R^3}\left(\dfrac{r^2}{3} - R^2\right), & \text{if } r < R \\ \dfrac{e^2Z}{4\pi\epsilon_0 r}, & \text{if } r \geq 0, \end{cases} \tag{3.1}$$

where R is the cluster radius and Z the total charge of the ionic background. For neutral clusters $Z = N$ and $R = N^{1/3}r_s$. Note that inside the jellium sphere the potential is harmonic. In a harmonic potential the center of mass motion separates out from the internal motion of the electrons. In large clusters the electrons are mainly inside the harmonic well and consequently the center of mass motion is the dominating collective resonance, *i.e.* the plasmon [2,19]. In the limit of larce clusters the resonance frequency is then independent of the electron-electron interactions. However, in studying the single particle motion the electron-electron interaction has to be taken into account. In the Kohn–Sham method this means that the electrons move in an effective potential (see above) which also include the electrostatic potential of the electrons and the exchange-correlation contributions. In a neutral jellium cluster the inside of the cluster will be electrically neutral and consequently the effective potential will be nearly constant inside the cluster and the potential resembles more a potential well with a rounded edge than a harmonic oscillator potential. The effective potential can be approximated with the Woods–Saxon potential

$$V(r) = -\frac{V_0}{1 + e^{\alpha(r-R)}}, \tag{3.2}$$

where V_0 is the depth of the potential and α a parameter determining the softness of the surface. Physically this softness comes from spilling out of electrons from the jellium sphere. There are several other ways to construct a model potential for stydying the electronic structure in spherical clusters. Mansikka-aho *et al.* [20] suggested the form

$$V(r) = -\frac{V_0}{1 + (r/R)^\alpha} \cdot \tag{3.3}$$

The advantage of this potential is that it can be smoothly transformed from a square well ($\alpha \to 0$) to a potential with a Coulomb tail ($\alpha = 1$).

The self-consistent solution of the Kohn–Sham equations for jellium clusters can be found only numerically. First this was done by Martins *et al.* [17] using a parametrized potential and later fully self-consistently by many others [21–24]. The results for the electronic structure, ionization potential and stability are not very sensitive to the degree of the self-consistency and even

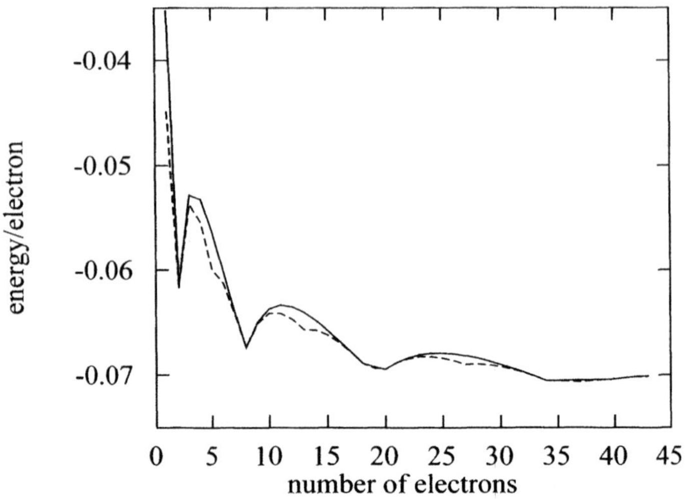

Fig. 1. Total energy per electron as a function of the number of electrons of a spherical jellium cluster ($r_s = 4$). The solid line is the LDA result and The dashed line the LSDA results showing the effect of the Hund's rule.

the simple model potentials produce very well the essential features of the electronic shell structure.

Figure 1 shows the total energy per electron of a jellium cluster and Figure 2 its second derivative $\Delta_2 E = E(N-1) - 2E(N) + E(N+1)$. The magic numbers 2, 8, 20, 34, and 40 are seen in the total energy of the LDA as minima and in the second derivative as maxima. The open shell clusters represent degenerate ground states. In these cases, the Jahn–Teller theorem dictates that if there is a mechanism to remove the degeneracy, the total energy can be decreased. In metal clusters the shape deformation is the most important mechanism (it will be discussed in the following sections). However, there is another mechanism to reduce the degeneracy. This is the first Hund's rule which suggests that in an open shell the z-component of the spin will be at maximum. In the spin-density functional formalism (LSDA) this means that the single particle energy levels of the majority spin will be pulled down and those of the minority spins will be pushed up, thus reducing the degeneracy. In the case of a half-filled shell the degeneracy totally disappears since the spin-up electron shell will be full and the spin-down electron shell empty. Indeed, if we use LSDA in the spherical jellium model the $\Delta_2 E$ will have clear maxima also at half-filled shells as shown in Figure 2. In Figure 1 we see that for the nonmagic clusters the LDSA gives lower total energies than the LDA.

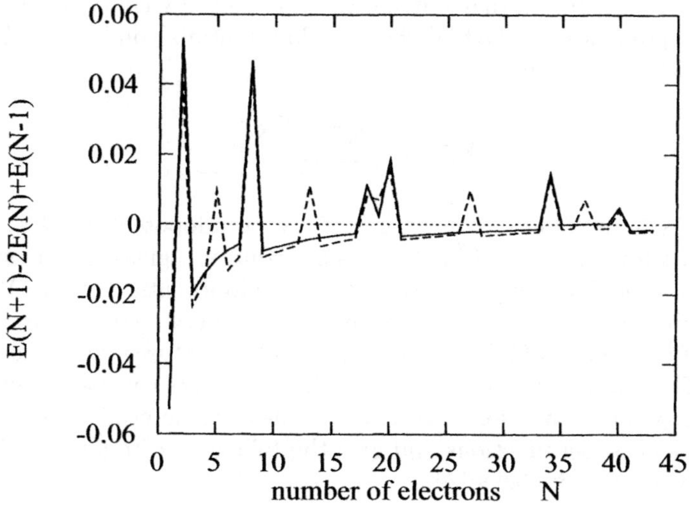

Fig. 2. The second derivative of the total energy as a function of the number of electrons of the spherical jellium cluster. The solid line is the LDA result and the dashed line the LSDA results showing the effect of the Hund's rule.

In the case of spherical clusters the open electron shell would also mean a large magnetic moment associated with the total spin, as first pointed out by Geguzin [25] for a pseudopotential model and by Hintermann and Manninen for jellium clusters [21]. However, later experiments and theoretical considerations have shown that the shape deformation is much more effective in reducing the degeneracy of the open electronic shells. There is no evidence of magnetism of *free* alkali metal clusters. Nevertheless, there are suggestions that clusters confined in spherical cavities in zeolite could prevent the deformation and lead to magnetism according to the Hund's rule.

4 Effect of the lattice

So far we have assumed that the electrons form an interacting electron gas but that they do not feel the underlaying lattice. The lattice can be taken into accout by using perturbation theory and proper pseudopotentials. It is well known from the solid state physics that model pseudopotentials can reproduce the band structure of crystalline metals as well as describe their cohesive erengy. Consider first the effect of the pseudopotentials on the shell structure of alkali metal clusters. The band structure theory for nearly free electron metals shows that outside the immediate vicinity of the Brillouin zone boundaries the effect of the pseudopotentials come from the second order perturbation theory

$$\epsilon_{\mathbf{k}} - \epsilon_{\mathbf{k}}^{0} = \sum_{\mathbf{G}} \frac{|U_{\mathbf{G}}|^2}{\epsilon_{\mathbf{k}}^{0} - \epsilon_{\mathbf{k}-\mathbf{G}}^{0}}, \tag{4.1}$$

where $U_{\mathbf{G}}$ is the Fourier component of the pseudopotential, $\mathbf{G}$ a reciprocal lattice vector and $\epsilon_{\mathbf{k}}^{0} = \hbar^2 k^2/2m$. This equation gives us an estimate of the crystal field spitting caused by the pseudopotential lattice. When this energy difference becomes comparable to the energy gap between adjacent shells of the spherical jellium model the shell structure effects should disappear. The energy gaps in spherical jellium clusters can be estimated from the energy levels in a spherical cavity. They are determined by the zeroes of spherical bessel functions and give the following estimate for the energy gap between major shells [26]

$$\Delta E = \epsilon_{\mathrm{F}} \left(\frac{9}{4\pi N} \right)^{1/3}, \tag{4.2}$$

where N is the number of electrons and ϵ_{F} the fermi energy. The energy gap will be of the same order than the crystal field splitting when the cluster has of the oder of 10^6 valence electrons. This number is far larger than the size region where the shell structure or supershell structure is observed. In fact, it turns out that the surface effects cause much stronger disturbance to the shell structure than the pseudopotential lattice inside the cluster.

Three different properties of the surface have effects on the electronic shell structure: softness of the surface potential, surface roughness, and faceting. The softness can be described by a model potentials (9) or (10) and has found to have rather small effects on the shell structure of small clusters. Only extreme softness (which will prevent classical triangular orbits of electrons) can drastically change the shell structure [20, 27]: magic numbers would still exist but with different electron numbers.

The surface roughness is related to the crystal field splitting caused by the pseudopotentials inside the metal. However, since the periodicity of the lattice is broken at the surface the effect is much stronger. Several different models [28–32] have been applied for studying the effect of the surface roughness. (Simplest way to study it is to use the tight binding model described in the next section.) The roughness widens the discrete degenerate electron shells and make the density of state smoother. The shell structure is still there but it is weaker. The surface roughness also increases the total energy of the cluster (this is the reason why a simple pseudopotential correction to the spherical jellium model fails).

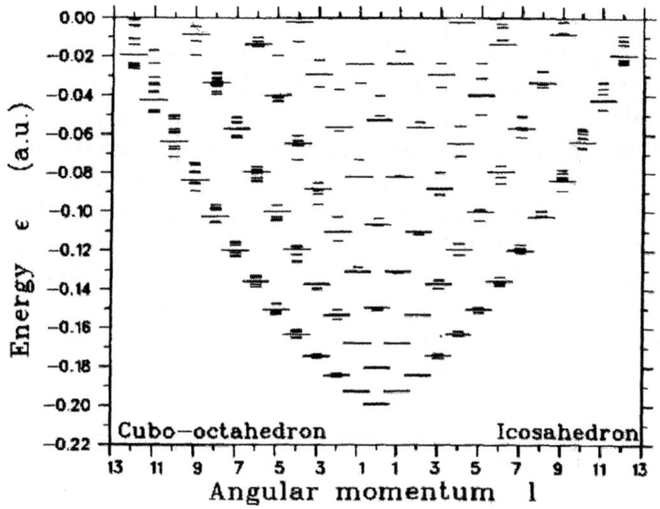

Fig. 3. Energy levels in a potential well corresponding to a 309 atom sodium cluster. The long lines are results for a spherical well, the short lines on the left-hand side are results for a cuboctahedral well and the short lines on the right-hand side are results for an icosahedral well. For a nonspherical cluster the angular momentum is only approximately correct. From reference [37].

The surface of a crystalline material tends to be faceted, consisting of only low-index facets (in FCC lattice [111], [110], [100]). In finite clusters the minimization of the surface energy leads to so-called Wulff construction [33] which is a regular polyhedron. In small clusters also noncrystalline faceted clusters like icosahedra and decahedra are possible. Indeed, experiments have shown that large sodium clusters are icosahedra [34] while aluminium clusters seem to have octahedral shapes [35]. Generally, it seems that when the surface faceting determines the geometry, the electronic shell structure does not any more have a marked effect of the cluster stability. It is then impontant to know the effect of the faceting to the electronic shell structure. Mansikka-aho *et al.* [36, 37] have used two different models to study the effect of the icosahedral shape and Wulff construction to the electronic shell structure. The results show that the Wulff construction in in an FCC lattice destroys the shell structure when the cluster has about 300 electrons while the icosahedral shape still has the shell structure when the cluster has about 1000 electrons. Figure 3 shows the effect of the surface faceting on the electronic structure in the cases of icosahedral and cuboctahedral shapes.

5 Tight-binding model

Before studying further the effect of the surface and deformation on the electronic structure we will introduce another simple model. The tight binding model is often considered of being an opposite to the nearly free electron model. While in the latter the basis functions are delocalized plane waves, in the tight binding model one starts from localized orbitals. The simplest tight binding model assumes only one s-electron per atom, neglects the differential overlap between orbitals and assumes coupling only between nearest neighbours. This model is frequently called the Hückel model or the CNDO model (complete neglect of differential overlap).

The hamiltonian to be diagonalized is

$$H_{ij} = \begin{cases} \alpha & \text{if } i = j \\ -\beta & \text{if } i \text{ and } j \text{ nearest neighbours} \\ 0 & \text{otherwise} \end{cases} \tag{5.1}$$

where α is the energy eigenvalue at a free atom and β so-called hopping integral between nearest neighbours. Without restrictions we can use Hückel units by taking $\alpha = 0$ and $\beta = 1$. With this simple Hamiltonian it is easy to solve the energy eigenvalues for infinite lattices with periodic boundary conditions or for finite clusters with up to tens of thousands of atoms. The total energy of the simple tight binding model can be defined as the sum of the occupied energy levels. In determining the geometry the bond length is usually kept fixed and the atoms are not allowed to be closer to each other than the bond length (hard spheres).

Even if the model is so simple, it has turned out to be extremely difficult to find the optimum geometry (which minimizes the total energy) as soon as the cluster has more than 10 atoms. Wang *et al.* [38, 39] computed the ground state geometries for Hückel clusters up to 9 atoms by testing all possible topologies. It is interesting to note that the ground state geometries obtained with the Hückel model are very similar to those obtained with *ab initio* calculations [3, 40] for alkali metals. For larger clusters simulated annealing technique [41] and genetic algortithms [42] have been used to look for the ground state geometry. However, the Hückel model seems to be a much more difficult optimization problem than for example finding the ground state geometries for Lennard–Jones clusters [43].

In studying large spherical Hückel clusters Lindsay *et al.* [44] observed an electronic shell structure, similar to that of the jellium model. These clusters were cutted off from cubic lattices and contained hundreds of atoms.

For understanding these surprising results two different considerations have have to be made. We will first show the equivalence between the Hückel model and the free electron jellium model in spherical clusters.

The understanding of the insensitivity of the overall shape of small clusters on the theoretical approach is slightly more complicated and will be discussed in the next section. Consider now the single particle Schrödinger equation in an infinite jellium. The potential is constant and thus the solution is a free electron plane wave. However, lets assume that we do not know that, but instead try to solve the equation numerically in a three-dimensional cubic grid $\mathbf{r}_i$ with a lattice constant a. The Schrödinger equation then becomes

$$-\frac{\hbar^2}{2m}\nabla^2\psi(\mathbf{r}_i) \approx -\frac{\hbar^2}{2ma^2}\left[\sum_j^{n.n.}\psi(\mathbf{r}_j) - c\psi(\mathbf{r}_i)\right] = \epsilon\psi(\mathbf{r}_i), \tag{5.2}$$

where the sum goes over the nearest neighbours and c is the coordination number (*i.e.* the number of nearest neighbours). It is now obvious that the solution of the group of equations (5.2) is equivalent of finding the eigenvalues and eigenvectors of the Hamiltonian (5.1) with

$$\alpha = \frac{\hbar^2 c}{2ma^2}, \qquad \beta = \frac{\hbar^2}{2ma^2}\,. \tag{5.3}$$

The tight binding model in a cubic grid is thus an approximation to the free electron model. This explains the similar shell structure at low energies. We should notice, however, that the "numerical" results are accurate only when the variation of the wave function is small within the distance a. This means that if a is a true lattice constant the Hückel model is only accurate close to the bottom of the conduction band. Another way to state the similarity of the Hückel model and the nearly free electron model is that in the case of an isotropic effective mass the bottom of the energy band always looks like a free electron band.

The numerical solution of the Hückel model has been used to study the shell structure of nonsherical clusters like icosahedra [36], cuboctahedra [36] and tetrahedra [45] and effect of the surface roughness on the shell structure and level spacing statistics [28, 29].

6 Shape deformation

Computations of the atomic structure of metal clusters have shown that small clusters prefer shapes that generally are not closed packed structures of hard or soft spheres as would be a result of a pairwise interaction (*e.g.* Lennard–Jones clusters). For small alkali metal clusters the basic geometries seem to be quite insensitive to to accuracy of the treatment of the many-body problem of electrons and ions: *ab initio* quantum chemical methods (CI), Kohn–Sham method with pseudopotentials, and even the simple

Hückel model gives quite similar planar structures for sodium clusters up to 5 or 6 atoms and nearly the same structures also for clusters with 7 or 8 atoms. This suggests that there should be a simple single-particle explanation which determines the basic structures.

Experimentally, the observation of the geometries is not straightforward and the information of geometries of small clusters comes only from indirect measurements. The photoabsorption and photoelectron spectra seem to be sensitive to the detailed geometry of the cluster, but both methods need a comparison of theoretical and experimental results for determining the structure [3, 46]. A simple theory of photoelectron spectroscopy is shortly explained in Section 9. In large clusters with more than, say, 40 atoms, there is only very limited experimental knowledge of the cluster shapes (expect in some cases of large clusters where diffraction measurents have confirmed the clystal lattice [48], or the mass spectrum had revealed the growth pattern [34, 35]). Nevertheless, the splitting of the plasmon resonance can be explained by an overall prolate or oblate geometry [49] and the intensity variation of the mass spectrum has been explained with help of deformation [50, 51].

In the jellium model the deformation can be understood as a Jahn–Teller effect, which tries to reduce the degeneracy of an open shell structure. First ellipsoidal deformation with constant background density was considered [52]. This model already gives good understanding of the basic features in the small cluster limit. It predicts correctly that in the beginning of a shell filling the shape is prolate and at the end of the shell filling it is oblate. Also, it explaines the odd-even staggering observed in the ionization potential of small alkali metal clusters. The simple explanation of the deformation in the first p-shell is the following: if only one or two electrons are in the p-shell they can occupy the same, say, p_x state along the elongated axis of the prolate ellipsoid. The 4th asd 5th electron occupy the p_y-state and the cluster prefers an oblate shape. The analysis becomes more complicated in filling higher angular momentum shells but, nevertheless, it is the single particle picture which explains the deformation also in large clusters [53,54].

The deformed jellium model can be improved by relaxing the requirement of constant background density. This makes also the numerical computation of the Coulomb energy and potential easier. In the ultimate limit the lowest energy is obtained by allowing the background density at every point to be the same as the electron density [55, 56]. In this so-called ultimate jellium model the background density of the bulk has the value corresponding to the minimum energy of the homogeneous electron gas. The cluster shapes are entirely determined by the electrons. The results for small clusters are shown in Figure 4. The shapes are more complicated than just ellipsoids. Nevertheless, it is important that these shapes are not

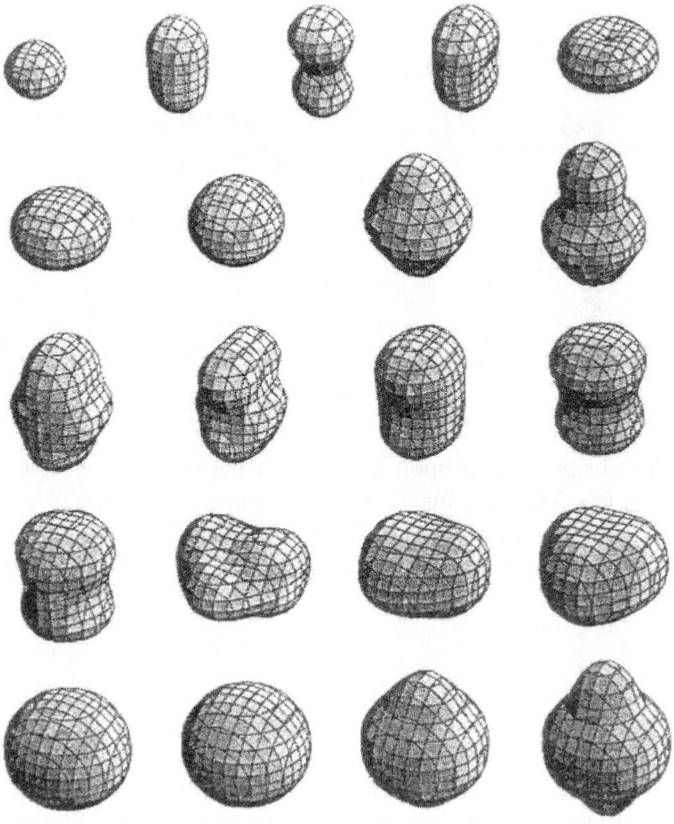

Fig. 4. Ground state shapes of clusters with 2 to 22 electrons (with increasing order) calculated with the ultimate jellium model. A constant density contour corresponding to about 40% of the bulk density is shown.

at all sensitive to how the exchange-correlation potential is approximatied in the Kohn–Sham theory. In fact, essentially the same shapes are obtained if parametrization for nuclear matter is used [57].

Häkkinen *et al.* [58] have studied the universality of the shapes of fermion clusters in the local density approximation of the density functional theory. Indeed, for small cluster the multipole expansion of the shape shows that the shape parameters are nearly independent of the interparticle interactions. This allows accurate calculation of internal nuclear quadrupole moments, eventhough the equation of state of the nuclear matter is not well known. The results are shown in Figure 5. It should be noted that the simple model only applies for nuclei with the same number of protons and neutrons.

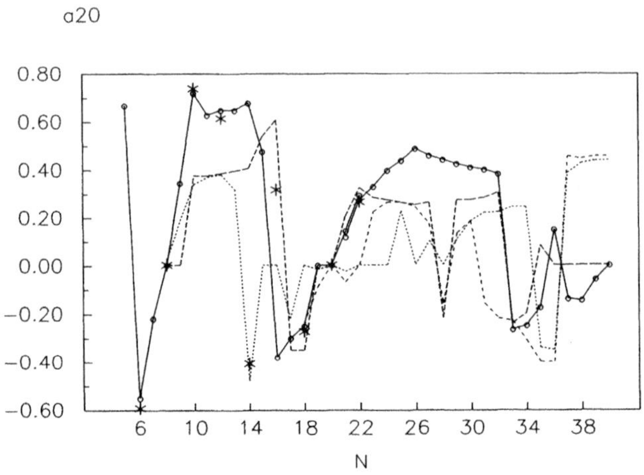

Fig. 5. The internal quadrupole moments of nuclei shown as a deformation parameter a_{20} (see [56]) as a function of the number of protons. The experimental results are shown a stars. The results of the universal model are shown as black dots connected with a solid line. The other lines show results of other models. From [73].

The reduction of the total energy due to the defromation is quite dramatic. In Figure 6 we show the total energy per electron of the ultimate jellium model as compared to the spherical jellium model for sodium. The energy does not any more increase in between the magic numbers and there is a strong odd-even staggering. It is also interesting to note that the total energy of the ultimate jellium model is not very different from the conventional jellium model for the same density (the bulk density of the ultimate jellium model is $r_s = 4.2$ while the spherical jellium result in Fig. 6 is for $r_s = 4.0$).

The key quantity for determining the shape in the single particle picture is the sum of the single particle eigenvalues. The eigenvalues depend on the shape more sensitively than the integrated parts of the total energy functional of the density functional theory [61]. The coupling of the eigenvalue spectrum to the shape was first utilized in the Nilsson model [59] for nuclei (as explained by Mottelson in the 1957 Les Houches lectures [60]) and with the same model by Clemenger [53] for clusters. The Nilsson model is based on the harmonic oscillator where the energy eigenvalues have a simple analytic expression. To demonstrate that the shape is insensitive to the details of the single particle potential we write the following expression for

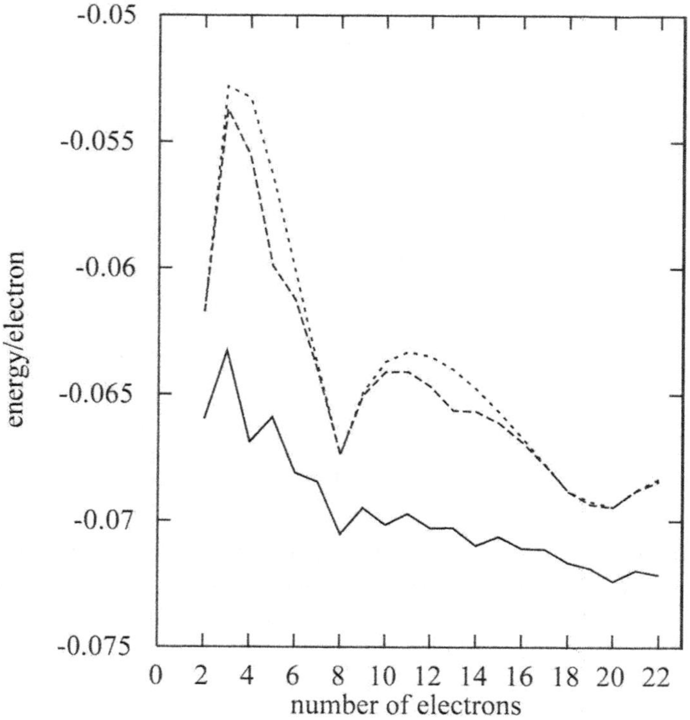

Fig. 6. Energy per electron for the ultimate jellium model (solid line) compared to the spherical jellium model with LDA (dotted line) and LSDA (dashed line).

the eigenvalues [61]

$$\epsilon(n_1, n_2, n_3) = \omega_1 n_1^\gamma + \omega_2 n_2^\gamma + \omega_3 n_3^\gamma + (\gamma/2 - 1)(\omega_1 + \omega_2 + \omega_3), \qquad (6.1)$$

where n_i are positive integers (zero not included) and γ is a parameter which interpolates the eigenvalues between the harmonic oscillator $\gamma = 1$ and a rectangular box $\gamma = 2$. We use a volume conservation $\omega_1\omega_2\omega_3 = \omega_0^3$ and assume an "axial" deformation $\omega_1 = \omega_2$. A deformation constant δ can then be defined as

$$\omega_3 = \omega_0(1 - \xi\delta)^\gamma \qquad \omega_1 = \omega_2 = \frac{\omega_0}{(1 - \xi\delta)^{\gamma/2}} \cdot \qquad (6.2)$$

Here the exponent γ guarantees the correct single particle energy and the second equation follows from the volume conservation. The coefficient $\xi = 2(0.51 + 0.49\gamma)/3$ is a purely goemetrical factor introduced to take into account the δ-dependence of the moment of inertia. In the case of the

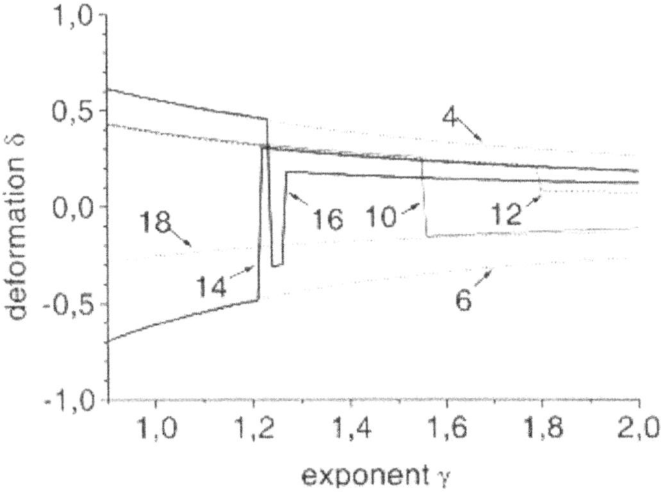

Fig. 7. Deformation as a function of the parameter γ of the single particle spectrum. $\gamma = 1$ corresponds to a harmonic oscillator and $\gamma = 2$ a rectangular box. Different lines correspond to the electron numbers indicated.

harmonic oscillator the deformation parameter is the same as frequently used in nuclear physics [62].

Figure 7 shows the dependence of the optimal deformation parameter on the single particle spectrum, *i.e.* parameter γ, for even cluster from 4 to 18 particles. In the cases of 8 and 20 particles the cluster is not deformed and $\delta = 0$. We assume that each level can occupy two electrons due to the spin and that the total energy is the sum of the occupied single particle levels. We can see that the 4 electron cluster is prolate $\delta > 0$ and the 6 electron cluster is oblate $\delta < 0$ for all values of γ. In filling the d-shell the situation becomes more complicated: the clusters can flip between the prolate and oblate isomers when γ is increased. This means the there are two isomers which are quite close in energy. This situation is known from the Nilsson model. However, it is more important to notice that the deformation parameter changes only slowly as a function of the parameter γ which describes the single particle spectrum. This is the key for understanding the insensitivity of the shapes of small clusters on the details of the system in question. Finally, we whould mention that the Nilsson model can be extended to a quantitative model by the Strutinsky shell correction method [63], which has been suggesfully used for metal clusters [54, 64, 65].

7 Tetrahedral and triangular shapes

Two-dimensional triangular and three-dimensional tetrahedral clustes form a special class of deformed clusters. Electrons confined in a triangular cavity have a simple shell structure that can be solved analytically [66]. Similarly the periodic orbit theory shows that there are only two important periodic orbits in a triangle. The regular shell structure is accompanied with a regular supershell structure [67]. The shell closings are the same as in the 2D harmonic oscillator. In the case of the 3D tetrahedron, the situation is not so simple: an analytic solution of the energy levels (in a cavity) do not exists and there are no nontrivial periodic orbits. However, the first shell fillings happen at the same electron numbers as in a 3D harmonic oscillator, *i.e.* 8, 20, 40, 70, 112 [45].

An octupole deformation will transform a sphere into a tetrahedron. Hamamoto *et al.* [68] have shown that a single-particle picture, related to that of the Nilsson model, indeed will lead to an octupole deformation with particle numbers 40, 70 and 112. This has been confirmed with the ultimate jellium model [45], which also shows that the tetrahedral deformation has an important role in stabilizing 40 as a strong magic number in the jellium model.

The triangular and tetrahedral clusters have an interesting possibility of making "double magic" clusters. Complete tetrahedra are closed packed FCC structures having 4, 10, 20, 35, 56 etc. atoms [69]. When multiplied by two these numbers equal to the shell closings in tetrahedral cavity. A tetrahedron consisting of divalent metal atoms has then a closed shell of atoms and a closed shell of electrons. *Ab initio* electronic structure calculations show that indeed a 35 atom magnesium tetrahedron has a clear electronic shell structure [45]. Similarly, in two-dimensional clusters the triangles of divalent atoms would be "doubly magic". Nevertheless, the sharp corners of tetrahedra and triangles are likely to make them energetically less favourable than more rounded geometries. The existence of double magic triangular clusters is still uncertain.

8 Odd–even staggering in metal clusters

The ionization potential of alkali metal clusters as a function of the number of atoms show a clear odd-even staggering. The ionization potential of an even cluster is larger than that of the neighbouring odd clusters [50]. Similar odd-even staggering is also seen in the abundances of clusters in the mass spectra and in the measured binding energies. In the case of small clusters ab initio calculations of quantum chemistry [3] give a quantitative description and explanation to the observed odd-even staggering. Nevertheless, the similarity of the odd-even staggering between different metals and its similarity to that observed in nuclear physics calls for a simple model for a qualitative explanation.

According to the previous section the deformation tries to remove the degeneracy of the many-body state arising of a open electronic shell. The reason for the odd-even staggering is the spin-degeneracy, which the deformation can not remove. Simple models, like the Clemenger–Nilson model, already explains the odd-even staggering [53]. In the limit of small clusters the odd-even staggering shows similar universality as the deformation [58]: it becomes nearly independent of the system in question and the odd-even staggering is quantitatively similar (in properly scaled energy units [58]) in clusters and in small nuclei.

In atomic clusters the detailed atomic geometry causes that the odd-even staggering is not as smooth function of the particle number as in the case of nuclei. However, at high temperatures where the clusters will sample a large collection of isomers, the average odd-even staggering of the Fermi energy will be smoother. The effect of the temperature on the odd-even staggering has been studied in [70] using several simple models.

Qunatitatively the odd-even staggering of the total energy can be determined as the absolute value of the second derivative of the total energy as a function of the number of electrons. Figure 8 shows the odd-even staggering calculated using the ultimate jellium model with the parameters suitable for nuclear matter. The figure also shows the experimental results for nuclei. For nuclei with less than about 20 protons and neutrons the deformation counts most of the odd-even staggering while in larger clusters the experimental staggering is much stronger. Also, in nuclei the odd-even staggering is a much smoother function of the particle number than in the results of the ultimate jellium model.

Figure 8 also show the experimental results for atomic clusters [71]. At smallest sizes the clusters show nearly as strong staggering as the nuclei, but the staggering gets weaker when the cluster size increases. Part of this rapic decrese might be an effect of the finite temperature of the clusters.

Finally, it should be noted that in nuclear physics the pairing of nucleons in accordance with the BCS-theory [62] seems to be an important element in the observed odd-even staggering. This is especially the case in large nuclei. From Figure 8 we can see that when the nucleus has more than about 20 protons (and neutrons) the deformation can account in average only about half of the observed odd-even staggering of the total energy. In the case of metal clusters the binding energies of Cooper pairs are orders of magnitude smaller than the observed odd-even staggering in the ionization potential. Consequently, in metal clusters the odd-even staggering is solely a deformation effect.

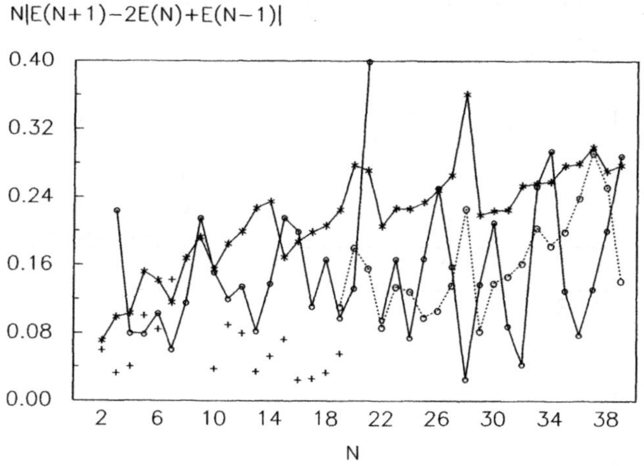

Fig. 8. Odd-even staggering of the total energy. Experimental results for nuclei are shown as stars, the results of the ultimate jellium model are shown as black dots. The experimental results for atomic clusters are shown as plus-signs. The open circles show theoretical results of from [72]. N is the number of protons and neutrons in the nucleus or the number of electrons in the case of atomic clusters.

9 *Ab initio* electronic structure: Shape and photoabsorption

Above we have considered clusters mainly as islands of electrons and the ions are used only to provide a smooth background to confine the electrons. The treatment of the true ionic sceleton together with the electrons can be done either using a perturbation theory or so-called *ab initio* computations. The *ab initio* calculations can be made with several different degrees of approximations. The traditional physical chemistry approach is to use Hatree–Fock or Configuration Interaction (CI) methods. The latter approach has the advantage of being essentially exact for small electron numbers and of providing the electronic excitation spectrum which often can be directly compared with experiments [3].

Another approach is to use the density functional Kohn–Sham method together with proper pseudopotentials. Here the advantage is that large electron or atom numbers, up to thousands, can be treated. In any *ab initio* method the atomic positions can be determined by minimizing the total energy. The Born–Oppenheimer approximation allows us to use classical molecular dynamics and, for example, to do simulated annealing for looking the ground state geometry. This method is often called *ab initio* molecular dynamics or "quantum molecular dynamics" since the forces are determined by solving the electronic structure. This can be done either adiabatically [74]

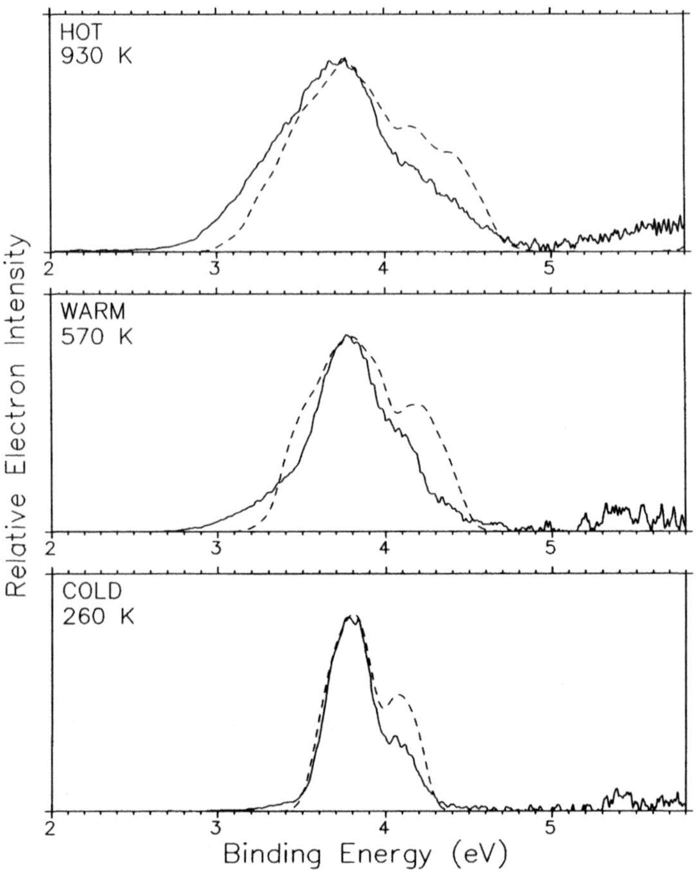

Fig. 9. Comparison of measured (solid lines) and calculated (dashed lines) photoemission spectra of Al_{13}^- at three different temperatures. From reference [46].

or by letting the electron wave functions also to be as dynamical variables as done in the Car–Parrinello method [75].

As an example we will study the photoionization of aluminium clusters. In this method a laser is used to ionize the cluster and the kinetic energy of the outcoming electron is measured. The kinetic energy reflects the binding energy of the electron in question and, in the single-electron picture, the density of electron levels can be measured. In practice, if mono-size clusters are studied the measurement has to be made for cluster ions or anions.

The Koopmans' theorem for Hartree–Fock approximation suggests that the kinetic energy of the photoelectron could be approximated by the single particle eigenvalue of the electron. In the case of the Kohn–Sham method

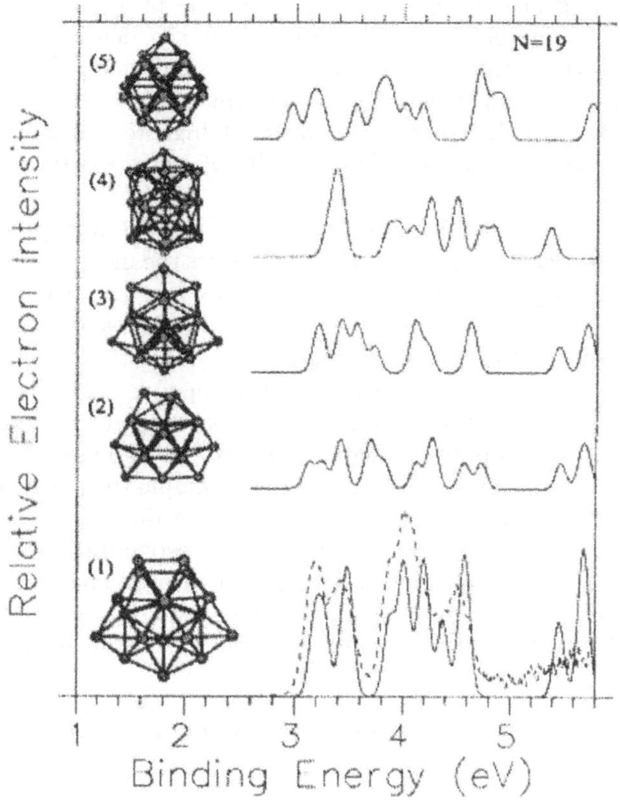

Fig. 10. Photoemission spectra of Al_{19}^- isomers. Solid lines are calculated spectra for isomers shown. The dashed line is the experimental spectrum. The lowest energy isomer gives the best agreement with the experiment. From reference [47].

it can be shown that the Koopmans' theorem does not give the correct ionization potential, but the single particle eigenvalue has to be corrected by the asymptotic limit of the exchange-correlation potential [46]. Assuming that the same is true also for the detachment of electrons from the deeper single particle levels, we can approximate that the photoelectron energy as

$$E_{\text{kin}} = h\nu + \epsilon_i - v_{\text{xc}}^{\infty}, \tag{9.1}$$

where $h\nu$ is the photon energy, ϵ_i the energy energy eigenvalue of the electron in the cluster, and v_{xc}^{∞} the asumptotic limit of the exchange-correlation potential. Model calculations have shown that v_{xc}^{∞} is nearly independent of the energy eigenvalue in question and can be approximated by a constant

[47]. Assuming further that the photoemission propability is equal for all electrons, the photoemission spectrum gives the density of states of the cluster. At zero temperature this consists of discrete lines. At a finite temperature the ionic motion moves the single particle levels up and down giving a width to the peaks. Moreover, at high temperatures the cluster will sample several isomers and the density of states will be an average over the isomers.

Ab initio molecular dynamics provises a tool to average the density of states at a desired temperature. Figure 9 shows the measured photoemission spectrum for Al_{13}^- at three different temperatures compared to calculated densities of states. Clearly this type of experiments and computations can give information of the cluster temperature.

The electronic structure of the cluster depends on the ionic positions. Consequently, the photoemission spectra measured at low temperatures together with the computed density of states can be used to determine the cluster goemetries. Figure 10 shows as an example the photoemission spectrum of Al_{19}^- compared to calculated spectra for different geometries shown in the figure. In this case only the isomer corresponding to the lowest total energy gives a good agreement with the measured spectrum. This gives confidence that properties of small clusters can be rather reliably computed using the Kohn–Sham method and *ab initio* molecular dynamics.

10 Quantum dots: Hund's rule and spin-density waves

In semiconductor heterostructures the conduction electron gas can be confined to move in a two-dimensional (2D) layer. There are several ways to further confine a finite number of electrons to form a quantum dot [76, 77]. Typically, in a quantum dot the electrons move in a nearly harmonic potential which can have a circular or elliptic symmetry. The electronic properties of the quantum dot are most often measured by studying the electron transport through the dot. We will not go into the details how the quantum dot is manufactured or the measurements done, but concentrate on the study of the interesting quantum mechanics of a small number of electrons in a 2D harmonic oscillator. However, it should be emphasized that the electrons in question are the conduction electrons of the semiconductor. They are characterized by a small effective mass and very low density. An external magnetic field can fully polarize the electron gas and lead to new many-body effects as most clearly manifested in the quantum Hall effect.

Noninteracting electrons in a quantum dot have a shell structure of the 2D harmonic oscillator. The shell fillings (magic numbers) occur at electron numbers 2, 6, 12, 20 etc. The electron-electron interaction will reduce the large degeneracies of the harmonic oscillator potential due to the fact that the effective single particle potential is not harmonic. The situation is similar to the spherical jellium model for clusters: the electrons move in a harmonic well, but the effective mean field potential is not harmonic. Nevertheless, the harmonic confinement means that, like in metal clusters, a collective plasmon excitation corresponding to the center of mass motion is the dominating dipole excitation. Again, the density functional Kohn–Sham formalism can be used to study the electronic structure. The only difference from the jellium model is that in the case of quantum dots the system is two-dimensional and consequently in the local density approximation the exchange-correlation potential is different from that of the 3D case [78].

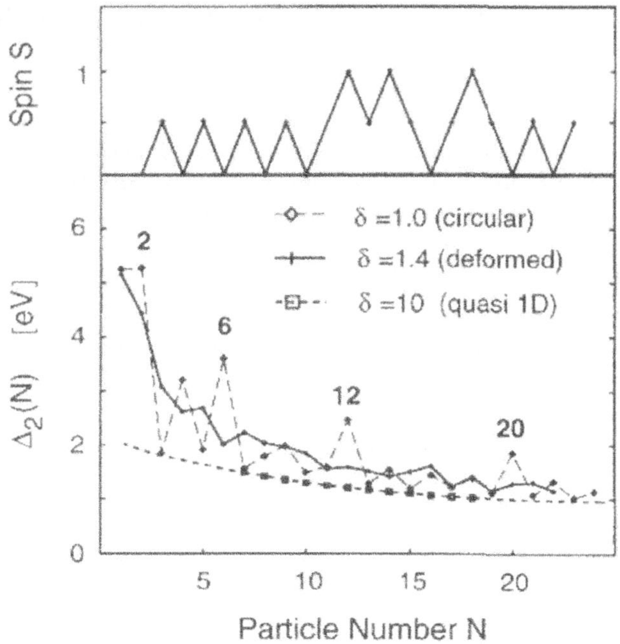

Fig. 11. Effect of the deformation on the second derivative of the total energy in ellipsoidal quantum dots (lower panel). The dashed line shows the result for the circular dots with magic numbers 2, 6, 12 and 20. Note the maxima at 4 and 9 caused by the Hund's rule. The solid line shows the result for an ellipsoidal dot and dashed line for a quasi-one-dimensional wire. The deformation reduces the shell structure as in metal clusters. The upper panel shows the total spin of the ellipsoidal dot. (In spherical dots the spin is determined by the Hund's rule, being zero at magic numbers and at maximum at half-filled shells).

In the case of an open shell quantum dot the spin-density functional theory should be used to allow the minimization of the total energy with respect to the total spin. If this is done the shell fillings obey the Hund's first rule in agreement with the experimental findings [79]. Figure 11 shows the second derivative of the total energy as a function of the electron number in a circular and deformed quantum dots. The magic numbers corresponding the full shells are clearly seen as well as the maxima corresponding to the half-filled shells at electron numbers 4, 9 and 16. The Hund's rule is caused by the breaking the spin-symmetry which separates the spin-up and spin-down energy levels. In the case of a half-filled shell this opens a gap at the Fermi level and reduces the total energy in a similar fashion as in the spherical jellium model for metal clusters. In reality the metal clusters can spontaneously deform and the Hund's rule does not have any role, while in quantum dots the shape is fixed and the Hund's rule can be observed experimentally.

In a large quantum dot there is, however, another mechanism to increase the gap at the Fermi level. This is a static spin-density wave: the total spin is zero but the spin-polarization is an oscillating function. The LSDA calculations show that for example the quantum dots with 24 and 34 electrons have a spin-density wave in their ground states when the electron density is low enough [80].

The spin-density waves are even more clearly seen in the quasi-one-dimensional systems such as quantum wires and quantum rings. In an infinite one-dimensional wire it can be understood as a manifestation of the spin-Peierls transition [81]. In a 1D metal the Fermi surface consists of only two points. Thus any periodic potential with the periodicity determined by the Fermi wave length will open a gap at the Fermi surface and reduce the total energy. The spin-density wave will provide a periodic potential (opposite phase for spin-up and spin-down electrons) which opens the gap [82].

Figure 12 shows the spin-density wave for a ring of 12 electrons calculated using the local density approximation. The advantage of the local density approximation is that the existence and origin of the spin-density wave is easy to determine. The Kohn–Sham theory as a mean field theory can brake the symmetry of the system and reflect the internal symmetry of the many-body state.

In principle, in circular systems the symmetry of the Hamiltonian requires that the total electron density and spin-density have also the circular symmetry. The existence of the spin-density wave can then be seen only in the internal structure of the many-body wave function. This can be done by studying correlation functions. However, since the Pauli exclusion principle already dictates certain correlation between spin-up and spin-down

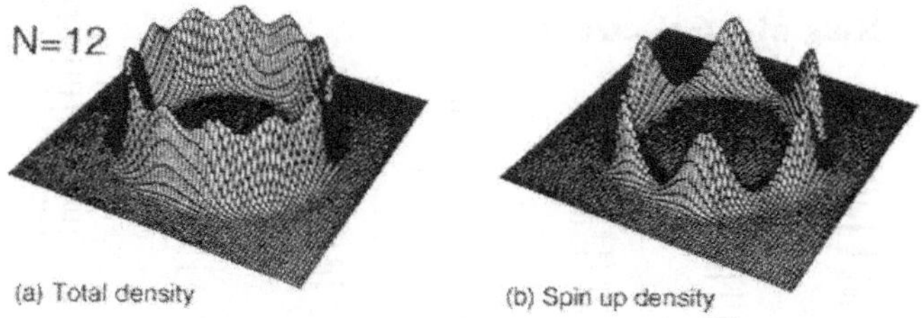

Fig. 12. The total electron density and the spin-up electron density in a narrow ring of 12 electrons calculated using the LSDA approximation. The spin-down density is in the opposite phase than the spin-up density. Together they form an antiferromagnetic ring of electrons.

electrons, it is not easy to see the onset of a spin-density wave state by lookling, for example, at the pair correlation function.

In the case of quasi-1D-rings the existence of the internal spin-density wave can be most easily seen by studying the excitation spectum of the many-body state [83]. Figure 13 shows the rotational spectrum of a six-electron ring calculated using an essentially exact CI method. The spectrum is characterized by a low-energy band which has a six-fold symmetry: the energy differencies and spin-assignments of the $M = 0$ and $M = 6$ are equal, similarly $M = 1$ and $M = 2$ states are similar to those with $M = 5$ and $M = 4$. This spectrum can be explained quantitatively with a model Hamiltonian

$$H = J \sum_{ij} \mathbf{S}_i \cdot \mathbf{S}_j + \frac{1}{2I} M^2 + \sum_{\nu} \hbar\omega_\nu, \tag{10.1}$$

where the first term is an antiferromagnetic Heisenberg Hamiltonian [5] (with coupling constant J), the second term describes rigid rotations of a ring of electrons (with moment of inertia I), and the last term describes vibrations of localized electrons.

The lowest band of Figure 13 corresponds to the vibrational ground state. The spin-structure of the band is determined by the Heisenberg model and the overall rise with the angular momentum comes from the rigid rotations. This result shows that the electrons in a quasi-1D-ring are localized in an antiferromagnetic chain. Consequently, the internal structure indeed has the spin-density wave suggested by the simple local density approximation (Fig. 12). This example shows that the study of the internal structure of the complicated many-body wave function can indeed be cumbersome and the simple models often give a more transparent picture of

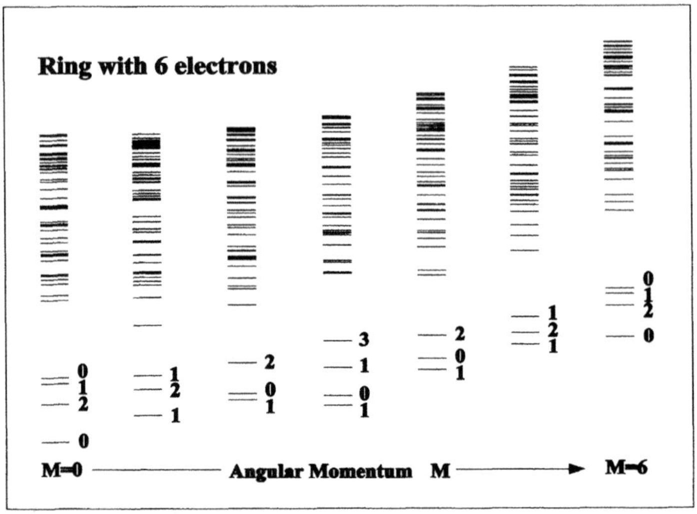

Fig. 13. Exact many-body spectrum of a quantum ring with six electrons. The numbers indicate the total spins of the lowest states.

the key properties of the system. Note that it is the structure of the model Hamiltonian that tells us that the system is an antiferromagnetic chain of localized electrons. By solving the density or the spin-density of coming from the model Hamiltonian does not leed to anything since both are circularly symmetric. The pair-correlation function does not tell us much either, since it is known that the correlation in a 1D Heisenberg model decays as $1/r$, r being the distance between the spins.

11 Deformation in quantum dots

The main difference in studying the deformation of metal clusters and quantum dots is that while in the metal clusters the shape is determined by the electronic structure due to the Jahn–Teller effect in quantum dots the shape is determined by the external confinement caused by external electrodes or the actual shape of the dot structure. In quantum dots the electronic structure can then in principle be freely chosen. Nevertheless, the external potential in a quantum dot is always nearly harmonic, at least at the bottom of the well. In the case of a small number of electrons the most important deformation is then ellipsoidal. The simple Nilsson diagram [59] can be used to study the effect of deformation to the shell structure. Since the deformation redistributes the energy levels and opens gaps at different places than in the circular case, the magnetism (Hund's rule) and magic numbers depend on the deformation, as illustrated in Figure 11. For example, the magnetic moment of the quantum dot can be changed by deformation caused by external electrodes. Reimann *et al.* [84] have studied in detail the effect of the deformation on the magnetism in quantum dots.

In layered semiconductors it is possible to construct a structure where an electron quantum dot is on top a quantum dot of holes. In this case the electrons and holes can bind each other without external confinement. The shape is then free to deform like in the case of metal clusters. The structure of the electron and hole droplets will depend on the distance between the layers and on the number of particles. If the distance between the electron and hole layers is small compared to the dot size we can approximate that the electrons and holes are in the same layer in calculating the electronic structure. The effect of the separating layer is then only in preventing recombination. Futhremore, if the number of electrons and holes is the same and they have the same effective mass, we end up in a model which is a two-dimensional version of the ultimate jellium model described earlier.

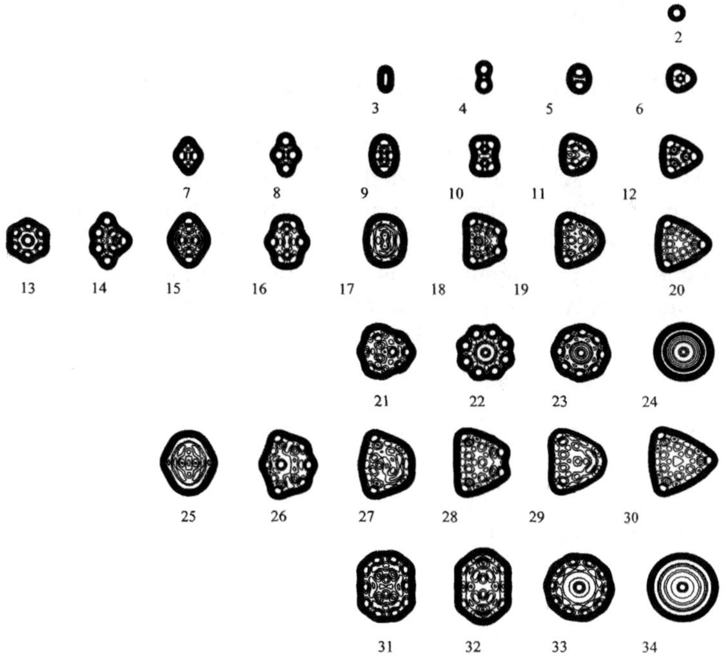

Fig. 14. Contours of the ground state densities of 2D electron-hole clusters calculated using the ultimate jellium model. The right column has the magic clusters. From reference [85].

In the ultimate model the electron and hole densities (which are equal) are determined by the properties of the 2D electron gas: There is no average Coulomb interaction and the cluster is free to deform. Figure 14 shows the optimal shapes of such a model [85]. It is seen that the magic clusters have triangular shapes in addition to the spherical ones. This is related to the tendency of octupole deformation of some magic clusters in 3D case discussed earlier. Moreover, the freely deformed shapes in 2D seem to have a much richer variety of shapes than in the 3D case. In this model the quantum dot is always nonmagnetic, since the deformation can more effectively open a gap at the Fermi level than the magnetization. The situation is the same as in the 3D case.

Using the Kohn–Sham method it is straighforward to improve the model of layered electron/hole dots. The actual distance between the dots as well as the interdot Coulomb potential can be taken into account, as well as the necessary external confinement. However, in the case that the external confinement is weak in comparison to the interdot Coulomb potential the spontaneous deformation of the electron and hole clouds in the quantum dot should be possible. In principle the same should happen in 3D droplets of electron-hole plasma [55].

12 Localization of electrons in a strong magnetic field

The external magnetic field is described in quantum mechanics by a vector potential ($\mathbf{B} = \nabla \times \mathbf{A}$) which is added to the momemtum operator (minimal substitution). In the density functional Kohn–Sham formalism part of the kinetic energy of the interacting electrons is hidden in the exchange-correlation energy. Consequently, it is not enough to include the magnetic field contribution only to the kinetic energy operator of the single particle Hamiltonian, but also in the effective potential. This leads to so-called current-density functional formalism developed in detail by Vignale and Rasolt [86].

The current-density-functional method is rather complicated due to the diffucult coupling between the exchange-correlation energy and the vector potential. We do not desctribe the formalism here (see Refs. [86,87]), but only state that the single particle Kohn–Sham equation now is

$$\left[\frac{1}{2m}\left(-i\hbar\nabla + \mathbf{A}(\mathbf{r})\right)^2 + V_{\text{eff}}^{\sigma}(\mathbf{r})\right]\psi_{i\sigma} = \epsilon_{i\sigma}\psi_{i\sigma}, \tag{12.1}$$

where vector potential includes a contribution coming from the electron currents and the effective potential depends on the vector potential. V_{eff} also includes the direct interaction of the external magnetic field with the electron spin, so-called Zeeman term. As in the normal Kohn–Sham method the equations can be solved iteratively. The results will give the total spin, electron and spin densities and the internal currents.

The actual numerical solution of the equations is quite demanding. However, some general features can be understood by looking at the noninteracting electrons in a 2D harmonic oscillator in the presence of a perpendicular magnetic field. This problem can be solved analytically and the single particle levels form so-called Darwin–Fock basis and energy levels

$$\epsilon_{nL} = (2n + 1 + |L|)\hbar\sqrt{\frac{1}{4}\omega_c^2 + \omega_0^2} - \frac{1}{2}L\hbar\omega_c + g^*\mu_B S_z B, \tag{12.2}$$

where L is the orbital angular momentum and ω_c and ω_0 the cyclotron and confinement frequencies, respectively. The last term takes into account the Zeeman splitting (g^* is a reduced Landé factor of the semiconductor). The cyclotron frequency is proportional to the magnetic field. In a high magnetic field only the lowest Landau band is filled ($n = 0$) and each harmonic oscillator state will have only one electron corresponding to the maximum orbital angular momentum. The electron density distribution will be homogeneous, so-called maximum density droplet. This happens also in the case of the interacting electrons in the quantum dot. The only difference is that in this case the single particle states are not exactly the Darwin–Fock states but solutions of the circular problem with a more complicated effective potential.

The Zeeman splitting can be so small that a maximum density droplet is first formed with both spin states equally populated. When the field increases the droplet with spin-up states starts to shrink and that of the spin-down states expand until all electrons populate the spin-down states. Up to this point the interacting electron system behaves quite similarly with that of the noninteracting electrons, the only differences being in the order the single particle levels are filled before the maximum density droplet state is reached.

The electron-electron interactions change the situation drastically when the magnetic field is further increased. In the noninteracting case the maximum density droplet continues becoming smaller in size. In the interacting case the Coulomb repulsion prevents the shrinking of the droplet forcing electrons from small angular momentum values to move to higher angular momentum values. In the density this is seen as a ring of electrons separating from the maximum density droplet [88]. The current-spin-density formalism gives a result that this ring consists of localized electrons as shown in Figure 15. Each maxima at the ring corresponds to a single current vortex consisting of one electron. Increasing the magnetic field eventually makes all the electrons to localize into single "vortices".

The localization of electrons in the density functional model is an other example of an internal symmetry breaking of the system. Again, as in the

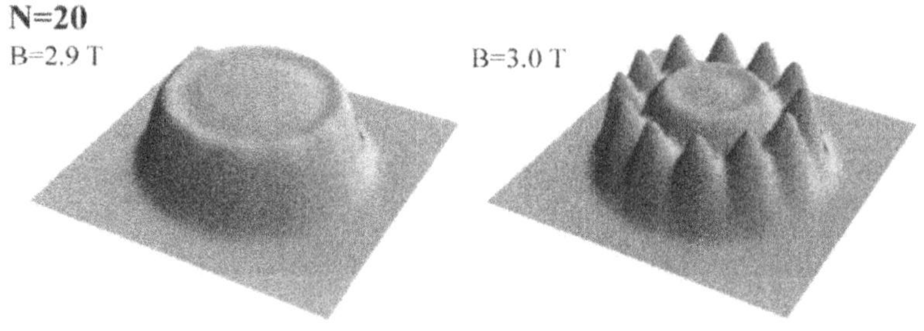

Fig. 15. Electron density in a circular quantum dot in a high magnetic field. The Maximum density droplet is shown in the left. Increasing the field separates a ring of electrons as shown in the right.

cases of the deformation of clusters or spin-density waves in quantum dots, the symmetry of the Hamiltonian requires that in the laboratory frame of reference the total density should have a circular symmetry. The density functional theory, as a mean field theory, can reduce the energy by breaking the symmetry of the effective single-particel potential. Similar localization of electrons in high magnetic fields is obtained also in the unrestricted Hartree–Fock calculations [89].

Exact many-body calculations, like CI, naturally give a symmetric total electron density, and correlation functions have to be studied to see the internal localization of electrons [90]. Another possibility would be to study the excitation spectrum of the many-body states as done above in the case of a quantum ring where the antiferromagnetic Heisenberg model explained the electron localization. In the presense of a magnetic field, however, the computations are less accurate and the analysis of the rotational spectra might not be as simple. However, there is another way to reveal the electron localization. By reducing the symmetry of the external confinement from the circular symmetry, the symmetry broken solutions become allowed. Koskinen *et al.* [91] have performed "exact" CI calculations for ellipsoidal quantum dots in a high magnetic field and indeed obtained the localization of electrons. The results are shown in Figure 16.

The above discussion and results show that the local current-density approximation can quite well describe the electrons in a magnetic field and give information of the internal structure of the many-body wave function which is not always so easy to see from the exact result (even if it is available). The basic reason for this is that the system is dominated by the single (quasi) particle picture. This should brake down in even stronger magnetic fields where the fractional quantum Hall regime is reached. In this case the many-body wave function can be approximated by the Laughlin wave

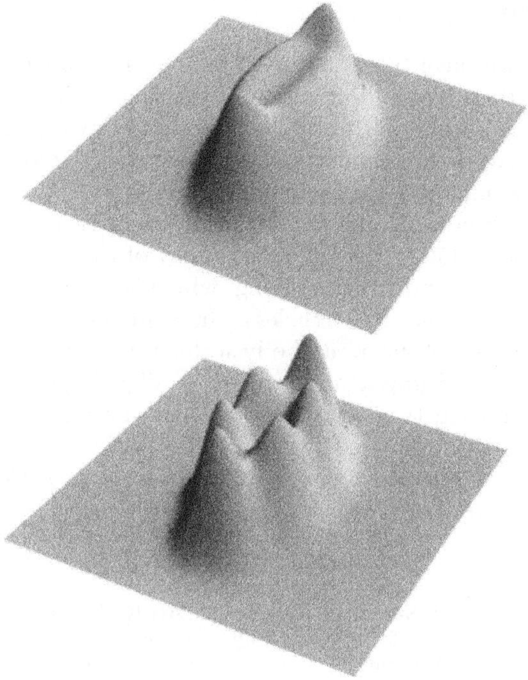

Fig. 16. Exact electron densities in ellipsoidal quantum dots containing six electrons. The upper panel shows the maximum density droplet and the lower panel the localization of electrons in a high magnetic field.

function [92]

$$\Psi = \prod_{i<j} (z_i - z_j)^q \exp\left(-\sum_k |z_k|^2/2 \right) , \qquad (12.3)$$

where $z = x + iy$ and the x and y-coordinates are in the units of the magnetic length. The exponent q has to be odd (for fermions). The fractional quantum Hall regime is reached when $q = 3$.

It is interesting to note that for $q = 1$ the Laughlin wave function is just the maximum density droplet (for noninteracting particles), *i.e.* a Slater determinant of single particle states. For $q = 3$ it is a rather complicated superposition of Slater determinants made of single-particle states of the lowest Landau band. It seems unlikely that the local density approximation could describe the Laughlin state properly. Nevertheless, suggestions has been made how to include the fractional quantum Hall regime into the exchange-correlation functional.

13 Conclusions

In this chapter we have tried to illustrate that simple single-particle models can give good understanding of many features of finite electron systems like metal clusters and quantum dots. The underlaying theory is the density functional Kohn–Sham method which formally folds the many-body problem into a single particle problem.

In the Kohn–Sham formalism all the quasi-particles move in the same effective potential. Due to the self-consistency of this potential the method allows a spontaneus symmetry breaking, which is sometimes viewed only as an artifact of the model. Nevertheless, the symmetry breaking can reflect the internal structure of the many-body state and in many cases gives a more transparent picture of important properties, like shape deformation, spin-density waves and electron localization, than the exact many-body wave function.

The many-body physics of small qauntum systems is rich. There is still need for exact many-body calculations, approximative methods as well as for simple models.

The author wihes to thank Stephanie Reimann, Matti Koskinen, Ben Mottelson, Pertti Lipas and Jere Kolehmainen for many helpful and exciting discussions. This work has been supported by the Academy of Finland under the Finnish Centre of Excellence Programme 2000-2005 (Project No. 44875, Nuclear and Condensed Matter Programme at JYFL).

References

[1] W.D. Knight, K. Clemenger, W.A. de Heer, W.A. Saunders, M.Y. Chou and M. Cohen, *Phys. Rev. Lett.* **52** (1984) 2141.

[2] M. Brack, *Rev. Mod. Phys.* **65** (1993) 677.

[3] V. Bonačić Koutecký, P. Fantucki and J. Koutecký, *Chem. Rev.* **91** (1991) 1035.

[4] W. Kohn and L.J. Sham, *Phys. Rev. A* **140** (1965) 1133.

[5] N.W. Ashcroft and N.D. Mermin, *Solid State Physics* (Holt, Reinhard and Winston, 1976).

[6] R.O. Jones and O. Gunnarsson, *Rev. Mod. Phys.* **61** (1989) 689.

[7] S.H. Vosko, L. Wilk and M. Nusair, *Can. J. Phys.* **58** (1980) 1200.

[8] J.P. Perdew and A. Zunger, *Phys. Rev. B* **23** (1981) 5048.

[9] N.D. Lang and W. Kohn, *Phys. Rev. B* **1** (1970) 4555.

[10] N.D. Lang, *Solid State Phys.* **28** (1973) 225.

[11] R. Monnier and J.P. Perdew, *Phys. Rev. B* **17** (1978) 2595.

[12] M. Manninen and R.M. Nieminen, *J. Phys. F: Metal Phys.* **8** (1978) 2243.

[13] J.P. Perdew, H.Q. Tran and E.D. Smith, *Phys. Rev. B* **42** (1990) 11627.

[14] H.B. Shore and J.H. Rose, *Phys. Rev. Lett.* **66** (1991) 2519.

[15] P. Ziesche, J.P. Perdew and C. Fiolhais, *Phys. Rev. B* **49** (1994) 7916.

[16] M. Manninen, R. Nieminen, P. Hautojärvi and J. Arponen, *Phys. Rev. B* **12** (1975) 4012.

[17] J. Martins and J. Buttet, *Surf. Sci.* **106** (1981) 261.
[18] M. Koskinen and M. Manninen, *Phys. Rev. B* **54** (1996) 14796.
[19] M. Koskinen, M. Manninen and P.O. Lipas, *Phys. Rev. B* **49** (1994) 8418.
[20] J. Mansikka-aho, M. Manninen and E. Hammarén, *Z. Phys. D* **31** (1994) 253.
[21] A. Hintermann and M. Manninen, *Phys. Rev. B* **27** (1983) 7263.
[22] W. Ekardt, *Phys. Rev. B* **29** (1984) 1558 (1984).
[23] D.E. Beck, *Phys. Rev. B* **30**, 6935 (1984).
[24] M.Y. Chou, A. Cleland and M.L. Cohen, *Solid State Commun.* **52** (1984) 645.
[25] I.I. Geguzin, *Zh. Eksp. Teor. Fiz. Pis'ma Red* **35** (1981) 584 [*JETP Lett.* **33** (1981) 568].
[26] M. Manninen and P. Jena, *Europhys. Lett.* **14** (1991) 643.
[27] J. Mansikka-aho, M. Manninen and H. Nishioka, *Phys. Rev. B* **48** (1993) 1873.
[28] N. Pavloff and M.S. Hansen, *Z. Phys. D* **24** (1992) 57.
[29] J. Mansikka-aho, M. Manninen and E. Hammarén, *Phys. Rev. B* **47** (1993) 10675.
[30] V.M. Akulin, *Phys. Rev. B* **48** (1993) 3532.
[31] C. Brechignac, Ph. Cahusac, J. Leyginer, A. Sarfati and V.M. Akulin, *Phys. Rev. A* **51** (1995) 3902.
[32] N. Pavloff, *J. Phys. A* **28** (1995) 4123.
[33] A. Zangwill, *Physics at surfaces* (Cambridge University Press, 1988).
[34] H. Göhlich, T. Lange, T. Bergmann and T.P. Martin, *Phys. Rev. Lett.* **65** (1990) 748.
[35] T.P. Martin, U. Näher and H. Schaber, *Chem Phys. Lett.* **199** (1992) 470.
[36] J. Mansikka-aho, M. Manninen and E. Hammarén, *Z. Phys. D* **21** (1991) 271.
[37] J. Mansikka-aho, E. Hammarén and M. Manninen, *Phys. Rev. B* **46** (1992) 12649.
[38] Y. Wang, T.F. George and D.M. Lindsay, *J. Chem. Phys.* **86** (1987) 3493.
[39] D.M. Lindsay, Y. Wang and T.F. George, *J. Phys. Chem.* **86** (1987) 3500.
[40] J.L. Martins, J. Buttet and R. Car, *Phys. Rev. Lett.* **53** (1984) 655.
[41] A. Yoshida, T. Døssing and M. Manninen, *J. Chem. Phys.* **101** (1994) 3041.
[42] R. Poteau and G.M. Partos, *Eur. Phys. J.* **9** (1999) 235.
[43] J.P.K. Doye and D.J. Wales, *Z. Phys. D* **40** (1997) 194.
[44] D.M. Lindsay, Y. Wang and T.F. George, *J. Cluster Sci.* **1** (1990) 107.
[45] S.M. Reimann, M. Koskinen, H. Häkkinen, P.E. Lindelof and M. Manninen, *Phys. Rev. B* **56** (1997) 12147.
[46] J. Akola, M. Manninen, H. Häkkinen, U. Landman, X. Li and L.-S. Wang, *Phys. Rev. B* **60** (1999) R11297.
[47] J. Akola, M. Manninen, H. Häkkinen, U. Landman, X. Li and L.-S. Wang, *Phys. Rev. B* **62** (2000) 13216.
[48] D. Reinhard, B.D. Hall, D. Ugarte and R. Monot, *Phys. Rev. B* **55** (1997) 7868.
[49] J. Borggreen, P. Chowdhury, N. Kebaili, L. Lundsberg-Nielsen, K. Lützenkirchen, M.B. Nielsen, J. Pedersen and H.D. Rasmussen, *Phys. Rev. B* **48** (1993) 17507.
[50] W. de Heer, *Rev. Mod. Phys.* **65** (1993) 611.
[51] S. Bjørnholm and J. Borggreen, *Phil. Mag. B* **79** (1999) 1321.
[52] W. Ekard and Z. Penzar, *Phys. Rev. B* **38** (1988) 4273.
[53] K. Clemenger, *Phys. Rev. B* **32** (1985) 1359.
[54] S. Frauendorf and V.V. Pashkevich, *Z. Phys. D* **26** (1993) S98.
[55] M. Manninen, *Phys. Rev. B* **34** (1986) 6886.

[56] M. Koskinen, P.O. Lipas and M. Manninen, *Z. Phys. D* **35** (1995) 285.

[57] M. Koskinen, P.O. Lipas and M. Manninen, *Nucl. Phys. A* **591**, 421.

[58] H. Häkkinen, J. Kolehmainen, M. Koskinen, P.O. Lipas and M. Manninen, *Phys. Rev. Lett.* **78** (1997) 1034.

[59] S.G. Nilsson, *Mat.-Fys. Medd. Dansk Vidensk. Selsk.* **29** (1955) 16.

[60] B.R. Mottelson, in *The Many-Body Problem, Les Houches Lectures 1957* (Dunod, Paris, 1959).

[61] M. Manninen, J. Kolehmainen, M. Koskinen and S.M. Reimann, *Phys. Scr.* **T88** (2000) 90.

[62] A. Bohr and B.R. Mottelson, *Nuclear structure II* (Benjamin, New York, 1975).

[63] V.M. Strutinsky, *Nucl. Phys. A* **95** (1967) 420.

[64] S.M. Reimann, S. Frauendorf and M. Brack, *Z. Phys. D* **34** (1995) 125.

[65] C. Yannouleas and U. Landman, *Phys. Rev. B* **51** (1995) 1902.

[66] F.E. Borgnis and C.H. Papas, in *Encyclopaedia of Physics*, ed. by S. Flügge (Springer, Berlin, 1958), Vol. XVI, p.285.

[67] M. Brack, J. Blaschke, S.C. Creagh, A.G. Magner, P. Meier and S.M. Reimann, *Z. Phys. D* **40** (1997) 276.

[68] I. Hamamoto, B.R. Mottelson, H. Xie and X.Z. Zhang, *Z. Phys. D* **21** (1991) 163.

[69] T.P. Martin, *Phys. Reports* **273** (1996) 199.

[70] M. Manninen, J. Mansikka-aho, H. Nishioka and Y.Z. Takahashi, *Z. Phys. D* **31** (1994) 259.

[71] C. Brechignac, Ph. Cahuzac, J. Leygnier and J. Weiner, *J. Chem. Phys.* **90** (1989) 1492.

[72] P, Møller, J.R. Nix, W.D. Myers and J. Swiatecki, *Atom Data Nucl. Data Tab.* **59** (1995) 185.

[73] J. Kolehmainen, M. Koskinen, H. Hk̈kinen, M. Manninen and S. Reimann, *Czech. J. Phys.* **48** (1998) 679.

[74] R.N. Barnett and U. Landman, *Phys. Rev. B* **48** (1993) 2081.

[75] R. Car and M. Parrinello, *Phys. Rev. Lett.* **55** (1985) 2471.

[76] P. Harrison, *Quantum wells, wires and dots* (Wiley, New York, 1999).

[77] T. Chakraborty, *Quantum dots: a survay of the properties of artificial atoms* (Elsevier, Amsterdam, 1999).

[78] B. Tanatar and D.M. Ceperley, *Phys. Rev. B* **39** (1989) 5005.

[79] S. Tarucha, D.G. Austing, T. Honda, R.J. van der Hage and L.P. Kouwenhoven, *Phys. Rev. Lett.* **77** (1996) 3613.

[80] M. Koskinen, M. Manninen and S.M. Reimann, *Phys. Rev. Lett.* **79** (1997) 1389.

[81] S.M. Reimann, M. Koskinen and M. Manninen, *Phys. Rev. B* **59** (1999) 1613.

[82] J.R. Hook and H.E. Hall, *Solid State Physics* (Wiley, Chichester, 1991).

[83] M. Koskinen, M. Manninen, R.B. Mottelson and S.M. Reimann, *Phys. Rev. B* **63** (2001) 205323.

[84] D.G. Austing, S. Sasaki, S. Tarucha, S.M. Reimann, M. Koskinen and M. Manninen, *Phys. Rev. B* **60** (1999) 11514.

[85] S.M. Reimann, M. Koskinen, J. Helgesson, P.E. Lindelof and M. Manninen, *Phys. Rev. B* **58** (1998) 8111.

[86] G. Vignale and M. Rasolt, *Phys. Rev. B* **37** (1988) 10685.

[87] M. Koskinen, J. Kolehmainen, S.M. Reimann, J. Toivanen and M. Manninen, *Eur. Phys. J.* **9** (1999) 487.

[88] C. de Chamon and X.G. Wen, *Phys. Rev. B* **49** (1994) 8227.
[89] H.-M. Müller and S.E. Koonin, *Phys. Rev. B* **54** (1996) 14532.
[90] P.A. Maksym, *Phys. Rev. B* **53** (1996) 10871.
[91] M. Koskinen, M. Manninen and S.M. Reimann, to be published.
[92] M.P. Marder, *Condensed Matter Physics* (Wiley, New York, 2000).

COURSE 8

THEORY OF CLUSTER MAGNETISM

G.M. PASTOR

*Laboratoire de Physique Quantique,
CNRS, Toulouse, France*

Contents

THEORY OF CLUSTER MAGNETISM

G.M. Pastor

1 Introduction

Magnetism is a major area in fundamental and applied research, a subject of main concern in human knowledge since the antiquity, which remains, even nowadays, quite elusive. The principal sources of magnetism in atoms, clusters and solids are the currents associated to the electronic motion (orbital magnetism) and the intrinsic angular momentum of the electrons (spin magnetism). These properties are of quantum mechanical nature and cannot be regarded as the superposition of independent individual-electron contributions. They result from the behavior of the ensemble of electrons in the relevant low-energy scale (valence electrons). Consequently, the magnetic properties of matter are very sensitive to the details of the electronic correlations and to temperature.

Finite clusters constitute a new state of matter with its own fascinating phenomena. From this point of view, the properties of atoms and solids may be regarded as limiting cases of a much richer and complex dependence on cluster size. One of the main goals of the theory of metal clusters is to understand how the properties of localized, atomic electrons change as they become part of a finite cluster and start to delocalize, and how bulk-like behavior is attained. The study of magnetism constitutes one of its major challenges. In isolated atoms almost all elements show a non-vanishing magnetic moment given by Hund's rules, while in the solid state only a few of them (some transition metals of the Fe group, the lanthanides and actinides) preserve a non-vanishing magnetization. The latter is a consequence of the delocalization of the valence electrons which favors equal occupation of states having opposite projections of spin and orbital moments. In this context the magnetism of transition-metal (TM) clusters is of fundamental interest, since atomic and bulk behaviors are intrinsically different. Atomic magnetism is due to electrons which occupy localized orbitals, while in TM solids the electrons responsible for magnetism are itinerant, conducting *d*-electrons. Therefore, the magnetic properties of TM's are very sensitive to system size and to the local environment of the atoms. Clusters of these elements show a wide variety of very interesting magnetic phenomena.

The purpose of these lecture notes is to introduce the main concepts in the theory of cluster magnetism giving emphasis to the properties of transition metals. To get started, the next section recalls the basic physical pictures of localized (atomic) and itinerant (band) magnetism. In Section 3 representative experimental results on free and deposited transition-metal clusters are briefly reviewed. Sections 4–6 present various theoretical methods to determine the electronic and magnetic properties of clusters as well as the main results derived from them. In Section 4 the size and environment dependence of $T = 0$ properties, such as spin moments, orbital moments, magnetic anisotropy and electronic structure, are investigated in the framework of a mean-field approach. Electron-correlation effects are analyzed explicitly in Section 5. Section 6 is focused on temperature dependent properties – average magnetization per atom, magnetic susceptibility, specific heat, etc. – and on their relation to the many-body excitation-spectrum. Finally, we conclude in Section 7 by pointing out some future research directions.

2 Background on atomic and solid-state properties

2.1 Localized electron magnetism

In the presence of an external magnetic field $\vec{H} = \vec{\nabla} \times \vec{A}(\vec{\nabla} \cdot \vec{A} = 0)$ the Hamiltonian of a non relativistic electron is modified in two major ways. First, as in classical mechanics, the linear momentum $\vec{p}$ is replaced by $\vec{p} + (e/c)\vec{A}$ $(e > 0)$. The kinetic energy can then be written as

$$T = \frac{1}{2m} p^2 + \frac{e\hbar}{mc} \vec{H} \cdot \vec{\ell} + \frac{e^2}{2mc^2} \frac{(\vec{H} \times \vec{r})^2}{4}, \tag{2.1}$$

where $\vec{\ell} = \vec{r} \times \vec{p}$ is the orbital angular momentum operator. Second, since the electron has an intrinsic angular momentum, we must take into account its direct interaction with the magnetic field given by

$$\Delta\mathcal{H}_{\text{spin}} = g_0 \, \mu_{\text{B}} \, \vec{s} \cdot \vec{H}, \tag{2.2}$$

where $\vec{s} = \vec{\sigma}/2$ is the spin operator, $\mu_{\text{B}} = e\hbar/2mc$ the Bohr magneton and $g_0 = 2.0023 \simeq 2$ the electronic g-factor.

For a many-electron system (atom or solid) the Hamiltonian reads

$$\mathcal{H} = \mathcal{H}_0 + \mu_{\text{B}} \vec{H} \cdot (\vec{L} + g_0 \vec{S}) + \frac{e^2}{8mc^2} \sum_i (\vec{H} \times \vec{r}_i)^2. \tag{2.3}$$

$\vec{L} = \sum_i \vec{\ell}_i$ refers to the total angular momentum, $\vec{S} = \sum_i \vec{s}_i$ to the total spin, and $\mathcal{H}_0$ to the Hamiltonian in the absence of magnetic field. The last term in equation (2.3) is a repulsive potential that gives a diamagnetic contribution to the magnetic susceptibility χ [1]. The second term, the Zeeman interaction, gives a paramagnetic contribution to χ that dominates over the diamagnetic term as soon as $\langle \vec{L} + 2\vec{S} \rangle \neq 0$. This is the case of almost all open-shell systems we are most interested in, *e.g.*, TM or rare-earth atoms or ions. Notice that both orbital and spin contributions add up to form $\vec{\mu}_{\rm at} = -\mu_{\rm B}(\vec{L} + 2\vec{S})$ which can be regarded as the atomic intrinsic magnetic moment. Consequently, the low-lying eigenvalues of $\vec{L}$, $\vec{S}$ and the total angular momentum $\vec{J} = \vec{L} + \vec{S}$ are of central importance to the magnetic response. We shall therefore recall Hund's rules which summarize the main results of the elementary theory of atomic multiplets [2]. These properties are important for understanding the sources of localized magnetism in isolated atoms or ions diluted in insulators, and also as a reference for comparison with the behavior of itinerant-electron magnetism in solids and as a function of cluster size.

2.1.1 Magnetic configurations of atoms: Hund's rules

Consider an atom or ion with only one partially filled shell (for example, a TM or rare-earth atom) which is characterized by a given principal quantum number n and angular momentum ℓ. Since closed-shell states are non degenerate and have $L = S = J = 0$ we need only to consider the electrons in the open shell. We have therefore $2(2\ell + 1)$ one-electron states which can be characterized by the orbital momentum ℓ of the shell, the azimuthal quantum number $m\,(-\ell \leq m \leq \ell)$ and the spin $\sigma = \uparrow, \downarrow$. Within the open shell the single-particle Hamiltonian of the electrons in a central field (ignoring electron-electron and spin-orbit interactions) is given by

$$\mathcal{H}_0 = \mathcal{E}^0_{n\ell} \sum_{m,\sigma} \hat{c}^\dagger_{m\sigma} \hat{c}_{m\sigma} = \mathcal{E}^0_{n\ell} \sum_{m,\sigma} \hat{n}_{m\sigma} \qquad -\ell \leq m \leq \ell\,,\ \sigma = \uparrow, \downarrow \tag{2.4}$$

where

$$\mathcal{E}^0_{n\ell} = \int R^*_{n\ell}(r) \left[-\frac{\hbar^2}{2m} \nabla^2 + V_{\rm core}(r) \right] R_{n\ell}(r) r^2 {\rm d}r \tag{2.5}$$

is the average of kinetic and electron-core energies in the shell $n\ell$, and $\hat{c}^\dagger_{m\sigma}$ ($\hat{c}_{m\sigma}$) is the creation (annihilation) operator for an electron of spin σ at orbital $u_{n\ell m} = R_{n\ell}(r)\, Y_{\ell m}(\theta, \phi)$. Since $\mathcal{E}^0_{n\ell}$ is independent of m and σ, a shell having $\nu = \nu_\uparrow + \nu_\downarrow$ electrons would be $\binom{2\ell+1}{\nu_\uparrow}\binom{2\ell+1}{\nu_\downarrow}$-fold degenerate.

Part of this degeneracy is removed by the electron-electron interaction

$$\mathcal{H}_{\rm ee} = \frac{1}{2} \sum_{\substack{ijrt \\ \sigma\sigma'}} V_{ijrt}\, \hat{c}^\dagger_{i\sigma}\, \hat{c}^\dagger_{j\sigma'}\, \hat{c}_{t\sigma'}\, \hat{c}_{r\sigma}, \tag{2.6}$$

where

$$V_{ijrt} = \int u_i^*(\vec{r}_1)\, u_j^*(\vec{r}_2)\, \frac{\mathrm{e}^2}{|\vec{r}_1 - \vec{r}_2|}\, u_r(\vec{r}_1)\, u_t(\vec{r}_2)\, \mathrm{d}^3 r_1\, \mathrm{d}^3 r_2 \tag{2.7}$$

are the Coulomb integrals ($i \equiv nlm$). The dominant V_{ijrt} are those which involve only one pair of distinct electrons. Thus, one may simplify equation (2.6) by considering only the direct Coulomb integrals $U_{mm'} = V_{mm'mm'}$ that represent the electrostatic repulsion between 2 electrons occupying orbitals m and m', and the exchange integrals $J_{mm'} = V_{mm'm'm}$ which are all positive [2]. $J_{mm'}$ measures the energy difference between the singlet and triplet states that can be formed with two electrons occupying orbitals m and m'. Typical values for TM atoms are $U_{mm'} \simeq 10$ eV and $J_{mm'} \simeq 0.5$–1.0 eV. Keeping the terms ($i = r$, $j = t$) and ($i = t$, $j = r$), $\mathcal{H}_{\rm ee}$ takes the form

$$\mathcal{H}_{\rm ee} = \frac{1}{2} \sum_{\substack{mm' \\ \sigma\sigma' \\ (m\sigma)\neq(m'\sigma')}} U^{\sigma\sigma'}_{mm'} \hat{n}_{m\sigma} \hat{n}_{m'\sigma'} - \frac{1}{2} \sum_{m\neq m'} J_{mm'} (\hat{s}^+_m \hat{s}^-_{m'} + \hat{s}^-_m \hat{s}^+_{m'}), \tag{2.8}$$

where $\hat{n}_{m\sigma} = \hat{c}^\dagger_{m\sigma}\hat{c}_{m\sigma}$ and $\hat{s}^+_m = c^\dagger_{m\uparrow}\hat{c}_{m\downarrow}$ [$\hat{s}^-_m = (\hat{s}^+_m)^\dagger$]. $U^{\uparrow\uparrow}_{mm'} = U^{\downarrow\downarrow}_{mm'} = U_{mm'} - J_{mm'}$ and $U^{\uparrow\downarrow}_{mm'} = U^{\downarrow\uparrow}_{mm'} = U_{mm'}$ represent the effective Coulomb repulsion between electrons having parallel and unparallel spins.

Hund's first rule states that the ground-state configuration of open-shell atoms has the largest possible total spin S. From equation (2.8) one observes that it is energetically more favorable to set all the electrons in the open shell with the same spin due to the exchange interaction $J_{mm'}$. The ground-state spin S of an atom having ν electrons in a partially filled shell is therefore $S = \nu/2$ for $\nu \leq 2\ell + 1$, and $S = 2\ell + 1 - \nu/2$ for $2\ell + 1 \leq \nu \leq 2(2\ell + 1)$. The orbital dependence of $U_{mm'}$ and $J_{mm'}$ is not essential in this case, as can be easily verified by setting $U_{mm'} = U$ and $J_{mm'} = J$. However, if one assumes $J_{mm'} = J$ and $U_{mm'} = U$, an important degeneracy remains, since all $\binom{2\ell+1}{\nu_\uparrow}$ states having maximal $S_z = \nu_\uparrow/2$ ($\nu_\uparrow \leq 2\ell + 1$ and $\nu_\downarrow = 0$) would have the same energy $E(S{=}\nu_\uparrow/2) = (U - J)\nu_\uparrow(\nu_\uparrow - 1)/2$.

Hund's second rule states that the lowest-energy configuration among those having maximal S (1st rule) is the one with the largest total orbital momentum L. This is a consequence of the orbital dependence of

$U_{mm'}$ and $J_{mm'}$. In fact, the Coulomb repulsion energies among electrons with the largest possible m are weaker, since the corresponding electronic densities $|\psi_{\ell m}(\theta,\phi)|^2$ are more distant in average [2]. For example, for d-electrons the deviations $\Delta U_{mm'}$ of $U^{\sigma\sigma}_{mm'} = U_{mm'} - J_{mm'}$ from the average value are approximately $\Delta U_{2,1} = -58\,u$, $\Delta U_{2,0} = -58\,u$, $\Delta U_{2,-1} = -4\,u$, $\Delta U_{2,-2} = 50\,u$, $\Delta U_{1,0} = 23\,u$ and $\Delta U_{1,-1} = -31\,u$, where u is a constant related to a radial Coulomb integral ($u = F^2/441$ and $F^2 \sim 49J$ [2]). The deviations of $U_{mm'}$ and $J_{mm'}$ from the averages U and J are typically of the order of $J/10 \simeq 50-100$ meV. Taking into account the 1st rule and the exclusion principle we have $L = \nu(2l - \nu + 1)/2$ for $\nu \leq 2\ell + 1$, and $L = (\nu - 2l - 1)(4l - \nu + 2)/2$ for $2\ell + 1 \leq \nu \leq 2(2\ell + 1)$. For example, for $\nu = 2$ ($\nu = 7$) d electrons the lowest energy configuration has $S = 1$ ($S = 3/2$) and $L = 3$.

Hund's third rule states that the lowest-energy multiplet has a total angular momentum $J = |L - S|$ for $\nu \leq 2\ell + 1$, and $J = L + S$ for $\nu \geq 2\ell + 1$. This is the result of spin-orbit interactions. In fact, the non-relativistic Hamiltonian commutes with $\vec{L}$ and $\vec{S}$. Therefore, the $(2S + 1)(2L + 1)$ ground-state degeneracy cannot be removed at the non-relativistic level. For not too heavy elements (TM's and even lanthanides) L and S can still be considered as good quantum numbers. The major non-scalar relativistic effects are then taken into account by the spin-orbit Hamiltonian

$$\mathcal{H}_{\rm so} = A\,\vec{L}\cdot\vec{S}, \tag{2.9}$$

where $A > 0$ for $0 < \nu < 2\ell + 1$ and $A < 0$ for $2\ell + 1 < \nu < 2(2\ell + 1)$ [3]. Consequently, $\mathcal{H}_{\rm so}$ yields antiparallel alignment of L and S for shells less than half filled ($J = |L - S|$) and parallel alignment of L and S for shells more than half filled ($J = L + S$).

2.1.2 Magnetic susceptibility of open-shell ions in insulators

At low temperatures the free energy $F = -k_{\rm B}T \ln Z$ of an ion in an insulator can be calculated by considering its $2J + 1$ lowest-multiplet states ($J \neq 0$). From equation (2.3) the field dependence of the energy of $|JLSJ_z\rangle$ is given by

$$E_{JLSJ_z} = E_0(JLS) + \mu_{\rm B}H\langle JLSJ_z|L_z + g_0S_z|JLSJ_z\rangle\,\cdot$$

Applying the Wigner–Eckardt theorem within the JLS multiplet we have

$$\langle JLSJ_z|L_z + g_0S_z|JLSJ_z\rangle = g(JLS)\,\langle JLSJ_z|J_z|JLSJ_z\rangle,$$

where

$$g(JLS) = \frac{3}{2} + \frac{1}{2}\left[\frac{S(S+1) - L(L+1)}{J(J+1)}\right]$$

is the gyromagnetic Landé factor ($g_0 = 2$). The magnetic field splits the $2J+1$ states uniformly by $g(JLS)\,\mu_B\,H$. The magnetization is then given by

$$M = -\frac{\partial F}{\partial H} = g\mu_B J\, B_J\left(\frac{g\mu_B J H}{k_B T}\right), \tag{2.10}$$

where $B_J(x)$ is the Brillouin function, and the zero-field magnetic susceptibility by

$$\chi = -\frac{\partial^2 F}{\partial H^2}\bigg|_{H=0} = \frac{g^2\mu_B^2}{3}\frac{J(J+1)}{k_B T}. \tag{2.11}$$

This is Curie's law which describes the magnetic response of impurities that keep a localized permanent moment, for example, rare-earth impurities in insulators. In order to compare with experiments it is customary to write

$$\chi = \frac{1}{N_a}\frac{1}{3}\mu_B^2\frac{p^2}{k_B T}, \tag{2.12}$$

where p is the effective Bohr magneton number. From equation (2.11) one observes that the theoretical free-ion value of p is $p_{th} = g(JLS)\sqrt{J(J+1)}$.

Extensive experimental studies show that in the case of rare-earth in insulators Curie's law is followed very closely with $J = |L \pm S|$, *i.e.*, the effective moment is the result of the vector sum of spin and orbital contributions as in the free ion [1]. This is a consequence of the fact that the $4f$ valence orbitals are very localized.

In the case of TM impurities, Curie's law is still followed very closely in what the $1/T$ dependence is concerned, but the measured effective moments p deviate considerably from $p_{th} = g(JLS)\sqrt{J(J+1)}$ [1]. The d-electron states in TM's are more delocalized and thus more affected by the atomic environment than the $4f$ orbitals. In TM ions the experimental p no longer follows the free-ion total moment $J = |L \pm S|$ but rather $p \simeq 2\sqrt{S(S+1)}$, which corresponds to $J \simeq S$. One says that the orbital moment vanishes or that it is "quenched" under the action of the crystal-field. Still, L^2 remains a good quantum number, *i.e.*, the $3d$ states preserve their identity. For TM ions the interaction with the atoms surrounding the ion are more important than the weaker spin-orbit energy. For the localized f states of the heavier rare-earths the opposite trend holds. The non-spherical potential created by the atoms surrounding the magnetic impurity, known as crystal field, induces transitions between the different $m = \pm 1$ and $m = \pm 2$ d orbitals, and may also favor energetically the occupation of the $m = 0$ orbital yielding $\langle L_z \rangle \simeq 0$. However, notice that in many cases the spin-orbit interactions still induce small orbital moments $\langle L_z \rangle$ which add up to the dominant spin moments. For example, one observes that p_{exp} is somewhat larger than $p_{th}(J = S)$ for Fe, Co and Ni [1].

2.1.3 Interaction between local moments: Heisenberg model

Consider a system composed by N local moments, for example, a periodic array, a finite cluster of magnetic atoms, or an alloy with non-vanishing concentration of magnetic impurities. As long the magnetic moments preserve their local nature, the quantum mechanical state of ion i is characterized by the vector operator $\vec{S}_i$, that can be the total moment or the spin operator depending on the role of L. The simplest, lowest-order rotational invariant that can be formed with two vector is their scalar product. Thus, the interaction energy between two local spins can be described quite generally by the Heisenberg model

$$\mathcal{H} = -\sum_{i<j} J_{ij}\, \vec{S}_i \cdot \vec{S}_j. \qquad (2.13)$$

The interatomic exchange coupling constants J_{ij} depend on the microscopic origin of the interactions and may be positive, which favors ferromagnetic order (parallel spins), or negative, which favors antiferromagnetism (antiparallel spins). The general symmetry considerations behind the form of the Heisenberg interaction explains the success of the model for a variety of different physical situations. One may distinguish, for example, direct exchange interactions yielding usually $J_{ij} < 0$, superexchange or RKKY indirect exchange in which the local spins couple through their interactions with the surrounding insulating or metallic electrons, and itinerant exchange as found when the electrons forming the spins moments $\vec{S}_i$ are delocalized conduction electrons (metallic magnetism).

In order to discuss a few important properties of ferromagnets we consider the mean-field approximation to equation (2.13). We focus on a particular site j and approximate the interaction with the surrounding sites i using the average value of the magnetization $\langle \vec{S}_i \rangle$. The resulting mean-field Hamiltonian reads

$$\mathcal{H}_{\mathrm{MF}} = -\sum_j \vec{S}_j \cdot \sum_i J_{ij}\, \langle \vec{S}_i \rangle \,.$$

Notice that this is a quite serious approximation, since it neglects the exchange couplings $S_i^+ S_j^- + S_i^- S_j^+$ responsible of spin waves as well as short-range magnetic correlations ($\langle \vec{S}_i \cdot \vec{S}_j \rangle \neq \langle \vec{S}_i \rangle \cdot \langle \vec{S}_j \rangle$). Within this approach the effect of spin interactions can be regarded as an effective local magnetic field

$$H_j^{\mathrm{eff}} = \frac{1}{g\mu_{\mathrm{B}}} \sum_i J_{ij} \langle \vec{S}_i \rangle = \frac{1}{g\mu_{\mathrm{B}}} z_j J \langle \vec{S}_z \rangle, \qquad (2.14)$$

where z_j is the local coordination number at atom j ($J_{ij} = J$ only for NN's).

The magnetization M is obtained by replacing H by $H + H^{\text{eff}}$ in equation (2.10):

$$M = g\mu_{\text{B}}\langle S_z\rangle = g\mu_{\text{B}} S B_S \left(\frac{g\mu_{\text{B}} SH + zJ\langle S_z\rangle S}{k_{\text{B}}T}\right). \tag{2.15}$$

The physical picture of the temperature dependence of the magnetization is the following. At low temperatures, FM order dominates since parallel ordering of local spins yields the lowest energy. As the temperature increases the spins start to disorder in order to gain the entropy of mixing the different spin configurations. Temperature-induced spin flips reduce $\langle S_z\rangle$ and tend to average out the exchange interactions which sustain long-range magnetic order ($H^{\text{eff}} \propto \langle S_z\rangle$). $\langle S_z\rangle(T)$ decreases with increasing T vanishing at the Curie temperature T_{C}. If short-range magnetic order is neglected, all $(2S+1)^N$ spin configurations are equally probable for $T > T_{\text{C}}$, once $\langle S_z\rangle = 0$. The entropy of spin mixing, $\mathcal{S} = k_{\text{B}} N \ln(2S+1)$, is therefore an important contribution to the entropy of a ferromagnetic system.

In mean-field approximation the Curie temperature

$$T_{\text{C}} = \frac{zJS(S+1)}{3k_{\text{B}}} \tag{2.16}$$

can be regarded as proportional to the energy required to flip a spin from $S_z = S$ to $S_z = -S$ in the field H^{eff} generated by z surrounding spins having $S_z = S$. Differentiation of equation (2.15) yields the magnetic susceptibility above T_{C} which follows the Curie–Weiss law

$$\chi = g\mu_{\text{B}} \left.\frac{\partial\langle S_z\rangle}{\partial H}\right|_{H=0} = \frac{g^2\mu_{\text{B}}^2\, S(S+1)}{3k_{\text{B}}} \frac{1}{T - T_{\text{C}}}. \tag{2.17}$$

The divergency of χ at T_{C} indicates the onset of long-range magnetic order. In real systems one observes that close to the phase transition $\chi \propto (T - T_{\text{C}})^{-\gamma}$ with a critical exponent $\gamma \neq 1$.

This closes our discussion of localized-electron magnetism. We have seen that the atomic like total moments $J = |L \pm S|$, preserve their identity in the case of localized $4f$ electrons and that the crystalline environment modifies appreciably the magnetic behavior of TM ions, even in the case of an insulating matrix. The magnetic interactions between local moments have been described by the Heisenberg Hamiltonian, where the temperature dependence of the magnetization results from fluctuations of local degrees of freedom. Although successful in several respects, the local picture alone is certainly inappropriate for describing magnetic metals where the electrons are delocalized throughout the entire system (cluster or solid). The main features of itinerant magnetism are discussed in the following section.

2.2 *Stoner model of itinerant magnetism*

The 3*d* electrons in TM solids are itinerant electrons that form bands and that contribute to the conductivity and to the low-T linear specific heat. It is therefore of considerable interest to discuss the sources of a ferromagnetic instability in a metallic system, particularly in view of the study of TM clusters. Stoner proposed a simple model which illustrates the effect of exchange interactions on itinerant electrons in a clear yet oversimplified way. The interaction energy of the electron gas is approximated by considering two different Coulomb repulsion strengths. The first one, denoted by U, is the average Coulomb repulsion between electrons having different spins and the second one, denoted by $U-J$, is the average repulsion between electrons having the same spin, which is reduced by the exchange integral J. For a given number of electrons per atom $\nu = n_\uparrow + n_\downarrow$ and spin polarization $n_\uparrow - n_\downarrow$, the Coulomb interaction energy per atom is given by

$$E_{\rm C} = \frac{1}{2}\left(U - \frac{J}{2}\right)(n_\uparrow + n_\downarrow)^2 - \frac{J}{4}(n_\uparrow - n_\downarrow)^2 . \tag{2.18}$$

Exchange interactions tend to stabilize a ferromagnetic solution, *i.e.*, $n_\uparrow - n_\downarrow \neq 0$. This tendency to maximize $\langle S_z \rangle = \frac{1}{2}(n_\uparrow - n_\downarrow)$, that in the isolated atom leads to Hund's 1st rule, is counteracted by the kinetic energy $E_{\rm K}$ associated to electron delocalization throughout the cluster or solid, which favors equal filling of spin up and down states. Notice that in the solid the spin of the atomic *d*-shell is no longer well defined as in the isolated atom. Nevertheless, the *d*-electron states remain the fundamental building blocks of the magnetic bands.

In order to calculate $E_{\rm K}$ for given $n_\uparrow$ and $n_\downarrow$, we occupy the lowest-energy k states having $\varepsilon_k < \varepsilon_{\rm F} - \Delta$ for spin down, and $\varepsilon_k < \varepsilon_{\rm F} + \Delta$ for up spin. Assuming that the paramagnetic density of states (DOS) per spin $\mathcal{N}_0(\varepsilon) = (1/N_{\rm a}) \sum_k \delta(\varepsilon - \varepsilon_k)$ depends weakly on ε for $|\varepsilon - \varepsilon_{\rm F}| \leq \Delta$, we have $n_\uparrow - n_\downarrow = 2\Delta \mathcal{N}_0(\varepsilon_{\rm F})$ and

$$E_{\rm K} = E_{\rm K}^0 + \mathcal{N}_0(\varepsilon_{\rm F})\Delta^2 = E_{\rm K}^0 + \frac{(n_\uparrow - n_\downarrow)^2}{4\mathcal{N}_0(\varepsilon_{\rm F})} . \tag{2.19}$$

The higher the DOS at $\varepsilon_{\rm F}$ is – *i.e.*, the narrower the band is or the closer we are to the atomic limit – the smaller the kinetic-energy increase upon spin polarization. From equations (2.18) and (2.19) one obtains that a ferromagnetic instability occurs when

$$J > \frac{1}{\mathcal{N}_0(\varepsilon_{\rm F})} . \tag{2.20}$$

This is known as Stoner's criterion for band ferromagnetism. Despite the limitations of the involved approximations (mean-field treatment of $E_{\rm C}$, rigid-band calculation of E_K, etc.) the Stoner criterion already reflects the fundamental interplay between kinetic energy and Coulomb energy which is so important in itinerant-electron magnetism.

The self-consistent value of the magnetization $M = \mu_B(n_\uparrow - n_\downarrow)$ results from minimizing the total energy

$$\begin{aligned} E &= E_C + E_K \\ &= E_0 + \frac{J}{4}\left(\int_{-\Delta}^{\Delta} \mathcal{N}_0(\varepsilon)\mathrm{d}\varepsilon\right)^2 + \int_0^{\Delta} \left[\mathcal{N}_0(\varepsilon) - \mathcal{N}_0(-\varepsilon)\right] \varepsilon \mathrm{d}\varepsilon \end{aligned}$$

with respect to Δ ($\varepsilon_F = 0$). In the solid, depending on the shape of the density of states and on the value of J, $M(T = 0)$ may take any value from the non-magnetic case ($n_\uparrow = n_\downarrow$), over non-saturated ferromagnetic solutions ($0 < n_\uparrow - n_\downarrow < \nu$), up to saturated ferromagnetism ($n_\uparrow = \nu$, $n_\downarrow = 0$). This is a consequence of the delocalization of the electronic states and of the continuum DOS of the macroscopic metal. As we shall see, in the case of finite clusters M increases with J showing discrete steps $\Delta M = 2\ell/N_a$ with integer ℓ.

The explanation of non-integer values of the local moments at $T = 0$ is the mayor goal of Stoner's theory and of the various mean-field approaches to ferromagnetism that were inspired by it. Moreover, extensions of the Stoner theory describe correctly the environment dependence of TM magnetism in alloys, close to surfaces and clusters. In fact, we shall use such an approach to explain novel magnetic properties of clusters at $T = 0$, such as the enhancement of spin and orbital moments in Fe, Co and Ni clusters, the onset of magnetism in clusters of 4*d* TM's such as Rh_N ($N \leq 55$) which is non magnetic in the bulk, or the size dependence of magnetic anisotropy energies.

However, the straightforward extension of Stoner's theory to finite temperatures – which would consist in taking into account the Fermi function in the calculation of the kinetic energy – fails completely to describe the behavior of TM solids, surfaces and clusters at $T > 0$. First, the magnetic susceptibility above T_C does not follow the Curie law as observed in all ferromagnetic TM's. Second, at low T, $M(T)$ decreases too slowly with increasing T: $M(0) - M(T) \propto e^{-\Delta/k_B T}$ instead of $M(0) - M(T) \propto T^{3/2}$. Third, the calculated Curie temperatures are of the order of the $T = 0$ exchange-splitting $\Delta\varepsilon_X \sim J \simeq 1$ eV $\simeq 10^4$ K, *i.e.*, an order of magnitude larger than those observed in experiments ($T_C^{exp} \sim 1000$ K in TM's). Finally, the entropy around T_C is far too small, which indicates that the Stoner model fails to reproduce the mechanism for the disappearance of long-range magnetic order. Rather than the simplified form of the model Hamiltonian, the main reason for these drawbacks is the mean-field treatment of electron-electron interactions. In other words, Stoner's approach does not take into account the fluctuations of local spin degrees of freedom and is thus unable to discern between the rather large exchange energy associated to the formation of local moments (Hund's first rule) and the much smaller energy required to produce spatial fluctuations of the orientation of local spin polarizations (spin waves). Therefore, we shall not stop here to discuss temperature dependent properties in the framework of Stoner's theory. The interested reader should refer to [1] and [5]. We shall return to the problem of spin fluctuations in itinerant magnetism in Section 6.1, where we discuss the finite-temperature magnetic properties of TM clusters using a functional integral approach.

2.3 *Localized and itinerant aspects of magnetism in solids*

The localized and itinerant models of magnetism have diametrically opposed starting points. The former starts from electrons occupying atomic like states (localized in real space), while the latter starts from Bloch-like electrons which are spread out throughout the system (localized in reciprocal k space). They may be regarded as the two limiting cases obtained by varying the relative strength of the local Coulomb interactions (U and J) as compared to the kinetic energy of electron delocalization (bandwidth W). The local moment picture fits at best to $4f$ or $3d$ magnetic impurities in insulators, while the band picture of magnetism is a good starting point for metallic ferromagnets having small magnetizations ($\sim 0.1\mu_B$) and low Curie temperatures (~ 10 K). However, none of these approaches yields a satisfactory description of transition-metal solids. Some of the properties of TM's can be understood within the local moment picture but not within Stoner's model, and *vice versa*. For example, the non-integer values of the $T = 0$ magnetizations [$\mu_b(\text{Fe}) = 2.2\mu_B$, $\mu_b(\text{Co}) = 1.7\mu_B$ and $\mu_b(\text{Ni}) = 0.6\mu_B$] and the strong environment dependence of the local magnetic moments close to surfaces, in films or alloys are well described by the band approach to magnetism. These are characteristics of itinerant magnetism which contrast with the environment independent integer or half-integer values of the atomic-like angular momentum J or spin S. On the other side, the $T^{3/2}$ low-temperature decrease of $M(T)$ indicates the presence of low-energy spin-wave excitations. Neutron scattering experiments show the existence of local magnetic moments with an appreciable degree of short-range magnetic order (SRMO) at temperatures above T_C. A Curie-Weiss susceptibility $\chi \sim 1/(T - T_C)$ is observed for $T > T_C$, and the measured Curie temperatures are of the order of 10^3 K [$T_C(\text{Fe}) = 1043\ \text{K}, T_C(\text{Co}) = 1388\ \text{K}, T_C(\text{Ni}) = 627\ \text{K}$]. None of these properties can be understood using a straightforward finite-temperature extension of the band approach. In fact, Stoner's theory predicts erroneously *i)* low-temperature spin-flip excitations across a gap of the order of 1 eV, *ii)* $k_B T_C \simeq JM(T = 0) \simeq 10^4$ K, *iii)* the disappearance of all local magnetic moments at $T = T_C$, and *iv)* no Curie–Weiss susceptibility at $T > T_C$.

Summarizing, the mean-field approach to band magnetism is qualitatively correct at $T = 0$ but fails at $T > 0$, whereas the local moment approach – obviously inappropriate for metals – fails at $T = 0$ but is qualitatively correct as a picture for the finite-temperature excitations. These properties suggest the existence of local moments in magnetic TM's. The duality between itinerant and local behaviors is an essential feature of TM magnetism which reflects the importance of electron correlations in these systems. A theoretical framework which unifies these two approaches was developed in the 70's by applying functional-integral methods to single-band and d-band Hubbard models [6, 7]. This is known as spin-fluctuation theory of itinerant-electron magnetism and shall be discussed later on in the context of finite-temperature properties of clusters.

3 Experiments on magnetic clusters

Theorists were the first to study the magnetism of clusters. Salahub *et al.* determined the magnetic moments of small $3d$ TM clusters using the X-α approximation [8]. It was predicted that the magnetic moment per atom $\bar{\mu}_N$ in small bcc-like Fe_N clusters should be $\bar{\mu}_9 = 2.89\mu_B$ and $\bar{\mu}_{15} = 2.67\mu_B$, values that are considerably larger than the bulk moment $\mu_b = 2.2\mu_B$. These results were confirmed qualitatively by the local-spin-density calculations of Lee, Callaway and coworkers who obtained $\bar{\mu}_9 = 2.89\mu_B$ and $\bar{\mu}_{15} = 2.93\mu_B$ [9]. An alternative to *ab initio* calculations was provided by the self-consistent tight-binding (SCTB) studies on Cr, Fe and Ni clusters ($N \leq 51$) [10]. This method yields results in good agreement with *ab initio* calculations [$\bar{\mu}(Fe_N) \simeq 3.0\mu_B$] and, taking advantage of the flexibility of the parameterized minimal-basis approximation, it explored a variety of cluster structures and sizes that remain even nowadays unaccessible to first principles techniques. The main predictions of these theoretical studies [8–10] were confirmed a couple of years later by Stern–Gerlasch measurements. The theoretical research on cluster magnetism was considerably boosted by these remarkable experimental findings [11, 12]. An important number of calculations on magnetic TM clusters were then performed within the framework of the local spin density approximation, self-consistent tight-binding methods and many-body model Hamiltonians [13–16].

The first experiments on the magnetic properties of *free* TM clusters were performed by Cox *et al.* [17]. As in all the other gas-phase experimental studies that followed, the Stern–Gerlasch (SG) deflection is used to determine the magnetic properties. A typical experimental arrangement – see Figure 1 of [17] – involves three main steps: *i)* A beam of neutral clusters containing a more or less broad distribution of sizes is produced, usually by using a laser vaporization source. *ii)* The neutral clusters pass through a SG magnet where they are deflected along the field-gradient direction. The deflection depends on the value of the projection of the net cluster magnetization onto the magnetic field H. Clusters which magnetization is parallel (antiparallel) to H are deflected in the direction of increasing (decreasing) field. *iii)* Finally, the clusters are ionized in order to be detected after mass selection.

Cox *et al.* observed that the beam intensity at the beam axis is depleted upon switching on the SG magnetic field. This was the first experimental indication that small Fe_N clusters are magnetic ($2 \leq N \leq 17$) [17]. The measurement were interpreted assuming that the magnetic clusters should be deflected in the external magnetic field H essentially like atoms carrying a large magnetic moment $N\bar{\mu}_N$. This would correspond to an equal probability in the direction of increasing and decreasing field, since it is *a priori* equally probable to find $\bar{\mu}_N$ parallel or antiparallel to H. However, later experiments [11, 12] revealed that this assumption was incorrect and that performing only on-axis detection (ionization) is a strong limitation. Measuring only the on-axis depletion factor is therefore insufficient for determining $\bar{\mu}_N$.

The next important experimental progress was achieved by de Heer *et al.* [11]. They performed SG experiments on Fe_N ($15 \leq N \leq 650$) and measured the cluster intensity all along the direction of the field gradient, which is perpendicular to the beam axis. In this way they discovered that the clusters deflect dominantly in the direction of *increasing* field, showing a somewhat broad spatial distribution (see Fig. 1 of Ref. [11]). This implies that the magnetization of *isolated* Fe clusters tends to align parallel to an external magnetic field, a remarkable effect which was not expected [11]. In this work, the average deflection D was related directly to the average magnetic moment per atom $\bar{\mu}$ by using the relation $\bar{\mu} \propto Dmv^2/H$. However, it is actually the average magnetization of the cluster *ensemble* including relaxation effects that should be related to D. The values of $\bar{\mu}$ derived in this way increase with increasing magnetic field H and are much *smaller* than the bulk atomic moments μ_b. De Heer *et al.* already recognized that the cluster magnetic moments determined in such a way are a lower bound to the average magnetic moments of the cluster [8–10]. Due to the assumption $\bar{\mu} \propto Dmv^2/H$ the comparison between theory and experiment seemed quite controversial, also concerning the temperature dependence of D. It was not until the work of Bucher *et al.* [12] that the relation between the measured deflections D and the intrinsic cluster moment $\bar{\mu}_N$ became clear.

Bucher *et al.* performed experiments on Co_N clusters ($20 \leq N \leq 200$) and showed that the observed small average deflections could be interpreted as the result of relaxation of the magnetization of superparamagnetic clusters in the direction of H [12]. Theoretical analysis by Khanna and Linderodth [18] and by Jensen *et al.* [19] supported this interpretation. Superparamagnetic relaxation is a well-established phenomenon in the case of supported magnetic nanoparticles at temperatures above their blocking temperature. While the later is a consequence of thermal average of the orientation of magnetic moments in contact with a macroscopic environment, the mechanism of spin relaxation in free clusters is not yet completely understood [20]. Assuming that the clusters undergo superparamagnetic relaxation, Bucher *et al.* inferred intrinsic magnetic moments per atom which are larger than the bulk moment $[\bar{\mu}(\mathrm{Co}_N) = (2.1 \pm 0.2)\mu_\mathrm{B}]$ in agreement with existing calculations [8–10]. The superparamagnetic regime opened the possibility of a series of systematic experimental studies [21–26]. Rh_N clusters were found to have rather large magnetizations [23]. This was the first experimental observation of a transition from non-magnetic to magnetic behavior upon reduction of system size. Theoretical results already indicated this possibility [14, 27, 28]. Experiments on clusters of other non-magnetic TM's (V_N, Cr_N and Pd_N) have yielded no measurable SG deflection so far [22].

Billas *et al.* determined the magnetization of Fe_N clusters as a function of size and temperature for $25 \leq N \leq 700$ and $100\ \mathrm{K} \leq T \leq 900\ \mathrm{K}$ [24]. They obtained that the low-temperature average moment $\bar{\mu}_N$ increases with decreasing cluster size, exhibiting some oscillations as a function of N and reaching a value of about $3\mu_\mathrm{B}$ for the smallest sizes ($T \simeq 100$ K). A similar behavior is also observed for Co and Ni clusters. Stern–Gerlasch measurements on Ni_N clusters have been also reported in [26]. However, there are still some quantitative discrepancies between the $\bar{\mu}_N$ of Ni_N obtained in different experiments [25, 26].

The experimental results for the temperature dependence of the average magnetization per atom $\bar{\mu}_N(T)$ are far from simple and depend significantly on the considered TM [25]. In Ni_N the magnetization curves are qualitatively similar to the bulk, except of course for an important finite-size broadening of the transition around the Curie temperature T_C. Experiments on Co_N show that the magnetization per atom is about $0.1-0.5\mu_\mathrm{B}$ larger than the bulk $\mu_\mathrm{b}(T)$ for $50 \leq N \leq 600$ and $100\ \mathrm{K} \leq T \leq 1000\ \mathrm{K}$ [25]. At low temperatures, $100\ \mathrm{K} \leq T \leq 500\ \mathrm{K}$, the magnetization of Co_N is found to increase slightly with T. This is an unexpected effect that is not observed in the solid. Note that these temperatures are significantly lower than the bulk Curie temperature $T_\mathrm{C}(\mathrm{Co}) = 1388$ K. In Fe clusters the temperature dependence is qualitatively different from that of Ni or Co clusters. For $250 \leq N \leq 600$ one observes a rapid, almost linear decrease of the magnetization with increasing T ($T \leq 500-600$ K). For $T \geq 300$ K, $\bar{\mu}_N(T)$ is smaller than in bulk $M(T)$ although at $T = 0$ it was larger

[T_C(Fe-bulk) $= 1043$ K]. As the cluster size increases ($250 \leq N \leq 600$) $\bar{\mu}_N(T)$ decreases further making the difference between cluster and bulk magnetizations even larger [25]. One expects that this trend should change at larger sizes, although no experimental evidence is available so far.

In addition to the values of the intrinsic magnetic moments of isolated clusters, the Stern–Gerlasch beam experiments yielded remarkable results for the magnetic behavior of a cluster ensemble [11, 12]. Several theoretical studies were motivated by the asymmetric magnetic deflection and by the magnetic-field dependence of the average magnetization of the cluster ensemble which shows strong deviations from Langevin behavior [18–20]. Assuming superparamagnetic relaxation, the experimental findings could be analyzed consistent with electronic calculations of $\bar{\mu}_N$. However, a microscopic understanding of the remarkable spin relaxation process in an isolated cluster deserves further study. The experiments on Fe_N in a magnetic field also revealed the breakdown of the superparamagnetic regime in the case of supersonically cooled clusters (see Fig. 1 of Ref. [24]). This has been interpreted as the result of a resonant coupling between the Zeeman levels that is induced by the rotations of the cluster in the magnetic field [20, 24].

Experiments on clusters embedded in a macroscopic matrix or deposited on surfaces are of considerable interest in view of applications in storage and recording technologies. From a fundamental standpoint, they provide complementary information to studies of free clusters with several experimental advantages, such as a precise control of the sample temperature including the possibility of cooling down to very low temperatures, both of which are difficult in the gas phase. Large densities of clusters in colloids, clusters codeposited in a matrix or deposited on surfaces can be produced which can be investigated by powerful condensed-matter techniques such as magnetization measurements, Mössbauer spectroscopy [29] and X-ray magnetic circular dichroism [30, 31]. However, note that in these cases the hybridizations, distortions and bond-length changes induced by the interactions with the support often result in magnetic properties that may differ considerably from those of free clusters.

The study of supported clusters emphasizes the close relations between the magnetic properties clusters, nanocrystals, nano-structured materials, thin films and surfaces. For example, experiments on Co nanoparticles (1.5–2 nm diameter) in colloidal solutions [32] show an enhancement of the mean magnetic moment per atom of about 20% with respect to the solid and magnetic anisotropies that are much larger than in the bulk [29]. For 2-layer-high Co islands deposited on Au(111) [30] one also observes an enhancement of the magnetic anisotropy energy per atom and of the orbital magnetic moments as the cluster size is reduced. Fe clusters deposited on a graphite surface (HOPG) with sizes ranging from 1–5 nm (40–5000 atoms) present orbital moments $\langle L \rangle$ that are twice as large as in the bulk ($L_b \simeq 0.1\,\mu_B$) [31]. These results are in agreement with theoretical predictions on small clusters [33].

4 Ground-state magnetic properties of transition-metal clusters

In this section we introduce the electronic models relevant for studying the magnetic properties of low-dimensional TM systems. The ground-state properties of clusters are investigated by using a realistic *spd*-band model Hamiltonian and a mean-field treatment of electron-electron interactions. The size and structural dependence of several magnetic properties are analyzed and the microscopic origin of experimental observations is discussed. A rigorous treatment of electron-correlation effects is developed in Section 5.

4.1 Model Hamiltonians

We consider the non-relativistic Schrödinger equation for the valence s, p and d electrons and we expand the field operator $\hat{\psi}_\sigma(r) = \sum_{i\alpha} \phi_{i\alpha}(r)\, \hat{c}_{i\alpha\sigma}$ in a set of orbitals $\phi_{i\alpha}(r)$ centered at each atom i. $\phi_{i\alpha}(r) = \phi_\alpha(\vec{r} - \vec{R}_i)$ where $\vec{R}_i$ refer to the position of atom i, and α denotes the different s, p and d orbitals. The Hamiltonian can be written as

$$\mathcal{H} = \mathcal{H}_0 + \hat{V}, \tag{4.1}$$

$$\mathcal{H}_0 = \sum_{\substack{i\alpha \\ j\beta}} t_{ij}^{\alpha\beta} \hat{c}_{i\alpha\sigma}^\dagger \hat{c}_{j\beta\sigma}, \tag{4.2}$$

$$\hat{V} = \frac{1}{2} \sum_{\sigma\sigma'} \sum_{ijkl} \sum_{\alpha\beta\gamma\delta} V_{ijkl}^{\alpha\beta\gamma\delta} \hat{c}_{i\alpha\sigma}^\dagger \hat{c}_{j\beta\sigma'}^\dagger \hat{c}_{k\gamma\sigma'} \hat{c}_{l\delta\sigma}, \tag{4.3}$$

where

$$t_{ij}^{\alpha\beta} = \int \phi_\alpha^*(\vec{r} - \vec{R}_i) \left[-\frac{\hbar^2}{2m} \nabla^2 + V_{\text{ext}} \right] \phi_\beta(\vec{r} - \vec{R}_j) \mathrm{d}^3 r$$

is the transfer or hopping integral between the β-th orbital at atom j and the αth orbital at atom i. $\varepsilon_{i\alpha}^0 = t_{ii}^{\alpha\alpha}$ refers to the single-particle energy level of the orbital $i\alpha$.

The Coulomb integrals

$$V_{ijkl}^{\alpha\beta\gamma\delta} = \int \mathrm{d}^3 r \mathrm{d}^3 r' \phi_\alpha^*(\vec{r} - \vec{R}_i) \phi_\beta^*(\vec{r} - \vec{R}_j) \frac{e^2}{|r - r'|} \phi_\gamma(\vec{r} - \vec{R}_k) \phi_\delta(\vec{r} - \vec{R}_l)$$

represent transitions of a pair of electrons from at the orbitals $k\gamma$ and $l\delta$ to the orbitals $i\alpha$ and $j\beta$.

The intra-atomic terms in $\hat{V}$ are the dominant ones for the magnetic properties. As already discussed, they are responsible for the Hund-rule correlations that lead to the formation of local magnetic moments in the open d shell. Therefore, in first approximation, we may treat explicitly only the intra-atomic Coulomb terms ($i = j = k = l$) by including the interatomic Coulomb contributions as a mean-field correction to the single-particle potential $V_{\text{ext}} = V_{\text{core}} + V_{\text{inter}}$. In this way the interatomic terms screen quite efficiently the potential generated by the ions beyond the 1st or 2nd nearest-neighbor shell. An explicit treatment of interatomic contributions becomes important for heterogeneous clusters such as TM oxides where site-dependent charge transfers occur.

The Coulomb integrals $V_{ijkl}^{\alpha\beta\gamma\delta}$ are roughly proportional to the product of the overlap between $\phi_{i\alpha}$ and $\phi_{l\delta}$, and between $\phi_{j\beta}$ and $\phi_{k\gamma}$. Thus, we may simplify the interaction Hamiltonian further by retaining explicitly only the terms involving at most two different orbitals, namely, the direct terms having $i\alpha = l\delta$ and $j\beta = k\gamma$, and the exchange terms having $i\alpha = k\gamma$ and $j\beta = l\delta$ (two-center approximation). Thus, the interaction term takes the form

$$\hat{V} = \frac{1}{2}\sum_{i\sigma}\left\{\sum_{\alpha\beta} U_{\alpha\beta}\hat{n}_{i\alpha\sigma}\hat{n}_{i\beta\bar{\sigma}} + \sum_{\alpha\neq\beta}\left(U_{\alpha\beta} - J_{\alpha\beta}\right)\hat{n}_{i\alpha\sigma}\hat{n}_{i\beta\sigma}\right.$$
$$\left. - \sum_{\alpha\neq\beta} J_{\alpha\beta}\hat{c}^{\dagger}_{i\alpha\sigma}\hat{c}_{i\alpha\bar{\sigma}}\hat{c}^{\dagger}_{i\beta\bar{\sigma}}\hat{c}_{i\beta\sigma}\right\}, \tag{4.4}$$

where $U_{\alpha\beta} = U_{iiii}^{\alpha\beta\beta\alpha}$ and $J_{\alpha\beta} = U_{iiii}^{\alpha\beta\alpha\beta}$ are, respectively, the direct and exchange Coulomb integrals between the orbitals α and β of atom i. For homogeneous clusters they are independent of i. $J_{\alpha\beta}$ is responsible for Hund's first rule in the atom. With present computer facilities it is possible to perform mean-field ground-state calculations on small clusters using complex multiband models that treat all s, p and d valence electrons explicitly (*e.g.*, Hartree–Fock or density functional theory in the local spin density approximation). The situation changes completely as soon as electron correlation effects or finite-temperature properties are addressed. In this case it is necessary to simplify the valence-electron dynamics further by focusing on the d orbitals responsible for magnetism. Comparative studies between *spd* and d models at $T = 0$ show, as expected, that the magnetic properties of TM systems are largely dominated by the d valence electrons. Therefore, one often restricts α and β in equations (4.2) and (4.4) to the 3d states. Replacing for simplicity $\varepsilon^0_{i\alpha}$, $U_{\alpha\beta}$ and $J_{\alpha\beta}$ by their average values ε^0_d, U and

J, one obtains

$$\mathcal{H}_0 = \sum_{i\alpha\sigma} \varepsilon_d^0 \hat{n}_{i\alpha\sigma} + \sum_{\substack{i\neq j \\ \alpha\beta\sigma}} t_{ij}^{\alpha\beta} \hat{c}_{i\alpha\sigma}^{\dagger} \hat{c}_{j\beta\sigma} \tag{4.5}$$

and

$$\hat{V} = \frac{1}{2} \sum_{\substack{i,\alpha\neq\beta \\ \sigma\sigma'}} U_{\sigma\sigma'} \hat{n}_{i\alpha\sigma} \hat{n}_{i\beta\sigma'} - \frac{1}{2} J \sum_{i,\alpha\neq\beta} \left(\hat{s}_{i\alpha}^{+} \hat{s}_{i\beta}^{-} + \hat{s}_{i\alpha}^{-} \hat{s}_{i\beta}^{+} \right), \tag{4.6}$$

where $U_{\sigma\bar{\sigma}} = U$, $U_{\sigma\sigma} = U - J$, and $\hat{s}_{i\alpha}^{+} = \hat{c}_{i\alpha\uparrow}^{\dagger} \hat{c}_{i\alpha\downarrow}$. Notice that the self-interaction term $\alpha = \beta$ has been included in the first sum by replacing ε_d^0 by $\varepsilon_d^0 - U_{\sigma\sigma}/2$ in $\mathcal{H}_0$. This model will be used in the following to investigate the size and structural dependence of a wide variety of magnetic properties of clusters.

4.2 Mean-field approximation

Introduce c-numbers $\nu_{i\alpha\sigma}$ and $\zeta_{i\alpha}^{+}$ $\left[\zeta_{i\alpha}^{+} \in \mathbb{C}, (\zeta_{i\alpha}^{+})^{*} = \zeta_{i\alpha}^{-}\right]$ and rewrite $\hat{V}$ (Eq. (4.6)) using

$$\begin{aligned} \hat{n}_{i\alpha\sigma} \hat{n}_{i\beta\sigma'} &= \nu_{i\alpha\sigma} \hat{n}_{i\beta\sigma'} + \nu_{i\beta\sigma'} \hat{n}_{i\alpha\sigma} - \nu_{i\alpha\sigma} \nu_{i\beta\sigma'} \\ &\quad + (\hat{n}_{i\alpha\sigma} - \nu_{i\alpha\sigma})(\hat{n}_{i\beta\sigma'} - \nu_{i\beta\sigma'}) \end{aligned}$$

and

$$\hat{s}_{i\alpha}^{+} \hat{s}_{i\beta}^{-} = \zeta_{i\alpha}^{+} \hat{s}_{i\beta}^{-} + \zeta_{i\beta}^{-} \hat{s}_{i\alpha}^{+} - \zeta_{i\alpha}^{+} \zeta_{i\beta}^{-} + (\hat{s}_{i\alpha}^{+} - \zeta_{i\alpha}^{+})(\hat{s}_{i\beta}^{-} - \zeta_{i\beta}^{-})$$

as

$$\begin{aligned} \hat{V} &= \hat{V}_1 + \hat{V}_2, \\ \hat{V}_1 &= \sum_{i\alpha\sigma} \left(\sum_{\beta\sigma'} U_{\alpha\beta}^{\sigma\sigma'} \nu_{i\beta\sigma'} \right) \hat{n}_{i\alpha\sigma} \\ &\quad - \sum_{i\alpha} \left[\left(\sum_{\beta\neq\alpha} J_{\alpha\beta} \zeta_{i\beta}^{-} \right) \hat{s}_{i\alpha}^{+} + \left(\sum_{\beta\neq\alpha} J_{\alpha\beta} \zeta_{i\beta}^{+} \right) \hat{s}_{i\alpha}^{-} \right] \\ &\quad - \frac{1}{2} \sum_{\alpha\beta} U_{\alpha\beta}^{\sigma\sigma'} \nu_{i\alpha\sigma} \nu_{i\beta\sigma'} + \sum_{\alpha\neq\beta} J_{\alpha\beta} \zeta_{i\alpha}^{+} \zeta_{i\beta}^{-}, \\ \hat{V}_2 &= \frac{1}{2} \sum_{\substack{i\alpha\beta \\ \sigma\sigma'}} U_{\alpha\beta}^{\sigma\sigma'} \left(\hat{n}_{i\alpha\sigma} - \nu_{i\alpha\sigma} \right) \left(\hat{n}_{i\beta\sigma'} - \nu_{i\beta\sigma'} \right) \\ &\quad - \frac{1}{2} \sum_{i,\alpha\neq\beta} J_{\alpha\beta} \left[\left(\hat{s}_{i\alpha}^{+} - \zeta_{i\alpha}^{+} \right) \left(\hat{s}_{i\beta}^{-} - \zeta_{i\beta}^{-} \right) + \text{h.c.} \right]. \end{aligned}$$

The mean-field approximation is obtained by neglecting the fluctuation (or correlation) term $\hat{V}_2$ and by choosing $\nu_{i\alpha\sigma}$ and $\zeta^+_{i\alpha}$ such that the ground-state energy of $\mathcal{H}_1 = \mathcal{H}_0 + \hat{V}_1$ is minimized. The conditions

$$\frac{\partial\langle\mathcal{H}_1\rangle}{\partial\nu_{i\alpha\sigma}} = 0 \qquad \text{and} \qquad \frac{\partial\langle\mathcal{H}_1\rangle}{\partial\zeta^+_{i\alpha}} = 0$$

yield the usual selfconsistent equations

$$\nu_{i\alpha\sigma} = \langle\hat{n}_{i\alpha\sigma}\rangle \tag{4.7}$$

$$\zeta^+_{i\alpha} = \langle\hat{s}^+_{i\alpha}\rangle \tag{4.8}$$

where $\langle\cdots\rangle$ implies average over the ground state of $\mathcal{H}_1$. For simplicity we assume in the following that the magnetic moments are collinear, *i.e.*, $\zeta^+_{i\alpha} = 0$, $\forall\, i\alpha$. The collinear state is always a selfconsistent solution since, if $\zeta^+_{i\alpha} = 0$, the eigenstates of $\mathcal{H}_1$ are products of spin-up and spin-down states, and then $\langle\hat{s}^+_{i\alpha}\rangle = 0$. However, it is sometimes possible to find lower-energy solutions having non-collinear arrangements of the local spin polarizations. This is the case when magnetic frustrations are present, for example, in antiferromagnetic compact clusters (see Sect. 5.4 and Refs. [34–36]).

In the mean-field collinear approximation the Hamiltonian is given by

$$\mathcal{H} = \sum_{i\alpha\sigma} \varepsilon_{i\alpha\sigma}\hat{n}_{i\alpha\sigma} + \sum_{\substack{i\neq j\\ \alpha\beta\sigma}} t^{\alpha\beta}_{ij}\hat{c}^\dagger_{i\alpha\sigma}\hat{c}_{j\beta\sigma} \tag{4.9}$$

with

$$\varepsilon_{i\alpha\sigma} = \varepsilon^0_{i\alpha} + \sum_\beta (U_{\alpha\beta} - J_{\alpha\beta}/2)\,\nu_{i\beta} - \frac{\sigma}{2}\sum_\beta J_{\alpha\beta}\mu_{i\beta}. \tag{4.10}$$

Here, $\nu_{i\beta} = \langle\hat{n}_{i\beta\uparrow}\rangle + \langle\hat{n}_{i\beta\downarrow}\rangle$ and $\mu_{i\beta} = \langle\hat{n}_{i\beta\uparrow}\rangle - \langle\hat{n}_{i\beta\downarrow}\rangle$ refer, respectively, to the average occupation and spin polarization of orbital $i\beta$.

The average occupation numbers $\langle\hat{n}_{i\alpha\sigma}\rangle$ are determined self-consistently from

$$\langle\hat{n}_{i\alpha\sigma}\rangle = \int_{-\infty}^{\varepsilon_F} \mathcal{N}_{i\alpha\sigma}(\varepsilon)\mathrm{d}\varepsilon,$$

where

$$\mathcal{N}_{i\alpha\sigma}(\varepsilon) = -\frac{1}{\pi}\mathrm{Im}\left\{\mathrm{G}_{i\alpha\sigma,i\alpha\sigma}(\varepsilon)\right\}$$

is the local density of states (LDOS) per spin orbital and $\mathrm{G}_{i\alpha\sigma,i\alpha\sigma}(\varepsilon)$ the local Green's function $[\hat{\mathrm{G}} = (\varepsilon - \mathcal{H})^{-1}]$. A particularly efficient way of determining $\mathcal{N}_{i\alpha\sigma}(\varepsilon)$ in systems lacking translational symmetry (free and deposited clusters, surfaces, thin films, etc.) is the Haydock–Heine–Kelly's recursion method [37], which expresses $G_{i\alpha\sigma,i\alpha\sigma}(\varepsilon)$ as a local expansion around each atom i of the cluster. This local approach is physically very appealing since it stresses the role of the local atomic environment and its changes as function of cluster size or as we move from the interior to the surface of the cluster.

4.3 *Second-moment approximation: Enhancement of local and average spin magnetic moments*

The lowest-order local approximation to $\mathcal{N}_{i\alpha\sigma}(\varepsilon)$ is the second-order approximation which is obtained by taking into account only the contributions to $G_{i\alpha\sigma,i\alpha\sigma}(\varepsilon)$ from the nearest neighbors of atom i. Dropping the orbital dependence ($U_{\alpha\beta} = U, J_{\alpha\beta} = J$) and averaging over α one obtains

$$\mathcal{N}^{(2)}_{i\sigma}(\varepsilon) = \frac{10}{\pi w_i}\left\{1 - \frac{(\varepsilon - \varepsilon_{i\sigma})^2}{w_i^2}\right\}^{1/2} \tag{4.11}$$

where $w_i^2 = (z_i/z_b)w_b^2$ is the second moment of the local DOS[1]. $w_i = \sqrt{z_i/z_\mathrm{b}}\,w_\mathrm{b}$ plays the role of an effective local band-width that depends on the local coordination number z_i at the cluster atom i. z_b refers to the coordination number in the bulk ($z_\mathrm{b} = 12$ for an FCC lattice, for example) and w_b to the bulk band-width.

As illustrated in Figure 1, the reduction of local coordination number results in a reduction of the local effective band width and thus in an enhanced paramagnetic density of states $\mathcal{N}_{i0}(\varepsilon_\mathrm{F})$ at the Fermi energy. The resulting self-consistent magnetic moments are given in Figure 2. This very simple approximation already explains qualitatively various major trends in low-dimensional TM magnetism:

i) the local spin magnetic moments μ_i increase as the local coordination number decreases since the kinetic (promotion) energy is smaller for larger $\mathcal{N}_{i\sigma}(\varepsilon)$ (Stoner's criterion). Thus, open surfaces show larger local moments than closed ones [38];

ii) the average spin moment per atom $\bar{\mu}_N$ of an N-atom cluster increases with decreasing N since $\langle z_i\rangle$ decreases as the cluster size is reduced. Moreover, the enhancement of $\bar{\mu}_N$ originates at the cluster surface where z_i is smallest;

---

[1] The second moment of $\mathcal{N}_{i\alpha\sigma}(\varepsilon)$ is $w^2_{i\alpha\sigma} = \int_{-\infty}^{\infty}(\varepsilon - \varepsilon_{i\alpha\sigma})^2\mathcal{N}_{i\alpha\sigma}(\varepsilon)d\varepsilon$. Setting $\varepsilon_{i\alpha\sigma} = 0$, we have $w^2_{i\alpha\sigma} = \sum_k |\langle k|i\alpha\sigma\rangle|^2\,\varepsilon_k^2 = \langle i\alpha\sigma|\mathcal{H}^2|i\alpha\sigma\rangle = \sum_{j\beta}|t_{ij}^{\alpha\beta}|^2 \simeq 5z_i\bar{t}^2 = (z_i/z_\mathrm{b})w_\mathrm{b}^2$, where w_b^2 and z_b are the second moment and local coordination number in the bulk.

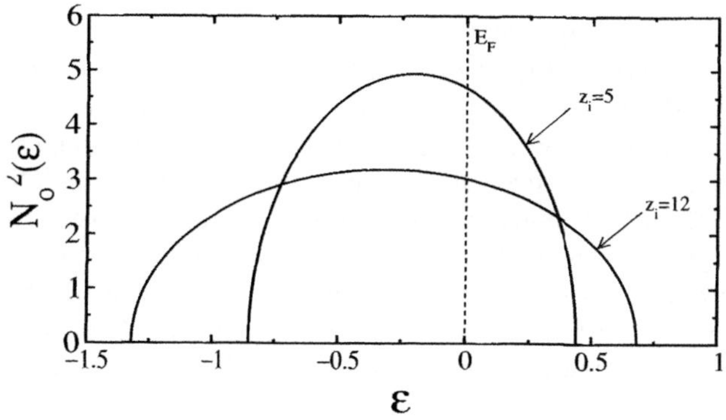

Fig. 1. Local density of states $\mathcal{N}_{i0}^{(2)}(\varepsilon)$ in the second-moment approximation for $z_b = 12$ and $z_i = 5$ (Eq. (4.11)). The Fermi energy $\varepsilon_F = 0$ corresponds to Fe ($n_d = 7$).

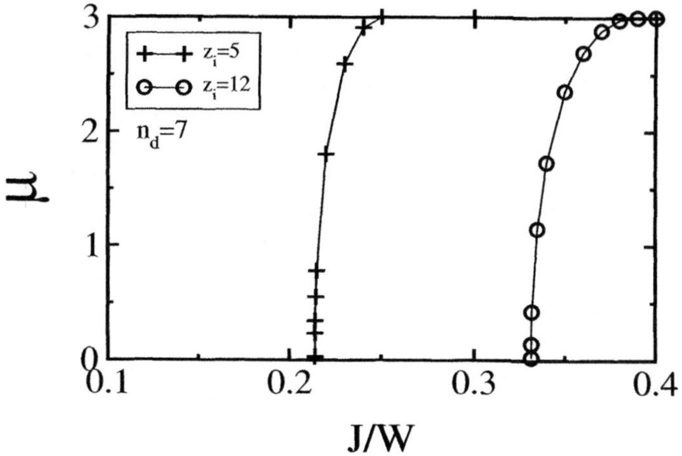

Fig. 2. Selfconsistent magnetic moments as a function of J/w_b as obtained by using the second-moment approximation to the local density of states.

iii) bond-length contraction results in a reduction of μ_i and $\bar{\mu}_N$ since for shorter interatomic distances the hopping integrals and thus w_b increase;

iv) it is in principle possible that some TM's which are non-magnetic in the bulk could develop non-vanishing magnetizations if the system dimensions are reduced. This has been observed experimentally in the case of Rh_N clusters [23].

Of course, such a simple structureless LDOS cannot explain many other important properties. Equation (4.11) fails to reproduce that in clusters $\bar{\mu}_N$ can only adopt discrete values due to the discreteness of the electronic spectrum in a finite system. The same holds for the changes in the local spin moments observed even for atoms having a complete NN shell below the cluster surface. Equation (4.11) is also unable to distinguish between different magnetic orders which may depend on the cluster structure. These are just a few examples. Nevertheless, despite these drawbacks, the 2nd moment approximation remains an illustrative start point for a more rigorous analysis of cluster magnetic properties from a local point of view.

4.4 Spin magnetic moments and magnetic order

The local magnetic moments $\mu(i)$ at the different atoms i, the magnetic order and average magnetic moment per atom $\bar{\mu}_N$ of the cluster are determined using the self-consistent tight-binding (SCTB) model given by equations (4.9) and (4.10). The element-specific parameters required for the calculations are estimated as follows. The interatomic hopping integrals $t_{ij}^{\alpha\beta}$ for d electrons are given by the canonical two-center approximation taking into account the dependence on interatomic distances R_{ij}: $dd(\sigma, \pi, \delta) = (-6, 4, -1)(W_d/2.5)(R_0/R_{ij})^5$, where R_0 is the bulk NN distance. This involves one scaling parameter, namely, the bulk d band-width W_d (for example, W_d(Fe) $= 6.0$ eV and W_d(Rh) $= 7.4$ eV). Other more sophisticated choices may be considered, for instance, $t_{ij}^{\alpha\beta}$ can be fitted to local-spin-density band-structure calculations for the solid. However, this does not affect the main physical conclusions. The direct Coulomb integrals U are estimated from atomic data, *e.g.*, U_{dd}(Fe) $= 6.0$ eV and U_{dd}(Rh) $= 7.8$ eV. These values yield a density distribution close to local charge neutrality so that the results of SCTB calculations are not very sensitive to the precise value of U [10]. The main parameter for the determination of magnetic properties is the d-electron exchange integral J_{dd}, which is usually fitted to the bulk moment μ_b at $T = 0$ (*e.g.*, J_{dd}(Fe) $= 0.70$ eV). For non-magnetic TM's such as Rh the exchange integrals are obtained from local-spin-density (LSD) Stoner theory [39] by taking into account a reduction of 20% in order to account for correlation effects [40]. In the case of Rh this yields J_{dd}(Rh) $= 0.48$ eV [41]. Applying the same procedure to Fe one obtains J_{dd}(Fe) $= 0.71$ eV, which is very close to the value derived from the bulk magnetization.

4.4.1 Free clusters: Surface effects

The onset of magnetism in clusters and its relation to the electronic structure can be analyzed systematically by determining the magnetic properties

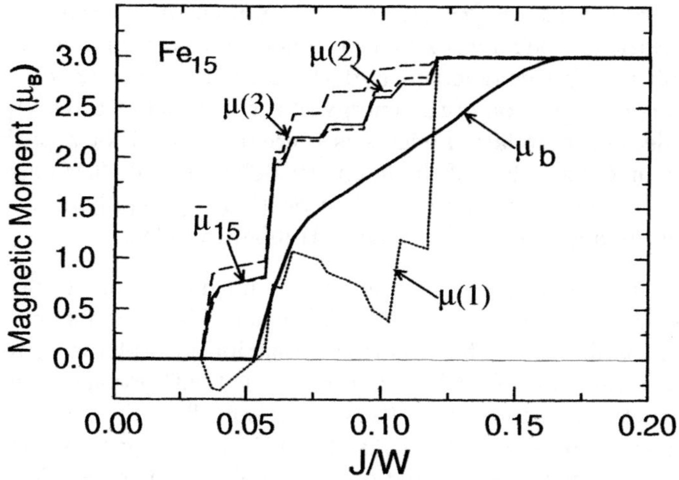

Fig. 3. Local magnetic moments $\mu(i)$ and average magnetic moment per atom $\bar{\mu}_{15}$ (full curve) of Fe_{15} with bcc-like structure as a function of the ratio between the d-electron exchange integral J and the bulk d band-width W (see Ref. [10]). $i = 1$ refers to the central atom, $i = 2$ to its 1st NN's and $i = 3$ to the outermost shell of the 2nd NN's of the central atom. The corresponding bulk magnetization μ_b is also given.

as a function of the ratio between the intra-atomic d-electron exchange integral J and the band width W of the corresponding solid. In Figure 3 the local magnetic moments $\mu(i)$ and the average magnetic moment per atom $\bar{\mu}_N$ of a bcc-like Fe_{15} cluster are given as a function J/W. The variations of J/W can be regarded as resulting from changes in the interatomic distances R_{ij} (*e.g.*, for TM's, $W \propto t_{ij} \propto R_{ij}^{-5}$) or from changes in the spatial extension of the d orbitals for different elements within the same group.

One observes that magnetism sets in for $J > J_c$, where the critical value J_c depends on size and structure. In clusters J_c is smaller than the bulk Stoner value $J_c = 1/\mathcal{N}_0(\varepsilon_F)$. The cluster magnetic moments are thus larger than the bulk magnetization. This is a consequence of the reduction of the local coordination numbers as the size of the system decreases [10] in agreement with cluster experiments and well-known surface properties [12, 24, 25]. Larger clusters show a similar behavior.

A particularly interesting result is that the clusters can be magnetic for values of J/W for which the macroscopic solid is non-magnetic. A transition form paramagnetic to ferromagnetic behavior is therefore possible by reducing the system size. Indeed, this phenomenon has been experimentally observed in the case of Rh clusters [23]. For more recent specific calculations on Rh_N see [28, 41, 42]. At this point it is important to recall that the Hartree–Fock approximation tends to overestimate systematically the stability of ferromagnetism and the formation of magnetic moments due to an overestimation of the energy of non-magnetic states. Taking into account electron correlation effects is therefore crucial for a profound and systematic understanding of the magnetic behavior of $4d$ TM clusters. This problem is discussed in more detail in Section 5 by applying exact diagonalization methods to the single-band Hubbard-model.

As J is increased beyond J_c, the average magnetic moment per atom $\bar{\mu}_N$ increases by discrete steps $\Delta\bar{\mu}_N = 2l/N$, where l is the number of electrons whose spin is flipped. The resulting redistributions of the spin-polarized density as a function of J/W are accompanied with strong changes in the local magnetic moments $\mu(i)$. Close to a spin flip, the local densities of electronic states present large peaks at the Fermi energy ε_F, which indicates that the states changing occupation are highly degenerate. The electronic structure is thus very sensitive to the interatomic distances (see Figs. 3 and 4 of Ref. [14]).

Concerning the magnetic order within the cluster one finds that bcc-like clusters are in general ferromagnetic [$\mu(i) > 0, \forall i$]. However, one sometimes observes, in a limited range of small values of J/W, that the local magnetic moments at the center of the cluster point opposite to the average magnetization. In fact, for unsaturated magnetism, $\mu(i)$ depends very strongly on the atomic position i. In particular for Rh_N this results in a remarkable size dependence of local and average magnetic moments [42]. The local spin moments tend to increase as we go from the center to the surface of the cluster. Therefore, the enhancement of the average spin magnetization of the cluster is dominated by the surface contributions. As already discussed, this can be qualitatively understood as a consequence of the reduction of the local coordination number and of the effective local d-band width [10].

The structural dependence of the magnetic properties can be illustrated by comparing the behaviors of clusters having fcc- and bcc-like structures. In contrast to bcc clusters, fcc Fe clusters present antiferromagnetic-like order, and thus much smaller average moments $\bar{\mu}_N$ [10, 14]. The presence of antiferromagnetic order in compact structures raises quite naturally the question of magnetic frustrations and possible non-collinear spin arrangements in finite systems. Such more complicated magnetic structures have been explored in the framework of the single-band Hubbard model [35] (see Sect. 5.4) and are expected to be important in Cr_N as well as in mixed clusters like $(Cr_xFe_{1-x})_N$ [43]. The remarkable role played in free clusters by the surface atoms let us expect a very interesting physics at the interface of embedded clusters.

4.4.2 Embedded clusters: Interface effects

The magnetic properties of clusters are a subject of considerable interest in materials science in view of possible applications in magnetic recording and storage technologies. For many practical applications involving clusters, these are not isolated but embedded in a matrix or deposited on a surface. It is therefore important to extend our analysis of cluster magnetism to situations where the clusters are in contact with a macroscopic environment. The studies on free and embedded clusters yield complementary informations which contribute to the characterization of the various specific behavior of these novel materials.

In this section we consider the magnetic properties of Fe clusters embedded in a Cr matrix. These materials are very interesting since they show different magnetic behaviors which are very sensitive to the structural and chemical environment of the atoms. The competition between the antiferromagnetic order of the Cr matrix and the tendency of Fe_N clusters to order ferromagnetically leads to particularly strong cluster-matrix magnetic interactions. Therefore, one expects to encounter a variety of novel magnetic effects. The problem is to some extent similar to that of Fe/Cr multilayers, which deserved considerable attention in past years.

In Figure 4 results are given for the local magnetic moments μ_i of Fe_N clusters embedded in a Cr matrix as substitutional impurities [44, 45]. The main conclusions are summarized below:

i) the magnetic moment of the atomic impurity, $\mu(1) = 0.26\mu_B$, is drastically reduced with respect to the bulk magnetization $\mu_b = 2.2\mu_B$ of pure Fe. This can be interpreted as a consequence of the smaller exchange splittings of the Cr atoms with which the Fe-impurity orbitals hybridize strongly. Moreover, the local magnetic moment at the Fe impurity is antiparallel to its NN's, *i.e.*, the magnetic coupling is consistent with the spin-density-wave state of the matrix;

ii) as long as the size of the embedded cluster is very small ($N \leq 4$), the magnetic order *within* the Fe_N cluster is antiferromagnetic with strongly reduced local magnetic moments $\mu(i)$. In other words, very small Fe clusters adopt the magnetic order induced by the spin-density wave of the matrix. This contrasts with the behavior observed for free clusters which show ferromagnetic order and $\mu(i) > \mu_b$ (see Sect. 4.4.1);

iii) for $N \geq 6$ one observes the expected transition from antiferromagnetic to ferromagnetic order within Fe_N. This is followed by a considerable increase of the local magnetic moments at the Fe atoms, although

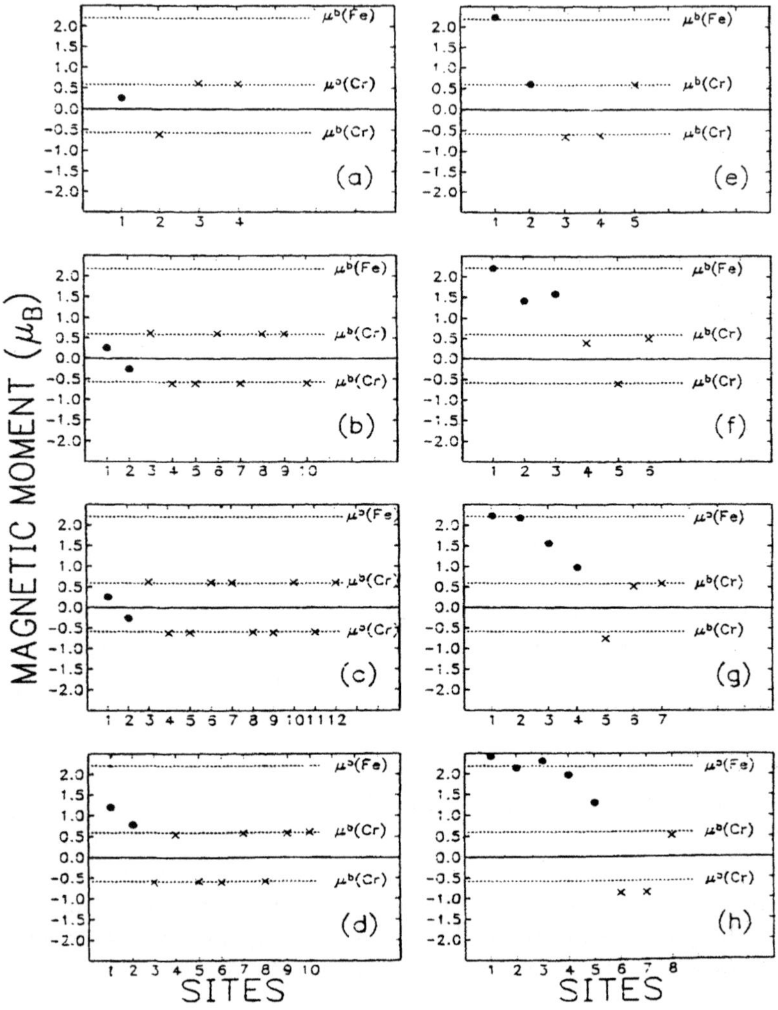

Fig. 4. Spatial distribution of the local magnetic moments $\mu(i)$ for Fe_N clusters ($N \leq 51$) embedded in an antiferromagnetic Cr matrix. The non-equivalent atomic sites i are ordered by increasing distance to the central Fe-atom. Dots correspond to Fe atoms and crosses to Cr atoms. Notice the transition from antiferromagnetic to ferromagnetic order within Fe_N for $N = 4$–6, and the decrease of $\mu(i)$ at the Fe atoms close to the Fe-Cr interface.

close to the interface with the matrix the $\mu(i)$ are still smaller than μ_b (see Fig. 4);

iv) at the interface there is a strong tendency to antiferromagnetic coupling between Fe and Cr moments. The magnetic order among Cr atoms, given by the sign of $\mu(i)$, is not affected by the presence of the Fe cluster, even in cases where the shape of the cluster prevents a perfect antiferromagnetic coupling at the interface. However, one observes important quantitative changes in $|\mu(i)|$ at the Cr atoms close to the cluster;

v) the magnetic moments $\mu_{\mathrm{Fe}}(i)$ of the Fe atoms are very sensitive to the local atomic environment. $\mu_{\mathrm{Fe}}(i)$ is found to be roughly proportional to the number of Fe atoms $z_{\mathrm{Fe}}(i)$ found in the 1st NN shell of atom i [$\mu_{\mathrm{Fe}}(i) = 0.61\mu_{\mathrm{B}}$, $0.99\mu_{\mathrm{B}}$, $1.29\mu_{\mathrm{B}}$ and 1.42–$1.59\mu_{\mathrm{B}}$ for $z_{\mathrm{Fe}}(i) = 1$–4, respectively]. For larger clusters ($N \geq 9$), the magnetic moment at the center of the cluster is quite close to the Fe-bulk value. However, the average magnetization per atom is always smaller than $\mu_{\mathrm{b}}(\mathrm{Fe})$ due to the contributions of interface Fe atoms. The trend is thus opposite to that of free clusters or infinite surfaces;

vi) the electronic densities of states reflect very clearly the cluster-matrix hybridizations and the transition from antiferromagnetic to ferromagnetic order within Fe_N. Small Fe_N clusters in Cr ($N \leq 4$) present an electronic structure which is quite similar to that of pure Cr. Larger Fe clusters ($N \geq 9$) show a ferromagnetic-like exchange splitting and the first signs of convergence towards Fe-bulk behavior. Nonetheless, the changes induced by the cluster-matrix interactions remain significant [44,45].

Fe clusters embedded in a Cr matrix have many properties in common with Fe/Cr multilayers. The antiferromagnetic order in the Cr even close the interfaces, the antiferromagnetic coupling between Fe and Cr moments at the interface, the strong reduction of the local magnetic moments at Fe atoms close to the interface, the possibility of a slight enhancement of the Fe moment beyond the bulk value at the center of a cluster or film, the strong changes of $\mu(i)$ associated to frustrations, are all characteristics shared by Fe-Cr systems both in the form of embedded clusters or multilayer structures [44]. One concludes that, at least in this case, the immediate local environment of the atoms gives the dominant contribution to the magnetic behavior. Nevertheless, for quantitative predictions a detailed determination of the electronic structure and the geometrical structure at a larger length scale are certainly important.

The results discussed in this section have shown that magnetic properties of Fe particles embedded in a matrix are qualitatively different form those of free (unsupported) clusters. For Fe_N in Cr or V the changes are so dramatic because the magnetic clusters interact very strongly with the TM matrix.

In the case of inert matrices or colloidal solutions the properties of embedded clusters often resemble those of free clusters in many respects. Clusters on surfaces experience both the reduction of the local coordination number, as free clusters, and the cluster-substrate interactions, as embedded clusters. A variety of very interesting magnetic behaviors results from these competing effects. Research in this direction is a subject of mayor current interest.

4.5 *Magnetic anisotropy and orbital magnetism*

The electronic and magnetic properties of ferromagnets depend to some extent on the orientation of the magnetization with respect to the crystal structure and to the external shape of the system. The anisotropy of the ground-state energy (or of the free energy at $T > 0$) is one of the main characteristics of a magnetic material. It determines the low-temperature orientation of the magnetization with respect to the geometry of the system and the stability of the magnetization direction in the case of single-domain particles. The magnetic anisotropy energy (MAE) – defined as the energy difference involved in changing the magnetization direction from the low-energy direction or easy axis to a high-energy direction or hard axis – is a property of crucial importance in technological applications (*e.g.*, magnetic recording or memory devices) where the magnetization must be pinned to a given direction in space. In the case of clusters, the magnetic anisotropy deserves a special attention, not only from a purely theoretical point of view but also because of its implications in cluster-beam Stern–Gerlasch experiments [11, 12]. In this section we consider the main relativistic corrections at the origin of magnetic anisotropy in transition metals and we discuss the size and structural dependence of the MAE and orbital magnetic moments L. In particular we investigate the transition from large atomic-like L, given by Hund's second rule, to bulk-like quenching of L as a function of cluster size.

4.5.1 Relativistic corrections

So far we have discussed the magnetic properties of clusters at a purely non-relativistic (NR) level taking into account of course that the electrons are spin-1/2 fermions. In non-relativistic quantum mechanics the total spin operator $\vec{S} = \sum_i \vec{S}_i$ commutes with the Hamiltonian operator and therefore the direction of the magnetization relative to the cluster structure plays no role in the electronic properties. Thus, in the absence of an external magnetic field the NR ground-state and excited-state energies are independent of the direction of the magnetization. The magneto-anisotropic behavior of magnetic materials is the result of relativistic corrections to the single-electron dynamics (Dirac equation) and to the electron-electron interaction

(Breit interaction) (see, for instance, Ref. [46]). The leading contributions are of second order in the ratio v/c between the velocity of the electrons and the velocity of light.

The relativistic corrections to the electron-electron interaction have two origins. First, the electrons are *moving* charges and therefore each electron interacts with the magnetic field generated by the current and by the spins of the other electrons. Second, the electromagnetic interactions are mediated by photons which travel at the speed of light and are thus affected by retardation effects. In the Hartree approximation to the Breit interaction, the magnetic dipole-dipole energy is given by [46, 47]

$$\begin{aligned} E_{\mathrm{DD}} = \frac{\mu_{\mathrm{B}}^2}{2} \int \mathrm{d}^3r\mathrm{d}^3r' \Bigg(& \frac{\vec{m}(r)\cdot\vec{m}(r')}{|r-r'|^3} \\ & -3\frac{\left[\vec{m}(r)\cdot(\vec{r}-\vec{r'})\right]\left[\vec{m}(r')\cdot(\vec{r}-\vec{r'})\right]}{|r-r'|^5}\Bigg) \end{aligned} \tag{4.12}$$

where $\vec{m}(r) = \langle\hat{\vec{m}}(r)\rangle$ is the magnetization density. For transition metals we may replace equation (4.12) by the sum of the dipole-dipole interactions between the local magnetic moments $\vec{\mu}_i$ at each atom i:

$$E_{\mathrm{DD}} = \frac{\mu_{\mathrm{B}}^2}{2} \sum_{i\neq j} \left[\frac{\vec{\mu}_i\cdot\vec{\mu}_j}{R_{ij}^3} - 3\frac{(\vec{R}_{ij}\cdot\vec{\mu}_i)(\vec{R}_{ij}-\vec{\mu}_j)}{R_{ij}^5} \right].$$

The dipole-dipole interaction energy between pairs local moments is very small even at NN distances ($E_{\mathrm{DD}} \simeq 3\mu$ eV for $R_0 \simeq 2.5$ Å and $\mu \simeq 1\mu_{\mathrm{B}}$). However, E_{DD} decreases slowly as a function of distance ($E_{\mathrm{DD}} \sim 1/R_{ij}^3$) and the summation over pairs ij converges very slowly. Therefore, E_{DD} depends on the shape of the nanoparticle. In the case of small clusters the main contribution to the magnetic anisotropy energy results from spin-orbit interactions.

The other relativistic corrections to the Schrödinger equation come from the single-electron dynamics which is ruled by the Dirac equation. The non-relativistic limit of the Dirac equation – including the terms up to the order $(v/c)^2$ – is the Pauli Hamiltonian $\mathcal{H}_{\mathrm{Pauli}} = \mathcal{H}_{\mathrm{NR}} + \mathcal{H}_{\mathrm{SR}} + \mathcal{H}_{\mathrm{SO}}$ [46]. One distinguishes, on the one side, the non-relativistic (Schrödinger) term $\mathcal{H}_{\mathrm{NR}}$ and the scalar relativistic corrections $\mathcal{H}_{\mathrm{SR}}$, usually included in the external (pseudo)potential that defines the hopping integrals (Eq. (4.2)). On the other side, one finds the spin-orbit interactions

$$\mathcal{H}_{\mathrm{SO}} = \frac{e\hbar}{4m^2c^2}\vec{\sigma}\cdot\left(\vec{E}\times\vec{p}\right) \tag{4.13}$$

which are qualitatively important since they modify the symmetry of the wave function. The spin-orbit coupling represents the interaction of the magnetic moment of the electron with the magnetic field that results from the electronic motion relative to the lattice potential $V(\vec{r})$. In other words, the magnetic moment of the electron interacts with the magnetic field that is generated by $V(\vec{r})$ as seen in the reference frame of the electron. Since the electric field is strongest close to the nuclei, the SO effects are most important for the heaviest elements and for the most localized orbitals (core-electron states and d or f valence orbitals). Approximating V by its spherical average, one may write

$$\mathcal{H}_{\mathrm{SO}} = \frac{\hbar^2}{2m^2c^2r}\frac{\partial V}{\partial r}\vec{\ell}\cdot\vec{s} = \alpha(r)\vec{\ell}\cdot\vec{s}.$$

For the d-electrons in transition metals it is a good approximation to consider only the intra-atomic terms. Thus, $\mathcal{H}_{\mathrm{SO}}$ may be expressed as

$$\mathcal{H}_{\mathrm{SO}} = \xi \sum_{\substack{i\,m,m\\ \sigma\sigma'}} (\vec{\ell}\cdot\vec{s})_{m\sigma,m'\sigma'}\hat{c}^{\dagger}_{im\sigma}\,\hat{c}_{im'\sigma'},$$

where $\xi = \int R^{*}_{3d}(r)\,\alpha(r)\,R_{3d}(r)\,r^2\,dr$ is the spin-orbit coupling constant ($\xi > 0$), and $(\vec{\ell}\cdot\vec{s})_{m\sigma,m'\sigma'}$ are the matrix elements of $\vec{\ell}\cdot\vec{s}$ among the d states.

A simple physical picture for the MAE can be derived in the limit of saturated magnetic moments with large exchange splitting $\varepsilon_{id\uparrow} - \varepsilon_{id\downarrow} = J\mu$ [48]. In this case, at lowest-order perturbation theory, we may neglect the terms that mix spin-up and spin-down states. Thus, $\mathcal{H}_{\mathrm{SO}} \simeq (\xi/2)\sum_i(\hat{\ell}_{iz\uparrow} - \hat{\ell}_{iz\downarrow})$ can be regarded as a magnetic field acting on the orbital moment $\hat{\ell}_{iz\sigma}$. Moreover, $\langle\mathcal{H}_{\mathrm{SO}}\rangle/N_{\mathrm{a}} = (\xi/2)\left(\langle L_{z\uparrow}\rangle - \langle L_{z\downarrow}\rangle\right)$ where $\langle L_{z\sigma}\rangle$ is the average orbital moment per atom. Taking the up spins as the majority ones, we have $\langle L_{z\downarrow}\rangle \simeq 0$ for $n_d < 5$, and $\langle L_{z\uparrow}\rangle \simeq 0$ for $n_d > 5$. Thus, $\langle\mathcal{H}_{\mathrm{SO}}\rangle/N_{\mathrm{a}} \simeq \pm(\xi/2)\langle L_z\rangle$ where the + (−) sign corresponds to $n_d < 5$ ($n_d > 5$). As in the atom, antiparallel (parallel) alignment of $\langle L_z\rangle$ and $\langle S_z\rangle$ is favored for $n_d < 5$ ($n_d > 5$). The lowest-energy magnetization direction (easy axis) is the one yielding the largest orbital moment. The spin magnetization is "turned" to the direction yielding the largest $\langle L\rangle$.

4.5.2 Magnetic anisotropy of small clusters

In Figure 5 results are given for the MAE ΔE, the orbital angular momentum $\langle L_\delta\rangle$ along the magnetization direction δ ($\delta = x, y, z$), and the average

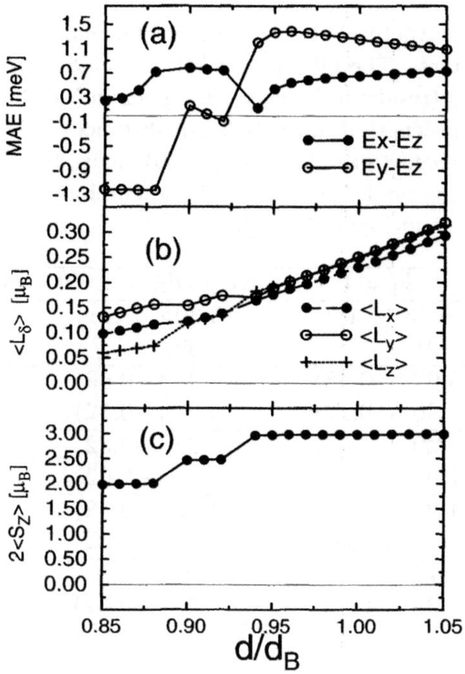

Fig. 5. (a) Magnetic anisotropy energy (MAE), (b) orbital magnetic moment $\langle L_\delta \rangle$ and (c) spin magnetic moment $\langle S_z \rangle$ of Fe_4 with rhombohedral structure as a function of the bond length d (d_B = bulk NN distance). The magnetization direction $\delta = x$ is along the middle bond, the direction $\delta = y$ is perpendicular to x and within the plane of the cluster, and $\delta = z$ is perpendicular to x and y.

spin projection $\langle S_z \rangle$ of an Fe_4 cluster with rhombohedral structure [33]. The two-center hopping integrals and the intra-atomic Coulomb integrals used in the calculations are the same as in previous sections. The value of the spin-orbit coupling constant corresponding to Fe is $\xi = 0.05$ eV [33]. The results are given as a function of the bond length d in order to analyze the role of cluster relaxation and to infer the possible coupling of the magnetization direction to vibrations and distortions. Only $\langle S_z \rangle$ is shown, since the magnitude of the spin magnetization $|\langle \vec{S} \rangle|$ depends very weakly on the considered direction (typically, $|\langle S_z \rangle - \langle S_x \rangle| \sim 10^{-3}$–$10^{-4}$). Comparing Figures 5a and 5c it is clear that the variations of the MAE are related to the variations of $\langle S_z \rangle$ and to the resulting changes in the electronic spectrum. For large values of d/d_B the spin magnetic moments are saturated, *i.e.*, $\langle S_z \rangle \simeq (10 - \nu_d)/2 = 3/2$. When d/d_B decreases, discrete changes in the spin polarization occur and non-saturated moments are obtained (see

Table 1. Size and structural dependence of the magnetic anisotropy energy (MAE) of Fe_N clusters. The off-plane MAE $\Delta E = E_x - E_z$ and the in-plane MAE $\Delta E = E_x - E_y$ (results in brackets) are given in MeV for different values of the inter-atomic bond-length d (d_B = bulk NN distance). Different structures are considered for each cluster size: (a) triangle, (b) chain, (c) square, (d) rhombus, (e) trust, (f) square pyramid, (g) triangle, (h) square bipyramid, (i) hexagon, and (j) pentagonal bipyramid (see Ref. [33]).

N Struct.	$\Delta E(d/d_B = 1.05)$	$\Delta E(d/d_B = 1.00)$	$\Delta E(d/d_B = 0.90)$
3 (a)	5.13 (1.43)	5.30 (−1.01)	−0.11 (−0.10)
(b)	2.01	1.69	1.25
4 (c)	5.02 (−0.48)	5.21 (−0.59)	−1.03 (−0.11)
(d)	0.30 (−0.12)	0.27 (−0.12)	0.33 (0.31)
5 (e)	0.29 (−1.01)	0.28 (−0.82)	−0.10 (−0.75)
(f)	1.09 (−0.01)	0.88 (−0.03)	−0.07 (0.04)
6 (g)	−1.25 (−0.76)	−1.12 (−0.62)	0.33 (0.29)
(h)	4.66 (0.04)	4.82 (−0.02)	−0.07 (−0.23)
7 (i)	1.78 (0.40)	1.91 (0.09)	2.22 (0.30)
(j)	4.32 (−0.03)	4.52 (−0.02)	−0.39 (0.00)

Sect. 4.4.1). For constant values of $\langle S_z \rangle$, the MAE and orbital momentum $\langle L_\delta \rangle$ vary continuously, since the electronic spectrum and the local magnetic moments are continuous functions of d/d_B. However, this is not the case when $\langle S_z \rangle$ changes, since a strong and discontinuous redistribution of the spin-polarized density takes place. Important changes in the energy-level structure occur around ε_F that modify the details of the SO mixing. The resulting changes in the electronic energy depend on the explicit form of $\mathcal{H}_{SO}$ and therefore on the direction of the magnetization (see Fig. 5). Consequently, very significant and discontinuous variations of the MAE are observed which may even lead to a change of sign of the MAE. Notice, that the rhombohedral Fe_4 cluster presents a remarkable in-plane anisotropy $(E_x - E_y) \simeq 0.4$ eV. This indicates that uniaxial anisotropy models are not directly applicable to clusters that have strongly reduced symmetry. A similar situation is found for other clusters and band-fillings [33].

In Table 1 results are given for the MAE of Fe clusters with different structures and interatomic distances [33]. The following general trends are obtained:

i) the MAE is much larger in small clusters than in the corresponding crystalline solids. In fact, the anisotropy energy ΔE is often even larger than in thin films. For instance, $\Delta E \sim 4$–5 MeV is a typical value for clusters (see Tab. 1). This is in agreement with experiments on free clusters and supported Fe nanoparticles [29];

ii) ΔE depends much more sensitively than the spin moments on the geometrical structure of the cluster. Indeed, changes of sign in ΔE are found as a function of the interatomic distance d even in situations where the magnetic moments are saturated and consequently do not depend on the cluster structure;

iii) the in-plane MAE's are considerably important in general. In some cases they are even larger than the usually considered off-plane anisotropy. The in-plane MAE is of course largest for low-symmetry structures and decreases, though not monotonically, as the angle between non-equivalent x and y directions decreases. Experiments support these conclusions.

4.5.3 Enhancement of orbital magnetism

In atoms, Hund's rules predict maximum orbital angular moment L compatible with maximum spin multiplicity. In transition-metal (TM) solids, electron delocalization and band formation result in the almost complete disappearance or quenching of $\langle L \rangle$. Such intrinsic differences between atomic and bulk behaviors are characteristic of systems developing itinerant-electron magnetism. Consequently, the orbital magnetism in TM clusters is expected to show novel size-dependent effects that are important both from a fundamental standpoint, and in view of applications of cluster-based magnetic materials.

The previous general considerations indicate that $\langle L \rangle$ should be very sensitive to the local atomic environment. In fact, as already discussed in Section 3, recent experiments show that the orbital moments at TM surfaces, thin films, and nanostructures are typically 100–200% larger than in the corresponding solids [49,50]. Moreover, a size-dependent enhancement of $\langle L \rangle$ should have direct consequences on the results for the average magnetic moments per atom and on the comparison with experiment [24,26]. It is therefore important to investigate the dependence of orbital magnetism on variables such as size, structure, local atomic environment, and d-band filling.

In order to analyze the crossover from atomic L to bulk-like quenching of $\langle L \rangle$ with increasing size we focus on Ni_N clusters. Besides its experimental relevance, Ni is a very interesting system since the differences between atomic and bulk orbital moments are dramatic ($L = L_\mathrm{a} = 2\mu_\mathrm{B}$ in the atomic s^1d^9 configuration, and $\langle L \rangle = L_\mathrm{b} = 0.05\mu_\mathrm{B}$ in the solid). Moreover, the spin moments in Ni_N are the smallest among the ferromagnetic $3d$

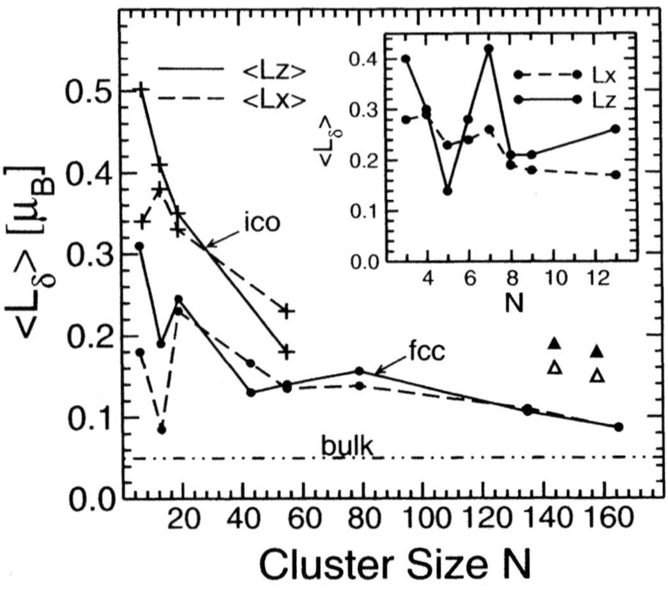

Fig. 6. Average orbital magnetic moment per atom $\langle L_\delta \rangle = |\sum_{i=1}^{N} \vec{L}_\delta(i)|/N$ of Ni_N clusters. Results are given for fcc-like (dots) and icosahedral-like structures (crosses). The magnetization direction δ is taken along a principal C_n symmetry axis ($\delta = z$, full lines) or along a NN bond perpendicular to z ($\delta = x$, dashed lines). Full (open) triangles refer to coin-like bilayer clusters with perpendicular (in-plane) magnetization. Results for small clusters are given in the inset. The lines are a guide for the eye [52].

TM's, and therefore the contribution of $\langle L \rangle$ to the total magnetic moment $\langle \vec{M} \rangle = 2\langle \vec{S} \rangle + \langle \vec{L} \rangle$ should be particularly significant.

In Figure 6 results are given for the average orbital moment per atom

$$\langle L_\delta \rangle = |\sum_{i=1}^{N} \vec{L}_\delta(i)|/N$$

of Ni_N clusters having $N \leq 169$, and face-centered-cubic (fcc) or icosahedral-like geometries. $\delta = x, y$, and z refers to different orientations on the magnetization. On the one side, we observe that the reduction of system size causes a remarkable enhancement of $\langle L_\delta \rangle$ with respect to the bulk. Values about eight times larger than $L_b = 0.05\mu_B$ are not uncommon. On the other side, comparison with the atomic result $L_a = 2\mu_B$ shows that the largest part of the quenching of L takes place already at the smallest clusters, as soon as full rotational symmetry is lost. For example, for Ni_3

(triangle) we obtain $\langle L_z \rangle = 0.40\mu_B$, and for Ni_4 (tetrahedron) $\langle L_z \rangle = 0.30\mu_B$. Concerning the size dependence, one observes that $\langle L_\delta \rangle$ decreases with increasing N showing some oscillations as the bulk limit is approached (see Fig. 6). Notice that an important enhancement of $\langle L_\delta \rangle$, about 100%, is still present even for the largest considered sizes ($\langle L_\delta \rangle / L_b = 1.8$ for $N = 165$). In small clusters, $\langle L_\delta \rangle / L_b$ ranges from $\langle L_\delta \rangle / L_b \simeq 3$ for $N = 50{-}80$, to $\langle L_\delta \rangle / L_b \simeq 6$ for $N = 10{-}20$. Large $\langle L_\delta \rangle$ are also found for $N \leq 10$, as shown in the inset figure. The results correspond here to the optimized cluster geometries determined in the framework of the local spin density approximation [51]. One concludes that orbital magnetism is the source of an important contribution to the size-dependent magnetic properties of Ni clusters.

In Figure 6 results are also given for icosahedral clusters. In this case $\langle L_\delta \rangle$ decreases with increasing N as for the fcc geometries but significant quantitative differences are observed. The five-fold symmetric structures yield in general larger $\langle L_\delta \rangle$ than the cubic symmetric ones. Strong structural dependences are also found in smaller clusters. Notice that the results for $N \geq 13$ given in Figure 6 concern mainly highly-symmetric structures with nearly spherical shape and closed nearest neighbors (NN) shells. Clusters with lower symmetry usually present a richer size dependence and larger orbital moments, particularly when the number of surface atoms is larger. This could affect the convergence to bulk-like quenching for large N. Remarkable structural effects are expected for clusters deposited on surfaces, whose shape can be experimentally tuned to some extent by changing the landing velocities [50]. In order to illustrate this effect we present results for two-layer-thick coin-like Ni_N, as indicated by the triangles in Figure 6 ($N = 144$ and 158). In these two-dimensional islands, $\langle L_\delta \rangle$ is much larger than for three-dimensional fcc clusters of comparable size. The enhancement with respect to the bulk is here about 200%. Similar trends are found for other TM's, in qualitative agreement with recent experiments on deposited Fe particles [50]. Structure and shape have a strong influence on the orbital moments of TM nanoparticles.

The enhancement of $\langle L_\delta \rangle$ can be related to the changes in the local environment of the atoms. First, the reduction of local coordination number with decreasing N causes a reduction of the effective d-band width and an decrease of the average energy-level spacings. Consequently, the local spin polarizations $\langle S_z(i) \rangle$ are increased, which induces larger orbital moments by means of the spin-orbit interactions. Second, orbital polarization (OP) contributions resulting from the orbital dependence of the Coulomb integrals $U_{mm'}$ and $J_{mm'}$ amplify this effect ($U_{mm'} \simeq U - m m' B$ with $B \simeq 0.1$ eV). This amounts typically to 30–50% of the enhancement of $\langle L_\delta \rangle$. Third, the reduction of the effective band-width and the presence of degeneracies in the single-particle spectrum favor a more effective spin-orbit mixing, which enhances $\langle L_\delta(i)\rangle$ even in situations where the spin polarizations are saturated ($2\langle S_z\rangle \simeq 10 - n_d$). Notice that the changes in $\langle S_z\rangle$ and $\langle L_\delta\rangle$ involve different energy scales. Although the exchange energies dominate over SO and OP energies ($J \simeq 1$ eV, $B \simeq \xi \simeq 0.1$ eV), the occupations of the different m orbitals remain very sensitive to SO and OP interactions, since they may vary without altering the spin polarizations. Compare, for example, icosahedral and fcc clusters in Figure 6. Finally, there are less predictable effects related to the details of the electronic structure and its dependence on cluster geometry (*e.g.*, the presence of high-symmetry axes, changes in bond-length, etc.) which may affect the orbital moments depending on the d-band filling. A similar analysis holds for transition-metal surfaces and thin films [53, 54].

The environment dependence of the local orbital moments $L_\delta(i)$ provides further insight on the enhancement of $\langle L_\delta\rangle$ in small clusters, and on the development of bulk-like quenching for large N. One observes that $L_\delta(i)$ generally increases, with some oscillations, as we move from the center to the surface of the cluster [52]. Thus, the enhancement of $\langle L_\delta\rangle$ is driven by the cluster surface, as it is the case for the spin moments. For small N ($N \leq 55$) the enhancement of $L_\delta(i)$ concerns practically all atoms, including those with a complete NN shell, well below the surface. For larger sizes ($N \geq 135$) bulk-like quenching starts to set in at the interior of the cluster, leaving significantly enhanced orbital moments only at a few outermost shells. As expected, one approaches the behavior found for TM surfaces, where enhanced $L_\delta(i)$ are restricted to a few uppermost layers [54]. A droplet model, *i.e.*, $(\langle L_\delta\rangle - L_b) \propto N^{-1/3}$, is in fact a good first approximation for $N \geq 150$. Notice, however, that the convergence to L_b is rather slow, even for $N \simeq 200$. The fact that $L_\delta(i) \simeq L_b$ for inner atoms indicates that cubic point-group symmetry – which is absent in all but the central atom – is not essential for reaching almost bulk-like quenching. Instead, a local picture, namely, the recovery of bulk-like local atomic environment, is a more appropriate interpretation.

Finally, let us point out that the contribution of the orbital moments – that in Ni align parallel to the spin moments – increases the value of the total magnetic moment $\langle L\rangle + 2\langle S\rangle$ predicted by theory, thereby improving the agreement with experiment. In fact, despite some quantitative differences among the results obtained by different groups, available spin-only calculations yield ground-state spin magnetizations that underestimate systematically the experimental $\bar{\mu}_N$ by about 0.3–0.6μ_B for $N \leq 13$ [51]. The largest part of this discrepancy is removed by including the enhanced orbital contributions.

5 Electron-correlation effects on cluster magnetism

A deeper understanding of the electronic correlations underlying the magnetic properties of clusters is not only important from a fundamental point of view, but it also crucial in investigations of more delicate properties such as excitation spectra, magnetic behavior at finite temperatures or even the geometry of magnetic clusters. It is therefore necessary to go beyond simple mean-field approximations by including charge and spin fluctuations explicitly. The simultaneous determination of electron-correlation effects, magnetism and cluster geometry is a very difficult task. In fact, most theoretical studies performed so far have attempted to deal with only part of these aspects of the problem at a time. The localized character of the atomic-like $3d$ orbitals, and the complicated dependence of the magnetic moments and magnetic order on cluster geometry, have prevented the development of rigorous first-principles treatments of electron correlations.

5.1 The Hubbard model

The mean-field approximation discussed in Section 4.1 could be improved systematically, for example by treating the residual interactions by perturbation methods or by using Gutzwiller or Jastrow variational *Ansätze* [55, 56]. However, a general implementation of such calculations for TM clusters would be very demanding, particularly if the structure is arbitrary, *i.e.*, lacking of any symmetry. An alternative approach which has provided numerous significant results is to consider a simpler electronic model which allows an exact or at least very accurate solution of the many-body problem and which at the same time contains enough complexity to be able to shed light on the physics of real systems. With this in mind, we simplify the model given by equations (4.5) and (4.6) by droping the d-band degeneracy. One obtains then the single-band Hubbard Hamiltonian [57]

$$H = -t \sum_{\langle i,j\rangle,\sigma} \hat{c}^{\dagger}_{i\sigma}\hat{c}_{j\sigma} + U \sum_{i} \hat{n}_{i\uparrow}\hat{n}_{i\downarrow}, \tag{5.1}$$

where $\hat{c}^{\dagger}_{i\sigma}$ ($\hat{c}_{i\sigma}$) refers to the creation (annihilation) operator for an electron at site i with spin σ, and $\hat{n}_{i\sigma} = \hat{c}^{\dagger}_{i\sigma}\hat{c}_{i\sigma}$ to the corresponding number operator. The first term is the kinetic-energy operator, which describes electronic hoppings between nearest neighbor (NN) sites i and j leading to electron delocalization and bond formation ($t > 0$). The second term takes into account the intra-atomic Coulomb repulsion, which is the dominant contribution from the electron-electron interaction ($U > 0$) [57]. Equation (5.1) can be considered as a minimum model for describing itinerant electrons in a lattice. At low energies the electronic properties result from a delicate balance between the tendency to delocalize the valence electrons in order to reduce their kinetic energy, and the effect of the Coulomb repulsions associated to local charge fluctuations. The relative importance of these contributions depends strongly on the electron density or number of electrons ν, on the total spin S and on the ratio U/t.

The Hubbard model for small clusters is solved numerically by expanding its eigenfunctions $|\Psi_l\rangle = \sum_m \alpha_{lm}|\Phi_m\rangle$ in a complete set of basis states $|\Phi_m\rangle = \left[\prod_{i\sigma}(\hat{c}^\dagger_{i\sigma})^{n^m_{i\sigma}}\right]|vac\rangle$ which have definite occupation numbers $n^m_{i\sigma}$ at all orbitals $i\sigma$ ($\hat{n}_{i\sigma}|\Phi_m\rangle = n^m_{i\sigma}|\Phi_m\rangle$ with $n^m_{i\sigma} = 0$ or 1). The values of $n^m_{i\sigma}$ satisfy the conservation of the number of electrons $\nu = \nu_\uparrow + \nu_\downarrow$ and of the z component of the total spin $S_z = (\nu_\uparrow - \nu_\downarrow)/2$, where $\nu_\sigma = \sum_i n^m_{i\sigma}$. Taking into account all possible electronic configurations may imply a considerable numerical effort which in practice sets a drastic limit to the size of the clusters under study. For example, at half-band filling and minimal S_z the dimension of the Hilbert space $D = \binom{N}{\nu_\uparrow}\binom{N}{\nu_\downarrow}$ is $D = 853\,776$, $2\,944\,656$ and $11\,778\,624$, for $N = 12$, 13 and 14 respectively. For not too large clusters, the expansion coefficients α_{lm} corresponding to the ground state ($l = 0$) and low-lying excited states are determined by sparse-matrix diagonalization procedures such as the Lanczos iterative method [58]. Several cluster properties can be calculated in terms of the coefficients α_{lm} of the eigenstate $|\Psi_l\rangle$ by simple operations on the basis states $|\Phi_m\rangle$, for example, the spin-density distribution

$$\langle \hat{n}_{i\sigma}\rangle_l = \sum_m |\alpha_{lm}|^2 n^m_{i\sigma}, \tag{5.2}$$

the density-density correlation functions

$$\langle \hat{n}_{i\sigma}\hat{n}_{j\sigma'}\rangle_l = \sum_m |\alpha_{lm}|^2 n^m_{i\sigma} n^m_{j\sigma'}, \tag{5.3}$$

and the spin correlation functions

$$\langle \hat{S}^z_i \hat{S}^z_j\rangle_l = \frac{1}{4}\sum_m |\alpha_{lm}|^2 (n^m_{i\uparrow} - n^m_{i\downarrow})(n^m_{j\uparrow} - n^m_{j\downarrow}). \tag{5.4}$$

Within a controlled accuracy, the results are thus exact in the framework of the Hubbard model.

5.2 Geometry optimization in graph space

The Hubbard model considers a single s-like orbital per site and NN hoppings. The hopping integrals t_{ij} between sites i and j take only two possible values, namely, $t_{ij} = -t$ if $R_{ij} = R_0$, and $t_{ij} = 0$ if $R_{ij} > R_0$, where R_{ij} refers to the interatomic distance and R_0 to the NN distance. Consequently, only the topological aspect of the structure is relevant for the electronic properties. In other words, defining the cluster structure is equivalent to defining for each atom i those atoms j which are connected to i by a hopping element t. This implies that the set of all possible non-equivalent cluster structures is a subset of the set of *graphs* with N vertices [59]. For example, for $N = 3$ there are two different graphs (triangle and linear chain) and for $N = 4$ these are six (tetrahedron, rhombus, square, star, etc.). Taking into account only NN hoppings with fixed bond length results in a discretization of the configurational space so that the geometry optimization can be performed within graph space. Notice, however, that the number of graphs n_g or site configurations to be considered increases extremely rapidly with N.

For the study of clusters one must consider only those graphs which can be represented as a true *structure* in real space. A *graph* is acceptable as a cluster *structure* if a set of atomic coordinates $\vec{R}_i$ ($i = 1, \ldots, N$) exists, such that the interatomic distances R_{ij} satisfy the conditions $R_{ij} = R_0$ if the sites i and j are connected in the graph (*i.e.*, if $t_{ij} = -t$) and $R_{ij} > R_0$ otherwise (*i.e.*, if $t_{ij} = 0$). While for $N \leq 4$ all graphs are possible cluster structures, this is no longer true for $N \geq 5$. For example, for $N = 5$ it is not possible to have all atoms being NN's from each other. With increasing cluster size it becomes increasingly difficult to identify the graphs which are not representable as a structure in $D \leq 3$ dimensions. This problem can be solved with a computer algorithm based on the optimization of objective functions under different boundary conditions [16].

5.3 Ground-state structure and total spin

The results for the most stable structure and the corresponding total spin S as a function of U/t and ν are summarized in the form of "phase" or structural diagrams as the ones shown in Figure 7. The main conclusions and trends derived from the calculations for $N \leq 8$ are summarized below.

For low electron or hole concentration (*i.e.*, $\nu/N \leq 0.4$–0.6 and $2 - \nu/N \leq 0.3$–0.6) the optimal cluster structure is independent of U/t, *i.e.*, the structure which yields the minimal kinetic energy (uncorrelated limit) remains the most stable one, irrespectively of the strength of the Coulomb interactions. Furthermore, no magnetic transitions are observed: the ground-state is always a singlet or a doublet (see Fig. 7). This indicates that, for low carrier concentration, the Coulomb interactions are very efficiently suppressed by the correlations, so that the magnetic and geometric structure of the clusters are dominated by the kinetic term. This physically plausible conclusion is further supported by analytical results for the case of two carriers. Nonetheless, the fact that this holds for finite values of ν/N and $U/t \to +\infty$, seems not obvious *a priori*.

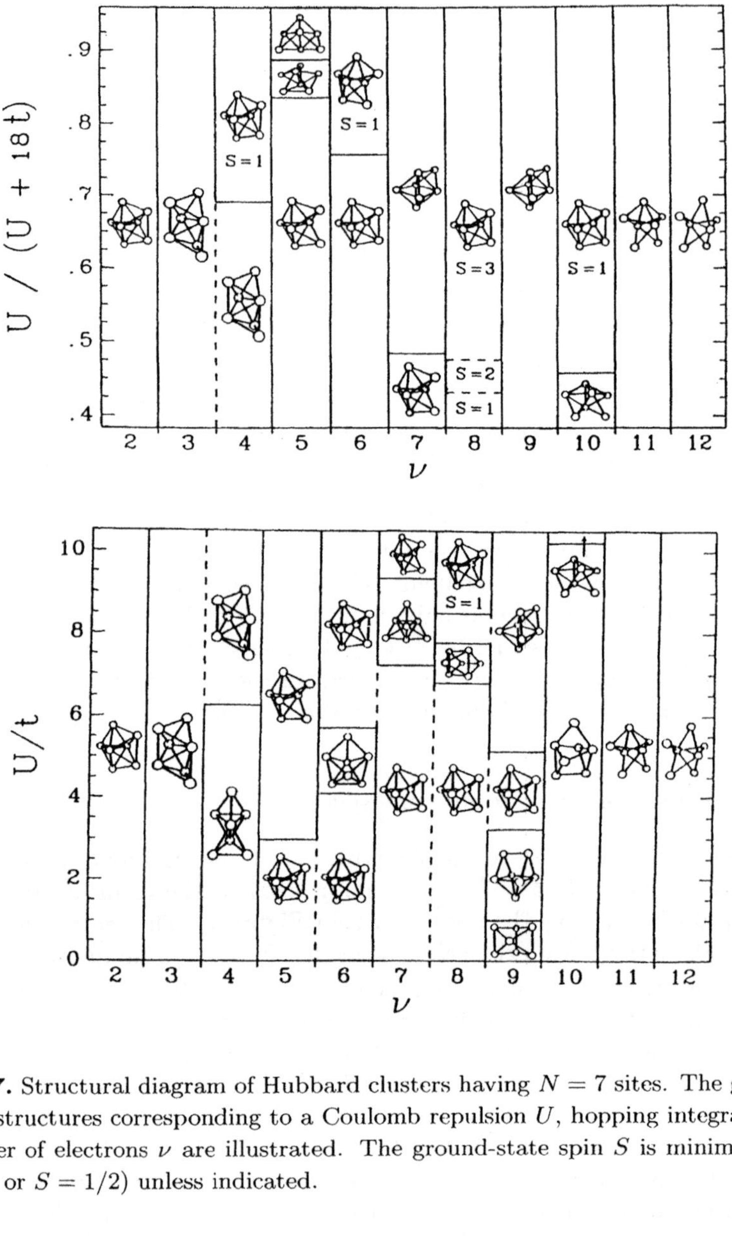

Fig. 7. Structural diagram of Hubbard clusters having $N = 7$ sites. The ground-state structures corresponding to a Coulomb repulsion U, hopping integral t and number of electrons ν are illustrated. The ground-state spin S is minimal (*i.e.*, $S = 0$ or $S = 1/2$) unless indicated.

For small ν we obtain compact structures having maximal average coordination number $\bar{z}$ ($t_{ij} = -t < 0$). These are all substructures of the icosahedron and have the largest possible number of *triangular* loops.

In contrast, for large ν (small $\nu_h = 2N - \nu$) open structures are found. In particular for $\nu_h = 2$ we obtain bipartite structures, which have the largest possible number of *square* loops. This can be qualitatively understood in terms of the single-particle spectrum. In the first case (small ν) the largest stability is obtained for the largest band-width for bonding (negative-energy) states ($\varepsilon_b \leq -\bar{z}t$), while in the second case (small ν_h) it is obtained for the largest band-width for antibonding (positive-energy) states, *i.e.*, for the most compact bipartite structure.

A much more interesting interplay between electronic correlations, magnetism and cluster structure is observed around half-band filling (*i.e.*, $|\nu/N - 1| \leq 0.2$–0.4). Here, several structural transitions are found as a function of U/t (see Fig. 7). Starting from the ground-state structures for $U = 0$, one observes that as U is increased, first one or more of the weakest cluster bonds are broken. This change to more open structures occurs for $U/t \simeq 1$–4 and is most often seen for $\nu < N$, since in this case the $U = 0$ structures are more compact, while for $\nu > N$ the structures are rather open already for $U \to 0$ (see Fig. 7). As U is further increased ($U/t > 5$–6) it is energetically more advantageous to create additional new bonds. Higher coordination gives the strongly correlated electrons more possibilities for performing a mutually avoiding motion that lowers the kinetic energy. Moreover, these compact structures are in general more symmetric and the electron-density distribution is more uniform, which also contributes to lower the Coulomb-repulsion energy.

The structural changes at larger U are often accompanied by strong changes in the magnetic behavior. One may indeed say that these structural changes are driven by magnetism [60]. For half-band filling ($\nu = N$) the optimal structures show minimal total spin S and strong antiferromagnetic correlations. None of the structures having a (unsaturated) ferromagnetic ground-state [61] were found to be the most stable ones for any value of U/t. The optimal antiferromagnetic structures are *non-bipartite*. For instance, the rhombus is more stable than the square for $N = \nu = 4$ (see also Fig. 7). The bonds that are frustrated in a static picture of antiferromagnetism, yield an appreciable energy lowering when quantum fluctuations are taken into account. Therefore, Hubbard clusters with one electron per site and large U/t can be best seen as frustrated quantum antiferromagnets.

For all studied cluster sizes ($N \leq 8$), the most stable structures for $\nu = N + 1$ show ferromagnetism for large U (typically $U/t > 4$–14). This is in agreement with Nagaoka's theorem [62]. For the smaller clusters, *i.e.*, $N = 3$, 4 and 6, this is the only case where the optimal structures are ferromagnetic. For $N = 5$ unsaturated ferromagnetism ($S = 1$) is also found for $\nu = 4$, though at large values of U ($U/t > 30$). For larger clusters, ferromagnetism extends more and more throughout the (U/t)–(ν/N)

phase diagram. Clusters with $N = 7$ ($N = 8$) are ferromagnetic for $\nu = 4$, 6, 8 and 10 ($\nu = 9$–12). The tendency towards ferromagnetism is much stronger above half-band filling than below. This is qualitatively in agreement with experiments on $3d$-TM clusters. In fact one observes that the magnetic moments per atom μ in V and Cr clusters are very small if not zero ($\mu < 0.6$–$0.8\mu_B$) [22], while Fe and Co clusters show large magnetizations [12, 24, 25]. Finally, let us remark that the appearance of ferromagnetism is much less frequent than what one would expect from mean-field Hartree–Fock arguments (Stoner criterion). This reflects the importance of correlations in low-dimensional systems, as it will be discussed in the next section [60, 63]. It should be nonetheless noted that the Hubbard model for clusters probably exaggerates the effects of quantum fluctuations, since it is one of the most extremely low-dimensional systems one can consider. Improvements on the model, either by including several bands or non-local interactions, should tend to weaken such strong fluctuation effects.

The Hubbard model is in accordance with nature in a further interesting aspect. While the number of possible site configurations or graphs n_g increases in an explosive way with the number of atoms N, the number of structures n_o which are optimal for some value of the parameters U/t and ν/N remains a handful. For example, for $N = 7$ (8), $n_g = 853$ (11 117) while $n_o = 18$ (23). As the cluster size increases, some growth patterns start to dominate and the same or very similar structures cover larger and larger regions of the phase diagram (see Fig. 7). The situation tends to what one observes in the macroscopic limit (solid state) where – regardless of the infinite number of topologically different atomic arrangements – the equilibrium lattice structures of the elements are only a few (fcc, hcp, bcc, etc.).

5.4 Comparison with non-collinear Hartree–Fock

The unrestricted Hartree–Fock (UHF) approximation developed in Section 4.2 in the context of *spd*-band models can also be applied in an straightforward way to the single-band Hubbard model [35]. Comparison between mean-field and exact results allows to quantify the role of electron correlations on various magnetic and structural properties.

For the Hubbard model the UHF Hamiltonian reads

$$H_{\text{UHF}} = -t \sum_{\langle lm \rangle \sigma} c^\dagger_{l\sigma} c_{m\sigma} + U \sum_{l\sigma} \left(\rho_{l\sigma,l\sigma} c^\dagger_{l\bar{\sigma}} c_{l\bar{\sigma}} - \rho_{l\sigma,l\bar{\sigma}} c^\dagger_{l\bar{\sigma}} c_{l\sigma} \right), \tag{5.5}$$

from which a single-determinant approximation $|\text{UHF}\rangle$ to the ground state is obtained. $\rho_{l\sigma,l\sigma'} = \langle c^\dagger_{l\sigma} c_{l\sigma'} \rangle$ are the matrix elements of the density matrix, where $\langle ... \rangle = \langle \text{UHF}|...|\text{UHF}\rangle$ implies self-consistency. The distribution of the electron density is given by

$$\langle n_l \rangle = \rho_{l\uparrow,l\uparrow} + \rho_{l\downarrow,l\downarrow}, \tag{5.6}$$

and the spin polarization $\langle \vec{S}_l \rangle = \langle (S_l^x, S_l^y, S_l^z) \rangle$ by

$$\begin{aligned} \langle S_l^x \rangle &= (\rho_{l\uparrow,l\downarrow} + \rho_{l\downarrow,l\uparrow})/2, \\ \langle S_l^y \rangle &= -i\ (\rho_{l\uparrow,l\downarrow} - \rho_{l\downarrow,l\uparrow})/2, \\ \langle S_l^z \rangle &= (\rho_{l\uparrow,l\uparrow} - \rho_{l\downarrow,l\downarrow})/2. \end{aligned} \tag{5.7}$$

Notice that the local magnetic moments $\langle \vec{S}_l \rangle$ are collinear only if $\rho_{l\sigma,l\bar{\sigma}} = 0$, $\forall l$. In Figure 8 the selfconsistent spin arrangements obtained in fcc clusters with $\nu = N = 13$ are shown for representative values of U/t.

In Figure 9 UHF and exact results are given for the local moments $\mu_l^2 = \langle S_l^2 \rangle$ and spin correlations $\langle \vec{S}_l \cdot \vec{S}_m \rangle$ (fcc-like 13-atom cluster). The UHF values for μ_l^2 are quantitatively not far from the exact ones. The uncorrelated limit ($U = 0$) is reproduced and a fairly good agreement is also obtained in the large U/t regime. Main trends such as the larger μ_l^2 at the cluster surface for small U/t and the reduction of the difference between surface and inner moments for large U/t come out correctly. However, UHF underestimates the increase of μ_l^2 for $U/t < 3.7$ and anticipates the tendency to localization with increasing U/t. This results in a much more rapid crossover from weak to strong interacting regimes and in significant inaccuracies for intermediate U/t.

The quantitative discrepancies are more important in the case of spin correlation functions $\langle \vec{S}_l \cdot \vec{S}_m \rangle$ particularly for large U/t (Fig. 9b). Here we find that UHF underestimates the strength of AF spin correlations γ_{11} at the surface. This can be interpreted as a consequence of the formation of permanent local moments that blocks quantum spin fluctuations along the transversal directions. Still, in both UHF and exact calculations, the increase of AF correlations at the surface (increase of $|\gamma_{11}|$) is done at the expense of a decrease of the spin correlations with the central atom (decrease of $|\gamma_{01}|$). AF correlations among the larger surface magnetic moments dominate, while the AF couplings between central and surface atoms tend to cancel out. The discrepancies between UHF and exact results for $N = 13$ show the limits of mean-field and give us an approximate idea of the corrections to be expected in correlated calculations on larger clusters. As expected, UHF yields better results for properties like the local moments, that are related to the spin density distribution, than for the correlation functions. Nevertheless, since the trends given by UHF are qualitatively correct, one can be reasonably confident on the validity of the conclusions derived using mean-field.

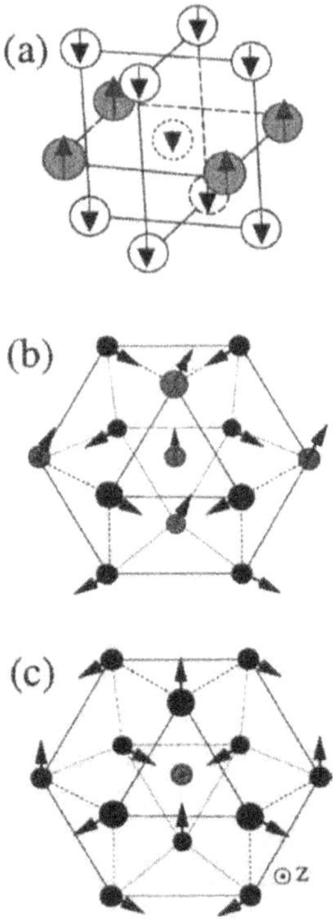

Fig. 8. Illustration of the UHF magnetic order in fcc-like clusters having $N = 13$ atoms, $\nu = 13$ electrons and representative values of U/t (Hubbard model). (a) $U/t < 3.7$: collinear spin arrangement; (b) $4.8 \leq U/t < 5.1$: all local spin polarizations $\langle \vec{S}_l \rangle$ within the (111) plane; (c) $U/t \geq 5.1$: non-collinear spin arrangement in 3 dimensions. The arrows indicate here the projection of $\langle \vec{S}_l \rangle$ on to the (111) plane.

The determination of the cluster structure is a central problem in cluster research in particular for magnetism since, as we have seen, structure and magnetic behavior are interrelated. Most calculations of cluster structures are based upon mean-field approximations. Therefore, it is very interesting to evaluate the role of electronic correlations on cluster structure. In order to explore this problem we determine the relative stability of a few representative cluster structures in the framework of the UHF approximation and we compare the results with the exact solution [16]. Four different symmetries are considered: icosahedral, face centered cubic (fcc), hexagonal close packed (hcp), and body centered cubic (bcc) clusters. The magnetic and structural diagrams for $N = 13$ are shown in Figures 10 and 11.

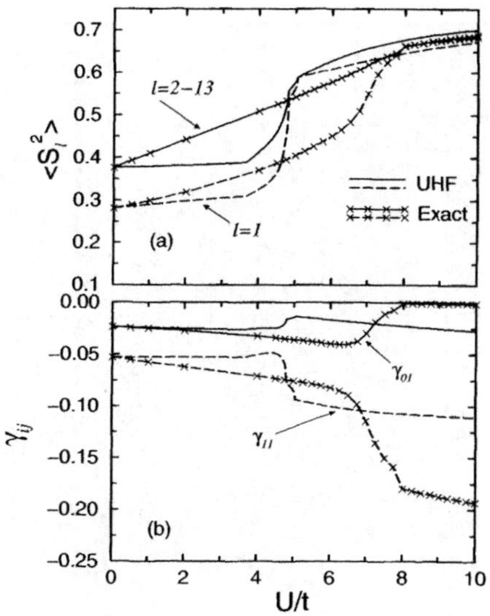

Fig. 9. UHF and exact results for an fcc-like 13-atom cluster ($\nu = N = 13$). (a) Local magnetic moments $\langle S_l^2 \rangle$ at the central site ($l = 1$, dashed) and at the cluster surface ($l = 2$–13, solid). (b) Spin correlation functions $\langle \vec{S}_l \cdot \vec{S}_m \rangle$ between the central site and its first NN's at the surface (γ_{01}, solid) and between NN's at the surface shell (γ_{11}, dashed). Curves with (without) crosses correspond to exact (UHF) results.

A qualitative description of the magnetic order obtained in the UHF calculations is indicated by the different shadings. One may distinguish three different *collinear* spin arrangements: non-magnetic solutions (NM), non-saturated or weak ferromagnetic solutions (WFM) and saturated ferromagnetic solutions (SFM). The NM states are paramagnetic with minimal total moment $\mu_T = 0$ or $1/2$. Concerning the *noncollinear* spin arrangements we distinguish two cases: noncollinear nonmagnetic states (NC) in which non-vanishing (eventually large) local moments μ_l sum up to an approximately minimal total moment $\mu_T < 1$, and noncollinear ferromagnetic states (NCFM) that show a net magnetization $\mu_T \geq 1$. The NC states include all sort of frustrated antiferromagnetic-like spin structures, for example, those illustrated in Figure 8.

For small U/t ($U/t < 10$) the UHF results are in agreement with the exact calculations (compare Figs. 10 and 11). The icosahedral cluster yields the lowest energy for low carrier concentrations ($\nu \leq 6$). As already discussed, the largest coordination numbers are favored for small ν since the kinetic energy dominates stabilizing the structure with the largest

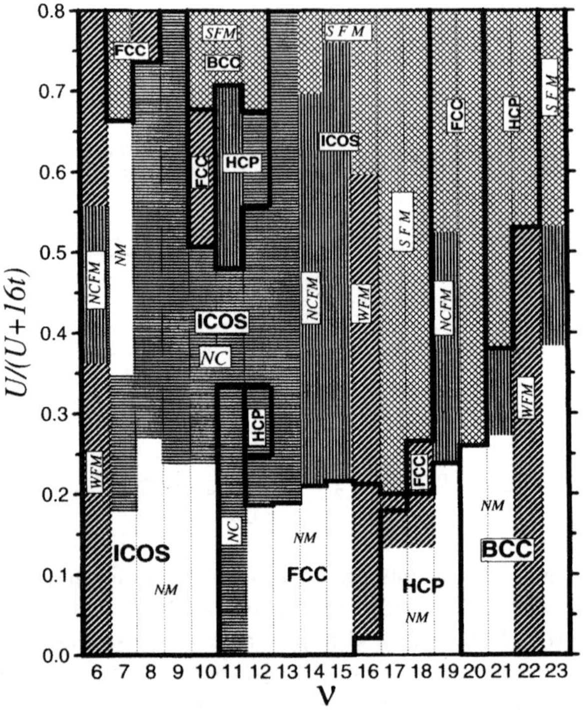

Fig. 10. Magnetic and structural diagram of clusters having $N = 13$ atoms as obtained by using the Hubbard model in the unrestricted Hartree–Fock (UHF) approximation. Four different types of structures are considered: icosahedral (ICOS), face centered cubic (FCC), hexagonal close packed (HCP), and body centered cubic (BCC). The corresponding UHF magnetic orders are indicated by different shading patterns.

bandwidth [16]. As ν is increased several structural transitions occur. For small U/t, both exact and UHF calculations present the same structural changes: from icosahedral to fcc structure at $\nu = 11$, from fcc to hcp at $\nu = 17$, and from hcp to bcc at $\nu = 20$. At larger U/t ($U/t > 12$) the interplay between the kinetic and Coulomb energies introduces correlations that cannot be accounted for within UHF and that are important to the magnetic and structural properties. A main source of discrepancy is the too strong tendency of UHF to yield SFM ground states, particularly above half-band filling. Consequently, the optimal structure is missed rather frequently in the limit of large U/t. For example, for $\nu = 19$ and $U/t > 18$, UHF predicts the fcc structure with a SFM ground state, while in the exact calculation the icosahedral cluster with $S = 1/2$ is the optimum. Similar

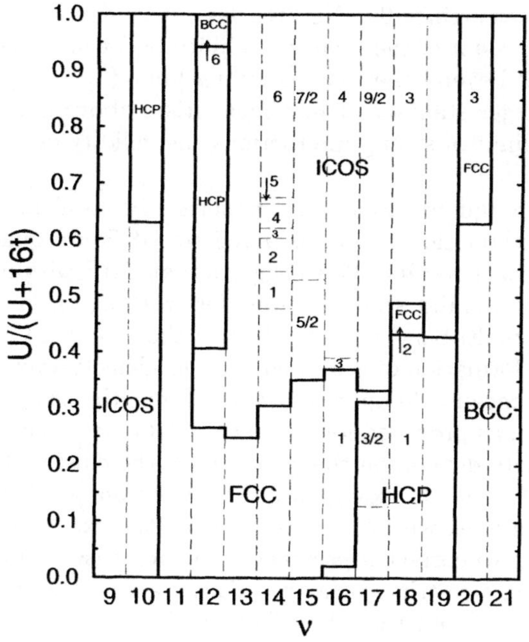

Fig. 11. Exact calculated structural and magnetic diagram of Hubbard clusters having $N = 13$ atoms as obtained by considering the same four structures as in Figure 10. The corresponding ground-state spin S is minimal ($S = 0$ or $1/2$) unless explicitly indicated. Broken lines separate regions having the same structure but different S. For $\nu \leq 9$ the icosahedron yields the lowest energy for all U/t, while for $\nu \geq 21$ the bcc structure is the most stable.

drawbacks are seen for other band fillings such as $\nu = 7$, 10, 21, and 22 (large U/t). Still, in the event that the true ground state does show strong ferromagnetism, UHF succeeds since the ground state is the superposition of single-particle states. Examples of this kind are $\nu = 12$, 14 and 20 for large U/t. These are rather exceptions since the presence of a FM ground state in the exact solution is in general much less frequent than predicted by UHF. Around half-band filling UHF reproduces qualitatively well the transition from fcc or hcp to icosahedral structure with increasing U/t as well as nontrivial re-entrant effects ($\nu = 13-18$). However, UHF underestimates the value of U/t at which the structural changes occur (see Figs. 11 and 10).

Summarizing, UHF fails to reproduce the exact phase diagram in detail, particularly in some of the most interesting AF or weak FM regimes. Structural transition are sometimes missing, and in other cases changes of structures appear artificially. Nevertheless, it is also fair to say that UHF yields a good account of the relative stability between the considered structures, even well beyond the weak coupling limit (up to $U/t \simeq 16$). It also explains the larger stability of ferromagnetism above half-band filling, although in the limit of strong interactions the validity of UHF often breaks down.

Despite some quantitative discrepancies, the local magnetic moments and the spin correlation functions given by UHF are qualitatively in accord with the exact results. This suggests that UHF with noncollinear spin arrangements is a valid starting point for working on larger clusters and more complex multi-band models. Notice that the breaking of spin symmetry and the formation of a permanent spin-density-wave states are often artifacts of the mean-field approximation. While the broken spin-symmetry can be restored, as proposed for instance in [56], the success of such an approach is likely to depend on the geometry of the cluster. In low-symmetry structures UHF tends to exaggerate the formation of inhomogeneous spin-density distributions, which may be far from the exact solution and which may be difficult to correct *a posteriori*. In such cases a Jastrow-like variational Ansatz on a restricted Hartree–Fock state is often more appropriate. Finally, let us point out that the effects of charge fluctuations and correlations are probably more drastic in the single-band Hubbard model than in d-band Hamiltonians, more appropriate for the description of TM's.

6 Finite-temperature magnetic properties of clusters

The temperature dependence of properties reflecting phase transitions in finite systems are intrinsically different from the critical phenomena of the solid state [64]. In a ferromagnet, for example, the divergencies in the specific heat $C_p(T)$ and magnetic susceptibility $\chi(T)$ at the Curie temperature T_C disappear, since the magnetic fluctuations at long wave-lengths are suppressed by the finite size of the cluster. Instead, these properties present a peak at a characteristic temperature $T_C(N)$ with a size dependent width. In addition, the peaks in $C_p(T)$ and $\chi(T)$ occur in general at somewhat different T. Therefore, the "Curie" temperature $T_C(N)$ derived from different properties usually differ to some extent.

General trends in the size dependence of $T_C(N)$ seem difficult to infer *a priori*. On the one side, taking into account the enhancement of the local magnetic moments $\mu(i)$ and of the d-level exchange splittings $\Delta\varepsilon_x(i) = \varepsilon_{i\downarrow} - \varepsilon_{i\uparrow}$, one could expect that $T_C(N)$ should be larger in small clusters than in the bulk. However, on the other side, it should be energetically easier to disorder the local magnetic moments in a cluster (*e.g.*, by flipping or canting $\mu(i)$) since the local coordination numbers are smaller.

If the later effect dominates, $T_C(N)$ should decrease with decreasing N. In addition, recent model calculations [16] indicate that structural changes or fluctuations are likely to affect the temperature dependence of the magnetization. This is particularly relevant for systems like Fe_N and Rh_N which are remarkably structural dependent already at $T = 0$. In order to derive reliable conclusions concerning the size dependence of $T_C(N)$, the theory must take into account both the fluctuations of the magnetic moments and the itinerant character of the d electron states. Simple spin Hamiltonians, for example based on the Heisenberg or Ising model, are not expected to be very predictive, at least until they incorporate the electronic effects responsible for the size dependence of the interactions between the magnetic moments. In fact, previous studies on TM surfaces have already shown that the effective exchange interactions J_{ij} between NN moments $\mu(i)$ and $\mu(j)$ depend strongly on the local coordination numbers of sites i and j [65]. A similar behavior is also found in clusters, as it will be discussed below.

6.1 Spin-fluctuation theory of cluster magnetism

A complete understanding of the finite-temperature properties of magnetic TM's remains an open problem even in periodic solids. Stoner's mean-field theory gives a qualitatively correct description of the electronic structure and magnetization at $T = 0$ but fails completely at finite T. Considerable progress has been made in solids by using d-band model Hamiltonians and functional integral theory [6, 7, 66]. This formalism incorporates both the itinerant character of the d states and the local spin excitations (spin fluctuations) relevant at finite temperatures. The purpose of this section is to extend this theory to finite clusters.

Consider the tight-binding Hubbard Hamiltonian for d electrons given by equations (4.5) and (4.6). For calculating the canonical partition function Z it is useful to rewrite the interaction term $\widehat{V}$ as a sum of squared operators. Among several possibilities we choose, as proposed by Hasegawa [67], the local electron number operator $\widehat{N}_i = \sum_\alpha (\widehat{n}_{i\alpha\uparrow} + \widehat{n}_{i\alpha\downarrow})$ and the local z-component of the spin operator $\widehat{S}_{iz} = \frac{1}{2}\sum_\alpha (\widehat{n}_{i\alpha\uparrow} - \widehat{n}_{i\alpha\downarrow})$ at cluster site i, which retain explicitly the main physical variables of the problem. The transversal exchange terms are neglected for simplicity. In this way one obtains

$$\widehat{V} = \sum_i \left(\frac{U}{2} \widehat{N}_i^2 - J \widehat{S}_{iz}^2 \right), \tag{6.1}$$

where $U = (U_{\uparrow\downarrow} + U_{\uparrow\uparrow})/2$ refers to the average direct Coulomb interaction between d-electrons and $J = U_{\uparrow\downarrow} - U_{\uparrow\uparrow}$ to the exchange integral (see Sect. 4.1). The Stratonovich-Hubbard identity can then be used to express

Z as a functional integral [68]. Neglecting irrelevant multiplicative factors, one obtains

$$Z \propto \prod_{i,t} \int_{-\infty}^{+\infty} d\eta_i(t) d\xi_i(t) \exp\left\{-\frac{\beta}{2}\left(U\eta_i(t)^2 + \frac{J}{2}\xi_i(t)^2\right)\right\}$$
$$\times \operatorname{Tr}\left[\widehat{T} \exp\left\{-\beta \int_0^1 dt \widehat{H}'(t)\right\}\right],$$

where

$$\widehat{H}'(t) = \widehat{H}_\circ(t) + U \sum_j i\eta_j(t)\widehat{N}_j(t) - J \sum_j \xi_j(t)\widehat{S}_{jz}(t). \tag{6.2}$$

The integration variables $\xi_i(t)$ and $\eta_i(t)$ are interpreted as time-dependent exchange and charge fields at cluster site i, since they are coupled to the spin operator $\widehat{S}_{iz}$ and charge operator $\widehat{N}_i$, respectively. Z is given by a functional integral over the fields $\xi_i(t)$ and $\eta_i(t)$ of the partition function Z' of noninteracting electrons moving under the action of these arbitrary time-dependent fields $[\vec{\xi} = (\xi_1, ..., \xi_N)$ and $\vec{\eta} = (\eta_1, ..., \eta_N)]$. The operator $\widehat{T}$ ensures the conventional time ordering [69]. One may attempt to take into account the time dependence systematically by developing $\xi_i(t)$ and $\eta_i(t)$ in its Fourier components. The first term of such an expansion yields the so called *static approximation* which neglects the time dependences. The functional integral reduces then to an ordinary one:

$$Z \propto \int d\vec{\eta} d\vec{\xi}\, Z'(\vec{\xi}, \vec{\eta}), \tag{6.3}$$

where

$$Z'(\vec{\xi}, \vec{\eta}) = \exp\{-\beta F'(\vec{\xi}, \vec{\eta})\} \tag{6.4}$$
$$= \exp\left\{-\frac{\beta}{2}\sum_i \left(U\eta_i^2 + \frac{J}{2}\xi_i^2\right)\right\} \operatorname{Tr}\left[\exp\left\{-\beta\left(\widehat{H}' - \mu\widehat{N}\right)\right\}\right]. \tag{6.5}$$

The static approximation is exact in the highly correlated limit ($t_{ij}^{\alpha\beta} = 0$, $\forall i \neq j$) where no charge fluctuations are present, and in the non-interacting limit ($U_{\sigma\sigma'} = 0$). For nontrivial cases it describes the dynamics of the electrons as independent particles moving in a random alloy with energy levels

$$\varepsilon'_{i\sigma} = \varepsilon_i^0 + Ui\eta_i - \sigma\frac{J}{2}\xi_i. \tag{6.6}$$

The thermodynamic properties of the system are given by a statistical average over all possible distributions of $\varepsilon'_{i\sigma}$ throughout the cluster. For $T \to 0$ the field configuration ξ_i^0, η_i^0 that minimizes the free energy $F'(\vec{\xi}, \vec{\eta})$ dominates. This corresponds to the self-consistent equations

$$\left.\frac{\partial F'}{\partial \xi_i}\right|_0 = \frac{J}{2}(\xi_i^0 - 2\langle \widehat{S}_{iz}\rangle) = 0, \tag{6.7}$$

$$\left.\frac{\partial F'}{\partial \eta_i}\right|_0 = -iU(i\eta_i^0 - \langle \widehat{N}_i\rangle) = 0, \tag{6.8}$$

where $\langle \ldots \rangle$ indicates ground-state average. Note that equations (6.7) and (6.8) are the usual Hartree–Fock equations (4.7) and (4.8) that were used in Section 4.4 for calculating the magnetic properties of clusters and nanostructures at $T = 0$. The static approximation is thus a suitable extension of Hartree–Fock theory that includes spin fluctuations at finite temperatures. Notice that at $T > 0$ the fields ξ_i, η_i fluctuate around ξ_i^0, η_i^0 and other possible local minima.

If one is mainly interested in the magnetic properties it is reasonable to neglect the thermal fluctuations of the charge fields η_i. For each exchange field configuration $\vec{\xi}$, we set $\eta_i = \overline{\eta}_i(\vec{\xi})$ where $\overline{\eta}$ is the saddle point of $F(\vec{\xi}, \vec{\eta})$ given by

$$\left.\frac{\partial F'}{\partial \eta_i}\right|_{\overline{\eta}_i} = -iU(i\overline{\eta}_i - \langle \widehat{N}_i\rangle) = 0. \tag{6.9}$$

Physically, this means that the charge distribution $\langle \widehat{N}_i\rangle$ is calculated self-consistently for each $\vec{\xi}$. Since the η_i are implicit functions of $\vec{\xi}$,

$$Z \propto \int \mathrm{d}\vec{\xi} Z'(\vec{\xi}) = \int \mathrm{d}\vec{\xi} \exp\{-\beta F'(\vec{\xi})\} \tag{6.10}$$

depends only on the relevant exchange variables ξ_i. The integrand in equation (6.10) can be interpreted as proportional to the probability $P(\vec{\xi})$ for a given exchange field configuration $\vec{\xi}$. All thermodynamic quantities are given by an average over all possible $\vec{\xi}$ with $\exp\{-\beta F'(\vec{\xi})\}$ as weighting factor. For example, the local magnetic moments are obtained from

$$\mu_i(T) = \frac{1}{Z}\int \mathrm{d}\vec{\xi} 2\langle S_{iz}\rangle \mathrm{e}^{-\beta F'(\vec{\xi})} = \frac{1}{Z}\int \mathrm{d}\vec{\xi} \xi_i \mathrm{e}^{-\beta F'(\vec{\xi})}. \tag{6.11}$$

This equation justifies the intuitive association between the fluctuations of the local moments $2\langle \hat{S}_{iz}\rangle$ and those of the exchange field ξ_i. The cluster magnetization per atom is given by the average of the local magnetizations $\mu_i(T)$.

In order to discuss the physical behavior of the free energy $F'(\vec{\xi})$ from which the magnetic properties are derived, and in particular the spin excitations for different cluster sites i, we consider the probability distribution function $P_i(\xi)$ of having $\xi_i = \xi$ at atom i, which is given by

$$P_i(\xi) = \frac{1}{Z} \int \prod_{l \neq i} \mathrm{d}\xi_l \exp\{-\beta F'(\xi_1, \ldots, \xi_{i-1}, \xi, \xi_{i+1}, \ldots, \xi_N)\} \tag{6.12}$$

$$= \frac{1}{Z} \exp\{-\beta F_i'(\xi)\} \cdot \tag{6.13}$$

For very low temperatures the integration in equation (6.13) may be simplified by setting $\xi_l = \xi_l^0$ for $l \neq i$ (see Eqs. (6.7) and (6.8)). The free-energy difference $\Delta F_i'(\xi) = F_i'(\xi) - F'(T=0)$ obtained in this way,

$$\Delta F_i'(\xi)\Big|_{T\to 0} = F'(\xi_1^0, \ldots, \xi_{i-1}^0, \xi, \xi_{i+1}^0, \ldots, \xi_N^0) - F'(\xi_1^0, \ldots, \xi_i^0, \ldots, \xi_N^0), \tag{6.14}$$

represents the energy involved in an exchange-field fluctuation ξ above the Hartree–Fock ground state.

6.2 Environment dependence of spin fluctuation energies

We consider the low-temperature limit of the local free-energy $F_i'(\xi)$ and compare the behavior of $\Delta F_i'(\xi)$ for different atoms i within the ferromagnetic cluster. This provides useful information on the stability of the local magnetizations and its environment dependence. In Figures 12 and 13 results are given for $\Delta F_i'(\xi)$ in Fe_9 and Fe_{15} clusters with bcc-like structure. First of all notice that $\Delta F_i'(\xi) = F_i'(\xi) - F_i'(\xi = \mu_i^0) > 0$ which indicates, as expected, that the ferromagnetic order is stable at low temperatures. The surface atoms of Fe_9, which have the largest local magnetic moments μ_i^0 at $T = 0$, show two minima located at $\xi^+ = \mu_i^0$ and $\xi^- \simeq -\mu_i^0$. This double-minimum structure in $F_i'(\xi)$ indicates that flips of the surface magnetic moments keeping their amplitude approximately constant are the dominant magnetic excitations. In contrast at the central atom, which has a much smaller μ_i^0, $F_i'(\xi)$ has a single minimum. Here the amplitude fluctuations of ξ dominate. At the surface of the cluster only small fluctuations of ξ are possible with $\Delta F_i'(\xi)$ smaller than the energy $\Delta F_i'(\xi^-) = F_i'(\xi^-) - F_i'(\xi^+)$ required to flip a local magnetic moment. The probability $P_i(\xi) \propto \exp\{-\beta \Delta F_i'(\xi)\}$ has 2 sharp maxima at $\xi \simeq \xi^+$ and $\xi \simeq \xi^-$ with $P_i(\xi^-) \gg P_i(\xi = 0)$. The fact that very small clusters and particularly cluster atoms having small local coordination numbers z_i show such a Heisenberg- or Ising-like behavior is not surprising, since the kinetic-energy loss caused by flipping a local magnetic moment $[\xi \simeq \mu_i \to \xi \simeq -\mu_i]$

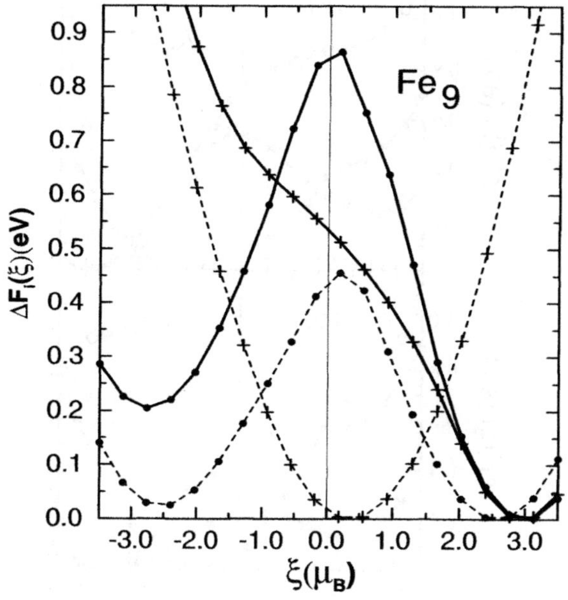

Fig. 12. Local spin-fluctuation energy $\Delta F_i'(\xi) = F_i'(\xi) - F_i'(T=0)$ as a function the exchange field ξ at different atomic sites i of Fe_9 with bcc-like structure. Dots refer to the central atom and crosses to one of the eight surface atoms (NN of the central one). Results are given for $d/d_B = 1.0$ (full curves) and $d/d_B = 0.92$ (dashed curves), where d refers to the NN distance (d_B = bulk NN distance).

is smaller when z_i is smaller ($E_K \propto \sqrt{z_i}$). At the same time, the exchange energy $\Delta E_x = (J/4)\sum_i \mu_i^2$, being a local property, is much less affected by the change of sign of ξ. Even Ni, that in the solid state has a single minimum in $F_i'(\xi)$ and is therefore dominated by amplitude fluctuations of ξ [66], tends to show a double minimum structure in $F_i'(\xi)$ for sufficiently small N. These results suggest the existence of a transition or crossover from Heisenberg-like to itinerant-like behavior with increasing cluster size. Physically this can be interpreted as a consequence of the competition between the Coulomb interaction energy, which is relatively more important in small clusters, and the kinetic d-band energy, which is most important in the bulk.

It is also interesting to determine how $\Delta F_i'(\xi)$ depends on the interatomic distances in order to infer the effects of structural distortions on the

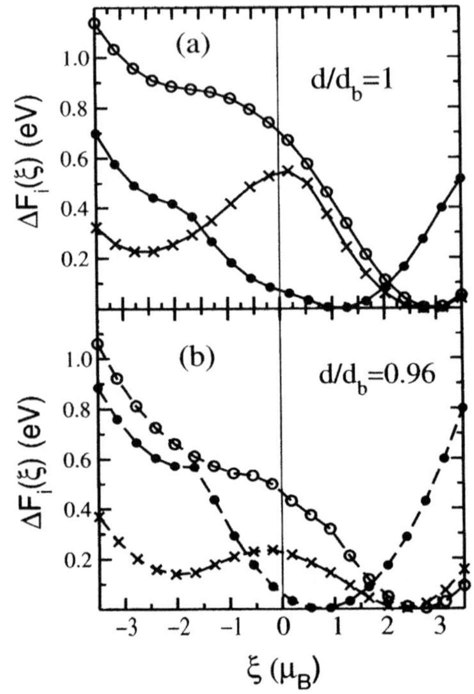

Fig. 13. Local spin-fluctuation energy $\Delta F_i'(\xi) = F_i'(\xi) - F_i'(T=0)$ as a function the exchange field ξ at different atomic sites i of Fe_{15} with bcc-like structure. Dots refer to the central atom, crosses to the first NN shell, and open circles to the outermost shell. Results are given for (a) $d/d_b = 1.0$ in full curves and (b) $d/d_b = 0.96$ in dashed curves, where d refers to the NN distance (d_b = bulk NN distance).

cluster magnetization curves. If the free-energy is optimized by changing the NN distances and by keeping the cluster symmetry unchanged (uniform relaxation) usually a bond-length contraction is obtained ($d < d_B$) [10]. Therefore, we also show in Figure 12 results for Fe_9 using $d/d_B = 0.92$. In this case one finds no qualitative, but strong quantitative changes in $F_i'(\xi)$ as a function of ξ. Besides the shift of the position of the minimum at ξ^+ reflecting the reduction of the local magnetic moments μ_i^0 at $T = 0$, one observes a remarkable reduction (about a factor 10) of the free energy $\Delta F_i'(\xi^-) = F_i'(\xi^-) - F_i'(\xi^+)$ required to flip a local magnetic moment at the surface atoms. A similar large reduction of the cluster "Curie" temperature $T_C(N)$ is expected to occur upon relaxation, since in first approximation

$T_C(N) \propto \Delta F_i'(\xi^-)$. These results show once more the strong sensitivity of the magnetic properties of $3d$ TM clusters to changes in the local environment. As it will be discussed in the next section, recent calculations including correlations effects exactly within the single-band Hubbard model have also revealed the importance of structural fluctuations for the temperature dependence of the magnetic properties of clusters. As in the unrelaxed case, $F_i'(\xi)$ depends strongly on the position of the atom i within the cluster. Similar effects are also observed for other sizes and structures, for example, for bcc-like Fe_{15} as shown in Figure 13. Comparing $\Delta F_i'(\xi^-)$ at different atomic sites i one observes that $\Delta F_i'(\xi^-)$ does not scale simply with the local coordination number z_i. Thus, the effective exchange coupling between local magnetic moments depends on the local environment, as was found at the surfaces of macroscopic TM's [65].

The discussion of $F'(\vec{\xi})$ as a function of a single exchange variable $\xi_i = \xi$, does not mean that the local magnetic moment on different atoms fluctuate independently of each other. Short-range magnetic correlations are known to be present in TM solids, surfaces and clusters even for $T > T_C$ [70]. They can be important for determining the cluster magnetic properties since they introduce additional couplings between sites with different local environments. In the framework of functional-integral theory, short-range magnetic order results in correlations between the exchange fields at neighboring sites. For example, in the ferromagnetic case $\langle \xi_i \xi_j \rangle > \langle \xi_i \rangle \langle \xi_j \rangle$, for i and j being nearest neighbors. Particularly interesting finite-size effects should occur when the cluster size becomes of the order of spin-spin correlation length. In this case, the entropy increase leading to a reduction of the ferromagnetic order necessarily requires breaking the energetically favorable short-range magnetic correlations.

6.3 *Role of electron correlations and structural fluctuations*

Stern–Gerlasch deflection experiments in free transition-metal (TM) clusters show that temperature dependence of $\bar{\mu}_N$ derived from experiment depends qualitatively on the considered TM [24]. The theoretical study of finite-temperature magnetic properties is a difficult problem due to the strong interdependence of electron correlations, magnetism and cluster structure. On the one side, as in solids, one requires an accurate treatment of electron correlations in order to determine the electronic excitations, particularly when spin fluctuations or changes in the total spin are involved. On the other side, in clusters one should take into account possible temperature-induced changes or fluctuations of the geometry on which the magnetic properties of itinerant electrons are known to depend strongly.

In order to investigate this problem we consider the Hubbard Hamiltonian

$$H = -t \sum_{\substack{<i \neq j> \\ \sigma}} \hat{c}^{\dagger}_{i\sigma} \hat{c}_{j\sigma} + U \sum_{i} \hat{n}_{i\downarrow} \hat{n}_{i\uparrow}. \tag{6.15}$$

For small clusters the eigenstates of equation (6.15) can be calculated exactly by numerical methods [71]. The finite-temperature properties are then derived from the canonical partition function

$$Z = \sum_{g} \sum_{l} \mathrm{e}^{-\beta \varepsilon_l (g)}, \tag{6.16}$$

over electronic and structural degrees of freedom. In particular the average total spin $\langle S \rangle$ is obtained from

$$\langle \hat{S}^2 \rangle = \frac{3}{Z} \sum_{g} \sum_{l} \langle \hat{S}_z^2 \rangle \mathrm{e}^{-\beta \varepsilon_l (g)}. \tag{6.17}$$

In equations (6.16) and (6.17) $\varepsilon_l(g)$ stands for the lth eigenenergy corresponding to a cluster structure or geometry g. $\beta = 1/k_\mathrm{B}T$, where T is the temperature of the cluster source that defines the macroscopic thermal bath with which the small clusters are in equilibrium before expansion. Thermal average refers then to the ensemble of clusters in the beam. Keeping ν and N fixed (canonical ensemble) corresponds to the experimental situation in charge and size selected beams [24]. In order to sample all relevant cluster geometries we recall that in the Hubbard model only the topological aspect of the structure is relevant to the electronic properties. Taking into account only NN hoppings with fixed bond lengths results in a discretization of the configurational space so that the sampling of cluster geometries can be performed within graph space (see Sect. 5.2).

In Figure 14 the temperature dependence of the local moments $\langle \hat{S}_i^2 \rangle$ and of the NN spin-correlation functions $\gamma_{ij} = \langle \vec{S}_i \cdot \vec{S}_j \rangle$ are shown for clusters having $N = 6$ atoms and $\nu = 7$ electrons. $U/t = 64$ is representative of the limit of strong electron correlations for which the ground state is ferromagnetic (FM) with saturated moment $S = 5/2$. For simplicity we first assume that the structure is independent of T. The considered cluster geometry, shown in the inset, is the most stable one at $T = 0$. The effects of structural fluctuations are discussed below. Figure 14 shows that $\gamma_{ij} > 0$ for all T, which indicates FM order, and that γ_{ij} decreases with increasing T vanishing in the high-temperature limit. The crossover from the low-temperature ferromagnetically-ordered state to the high-temperature disordered state is significantly broadened by the finite size of the cluster. However, a characteristic temperature scale $T_\mathrm{C}(N)$ may be identified above

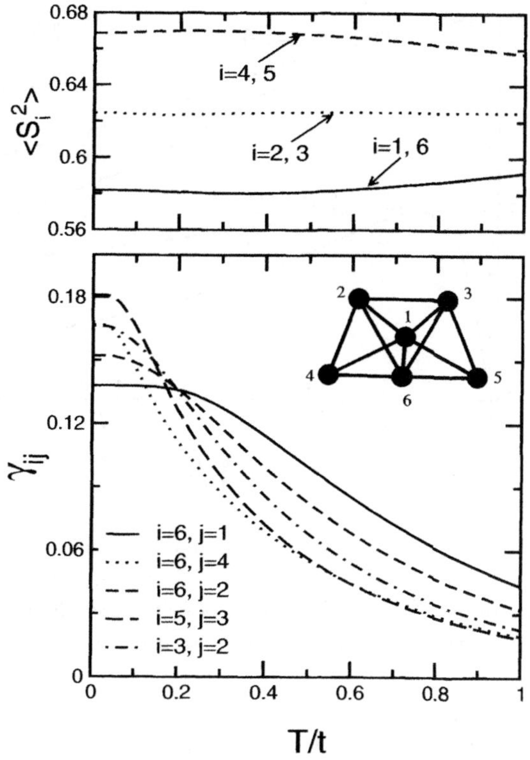

Fig. 14. Temperature dependence of the local magnetic moments $\langle S_i^2 \rangle$ and spin correlation functions $\gamma_{ij} = \langle \vec{S}_i \cdot \vec{S}_j \rangle$ between different NN sites i and j in a cluster having $N = 6$ atoms, $\nu = 7$ electrons, and $U/t = 64$. The results correspond to the most stable structure at $T = 0$ as illustrated in the inset.

which the FM correlations are strongly reduced by thermal spin fluctuations. In contrast, the local moments $\langle \hat{S}_i^2 \rangle$ remain essentially unaffected well above T_C. Taking the inflection point in $\gamma_{ij}(T)$ as a measure of $T_C(N)$ one obtains $T_C/t \simeq 0.1$–0.2. This corresponds to about 1/100 of the bulk band width W_b or to 500–600 K if parameters appropriate for TM's are used ($W_b = 16t$ for the fcc structure and $W_b = 5$–6 eV for TM d-bands). Around this temperature the total average magnetization $\langle S \rangle$ is significantly reduced with respect to the low-temperature value $S = 5/2$ showing also an inflection point (see Fig. 15 for $U/t = 64$). Furthermore, other physical properties also show remarkable changes at $T \simeq T_C = 0.15t$. For example, the magnetic susceptibility $\chi(T)$ has a Curie–Weiss like behavior of the form $\chi \propto 1/(T - T_C)^{-1}$ for $T > T_C$, and the specific heat $C_p(T)$ shows a peak at

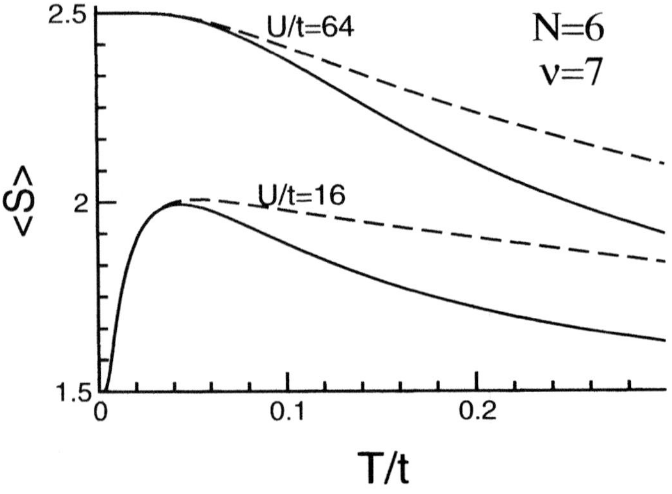

Fig. 15. Temperature dependence of the average total spin $\langle S\rangle$ of clusters having $N = 6$ atoms and $\nu = 7$ electrons $[\langle S\rangle(\langle S\rangle + 1) = \langle \hat{S}^2\rangle]$. The results are obtained by taking into account either the electronic excitations alone (dashed lines), or both electronic excitations and structural fluctuations (solid lines).

$T \simeq T_C$ due to magnetic excitations. These results can be interpreted as a precursor of a magnetic phase transition in the infinite solid.

$\langle \hat{S}_i^2\rangle$ and $\gamma_{ij}(T)$ present a very rich environment dependence. The local moments decrease with increasing local coordination number z_i, as already observed in mean-field calculations. For some bonds (*e.g.*, $(i,j) = (5,3)$) the spin correlations are largest at low T but then decrease quite rapidly with increasing T. This corresponds to bonds with the smallest local coordination numbers. On other bonds with larger z_i (*e.g.*, $(i,j) = (6,1)$) the short-range magnetic order resists better the effects of temperature fluctuations, despite the fact that the local moments and the ground-state spin correlations are somewhat weaker.

The temperature dependence of the average total magnetic moment $\langle S\rangle$ is shown in Figure 15 $[\langle S\rangle(\langle S\rangle + 1) = \langle \hat{S}^2\rangle]$. For large U/t (saturated FM ground state) $\langle S\rangle$ decreases monotonically with increasing T tending to a remnant value $\langle S\rangle_\infty > 0$ at high temperatures ($T > t$). $\langle S\rangle_\infty$ corresponds to an equally probable occupation of all electronic states within the first Hubbard band, *i.e.*, for a minimal number of double occupations ($k_B T \ll U$). In the present case $\langle S\rangle_\infty = 1.5$. The classical analogous of $\langle S\rangle_\infty$ is the average of N random spins which does not vanish in a finite system ($\langle \mu_\infty\rangle = \mu_0/\sqrt{N}$). The difference $\langle S\rangle - \langle S\rangle_\infty$ is a measure of the importance of short-range magnetic order [70]. For smaller values of U/t corresponding to a non-saturated FM ground-state – for example, $S = 3/2$ for $U/t = 16$ – a completely different, non-monotonous temperature dependence of $\langle S \rangle$ is found. Here, one observes first an increase of $\langle S \rangle$ with increasing T, for $T < 0.05t$, followed by a decrease towards the high-temperature limit $\langle S \rangle_\infty$. The increase of $\langle S \rangle$ at low T results from populating low-lying excited states which have higher S than the ground state [$E(S = 5/2) - E(S = 3/2) = 0.02t$ for $U/t = 16$]. If the temperature is further increased, $\langle S \rangle$ decreases since the FM correlations are destroyed in a similar way as for $U/t = 64$ (see Fig. 14). It is interesting to observe that a weak increase of the average magnetization per atom with increasing T has been experimentally observed in large Co_N clusters ($N \simeq 50$–600 and $T \simeq 400$) [24]. The temperature dependence of $\langle S \rangle$ in non-saturated FM clusters is very sensitive to the value of U/t. If U/t is increased beyond $U/t = 16$, the $S = 5/2$ states are stabilized with respect to the quartet and the maximum in $\langle S \rangle(T)$ shifts to lower temperatures. For sufficiently large U/t the sextet becomes the ground state. In contrast, at smaller U/t (*e.g.*, $U/t = 12$) the excitation energies to high-spin states increase and the low-lying excitations correspond to minimal S. Consequently, $\langle S \rangle$ decreases rapidly at low T showing eventually a minimum at intermediate T. The strong sensitivity of the spin excitation spectra on U/t reflects the importance of electron correlations to the finite temperature behavior.

The solid curves in Figure 15 are obtained by including temperature-induced changes of structure (see Eqs. (6.16) and (6.17)). Structural fluctuations play no role at low temperatures where the ground-state structure dominates ($T/t < 0.05$). However, at higher T they contribute to a more rapid decrease of $\langle S \rangle$. In order to analyze this effect one may compute the temperature dependence of the average coordination number $\langle z \rangle$. With increasing T, $\langle z \rangle$ decreases since breaking NN bonds becomes increasingly probable. The reduction of $\langle z \rangle$ is more important for $U/t = 16$ than for $U/t = 64$ since ferromagnetism tends to stabilize the more compact structures [16]. Consequently, the effect of structural fluctuations on $\langle S \rangle(T)$ is somewhat stronger for moderate U/t, where the ground-state magnetic moments are not saturated. The reduction of $\langle S \rangle$ as a result of structural fluctuations may be qualitatively understood as follows. In open structures the band width for bonding (negative energy) states is smaller. Therefore, the eigenstates with maximal spin, whose energy is dominated by the minority electrons, are comparatively less stable than lower spin states. In other words, the spin-flip energy $\Delta E = E(S_{\max} - 1) - E(S_{\max})$ tends to decrease with decreasing coordination number. Thus, $\langle S \rangle(T)$ is reduced as more open structures are populated. Notice, however, that more complex structural effects may be expected for other ν/N and U/t. For example, one may find structures having a higher energy *and* a higher ground-state spin S than the optimal structure at $T = 0$ [16]. In such cases taking into account structural fluctuations could tend to increase $\langle S \rangle(T)$. Summarizing, thermal-induced structural fluctuations play an important role in the temperature dependence of cluster magnetic properties.

7 Conclusion

The different problems discussed in this chapter reflect the remarkable activity in the field of cluster magnetism and suggest possible future studies. We have seen, for example, that the magnetic moments in free $3d$ transition-metal clusters are enhanced with respect to the corresponding solids. This has been related to the reduction of the local coordination number and to the resulting changes in the electronic structure. In contrast, when the same clusters are embedded in a non-magnetic metallic matrix their magnetic moments are *smaller* than the bulk magnetization per atom, a consequence of strong cluster-matrix hybridizations. In the case of magnetic clusters deposited on surfaces the interplay between reduction of local coordination number and cluster-substrate hybridization effects should reveal a wide variety of interesting behaviors. The properties of clusters, islands and dots on surfaces as a function of size and structure motivate a considerable experimental and theoretical activity, particularly in view of developing new magnetic materials for applications in high-density recording and memory devices.

Electron-correlation phenomena and their temperature dependence are subjects of fundamental importance in cluster research where many open questions remain to be elucidated. Methodological improvements on both theoretical and experimental methods are worthwhile and necessary. For instance, a systematic theoretical study of the photoemission spectra of TM clusters would allow to relate the size dependence of excitation spectra and magnetic properties, thus providing valuable information that would complement the results of Stern–Gerlasch deflection experiments. The determination of electronic excitations motivates the development of an improved treatment of many-body effects. Rigorous calculations using realistic d-band Hamiltonians and exact diagonalization methods on small clusters could serve as a basis for testing approximate methods applicable to larger systems. The analysis of the evolution of the many-body electronic structure as a function of size would also improve our understanding of fundamental properties such as the spin and orbital magnetic moments, the magnetic anisotropy and the transition from localized to itinerant magnetism.

The enhancement of the local magnetic moments in low-dimensional systems and the presence of significant short-range magnetic order above the

Curie temperature $T_C(N)$ are two characteristics of TM magnetism which play a significant role for determining the finite temperature behavior. In the case of clusters, changes or fluctuations of structure can also be very important. The spin-fluctuation theory developed in this chapter is an appropriate means of addressing the main issues of this problem, since the local fluctuations of the magnetic moments and the itinerant character of the d-electron states are taken into account at the same electronic level. Presently, more systematic studies of finite-temperature properties are necessary, in particular concerning the size dependence of the cluster Curie temperature $T_C(N)$. Monte-Carlo algorithms could be applied to perform the functional integration over exchange fields, including contributions beyond the usual single-site approximations. For large clusters and cluster-based nanostructures simpler spin models could be formulated by deriving effective spin-spin interactions from the electronic calculations. In this context the non-adiabatic relaxation dynamics of spin excitations deserves special attention. Small clusters would be also useful as test models for investigating non-equilibrium magnetic processes such as the dynamics of recording within a microscopic theory.

In conclusion, cluster magnetism offers a challenging research area that will continue to reveal a series of novel phenomena, fascinating from the fundamental standpoint and potentially important for technological applications.

It is a pleasure to thank the organizing committee and staff of the *Les Houches Summer School of Theoretical Physiscs* for their kind hospitality and support. Numerous fruitful discussions with Prof. J. Dorantes–Dávila are most gratefully acknowledged.

References

[1] N.W. Ashcroft and N.D. Mermin, *Solid State Physics* (Saunders College, 1976).

[2] J.C. Slater, *Quantum Theory of Atomic Structure*, Vol. 1 (McGraw Hill, 1960).

[3] L.D. Landau and E.M. Lifshitz, *Quantum Mechanics*, Vol. 3 (Pergamon Press, 1965).

[4] C. Kittel, *Introduction to Solid State Physics*, 5th edition (Wiley, New York, 1976).

[5] T. Moriya, *Spin fluctuations in itinerant electron magnetism*, Solid State Sciences, Vol. 56 (Springer, Berlin, 1985).

[6] J. Hubbard, *Phys. Rev. B* **19** (1979) 2626; *ibid.* **20** (1979) 4584; H. Hasegawa, *J. Phys. Soc. Jpn.* **49** (1980) 178; *ibid.* **49** (1980) 963.

[7] *Electron Correlation and Magnetism in Narrow-Band Systems*, edited by T. Moriya, Springer Series in Solid State Sciences, Vol. 29 (Springer, Heidelberg, 1981).

[8] D.R. Salahub and R.P. Messmer, *Surf. Sci.* **106** (1981) 415; C.Y. Yang, K.H. Johnson, D.R. Salahub, J. Kaspar and R.P. Messmer, *Phys. Rev. B* **24** (1981) 5673.

[9] K. Lee, J. Callaway and S. Dhar, *Phys. Rev. B* **30** (1985) 1724; K. Lee, J. Callaway, K. Wong, R. Tang and A. Ziegler, *ibid.* **31** (1985) 1796; K. Lee and J. Callaway, *ibid.* **48** (1993) 15358.

[10] G.M. Pastor, J. Dorantes–Dávila and K.H. Bennemann, *Physica B* **149** (1988) 22; *Phys. Rev. B* **40** (1989) 7642.

[11] W.A. de Heer, P. Milani and A. Châtelain, *Phys. Rev. Lett.* **65** (1990) 488.

[12] J.P. Bucher, D.C. Douglass and L.A. Bloomfield, *Phys. Rev. Lett.* **66** (1991) 3052.

[13] See, for example, B.I. Dunlap, *Z. Phys. D* **19** (1991) 255; J.L. Chen, C.S. Wang, K.A. Jackson and M.R. Perderson, *Phys. Rev. B* **44** (1991) 6558; M. Castro and D.R. Salahub, *Phys. Rev. B* **49** (1994) 11842; P. Ballone and R.O. Jones, *Chem. Phys. Lett.* **233** (1995) 632.

[14] J. Dorantes–Dávila, H. Dreyssé and G.M. Pastor, *Phys. Rev. B* **46** (1992) 10432.

[15] A. Vega, J. Dorantes–Dávila, L.C. Balbás and G.M. Pastor, *Phys. Rev. B* **47** (1993) 4742.

[16] G.M. Pastor, R. Hirsch and B. Mühlschlegel, *Phys. Rev. Lett.* **72** (1994) 3879; *Phys. Rev. B* **53** (1996) 10382; F. López–Urías and G.M. Pastor, *Phys. Rev. B* **59** (1999) 5223.

[17] D.M. Cox, D.J. Trevor, R.L. Whetten, E.A. Rohlfing and A. Kaldor, *Phys. Rev. B* **32** (1985) 7290.

[18] S.N. Khanna and S. Linderoth, *Phys. Rev. Lett.* **67** (1991) 742.

[19] P.J. Jensen, S. Mukherjee and K.H. Bennemann, *Z. Phys. D* **21** (1991) 349.

[20] P.J. Jensen and K.H. Bennemann, *Z. Phys. D* **26** (1993) 246; A. Maiti and L.M. Falicov, *Phys. Rev. B* **48** (1993) 13596; N. Hamamoto, N. Onishi and G. Bertsch, *Phys. Rev. B* **61** (2000) 1336.

[21] D.C. Douglass, A.J. Cox, J.P. Bucher and L.A. Bloomfield, *Phys. Rev. B* **47** (1993) 12874.

[22] D.C. Douglass, J.P. Bucher and L.A. Bloomfield, *Phys. Rev. B* **45** (1992) 6341.

[23] A.J. Cox, J.G. Louderback and L.A. Bloomfield, *Phys. Rev. Lett.* **71** (1993) 923; A.J. Cox, J.G. Louderback, S.E. Apsel and L.A. Bloomfield, *Phys. Rev. B* **49** (1994) 12295.

[24] I.M.L. Billas, J.A. Becker, A. Châtelain and W.A. de Heer, *Phys. Rev. Lett.* **71** (1993) 4067.

[25] I.M.L. Billas, A. Châtelain and W.A. de Heer, *Science* **265** (1994) 1682.

[26] S.E. Apsel, J.W. Emert, J. Deng and L.A. Bloomfield, *Phys. Rev. Lett.* **76** (1996) 1441.

[27] R. Galicia, *Rev. Mex. Fis.* **32** (1985) 51.

[28] B.V. Reddy, S.N. Khanna and B.I. Dunlap, *Phys. Rev. Lett.* **70** (1993) 3323.

[29] F. Bodker, S. Morup and S. Linderoth, *Phys. Rev. Lett.* **72** (1994) 282.

[30] H.A. Dürr *et al.*, *Phys. Rev. B* **59** (1999) R701.

[31] K.W. Edmonds *et al.*, *Phys. Rev. B* **60** (1999) 472.

[32] M. Respaud *et al.*, *Phys. Rev. B* **57** (1998) 2925.

[33] G.M. Pastor, J. Dorantes–Dávila, S. Pick and H. Dreyssé, *Phys. Rev. Lett.* **75** (1995) 326.

[34] T. Oda, A. Pasquarello and R. Car, *Phys. Rev. Lett.* **80** (1998) 3622.

[35] M.A. Ojeda, J. Dorantes–Dávila and G.M. Pastor, *Phys. Rev. B* **60** (1999) 6121; *ibid.* **60** (1999) 9122.

[36] C. Kohl and G.F. Bertsch, *Phys. Rev. B* (1999).

[37] R. Haydock, *Solid State Physics* (Academic, New York, 1980), Vol. 35, p. 216.

[38] L.M. Falicov and G.A. Somorjai, *Proc. Natl. Acad. Sci. USA* **82** (1985) 2207.

[39] N.E. Christensen, O. Gunnarsson, O. Jepsen and O.K. Andersen, *J. Phys. France* **49** (1988) C8-17; O.K. Andersen, O. Jepsen and D. Glötzel, *Highlights of Condensed Matter Theory*, edited by F. Bassani, F. Fumi and M.P. Tosi (North Holland, Amsterdam, 1985), p. 59.

[40] G. Stollhoff, A.M. Oles and V. Heine, *Phys. Rev. B* **41** (1990) 7028.

[41] P. Villaseñor–González, J. Dorantes–Dávila, H. Dreyssé and G.M. Pastor, *Phys. Rev. B* **55** (1997) 15084.

[42] Y. Jinlong, F. Toigo and W. Kelin, *Phys. Rev. B* **50** (1994) 7915; Z.-Q. Li, J.-Z. Yu, K. Ohno and Y. Kawazoe, *J. Phys. Cond. Matt.* **7** (1995) 47; B. Piveteau, M-C. Desjonquéres, A.M. Olés and D. Spanjaard, *Phys. Rev. B* **53** (1996) 9251.

[43] A. Vega, J. Dorantes–Dávila, G.M. Pastor and L.C. Balbás, *Z. Phys. D* **19** (1991) 263.

[44] A. Vega, L.C. Balbás, J. Dorantes–Dávila and G.M. Pastor, *Phys. Rev. B* **50** (1994) 3899; *Comput. Mater. Sci.* **2** (1994) 463.

[45] P. Alvarado, J. Dorantes–Dávila and G.M. Pastor, *Phys. Rev. B* **58** (1998) 12216.

[46] P. Strange, *Relativistic Quantum Mechanics* (Cambridge University Press, 1998).

[47] H.J.F. Jansen, *Phys. Rev. B* **38** (1998) 8022.

[48] P. Bruno, *Phys. Rev. B* **39** (1989) 865.

[49] J. Trygg *et al.*, *Phys. Rev. Lett.* **75** (1995) 2871; M. Tischer *et al.*, *ibid.* (1995) 1602; D. Weller *et al.*, *ibid.* (1995) 3752; A.N. Anisimov *et al.*, *ibid.* **82** (1999) 2390; H.A. Dürr *et al.*, *Phys. Rev. B* **58** (1998) R11853.

[50] H.A. Dürr *et al.*, *Phys. Rev. B* **59** (1999) R701; K.W. Edmonds *et al.*, *ibid.* **60** (1999) 472.

[51] M. Castro, Ch. Jamorski and D. Salahub, *Chem. Phys. Lett.* **271** (1997) 133; B.V. Reddy *et al.*, *J. Phys. Chem. A* **102** (1998) 1748; F.A. Reuse and S. Khanna, *Eur. Phys. J. D* **6** (1999) 77.

[52] R. Guirado–López, J. Dorantes–Dávila and G.M. Pastor, to be published.

[53] O. Hjortstam *et al.*, *Phys. Rev. B* **53** (1996) 9204.

[54] J.L. Rodriguez–López, J. Dorantes–Dávila and G.M. Pastor, *Phys. Rev. B* **57** (1998) 1040.

[55] P. Fulde, *Electron correlations in atoms, molecules and solids* (Springer, Berlin, 1990).

[56] S.L. Reindl and G.M. Pastor, *Phys. Rev. B* **47** (1993) 4680.

[57] J. Hubbard, *Proc. R. Soc. London* **A276** (1963) 238; **A281** (1964) 401; J. Kanamori, *Prog. Theo. Phys.* **30** (1963) 275; M.C. Gutzwiller, *Phys. Rev. Lett.* **10** (1963) 159.

[58] C. Lanczos, *J. Res. Nat. Bur. Stand.* **45** (1950) 255; B.N. Parlett, *The Symmetric Eigenvalue Problem* (Prentice-Hall, Engelwood Cliffs, 1980); J.K. Collum and R.A. Willoughby, *Lanczos Algorithms for large Symmetric Eigenvalue Computations* (Boston, 1985), Vol. I.

[59] Y. Wang, T.F. George, D.M. Lindsay and A.C. Beri, *J. Chem. Phys.* **86** (1987) 3493.

[60] J. Callaway, D.P. Chen and R. Tang, *Z. Phys D* **3** (1986) 91; *Phys. Rev. B* **35** (1987) 3705.

[61] E.H. Lieb, *Phys. Rev. Lett.* **62** (1989) 1201.

[62] Y. Nagaoka, *Solid State Commun.* **3** (1965) 409; D.J. Thouless, *Proc. Phys. Soc. London* **86** (1965) 893; Y. Nagaoka, *Phys. Rev.* **147** (1966) 392; H. Tasaki, *Phys. Rev. B* **40** (1989) 9192.

[63] L.M. Falicov and R.H. Victora, *Phys. Rev. B* **30** (1984) 1695; Y. Ishii and S. Sugano, *J. Phys. Soc. Jpn.* **53** (1984) 3895.

[64] P.G. Watson, *Phase Transitions and Critical Phenomena*, Vol. 2, edited by C. Domb and M.S. Green (Academic, London, 1972), p. 101; M.N. Barber, *ibid.* Vol. 8, edited by C. Domb and J.L. Lebowitz (Academic, London, 1983), p. 145.

[65] J. Dorantes–Dávila, G.M. Pastor and K.H. Bennemann, *Solid State Commun.* **59** (1986) 159; *ibid.* **60** (1986) 465.

[66] Y. Kakehashi, *J. Phys. Soc. Jpn.* **50** (1981) 2251.

[67] H. Hasegawa, *J. Phys. F.: Met. Phys.* **16** (1986) 347; *ibid.* (1986) 1555.

[68] J. Hubbard, *Phys. Rev. Lett.* **3** (1959) 77.

[69] R.P. Feynman, *Phys. Rev.* **84** (1951) 108.

[70] G.M. Pastor and J. Dorantes–Dávila, *Phys. Rev. B* **52** (1995) 13799.

[71] F. López–Urías and G.M. Pastor, *J. Appl. Phys.* **87** (2000) 4909; to be published.

COURSE 9

ELECTRON SCATTERING ON METAL CLUSTERS AND FULLERENES

A.V. SOLOV'YOV

Institute for Theoretical Physics,
60054 Frankfurt am Main, Germany

Contents

ELECTRON SCATTERING ON METAL CLUSTERS AND FULLERENES

A.V. Solov'yov

Abstract

This paper gives a survey of physical phenomena manifesting themselves in electron scattering on atomic clusters. The main emphasis is made on electron scattering on fullerenes and metal clusters, however some results are applicable to other types of clusters as well. This work is addressed to theoretical aspects of electron-cluster scattering, however some experimental results are also discussed. It is demonstrated that the electron diffraction plays important role in the formation of both elastic and inelastic electron scattering cross sections. It is elucidated the essential role of the multipole surface and volume plasmon excitations in the formation of electron energy loss spectra on clusters (differential and total, above and below ionization potential) as well as the total inelastic scattering cross sections. Particular attention is paid to the elucidation of the role of the polarization interaction in low energy electron-cluster collisions. This problem is considered for electron attachment to metallic clusters and the plasmon enhanced photon emission. Finally, mechanisms of electron excitation widths formation and relaxation of electron excitations in metal clusters and fullerenes are discussed.

1 Introduction

Clusters have been recognized as new physical objects with their own properties relatively recently. This became clear after such experimental successes as the discovery of electron shell structure in metal clusters [1,2], observation of plasmon resonances in metal clusters [3–5] and fullerenes [6,7], formation of singly and doubly charged negative cluster ions [8] and many

On leave from A.F. Ioffe Physical-Technical Institute, Russian Academy of Sciences, 194021 St. Petersburg, Russia.

more others. Complete review of the field can be found in review papers and books, see *e.g.* [9–14].

The novelty of cluster physics is greatly connected with the fact that cluster properties explain the transition from single atoms or molecules to the solid state. New possibilities in experiment made it possible to study this transition. Increasing cluster size, one can observe the appearance of the new physical features in the system, such as plasmon excitations, electron conduction band formation, superconductivity and super liquidity, phase transitions, fissioning and many more others. Most of these many-body phenomena exist in solid state and are absent in a single atom.

Clusters can naturally exist in all forms of matter: solid state, liquid, gas and plasmas and often collision processes involving atomic clusters play an important role in understanding properties of various physical systems. For example, positive and negative cluster ions do exist in plasmas and understanding of mechanisms of their creation and decay is essential for the description plasma properties and its diagnostics.

Properties of clusters can be studied by means of photon, electron and ion scattering. These methods are the traditional tools for probing properties and internal structure of various physical objects. There are many different types of clusters, such as metallic clusters, fullerenes, molecular clusters, semiconductor clusters, organic clusters, quantum dots, positively and negatively charged clusters, which can be studied by these means. This makes the number of various physical situations to be examined quite large. Therefore when considering electron cluster collisions one needs primarily to focus on certain specific cases and try to recognize physical phenomena relevant for various types of clusters.

Following this line we consider electron collisions with metal clusters and fullerenes, being in a gas phase, and focus on the following physical problems: manifestation of electron diffraction both in elastic and inelastic collisions [15–18], the role of surface and volume plasmon excitations in the formation of electron energy loss spectra (differential and total, above and below ionization potential) as well as the total inelastic scattering cross sections [15–19], importance of the polarization effects in electron attachment process and photon emission processes [20–30]. We also discuss briefly mechanisms of electron excitation width formation and relaxation of electron excitations in metal clusters [31–33]. The choice of these problems is partially made, because of their links with experimental efforts performed in the field.

The outlined problems can be tackled using various approaches. In this lecture solutions of the outlined problems are given on the basis of the consistent many-body theory developed with the Hartree–Fock jellium model wave functions. Many electron correlations in the system are taken into account, where it is necessary, using the random phase approximation with exchange (RPAE) and the Dyson equation method. These approaches are very well known in atomic and nuclear physics. Their effective use for clusters is one of the great advantages of the jellium model, which we briefly discuss as a basis for the electron scattering theory on metal clusters and fullerens.

In this paper the atomic system of units, $\hbar = |e| = m_e = 1$, is used.

2 Jellium model: Cluster electron wave functions

Metallic clusters are characterized by the property that their valence electrons are fully delocalized. To some extent this feature is also valid for fullerenes, where the delocalization of electrons take place on the surface in the vicinity of the fullerene's cage. When considering electron collisions involving metal clusters and fullerens, often, namely the valence delocalized electrons play the most important role in the formation of the cross sections of various collision processes. Therefore, an adequate description of such processes is possible to achieve on the basis of the jellium model.

Let us briefly discuss the main postulates and the results of the jellium model essential for our further consideration. Under different experimental conditions, the detailed ionic structure has been found not to affect the properties of alkali and other simple metal clusters very much (see *e.g.* [9–11] for a review). This behavior suggests the validity of a jellium model, defined by a Hamiltonian which treats the electrons in the usual quantum mechanical way, but approximates the ionic density $n(r)$ as a uniform positively charged background: $n(r) = \frac{N}{V}\Theta(R-r)$, where $V = \frac{4}{3}\pi R^3$ is the cluster volume, $R = r_s N^{1/3}$ is the cluster radius, r_s is the Wigner–Seitz radius, N is the number of atoms in a cluster, $\Theta(R-r)$ is the theta-function.

The mean filed of ions in this approximation is equal to

$$U(r) = \begin{cases} -\frac{N}{2R}\left(3 - \left(\frac{r}{R}\right)^2\right), \; r < R \\ -\frac{N}{r}, \; r > R. \end{cases} \tag{2.1}$$

This model leads to a description of the electron density in terms of single electron wave functions that extend over the entire cluster. The important consequence of this approach is the electronic shell structure in free alkali clusters, which was described theoretically and observed experimentally in [1, 2].

Initially, jellium calculations for metal clusters were based on the density functional formalism using pseudopotentials for description of the electron relaxation effects and lattice structure [34]. Fully self-consistent calculations for the spherical jellium metal clusters have been performed in the framework of the spin-density-functional formalism [35] and the Kohn–Sham formalism for self-consistent determination of the electron wave functions [2, 36].

Later the spherical jellium model for metal clusters based on the Hartree–Fock scheme of the self-consistent determination of the electron wave functions has been developed [37–39]. In spite of the significant differences in the theoretical justification of the many-body theory based on the use Hartree–Fock electron wave functions and the local density approximation, the predictions of two theories are rather similar [40].

Shortly after the discovery of electronic shell structure in free alkali clusters [1, 2] it was realized that the detailed N-dependencies of ionization potentials and other characteristics of small metal clusters can be understood as a consequence of non-spherical cluster shapes [41] by analogy with the nuclear shell model. A direct evidence for cluster deformation was achieved in experiments on photoabsorption, where splitting of plasmon resonances caused by the cluster deformation had been observed (see [9–11] and references therein).

Kohn–Sham calculations for spherical [2, 36] metal clusters have been generalized for spheroidal [42,43] and more general axial shapes of light clusters [44–50]. Light clusters of arbitrary shapes have been studied by means of the ultimate jellium model [45, 46]. It has been shown that alkali-metal clusters have similar shapes with small atomic nuclei [51]. In [33] shapes of deformed axially symmetric metal clusters have been studied within the framework of the Hartree–Fock jellium model.

Recently, the dynamic jellium model for metal clusters, which treats simultaneously the vibration modes of the ionic jellium background, the quantized electron motion and the interaction between the electronic and the ionic subsystems, has been suggested [31–33]. Quantized electron motion has been described on the basis of the Hartree–Fock approach. This model allowed to describe the widths of electron excitations and relaxation times of electron excitations in metal clusters beyond the adiabatic approximation.

Usually, the Hartee-Fock approximation serves as a basis for a systematic development of many-body (*i.e.* many-electron) theory. In general, the Hartree–Fock method itself combined with such methods as the random phase approximation with exchange (RPAE) [52], the Dyson equation method [53] and the consistent many-body theory provides one of the most fundamental and powerful approaches for the description many body phenomena in atomic, molecular and cluster physics. In cluster physics such program has been fulfilled in full for spherically symmetric metal clusters in a number of papers cited above. For deformed metal clusters analysis of many-electron correlations on the basis of the Hartree–Fock jellium model has not been performed so far and is a subject for further efforts.

Finally let us stress that many-body theory developed with the Hartree–Fock jellium model wave functions allows one to construct models on the basis of fundamental physical principles, which can then be refined by extending the quality of the approximations. Alternatively, one can construct pseudo-potentials, exchange-correlation functionals, etc., which of course do exist in principle as unique quantities but are not actually understood, so that they serve as a much less satisfactory basis for achieving a physical interpretation.

3 Diffraction of fast electrons on clusters: Theory and experiment

The jellium model of metal clusters and fullerenes can be examined in electron elastic scattering of fast electrons on metal clusters and fullerenes. Indeed, the jellium model implies that there is a rigid border in the ionic density distribution of a cluster. The presence of a surface in a cluster results in the specific oscillatory behaviour of the electron elastic scattering cross sections, which can be interpreted in terms of electron difraction of the cluster surface [15, 18].

The detailed theoretical treatment of the diffraction phenomena arising in electron scattering on metal clusters and fullerenes has been given in [15–17]. Experimentally, diffraction in electron elastic scattering cross sections on fullerenes in the gas phase has been observed for the first time in [18].

Let us explain the physical nature of the diffraction phenomena arising in elastic electron-cluster scattering on the example of fast electron scattering on the fullerene C_{60}. Due to the spherical-like shape of the C_{60} molecule, the charge densities of electrons and ions near the surface of the fullerene are much higher than in the outer region. These densities are characterized by the radius of the fullerene R and the width of the fullerene shell, $a \ll R$. The de Broglie wave length of a fast electron in collision with fullerene is small compared to R. At the same time the characteristic scattering length, $1/q$, where q is the transferred momentum can be larger than the average inter-atomic distance in C_{60}, which is approximately as large as a. Under these conditions the electron scattering amplitude becomes determined by the charge density of the entire fullerene rather than the charge density of a single carbon atom. In this limit, the scattering amplitude and the corresponding cross section possesses a series of diffraction maxima and minima. The positions of these minima and maxima are mainly determined by the radius of the molecule. In this case the process of electron elastic scattering on C_{60} becomes qualitatively similar to the diffraction of an electron at a shell. The similar behaviour occurs for other type of clusters as well.

The cross section of elastic scattering of a fast electron on C_{60} in the Born approximation (see *e.g.* [54]) reads as

$$\frac{d\sigma}{d\Omega} = \frac{4}{q^4} F(q)^2. \tag{3.1}$$

Here $F(q)$ is the form-factor of C_{60}, $q = |\mathbf{p} - \mathbf{p}'|$ is the momentum transfer and $\mathbf{p}$, $\mathbf{p}'$ are the momenta of the electron in the initial and the final states respectively.

For the description of the fullerene's form factor let us express $F(q)$ as product of the form factor of the atomic concentration $n(q)$ and the form factor of a single carbon atom $F_{\rm A}(q)$:

$$F(q) = F_{\rm A}(q) \sum_j \exp(i\mathbf{q}\mathbf{r}_j) = F_{\rm A}(q) n(q), \tag{3.2}$$

where the summation is performed over all coordinates, $\mathbf{r}_j$, of carbon atoms in the fullerene. The applicability of this approximation has been examined in [18] for metal clusters and fullerenes. The form-factor of an isolated carbon atom, $F_{\rm A}(q)$, calculated in the Hartree–Fock approximation decreases as a function of q on the scale $q \sim 1$. The form-factor of atomic concentration, $n(q)$, characterizing the geometry of C_{60}, varies more rapidly. Indeed, for the momenta transfer smaller than an atomic unit, $q \ll 1$, the model of the homogeneous distribution of carbon atoms over the fullerene sphere can be used. In this model $n(q) = N \sin(qR)/qR$ oscillates with the period $q = 2\pi/R \ll 1$. Namely these oscillations form the diffraction pattern of the differential cross Section (3.1). The angular dependence of the cross Section (3.1) for the projectile electron energy $\varepsilon = 809$ eV is shown in Figure 1 by solid line. It possesses a series of diffraction maxima and minima [18]. Experimental data points derived in [18] in the two sets of measurements are shown in Figure 1 by open and solid circles. The cross section dependence obtained theoretically is shown in Figure 1 by a solid line. Experimental data have been normalized to the theoretical cross section at the second diffraction maximum ($\theta = 5^{\rm o}$). Figure 1 shows quite good agreement of the experimental and the theoretical results in position of the first and the second maxima. The entire pattern of the differential cross section obtained theoretically is very similar to that from the experiment.

In the vicinity of diffraction maxima at $\theta < 10^{\rm o}$ the cross section greatly exceeds the elastic scattering cross section on the equivalent number of isolated atoms because of the coherent interaction of the projectile electron with the fullerene sphere. In the region $\theta > 10^{\rm o}$, where $q > 1$, the projectile electron scatters on individual carbon atoms of the fullerene rather than on

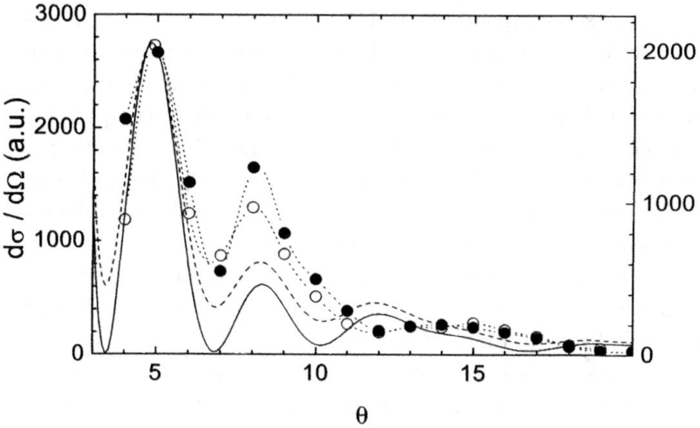

Fig. 1. Experimental (full and open circles) and theoretical (solid curve) angular dependencies of the differential elastic scattering cross section in collision of 809 eV electron with the C_{60} molecule [18]. Full and open circles correspond to the two independent sets of measurements. Dashed line is the differential cross section for the mixture containing 60% of C_{60} and 40% of equivalent isolated carbon atoms. The scale for this curve is given in the right hand side of the figure.

the entire fullerene sphere. Therefore diffraction features of the cross section in the region $\theta < 10^\circ$ are much more pronounced than in the region $\theta > 10^\circ$.

Note that in the region $\theta < 10^\circ$, where $q < 1$, theoretical cross section has zeros while experimental one does not. The presence of zeros at $q \approx \pi k/R < 1$ where k is integer, in theoretical curve is the consequence of the coherent scattering of electron on the fullerene sphere. However in experiment, zeros in the cross section can disappear because of various reasons. For example, this can occur due to the presence of carbon atoms or some other impurities in the gas cell. Figure 1 shows the differential cross section for the mixture containing 60% of C_{60} and 40% of isolated carbon atoms by dashed line. The differential electron elastic scattering cross section on single carbon atoms does not have diffraction oscillations and thus it forms the smooth background removing zeroes in the angular dependence of the cross section.

Finally, let us note that similar considerations are applicable to other types of clusters as well. For metal clusters they have been performed in [15].

4 Elements of many-body theory

Metal clusters and fullerenes possess prominent dynamic properties. Due to the presence of the highly movable delocalized valence electrons in the system, these clusters are higly polarizable. Collective oscillations of the delocalized electrons can be excited in a cluster by a photon or a projectile charged particle (electron or ion). These collective excitations are known as plasmons. Polarization effects are especially strong, when considering low-energy electron cluster collisions, when the velocity of the projectile electron is comparable with the energy of the delocalized cluster electrons. Multipole plasmon excitations is easier to study in the fast-electron cluster collisions. Note that the dipole plasmon excitation mode can also be effectively probed in the photoabsorption or photoionization process.

The dynamic properties of metal clusters are mainly determined by the delocalized cluster electrons and thus can be very well described on the basis of the jellium model. Therefore, let us consider here the essential elements of the many-body theory developed on the basis of the Hartree–Fock jellium model, which are necessary for the understanding of electron inelastic scattering processes described in further sections.

For this purpose let us analyze the transition amplitude M_{fi} defined as follows

$$M_{fi} = \left\langle \Psi_f(\mathbf{r}')\Psi_n(\mathbf{r}) \left| \frac{1}{|\mathbf{r}-\mathbf{r}'|} \right| \Psi_i(\mathbf{r}')\Psi_m(\mathbf{r}) \right\rangle . \tag{4.1}$$

This amplitude describes the transition of the projectile electron from the state Ψ_i to the state Ψ_f with simultaneous excitation of the target electron from the state Ψ_m to Ψ_n. This transition is caused by the Coulomb interaction between the electrons. We assume that the final states Ψ_n and Ψ_f of the electrons can either belong to continuous or discrete spectrum.

Such an amplitude or similar arises, when considering electron-cluster collisions on the basis of the perturbation theory. Depending on the final states of the particles this amplitude can describe either inelastic electron-cluster collision or electron attachment process.

The bound states for the extra electron in the field of the positive cluster can be calculated within the frozen-core Hartree–Fock approximation and by accounting independently for the cluster dynamic polarization potential. Bound states in the system of extra electron plus neutral target cluster do not exist if consider the system in the HF approximation. They appear however, when accounting for the polarization interaction between the electron and the cluster. The negative ion wave functions Ψ_n and the energies ε_n can be obtained as Dyson's equation with the non-local polarization potential [53]:

$$\hat{H}^{(0)}\Psi_n(\mathbf{r}) + \int \Sigma_{\varepsilon_n}(\mathbf{r},\mathbf{r}')\,\Psi_n(\mathbf{r}')\,d\mathbf{r}' = \varepsilon_n\Psi_n(\mathbf{r}). \tag{4.2}$$

Here, $\hat{H}^{(0)}$ is the static single-particle Hamiltonian of the cluster, $\Sigma_{\mathrm{E}}(\mathbf{r},\mathbf{r}')$ is the energy-dependent non-local potential, which is equal to the irreducible self-energy part of the single-electron Green's function of the system cluster + electron. $\Sigma_{\mathrm{E}}(\mathbf{r},\mathbf{r}')$ can be represented diagrammatically as a series on the inter-electron correlation interaction (see [26,53]). It is natural to calculate $\Sigma_{\mathrm{E}}(\mathbf{r},\mathbf{r}')$ and solve equation (4.2) by using the eigen single-particle wave functions ψ_i of the Hartree–Fock $\hat{H}^{(0)}$ Hamiltonian:

$$\hat{H}^{(0)}\psi_i(\mathbf{r}) = \epsilon_i\psi_i(\mathbf{r}) \tag{4.3}$$

where

$$\begin{aligned}\hat{H}^{(0)}\psi_i(\mathbf{r}) &= \left(-\frac{\Delta}{2} - U(\mathbf{r}) + \sum_j^{occ}\int \psi_j(\mathbf{r}')\frac{1}{|\mathbf{r}-\mathbf{r}'|}\psi_j(\mathbf{r}')\,\mathrm{d}\mathbf{r}'\right)\psi_i(\mathbf{r}) \\ &\quad -\sum_j^{occ}\int \psi_j(\mathbf{r}')\frac{1}{|\mathbf{r}-\mathbf{r}'|}\psi_i(\mathbf{r}')\,\mathrm{d}\mathbf{r}'\psi_j(\mathbf{r}). \end{aligned} \tag{4.4}$$

Here, $U(\mathbf{r})$ is the potential of the positive cluster core defined in (2.1). The exchange interaction in equation (4.4) is taken into account explicitly, which makes the potential in (4.4) non-local contrary to the local density approximation in which the exchange correlation interaction is always local.

Let us note that when the collective electron excitations in a cluster become important, one should treat the Coulomb many-electron correlations properly in order to calculate the matrix element (4.1) or similar correctly. For this purpose, let us treat the matrix element (4.1) and excitation energies ω_{fi} in the RPAE scheme using the Hartree–Fock wave functions calculated within the jellium model as a basis. This method similar to the one used in the dipole case [37] for photo-absorption by metal clusters.

The transition amplitude $M_{fi} \equiv \langle f|\hat{G}|i\rangle$ describes the transition between the ground state Ψ_i and the excited state Ψ_f. It can be expressed in the RPAE as a linear combination of the forward-propagating, X^k_{mj}, and the back-propagating, Y^k_{mj}, amplitudes. This relationship reads:

$$\left\langle f\left|\hat{G}\right|i\right\rangle = \sum_{jm}\left(X^{fi}_{mj}\left\langle m\left|\hat{g}\right|j\right\rangle + Y^{fi}_{mj}\left\langle j\left|\hat{g}\right|m\right\rangle\right). \tag{4.5}$$

Here, the single particle operator $\hat{g}$ is equal to $\int\frac{1}{|\mathbf{r}-\mathbf{r}'|}\Psi_f(\mathbf{r}')\Psi_i(\mathbf{r}')\mathrm{d}\mathbf{r}'$ and the many-body operator $\hat{G}$ is defined as $\sum_a\int\frac{1}{|\mathbf{r}_a-\mathbf{r}'|}\Psi_n(\mathbf{r}')\Psi_p(\mathbf{r}')\mathrm{d}\mathbf{r}'$. The eigenstates $|j>, |m>$ in (4.5) are the solutions of the single particle Hartree–Fock equations. Systems of coupled non-linear integro-differential Hartree–Fock equations for the occupied orbitals in the ground state have been solved numerically for clusters of different size in [37].

The forward-propagating, X^k_{mj}, and the back-propagating, Y^k_{mj}, amplitudes are obtained by solving the RPAE equation:

$$\begin{pmatrix} A & B \\ B^* & A^* \end{pmatrix} \begin{pmatrix} X^k \\ Y^k \end{pmatrix} = \omega_k \begin{pmatrix} X^k \\ -Y^k \end{pmatrix}, \tag{4.6}$$

where matrices A and B are defined as follows

$$A_{mi,nj} = (\varepsilon_m - \varepsilon_i)\, \delta_{ij}\delta_{mn} + \langle mj\, |V|\, in\rangle\,,$$

$$B_{mi,nj} = \langle mn\, |V|\, ij\rangle\,,$$

and $\hat{V}$ is defined as

$$\langle \psi_m \psi_l | \hat{V} | \psi_n \psi_k \rangle =$$
$$\int \frac{\psi^*_m(\mathbf{r})\psi_n(\mathbf{r})\psi^*_l(\mathbf{r}')\psi_k(\mathbf{r}') - \psi^*_m(\mathbf{r})\psi_n(\mathbf{r}')\psi^*_l(\mathbf{r}')\psi_k(\mathbf{r})}{|\mathbf{r} - \mathbf{r}'|} \mathbf{d}\mathbf{r}\mathbf{d}\mathbf{r}'.$$

Note that matrices A and B represent the time-forward and the time-reversed diagrams respectively. Using standard methods, one can extract the angular dependencies from equation (4.6).

The numerical solution of the RPAE equation requires a complete single-particle basis. Numerically, the RPAE equation is easier to solve when using a discrete basis $|j>, |m>$. In this case the RPAE equation (4.6) reduces to the matrix equation of the eigen-value problem. The problem of transformation of the complete basis of single-particle states, which includes both the discrete and continuous spectra, to an equivalent complete basis of discrete states can be solved by the B-spline method [55].

5 Inelastic scattering of fast electrons on metal clusters

Let us now consider the inelastic scattering of fast electrons on metal clusters and fullerenes, using approaches and methods described in the previous section. This process is of interest because the many-electron collective excitations of various multipolarity provide significant contribution to the cross section as demonstrated in [15–18].

Plasmon excitations in metal clusters and fullerenes have been intensively studied during last years (see *e.g.* [9–12, 23, 56–58]). They were observed in photo-absorption experiments with metal clusters [3–5] and in photo-ionization studies with the fullerenes [6, 7]. These studies were continued by a large number of experimental and theoretical works (see reviews cited above). In photoionization experiments with metal clusters only dipole collective excitations have been investigated. The same concerns the energy loss experiments mainly performed with the fullerene films for various charged particles, see *e.g.* [59]. These experiments studied the region of relatively small electron scattering angles and, therefore, only dipole collective excitations in the target have been probed.

Electron collective modes with higher angular momenta can be excited in metal clusters and fullerenes by electron impact if the scattering angle of the electron is large enough [15–18]. The plasmon excitations manifest themselves as resonances in the electron energy loss spectra. Dipole plasmon resonances of the same physical nature as in the case of the photoabsorption, dominate the electron energy loss spectrum if the scattering angle of the electron, and thus its transferred momentum, is sufficiently small. With increasing scattering angle plasmon excitations with higher angular momenta become more probable. The actual number of multipoles coming into play depends on the cluster size.

Collisions of electrons with metal clusters and fullerenes have been studied both experimentally and theoretically during the last years. Collisions of fast electrons with fullerenes were experimentally studied in [60]. Theoretically this problem was tackled for metal clusters and fullerenes in [15] and independently for fullerenes in [61]. *Ab initio* many-body quantum treatment of inelastic scattering of fast electrons on metal clusters have been performed in [16, 17]. In these papers both differential and total cross sections have been calculated. The total inelastic cross sections in electron collisions with metal clusters have been measured in [27, 28]. In electron-fullerene collisions, the measurements of the total inelastic scattering cross sections have been performed in [62–65].

Now let us consider the cross section of the scattering process:

$$\mathrm{d}^4\sigma = \frac{2\pi}{v}\left|M_{fi}(\mathbf{p},\mathbf{p}')\right|^2 \delta\left(\frac{\mathbf{p}'^2}{2} + \varepsilon_f - \frac{\mathbf{p}^2}{2} - \varepsilon_i\right) \mathrm{d}f \frac{\mathrm{d}\mathbf{p}'}{(2\pi)^3} \tag{5.1}$$

where $\mathbf{p}$ and $\mathbf{p}'$ are the initial and the final momenta of the projectile electron, ε_i and ε_f are the energies of the initial and the final states of the cluster, the integration over $\mathrm{d}f$ implies the summation over the discrete spectrum and the integration over the continuous spectrum of the final states of the cluster.

Let us describe the initial and the final states of the electron by plane waves. This assumption is valid when the collision velocity is much larger than the characteristic velocity of the delocalized electrons in a cluster (for estimates see [15]).

The amplitude M_{fi} of the process in the plane wave Born approximation reads as

$$M_{fi} = \frac{4\pi}{q^2} \left\langle \Psi_f \left| \sum_a e^{i\mathbf{q}\mathbf{r}_a} \right| \Psi_i \right\rangle , \qquad (5.2)$$

where r_a are the coordinates of the delocalized electrons in the cluster, $\mathbf{q} = \mathbf{p} - \mathbf{p}'$. The wave functions Ψ_i, Ψ_f are the initial and the final states of the cluster. Note that namely the many-electron wave function Ψ_f carries all the information about the collective plasmon excitation in a cluster.

Now let us perform the multipole expansion of the exponents in (5.2) (see *e.g.* [66])

$$M_{fi}(\mathbf{q}) = \frac{(4\pi)^2}{q^2} \sum_{lm} i^l Y^*_{lm}(\mathbf{n_q}) \left\langle \Psi_f \left| \sum_a j_l(qr_a) Y_{lm}(\mathbf{n}_a) \right| \Psi_i \right\rangle . \qquad (5.3)$$

If we consider spherical clusters with the zero total angular momentum of the ground state Ψ_i, the contributions of different multipolarity do not interfere in the cross section (5.1). In this case, substituting (5.3) to (5.1) and performing simple transformations, we come to the following expression for the triply differential inelastic scattering cross section

$$\frac{d^3\sigma}{d\varepsilon' d\Omega} = \frac{16\pi p'}{pq^4} \sum_{lm} (2l+1) \sum_f \left| \left\langle \Psi_f \left| \sum_a j_l(qr_a) Y_{lm}(\mathbf{n}_a) \right| \Psi_i \right\rangle \right|^2$$

$$\delta\left(\frac{\mathbf{p}'^2}{2} + \varepsilon_f - \frac{\mathbf{p}^2}{2} - \varepsilon_i \right) . \qquad (5.4)$$

Here, $d\Omega$ is the solid angle of the scattered electron. The summation over f implies the summation over the discrete spectrum and the integration over the continuous spectrum of the final states of the cluster. We calculate the many electron wave functions Ψ_i, Ψ_f and the excitation energies ε_f and ε_i, using the Hartree–Fock jelium model. As soon as collective electron excitations in a cluster play the significant role, then in order to obtain the correct result when calculating the matrix elements in (5.4) one should properly take into account many-electron correlations. This problem can be solved in the RPAE described in the previous section.

Integrating the triply differential cross Section (5.4) over $d\Omega$, we derive the total differential energy loss spectrum

$$\frac{d\sigma}{d\varepsilon'} = \frac{32\pi^2}{p^2} \sum_f \int_{q_{min}}^{q_{max}} \frac{dq}{q^3} \sum_{lm} (2l+1) \left| \left\langle \Psi_f \left| \sum_a j_l(qr_a) Y_{lm}(\mathbf{n}_a) \right| \Psi_i \right\rangle \right|^2$$

$$\delta\left(\frac{\mathbf{p}'^2}{2}+\varepsilon_f-\frac{\mathbf{p}^2}{2}-\varepsilon_i\right). \tag{5.5}$$

Integrating this equation over the transferred energies of the electron, we come to the expression for the total cross section of inelastic scattering

$$\sigma=\frac{32\pi^2}{p^2}\sum_f\int_{q_{\min}}^{q_{\max}}\frac{dq}{q^3}\sum_{lm}(2l+1)\sum_f\left|\left\langle\Psi_f\left|\sum_a j_l(qr_a)Y_{lm}(\mathbf{n}_a)\right|\Psi_i\right\rangle\right|^2. \tag{5.6}$$

The minimum and the maximum transferred momenta in (5.5) and (5.6) are equal to $q_{\min}=p(1-\sqrt{1-\omega_{fi}/\varepsilon})$ and $q_{\max}=p(1+\sqrt{1-\omega_{fi}/\varepsilon})$. Here, ε_f and ε_i are the energies of the levels i and f respectively, ε is the energy of the projectile electron. The contributions of different multipolarity do not interfere in (5.4–5.6), which is the result of spherical symmetry of the target cluster.

6 Plasmon resonance approximation: Diffraction phenomena, comparison with experiment and RPAE

Besides complex numerical calculations of the cross section (5.4–5.6) one can derive rather simple approximate analytical results giving the distinct physical picture of the process [15]. Indeed, let us consider the behaviour of the inelastic cross sections in the vicinity of the giant collective resonance when surface plasmon excitations give the main contribution. In this case the interaction of the projectile with electrons in the surface layer of the width a near the surface of the cluster should play the most significant role in the inelastic scattering process. The width a of this layer is determined by the width of the region near of the surface of the cluster where oscillations of electron density mainly occur. This width is of the same order of magnitude as the size of a single atom. Oscillations of electron density take place mainly near the surface, because the electron density inside the cluster is well compensated by the oppositely charged density of the ionic background. Mathematically this means that theory has a small parameter, namely $a/R \ll 1$, which allows one make the simplification of final results. Indeed, in the case of the collective excitation the main contribution to the matrix elements in (5.3–5.6) arises from the $r_a \approx R$. This feature is known for the dipole matrix elements from the photoabsorption studies (see *e.g.* [58]).

The condition $r_a \approx R$ allows us to simplify the matrix elements in (5.3) and express them *via* the multipole matrix elements Q_{fi}^{lm} as

$$\left\langle \Psi_f \left| \sum_a j_l(qr_a) Y_{lm}(\mathbf{n}_a) \right| \Psi_i \right\rangle \simeq \sqrt{\frac{2l+1}{4\pi}} \frac{j_l(qR)}{R^l} Q_{fi}^{lm}, \tag{6.1}$$

where

$$Q_{fi}^{lm} = \sqrt{\frac{4\pi}{2l+1}} \left\langle \Psi_f \left| \sum_a r_a^l Y_{lm}(\mathbf{n}_a) \right| \Psi_i \right\rangle .$$

Substituting these equations to (5.3) and using the relationship

$$\int \left| Q_{fi}^{lm} \right|^2 \delta(\omega - \varepsilon_i - \varepsilon_f) \mathrm{d}f = \frac{1}{\pi} Im\alpha_l(\omega), \tag{6.2}$$

where $\alpha_l(\omega)$ is the multipole dynamic polarizability of the cluster we finally obtain the expression of the cross section *via* the imaginary parts of the multipole dynamic polarizabilities

$$\frac{\mathrm{d}^3\sigma}{\mathrm{d}\varepsilon' \mathrm{d}\Omega} = \frac{4p'}{\pi p q^4} \sum_l (2l+1)^2 \frac{j_l^2(qR)}{R^{2l-2}} Im\alpha_l(\Delta\varepsilon). \tag{6.3}$$

Here, $\Delta\varepsilon = \mathbf{p}^2/2 - \mathbf{p}'^2/2$.

Integrating the cross section (6.3) over $\mathrm{d}\Omega$, we derive the expression for the total spectrum of the electron energy loss

$$\frac{\mathrm{d}\sigma}{\mathrm{d}\varepsilon'} = \frac{8}{p^2} \sum_l \frac{(2l+1)^2}{R^{2l-2}} S_l(\Delta\varepsilon) Im\alpha_l(\Delta\varepsilon). \tag{6.4}$$

Here, $\Delta\varepsilon = \varepsilon - \varepsilon'$. We have also introduced functions $S_l(\Delta\varepsilon)$ as follows

$$S_l(\Delta\varepsilon) = \int_{q_{\min}R}^{q_{\max}R} \frac{\mathrm{d}x}{x^3} j_l^2(x). \tag{6.5}$$

The limits $q_{\min}$ and $q_{\max}$ are equal to $q_{\min} = p(1 - \sqrt{1 - \Delta\varepsilon/\varepsilon})$, $q_{\max} = p(1 + \sqrt{1 - \Delta\varepsilon/\varepsilon})$. In the region the most interesting for our consideration, one derives $\Delta\varepsilon \sim \omega_p$ and $\varepsilon \sim \omega_p^2 R^2$. Therefore, $\Delta\varepsilon/\varepsilon \ll 1$ and the limits of integration in (6.5) are approximately equal to $q_{\min} \approx \Delta\varepsilon/v$ and $q_{\max} \approx 2p$. In this case $q_{\max}R \gg 1$, the upper limit in (6.5) can be replaced by the infinity and S_l becomes a function of $\Delta\varepsilon R/v \ll 1$.

Equations (6.4–6.5) establish the connection between the inelastic scattering cross section and the imaginary parts of the multipole dynamic polarizabilities. These polarizabilities have a resonance behavior in the region of frequencies where collective electron modes in a cluster can be excited. In the plasmon resonance approximation (see *e.g.* [15,58]) the polarizability $\alpha_l(\omega)$ can be written as

$$\alpha_l(\omega) = R^{2l+1} \frac{\omega_l^2}{\omega_l^2 - \omega^2 - i\omega\Gamma_l}, \tag{6.6}$$

where ω_l is the resonance frequency of the plasmon excitation with the angular momentum l (we have omitted index p when defined ω_l), which for metal clusters according to the Mie theory (see *e.g.* [58]) is equal to $\omega_l = \sqrt{\frac{3lN_e}{(2l+1)R^3}}$ and $\omega_l = \sqrt{\frac{l(l+1)N_e}{(2l+1)R^3}}$ for fullerenes [15,61]. Here, N_e is the number of delocalized electrons in the cluster. The parameter Γ_l in (6.6) is the width of the plasmon resonance with the angular momentum l.

These formulae demonstrate that due to the resonance behavior of the polarizability, the differential inelastic scattering cross section should also exhibit resonances, if the transferred energy lies in the range characteristic for plasmon excitations.

Integrating (6.4) over ε', we derive the plasmon contribution to the total inelastic cross section. Assuming that the plasmon resonance is narrow enough and using the pole approximation (6.6), we obtain the following expression for the plasmon contribution to the inelastic cross section

$$\sigma = \frac{4\pi e^2 R^3}{v^2} \sum_l (2l+1)^2 \omega_l S_l(\omega_l). \tag{6.7}$$

Note that excitations with large enough angular momenta l have a single particle nature rather than collective character. It follows, for instance, from the fact that with increasing l the wave length of the surface plasmon mode, $2\pi R/l$, becomes smaller than the characteristic wave length of the delocalized electrons on the surface of the cluster, $2\pi/\sqrt{2\varepsilon}$, where ε is the characteristic kinetic electron energy in the cluster. Another words, excitations with the angular momenta comparable or larger than the characteristic electron angular momenta of the ground state should have a single particle character rather than collective nature. Therefore analyzing contribution of the collective modes, which have prominent resonance character, we should restrict the consideration only by relatively low angular momenta. For example, according to the jellium model the maximum angular momentum of the delocalized electrons in the Na_{40} cluster is equal to 4. Therefore, only the dipole, the quadrupole and the octupole collective modes can be expected in this case. This means that only the first three terms in (6.6) should be considered. With increasing cluster size the number of collective modes grows as R.

The plasmon resonance approximation and its classical (see [67,68]) and quasi-classical (see [69,70]) modifications can be used for the estimation of the cross sections and investigation their dependence on various parameters even for the large clusters having hundreds and thousands of atoms, when *ab initio* calculations are hardly possible. For the small cluster systems its validity can be verified by performing the *ab initio* quantum mechanical calculation of cross sections. Such calculation for the electron collision with the Na_{20}, Na_{40}, Na_{58} and Na_{92} clusters have been performed in [16, 17]. In both approaches we have described the collision process in the Born approximation, which is applicable when the energy of the projectile electron surpasses typical energy of the delocalized electrons. Comparison of the two approaches demonstrates that collective excitations of electrons in a cluster provide dominating contribution to inelastic scattering cross sections and explain their most important features.

The resonance plasmon approximation provides a simple criterion for the estimation of the relative importance of various plasmon modes. Indeed, according to the resonance plasmon approximation the minima and the maxima of the contribution of the plasmon mode with the angular momentum l are determined by the minima and the maxima of the diffraction factor $j_l^2(qR)$ as it follows from (5.6). This factor shows that diffraction phenomena arise also in electron inelastic scattering on clusters. The main maximum of the partial contribution with the angular momentum l arises when $qR \sim l$. This condition reflects a simple fact that the probability of the excitation of the collective plasmon mode is maximum, when the characteristic collision distance is about the wave length of plasmon. This results in the significant dependence of the profile of the spectrum on the angle of the scattered electron.

Let us consider now the behaviour of the cross section for the electron-fullerene C_{60} collision. In Figure 2 (left pannel) we compare the calculated differential cross section, $d\sigma/d\varepsilon' d\Omega$, (solid line) with the experimental electron energy-loss spectrum [60] shown by dots. The electron impact energy is equal to 1 keV and the electron scattering angle is equal to 1.5^o in Figure 2a (left pannel) and to 10° in Figure 2b (left pannel). By the dashed line we show the dominating dipole plasmon contribution (Fig. 2a, left pannel) and the quadrupole plasmon contribution (Fig. 2b, left pannel). The monopole term does not contribute much to the cross section, since there practically exists no monopole surface plasmon excitation mode in the system. In Figure 2c (left pannel) the dominating octupole contribution is shown by the dashed line while the guadrupole and the dipole contributions are shown by dotted and dashed-dotted lines respectively. Unfortunately, there exist no experimental data related to this scattering angle.

Figures 2a, b (left pannel) show quite reasonable agreement of theoretical results with the experimental data. At low momentum transfer ($\theta = 1.5^o$) the dipole excitation dominates in the energy loss spectrum (Fig. 2a, left

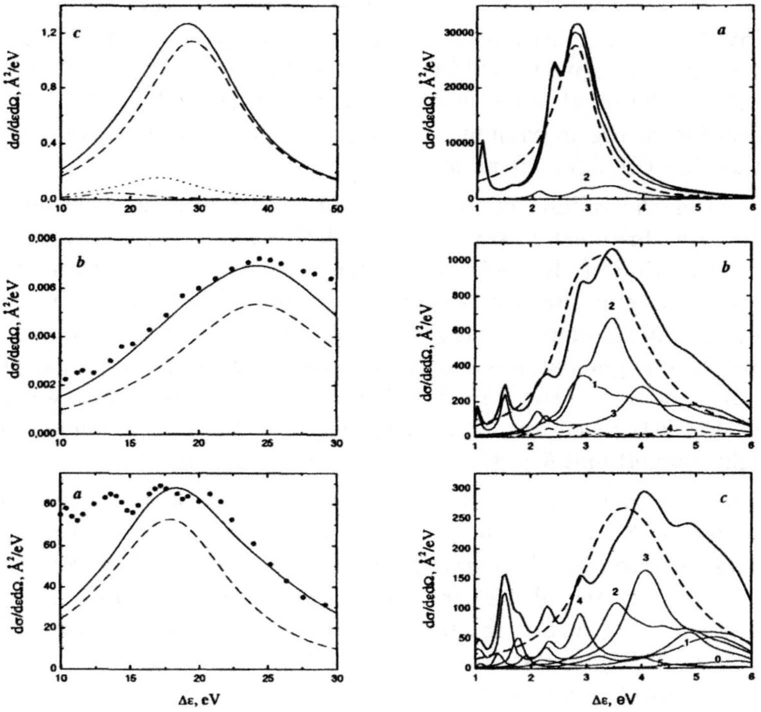

Fig. 2. Left pannel: the differential cross section $d\sigma/d\varepsilon' d\Omega$ (5.4) as a function of the transferred energy $\Delta\varepsilon$ calculated for the electron-fullerene C_{60} collision [17]. The impact electron energy is $\varepsilon = 1$ keV. The scattering angle is $\theta = 1.5°$ (a), $\theta = 10°$ (b) and $\theta = 5°$ (c). The results of the resonance plasmon approximation are shown by solid lines. Dots in figures (a) and (b) represent the experimental data from [60]. Dashed lines show the leading multipole (dipole (a), quadrupole (b) and octupole (c)) plasmon contribution to the spectrum. In Figure 1c, contributions of the qudrupole and the dipole plasmon modes are shown by dotted and dashed-dotted lines respectively. **Right pannel:** the differential cross section, $d\sigma/d\varepsilon' d\Omega$, (5.4) as a function of the transferred energy $\Delta\varepsilon$ calculated for the electron-Na_{40} collision [17]. The impact electron energy, $\varepsilon = 50$ eV. The electron scattering angle is $\theta = 1°$ (a), $\theta = 6°$ (b) and $\theta = 8°$ (c). Solid lines represent the results of the RPAE calculation with the Hartree–Fock jellium model basis wave functions. Thick solid line is the total energy loss spectrum. Thin solid lines marked with the angular momentum number represent various multipole contributions to the energy loss spectrum. By dashed line we plot the electron energy loss spectrum calculated in the resonance plasmon approximation.

pannel), while for $\theta = 10^\circ$ the quadrupole contribution (Fig. 2b, left pannel) provides the main contribution. As a result the position of the maximum of the energy loss spectrum shifts from the dipole plasmon frequency ω_1 at $\theta = 1.5^\circ$ to the quadrupole plasmon frequency ω_2 at $\theta = 10^\circ$. At $\theta = 5^\circ$, the position of the maximum is close to the octupole plasmon resonance frequency ω_3 (Fig. 2c, left pannel).

To check the validity of the pure resonance treatment of plasmon excitations in a cluster and establish the relative role of the collective modes in the formation of the inelastic scattering cross sections, let us compare the results obtained from the direct quantum calculations with those derived from the resonance treatment of plasmon excitations in a cluster. In Figures 2a–c (right pannel) the differential energy loss spectra, $\Delta\varepsilon d\sigma/d\varepsilon'$, (5.1) as a function of the transferred energy $\Delta\varepsilon$ are shown for the Na_{40} target cluster at the electron scattering angles $\theta = 1^\circ$ (a, right pannel), $\theta = 6^\circ$ (b, right pannel) and $\theta = 8^\circ$ (c, right pannel) respectively [17].

Figure 2a (right pannel) shows that at 1° the dipole plasmon excitation dominates in the electron energy loss spectrum. The quadrupole excitation provides relatively small contribution to the spectrum. The contributions of the monopole and all higher multipole excitations are almost negligible at this scattering angle. Figure 2b (right pannel) demonstrates that at 6° the quadrupole excitation becomes the leading excitation in the electron energy loss spectrum, shifting the maximum of the spectrum towards higher energies and also changing the profile of the resonance. The dipole and the octupole excitations provide also considerable contributions in a wide range of transferred energies broadening the spectrum. The dominance of quadrupole excitation is not as large as for the dipole excitation at 1°. At 9° the picture (Fig. 2c, right pannel) becomes more complex. In this case the octupole excitations provide the dominating contribution to the spectrum in the vicinity of the maximum of the energy loss spectrum at $\Delta\varepsilon \approx 4$ eV. Besides this region the dipole, quadrupole and even excitations with angular momentum 4 give comparable contributions to the energy loss spectrum and form rather broad structure. The monopole excitation and the excitations with angular momentum 5 and higher are almost negligible. With increasing scattering angle, excitations with the angular momentum 4 become more important. However, the corresponding spectrum does not possess a resonance behaviour, because it is mainly formed by single electron transitions.

Comparison of the results derived from the RPAE calculations with those obtained in the resonance plasmon approximation shows that, in spite of the simplicity, the plasmon resonance treatment is in quite good agreement with the consistent many-body quantum calculation. The main discrepancy between the two approaches arises from the single particle transitions omitted in the resonance plasmon approximation, but taken into account in the RPAE calculation. These transitions bring some structure to the final energy loss spectra manifesting themselves over the smooth resonance behaviour which is reproduced by the resonance plasmon approximation. At larger scattering angles plasmons with larger angular momenta can be excited. However, as we know, excitations with large enough angular momenta occur due to single particle transitions rather than due to collective excitations. Therefore, the agreement between the resonance plasmon approximation and the RPAE is better at small angles.

7 Surface and volume plasmon excitations in the formation of the electron energy loss spectrum

Now let us discuss the formation of the widths of plasmon resonances. Damping of the plasmon oscillations is connected with the decay of the collective electron excitations to the single-particle ones similar to the mechanism of Landau damping in infinite electron gas. Frequencies of the surface plasmon excitations in neutral metal clusters usually lie below the ionization threshold. Therefore single-particle excitations in the vicinity of the surface plasmon resonance have the discrete spectrum. In this case the width of a surface plasmon excitation caused by the Landau damping should be treated as the width of the distribution of the oscillator strengths in the vicinity of the resonance. The problem of the formation of the surface plasmon resonance widths in clusters has been studied during past years in a number of works [19,69,71,72].

Similar situation takes place for volume plasmon excitations in metal clusters [19]. In metal clusters the resonance frequencies of volume plasmons are above the ionization threshold. This means that the volume plasmon excitations are quasi-stable. They have the real channel of the Landau damping leading to the ionization of the cluster. Thus the process of inelastic scattering in the region of transferred energies above the ionization threshold can be described as follows. The projectile particle induces the oscillations of the electron density in the cluster. Oscillations of the electric field caused by the electron motion result in the ionization of the cluster. Note that the similar scenario takes place with damping of the surface plasmon resonances in fullerenes [6], which also decay *via* the autoionization channel.

The differential cross section of the electron inelastic scattering on metal clusters obtained in the plasmon resonance approximation with accounting for both surface and volume excitations [19] reads as

$$\frac{d^2\sigma}{d\varepsilon' d\Omega} = \frac{4p'R}{\pi p q^4}\sum_l (2l+1)^2 j_l^2(qR)\frac{\omega_l^2 \Delta\varepsilon \Gamma_{sl}}{(\Delta\varepsilon^2 - {\omega_l}^2)^2 + \Delta\varepsilon^2\ \Gamma_{sl}^2} \tag{7.1}$$
$$+\frac{2p'R^3}{\pi p q^2}\sum_l (2l+1)\frac{\omega_p^2 \Delta\varepsilon \Gamma_{vl}}{(\Delta\varepsilon^2 - \ {\omega_p}^2)^2 + \Delta\varepsilon^2\Gamma_{vl}^2}$$
$$\times\left(j_l^2(qR) - j_{l+1}(qR)j_{l-1}(qR) - \frac{2}{qR}j_{l+1}(qR)j_l(qR)\right).$$

Here $\omega_p = \sqrt{3N_e/\alpha}$ is the volume plasmon resonance frequency, $\omega_l = \sqrt{l/(2l+1)}\omega_p$ is the frequency of surface plasmon excitation with the angular momentum l, N_e is the number of delocalized electrons, α is the static polarizability of the cluster, Γ_{vl} and Γ_{sl} are the widths of the volume and surface plasmon resonances, which are defined below. Note that volume plasmon excitations with different angular momenta have the equal resonance frequency ω_p.

This cross section is totally determined by collective electron excitations in the cluster. The first and the second terms in (7.1) describe contributions of the surface and the volume plasmon excitations respectively. Note, that similar expression for the cross section have been also obtained in [70] for electron scattering on small metal particles by means of classical electrodynamics.

According to [19], the width of the surface plasmon resonance in the plasmon resonance approximation is equal to:

$$\Gamma_{sl} = \frac{4\pi\omega_l}{(2l+1)R}\sum_{\nu\mu}\left|\langle\psi_\mu\left|\varphi_{sl}(\mathbf{r})\right|\psi_\nu\rangle\right|^2\delta(\omega_l - \varepsilon_\mu + \varepsilon_\nu), \tag{7.2}$$

where $\varphi_{sl}(\mathbf{r}) = \left((r/R)^l\theta(R-r) + (R/r)^{(l+1)}\theta(r-R)\right)Y_{lm}(\mathbf{n})$. Note that the same expression was obtained earlier in [69, 71] using other methods. Evaluation of the expression (7.2) for sufficiently large clusters leads to the well known result for the Landau damping of the surface plasmon oscillations $\Gamma_{sl} = 3lv_F/R$ [71], where v_F is the velocity of the cluster electron on the Fermi surface.

In the plasmon resonance approximation one can also determine the autoionization width of the volume plasmon resonance [19], which is equal to:

$$\Gamma_{vl} = \frac{8\pi^2\omega_p}{q^2R^3}\frac{\sum_\nu\int\left|\langle\psi_\mu\left|\varphi_{vl}(\mathbf{r})\right|\psi_\nu\rangle\right|^2\delta(\omega_p - \varepsilon_\mu + \varepsilon_\nu)d\mu}{(j_l^2(qR) - j_{l+1}(qR)j_{l-1}(qR) - \frac{2}{qR}j_{l+1}(qR)j_l(qR))}, \tag{7.3}$$

where $\varphi_{vl}(\mathbf{r}) = \left(j_l(qr) - j_l(qR)(r/R)^l\right)\theta(R-r)Y_{lm}(\mathbf{n})$. Here the summation is performed over the occupied single- electron states ν and the integration is performed over the electronic states μ of the continuous spectrum.

The projectile particle excites simultaneously numerous modes of the volume plasmon. The sum of the potentials of all the modes gives the resulting potential $\varphi_{vl}(\mathbf{r})$. It is essential that all normal modes of the volume plasmon have the same resonance frequency ω_{p}, but the excitation probability for these modes depends on the kinematics of collision. This leads to the dependence of the volume plasmon potential $\varphi_{vl}(\mathbf{r})$ upon the transferred momentum. The oscillations of the volume plasmon potential result in the ionization of the cluster, which probability and the volume plasmon resonance width depend on transferred momentum q. However, the numerical analysis performed in [19] shows that the dependence of Γ_{vl} on q is rather weak in the region of $q \ll 1$, where collective electron oscillations mainly take place. Therefore the volume plasmon resonance width with the given l can be approximated by the limiting value following from (7.3) at $q = 0$:

$$\Gamma_{vl} = (2l+5)\frac{\pi^2\omega_{\mathrm{p}}}{R}\sum_{\nu}\int \left|\langle\psi_\mu\left|\varphi^0_{vl}(\mathbf{r})\right|\psi_\nu\rangle\right|^2 \delta(\omega_{\mathrm{p}} - \varepsilon_\mu + \varepsilon_\nu)\mathrm{d}\mu, \qquad (7.4)$$

where $\varphi^0_{vl}(\mathbf{r}) = (r/R)^l \left(1 - (r/R)^2\right)\theta(R-r)Y_{lm}(\mathbf{n})$.

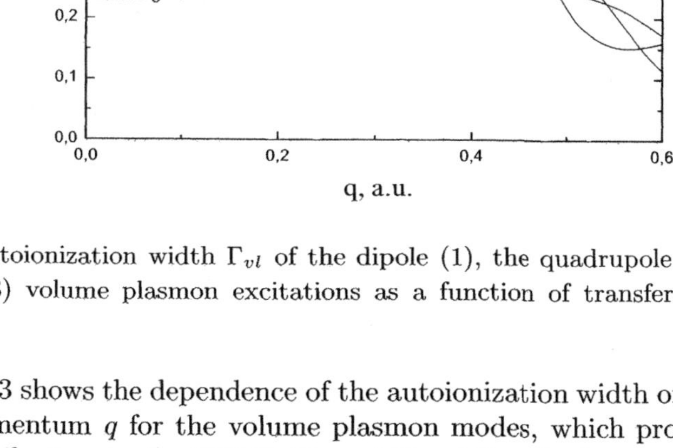

Fig. 3. Autoionization width Γ_{vl} of the dipole (1), the quadrupole (2) and the octupole (3) volume plasmon excitations as a function of transferred momentum q [19].

Figure 3 shows the dependence of the autoionization width on the transferred momentum q for the volume plasmon modes, which provide significant contribution to the EELS. The width of the dipole, the quadrupole and the octupole volume plasmon resonances has been calculated according to (7.3). The transferred momentum q plays the role of the wave vector for the volume plasmon excitations. All three plasmon modes have the similar

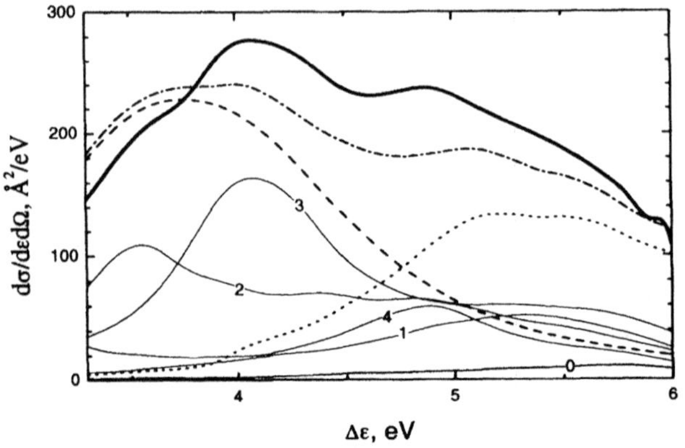

Fig. 4. Differential cross section, $d\sigma/d\varepsilon' d\Omega$ as a function of the transferred energy $\Delta\varepsilon$ calculated for the collision of 50 eV electron with the Na_{40} clusters [19]. The electron scattering angle is $\theta = 9°$. Solid lines represent the RPAE results. Thick solid line is the total energy loss spectrum. Thin solid lines marked by the corresponding angular momentum number represent various partial contributions to the energy loss spectrum. Contributions of the surface and the volume plasmons calculated in the plasmon resonance approximation (7.1) are shown by dashed and dotted lines respectively. Dashed-dotted line represents the sum of the surface and volume plasmon contributions to the EELS.

dependence of Γ_{vl} upon q. The width grows slowly in the region of small q and it decreases rapidly at larger q. In the latter region the probability of volume plasmon excitation by the incoming electron is correspondingly reduced. Note that the wave length of a collective electron oscillation should be larger than the inter-electronic distance in the cluster, *i.e.* plasmon wave vector should be smaller than the Fermi momentum of cluster electrons $q < 0.5$. In the region $q < 0.5$, where the latter condition is fulfilled, the dependence of Γ_{vl} upon q is rather weak. We can approximate the resonance width by following values $\Gamma_{v1} \simeq 0.5\omega_p$, $\Gamma_{v2} \simeq 0.3\omega_p$, $\Gamma_{v3} \simeq 0.23\omega_p$. Contrary to surface plasmons, the autoionization width of a volume plasmon decreases with the growth of the angular momentum.

Let us consider electron energy loss spectra in collision of an electron with the Na_{40} cluster in the region above the ionization potential, where volume plasmon modes become significant.

Figure 4 shows the EELS in the region above the ionization threshold $\Delta\varepsilon > 3.3$ eV calculated using the RPAE. The energy of the collision is equal to $\varepsilon = 50$ eV and the scattering angle $\theta = 9°$. Thick line corresponds to the total energy loss spectrum, while thin lines show various partial contributions. The angular momenta corresponding to these lines are marked by numbers from 0 to 4. The partial contribution to the EELS with $l < 3$ have the broad maximum in the vicinity of $\Delta\varepsilon \simeq 5.1$ eV. Comparison of the EELS calculated in the two different approaches proves our assumption that the peculiarity in the EELS in the vicinity of $\Delta\varepsilon \sim 5$ eV is connected with the volume plasmon excitation. Figure 4 demonstrates that collective excitations provide dominating contribution to the total EELS determining its pattern.

8 Polarization effects in low-energy electron cluster collision and the photon emission process

Our consideration in the previous sections has been mainly focused on the collisions of fast electrons with metal clusters and fullerenes, because the results of the fast electron-cluster collisions theory have the most straightforward and simple connection to the experiment.

However, the review of the electron-cluster collisions theory would not be complete if one says nothing about the low energy electron- cluster collisions. In the low energy electron cluster collision, the electron collision velocity is lower or comparable with the characteristic velocities of the cluster delocalized electrons. This criterion can be traced from the Born theory of electron-cluster collisions, which have very much in common with the Born theory of electron-atom collisions [54]. In [15], it was shown that the electron collisions with metal clusters in the region of collision energies below 3–5 eV should be treated as slow, while for fullerenes, this region extends up to 30 eV.

In the low energy electron-cluster collisions the role of the cluster polarization and exchange correlation effects increases dramatically. The polarization potential of electron-cluster interaction sometimes changes completely the qualitative picture of the collision. This for example takes place, when considering low energy electron elastic scattering on metal clusters. In this case the resonant structures can appear in the energy dependence of the electron elastic scattering cross section due to the presence of the bound or quasi-bound states in the system [73, 74]. The resonance structure turns out to be very sensitive to the choice of the approximations made for its description and has not been experimentally observed so far.

In [73, 74] consistent many-body theory based on the jellium model and the Dyson equation method has been used for the description of low energy elastic electron-cluster collisions. Calculation of the low energy elastic scattering cross section was also performed in the semiclassical approximation [75] and entirely quantum mechanically [76]. The influence of the choice of various LDA potentials on the elastic scattering of low-energy electrons (below 10 eV) was considered in [77]. Low energy electron-fullerene scattering was studied experimentally in [78, 79]. Collisions of low and intermediate energy electrons with metal clusters were experimentally investigated in [27–30]. The response function of fullerene and its relation to the inelastic scattering problem was considered in [80].

During the last years, considerable attention has been devoted both experimentally and theoretically to the problem of electron attachment to metal clusters and fullerenes. The electron attachment process is one of the mechanisms, which lead to the negative cluster ion formation in gases and plasmas and thus it attracts the interest of numerous researches. For fullerenes, the experimental observations of electron attachment have been performed in [78, 79]. For metal clusters, the electron attachment problem has been the subject of the intensive experimental [27–30] and theoretical investigations [20, 21, 24, 26] and is not yet completely understood. Let us further discuss this problem in more details.

The very simple picture of attachment is described in many textbooks, see *e.g.* [81]. Let us assume that there exists a Langevin attraction potential of the form

$$V(r) = -\frac{1}{2}\frac{\alpha e^2}{r^4} \tag{8.1}$$

outside the cluster radius. The constant α is the static polarizability of the cluster. One can then show that there is an orbiting cross section

$$\sigma = \pi \left\{ \frac{2\alpha e^2}{\varepsilon} \right\}^{1/2}, \tag{8.2}$$

which sets an upper limit bound to the attachment cross section (so called the Langevin limit). Here ε is the kinetic energy of the projectile electron. This simple treatment, if valid, would explain the behaviour of the cross section in the vicinity of the threshold.

It is known that metal clusters possess a high polarizabilty, see *e.g.* [9]. Hence, large capture cross sections are anticipated. However, simple attempts to account for attachment by using the static polarizabilty α are not in accordance with observation [27]. A recent review of electric polarizability effects in metal clusters is given in [29]. The great weakness of the Langevin model is the treatment of α as an approximate constant. In fact it possesses a complicated energy dependence, due to the dynamical polarizabilty of the metallic cluster.

The possibility of resonances in the capture cross section was considered, first, theoretically in [20, 21]. It was theoretically demonstrated that electrons of low energy can excite a collective plasmon resonance within the metal cluster in the electron attachment process as a result of a strong dipole deformation of the charge density of the cluster. Later this idea commented in the context of the measurements performed in [28], although no clear evidence of resonant behaviour have been found. In [28] there have been measured total inelastic scattering cross sections, which include attachment as only one of several possible contributing channels.

The resonant electron attachment mechanism was named in [20, 21] as polarizational. An important consequence of the polarization mechanism is that the low energy electron falls into the target and the probability of this process is enhanced. Since the process as a whole is resonant, the enhancement is greatest for energies rather close to the plasmon resonance in the dynamic polarizability of the cluster. This fact explains another reason for the interest to this process. Indeed, the predictions of resonantly enhanced electron capture by metal clusters have been based on the jellium model. While there is convincing evidence for the jellium picture through the occurrence of magic numbers, ellipsoidal structures in studies of the stability of metallic clusters and through the energy splittings of plasmon resonances, the range of validity of the jellium model remains uncertain and is subject to current discussions.

In the attachment process the electron losses its excess energy. Polarizational bremsstrahlung is one of possible channels of the electron energy loss [20–23, 25]. Energy of the electron can also be transferred to the excitations of the ionic background of the cluster [32], which may lead to increase of its vibrations and final fragmentation. In spite of the significant physical difference between various channels of the electron energy loss, they have one important common feature: they all go *via* the plasmon excitation. Therefore, calculating the total electron attachment cross section including all possible channels of the electron energy loss in the system, one obtains [24, 26] qualitatively similar dependence of the cross section as it was obtained initially for the radiative channel of electron energy loss [20].

In [20] the attachment cross section has been calculated within the jellium model in a scheme which holds best if the kinetic energy of the electrons is somewhat higher than the energy of the resonance. Also, it was assumed that the attached ion is created in the ground state. As a useful step in simplifying the calculation, a Kramers-Kronig transformation procedure was introduced to compute the polarizability from the absorption coefficient, thereby circumventing the need for full *ab initio* calculations.

Within this approximate scheme, it was found [20, 21] that the resonant attachment cross section dominates over the non-resonant by a factor of about 10^3–10^4 near resonance, and is therefore a very significant pathway for electrons of low enough energy.

In [24, 26], the earlier theoretical work on attachment was extended by including the following improvements: (a) there were included all possible

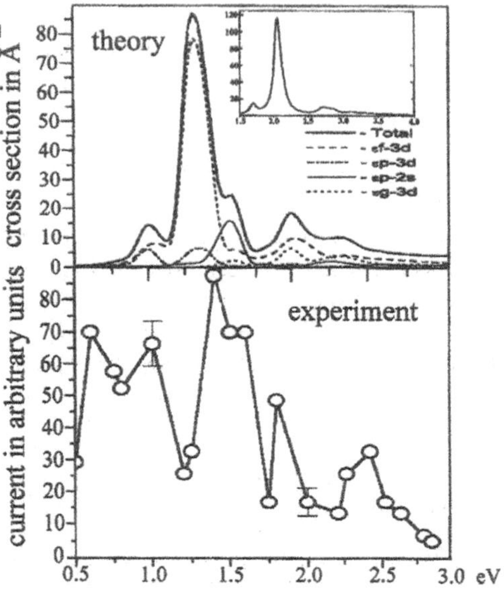

Fig. 5. A comparison between theory and experiment in the vicinity of the plasmon resonance. The theoretical curves from [26] and includes the photoabsorption spectrum. The experimental points are from [30].

channels of the electron attachment and calculated the total cross section of the process rather that analyzed a particular single channel, no assumption that the system can only return to its ground state had been made; (b) theoretical approximation allowed to treat electron energies not only in the resonance region, but actually throughout the range of interest; (c) there were performed an RPAE calculation of the dynamical polarizability and the corresponding electron attachment cross sections on the basis of the consistent many body theory with the use of the Hartree Fock jellium model wave function; (d) calculations were performed for both neutral and charged cluster targets; (e) the polarization effect on the incoming particle has been taken into account and collective excitations of different multipolarity in the target electron system are taken into account; (f) Dyson's equation was used to reduce the problem of the interaction of an extra electron with a many-electron target system to a quasi-one-particle problem in a similar way as it was done for negative atomic ions calculations [53].

An example of such calculation is shown on the upper plot of Figure 5. This plot represents the total and partial electron capture cross sections calculated for neutral potassium K_8 cluster. The insert demonstrates the photoabsorption spectrum of K_8. In [26] this calculation was performed in various approximations outlined in the previous paragraph. It was found that the resonance pattern in the electron capture cross section for the K_8 cluster turns out to be similar in various approaches, although for some other sodium and potassium clusters it is more sensitive to the approximations made [26]. The plasmon resonance in the electron capture cross section is shifted on the value of energy of the attached electron as compared to the photoabsorption case shown in the insert.

Experimental evidence for the resonance enhancement of the cross sections of electron attachment process has been recently obtained [30]. The experimental points from [30] are shown on the lower panel of Figure 5. Comparison of the upper and lower parts of Figure 5 shows that the reasonable agreement between the predictions of theory and the experimental results has been achieved, although more precise measurements would be desirable to resolve the more detailed structures in the electron attachment cross sections.

Note that in its present state, the theoretical description of metallic clusters does not allow such complex processes as resonantly enhanced attachment (in which electron correlations play a dominant role) to be computed with the inclusion of the ion core structure. Fortunately, it has been shown that the jellium approximation provides a good theoretical framework, capable of predicting many properties of metallic clusters. Within this framework, all the refinements of the RPAE become acceptable for the calculation of collective phenomena dominated by electron correlations.

It is also worth to mention that the plasmon resonance enhanced mechanism of electron attachment considered above is typical for metal clusters rather than for fullerenes. Although fullerenes have many similarities in the properties with metal clusters and also possess the plasmon resonances, the energies of these resonances are much higher ($\sim$7 eV and $\sim$19 eV) and thus cannot be reached at low kinetic energies of the projectile electron.

Finally, let us note that strong polarization effects arise also in the process of photon emission by an electron colliding with a cluster [21–23, 25]. Such a process is known as the polarization bremsstrahlung or polarization radiation (see *e.g.* [23]). In these papers it was demonstrated that the plasmon resonance structure manifests itself in the photon emission spectrum in collisions of electrons with metal clusters and fullerenes. This effect occurs because the cross section of this process is mainly determined by the dynamical polarizability of the cluster and is applicable to any polarizable system possessing a collective giant resonance.

9 How electron excitations in a cluster relax

In metal clusters, the plasmon resonances lie below the ionization thresholds, *i.e.* in the region of the discrete spectrum of electron excitations [11].

This fact rises an interesting physical problem about the eigen widths of those electron excitations, which possess large oscillator strengths and form the plasmon resonances. Knowledge of these widths is necessary for the complete description of the electron energy loss spectra, electron attachment, polarizational bremsstrahlung and photoabsorption cross sections in the vicinity of the plasmon resonances and the description of their dependence on the cluster temperature. Note, that experimentally the dependence of the plasmon resonance photoabsorption patterns of metal clusters on temperature has been studied in [82].

In metal clusters, the origination of the electron excitation widths is mainly connected with the dynamics of the ionic cluster core [31,32,83–87]. This is an example of effect, which has no analogy in atomic physics. Contrary, in fullerenes, the discrete transitions with the energies above the ionization threshold, which form mainly the plasmon resonance in the vicinity of 20 eV, possess the autoionization widths [6]. The latter mechanism of the line width broadening is well known in atomic physics [52] and we do not discuss it here.

Instead, let us focus on the influence of the dynamics of ions on the motion of delocalized electrons in metal clusters and discuss it on the basis of the dynamic jellium model suggested in [31] and further developed in [32,33]. This model generalizes the static jellium model [36–39], which treats the ionic background as frozen, by taking into account vibrations of the ionic background near the equilibrium point. The dynamic jellium model treats simultaneously the vibration modes of the ionic jellium background, the quantized electron motion and the interaction between the electronic and the ionic subsystems. In [31] the dynamical jellium model was applied for a consistent description of the physical phenomena arising from the oscillatory dynamics of ions.

The dynamic jellium model [31] allows to calculate widths of the electron excitations in metal clusters caused by the dynamics of ions and their temperature dependence accounting for two mechanisms of the electron excitation line broadening, namely adiabatic and non-dynamic ones.

The adiabatic mechanism is connected with the averaging of the electron excitation spectrum over the temperature fluctuation of the ionic background in a cluster. This phenomenon has been also studied earlier in a number of papers [31,32,83–87]. The adiabatic linewidth is equal to

$$\Gamma = \sqrt{\frac{4\ln 2}{m\Omega}\mathrm{cth}\left(\frac{\Omega}{2kT}\right)}\left|\hat{V}_{nn}\right| . \tag{9.1}$$

Here m and Ω are the mass and frequency corresponding to the generalized oscillatory mode considered, T is the cluster temperature, k is the Bolzmann constant, $\left|\hat{V}_{nn}\right|$ is the matrix element of the electron phonon coupling, calculated for surface and volume cluster vibration modes in [32].

The mechanism of dynamic or non-adiabatic electron excitation line broadening has been considered for the first time in [31, 32]. This mechanism originates from the real multiphonon transitions between the excited electron energy levels. Therefore the dynamic linewidths characterize the real lifetimes of the electronic excitations in a cluster. The analytic expression for the non-adiabatic width obtained in [31, 32] is rather cumbersome as compared to (9.1) and thus we do not present it here.

The adiabatic broadening mechanism explains the temperature dependence of the photoabsorption spectra in the vicinity of the plasmon resonance *via* the coupling of the dipole excitations in a cluster with the quadrupole deformation of the cluster surface. The photoabsorption spectra were calculated within the framework of deformed jellium model using either the plasmon pole approximation [83, 84] or the local density approximation [85–89].

In [31, 32] both the adiabatic and non-adiabatic linewidths of electron excitations in the vicinity of the plasmon resonance caused by coupling of electrons with various ionic vibration modes have been calculated. The non-adiabatic linewidths characterize the real lifetimes of cluster electron excitations. Naturally, the non-adiabatic widths turn out to be much smaller than the adiabatic ones due to the slow motion of ions in the cluster. However, the adiabatic linewidths do not completely mask the non-adiabatic ones, because the two types of widths manifest themselves differently. The adiabatic broadening determines the pattern of the photoabsorption spectrum in the linear regime. The non-adiabatic linewidths are important for the processes, in which the real lifetime of electron excitations and the electron-ion energy transfer are essential. The information about the non-adiabatic electron-phonon interactions in clusters is necessary for the description of electron inelastic scattering on clusters [15–17, 19], including the processes of electron attachment [20,24,26], the non-linear photo-absorption and bremsstrahlung [21–23, 25], the problem of cluster stability and fission. The non-adiabatic linewidths determined by the probability of multiphonon transitions are also essential for the treatment of the relaxation of electronic excitations in clusters and the energy transfer from the excited electrons to ions, which occurs after the impact- or photoexcitation of the cluster.

In [31, 32] there was investigated the role of the volume and the surface vibrations of the ionic cluster core in the formation of the electron excitation linewidths and demonstrated that the volume and surface vibrations provide comparable contributions to the adiabatic linewidths, but the surface vibrations are much more essential for the non-adiabatic multiphonon transitions than the volume ones. Calculations of adiabatic and non-adiabatic linewidths for sodium clusters and analysis of their temperature dependencies have been performed in [31, 32].

10 Concluding remarks

The have considered a number of problems arising in fast and slow electron cluster collisions. The choice of these particular problems was greatly influenced by the experimental efforts undertaken in the field. However, there are many more interesting problems in the field, which have been left aside in this lecture, but with no doubts deserve also careful theoretical and experimental consideration. Concluding this lecture I would like to mention some of them:

- further improvements of jellium model can be incorporated in theory, which allow accounting for the detailed ionic structure and the fragmentation channels of the system in various collision processes;
- it is interesting to elucidate collective nature of two electron excitations in a cluster and their manifestation in collision processes;
- both elastic and inelastic low energy electron-cluster collisions may become a good test for the many-body theories and the applicability of simple cluster jellium models;
- it is interesting to study consequences of the cluster deformations in various collision processes;
- molecular dynamics methods can be used for the description of electron – phonon coupling and electron excitation relaxation in atomic clusters;
- deeper insight in theory of various collision processes, in which non-adiabatic multiphonon transitions play essential role;
- generalization of the problems and application of the methods described in the lecture to other types of clusters and nonostructures.

I would like to express my gratitude to the organizers of the Summer School in Les Houches for the opportunity to deliver this lecture and also to the Volkswagen Foundation, Deutsche Forschungsgemeinschaft and INTAS for the support of this work.

References

[1] W.D. Knight, K. Clemenger, W.A. de Heer, W.A. Saunders, M.Y. Chou and M.L. Cohen, *Phys. Rev. Lett.* **52** (1984) 2141.

[2] W. Ekardt, *Phys. Rev. B* **29** (1984) 1558.

[3] C. Bréchignac, Ph. Cahuzac, F. Carlier and J. Leygnier, *Chem. Phys. Lett.* **164** (1989) 433.

[4] K. Selby, M. Vollmer, J. Masui, V. Kresin, W.A. de Heer and W.D. Knight, *Phys. Rev. B* **40** (1989) 5417.

[5] K. Selby, V. Kresin, J. Masui, M. Vollmer, W.A. de Heer, A. Scheidemann and W.D. Knight, *Phys. Rev. B* **43** (1991) 4565.

[6] G.F. Bertsch, A. Bulcac, D. Tomanek and Y. Wang, *Phys. Rev. Lett.* **67** (1992) 1991.

[7] I.V. Hertel, H. Steger, J. de Vries , B. Weisser, C. Menzel, B. Kamke and W. Kamke, *Phys. Rev. Lett.* **68** (1992) 784.

[8] A. Herlert, S. Krückeberg, L. Schweikhard, M. Vogel and C. Walther, *Phys. Scr. T* **80** (1999) 200.

[9] W.A. de Heer, *Rev. Mod. Phys.* **65** (1993) 611.

[10] M. Brack, *Rev. Mod. Phys.* **65** (1993) 677.

[11] C. Bréchignac and J.P. Connerade, *J. Phys. B Mol. Opt. Phys.* **27** (1994) 3795.

[12] H. Haberland, Clusters of Atoms and Molecules, Theory, Experiment and Clusters of Atoms, *Springer Series in Chemical Physics* **52** (Springer, Berlin, Heidelberg, New York, 1994).

[13] U. Näher, S. Bjørnholm, S. Frauendorf, F. Garcias and C. Guet, *Phys. Rep.* **285** (1997) 245.

[14] Ekardt W., *Metal Clusters* (Wiley, New York, 1999).

[15] L.G. Gerchikov, J.P. Connerade, A.V. Solov'yov and W. Greiner, *J. Phys. B: At. Mol. Opt. Phys.* **30** (1997) 4133.

[16] L.G. Gerchikov, A.N. Ipatov and A.V. Solov'yov, *J. Phys. B: At. Mol. Opt. Phys.* **30** (1997) 5939.

[17] L.G. Gerchikov, A.N. Ipatov, A.V. Solov'yov and W. Greiner, *J. Phys. B: At. Mol. Opt. Phys.* **31** (1998) 3065.

[18] L.G. Gerchikov, P.V. Efimov, V.M. Mikoushkin and A.V. Solov'yov, *Phys. Rev. Lett.* **81** (1998) 2707.

[19] L.G. Gerchikov, A.N. Ipatov, R.G. Polozkov and A.V. Solov'yov, *Phys. Rev. A* **62** (2000) 043201.

[20] J.P. Connerade and A.V. Solov'yov, *J. Phys. B: At. Mol. Opt. Phys.* **29** (1996) 365.

[21] J.P. Connerade and A.V. Solov'yov, *J. Phys. B: At. Mol. Opt. Phys.* **29** (1996) 3529.

[22] L.G. Gerchikov and A.V. Solov'yov, *Z. Phys. D: Atoms, Molecules, Clusters* **42** (1997) 279.

[23] A.V. Korol and A.V. Solov'yov, Topical Review, *J. Phys. B: At. Mol. Opt. Phys.* **30** (1997) 1105.

[24] J.P. Connerade, L.G. Gerchikov, A.N. Ipatov and A.V. Solov'yov, *J. Phys. B: At. Mol. Opt. Phys.* **31** (1998) L27.

[25] L.G. Gerchikov, A.N. Ipatov and A.V. Solov'yov, *J. Phys. B: At. Mol. Opt. Phys.* **31** (1998) 2331.

[26] J.P. Connerade, L.G. Gerchikov, A.N. Ipatov and A.V. Solov'yov, *J. Phys. B: At. Mol. Opt. Phys.* **32** (1999) 877.

[27] V. Kresin, A. Scheidemann and W.D. Knight, *Electron Collisions with Molecules, Clusters and Surfaces*, edited by H. Eberhardt and L.A. Morgan (Plenum, New York, 1994) p. 183.

[28] V. Kasperovich, G. Tikhonov, K. Wong, P. Brockhaus and V.V. Kresin, *Phys. Rev. A* **60** (1999) 3071.

[29] V.V. Kresin and C. Guet, *Philos. Mag.* **79** (1999) 1401.

[30] S. Sentürk, J.P. Connerade, D.D. Burgess and N.J. Mason, *J. Phys. B: At. Mol. Opt. Phys.* **33** (2000) 2763.

[31] L.G. Gerchikov, A.V. Solov'yov and W. Greiner, *Int. J. Mod. Phys. E* **8** (1999) 289.

[32] L.G. Gerchikov, A.N. Ipatov, A.V. Solov'yov and W. Greiner, *J. Phys. B: At. Mol. Opt. Phys.* **33** (2000) 4905.

[33] A.G. Lyalin, S.K. Semenov, N.A. Cherepkov, A.V. Solov'yov and W. Greiner, *J. Phys. B: At. Mol. Opt. Phys.* **33** (2000) 3653.

[34] J.L. Martins, R. Car and J. Buttet, *Surf. Sci.* **106** (1981) 265.

[35] A. Hintermann and M. Manninen, *Phys. Rev. B* **27** (1983) 7262.

[36] W. Ekardt, *Phys. Rev. B* **32** (1985) 1961.

[37] C. Guet and W.R. Johnson, *Phys. Rev. B* **45** (1992) 283.

[38] V.K. Ivanov, A.N. Ipatov, V.A. Kharchenko and M.L Zhizhin, *Pis'ma JETPh* **58** (1993) 649 (in Russian); *Phys. Rev. A* **50** (1994) 1459.

[39] V.K. Ivanov, A.N. Ipatov, *Correlations in clusters and related systems. New perspectives of the many-body problem*, edited by J.-P. Connerade (World Scientific Publishing, Singapure, 1996) p. 141; *Physics of clusters*, edited by V.D. Lakhno and G.N. Chuev (World Scientific Publishing, Singapure, 1996) p. 224.

[40] M. Madjet and C. Guet, *Phys. Rev. A* **51** (1995) 1327.

[41] K. Clemenger, *Phys. Rev. B* **32** (1985) 1359.

[42] W. Ekardt and Z. Penzar, *Phys. Rev. B* **38** (1988) 4273.

[43] W. Ekardt and Z. Penzar, *Phys. Rev. B* **43** (1991) 1322.

[44] B. Montag, Th. Hirschmann, J. Mayer, P.-G. Reinhard and M. Brack, *Phys. Rev. B* **52** (1995) 4775.

[45] M. Koskinen, P.O. Lipas and M. Manninen, *Europhys. Lett.* **30** (1995) 519.

[46] M. Koskinen, P.O. Lipas and M. Manninen, *Z. Phys. D: Atoms, Molecules, Clusters D* **35** (1995) 285.

[47] Th. Hirschmann, M. Brack and J. Mayer, "Atomic and Nuclear Clusters", Proceedings of the Second International Conference at Santorini, Greece (1993) p. 240.

[48] I. Hamamoto, B. Mottelson, H. Xie and X.Z. Zhang, *Z. Phys. D: Atoms, Molecules, Clusters D* **21** (1991) 163.

[49] S.M. Reimann, M. Koskinen, H. Häkkinen, P.E. Lindelof and M. Manninen, *Phys. Rev. B* **56** (1997) 12147.

[50] A. Rytkönen, H. Häkkinen and M. Manninen, *Phys. Rev. B* **80** (1998) 3940.

[51] H. Häkkinen, J. Kolehmainen, M. Koskinen, P.O. Lipas and M. Manninen, *Phys. Rev. Lett.* **78** (1997) 1034.

[52] M.Ya. Amusia, *Atomic photoeffect*, edited by P.G. Burke and H. Kleinpoppen (Plenum, New York, 1990).

[53] L.V. Chernysheva, G.F. Gribakin, V.K. Ivanov and M.Yu. Kuchiev, *J. Phys. B: At. Mol. Opt. Phys.* **21** (1988) L419.

[54] L.D. Landau and E.M. Lifshitz, *Quantum Mechanics* (Pergamon Press, London, 1965).

[55] C. de Boor, *A practical Guide to Splines* (Springer, New York, 1978).

[56] F. Alasia, R.A. Broglia, H.E. Roman, L.I. Serra, G. Colo and J.M. Pacheco, *J. Phys. B: At. Mol. Opt. Phys.* **27** (1994) L663.

[57] M. Madjet, C. Guet and W.R. Johnson, *Phys. Rev. A* **51** (1995) 1327.

[58] U. Kreibig and M. Vollmer, *Optical Properties of Metal Clusters* (Springer-Verlag, Berlin, Heidelberg, 1995).

[59] G. Gensterbum, J.J. Pireaux, P.A. Thirty, R. Candano, J.P. Vigneron, Ph. Lambin, A.A. Lucas and W. Kratschmer, *Phys. Rev. B* **45** (1992) 4297.
[60] J.W. Keller and M.A. Coplan, *Chem. Phys. Lett.* **193** (1992) 89.
[61] D.A. Gorokhov, R.A. Suris and V.V. Cheianov, *Phys. Lett. A* **223** (1996) 116.
[62] R. Völpel, G. Hofmann, M. Steidl, M. Stenke, M. Schlapp, R. Trassl and E. Salzborn, *Phys. Rev. Lett.* **71** (1993) 3439.
[63] D. Hathiramani, K. Aichele, W. Arnold, K. Huber and E. Salzborn, *Phys. Rev. Lett.* **85** (2000) 3604.
[64] P. Scheier, D. Hathiramani, W. Arnold, K. Huber and E. Salzborn, *Phys. Rev. Lett.* **84** (2000) 55.
[65] D. Hathiramani, P. Scheier, K. Aichele, W. Arnold, K. Huber and E. Salzborn, *Chem. Phys. Rev. Lett.* **319** (2000) 13.
[66] D.A. Varshalovich, A.N. Moskalev and V.K. Khersonskii *Quantum Theory of Angular Momentum* (World Scientific, 1988).
[67] F. Fujimoto and Komaki, *J. Phys. Soc. Japan* **25** (1968) 1679.
[68] U. Kreibig and Zacharias, *Z. Phys.* **231** (1974) 17.
[69] A.A. Lushnikov and A.J. Simonov, *Z. Physik* **270** (1974) 17.
[70] A.A. Lushnikov and A.J. Simonov, *Z. Physik B* **21** (1975) 357.
[71] C. Yannouleas and R.A. Broglia, *Ann. Phys.* **217** (1992) 105.
[72] C. Yannouleas, *Phys. Rev. B* **58** (1998) 6748.
[73] A.N. Ipatov, V.K. Ivanov, B.D. Agap'ev and W. Eckardt, *J. Phys. B: At. Mol. Opt. Phys.* **31** (1998) 225.
[74] P. Descourt, M. Farine and C. Guet, *J. Phys. B: At. Mol. Opt. Phys.* **33** (2000) 4565.
[75] B. Wasserman and W. Ekardt, *Z. Phys. D: Atoms, Molecules, Clusters* **19** (1991) 97.
[76] M. Bernath, O. Dragun, M.R. Spinella and H. Massmann, *Z. Phys. D: Atoms, Molecules, Clusters D* **33** (1995) 71.
[77] O.J. Kroneisen, H.J. Ludde and R.M. Dreizler, *Phys. Lett. A* **222** (1996) 405.
[78] M. Lezius, P. Scheier and T.D. Mark, *Chem. Phys. Lett.* **203** (1993) 232.
[79] J. Huang, H.S. Carman and R.N. Compton, **99** (1995) 1719.
[80] A. Bulgas and N. Ju, *Phys. Rev. B* **46** (1992) 4297.
[81] H.S.W. Massey, *Atomic and Molecular Collisions* (Taylor and Francis, London, 1979).
[82] Ch. Ellert, M. Schmidt, Ch. Schmitt, Th. Reiners and H. Haberland, *Phys. Rev. Lett.* **75** (1995) 1731.
[83] J.M. Pacheco and R.A. Broglia, *Phys. Rev. Lett.* **62** (1989) 1400.
[84] G.F. Bertsch and D. Tomanek, *Phys. Rev. B* **40** (1989) 2749.
[85] Z. Penzar, W. Ekardt and A. Rubio, *Phys. Rev. B* **42** (1990) 5040.
[86] B. Montag and T. Hirshmann, J. Mayer and P.-J. Reinhard, *Z. Phys. D* **32** (1994) 124.
[87] B. Montag and P.-J. Reinhard, *Phys. Rev. B* **51** (1995) 14686.
[88] Y. Wang, C. Lewenkopf, D. Tomanek, G. Bertsch and S. Saito, *Chem. Phys. Lett.* **205** (1993) 521.
[89] J.M. Pacheco and W.-D. Schone, *Phys. Rev. Lett.* **79** (1997) 4986.

COURSE 10

ENERGY LANDSCAPES

D.J. WALES

University Chemical Laboratories,
Lensfield Road, Cambridge CB2 1EW,
U.K.

Contents

ENERGY LANDSCAPES

D.J. Wales

1 Introduction

This overview of energy landscapes will focus on some rigorous properties of potential energy surfaces (PES's), especially the symmetry properties of steepest-descent paths, and on thermodynamics and cluster simulation. More detail concerning topics such as visualisation of global PES's and master equation dynamics may be found in a complementary review [1]. The present notes will also touch upon the Born–Oppenheimer approximation [2], tunnelling and global optimisation, but in rather less detail.

In classical mechanics the potential energy function, V, determines the structure, dynamics and thermodynamics of any system. Stable configurations occur at minima in V, defining the structure; the gradient of V gives (minus) the forces on the particles, which appear in the equations of motion; and the configuration integral, which determines the thermodynamics, is also a function of V. Analogous statements can be made in quantum mechanics within the Born–Oppenheimer approximation (Sect. 2), where the potential energy surface (PES) is the solution of the Schrödinger equation for frozen nuclear positions.

When we are interested in the structure and physical properties of a chemical system we often mean the properties of the global minimum on the PES, or the properties at zero Kelvin. However, the dynamics of a system at temperatures or energies where it can escape from the global minimum depend on larger regions of the surface. The topology and topography of these extended regions determine the precise behaviour.

When considering the wider features of the PES in this way, it has become common to refer to it as the "potential energy landscape", or just the "energy landscape". The latter expression is also used to refer to free energy, though this should be clear from the context.

As the size of the system increases, so does the dimensionality of the configuration space required to describe it (Sect. 3.1). In the late 1960's the first attempts were being made to predict the three-dimensional structure of globular proteins from their amino acid sequence. In 1964 Anfinsen had shown that some denatured proteins regained their native structure reliably on a laboratory time scale [3]. By coarse-graining configuration space, Levinthal realised that the number of possible conformations for a typical protein is astronomically large [4]. If they were searched at random on the time scale of a typical vibration the time required to find the native state would probably exceed the lifetime of the universe. For example, with just two conformations per amino acid residue, searched at a rate of 10^{12} s^{-1}, it would take $2^{100} \times 10^{-12}$ s to visit them all for a 100 residue protein, *i.e.* around 4×10^{10} years. This discrepancy has come to be known as Levinthal's paradox. It has now been largely resolved from a better understanding of energy landscapes [1].

1.1 Levinthal's paradox

The first attempts to address the paradox centred on the idea that not all the possible configurations have to be searched. Levinthal himself suggested that folding might occur *via* a specific pathway, initiated by the formation of a condensed nucleus and proceeding through a well-defined sequence of events [5]. Alternatively, it was also suggested that different regions of secondary structure could start to form simultaneously or that the initial random coil might collapse to a compact state or molten globule, which would then need to rearrange. Simulations of simple lattice models suggest that the initial collapse is indeed fast and insensitive to the residue sequence [6,7]. However, some proteins, such as cytochrome *c*, do not appear to follow this route [8].

It seems unlikely that a reduction in the search space is the solution to the paradox, since the number of states remaining would still be too large for a random search to succeed. However, such ideas also contain the implicit notion that, in some way, the search is not random. In terms of the energy landscape there are two reasons for this. Firstly, conformations have different statistical weights in the thermodynamic ensemble, and secondly, they are not arranged at random in configuration space.

Levinthal's analysis assumes that the energy landscape is flat, like a golf course with a single hole corresponding to the native state (Fig. 1) [9]. For a simple model that includes an energetic bias towards the native structure the search time can be dramatically reduced to physically meaningful scales [10, 11].

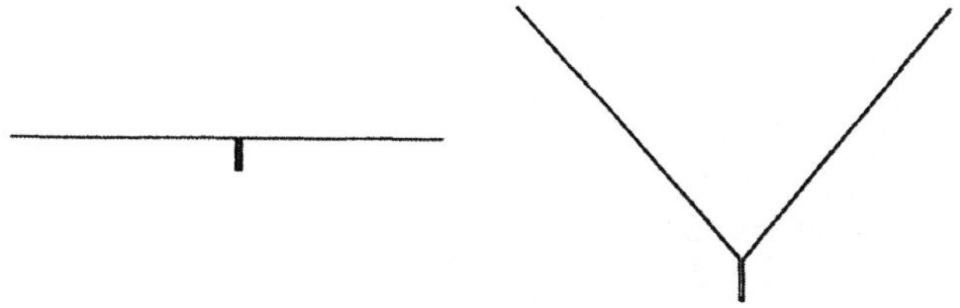

Fig. 1. Flat (left) and biased (right) golf courses.

Leopold, Montal, and Onuchic subsequently proposed that the landscape of a naturally occurring protein consists of a collection of convergent kinetic pathways that lead to a unique thermodynamically stable native state [12]. This landscape structure was termed a folding "funnel" because it focuses the manifold of misfolded states towards the correct target. The random search element of Levinthal's "paradox" is simply removed. In fact, we could ask similar questions about crystallisation, which may be an even more statistically improbable event, and how it is that buckminsterfullerene [13] is selected from a mass of vaporised small carbon chains when there are vast numbers of other possible isomers (Fig. 2)?

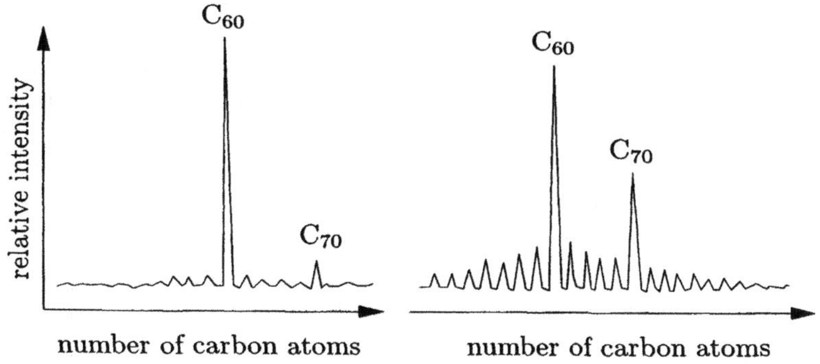

Fig. 2. Schematic time of flight mass spectra for carbon clusters prepared by laser vaporisation at high (left) and medium (right) buffer gas pressures.

Free energy funnels for proteins, which incorporate the entropy, have now been investigated for a variety of models [9, 14–16]. To do this it is necessary to express the free energy (usually Helmholtz) in terms of order parameters. The free energy landscape depends upon temperature and mass – the potential energy surface does not.

Often the dependence of the free energy, A, is expressed in terms of a single order parameter, Q, which is designed to somehow measure the similarity of the protein configuration to the native state. The function

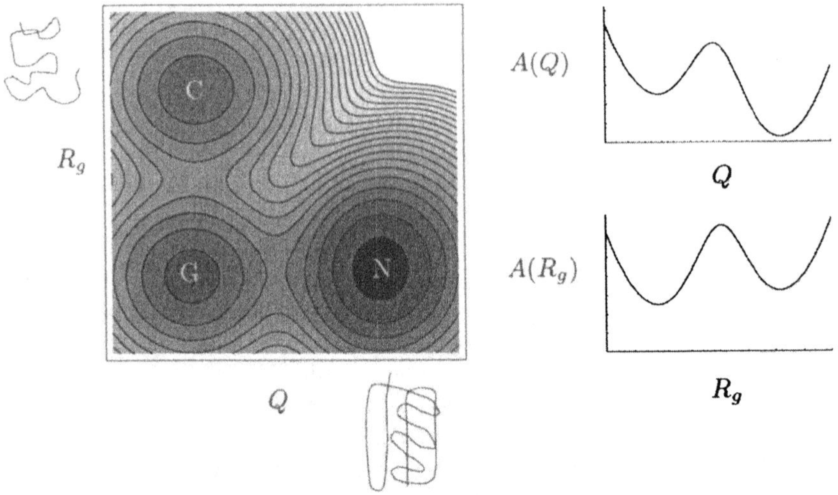

Fig. 3. Schematic protein free energy surface as a function of the radius of gyration, R_g, and the number of native contacts, Q. N = native state, C = "random coil", G = "molten globule".

$A(Q)$ may exhibit a double minimum over a particular temperature range, allowing a thermodynamic "transition state" to be defined between folded and unfolded states (Fig. 3). Some simulations suggest that the folding ability can be measured by the ratio of the folding temperature, T_f, below which the native state has the lowest free energy, to the "glass" transition temperature, T_g, where the kinetics exhibit a dramatic slowing down [7,17]. Folding is easiest for large $T_\mathrm{f}/T_\mathrm{g}$, since the native state is then statistically populated at temperatures where it is kinetically accessible. An important research goal is to understand why some amino acid sequences fold rapidly and reliably whilst others do not.

For a simple lattice model it was suggested that the necessary and sufficient condition for folding was a global potential energy minimum lying well below the other local minima [6,18]. This situation certainly increases the value of T_f, but the strength of the correlation between efficient folding and the energy gap separating the global minimum from the next lowest state has been questioned [19,20]. Relaxation from high-lying states must also depend on global properties of the potential energy surface, not just the region around the global minimum [21]. In fact, surfaces exist with pronounced potential energy minima that are kinetically inaccessible (Sect. 3) [1].

The energy of a protein depends on interactions between parts of the chain that come into contact as it folds. If the favourable interactions that define the native state involve bringing together other residues that interact unfavourably, the process is energetically frustrated. Geometric frustration arises if two similar structures can only be interconverted by disrupting contacts that have been made elsewhere. Both these effects lead to roughening of the potential energy landscape. Proteins that avoid these problems as far as possible are said to obey the principle of "minimum frustration" [9].

It is now generally believed, or at least assumed, that the native state of a protein is the global free energy minimum at the appropriate temperature (Anfinsen's thermodynamic principle) [9, 22]. To the extent that protein folding is a search for the global potential energy minimum, it is an example of the more general problem of global optimisation (Sect. 8), which is of great interest in fields as different as economics and microelectronics. The success of any global optimisation algorithm is inextricably linked to the structure of the landscape, and an understanding of these connections has recently been exploited to design more effective techniques (Sect. 8) [23–25].

1.2 "Strong" and "fragile" liquids

The potential energy landscape also plays an important role in determining the behaviour of both finite and bulk matter in the liquid state. Angell's scheme classifies liquids on a scale from "strong" to "fragile" using both thermodynamic criteria, such as the entropy relative to the crystal, and dynamic criteria, such as the viscosity [26]. Dynamic criteria are generally related to a reorganisation or structural relaxation time scale. In these terms strong liquids are more "resistant" to structural change as a function of temperature.

A strong liquid is characterised by a diffusion constant and viscosity that follow the Arrhenius form ($\propto \exp[A/T]$). These are often liquids with strongly bonded open network structures like SiO_2, BeF_2 and P_2O_5. Fragile liquids exhibit non-Arrhenius behaviour and tend to have more isotropic interactions (Fig. 4). Quantitative measures of strength and fragility have also been suggested [27].

Strong materials also have small heat capacity differences between the supercooled liquid and crystal; fragile materials exhibit large heat capacity differences. The apparent violation of the third law of thermodynamics by the extrapolated entropy of a supercooled liquid is known as the Kauzmann entropy paradox [28]. This possibility is avoided by the intervention of the glass transition where the viscosity increases rapidly and liquid-like diffusion ceases.

Whether we consider the viscosity or a general relaxation time, τ, the temperature dependence can often be described over a certain range by the

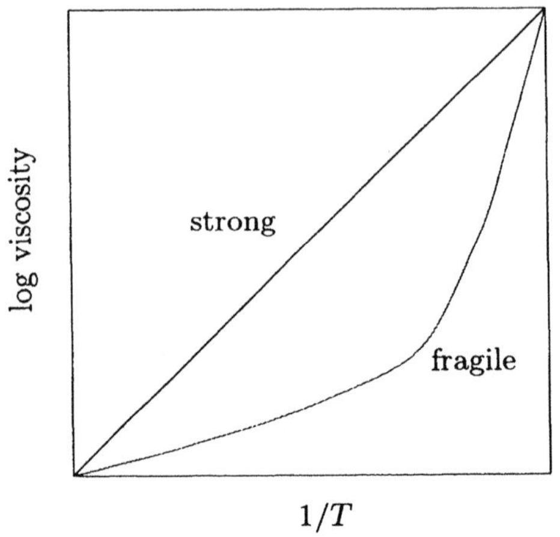

Fig. 4. Schematic behaviour of strong and fragile liquid viscosities.

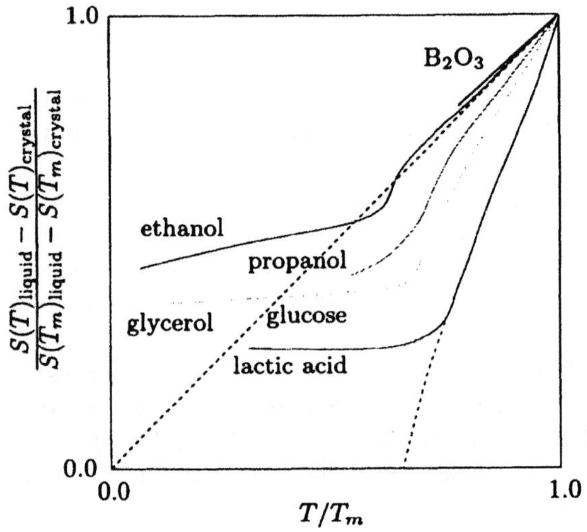

Fig. 5. Variation of the excess entropy with T/T_m for some strong and fragile glass formers.

empirical Vogel–Tammann–Fulcher (VTF) equation [29–31]:

$$\tau = \tau_0 \exp[DT_0/(T - T_0)], \tag{1.1}$$

where B, D and T_0 are fitted parameters. Arrhenius behaviour is recovered for small T_0 and large D. An alternative expression due to Ferry *et al.* often works equally well [32]:

$$\tau = \tau_0 \exp[-D/T^2]. \tag{1.2}$$

For strong liquids the relaxation behaviour of the property described by τ is usually found to fit a simple exponential form, $p(t) = p(0)e^{-t/\tau}$, where t is the time. For fragile liquids a variety of alternative functional forms have been suggested, such as the Kohlrausch–Williams–Watts stretched exponential [33], $p(t) = p(0)\exp[-a\,t^\theta]$. Debye relaxation corresponds to $\theta = 1$. Although the relation is by no means universal, increasing fragility is correlated with departure from Debye relaxation in a wide variety of liquids [34]. To define a general relaxation time we can use [35]

$$\int_0^\infty \frac{p(t) - p(\infty)}{p(0) - p(\infty)} \mathrm{d}t, \tag{1.3}$$

which gives the expected result for Debye relaxation.

As the temperature is lowered the absorption spectrum of a supercooled liquid or glass corresponding to dielectric relaxation often exhibits a peak that moves to lower frequency as the temperature decreases [36]. Once this primary or α peak has disappeared a weaker β absorption may be revealed which in turn moves to lower frequency as the temperature decreases. The α peak frequency as a function of temperature often exhibits non-Arrhenius behaviour and has been associated with slower molecular rearrangements.

Adam and Gibbs provided a heuristic connection between thermodynamics and dynamics by considering "cooperative relaxation" to obtain [37]

$$\tau = \tau_0 \exp(A\Delta\mu/TS_c), \tag{1.4}$$

where A is a constant, $\Delta\mu$ is the free energy barrier per molecule in the cooperative group and S_c is the excess configurational entropy of the liquid over the crystal. The VTF equation (1.1) can be recovered if $C_{p,\text{liquid}} - C_{p,\text{crystal}} \propto 1/T$.

In fact it is not always possible to classify liquids unambiguously as strong or fragile, and some strong liquids become fragile at higher densities. For example, the diffusion of both Si and O atoms in silica becomes strongly non-Arrhenius when the density is increased by 30% [38].

Glasses often (not always) exhibit other characteristic features such as a low temperature heat capacity that varies linearly with T. This phenomenon may be due to quantum mechanical tunnelling (Sect. 5) between nearly degenerate minima separated by tiny barriers [39–41]. Many glasses also exhibit an excess vibrational density of states at low frequency compared to the usual Debye dependence ($g(\nu) \propto \nu^2$). This feature is termed the "boson peak".

Much has been measured and fitted in the field of supercooled liquids and glasses. However, the fundamental basis for all glass phenomenology must lie in the underlying PES, as Goldstein realised as early as 1969 [42]. For example, Angell has proposed that fragile liquids have more local minima in a given energy range to account for the larger heat capacity, with low barriers between them to give a low value of D in the VTF equation (1.1) [38]. The slow α and fast β processes are generally presumed to be associated with mechanisms involving larger and smaller barriers [35].

2 The Born–Oppenheimer approximation

The Born–Oppenheimer approximation is a key element in the notion of a potential energy surface [2]. The Schrödinger equation for a molecule with n electrons, mass $m_{\rm e}$, and N nuclei, masses $M_{\rm s}$, is

$$\left[-\sum_{s=1}^{N}\frac{\hbar^2}{2M_{\rm s}}\nabla_s^2-\sum_{i=1}^{n}\frac{\hbar^2}{2m_{\rm e}}\nabla_i^2+V(\mathbf{x},\mathbf{X})\right]\psi(\mathbf{x},\mathbf{X})=E_{\rm exact}\psi(\mathbf{x},\mathbf{X}), \quad (2.1)$$

where $\mathbf{x}$ and $\mathbf{X}$ represent the electronic and nuclear coordinates. The potential energy is

$$V(\mathbf{x},\mathbf{X})=\frac{e^2}{4\pi\epsilon_0}\left[-\sum_{\text{all } i,t}\frac{Z_t}{r_{it}}+\sum_{i<j}\frac{1}{r_{ij}}+\sum_{t<s}\frac{Z_tZ_s}{r_{ts}}\right], \quad (2.2)$$

where Z_t is the atomic number (nuclear charge) of nucleus t and e is the modulus of the electronic charge. $r_{it}=|\mathbf{x}_i-\mathbf{X}_t|$ where $\mathbf{x}_i$ and $\mathbf{X}_t$ are the position vectors of electron i and nucleus t, etc.

It is more convenient to use atomic units where $e=1$, $m_{\rm e}=1$, $4\pi\epsilon_0=1$ and $\hbar=1$. The units of energy and length are then the hartree, $E_{\rm h}=e^2/4\pi\epsilon_0a_0$, and the Bohr radius, $a_0=4\pi\epsilon_0\hbar^2/m_{\rm e}e^2$.

We denote the total Hamiltonian operator in equation (2.1) by $\hat{\mathcal{H}}$, and the nuclear kinetic energy operator (the first term in Eq. (2.1)) by $\hat{T}_{\rm n}$. The Born–Oppenheimer electronic wavefunction is defined as the solution of

$$\left(\hat{\mathcal{H}}-\hat{T}_{\rm n}\right)\psi_{\rm e}=E_{\rm e}\psi_{\rm e}. \quad (2.3)$$

$\psi_{\rm e}$ is a function of the electronic coordinates $\mathbf{x}$, but only depends upon the nuclear positions $\mathbf{X}$ parametrically, because equation (2.3) is solved for some particular nuclear geometry. Hence we write $\psi_{\rm e}(\mathbf{x};\mathbf{X})$ and $E_{\rm e}(\mathbf{X})$ to show that different electronic wavefunctions and energies are obtained for different atomic structures.

The potential energy surface defines the variation of the electronic energy, $E_{\rm e}(\mathbf{X})$, with the nuclear geometry. In a diatomic molecule we simply consider $E_{\rm e}(R)$ as the bond-length, R, varies.

Assuming that $E_{\rm e}(\mathbf{X})$ defines an effective potential in which the nuclei move, the appropriate Schrödinger equation for the nuclear wavefunction, $\psi_{\rm n}(\mathbf{X})$, is

$$\left(\hat{T}_{\rm n} + E_{\rm e}(\mathbf{X})\right)\psi_{\rm n}(\mathbf{X}) = E_{\rm BO}\psi_{\rm n}(\mathbf{X}). \tag{2.4}$$

We now write $\psi(\mathbf{x},\mathbf{X}) = \psi_{\rm e}(\mathbf{x};\mathbf{X})\psi_{\rm n}(\mathbf{X})$, substitute this into equation (2.1) and neglect all the terms involving derivatives of $\psi_{\rm e}(\mathbf{x};\mathbf{X})$ with respect to nuclear coordinates, *i.e.* $\psi_{\rm n}\nabla_t^2\psi_{\rm e}$ and $\nabla_t\psi_{\rm n}\cdot\nabla_t\psi_{\rm e}$. A classical mechanical justification for this approximation is that the ratio of the mass of a proton to the mass of an electron is about 1836. Hence, nuclear velocities are expected to be very small in comparison with electronic velocities. Separating the variables gives (divide by $\psi_{\rm e}(\mathbf{x};\mathbf{X})\psi_{\rm n}(\mathbf{X})$)

$$-\sum_{i=1}^{n}\frac{\nabla_i^2\psi_{\rm e}(\mathbf{x};\mathbf{X})}{2\psi_{\rm e}(\mathbf{x};\mathbf{X})} + V(\mathbf{x},\mathbf{X}) = \sum_{t=1}^{N}\frac{\nabla_t^2\psi_{\rm n}(\mathbf{X})}{2M_t\psi_{\rm n}(\mathbf{X})} + E_{\rm BO} = E_{\rm e}(\mathbf{X}). \tag{2.5}$$

Hence we recover equations (2.3) and (2.4). The key point is that the latter two equations are much more tractable than equation (2.1).

Potential energy surfaces only exist within the Born–Oppenheimer approximation. If the approximation were exact then H–D would have no dipole moment, because the extra neutron in the frozen deuterium nucleus would not affect the electrons. This is why we usually assume that the (electronic) potential energy curve (or the effective force constant) of a diatomic molecule is unchanged by isotopic substitution. In fact H–D has a dipole moment of order 10^{-4} D (1 Debye = 3.336×10^{-30} Cm); for comparison, CH_3F has a dipole moment of 1.85 D. Hence the approximation is very good for H–D, and even better for heavier atoms.

2.1 *Normal modes*

Various approaches to the nuclear dynamics problem summarised in equation (2.4) are commonly employed. The simplest is normal mode analysis, which requires an orthogonal transformation of coordinates.

2.1.1 Orthogonal transformations

In general, we interpret the elements of an n-dimensional vector, $\mathbf{v} = (v_1, v_2, \ldots, v_n)$ as the components of a vector quantity in a particular basis of orthonormal (*i.e.* orthogonal and normalised) unit vectors, $\hat{\mathbf{e}}_\alpha$, defined by

$$\hat{\mathbf{e}}_\alpha = \frac{\partial \mathbf{r}}{\partial u_\alpha} \Big/ \left| \frac{\partial \mathbf{r}}{\partial u_\alpha} \right|, \tag{2.6}$$

where $\mathbf{r}$ is a position vector that depends upon the coordinate u_α. We must distinguish between the components, denoted by the row matrix $\mathbf{v}$, and the vector

$$\vec{\mathbf{v}} = \sum_{\alpha=1}^{n} v_\alpha \, \hat{\mathbf{e}}_\alpha. \tag{2.7}$$

If we change basis to a new set of orthonormal vectors, $\hat{\mathbf{e}}'_\alpha$, where

$$\hat{\mathbf{e}}'_\alpha = \sum_{\beta=1}^{n} C_{\beta\alpha} \, \hat{\mathbf{e}}_\beta, \tag{2.8}$$

$$\text{then} \quad \vec{\mathbf{v}}' = \vec{\mathbf{v}} = \sum_{\alpha=1}^{n} v'_\alpha \, \hat{\mathbf{e}}'_\alpha = \sum_{\alpha,\beta=1}^{n} v'_\alpha \, C_{\beta\alpha} \, \hat{\mathbf{e}}_\beta.$$

$$\text{So that} \quad v_\beta = \sum_{\alpha=1}^{n} v'_\alpha \, C_{\beta\alpha} \quad \text{or} \quad \mathbf{v} = \mathbf{v}' \, \mathbf{C}^T, \tag{2.9}$$

where superscript T denotes the transpose. The transformation matrix $\mathbf{C}$ therefore relates the components of a vector quantity in different bases.

If the components of $\mathbf{v}$ and $\mathbf{v}'$ correspond to different coordinates then the functions $f(\mathbf{v}) = f(\mathbf{v}'\mathbf{C}^T)$ must have the same value at any point. For example, the potential energy does not depend upon the coordinates, but the functional form may be different. If $V(x, y) = x^2 + 2y^2$ and $u = (x + y)/\sqrt{2}$, $v = (y - x)/\sqrt{2}$ then

$$\begin{aligned}
(u, v) &= (x, y)\mathbf{C}, \quad \text{where} \quad \mathbf{C} = \begin{pmatrix} 1/\sqrt{2} & -1/\sqrt{2} \\ 1/\sqrt{2} & 1/\sqrt{2} \end{pmatrix}, \\
V(u, v) &= 3(u^2 + v^2)/2 + uv, \\
\mathbf{g}(x, y) &= (\nabla_x V(x, y), \nabla_y V(x, y)) = (2x, 4y) \\
\text{and} \quad \vec{\mathbf{g}}(x, y) &= 2x\hat{\mathbf{e}}_x + 4y\hat{\mathbf{e}}_y, \\
\mathbf{g}(u, v) &= (\nabla_u V(u, v), \nabla_v V(u, v)) = (3u + v, u + 3v) \\
\text{and} \quad \vec{\mathbf{g}}(u, v) &= (3u + v)\hat{\mathbf{e}}_u + (u + 3v)\hat{\mathbf{e}}_v.
\end{aligned} \tag{2.10}$$

The components of $(x, y)\mathbf{C}$ are the values of the new coordinates (u, v) and the components of the gradient vectors are related by $\mathbf{g}(x, y)\,\mathbf{C} = \mathbf{g}(u, v)$, where $C_{\alpha\beta} = \hat{\mathbf{e}}_\alpha \cdot \hat{\mathbf{e}}_\beta$. The unit vectors in the (u, v) coordinate system are defined by

$$\hat{\mathbf{e}}_u = \frac{\partial \mathbf{r}}{\partial u} \Big/ \left| \frac{\partial \mathbf{r}}{\partial u} \right| = \left(\frac{\partial x}{\partial u} \hat{\mathbf{e}}_x + \frac{\partial y}{\partial u} \hat{\mathbf{e}}_y \right) \Big/ \left| \frac{\partial \mathbf{r}}{\partial u} \right| = \sqrt{\frac{1}{2}} (\hat{\mathbf{e}}_x + \hat{\mathbf{e}}_y)$$

$$\hat{\mathbf{e}}_v = \frac{\partial \mathbf{r}}{\partial v} / \left|\frac{\partial \mathbf{r}}{\partial v}\right| = \left(\frac{\partial x}{\partial v}\hat{\mathbf{e}}_x + \frac{\partial y}{\partial v}\hat{\mathbf{e}}_y\right) / \left|\frac{\partial \mathbf{r}}{\partial v}\right| = \sqrt{\frac{1}{2}}(\hat{\mathbf{e}}_y - \hat{\mathbf{e}}_x), \quad (2.11)$$

where $\mathbf{r} = x\hat{\mathbf{e}}_x + y\hat{\mathbf{e}}_y$. The gradient at any point is unaffected by the transformation:

$$\vec{\mathbf{g}}(u,v) = (3u+v)\hat{\mathbf{e}}_u + (u+3v)\hat{\mathbf{e}}_v = 2x\hat{\mathbf{e}}_x + 4y\hat{\mathbf{e}}_y = \vec{\mathbf{g}}(x,y), \quad (2.12)$$

but it is important to notice that the components of $\mathbf{g}(u,v)$ are different from those of $\mathbf{g}(x,y)$.

For transformations between orthonormal basis sets $\mathbf{C}$ is an orthogonal matrix and distances and scalar products are conserved. These results follow immediately from the conservation of the scalar product:

$$(\mathbf{v}_1')^T(\mathbf{v}_2') = (\mathbf{C}\,\mathbf{v}_1')^T(\mathbf{C}\,\mathbf{v}_2') = (\mathbf{v}_1)^T(\mathbf{v}_2). \quad (2.13)$$

2.1.2 The normal mode transformation

For motion near a local minimum we expand $E_{\rm e}(\mathbf{X})$ in a Taylor series:

$$E_{\rm e}(\mathbf{X}) \approx E_{\rm e}(\mathbf{X}^0) + \frac{1}{2}\sum_{\alpha,\beta=1}^{3N} \frac{\partial^2 E_{\rm e}(\mathbf{X}^0)}{\partial X_\alpha \partial X_\beta}(X_\alpha - X_\alpha^0)(X_\beta - X_\beta^0), \quad (2.14)$$

where the first derivatives vanish for equilibrium at $\mathbf{X}^0$.

A normal coordinate analysis then defines the fundamental vibrational modes of the molecule. Within the harmonic approximation the nuclear Hamiltonian in equation (2.4) can be written:

$$\hat{T}_{\rm n} + E_{\rm e}(\mathbf{X}) = -\sum_{t=1}^{N} \frac{\nabla_t^2}{2M_t} + \frac{1}{2}(\mathbf{X} - \mathbf{X}^0)^T \mathbf{H}(\mathbf{X}^0)(\mathbf{X} - \mathbf{X}^0), \quad (2.15)$$

where $H_{\alpha\beta}(\mathbf{X}^0) = \partial^2 E_{\rm e}(\mathbf{X}^0)/\partial X_\alpha \partial X_\beta$, and we have chosen the energy origin such that $E_{\rm e}(\mathbf{X}^0) = 0$.

Transforming to mass-weighted coordinates $\mathbf{q}_t = \mathbf{X}_t\sqrt{M_t}$, where $\mathbf{X}_t$ is the position vector of nucleus t, simplifies the kinetic energy operator to give

$$\hat{T}_{\rm n} + E_{\rm e}(\mathbf{q}) \approx -\sum_{\alpha=1}^{3N} \frac{1}{2}\frac{\partial^2}{\partial q_\alpha^2} + \frac{1}{2}(\mathbf{q} - \mathbf{q}^0)^T\, \overline{\mathbf{H}}(\mathbf{q}^0)\,(\mathbf{q} - \mathbf{q}^0), \quad (2.16)$$

where $\overline{H}_{\alpha\beta}(\mathbf{q}^0) = \partial^2 E_{\rm e}(\mathbf{q}^0)/\partial q_\alpha \partial q_\beta$, or $\overline{H}_{\alpha\beta} = H_{\alpha\beta}/\sqrt{M_\alpha M_\beta}$ where M_α is the nuclear mass corresponding to the component q_α. Both $\mathbf{H}$ and $\overline{\mathbf{H}}$ are commonly referred to as Hessian or second derivative matrices.

The symmetric $3N \times 3N$ matrix $\overline{\mathbf{H}}$ can be diagonalised using a matrix, $\mathbf{A}$, whose columns are the eigenvectors of $\overline{\mathbf{H}}$:

$$\sum_{\beta=1}^{3N} \overline{H}_{\alpha\beta}\, A_{\beta\gamma} = \omega_\gamma^2\, A_{\alpha\gamma}. \tag{2.17}$$

$A_{\beta\gamma}$ with β running from 1 to $3N$ corresponds to the γ'th eigenvector of $\overline{\mathbf{H}}$. Since $\overline{\mathbf{H}}$ is symmetric $\mathbf{A}$ is orthogonal and $\mathbf{A}\mathbf{A}^T = \mathbf{I}$ or $\sum_{\beta=1}^{3N} A_{\alpha\beta}A_{\beta\gamma}^T = \delta_{\alpha\gamma}$ and $\sum_{\beta=1}^{3N} A_{\beta\alpha}A_{\beta\gamma} = \delta_{\alpha\gamma}$.

We now define new coordinates, Q_γ, by the orthogonal transformation

$$q_\alpha = \sum_{\gamma=1}^{3N} A_{\alpha\gamma}\, Q_\gamma, \quad \text{so that} \quad Q_\beta = \sum_{\alpha=1}^{3N} A_{\alpha\beta}\, q_\alpha. \tag{2.18}$$

In terms of these new variables and moving the origin to the stationary point in question:

$$\begin{aligned}
\frac{1}{2}\sum_{\alpha,\beta=1}^{3N} \overline{H}_{\alpha\beta}\, q_\alpha\, q_\beta &= \frac{1}{2}\sum_{\alpha,\beta,\gamma,\delta=1}^{3N} \overline{H}_{\alpha\beta}\, A_{\alpha\gamma}\, A_{\beta\delta}\, Q_\gamma\, Q_\delta \\
&= \frac{1}{2}\sum_{\alpha,\gamma,\delta=1}^{3N} A_{\alpha\gamma}\, \omega_\delta^2\, A_{\alpha\delta}\, Q_\gamma\, Q_\delta = \frac{1}{2}\sum_{\delta=1}^{3N} \omega_\delta^2\, Q_\delta^2. \\
-\frac{1}{2}\sum_{\alpha=1}^{3N} \frac{\partial^2}{\partial q_\alpha^2} &= -\frac{1}{2}\sum_{\alpha,\beta,\gamma=1}^{3N} \frac{\partial^2}{\partial Q_\beta \partial Q_\gamma} A_{\alpha\beta}\, A_{\alpha\gamma} \\
&= -\frac{1}{2}\sum_{\beta=1}^{3N} \frac{\partial^2}{\partial Q_\beta^2}.
\end{aligned} \tag{2.19}$$

The overall transformation $Q_\beta = \sum_{\alpha=1}^{3N} A_{\alpha\beta}\, X_\alpha \sqrt{M_\alpha}$ therefore reduces the Hamiltonian to an uncoupled sum of $3N$ simple harmonic oscillator Hamiltonians, one for each normal coordinate:

$$\hat{T}_\mathrm{n} + E_\mathrm{e}(\mathbf{Q}) \approx \frac{1}{2}\sum_{\alpha=1}^{3N}\left[-\frac{\partial^2}{\partial Q_\alpha^2} + \omega_\alpha^2\, Q_\alpha^2\right]. \tag{2.20}$$

For such a separable Hamiltonian the wavefunction is a simple product and the energy is a simple sum. The corresponding classical Hamiltonian is

$$\frac{1}{2}\sum_{\alpha=1}^{3N}\left[\dot{Q}_\alpha^2 + \omega_\alpha^2\, Q_\alpha^2\right], \tag{2.21}$$

where $\dot{Q}_\alpha$ denotes the time derivative. The solutions have the form

$$Q_\gamma = c_\gamma \cos\left(\omega_\gamma t + \epsilon_\gamma\right), \tag{2.22}$$

and the general solution for the motion of any nuclear coordinate is

$$X_\alpha = \sum_{\gamma=1}^{3N} A_{\alpha\gamma}\, Q_\gamma / \sqrt{M_\alpha} = \sum_{\gamma=1}^{3N} A_{\alpha\gamma}\, c_\gamma \cos\left(\omega_\gamma t + \epsilon_\gamma\right) / \sqrt{M_\alpha}. \tag{2.23}$$

Motion corresponding to a single coordinate Q_γ is called a "normal mode". The displacements X_α oscillate in phase with frequency ω_γ and relative amplitudes $A_{\alpha\gamma}/\sqrt{M_\alpha}$. A general vibration is a superposition of all $3N$ modes. ω may be thought of as an angular frequency, related to a conventional frequency, ν, by $\omega = 2\pi\nu$. ω^2 also defines the curvature (second derivative) of the corresponding normal coordinate at any point. The normal mode approach is straightforward but limited by the validity of the second order Taylor expansion of the potential energy.

3 Describing the potential energy landscape

3.1 Introduction

For an N-atom system the potential energy is a $3N$-dimensional function. To refer to a potential energy hypersurface we must embed the function in a $3N+1$ dimensional space where the extra dimension corresponds to the "height" of the surface. The main problem with treating many-dimensional potential energy surfaces is that the number of stationary points grows rapidly with size. For a system composed of N_A atoms of type A, N_B atoms of type B, etc., the Hamiltonian is invariant to all permutations of equivalent nuclei and to inversion of all coordinates through a space-fixed origin. The number of permutation-inversion isomers of any given configuration could therefore be as large as

$$2 \times N_\mathrm{A}! \times N_\mathrm{B}! \times N_\mathrm{C}! \times \cdots, \tag{3.1}$$

but is reduced if the system possesses any point group symmetry elements other than the identity.

Following Pechukas [43] we associate any point group operation of a nuclear configuration $\mathbf{X}$ with a matrix $\mathbf{C}$ for which $\mathbf{C}\,\mathbf{X} = \mathbf{X}$. For proper point group operations (pure rotations and the identity) $\mathbf{C}$ can be written as a product of matrices associated with the mapping induced by the corresponding rotation, $\mathbf{R}$, and the permutation of nuclear labels required to reorder the coordinate vector, $\mathbf{P}$. We can show that any rotation or reflection commutes with any permutation by considering how these operations act on the atomic position vectors, $\mathbf{X}_i$, within the $3N$-dimensional vector $\mathbf{X}$. For example:

$$\mathbf{R}\begin{pmatrix}\mathbf{X}_1\\ \mathbf{X}_2\\ \vdots\\ \mathbf{X}_N\end{pmatrix}=\begin{pmatrix}\mathbf{r}\mathbf{X}_1\\ \mathbf{r}\mathbf{X}_2\\ \vdots\\ \mathbf{r}\mathbf{X}_N\end{pmatrix}\quad\text{and}\quad\mathbf{P}\begin{pmatrix}\mathbf{X}_1\\ \mathbf{X}_2\\ \vdots\\ \mathbf{X}_N\end{pmatrix}=\begin{pmatrix}\mathbf{X}_{P(1)}\\ \mathbf{X}_{P(2)}\\ \vdots\\ \mathbf{X}_{P(N)}\end{pmatrix},\qquad(3.2)$$

where $P(i)$ defines the mapping of atomic labels under the permutation operation, and $\mathbf{r}$ is the 3×3 transformation matrix corresponding to the proper operation. Hence

$$\mathbf{R}\,\mathbf{P}\,\mathbf{X}=\mathbf{P}\,\mathbf{R}\,\mathbf{X}=\begin{pmatrix}\mathbf{r}\mathbf{X}_{P(1)}\\ \mathbf{r}\mathbf{X}_{P(2)}\\ \vdots\\ \mathbf{r}\mathbf{X}_{P(N)}\end{pmatrix},\qquad(3.3)$$

and if $\mathbf{R}\,\mathbf{P}\,\mathbf{X}=\mathbf{X}$ then $\mathbf{r}\mathbf{X}_{P(\alpha)}=\mathbf{X}_\alpha$. For such configurations the permutation corresponding to $\mathbf{P}\,\mathbf{X}$ produces a labelled structure that can be superimposed on $\mathbf{X}$ by the pure rotation associated with the matrix $\mathbf{r}^{-1}$.

For improper point group operations we have $\mathbf{C}\,\mathbf{X}=\mathbf{P}\,\mathbf{S}\,\mathbf{X}=\mathbf{X}$ where the matrix corresponding to the improper symmetry element can be written as $\mathbf{S}=\mathbf{E}^*\mathbf{R}$ with $\mathbf{R}$ a pure rotation and $\mathbf{E}^*$ the inversion of the nuclear coordinates through the origin, *i.e.* $\mathbf{E}^*=-\mathbf{I}$. Note that $\mathbf{R}$ need not correspond to a point group operation. Hence $\mathbf{P}\,\mathbf{E}^*\,\mathbf{R}\,\mathbf{X}=\mathbf{X}$ and $\mathbf{P}\,\mathbf{E}^*$ produces a labelled structure that can be superimposed on $\mathbf{X}$ by the pure rotation associated with $\mathbf{r}^{-1}$.

For a (non-linear) nuclear configuration with point group of order h there are therefore h permutation and permutation-inversion operations that leave the configuration $\mathbf{X}$ unchanged, aside from possible reorientations. The number of distinct points or "permutational isomers" of $\mathbf{X}$ is therefore $n_{\mathbf{X}}=2N_{\mathrm{A}}!N_{\mathrm{B}}!N_{\mathrm{C}}!\cdots/h$. The ratio of the number of permutational isomers of any two configurations of the same molecule, A and B, is therefore $n_{\mathrm{A}}/n_{\mathrm{B}}=h_{\mathrm{B}}/h_{\mathrm{A}}$. The permutation and permutation-inversion operations defined above form a group, $\mathcal{G}_{\mathrm{RM}}$, known as the rigid molecule group [44]. The isomorphism between $\mathcal{G}_{\mathrm{RM}}$ and the usual rigid molecule point group, $\mathcal{G}_{\mathrm{pt}}$, is established by the explicit construction given above, and the elements of $\mathcal{G}_{\mathrm{RM}}$ may be obtained in general from the following prescription:

- for each proper operation, $\hat{\mathcal{R}}$, in $\mathcal{G}_{\mathrm{pt}}$ there is a permutation in $\mathcal{G}_{\mathrm{RM}}$ corresponding to the mapping of labelled atoms defined by the action of $\hat{\mathcal{R}}$ on the rigid molecule considered as a macroscopic object;

- for each improper operation, $\hat{\mathcal{S}}$, in $\mathcal{G}_{\text{pt}}$ there is a permutation-inversion in $\mathcal{G}_{\text{RM}}$ where the permutation corresponds to the mapping of labelled atoms defined this time by the action of $\hat{\mathcal{S}}$.

For example, all planar molecules have a σ_{h} reflection element that corresponds to the inversion, E^*, combined with the identity permutation, so $\mathcal{G}_{\text{RM}}$ contains E^*.

In the above analysis we have considered rotations and reflections that act on the nuclear framework as a rigid macroscopic object. This is the way that the point group is deduced. However, when the symmetries of vibrational and electronic states are assigned it is conventional to define point group operations that act upon vibrational displacements and electronic coordinates relative to a molecule fixed axis system [44, 45]. Bunker and Jensen distinguish the group composed of the latter operations from $\mathcal{G}_{\text{pt}}$ and call it the "molecular point group" [44]. In fact, it was Hougen who first established that the rotational energy levels of nonlinear molecules can be classified according to irreducible representations of the full molecular point group [46, 47].

There are no strict rules for the growth in the number of stationary points corresponding to different structures with system size. The evolution is system dependent, but empirical observations [48] and simple theories [49–51] suggest an exponential increase in the number of structurally distinct minima. Multiplying by the permutational factor therefore suggests that the number of minima or transition states for a homonuclear system is likely to increase as

$$2N!\exp\left(aN^b\right), \tag{3.4}$$

where $a > 0$ and b is of order unity.

In fact it is possible to derive relations between the numbers of stationary points of different Hessian indices, the Morse rules [52], and upper and lower bounds for the number of stationary points of a given index can sometimes be defined [53]. Topological analysis of potential energy surfaces has been discussed in some depth by Mezey [54–56].

4 Stationary points and pathways

The most interesting points of a potential energy surface are stationary (or critical) points where the gradient vanishes. Non-linear molecules in field-free space have six zero normal mode frequencies at a stationary point, corresponding to overall translations and rotations that do not affect the energy [57]; linear molecules have five. It is worth investigating the origin of these zero eigenvalues properly to appreciate the fundamental difference between the translational and rotational degrees of freedom [45].

4.1 Zero Hessian eigenvalues

The potential energy must be invariant to an overall displacement of any magnitude, Δ, in the x direction, which we denote by $\Delta\,\hat{\mathbf{e}}_x$ where

$$\hat{\mathbf{e}}_x = \sqrt{\frac{1}{N}}\begin{pmatrix}1\\0\\0\\1\\0\\\vdots\end{pmatrix}, \tag{4.1}$$

is the unit vector in the x direction. Taylor expansion gives

$$\begin{aligned} V(\mathbf{X}^0) &= V(\mathbf{X}^0 + \Delta\,\hat{\mathbf{e}}_x) \\ &= V(\mathbf{X}^0) + \Delta\,\mathbf{g}(\mathbf{X}^0)^T\,\hat{\mathbf{e}}_x + \frac{\Delta^2}{2}\hat{\mathbf{e}}_x^T\,\mathbf{H}(\mathbf{X}^0)\,\hat{\mathbf{e}}_x + \ldots, \end{aligned}$$

where $\mathbf{g}(\mathbf{X}^0) = \nabla V(\mathbf{X}^0)$, or $g_\alpha(\mathbf{X}^0) = \partial V(\mathbf{X}^0)/\partial X_\alpha$, and $H_{\alpha\beta}(\mathbf{X}^0) = \partial^2 V(\mathbf{X}^0)/\partial X_\alpha \partial X_\beta$.

The terms in each power of Δ must vanish for any $\mathbf{X}^0$, so

$$\mathbf{g}(\mathbf{X}^0)^T\,\hat{\mathbf{e}}_x = 0, \quad \text{and} \quad \hat{\mathbf{e}}_x^T\,\mathbf{H}(\mathbf{X}^0)\,\hat{\mathbf{e}}_x = \mathbf{0}. \tag{4.2}$$

The first condition gives a conservation law for the x component of the centre of mass linear momentum, and the second implies that $\hat{\mathbf{e}}_x$ is an eigenvector of $\mathbf{H}$ with eigenvalue zero. The corresponding eigenvector of $\overline{H}_{\alpha\beta} = H_{\alpha\beta}/\sqrt{M_\alpha M_\beta}$ is therefore $(\sqrt{M_1}, 0, 0, \sqrt{M_2}, 0\ldots)/\sqrt{M}$, where M is the total mass, and gives the components of the corresponding normal mode, Q_x^{trans}, in terms of the mass-weighted coordinates q_α. Hence

$$\begin{aligned} Q_x^{\mathrm{trans}} &= \left(\sqrt{M_1}\,q_{1x} + \sqrt{M_2}\,q_{2x} + \cdots + \sqrt{M_N}\,q_{Nx}\right)/\sqrt{M}, \\ &= \left(M_1 x_1 + M_2 x_2 + \cdots + M_N x_N\right)/\sqrt{M}, \end{aligned} \tag{4.3}$$

where x_t is the x coordinate of atom t. Q_x^{trans} is proportional to the x coordinate of the centre of mass; the factor of $1/\sqrt{M}$ appears due to normalisation. Analogous results apply for translation in the y and z directions.

Overall rotations, however, are more complicated. First we need the general formula for the rotation of the position vector of atom t, $\mathbf{X}_t^0$, through an angle α about an axis defined by the unit vector $\hat{\mathbf{n}}$:

$$\mathbf{X}_t = \mathbf{X}_t^0 \cos\alpha + \hat{\mathbf{n}}\left(\hat{\mathbf{n}}\cdot\mathbf{X}_t^0\right)(1-\cos\alpha) + \mathbf{X}_t^0 \wedge \hat{\mathbf{n}}\sin\alpha. \tag{4.4}$$

The displacement vector for small $|\alpha|$ and $\hat{\mathbf{n}} = (1, 0, 0)$ for a rotation about the x axis becomes

$$\begin{aligned}\mathbf{X}_t - \mathbf{X}_t^0 &= -\frac{\alpha^2}{2}\left(\mathbf{X}_t^0 - \hat{\mathbf{n}}\left(\hat{\mathbf{n}} \cdot \mathbf{X}_t^0\right)\right) + \alpha \mathbf{X}_t^0 \wedge \hat{\mathbf{n}} \\ &= \left(0, \alpha z_t - \frac{1}{2}\alpha^2 y_t^0, -\alpha y_t^0 - \frac{1}{2}\alpha^2 z_t^0\right), \end{aligned} \tag{4.5}$$

where z_t^0 is the z component of the position vector of atom t in the reference position, etc.

Denoting the $3N$-dimensional displacement corresponding to overall rotation about the x axis by

$$\mathbf{R}_x = \begin{pmatrix} 0 \\ \alpha z_1^0 - \frac{1}{2}\alpha^2 y_1^0 \\ -\alpha y_1^0 - \frac{1}{2}\alpha^2 z_1^0 \\ 0 \\ \alpha z_2^0 - \frac{1}{2}\alpha^2 y_2^0 \\ -\alpha y_2^0 - \frac{1}{2}\alpha^2 z_2^0 \\ \vdots \end{pmatrix}, \tag{4.6}$$

we now obtain

$$V(\mathbf{X}^0) = V(\mathbf{X}^0 + \mathbf{R}_x) = V(\mathbf{X}^0) + \mathbf{g}(\mathbf{X}^0)^T \mathbf{R}_x + \frac{1}{2}\mathbf{R}_x^T \mathbf{H}(\mathbf{X}^0)\mathbf{R}_x + \dots$$

Equating terms in α to zero gives a conservation law for the component of angular momentum about the x axis:

$$\begin{aligned}\mathbf{g}(\mathbf{X}^0)^T \begin{pmatrix} 0 \\ z_1^0 \\ -y_1^0 \\ 0 \\ z_2^0 \\ -y_2^0 \\ \vdots \end{pmatrix} &= 0, \\ \text{or} \quad \left[\nabla_1 V(\mathbf{X}^0) \wedge \mathbf{X}_1^0\right]_x + \left[\nabla_2 V(\mathbf{X}^0) \wedge \mathbf{X}_2^0\right]_x + \dots &= 0,\end{aligned} \tag{4.7}$$

where the left-hand-side is the x component of the torque on the system.

At a general point we cannot equate the terms in α^2 to zero and obtain an eigenvector of the second derivative matrix because there are also terms in the gradient. If the gradient does not vanish then overall rotations are coupled to the intramolecular vibrations and we obtain no further zero eigenvalues. However, at a stationary point

$$\mathbf{H}(\mathbf{X}^0)\begin{pmatrix} 0 \\ z_1^0 \\ -y_1^0 \\ 0 \\ z_2^0 \\ -y_2^0 \\ \vdots \end{pmatrix} = \mathbf{0}. \tag{4.8}$$

We therefore deduce that

$$\begin{aligned} Q_x^{\mathrm{rot}} &= \left(\sqrt{M_1}\left(z_1^0 q_{1y} - y_1^0 q_{1z}\right) + \sqrt{M_2}\left(z_2^0 q_{2y} - y_2^0 q_{2z}\right)\ldots\right)/\sqrt{I_{xx}}, \\ &= \left(M_1\left(z_1^0 y_1 - y_1^0 z_1\right) + M_2\left(z_2^0 y_2 - y_2^0 z_2\right)\ldots\right)/\sqrt{I_{xx}}, \end{aligned} \tag{4.9}$$

is a normal mode with zero eigenvalue, and analogous arguments apply to rotations about the y and z axes. The diagonal component of the inertia tensor, I_{xx}, appears due to normalisation. The above translational and rotational normal mode vectors are orthonormal if the molecule fixed coordinate axes coincide with the principal axes and the centre of mass lies at the origin.

The accuracy of the "zero" eigenvalues can be a useful measure of how well a geometry optimisation has converged.

4.2 Classification of stationary points

The characteristics of any stationary point are determined by the corresponding Hessian. The number of negative Hessian eigenvalues is termed the Hessian index, and this is usually the same as the order defined as the number of mutually orthogonal directions in which the energy is maximal [58]. Minima are stationary points with no negative Hessian eigenvalues. Note that a negative Hessian eigenvalue also corresponds to a negative force constant and an imaginary normal mode frequency. A displacement along such a coordinate lowers the energy.

The Murrell Laidler theorem [59] defines a transition state as a stationary point with a single imaginary normal mode frequency. Consider an index two saddle with two negative force constants, k_1 and k_2, for normal modes Q_1 and Q_2. With all the other coordinates fixed the change in potential energy for small displacements away from the saddle is:

$$\Delta V = \frac{1}{2}(k_1\, Q_1^2 + k_2\, Q_2^2) < 0. \tag{4.10}$$

Hence any displacement in this space lowers the energy. The saddle is therefore a "hill" in these two dimensions and we can always find a lower energy path by walking around it (Fig. 6).

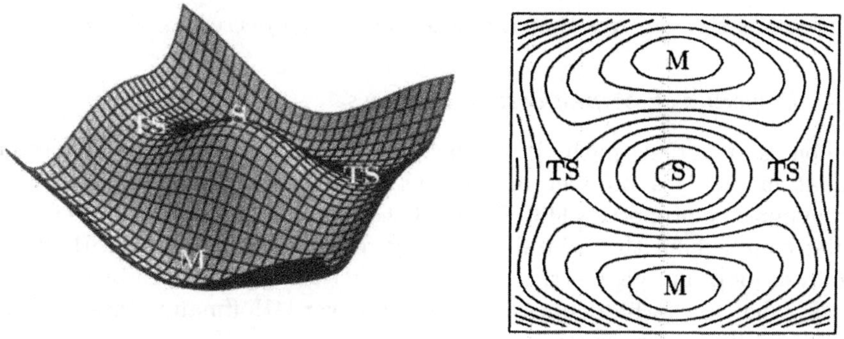

Fig. 6. Surface and contour plots illustrating the Murrell–Laidler theorem. M, TS and S denote minima, transition states and the index 2 saddle, respectively.

Hence if there is a path connecting two minima *via* a saddle point of index greater than one then it is (almost) always possible to find a lower energy path via one or more true transition states. Rate theory is therefore usually concerned with passage over true transition states.

4.3 Pathways

We can define a steepest-descent path or gradient line, $\mathbf{X}(s)$, in terms of a parameter s such that $|\mathrm{d}\mathbf{X}(s)/\mathrm{d}s| = 1$. $\mathbf{X}(s)$ as a function of s defines a continuous path through nuclear configuration space. Furthermore,

$$\mathbf{X}(s_0 + \mathrm{d}s) = \mathbf{X}(s_0) + \frac{\mathrm{d}\mathbf{X}(s_0)}{\mathrm{d}s}\mathrm{d}s + \dots \qquad (4.11)$$

Hence, for $\mathrm{d}s \to 0$ we have $\mathbf{X}(s_0 + \mathrm{d}s) - \mathbf{X}(s_0) = (\mathrm{d}\mathbf{X}(s_0)/\mathrm{d}s)\mathrm{d}s$ and the corresponding path length is $|\mathrm{d}\mathbf{X}(s_0)/\mathrm{d}s|\mathrm{d}s$. Since we have defined s such that $|\mathrm{d}\mathbf{X}(s_0)/\mathrm{d}s| = 1$ the integrated path length between $\mathbf{X}(s_1)$ and $\mathbf{X}(s_2)$ is

$$\int_{s_1}^{s_2} \left| \frac{\mathrm{d}\mathbf{X}(s)}{\mathrm{d}s} \right| \mathrm{d}s = s_2 - s_1. \qquad (4.12)$$

The parameter s therefore corresponds to the path length.

For a steepest-descent path the tangent to the curve must be antiparallel to the gradient vector, and the constraint $|\mathrm{d}\mathbf{X}(s_0)/\mathrm{d}s| = 1$ defines the constant of proportionality:

$$\frac{\mathrm{d}\mathbf{X}}{\mathrm{d}s} = -\frac{\mathbf{g}(\mathbf{X})}{|\mathbf{g}(\mathbf{X})|}. \tag{4.13}$$

The minus sign ensures that the potential energy decreases with s.

4.4 *Properties of steepest-descent pathways*

When a reaction path is defined in terms of the steepest-descent equation (4.13) a number of rigorous conditions can be deduced about its symmetry properties [43, 60–63]. These geometrical symmetry selection rules complement the orbital symmetry results of a Woodward–Hoffmann analysis [64]. The geometrical rules must be rigorously obeyed by a reaction path, whereas "forbiddenness" in the Woodward–Hoffmann sense implies a large but finite barrier, and can be overcome.

4.4.1 Uniqueness

If we assume that all the first and second derivatives of $V(\mathbf{X})$ exist and are continuous then every point in the nuclear configuration space, $\mathbf{X}$, lies on one and only one solution of (4.13), unless the gradient also vanishes. This result follows immediately from the uniqueness theorem of first order ordinary differential equations. Suppose we have two different paths, $\mathbf{X}(s)$ and $\mathbf{X}'(s)$, which coincide at some point s_0. Then at this point we have

$$\frac{\mathrm{d}\mathbf{X}(s_0)}{\mathrm{d}s} = -\frac{\mathbf{g}(\mathbf{X}(s_0))}{|\mathbf{g}(\mathbf{X}(s_0))|} = \frac{\mathrm{d}\mathbf{X}'(s_0)}{\mathrm{d}s}. \tag{4.14}$$

Hence the curves are coincident:

$$\begin{aligned}\mathbf{X}(s_0 + \mathrm{d}s) &= \mathbf{X}(s_0) + \frac{\mathrm{d}\mathbf{X}(s_0)}{\mathrm{d}s}\mathrm{d}s = \mathbf{X}'(s_0) + \frac{\mathrm{d}\mathbf{X}'(s_0)}{\mathrm{d}s}\mathrm{d}s \\ &= \mathbf{X}'(s_0 + \mathrm{d}s). \end{aligned} \tag{4.15}$$

A steepest-descent path is therefore uniquely specified by a single point that lies on it. This result is not true at a stationary point where $|\mathbf{g}(\mathbf{X})| = 0$ and equation (4.13) becomes undefined. In general there are infinitely many solutions to (4.13) that terminate at any given minimum.

4.4.2 Steepest-descent paths from a transition state

We will now prove that for a true transition state there are only two solutions to (4.13) with decreasing potential energy starting at $\mathbf{X}_{\mathrm{TS}}$. These paths set off in opposite directions along the eigenvector of $\mathbf{H}$ corresponding to negative curvature.

Consider an orthogonal transformation using the matrix $\mathbf{B}$ that satisfies

$$\sum_{\beta=1}^{3N} H_{\alpha\beta}B_{\beta\gamma} = \varepsilon_\gamma^2 B_{\alpha\gamma}. \tag{4.16}$$

$B_{\beta\gamma}$ with β running from 1 to $3N$ corresponds to the γ'th eigenvector of $\mathbf{H}$, just as $A_{\beta\gamma}$ corresponds to the γ'th eigenvector of $\overline{\mathbf{H}}$. Now define new coordinates $W_\alpha = \sum_{\beta=1}^{3N} B_{\beta\alpha}X_\beta$. Since $\mathbf{B}$ is an orthogonal matrix $X_\alpha = \sum_{\beta=1}^{3N} B_{\alpha\beta}W_\beta$. The potential energy in the Taylor expansion about a stationary point at the origin becomes

$$\begin{aligned}\frac{1}{2}\sum_{\alpha,\beta=1}^{3N} H_{\alpha\beta}\, X_\alpha\, X_\beta &= \frac{1}{2}\sum_{\alpha,\beta,\gamma,\delta=1}^{3N} H_{\alpha\beta}\, B_{\alpha\gamma}\, B_{\beta\delta}\, W_\gamma\, W_\delta \\ &= \frac{1}{2}\sum_{\alpha,\gamma,\delta=1}^{3N} B_{\alpha\gamma}\, \varepsilon_\delta^2\, B_{\alpha\delta}\, W_\gamma\, W_\delta = \frac{1}{2}\sum_{\delta=1}^{3N} \varepsilon_\delta^2\, W_\delta^2.\end{aligned} \tag{4.17}$$

$$\text{Since} \quad \frac{\partial}{\partial X_\alpha} = \sum_{\beta=1}^{3N} B_{\alpha\beta}\frac{\partial}{\partial W_\beta}, \tag{4.18}$$

the steepest-descent equation (4.13) at the stationary point in these new coordinates is equivalent to the $3N$ equations:

$$\frac{\mathrm{d}W_\alpha}{\mathrm{d}s} = -\frac{\varepsilon_\alpha^2 W_\alpha}{|\mathbf{g}|}. \tag{4.19}$$

In fact, the same paths are generated by

$$\frac{\mathrm{d}W_\alpha}{\mathrm{d}u} = -\varepsilon_\alpha^2 W_\alpha, \tag{4.20}$$

where $\mathrm{d}u = \mathrm{d}s/|\mathbf{g}|$, because if $\mathbf{W}(s_0) = \mathbf{W}(u_0)$ then

$$\begin{aligned}W_\alpha(s_0 + \mathrm{d}s) &= W_\alpha(s_0) + \frac{\mathrm{d}W_\alpha(s_0)}{\mathrm{d}s}\mathrm{d}s + \ldots \\ &= W_\alpha(s_0) - \frac{\epsilon_\alpha^2 W_\alpha(s_0)\mathrm{d}s}{|\mathbf{g}|} + \ldots \\ &= W_\alpha(u_0) + \frac{\mathrm{d}W_\alpha(u_0)}{\mathrm{d}u}\frac{\mathrm{d}s}{|\mathbf{g}|} + \ldots \\ &= W_\alpha(u_0 + \mathrm{d}u) \quad \text{for} \quad |\mathrm{d}s|, |\mathrm{d}u| \ll 1.\end{aligned} \tag{4.21}$$

Hence the solutions close to the stationary point (at $\mathbf{W} = \mathbf{0}$) can be written in parametric form as [43]

$$W_\alpha(u) = W_\alpha(0)\exp(-\varepsilon_\alpha^2 u). \tag{4.22}$$

Now suppose that the stationary point corresponding to $\mathbf{W} = \mathbf{0}$ is a transition state, with ε_1^2 the unique negative Hessian eigenvalue (or curvature). There are only two independent paths that approach the transition state as $u \to -\infty$: they have $W_\alpha = 0$ for $\alpha \neq 1$ and $W_1(0)$ either positive or negative. An infinite number of steepest-descent paths approach the transition state as $u \to \infty$, all with $W_1 = 0$. The energy rises if we step off the transition state in any of these directions. There are therefore only two distinct downhill steepest-descent paths leading from the transition state, and they begin parallel and antiparallel to the eigenvector corresponding to the unique negative Hessian eigenvalue (Fig. 7).

Fig. 7. Steepest descent paths (thick) and contours (thin) for the potential energy surface $V = -x^2 + 2y^2$.

In fact, further restrictions are needed to guarantee that there are only two steepest-descent paths leading down from a transition state. Zero Hessian eigenvalues other than those corresponding to overall translation and rotation can lead to all sorts of interesting surfaces. For example, a "monkey saddle" can occur for a cubic surface, as shown in Figure 8. The two-dimensional surface defined by $V = -\alpha x^2 - \beta y^3$ with $\alpha,\ \beta > 0$ has Hessian eigenvalues -2α and 0 at $x = y = 0$, but the energy is not invariant to displacements along the eigenvector corresponding to the zero eigenvalue [43]. The steepest descent paths are

$$x(u) = x(0)\mathrm{e}^{2\alpha u}, \qquad y(u) = 1/\left(1/y(0) - 3\beta u\right), \tag{4.23}$$

and an infinite number of steepest-descent paths approach the origin as $u \to -\infty$ in this case [43]. Real systems generally do not exhibit such exotic features: the appearance of an additional zero Hessian eigenvalue at a point where the gradient also vanishes exactly would be an improbable numerical accident.

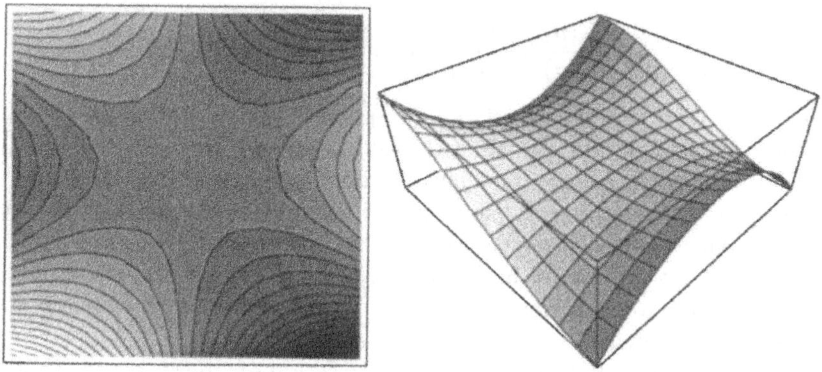

Fig. 8. Contour and surface plot for the monkey saddle exhibited by the function $x^3 - 3x^2y$.

There is also nothing to prevent one or both the downhill steepest-descent paths starting at a transition state from terminating at a stationary point that is not a minimum. Such features are probably quite common [65–70]; the addition of HF to ethene (Fig. 9) is an example [71]. A steepest-descent path links transition state 1 to the eclipsed form of fluoroethane, which is the transition state for internal rotation.

4.4.3 Principal directions

For any stationary point there are $2(3N-6)$ principal steepest-descent paths corresponding to positive and negative displacements along the $3N-6$

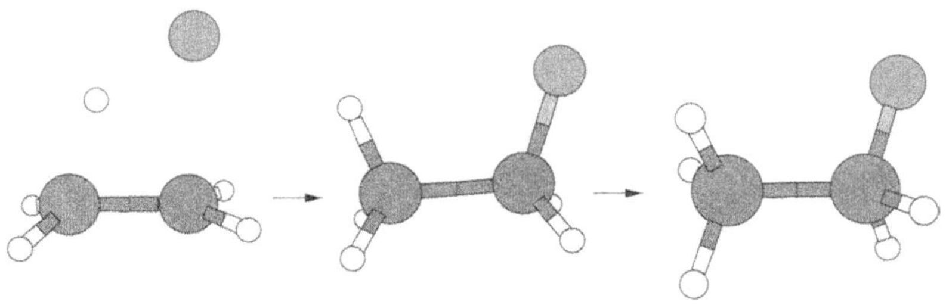

Fig. 9. HF addition to C_2H_4.

Hessian eigenvectors with non-zero eigenvalues. There is an infinite set of non-principal steepest-descent paths involving displacements along more than one principal direction. From equation (4.22) we see that

$$\frac{W_\alpha(u)}{W_\beta(u)} = \frac{W_\alpha(0)}{W_\beta(0)} \exp\left(-u(\varepsilon_\alpha^2 - \varepsilon_\beta^2)\right). \tag{4.24}$$

as $u \to \infty$ for a minimum all the components of $\mathbf{W}$ tend to zero except for the one corresponding to the smallest curvature. Steepest-descent paths therefore approach a minimum parallel to the principal direction with the smallest curvature for which they have a non-zero projection [43,72], so long as this eigenvalue is non-degenerate.

We have already noted in Section 4.4.2 that only two steepest-descent paths approach a transition state as $u \to -\infty$, while paths with $W_1 = 0$ approach the transition state when $u \to \infty$. The latter paths enter the transition state parallel to the principal direction of smallest curvature (ε^2) with which they have a non-zero overlap.

4.4.4 Birth and death of symmetry elements

Here we consider the symmetry group of a nuclear configuration $\mathbf{X}^0$, *i.e.* the point group of rotations and reflections, $\hat{R}$. The action of $\hat{R}$ mapping a general $3N$-dimensional vector $\mathbf{X}$ onto $\mathbf{Y}$ may be defined by

$$Y_\alpha = C_{\alpha\beta} X_\beta \quad \text{and} \quad X_\alpha = C_{\beta\alpha} Y_\beta. \tag{4.25}$$

For symmetry operations of $\mathbf{X}^0$

$$\mathbf{C}\,\mathbf{X}^0 = \mathbf{X}^0, \tag{4.26}$$

so the point group operations of configuration $\mathbf{X}^0$ correspond to orthogonal matrices for which $\mathbf{X}^0$ is an eigenvector with unit eigenvalue. Note that $\mathbf{C}$

includes not only the effect of the point group operation considered as a mapping of nuclear coordinates[1], but also a possible permutation of nuclear labels (see also Sect. 3.1) [43]. Using the chain rule gives

$$\frac{\partial V}{\partial Y_\alpha} = \sum_{\beta=1}^{3N} C_{\alpha\beta} \frac{\partial V}{\partial X_\beta}, \quad \text{or} \quad \mathbf{g}(\mathbf{Y}) = \mathbf{C}\,\mathbf{g}(\mathbf{X}). \tag{4.27}$$

$$\frac{\partial^2 V}{\partial Y_\alpha \partial Y_\beta} = \sum_{\gamma,\delta=1}^{3N} C_{\alpha\gamma} \frac{\partial^2 V}{\partial X_\gamma \partial X_\delta} C_{\beta\delta}, \quad \text{or} \quad \mathbf{H}(\mathbf{Y}) = \mathbf{C}\,\mathbf{H}(\mathbf{X})\,\mathbf{C}^T.$$

If $\mathbf{C}$ corresponds to a symmetry operation at $\mathbf{X}$ then $\mathbf{X} = \mathbf{Y}$ and $\mathbf{C}\,\mathbf{g}(\mathbf{X}) = \mathbf{g}(\mathbf{X})$. Hence the gradient vector transforms as the totally symmetric irreducible representation (IR) of the prevailing point group. For a small displacement $\Delta\,\mathbf{g}(\mathbf{X})$ from $\mathbf{X}$

$$\begin{aligned} \mathbf{C}(\mathbf{X} + \Delta\,\mathbf{g}(\mathbf{X})) &= \mathbf{C}\,\mathbf{X} + \Delta\,\mathbf{C}\,\mathbf{g}(\mathbf{X}) \\ &= \mathbf{Y} + \Delta\,\mathbf{g}(\mathbf{Y}). \end{aligned} \tag{4.28}$$

Hence, if points $\mathbf{X}$ and $\mathbf{Y}$ are related by the operation $\mathbf{C}$ then so are the steepest-descent paths passing through $\mathbf{X}$ and $\mathbf{Y}$. Furthermore, if $\mathbf{C}$ is a symmetry operation of the point $\mathbf{X}$ then the corresponding symmetry element is conserved along the steepest-descent pathway passing through $\mathbf{X}$ because $\mathbf{C}(\mathbf{X} + \Delta\,\mathbf{g}(\mathbf{X})) = \mathbf{X} + \Delta\,\mathbf{g}(\mathbf{X})$. However, additional symmetry elements may appear when a steepest-descent path reaches a stationary point [43].

The above proof shows that symmetry operations are conserved for infinitesimal displacements along the gradient. Hence a stationary point must have every symmetry operation of the steepest-descent paths that converge to it. However, we cannot apply the theorem the other way round because the gradient vanishes at any stationary point. Hence, stationary points may possess additional symmetry elements that are not conserved along connecting steepest-descent paths.

If we apply all the symmetry operations of a point $\mathbf{X}$, with matrix representations $\mathbf{R}_i$, to another point $\mathbf{Y}_1$ then the number of symmetry equivalent points generated, m, must be an integer divisor of the order, h_X, of the point group of $\mathbf{X}$, $\mathcal{G}_\mathrm{X}$. Suppose the point $\mathbf{Y}_1$ is invariant to s operations of $\mathcal{G}_\mathrm{X}$ comprising a subgroup, $\mathcal{G}_{\mathrm{Y}_1}$, with dimension s. Since we know there is at least one operation $\mathbf{R}_i \in \mathcal{G}_\mathrm{X}$ such that $\mathbf{R}_i\mathbf{Y}_1 = \mathbf{Y}_i$, it follows that all s operations in the coset $\{\mathbf{R}_i, \mathcal{G}_{\mathrm{Y}_1}\}$ map $\mathbf{Y}_1$ to $\mathbf{Y}_i$. There cannot be more

[1] In the conventional view, point group operations are considered to act upon vibronic coordinates.

than s operations because there is one operation that leaves $\mathbf{Y}_1$ invariant for each operation that maps $\mathbf{Y}_1$ to $\mathbf{Y}_i$. Hence there are precisely s operations mapping $\mathbf{Y}_1$ to each of the $\{\mathbf{Y}_1, \mathbf{Y}_2, \ldots, \mathbf{Y}_m\}$, so that $sm = h_X$, and m must be an integer divisor of h_X.

We proved in Section 4.4.1 that a transition state possesses precisely two downhill steepest-descent paths that correspond to positive or negative displacements along the unique Hessian eigenvector associated with negative curvature. Since there are only two such paths we deduce that a point group operation of the transition state must either leave both paths unchanged or interconvert them. Symmetry operations of the transition state are therefore either conserved along both downhill steepest-descent paths, or exchange them. *New symmetry operations may therefore appear at a transition state if they interconvert the two downhill steepest-descent paths.* Let $\mathbf{P}_1$ and $\mathbf{P}_2$ be two points on the two steepest descent paths leading from a transition state, which are interconverted by a symmetry operation of the transition state whose matrix representation is $\mathbf{R}$. Such operations must have even order, because any odd power of $\mathbf{R}$ applied to $\mathbf{P}_1$ must give $\mathbf{P}_2$, and hence cannot be the identity. Furthermore, $\mathbf{R}^{2n}\mathbf{P}_1 = \mathbf{P}_1$ for positive integer n, and hence all even powers of $\mathbf{R}$ correspond to symmetry elements that are conserved along both sides of the path. Similarly, all odd powers of $\mathbf{R}$ must interconvert the two sides. A new symmetry operation of the transition state must therefore also interconvert the stationary points ("product" and "reactant") at the two termini of the downhill steepest-descent paths. This situation can only arise if the pathway connects different permutational isomers of the same structure (a "degenerate" rearrangement [73]).

In the degenerate rearrangement of an icosahedral Ar_{55} cluster *via* a cuboctahedron six five-fold axes are created at the two minima and three four-fold axes at the transition state. The conserved symmetry elements are those of the point group T_h (order 24), whilst the minima belong to point group I_h (order 120) and the transition state is O_h (order 48).

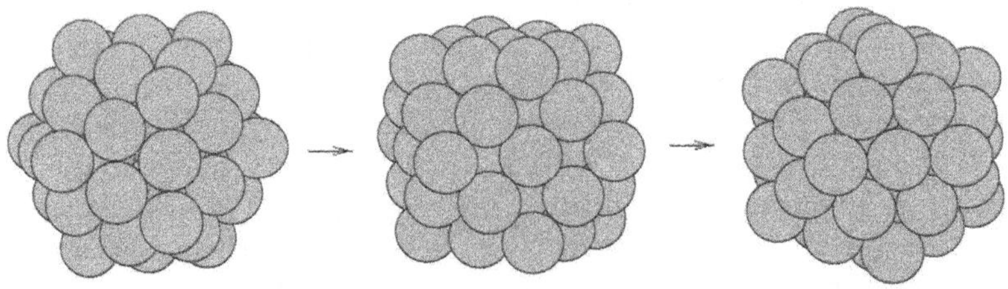

Fig. 10. Rearrangement of an Ar_{55} cluster.

Clearly if a reaction pathway links different structures (*i.e.* not permutational isomers) then no new symmetry elements can appear at the transition state. If the transition state does possess additional symmetry elements then it must have exactly twice as many symmetry operations as are conserved along the path, as illustrated above for Ar_{55}. If g symmetry elements are conserved along the steepest descent paths then the point group of the transition state also contains precisely g such elements, which form a subgroup of $\mathcal{G}_{ts}$, denoted $\mathcal{G}_{ts}^{+}$. If $\mathcal{G}_{ts}$ also contains an operation that interconverts the two paths, $\mathbf{R}$, then the product of $\mathbf{R}$ with any element conserved along the path must also interconvert the two sides. These operations must all be distinct and form a coset of the transition state point group [74] denoted $\mathbf{R}\mathcal{G}_{ts}^{+}$. Every element of $\mathcal{G}_{ts}$ belongs to $\mathcal{G}_{ts}^{+}$ or $\mathbf{R}\mathcal{G}_{ts}^{+}$, and the two sets are interconverted by any member of $\mathbf{R}\mathcal{G}_{ts}^{+}$. Hence the order of $\mathcal{G}_{ts}^{+}$ is $2g$.

4.5 *Classification of rearrangements*

We saw in Section 4.4.4 that the transition state for a degenerate rearrangement, which links permutational isomers of the same structure [73], can possess a symmetry element that maps one side of the path onto the other. This symmetry element is only present at the transition state.

Degenerate rearrangements are greatly outnumbered by non-degenerate pathways, but constitute an interesting minority. Transition state 2 in Figure 9 for the internal rotation of fluoroethane has the same point group as the minima it connects. However, the mirror plane in the transition state is not conserved along the path, and interconverts the corresponding minima. Degenerate rearrangements where the two sides of the path are related by a symmetry operation of the transition state are known as symmetric degenerate rearrangements [75]. For HF dimer this symmetry is evident in the plot of energy versus path length shown in Figure 11.

Asymmetric degenerate rearrangements are also possible, where the two minima have the same structure but the steepest-descent paths are not related by symmetry [75]. The energy profile is also asymmetric, and an example is shown for water trimer in Figure 12 [76].

The results of Section 4.4.4 may now be used to eliminate from consideration mechanisms whose pathways violate the symmetry results. For a non-degenerate rearrangement the stationary points at the termini of the steepest-descent paths descending from a transition state (the "products" (P) and "reactants" (R)) must possess all the symmetry elements of the transition state. Hence, (a) if the transition state is linear, the products and reactants are linear, (b) if the transition state is planar, the products and reactants are planar, (c) if the transition state is optically inactive, the products and reactants are optically inactive. Result (b) follows because a necessary and sufficient condition for planarity is that $E^{*} \in \mathcal{G}_{RM}$

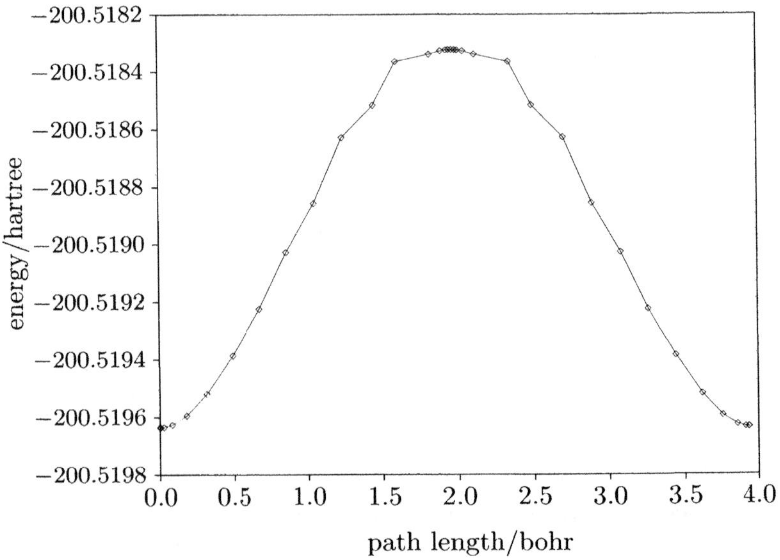

Fig. 11. DZP/MP2 energy profile for the symmetric degenerate rearrangement of HF dimer.

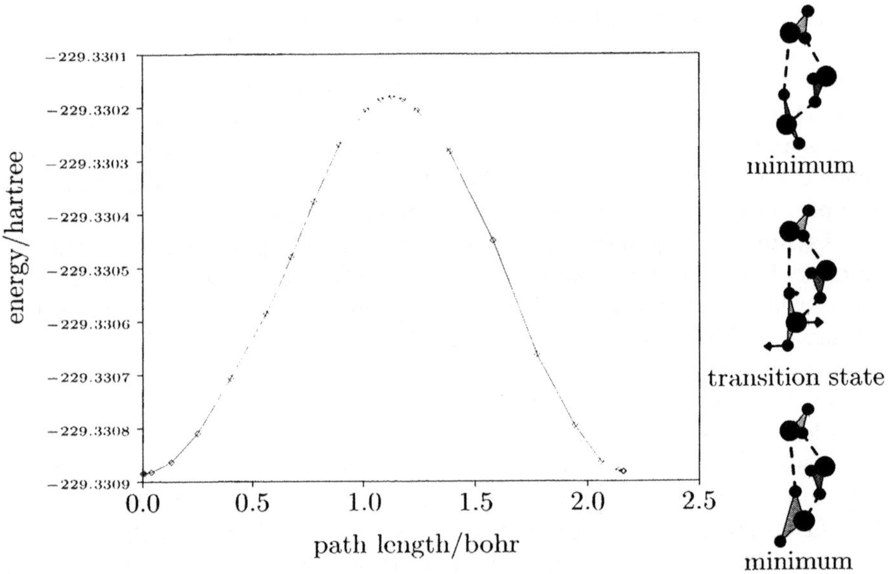

Fig. 12. DZP/MP2 energy profile for the asymmetric degenerate rearrangement of $(H_2O)_3$.

(see Sect. 3.1). Since the products and reactants may possess additional symmetry elements, an optically active transition state may link optically inactive products. However, the geometrical symmetries of the transition state for a non-degenerate rearrangement must also be present in the products and reactants, and must therefore form a subgroup in each of the point groups $\mathcal{G}_{\rm P}$ and $\mathcal{G}_{\rm R}$. The size of the largest common subgroup of $\mathcal{G}_{\rm P}$ and $\mathcal{G}_{\rm R}$ therefore sets an upper bound on the size of the transition state point group [43].

Pechukas has derived some further symmetry constraints that apply when one or both of the products and reactants is bimolecular [43]. For an association or dissociation reaction where the bimolecular species are different the transition state can possess at most a single rotation axis and a mirror plane that both contain the two centres of mass. These are the only symmetry elements that can exchange nuclei within two different separated molecules [43]. Additional transition state symmetries are possible for symmetric association or dissociation [43].

There have also been efforts to use symmetry in a predictive sense in analysing rearrangement mechanisms. A critique of earlier attempts is given by Bone *et al.*, along with a formulation in terms of the molecular symmetry group [77]. Given two permutational isomers of the same structure group theory can be used to determine the symmetry operations which could be generated at an intervening transition state [77–79]. These results have not yet been extensively exploited, perhaps partly because of the factorial growth in permutational isomers with system size, which makes calculations for all possible pairs of minima rather laborious.

4.6 The McIver–Stanton rules

McIver and Stanton proved a series of rules governing the properties of the "transition vector", which are either equivalent to or follow from the results of the previous section [62,63]. Here the transition vector corresponds to the Hessian eigenvector corresponding to the unique negative Hessian eigenvalue.

- Theorem I. The transition vector cannot transform according to a degenerate irreducible representation of the transition state point group;
- Theorem II. The transition vector must be antisymmetric under a transition state symmetry operation that converts reactants into products;
- Theorem III. The transition vector must be symmetric with respect to a symmetry operation that leaves either reactants or products unchanged;

- Theorem IV. If the transition vector for the reaction $R_1 \rightarrow P_1$ is symmetric under a symmetry operation $\hat{O}$ that converts reactants R_1 into the equivalent reactants R_2 and P_1 into P_2 then there exist lower energy transition states for the pathways $R_1 \rightarrow R_2$ and $P_1 \rightarrow P_2$ [case (a)]. If the transition vector is antisymmetric under $\hat{O}$ then there exists a lower energy transition state for the pathway $R_1 \rightarrow P_2$ [case(b)].

Theorem I is equivalent to the Murrell–Laidler theorem [59]: if the transition vector belongs to a degenerate irreducible representation of the transition state point group then there would have to be more than one negative Hessian eigenvalue. Theorem II is the case where a symmetry operation of the transition state maps one side of the path onto the other, while Theorem III applies to any symmetry operation that is conserved along the whole path.

Theorem IV will apply if a transition state links either (a) two transition states A and B for which $\hat{O}$ A=A and $\hat{O}$ B=B, or (b) the transition state mediates a symmetric degenerate rearrangement of permutational isomers A_1 and A_2 where $\hat{O}\,A_1=A_2$. In case (a) $\hat{O}\,R_1=R_2$ and $\hat{O}\,P_1=P_2$, so transition states A and B mediate symmetric degenerate rearrangements of the products and reactants. Theorem IVa is realized for the FNH_2F^- molecule in Figure 14. In case (b) the transition states A_1 and A_2 mediate the non-degenerate rearrangements of R_1 to P_1 and R_2 to P_2, respectively.

A number of corollaries are possible. For example, no structure can be a transition state if it possesses a C_3 rotation or other odd order operation that interconverts reactants and products [62, 63].

4.7 Coordinate transformations

The steepest-descent paths defined above by equation (4.13) have a rigorous definition. However, the literature contains references to the "minimum energy path" (MEP), "intrinsic reaction path" (IRP) and "intrinsic reaction coordinate" (IRC). Often authors simply refer to the "reaction coordinate" or the "reaction path" and MEP, IRP and SDP are used interchangeably. It is also common for these acronyms to refer to paths calculated not by solution of equation (4.13) but instead from mass-weighted coordinates defined by $q_\alpha = X_\alpha\sqrt{M_\alpha}$:

$$\frac{\mathrm{d}\mathbf{q}(s)}{\mathrm{d}s} = -\frac{\mathbf{g}(\mathbf{q}(s))}{|\mathbf{g}(\mathbf{q}(s))|}, \tag{4.29}$$

where M_α is the mass of the atom corresponding to nuclear coordinate α. The solutions are loosely referred to as steepest-descent paths in "mass-weighted coordinates".

In fact properties of a potential energy surface such as steepest-descent paths, principal directions and normal mode frequencies should be coordinate independent [80, 81]. If this were not the case then we would obtain different dynamic and thermodynamic properties from different sets of coordinates. However, if the metric tensor of the coordinate system does not correspond to the identity then extra care is needed to ensure that pathways and frequencies remain coordinate independent. For example, when second derivative matrices (Hessians) are constructed using ordinary partial derivatives of the energy the eigenvalues can exhibit coordinate dependence. This result does not mean that one particular coordinate system is correct and that others are not. Rather, it means that one should not use ordinary partial derivatives to construct the Hessian when the metric does not correspond to the identity matrix [82].

Relations between physical properties that are independent of the reference frame are obtained by taking proper account of metric tensors and using *covariant* derivatives [80, 81]. In general curvilinear coordinates, Z_α, the arc length, $\mathrm{d}s$, is obtained from

$$\mathrm{d}s^2 = \sum_{\alpha,\beta=1} G_{\alpha\beta}\,\mathrm{d}Z_\alpha\,\mathrm{d}Z_\beta = \sum_\alpha \mathrm{d}X_\alpha^2, \tag{4.30}$$

and the length of a vector, L, with components, v_α, is defined as

$$L^2 = \sum_{\alpha,\beta=1} G^{\alpha\beta}\,v_\alpha\,v_\beta. \tag{4.31}$$

$G_{\alpha\beta}$ and $G^{\alpha\beta}$ are the covariant and contravariant forms of the metric tensor for which[2] $\sum_{\beta=1}^{3N} G_{\alpha\beta}G^{\beta\gamma} = \delta_{\alpha\gamma}$. If $G_{\alpha\beta} = \delta_{\alpha\beta}$ we have a Euclidean space, otherwise the space is termed Riemannian.

To find $G_{\alpha\beta}$ for spherical polar coordinates:

$$\mathrm{d}s^2 = \mathrm{d}x^2 + \mathrm{d}y^2 + \mathrm{d}z^2 = \mathrm{d}r^2 + r^2\mathrm{d}\theta^2 + r^2\sin^2\theta\,\mathrm{d}\phi^2, \tag{4.32}$$

and so $G_{rr} = 1$, $G_{\theta\theta} = r^2$, $G_{\phi\phi} = r^2\sin^2\theta$ and all the other components are zero. In general

$$G_{\alpha\beta} = \sum_\gamma \left(\frac{\partial X_\gamma}{\partial Z_\alpha}\right)\left(\frac{\partial X_\gamma}{\partial Z_\beta}\right) \quad \text{and} \quad G^{\alpha\beta} = \sum_\gamma \left(\frac{\partial Z_\gamma}{\partial X_\alpha}\right)\left(\frac{\partial Z_\gamma}{\partial X_\beta}\right). \tag{4.33}$$

[2]Superscripts are conventionally used for contravariant components, and subscripts for covariant components. The differential coordinates $\mathrm{d}Z_1$, $\mathrm{d}Z_2, \ldots$ actually transform in a contravariant fashion, but subscripts are used here to maintain consistency with the rest of these notes.

The example of a free diatomic bound by a harmonic potential has been used in previous work to highlight the apparent dependence of normal mode frequencies on the coordinate system [83]. If the potential is $V = k(r - r_0)^2/2$ with $r^2 = (\mathbf{r}_1 - \mathbf{r}_2)^2$ then the Hessian eigenvalues are

$$0, \quad 0, \quad 0, \quad 2k(1 - r_0/r), \quad 2k(1 - r_0/r), \quad 2k, \tag{4.34}$$

where unit masses have been assumed for the two atoms. The three zero eigenvalues correspond to overall translation, while the two rotations are coupled to the bond vibration by a term linear in the gradient, as expected [45]. Only for a stationary point with $r = r_0$ do the two rotational modes exhibit zero eigenvalues.

The problem can be expressed in many other ways, for example in terms of

$$\mathbf{r}_1 \quad \text{and} \quad \mathbf{r}_1 + r(\sin\theta\cos\phi, \sin\theta\,\sin\phi, \cos\theta), \tag{4.35}$$

$$\text{or} \quad \mathbf{R} \pm r(\sin\theta\cos\phi, \sin\theta\,\sin\phi, \cos\theta)/2, \quad \mathbf{R} = (X, Y, Z), \tag{4.36}$$

where the corresponding coordinates are the components of $\mathbf{r}_1$ and r, θ and ϕ in the first case and X, Y, Z, r, θ and ϕ in the second case. Both of these coordinate systems give the same eigenvalue spectrum as equation (4.34) when the Hessian is calculated from the covariant derivatives

$$H_{ij} = \frac{\partial^2 V}{\partial q^i \partial q^j} - \sum_k \frac{\partial V}{\partial q^k} \Gamma^k_{ij}. \tag{4.37}$$

Γ^k_{ij} is a Christoffel symbol of the first kind,

$$\Gamma^k_{ij} = \frac{1}{2} \sum_l A^{kl} \left[\frac{\partial A_{jl}}{\partial q^i} + \frac{\partial A_{il}}{\partial q^j} - \frac{\partial A_{ij}}{\partial q^l} \right], \tag{4.38}$$

A_{ij} are components of the covariant metric tensor and A^{kl} are components of its inverse, the contravariant metric tensor. For example, with the coordinates $\{X, Y, Z, r, \theta, \phi\}$ the metric tensor is diagonal with components

$$A_{XX} = A_{YY} = A_{ZZ} = 2, \quad A_{rr} = \frac{1}{2}, \quad A_{\theta\theta} = \frac{r^2}{2}, \quad A_{\phi\phi} = \frac{1}{2} r^2 \sin^2\theta. \tag{4.39}$$

The generalised eigenvalue problem that accounts for the non-trivial metric tensor is [81]

$$\sum_j (H_{ij} - \lambda A_{ij}) c^j = 0, \tag{4.40}$$

and it is easily verified that solution yields the eigenvalues in (4.34) for both the coordinate systems defined above.

The use of covariant derivatives and the appropriate metric conserves all the Hessian eigenvalues for proper coordinate transformations, not only those associated with translation and rotation. The covariant first derivative is just the usual partial derivative. Coordinate dependence has been exploited in a number of studies concerning geometry optimisation [84–87], where the use of simple partial derivatives can produce more efficient convergence. This approach is legitimate, since the stationary points that are located are stationary points in any proper coordinates.

4.7.1 "Mass-weighted" steepest-descent paths

Coordinate independent steepest-descent paths can be defined by [81]:

$$\sum_{\beta=1}^{3N} G_{\alpha\beta}\frac{\mathrm{d}X_\beta(s)}{\mathrm{d}s} = -\frac{\dfrac{\partial V}{\partial X_\alpha}}{\sqrt{\displaystyle\sum_{\beta,\gamma=1}^{3N} G^{\beta\gamma}\frac{\partial V}{\partial X_\beta}\frac{\partial V}{\partial X_\gamma}}}. \tag{4.41}$$

For the mass-weighted coordinates, q_α, defined above, the metrics have non-vanishing components $G_{\alpha\alpha} = 1/M_\alpha$ and $G^{\alpha\alpha} = M^\alpha$. Steepest-descent paths should then be defined from equation (4.41) as

$$\frac{1}{M_\alpha}\frac{\mathrm{d}q_\alpha(s)}{\mathrm{d}s} = -\frac{\dfrac{\partial V}{\partial q_\alpha}}{\sqrt{\displaystyle\sum_{\beta=1}^{3N} M_\beta\left(\frac{\partial V}{\partial q_\beta}\right)^2}}. \tag{4.42}$$

Since $\partial/\partial q_\beta = (1/\sqrt{M_\beta})\partial/\partial X_\beta$ the above equation is equivalent to that for X_α.

Steepest-descent paths calculated from (4.41) are invariant to proper coordinate transformations $\mathbf{q} = \mathbf{C}\mathbf{X}$ where the matrix $\mathbf{C}$ is non-singular. Paths calculated from equation (4.29) are different, and were popularised by Fukui and coworkers because they correspond to the solutions of the classical equations of motion in the limit of zero kinetic energy [60, 88, 89].

We have previously defined the normal coordinates, Q_α, for which

$$q_\alpha = \sum_{\gamma=1}^{3N} A_{\alpha\gamma}\,Q_\gamma \quad \text{and} \quad Q_\beta = \sum_{\alpha=1}^{3N} A_{\alpha\beta}\,q_\alpha$$

$$\text{where} \quad \sum_{\beta=1}^{3N} \overline{H}_{\alpha\beta}\,A_{\beta\gamma} = \omega_\gamma^2\,A_{\alpha\gamma}, \tag{4.43}$$

$$\text{and} \quad \sum_{\alpha,\beta=1}^{3N} \overline{H}_{\alpha\beta}\, q_\alpha\, q_\beta = \sum_{\alpha=1}^{3N} \omega_\alpha^2\, Q_\alpha^2.$$

We can therefore repeat the analysis leading to equation (4.22) for the "mass-weighted" path to give

$$Q_\alpha(u) = Q_\alpha(0)\exp(-\omega_\alpha^2 u) \tag{4.44}$$

in the vicinity of a stationary point. Hence, the "mass-weighted" paths defined by (4.29) leave the transition state parallel or antiparallel to the normal mode corresponding to the unique imaginary frequency, Q_1.

Equation (4.43) shows that the displacements in terms of the mass-weighted coordinates q_β are parallel or antiparallel to the corresponding eigenvector of $\overline{\mathbf{H}}$. In contrast, true steepest-descent downhill pathways leave a transition state parallel or antiparallel to the unique eigenvector of $\mathbf{H}$ corresponding to the negative eigenvalue.

All the rules for conservation, creation and destruction of symmetry elements along true steepest-descent paths carry over to the "mass-weighted" paths. In principle, a surface could be constructed where the "mass-weighted" and true steepest-descent paths connect different stationary points. However, in practice the pathways usually do not look very different to the eye. For a homonuclear system they are, of course, the same.

4.7.2 Sylvester's law of inertia

The inertia of a matrix, i, is defined by i_+, i_- and i_0, the number of positive, negative and zero eigenvalues, respectively. Although the eigenvalues of $\overline{\mathbf{H}}$ are clearly different from those of $\mathbf{H}$ the numbers of positive, negative and zero eigenvalues are preserved. Hence the "mass-weighting" formulation cannot change the index of a stationary point, as we now prove.

In general, if two symmetric matrices $\mathbf{A}$ and $\mathbf{B}$ are related by

$$\mathbf{A} = \mathbf{S}\,\mathbf{B}\,\mathbf{S}^T, \tag{4.45}$$

where $\mathbf{S}$ is a non-singular matrix (*i.e.* $\mathbf{S}^{-1}$ exists) then $\mathbf{A}$ is said to be congruent to $\mathbf{B}$. Since $\mathbf{B} = \mathbf{S}^{-1}\,\mathbf{A}\,(\mathbf{S}^{-1})^T$, $\mathbf{B}$ is also congruent to $\mathbf{A}$. We will now prove that the inertia of $\mathbf{A}$, $i(\mathbf{A})$, is the same as the inertia of $\mathbf{B}$, $i(\mathbf{B})$ (Sylvester's theorem) [90].

We are interested in the case where $\mathbf{B} = \mathbf{H}$, $\mathbf{A} = \overline{\mathbf{H}}$ and $S_{\alpha\beta} = \delta_{\alpha\beta}/\sqrt{M_\alpha}$. First we show that $\mathbf{A}$ and $\mathbf{B}$ have the same number of zero eigenvalues. Let the $i_0(\mathbf{A})$ linearly independent eigenvectors of $\mathbf{A}$ with zero eigenvalues be denoted $\mathbf{a}_1,\ \mathbf{a}_2,\ldots,\ \mathbf{a}_{i_0(\mathbf{A})}$. Then we cannot find coefficients $c_1,\ c_2,\ldots,\ c_{i_0(\mathbf{A})}$, not all zero, satisfying

$$\sum_{\alpha=1}^{i_0(\mathbf{A})} c_\alpha \mathbf{a}_\alpha = \mathbf{0}. \tag{4.46}$$

Since $\mathbf{A}\,\mathbf{a}_\alpha = \mathbf{0}$ we must have $\mathbf{S}\,\mathbf{B}\,\mathbf{S}^T\mathbf{a}_\alpha = \mathbf{0}$ and $\mathbf{B}\,\mathbf{S}^T\mathbf{a}_\alpha = \mathbf{0}$. Hence the $i_0(\mathbf{A})$ vectors $\mathbf{S}^T\mathbf{a}_1^0,\ \mathbf{S}^T\,\mathbf{a}_2^0,\ldots,\ \mathbf{S}^T\,\mathbf{a}_{i_0(\mathbf{A})}^0$ are all eigenvectors of $\mathbf{B}$ with zero eigenvalues. Furthermore, multiplying equation (4.46) on the left by $\mathbf{S}^T$ we deduce that there are no coefficients, not all zero, satisfying

$$\sum_{\alpha=1}^{i_0(\mathbf{A})} c_\alpha\,\mathbf{S}^T\,\mathbf{a}_\alpha^0 = \mathbf{0}. \tag{4.47}$$

Hence the $i_0(\mathbf{A})$ vectors $\mathbf{S}^T\,\mathbf{a}_1^0,\ \mathbf{S}^T\,\mathbf{a}_2^0,\ldots,\ \mathbf{S}^T\,\mathbf{a}_{i_0(\mathbf{A})}^0$ are linearly independent and $i_0(\mathbf{B}) \geq i_0(\mathbf{A})$. However, we may interchange the roles of $\mathbf{A}$ and $\mathbf{B}$ to obtain $i_0(\mathbf{A}) \geq i_0(\mathbf{B})$, and so $i_0(\mathbf{A}) = i_0(\mathbf{B})$.

In fact we constructed eigenvectors corresponding to zero eigenvalues explicitly for the matrices $\mathbf{H}$ and $\overline{\mathbf{H}}$ in Section 4.1, and so we know that there must be equal numbers in this case.

Now denote the $i_+(\mathbf{A})$ linearly independent eigenvectors of $\mathbf{A}$ with positive eigenvalues by $\mathbf{a}_1^+,\ \mathbf{a}_2^+,\ldots,\ \mathbf{a}_{i_+(\mathbf{A})}^+$. Then we cannot find coefficients $c_1,\ c_2,\ldots,\ c_{i_+(\mathbf{A})}$, not all zero, satisfying

$$\sum_{\alpha=1}^{i_+(\mathbf{A})} c_\alpha\,\mathbf{a}_\alpha^+ = \mathbf{0}. \tag{4.48}$$

Let $\mathbf{x} = \sum_{\alpha=1}^{i_+(\mathbf{A})} c_\alpha\,\mathbf{a}_\alpha^+ \neq \mathbf{0}$. Then

$$\mathbf{x}^T\,\mathbf{A}\,\mathbf{x} = \sum_{\alpha=1}^{i_+(\mathbf{A})} c_\alpha^2\,\lambda_\alpha > 0. \tag{4.49}$$

Hence $\mathbf{x}^T\,\mathbf{S}\,\mathbf{B}\,\mathbf{S}^T\mathbf{x} > 0$, or $\mathbf{y}^T\,\mathbf{B}\,\mathbf{y} > 0$ where

$$\mathbf{y} = \sum_{\alpha=1}^{i_+(\mathbf{A})} c_\alpha\,\mathbf{S}^T\,\mathbf{a}_\alpha^+. \tag{4.50}$$

Hence there are at least $i_+(\mathbf{A})$ linearly independent eigenvectors of $\mathbf{B}$ with positive eigenvalues, and so $i_+(\mathbf{B}) \geq i_+(\mathbf{A})$. As above we may interchange the roles of $\mathbf{A}$ and $\mathbf{B}$ to give $i_+(\mathbf{A}) \geq i_+(\mathbf{B})$ and finally $i_+(\mathbf{A}) = i_+(\mathbf{B})$. This completes the proof, since $i_- = n - i_+ - i_0$ for an $n \times n$ matrix.

4.8 Branch points

The curvature along a steepest-descent path changes from negative to positive as we pass from the transition state to a minimum. Hence it must pass through zero somewhere along the path. The curvature associated with principal directions perpendicular to the steepest-descent path is normally positive, but it can change from positive to negative if the path leads to another transition state. The point at which the curvature along an orthogonal principal direction passes through zero is called a branch point or valley-ridge inflection point (Fig. 13) (see also Sect. 4.4.2).

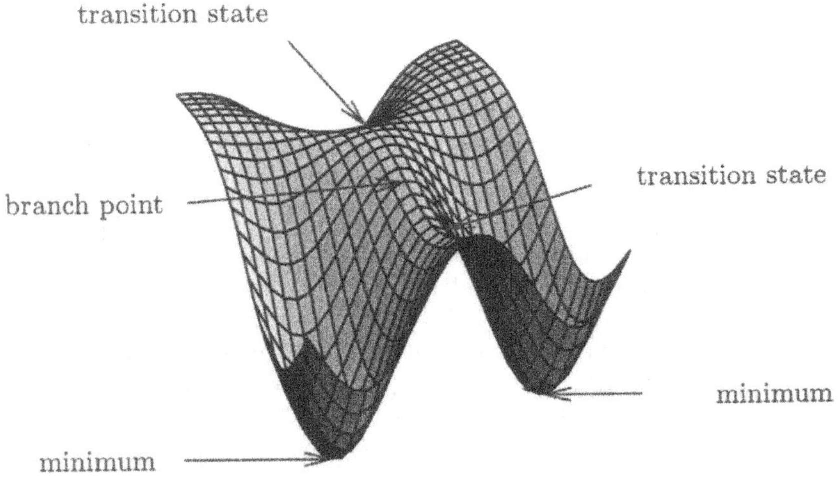

Fig. 13. Schematic view of a branch point.

Such features challenge our intuitive notion of a "minimum energy path" (MEP). The steepest-descent path, as defined above, continues along the ridge because the gradient has no component to send the path down either side of the branch point. Intuitively, we might feel that the MEP should depart from the steepest-descent path at such points. However, it is not possible to define such a path in terms of the solution of a differential equation, as we could for the steepest-descent path. Hence, if we define the "minimum energy path" as a steepest-descent path then it must go over the ridge [91]. To produce a definition more in keeping with intuition we would have to permit the path to bifurcate. If the system is infinitesimally perturbed from the true steepest-descent path then it will indeed follow such a trajectory. Unfortunately there does not seem to be a rigorous definition of the reaction path that allows this to happen. Hence, although branch points are sometimes referred to as "bifurcation" points, a bifurcation in the steepest-descent path cannot occur unless the branch point is also a stationary point [91]. Such a situation could only occur by accident in a realistic molecular system.

If a bifurcation point is defined as a point with two downhill steepest descent paths, then such a point must be a stationary point [58,91]. However, branch points are often referred to as bifurcation points in the literature.

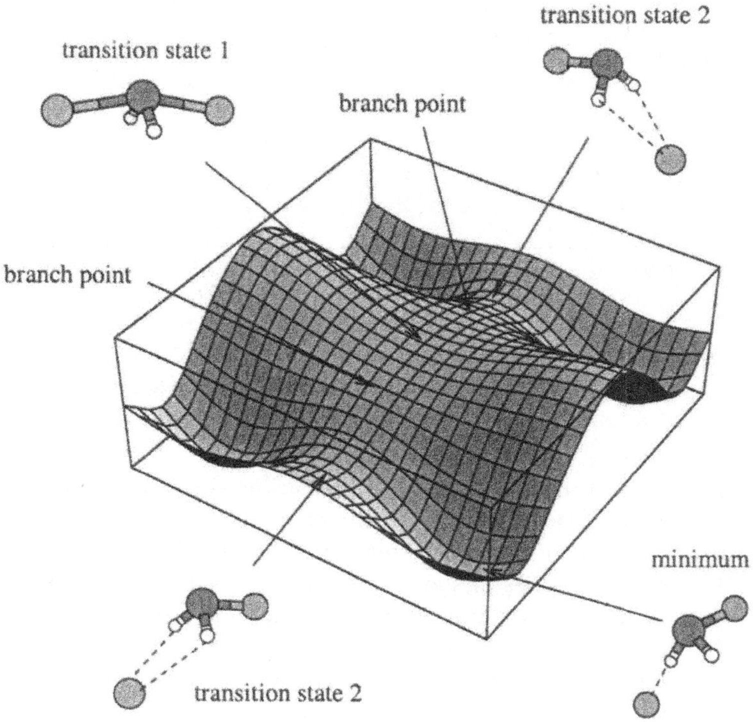

Fig. 14. Schematic potential energy surface for FNH_2F^-.

A branch point must occur whenever a steepest-descent path connects two transition states: the curvature along the initial reaction coordinate must become positive, whilst the curvature correlating with the reaction coordinate of the new transition state must change from positive to negative. There is a branch point between the two transition states involved in the addition of HF to ethene (Fig. 9) [71]. Another example is provided by rearrangements of FNH_2F^- (Fig. 14) [92]. For this system two equivalent

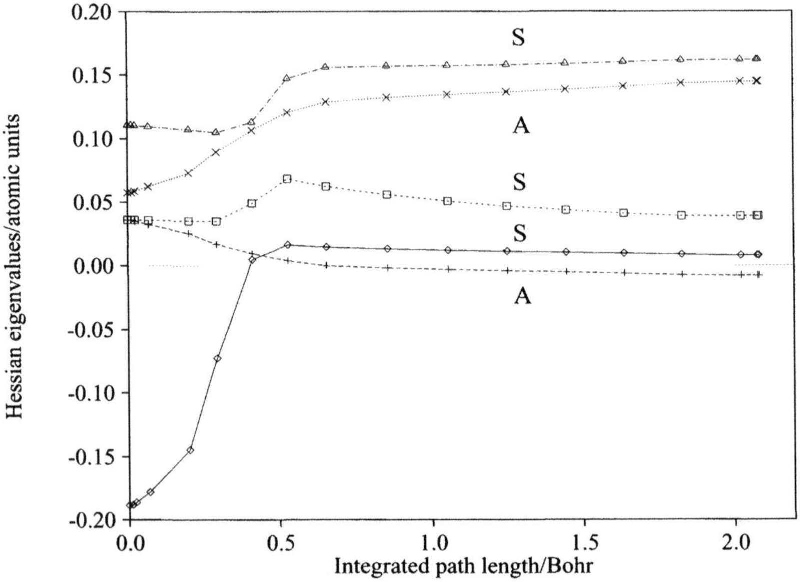

Fig. 15. Evolution of the lowest eigenvalues for FNH_2F^- along the path connecting the C_{2v} transition state (left) to a C_s transition state (right).

branch points exist between the C_{2v} transition state in the middle and the two transition states with C_s symmetry. C_s symmetry is preserved along the paths connecting these stationary points. The evolution of the lowest Hessian eigenvalues is shown for the $C_{2v} \rightarrow C_s$ path in Figure 15. The corresponding eigenvectors are S(ymmetric) and A(ntisymmetric) with respect to the conserved mirror plane. The motion perpendicular to the steepest-descent path at a branch point is usually associated with symmetry-breaking. For the above path the perpendicular direction whose curvature passes through zero at the branch point corresponds to loss of the conserved mirror plane. Notice also that the C_{2v} transition state has an additional symmetry element, namely a C_2 rotation axis.

Since the gradient vector transforms as the totally symmetric irreducible representation Γ_0 (Sect. 4.4.4), a branch point must occur whenever the eigenvalue associated with a Hessian eigenvector that does not transform according to Γ_0 passes through zero.

Ramquet *et al.* have recently argued that the location of valley-ridge inflection (or branch) points on a steepest descent path depends upon the coordinate system [58]. According to their definition, such points occur when the curvature vanishes in a direction, **d**, perpendicular to the steepest-descent path. Coordinate independent curvatures can be defined using covariant derivatives as [82, 93]

$$\frac{\sum_{ij} \mathrm{d}_i\, H_{ij}\, \mathrm{d}_j}{\sum_{ij} A_{ij}\, \mathrm{d}_i\, \mathrm{d}_j}. \tag{4.51}$$

Branch points defined according to this definition are also coordinate independent.

Ramquet *et al.* also note that the vanishing of a curvature perpendicular to a steepest-descent path need not correspond to a vanishing Hessian eigenvalue, and conclude that branch points need not possess a zero Hessian eigenvalue [58]. If we expand the direction, $\mathbf{d}$, in terms of the Hessian eigenvectors, $\hat{\mathbf{e}}_i$, where $\mathbf{H}\,\hat{\mathbf{e}}_i = \varepsilon_i^2\,\hat{\mathbf{e}}_i$ then

$$\mathbf{d}^T\,\mathbf{H}\,\mathbf{d} = \sum_{i=1}^{3N} c_i^2\,\varepsilon_i^2, \tag{4.52}$$

where $c_i = \mathbf{d}^T\,\hat{\mathbf{e}}_i$ and $\mathbf{d}$ is assumed to be normalised. The condition $\sum_{i=1}^{3N} c_i^2\,\varepsilon_i^2 = 0$ results when $\varepsilon_\alpha^2 = 0$ for a particular α, with $c_\alpha = 1$ and $c_{i\neq\alpha} = 0$. If one (or more) of the ε_i^2 are negative there can be solutions where $\mathbf{d}$ is not simply parallel to a Hessian eigenvector with a zero eigenvalue. An example occurs in Figure 16 for steepest-descent paths with broken symmetry lying close to the path that links TS1 and TS2.

Isotopic substitution has some interesting effects for surfaces with branch points [94]. Since the Born–Oppenheimer PES is independent of the atomic masses, the permutation of isotopes of the same element is also a symmetry operation. Hence, the coordinate independent steepest-descent paths defined by (4.41) are unchanged. However, the path corresponding to the intrinsic reaction coordinate (IRC), defined from (4.29), does not possess this additional symmetry, and deviates from the steepest-descent path obtained from (4.41) [94]. A branch point may therefore occur on the IRC in C_1 symmetry. A similar effect can be seen for the model surface shown in Figure 16. There is a branch point on the unique steepest-descent path linking TS1 and TS2, as expected, while nearby paths corresponding to broken symmetry also exhibit a perpendicular direction whose curvature changes from positive to negative.

5 Tunnelling

Tunnelling is the quantum mechanical situation where systems have a finite probability of being found in classically forbidden regions. The simplest example is a particle with total energy E_1 tunnelling through a barrier of height $E_2 > E_1$. The wavefunction in the barrier region is a decaying exponential and the transmission probability is largest when the energy

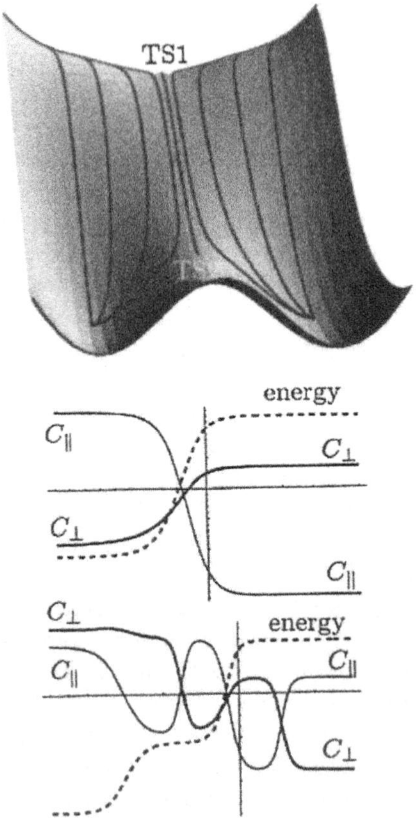

Fig. 16. Steepest descent paths and Hessian eigenvalues for a model potential energy surface. The top panel shows some paths superimposed on the surface itself, while the lower two panels show the curvatures parallel and perpendicular, $C_{\parallel}$ and $C_{\perp}$, to the path linking the two transition states (middle) and a path slightly displaced from it (bottom). The latter path avoids TS2, and finally converges to one of the two equivalent minima. It exhibits a branch point near to the one on the symmetry conserved path, and two ridge-valley inflection points: the first near TS1 where the path turns through 90°, and the other in the region between TS2 and the minimum.

difference and particle mass are small and the barrier is narrow. Hence tunnelling is only likely to give rise to observable effects for light atoms and short pathways with small barriers.

The classic example of tunnelling manifesting itself in spectroscopy is the inversion doubling of the ammonia molecule [95, 96]. The umbrella

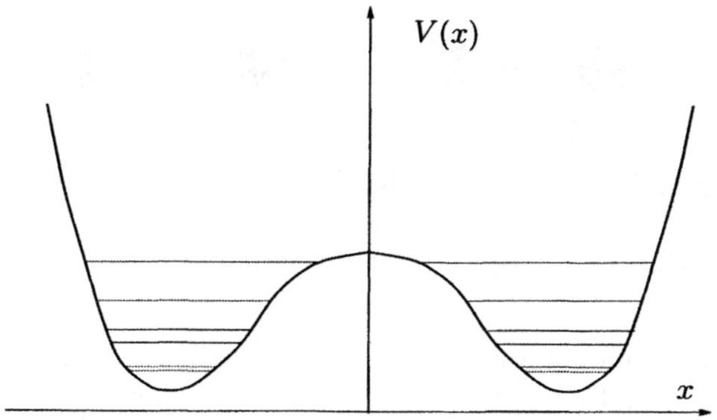

Fig. 17. Splitting of energy levels by tunnelling.

motion takes the system through the planar D_{3h} geometry from one C_{3v} minimum to a permutational isomer (Fig. 17). If the two potential wells were isolated they would support identical rovibronic energy levels. However, the barrier is sufficiently small for the localised wavefunctions in the two wells to interfere. The correct wavefunctions are therefore the sum and difference of the separate wavefunctions for the two wells and are delocalised.

Compare the interaction of atomic orbitals to give molecular orbitals. If two hydrogen atoms are far apart then we can treat each one as isolated but as they approach the $1s$ orbitals interact. The molecular orbitals are proportional to the sum and difference of the $1s$ functions, and the energy levels become $\alpha \pm \beta$ in the Hückel approximation, where α is the isolated energy level and β is the off-diagonal Hamiltonian matrix element between the two $1s$ orbitals. We can treat the double-well problem for ammonia in just the same way, forming a delocalised linear combination of localised wavefunctions, analogous to a linear combination of atomic orbitals. If we denote the localised wavefunctions by ψ_1 and ψ_2 and the matrix elements of the Hamiltonian by $\alpha = \langle\psi_1|\hat{\mathcal{H}}|\psi_1\rangle = \langle\psi_2|\hat{\mathcal{H}}|\psi_2\rangle$ and $\beta = \langle\psi_1|\hat{\mathcal{H}}|\psi_2\rangle = \langle\psi_2|\hat{\mathcal{H}}|\psi_1\rangle$ then the resulting secular equations are entirely analogous to the hydrogen molecule problem. The Hückel-type approximation may be justified for tunnelling problems because of the exponential decay of the wavefunctions in the barrier regions, although sometimes better approximations are needed [97].

For NH_3 the localised levels are split into pairs – hence the term inversion doubling. The splitting of the lowest level is $0.793\,\mathrm{cm}^{-1}$, and the splitting increases for higher energy states because the barrier is narrower and lower.

Tunnelling between minima with different structures (non-degenerate tunnelling) is generally assumed to lead to much smaller effects than

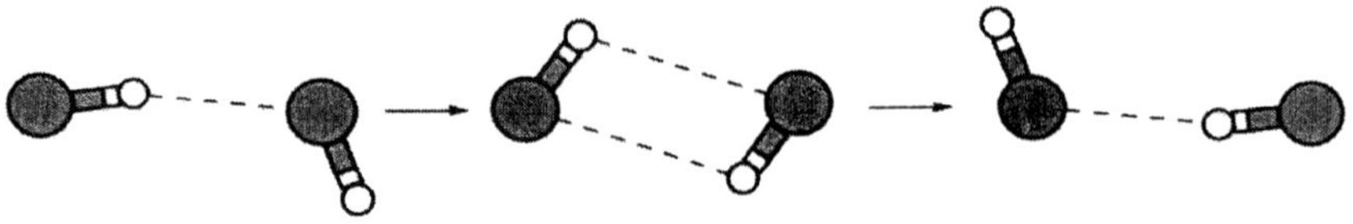

Fig. 18. Degenerate rearrangement of the HF dimer.

tunnelling between equivalent minima. The reason is basically that only the energy levels of permutational isomers of the same structure are in resonance.

5.1 Tunnelling in $(HF)_2$

The symmetric degenerate rearrangement of the $(HF)_2$ C_s minima *via* a C_{2h} transition state is shown in Figure 18 [98]. Note the appearance and disappearance of the C_2 symmetry element at the transition state. The secular determinant for each pair of degenerate wavefunctions is

$$\begin{vmatrix} \alpha - E & \beta \\ \beta & \alpha - E \end{vmatrix} = 0, \tag{5.1}$$

with solutions $E_{\pm} = \alpha \pm \beta$. The splitting is therefore 2β. The ground vibrational state of the dimer is subject to a tunnelling splitting of $0.66\,\text{cm}^{-1}$ [98]. However, this splitting decreases to 0.22 and $0.23\,\text{cm}^{-1}$ for the excited vibrational states corresponding to one quantum of vibration in the non-H-bonded and H-bonded HF monomers, respectively. The dependence of the splitting of low-frequency "floppy" modes upon excitation of high-frequency modes can often only be explained by detailed calculation [99–106].

5.2 Tunnelling in $(H_2O)_3$

New experiments in the far-infrared region have recently become possible that can probe the low frequency modes of a van der Waals complex and resolve rotational transitions [107–110]. Results for the water trimer, $(H_2O)_3$, have generated considerable theoretical activity in the last few years, and the spectra have now been assigned in some detail [111 117]. The lowest energy rearrangement mechanism of the cyclic global minimum is known as the "flip" (Fig. 19) [76, 118]. The global minimum is a "frustrated" structure because two "dangling" hydrogen atoms must be next to each other. The flip is an asymmetric degenerate rearrangement [75]. Each minimum can rearrange *via* two distinct flips, corresponding to either of the adjacent

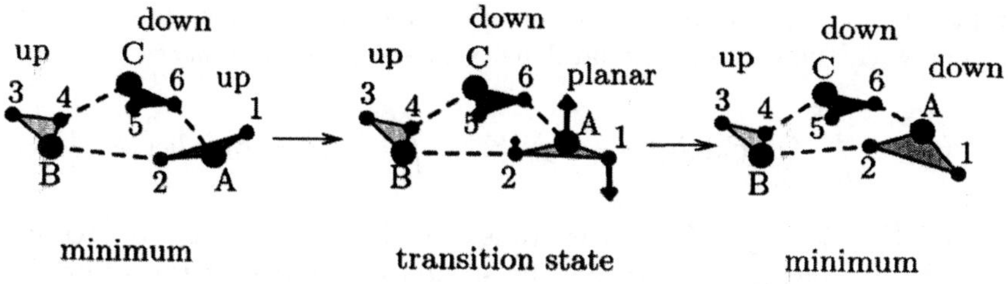

Fig. 19. Single flip rearrangement of $(H_2O)_3$.

"dangling" hydrogens. This process is facile, as no hydrogen bonds are made or broken, and is localised at a single hydrogen atom. The mechanism leads to a low effective mass, and, in combination with the low barrier, produces large tunnelling splittings on the order of 10 cm^{-1}.

It is not hard to show that three successive flips connect any permutational isomer to its enantiomer, while six flips can take us through a closed set of six such isomers [76]. The secular problem for the energy levels is analogous to the treatment of the π system of benzene in the Hückel approximation, with tunnelling levels at $\alpha \pm 2\beta$ (non-degenerate) and $\alpha \pm \beta$ (doubly degenerate), in good agreement with experiment and more accurate dynamical calculations [119, 120].

6 Global thermodynamics

6.1 The superposition approximation

$$H = E^0 + \frac{1}{2} \sum_{\alpha=1}^{3N-6} \left(\dot{Q}_\alpha^2 + \omega_\alpha^2 Q_\alpha^2 \right), \tag{6.1}$$

is the vibrational Hamiltonian in the harmonic approximation for a stationary point at the origin with normal coordinates Q_α. The total microcanonical (constant N, V and E) energy density of states for one minimum, $\Omega(E)$, may be obtained from the separate kinetic (momentum) and configurational densities of states, $\Omega_{\rm K}(E_{\rm K})$ and $\Omega_{\rm c}(E_{\rm c})$, respectively. $\Omega(E)$ is the differential of $G(E)$, the total phase volume with energy less than E, with respect to the total energy. Hence

$$G(E) = \int_0^E \Omega(E')\mathrm{d}E', \tag{6.2}$$

and the total phase volume is the convolution of either the momentum phase volume with the configurational density of states, or *vice versa:*

$$G(E) = \int_0^E \Omega_c(E_c) G_K(E - E_c) dE_c = \int_0^E G_c(E - E_K) \Omega_K(E_K) dE_K.$$

To find $G_c(E_c)$, $\sum_\alpha \omega_\alpha^2 Q_\alpha^2/2$ is transformed into a sum of squares using the substitution $\zeta_\alpha^2 = \omega_\alpha^2 Q_\alpha^2/2$. The standard formula for the volume of a hypersphere radius R in $\kappa = 3N - 6$ dimensions is $\pi^{\kappa/2} R^\kappa / \Gamma(\kappa/2+1)$ [121], and the Jacobian factor of $1/\prod_{s=1}^N m_s^{3/2}$, then gives

$$G_c(E_c) = \frac{(E_c - E^0)^{\kappa/2}}{(2\pi)^{\kappa/2} \Gamma(\kappa/2 + 1) \prod_{\alpha=1}^\kappa \nu_\alpha \prod_{s=1}^N m_s^{3/2}}, \tag{6.3}$$

where $\omega_\alpha = 2\pi\nu_\alpha$, Γ is the Gamma function and the mass of atom t is m_t. Hence

$$\Omega_c(E_c) = \frac{(E_c - E^0)^{\kappa/2-1}}{(2\pi)^{\kappa/2} \Gamma(\kappa/2) \prod_{s=1}^N m_s^{3/2}}. \tag{6.4}$$

Similarly,

$$\begin{aligned} G_K(E_K) &= \frac{(2\pi)^{\kappa/2} E_K^{\kappa/2} \prod_{s=1}^N m_s^{3/2}}{\Gamma(\kappa/2 + 1)}, \\ \text{and} \quad \Omega_K(E_K) &= \frac{(2\pi)^{\kappa/2} E_K^{\kappa/2-1} \prod_{s=1}^N m_s^{3/2}}{\Gamma(\kappa/2)}. \end{aligned} \tag{6.5}$$

Note that the Jacobian for the overall transformation of position and momentum variables to normal coordinates is unity. The total microcanonical phase volume may therefore be obtained as

$$\begin{aligned} G(E) &= \int_{E^0}^E \Omega_c(E_c - E^0) G_K(E - E_c) dE_c, \\ &= \frac{1}{\Gamma(\kappa/2)\Gamma(\kappa/2+1) \prod_j^\kappa \nu_j} \int_{E^0}^E (E_c - E^0)^{\kappa/2-1} (E - E_c)^{\kappa/2} dE_c, \\ &= \frac{(E - E^0)^\kappa}{\Gamma(\kappa + 1) \prod_{j=1}^\kappa \nu_j}, \end{aligned} \tag{6.6}$$

$$\text{and so} \quad \Omega(E) = \frac{(E - E^0)^{\kappa-1}}{\Gamma(\kappa) \prod_{j=1}^\kappa \nu_j}. \tag{6.7}$$

The above formula gives the harmonic approximation for the total energy density of vibrational states associated with a single minimum of a system with fixed centre of mass and fixed orientation with respect to space fixed axes. A factor of $1/h^\kappa$ gives the semiclassical approximation. Both $G(E)$ and $\Omega(E)$ vanish unless $E > E^0$, and this behaviour may be guaranteed by including a step function $\Theta(E - E^0)$ in (6.6) and (6.7).

In fact, equation (6.7) may be derived more quickly using the canonical partition function for a harmonic oscillator for each vibrational mode and taking an inverse Laplace transform. However, some of the intermediate results, such as $\Omega_\mathrm{K}(E_\mathrm{K})$ will be employed in later derivations.

To calculate the vibrational density of states for the whole system, all the minima need to be considered. In the superposition approximation we simply sum the density of states over all the minima low enough in energy to contribute [122]:

$$\Omega(E) = \sum_{E_s^0<E} \frac{n_s(E - E_s^0)^{\kappa-1}}{\Gamma(\kappa)\prod_{j=1}^{\kappa} h\nu_j^s}, \tag{6.8}$$

where the sum is over all the structurally distinct minima on the surface and n_s, the number of permutational isomers of minimum s, is given by $n_s = 2N!/h_s$ where h_s is the order of the point group of s (Sect. 3.1). The superposition approach was originally considered in the early days of cluster simulation [123–125], but has only recently become practical for larger clusters [122,126]. The first attempts to implement this decomposition of phase space over minima [127] were not very successful because of the difficulty in obtaining sufficiently large samples of minima, and because the effect of point group symmetry was omitted [125]. The latter mistake roughly doubles [1] the predicted melting point for LJ_{13}, the thirteen-atom cluster bound by the Lennard–Jones potential [128].

The canonical (constant N, V and T) partition function is obtained from Laplace transformation of the energy density of states, and hence the superposition approximation gives

$$Q(T) \;=\; \int \Omega(E)\mathrm{e}^{-\beta E}\mathrm{d}E \;=\; \sum_s \frac{n_s \mathrm{e}^{-\beta E_s^0}}{(\beta h \overline{\nu}_s)^\kappa}, \tag{6.9}$$

where $\beta = 1/kT$, k is the Boltzmann constant, and $\overline{\nu}_s$ is the geometric mean normal mode frequency of minimum s.

Thermodynamic properties can now be derived straightforwardly from $\Omega(E)$ and $Q(T)$. For example, the thermodynamic definition of temperature as $T = 1/(\partial S/\partial E)$ in the microcanonical ensemble gives:

$$\frac{1}{T} = \frac{k}{\Omega(E)} \sum_{E_s^0<E} \frac{n_s(E - E_s^0)^{\kappa-2}}{\Gamma(\kappa - 1)\prod_{j=1}^{\kappa} h\nu_j^s}, \tag{6.10}$$

while the average values of the internal and kinetic energy in the canonical ensemble are:

$$\begin{aligned}\langle E\rangle &= \kappa kT + \frac{1}{Q(T)}\sum_{s=1}^{N}\frac{n_s e^{-\beta E_s^0}E_s^0}{(\beta h\overline{\nu}_s)^\kappa},\\ \langle E_{\rm K}\rangle &= \frac{\kappa kT}{2}.\end{aligned} \tag{6.11}$$

The temperature and heat capacity in the microcanonical ensemble can also be expressed in terms of expectation values of the kinetic energy, using the probability distribution for $E_{\rm K}$ derived in (7.5).

An alternative microcanonical kinetic temperature, $T_{\rm K}$, may be defined in terms of the mean kinetic energy as $T_{\rm K} = 2\langle E_{\rm K}\rangle/k\kappa$. Since $G_{\rm K}(E_{\rm K}) = 2E_{\rm K}\Omega_{\rm K}(E_{\rm K})/\kappa$ (from (6.5)) it can also be shown that $\langle E_{\rm K}\rangle = \kappa G(E)/2\Omega(E)$, and hence that [129, 130]

$$T(E) = T_{\rm K}(E)\Big/\left[1 - k\left(\frac{\partial T_{\rm K}}{\partial E}\right)\right], \tag{6.12}$$

so that the thermodynamic and kinetic definitions of the microcanonical temperature differ by a leading term of order N^{-1}, as expected [131–133]. Umirzakov has shown that a van der Waals loop in $T_{\rm K}(E)$ necessitates a loop in $T(E)$, while the converse relation need not be true [130]. In fact, $T_{\rm K}$ corresponds to the temperature determined by the energy derivative of the entropy when it is defined as $S = k\ln G(E)$, because

$$\left(\frac{\partial\, k\ln G(E)}{\partial E}\right)_{N,V} = \frac{k\Omega(E)}{G(E)} = \frac{k\kappa}{2\langle E_{\rm K}\rangle} = \frac{1}{T_{\rm K}}. \tag{6.13}$$

The alternative definitions of entropy and temperature are identical in the thermodynamic limit [134].

Properties calculated from the superposition approximation are not equivalent to a weighted sum over minima. For example,

$$\begin{aligned}\sum_{E_s^0<E} P_s\ln\Omega_s(E) &= \ln\Omega(E) + \sum_{E_s^0<E} P_s\ln\left[\frac{\Omega_s(E)}{\Omega(E)}\right]\\ &= \ln\Omega(E) + \sum_{E_s^0<E} P_s\ln P_s.\end{aligned} \tag{6.14}$$

Hence the entropy calculated as a weighted sum differs from the true entropy by $-k\sum_{E_s^0<E} P_s\ln P_s > 0$, which is the entropy associated with the choice of minima in which the system may find itself.

6.2 Sample incompleteness

The difficulty with equation (6.8) is that for all but the very smallest clusters the sum involves an impractically large number of minima. For the LJ_{55} cluster there may be as many as 10^{21} terms. In such a case, as it is not possible to obtain a complete set of minima, a statistically representative sample is needed. A large set of minima can be obtained by systematic quenching from a high energy molecular dynamics (MD) trajectory. However, this gives a greater proportion of the low energy minima than of the high energy minima. Consequently, if the sample is used in equation (6.8) it is likely to underestimate the density of states due to the high energy minima, and so be inaccurate at high energies.

A method is needed that corrects for the incomplete nature of the sample of minima. This correction can be achieved by weighting the density of states for each known minimum by g_s, the number of minima of energy E_s^0 for which the minimum s is representative [122, 132]. Hence,

$$\Omega(E) = \sum_{E_s^0<E} \frac{g_s n_s (E - E_s^0)^{\kappa-1}}{\Gamma(\kappa)\prod_{j=1}^{\kappa} h\nu_j^s}, \tag{6.15}$$

where the sum is now over a representative sample of minima. The effect of g_s can be incorporated using the quench statistics. If the system is ergodic and the MD run is performed at constant energy E', the number of quenches to a minimum, γ, is assumed to be proportional to the density of states of the set of g_s minima, *i.e.* $\gamma(E')_s \propto g_s\Omega(E')_s$. Hence,

$$\begin{aligned} \Omega(E) &\propto \sum_{E_s^0<E} \gamma(E')_s \frac{\Omega(E)_s}{\Omega(E')_s} \\ &\propto \sum_{E_s^0<E} \gamma(E')_s \left(\frac{E - E_s^0}{E' - E_s^0}\right)^{\kappa-1}. \end{aligned} \tag{6.16}$$

This reweighting technique is analogous to the histogram Monte-Carlo approach [135–138], but instead of determining the configurational density of states from the canonical potential energy distribution, g, effectively a density of minima, is found from the microcanonical probability distribution for quenching to a minimum. If all the low energy minima are known equation (6.8) should be reasonably accurate at low energies. The proportionality constant in equation (6.16) can be found by matching it to the low energy form of equation (6.8) [132].

At low enough energy the term due to the global minimum dominates and the proportionality constant in (6.16), c, may be found by comparison with (6.8):

$$c = \frac{n_0(E' - E_0^0)^{\kappa-1}}{\gamma(E')_0 \Gamma(\kappa) \prod_{j=1}^{\kappa} h\nu_j^0}. \tag{6.17}$$

The harmonic superposition method has three main possible sources of error [122, 132]. The first is associated with the statistical accuracy of the quench frequencies. These errors can be eliminated in principle by having a long enough quench run to ensure ergodicity, and by choosing an appropriate energy for the MD run so that the relevant regions of phase space are all significantly sampled. When studying the thermodynamics of melting it is most appropriate to choose E' to lie in the coexistence region, so that quenches to solid-like, liquid-like and surface-melted states are frequent. Alternatively, a simulation that achieves better ergodicity could be used, such as simulated tempering [139]. The second possible source of error is the assumption that the phase space volumes for each minimum can be summed independently, *i.e.* the hyperellipsoids in phase space do not overlap. If overlap occurred $\Omega(E)$ would be overestimated. The third possible source of error is the harmonic approximation. Near the bottom of the well the harmonic assumption is reasonable, but as the energy is increased some parts of the well become increasingly flat. Consequently the harmonic approximation causes $\Omega(E)$ to be underestimated. Unfortunately it is also the most difficult to correct.

6.3 Thermodynamics and cluster simulation

There are some further subtleties involved in the interpretation of either molecular dynamics or Monte-Carlo simulations of clusters. If such simulations are run for long enough, then eventually atoms will evaporate at most energies or temperatures of interest. In practice some sort of constraint is applied, usually in the form of a container. Alternatively, one can check for evaporation periodically and reject or reverse such moves [140, 141]. Hence the densities of states and partition functions that we consider actually correspond to bound clusters. Fortunately, the effect of the container is relatively small for a reasonable range of container sizes [48]. However, such constraints influence the corresponding density of states and partition function, and hence thermodynamic and dynamic properties [142–145].

Constraints due to constant total linear and angular momenta, **P** and **L**, may be treated formally using the accessible total energy density of states:

$$\begin{aligned}\Omega(E, \mathbf{P}, \mathbf{L}) &= \int \delta\left[H(\mathbf{X}, \{\mathbf{p}_s\}) - E\right] \delta\left[\mathbf{P} - \sum_{s=1}^{N} \mathbf{p}_s\right] \delta\left[\mathbf{L} - \sum_{s=1}^{N} \mathbf{X}_s \wedge \mathbf{p}_s\right] \\ &\quad \mathrm{d}X_1 \mathrm{d}X_2 \ldots \mathrm{d}X_{3N} \mathrm{d}p_1 \mathrm{d}p_2 \ldots \mathrm{d}p_{3N},\end{aligned} \tag{6.18}$$

where $\mathbf{p}_s$ is the three-dimensional momentum vector for atom s. We can integrate over the momenta analytically [144]. A new derivation will be given here, denoting X_{sx} the x coordinate of atom s, etc., and defining the coordinates $\mathbf{P}^t$ and $\mathbf{P}^r$ with components:

$$
\begin{aligned}
P^t_X &= \sum_{s=1}^{N} \overline{p}_{sx}\sqrt{m_s}/\sqrt{M} \\
P^t_Y &= \sum_{s=1}^{N} \overline{p}_{sy}\sqrt{m_s}/\sqrt{M} \\
P^t_Z &= \sum_{s=1}^{N} \overline{p}_{sz}\sqrt{m_s}/\sqrt{M} \\
P^r_X &= \sum_{s=1}^{N} (\overline{X}_{sy}\overline{p}_{sz} - \overline{X}_{sz}\overline{p}_{sy})\sqrt{m_s}/\sqrt{I^0_x} \\
P^r_Y &= \sum_{s=1}^{N} (\overline{X}_{sz}\overline{p}_{sx} - \overline{X}_{sx}\overline{p}_{sz})\sqrt{m_s}/\sqrt{I^0_y} \\
P^r_X &= \sum_{s=1}^{N} (\overline{X}_{sx}\overline{p}_{sy} - \overline{X}_{sy}\overline{p}_{sx})\sqrt{m_s}/\sqrt{I^0_z},
\end{aligned}
\tag{6.19}
$$

where $\overline{\mathbf{p}}_s = \mathbf{p}_s/\sqrt{m_s}$, M is the total mass, I^0_x, I^0_y and I^0_z are the principal moments of inertia tensor with respect to the centre of mass, $\mathbf{R}$, we define $\overline{\mathbf{X}}_i = \mathbf{U}^T(\mathbf{X}_i - \mathbf{R})$, and

$$
\sum_{\beta\gamma} U^T_{\alpha\beta}\, I^0_{\beta\gamma}\, U_{\gamma\delta} = \delta_{\alpha\delta} I^0_\alpha \tag{6.20}
$$

$$
\text{with} \quad I^0_{\beta\gamma} = \sum_{s=1}^{N} m_s \left[(\mathbf{X}_s - \mathbf{R})^T(\mathbf{X}_s - \mathbf{R})\delta_{\beta\gamma} - (X_{s\beta} - R_\beta)(X_{s\gamma} - R_\gamma)\right].
$$

The coordinates in (6.19) define the first six rows of an orthogonal transformation matrix, $\mathbf{V}$, in the $3N$-dimensional momentum space:

$$\begin{pmatrix}
\sqrt{\frac{m_1}{M}} & 0 & 0 & 0 & \overline{X}_{1z}\sqrt{m_1} & -\overline{X}_{1y}\sqrt{m_1} & \cdots \\
0 & \sqrt{\frac{m_1}{M}} & 0 & -\overline{X}_{1z}\sqrt{m_1} & 0 & \overline{X}_{1x}\sqrt{m_1} & \cdots \\
0 & 0 & \sqrt{\frac{m_1}{M}} & \overline{X}_{1y}\sqrt{m_1} & -\overline{X}_{1x}\sqrt{m_1} & 0 & \cdots \\
\sqrt{\frac{m_2}{M}} & 0 & 0 & 0 & \overline{X}_{2z}\sqrt{m_2} & -\overline{X}_{2y}\sqrt{m_2} & \cdots \\
0 & \sqrt{\frac{m_2}{M}} & 0 & -\overline{X}_{2z}\sqrt{m_2} & 0 & \overline{X}_{2x}\sqrt{m_2} & \cdots \\
0 & 0 & \sqrt{\frac{m_2}{M}} & \overline{X}_{2y}\sqrt{m_2} & -\overline{X}_{2x}\sqrt{m_2} & 0 & \cdots \\
\vdots & \vdots & \vdots & \vdots & \vdots & \vdots & \cdots \\
\sqrt{\frac{m_N}{M}} & 0 & 0 & 0 & \overline{X}_{Nz}\sqrt{m_N} & -\overline{X}_{Ny}\sqrt{m_N} & \cdots \\
0 & \sqrt{\frac{m_N}{M}} & 0 & -\overline{X}_{Nz}\sqrt{m_N} & 0 & \overline{X}_{Nx}\sqrt{m_N} & \cdots \\
0 & 0 & \sqrt{\frac{m_N}{M}} & \overline{X}_{Ny}\sqrt{m_N} & -\overline{X}_{Nx}\sqrt{m_N} & 0 & \cdots
\end{pmatrix}$$

It is not difficult to show that these first six columns are orthonormal; for example,

$$\begin{aligned}
\sum_{i=1}^{3N} V_{i4}V_{i4} &= \sum_{s=1}^{N} m_s(\overline{X}_{sy}^2 + \overline{X}_{sz}^2)/I_x^0 = 1 \\
\sum_{i=1}^{3N} V_{i2}V_{i4} &= -\sum_{s=1}^{N} m_s\overline{X}_{sz}/\sqrt{M} = 0 \\
\sum_{i=1}^{3N} V_{i4}V_{i5} &= \sum_{s=1}^{N} m_s\overline{X}_{sx}\overline{X}_{sy}/\sqrt{I_x^0 I_y^0} = 0,
\end{aligned} \tag{6.21}$$

where the second and third lines follow from

$$\mathbf{U}^T\sum_{s=1}^{N} m_s(\mathbf{X}_s - R) = \mathbf{U}^T(M\mathbf{R} - M\mathbf{R}) = 0, \tag{6.22}$$

$$\text{and} \quad \sum_{\beta\gamma} U_{\alpha\beta}^T I_{\beta\gamma}^0 U_{\gamma\delta} = \delta_{\alpha\delta}I_\alpha^0 = \sum_{s=1}^{N} m_s\left[\overline{\mathbf{X}}_s^2\delta_{\alpha\delta} - \overline{\mathbf{X}}_{s\alpha}\overline{\mathbf{X}}_{s\delta}\right].$$

The Jacobian for the transformation from the $\{\mathbf{p}_s\}$ to the $\{\overline{\mathbf{p}}_s\}$ is $\prod_{s=1}^{N} m_s^{3/2}$. The kinetic energy is diagonal in terms of $\{\overline{\mathbf{p}}_s\}$, and this form is preserved by the orthogonal transformation specified by $\mathbf{V}$, for which the Jacobian is unity. The delta function for the centre of mass linear momentum may be rearranged as

$$\begin{aligned}
&\int \delta\left[\mathbf{P} - \sum_{s=1}^{N}\mathbf{p}_s\right] dP_X^t dP_Y^t dP_Z^t = \int \delta\left[\mathbf{P} - \mathbf{P}^t\sqrt{M}\right] dP_X^t dP_Y^t dP_Z^t \\
&= \int \delta\left[\mathbf{P}/\sqrt{M} - \mathbf{P}^t\right] /M^{3/2} dP_X^t dP_Y^t dP_Z^t,
\end{aligned} \tag{6.23}$$

because $\int \delta(ax)\mathrm{d}x = \int \delta(x)/|a|\mathrm{d}x$. Hence, integration over $\mathbf{P}^t$ fixes the centre of mass kinetic energy in H to $\mathbf{P}^2/2M$ and produces a factor of $M^{-3/2}$. Defining $\mathbf{L}^0 = \mathbf{R}\wedge\sum_{s=1}^N \mathbf{p}_s$ the delta function for the angular momentum becomes:

$$\delta\left[\mathbf{L}-\sum_{s=1}^{N}\mathbf{X}_s\wedge\mathbf{p}_s\right] = \delta\left[(\mathbf{L}-\mathbf{L}^0)-\sum_{s=1}^{N}(\mathbf{X}_s-\mathbf{R})\wedge\mathbf{p}_s\right] \tag{6.24}$$

$$= \delta\left[(\mathbf{L}-\mathbf{L}^0)-\mathbf{U}\mathbf{D}^{-1}\sum_{s=1}^{N}\mathbf{D}\mathbf{U}^T(\mathbf{X}_s-\mathbf{R})\wedge\mathbf{p}_s\right] \tag{6.25}$$

$$= \frac{|\mathbf{D}|}{|\mathbf{U}|}\delta\left[\mathbf{D}\mathbf{U}^T(\mathbf{L}-\mathbf{L}^0)-\mathbf{P}^r\right], \tag{6.26}$$

where $|\mathbf{D}|$ is the determinant of $\mathbf{D}$ and $\mathbf{D}^2 = \mathbf{U}^T(\mathbf{I}^0)^{-1}\mathbf{U}$, so that $D_{\alpha\beta} = \delta_{\alpha\beta}/\sqrt{I^0_\alpha}$ and $\sum_{s=1}^N \mathbf{D}\mathbf{U}^T(\mathbf{X}_s-\mathbf{R})\wedge\mathbf{p}_s = \mathbf{P}^r$. Integration over $\mathbf{P}^r$ therefore introduces a factor of $1/\sqrt{I^0_X I^0_Y I^0_Z} = 1/|\mathbf{I}^0|^{1/2}$ and fixes the contribution to the kinetic energy from angular momentum about the centre of mass to

$$\begin{aligned}(\mathbf{P}^r)^2/2 &= (\mathbf{L}-\mathbf{L}^0)^T\mathbf{U}\mathbf{D}^T\mathbf{D}\mathbf{U}^T(\mathbf{L}-\mathbf{L}^0)/2\\ &= (\mathbf{L}-\mathbf{L}^0)^T(\mathbf{I}^0)^{-1}(\mathbf{L}-\mathbf{L}^0)/2.\end{aligned} \tag{6.27}$$

It remains to integrate over the remaining $3N-6$ linear combinations of the $\overline{p}_{i\alpha}$, which we denote by $\zeta_7, \zeta_8, \ldots, \zeta_{3N}$. These linear combinations need not be specified so long as the overall transformation is orthogonal, and leaves the kinetic energy diagonal. Hence (6.18) reduces to

$$\Omega(E,\mathbf{P},\mathbf{L}) = \frac{\prod_{i=s}^N m_s^{3/2}}{M^{3/2}}$$

$$\times\int\delta\left[V(\mathbf{X})+\sum_{i=7}^{3N}\frac{\zeta_i^2}{2}+\frac{\mathbf{P}^2}{2M}+\frac{(\mathbf{L}-\mathbf{L}^0)^T(\mathbf{I}^0)^{-1}(\mathbf{L}-\mathbf{L}^0)}{2}-E\right]/|\mathbf{I}^0|^{1/2} \tag{6.28}$$

$$\mathrm{d}X_1\mathrm{d}X_2\ldots\mathrm{d}X_{3N}\mathrm{d}\zeta_7\mathrm{d}\zeta_8\ldots\mathrm{d}\zeta_{3N}. \tag{6.29}$$

Defining

$$\overline{E} = E - V(\mathbf{X}) - \frac{\mathbf{P}^2}{2M} - \frac{(\mathbf{L}-\mathbf{L}^0)^T(\mathbf{I}^0)^{-1}(\mathbf{L}-\mathbf{L}^0)}{2}, \tag{6.30}$$

$\overline{\zeta}_i = \zeta_i/\sqrt{2}$, and using the result [146] $\int\delta[f(x)]\mathrm{d}x = \int\delta[x-x_0]/|f'(x_0)|\mathrm{d}x$, where $f(x_0)=0$, we may write

$$\delta\left[\overline{\zeta}^2-\overline{E}\right] = \delta\left[\overline{\zeta}-\overline{E}^{1/2}\right]/2\overline{E}, \tag{6.31}$$

where $\overline{\zeta} = \sqrt{\sum_{i=7}^{3N} \overline{\zeta}_i^2}$. The standard formula for the surface area of a hypersphere radius R in κ dimensions is $\kappa\pi^{\kappa/2}R^{\kappa-1}/\Gamma(\kappa/2+1)$, and hence we finally obtain [144]

$$\Omega(E,\mathbf{P},\mathbf{L}) = \frac{(2\pi)^{\kappa/2}\prod_{s=1}^{N} m_s^{3/2}}{M^{3/2}\Gamma(\kappa/2)} \int \frac{\overline{E}^{\kappa/2-1}\Theta(\overline{E})}{|\mathbf{I}^0|^{1/2}} \mathrm{d}X_1\mathrm{d}X_2\ldots\mathrm{d}X_{3N}, \tag{6.32}$$

where Θ is the Heaviside step function, restricting the range of the integrand to $\overline{E} \geq 0$.

To describe the motion of a non-rigid molecule containing N atoms it is usual to use three centre of mass coordinates, three Euler angles and the $3N-6$ vibrational normal coordinates [44, 45, 57]. The orientation of the molecule relative to a stationary laboratory frame may be fixed using the Eckart conditions [44, 45, 57]:

$$\begin{aligned}
\sum_{s=1}^{N} m_s(X_{sx}^e X_{sy} - X_{sy}^e X_{sx}) &= 0, \\
\sum_{s=1}^{N} m_s(X_{sy}^e X_{sz} - X_{sz}^e X_{sy}) &= 0, \\
\sum_{s=1}^{N} m_s(X_{sz}^e X_{sx} - X_{sx}^e X_{sz}) &= 0,
\end{aligned} \tag{6.33}$$

where X_{sx}^e is the x component of atom s in the equilibrium reference geometry, etc. The above conditions provide an unambiguous way to specify the orientation of the "molecule fixed" (or "nuclear fixed" [44]) axes from the nuclear coordinates. From (4.9) we see that they are equivalent to setting the normal coordinates corresponding to infinitesimal rotations to zero. For the equilibrium nuclear configuration the Eckart conditions (6.33) are satisfied by the principal axes in which the moment of inertia tensor is diagonal. Otherwise (6.33) must be solved numerically to obtain the necessary Euler angles, and hence the orientation of the rotating molecule fixed axis system in which the vibrational angular momentum is minimised [44, 45, 57].

For simplicity, we now set the centre of mass to the origin, so that $\mathbf{L}_0 = \mathbf{0}$, and consider a non-translating, non-rotating cluster with $\mathbf{P} = \mathbf{L} = \mathbf{0}$. If we consider a normal mode expansion for the position coordinates about a local minimum, and assume that $|\mathbf{I}^0|^{1/2} = |\mathbf{I}_e^0|^{1/2}$, where $\mathbf{I}_e^0$ is the fixed equilibrium moment of inertia tensor, then (6.32) may be integrated over the vibrational coordinates as in Section 6.1 to obtain

$$\Omega(E,\mathbf{0},\mathbf{0}) = \frac{(2\pi)^{\kappa}E^{\kappa-1}}{\overline{\omega}^{\kappa}\Gamma(\kappa)M^{3/2}|\mathbf{I}_e^0|^{1/2}} \int \mathrm{d}\mathbf{Q}^{\mathrm{trans}}\mathrm{d}\mathbf{Q}^{\mathrm{rot}}, \tag{6.34}$$

where $\overline{\omega}$ is the geometric mean of the non-zero $\{\omega_\alpha\}$ from (6.1). To fix the centre of mass and orientation of the cluster we can transform the remaining six normal coordinates to the centre of mass coordinates, $\mathbf{R}$, and three Euler angles. This transformation requires a Jacobian factor, and then integration over delta functions in the centre of mass and Euler angle variables simply gives unity. Since $\mathbf{Q}^{\rm trans} = \sqrt{M}\mathbf{R}$ the Jacobian $|{\rm d}\mathbf{Q}^{\rm trans}/{\rm d}\mathbf{R}| = M^{3/2}$. Expressions for infinitesimal rotations about the principal axes may be obtained in terms of the change in the corresponding Euler angle and in terms of one of the ${\rm d}Q_\alpha^{\rm rot}$ from (4.9). Hence we obtain a further Jacobian factor of $|\mathbf{I}_e^0|^{1/2}$. The final result for a non-rotating cluster with fixed centre of mass in the normal mode approximation is:

$$\Omega(E,\mathbf{0},\mathbf{0}) = \frac{E^{\kappa-1}}{\overline{\nu}^\kappa \Gamma(\kappa)}, \tag{6.35}$$

in agreement with (6.7).

For a rotating cluster with $\mathbf{P}=\mathbf{0}$ (6.32) the kinetic and configurational densities of states may be factored [144] as

$$\Omega(E,\mathbf{0},\mathbf{L}) = \int_0^E \Omega_{\rm K}(E_{\rm K})\Omega_{\rm c}(E-E_{\rm K},\mathbf{L}){\rm d}E_{\rm K}, \tag{6.36}$$

where $\Omega_{\rm K}(E_{\rm K})$ is given by (6.5) and

$$\Omega_{\rm c}(E_{\rm c},\mathbf{L}) = \int \frac{\delta\left[E_{\rm c} - V(\mathbf{X}) - \mathbf{L}^T(\mathbf{I}^0)^{-1}\mathbf{L}\right)]}{M^{3/2}|\mathbf{I}^0|^{1/2}} {\rm d}X_1{\rm d}X_2\ldots{\rm d}X_{3N}. \tag{6.37}$$

Hence, even for a non-rotating cluster with $\mathbf{L}=\mathbf{0}$ there is a geometrical factor in the configurational density of states due to the term $1/|\mathbf{I}^0|^{1/2}$ [142–144].

6.4 *Example: Isomerisation dynamics of LJ_7*

Miller and Wales used frequent quenching from microcanonical MD simulations to calculate the distribution of lifetimes for the four geometrical isomers of the LJ_7 cluster [147]. The contribution of each isomer to the density of states was also obtained using Metropolis importance sampling of configuration space in the microcanonical ensemble. The occupation probabilities obtained from these contributions were found to be in good agreement with the MD quench statistics, in contrast to the results of Seko and Takatsuka, who concluded that the dynamics are non-ergodic in the melting region [148].

Although the MD and MC results of Figure 20 using (6.38) are very close, there is a perceptible systematic drift at higher energies because the

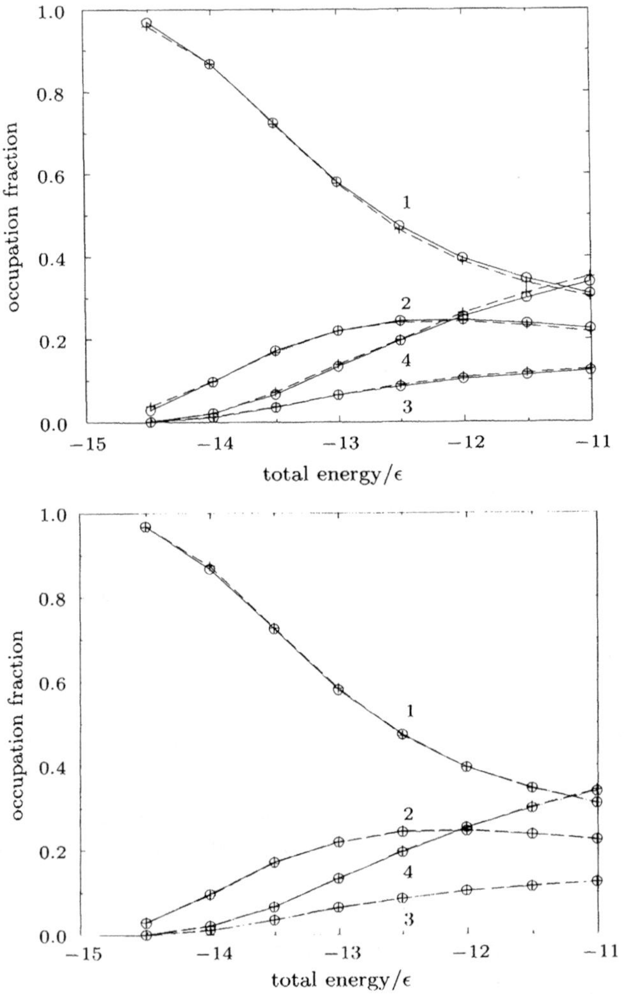

Fig. 20. Occupation fractions for each of the four minima of LJ_7 as a function of energy: top and bottom, results using (6.38) [147] and (6.39) [149], respectively.

moment of inertia factor was not included in the MC acceptance condition. Miller [149] has shown that the occupation probabilities from MC and MD simulations are in even better agreement when the MC acceptance probability is changed from the usual Metropolis condition [150] for microcanonical

MC [151]

$$\pi(\mathbf{X} \to \mathbf{X}') = \min\left(1, \frac{[E - V(\mathbf{X}')]^{\kappa/2-1}\,\Theta[E - V(\mathbf{X}')]}{[E - V(\mathbf{X})]^{\kappa/2-1}\,\Theta[E - V(\mathbf{X})]}\right), \tag{6.38}$$

to [144]

$$\pi(\mathbf{X} \to \mathbf{X}') = \min\left(1, \frac{[E - V(\mathbf{X}')]^{\kappa/2-1}\sqrt{|\mathbf{I}^0|}\,\Theta[E - V(\mathbf{X}')]}{[E - V(\mathbf{X})]^{\kappa/2-1}\sqrt{|\mathbf{I}^0|'}\,\Theta[E - V(\mathbf{X})]}\right). \tag{6.39}$$

The systematic drift observed with increasing energy is eliminated when the correct configuration space weighting is used (Fig. 20). The correction increases with energy because fluctuations in the moment of inertia increase. For larger clusters (with $\mathbf{L} = \mathbf{0}$) this effect would probably be even smaller, and so the uncorrected acceptance probability should give good results.

7 Finite size phase transitions

A first order phase transition in a bulk system occurs when the appropriate thermodynamic potential (*e.g.* free energy or entropy) exhibits a double minimum (or maximum) over some range of parameter space, with a barrier (or well) between the two extrema. For a solid-liquid transition the control parameter may be either temperature or pressure. If we consider conditions of constant N, P and T then the transition occurs where the chemical potentials of the solid and liquid phases are equal. Above and below the transition temperature one of the free energy minima corresponds to a metastable phase, and hysteresis (superheating or supercooling) may be observed in heating and cooling curves because of the free energy barrier between them [152]. For fixed N and P the two phases will coexist at the transition temperature, and in the $P - T$ plane there is a coexistence line in the phase diagram. The free energy derivative, and hence the entropy and other thermodynamic and structural properties, are discontinuous at the transition. The heat capacity is therefore singular at the transition temperature, since it is a second derivative property.

The melting transition of a finite system is different from bulk in several ways. For finite systems results obtained from different ensembles need only agree to order $\mathcal{O}(1/N)$ [131]. This ensemble dependence resulted in controversy and confusion in some earlier work [153], but was recognised in the context of astrophysics by Lynden–Bell and Lynden–Bell [154] and by Honeycutt and Anderson [155] in the context of clusters. Bixon and Jortner compared the caloric curves generated by simple model partition functions in the microcanonical and canonical ensembles, and noted that a negative slope is impossible in the latter ensemble because it is proportional to the mean squared energy fluctuation [156].

The singularities that characterise a bulk transition are rounded in finite systems, and the width of the transition is predicted [157, 158] to scale as $1/N$. Furthermore, the van Hove theorem [159], which forbids the appearance of van der Waals loops, or S-bends, in bulk material, does not apply. Accurate calculations of the total energy density of states using the histogram Monte Carlo approach [160] finally confirmed that such features are not the consequence of mean-field approximations [48, 161, 162]. In finite systems the energetic or entropic penalty for the formation of an interface can prevent solidlike and liquidlike forms from coexisting in contact, and instead it is common to observe a range of energy (or temperature) for which the system samples two separate forms [153]. This situation must be distinguished from the loops observed in approximate theories of the solid-liquid phase transition, which result from the neglect of the phase space corresponding to phase separation [163, 164].

7.1 Stability and van der Waals loops

Stability conditions are defined in terms of derivatives of the appropriate thermodynamic potential [165]. For the microcanonical ensemble the entropy, S, is maximised when $\partial T/\partial E > 0$. Some finite systems exhibit a loop in $\langle T(E)\rangle_\mu$, the microcanonical caloric curve. Some authors reserve the term "van der Waals loop" for loops that result from approximate theories, but here we will use "van der Waals loop" and "S-bend" interchangeably. The branches with positive slope are interpreted as stable solid-like and liquid-like forms (Fig. 21). The canonical caloric curve, $T(\langle E\rangle_{\mathrm{can}})$ cannot have a negative slope because $\partial E/\partial T = \left\langle (E - \langle E\rangle_{\mathrm{can}})^2\right\rangle_{\mathrm{can}}/kT^2$. Such phenomenology was certainly well understood by Hill in the context of ligand adsorption on proteins well before the first simulations of clusters [166–168].

More recently, order parameters have been used to provide a natural connection to bulk phase transitions and an operational criterion for phase coexistence [141, 164, 169]. Here we define a Landau free energy and entropy as

$$\begin{aligned} A_{\mathrm{L}}(Q) &= A(T) - kT\ln p_{\mathrm{can}}(Q), \\ S_{\mathrm{L}}(Q) &= S(E) + k\ln p_\mu(Q). \end{aligned} \tag{7.1}$$

A_{L} and S_{L} are examples of restricted thermodynamic potentials.

The canonical probability distribution of the total energy is:

$$p_{\mathrm{can}}(E;T) = \Omega(N,V,E)\exp(-E/kT)/Q(N,V,T)$$

so $$k\left(\frac{\partial \ln p_{\mathrm{can}}(E;T)}{\partial E}\right)_{N,V} = \left(\frac{\partial S}{\partial E}\right)_{N,V} - \frac{1}{T} = \frac{1}{\langle T(E)\rangle_\mu} - \frac{1}{T}. \tag{7.2}$$

General relations can be obtained for other conjugate variables and ensembles related by Legendre transforms [170].

The above results can be combined to deduce the following equivalent conditions related to the finite system analogue of the melting transition (Fig. 21):

- $p_{\rm can}(E)$ is bimodal for $T_f < T < T_m$;
- there is a loop in $\langle T(E)\rangle_\mu$, with turning points T_f, T_m;
- there are three energies where $\langle T(E)\rangle_\mu = T$ for $T_f < T < T_m$;
- $S(E)$ [or $\ln\Omega(E)$] must have two inflection points since $\partial S/\partial E$ must have the same slope at three points;
- the Landau free energy with $Q \equiv E$ has a double minimum as a function of E.

Physical considerations suggest that the energy density of states obeys power laws at high and low energies, so that $\partial^2 \ln\Omega(E)/\partial E^2$ is positive in both these limits [130]. Two inflection points are therefore guaranteed if $S(E)$ exhibits negative curvature.

Points E_1 and E_2 lying on a common tangent to $S(E)$ have $\langle T(E_1)\rangle_\mu = \langle T(E_2)\rangle_\mu = T^*$. Hence:

$$\int_{E_1}^{E_2}\left(\frac{1}{\langle T(E)\rangle_\mu} - \frac{1}{T^*}\right)\mathrm{d}E = S(E_1) - S(E_2) - \frac{(E_1 - E_2)}{T^*} = 0. \qquad (7.3)$$

This is the microcanonical analogue of the Maxwell tie-line construction.

In bulk systems the "convex intruder" [171] in $S(E)$ is forbidden by the van Hove theorem [159] and two phases coexist in contact. For a finite system the probability that the system is in a particular form decreases smoothly from 1 to 0 for $T_f < T < T_m$ (canonical) (Fig. 21) [172]. For a bulk transition the equilibrium constant is a step function, but metastable super heated or supercooled phases may still exist.

Mean field models, such as the van der Waals equation, wrongly predict the appearance of loops in the bulk limit because they neglect the phase space corresponding to solid and liquid in contact. We do not yet know at what size phase separation will begin to be seen in clusters, although for $(KCl)_{32}$ non-wetted structures have been found in which one part of the cluster has the rock-salt structure and the other part an amorphous liquid-like structure [173]. The results of a simple model for the total energy density of states, written as [163]

$$\Omega(E) = \Omega_{\rm solid}(E) + \Omega_{\rm liquid}(E) + \Omega_{\rm mix}(E), \qquad (7.4)$$

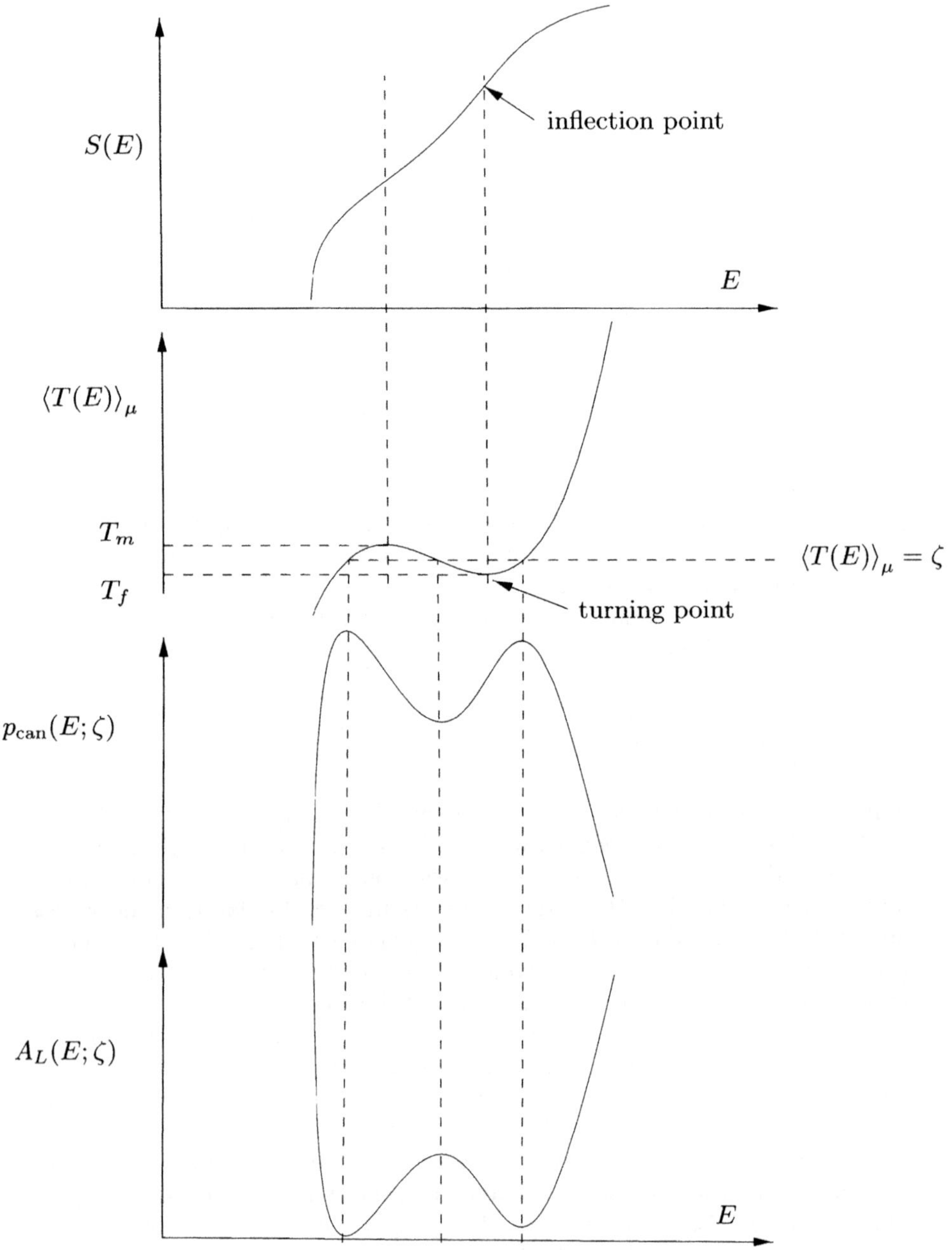

Fig. 21. Equivalent conditions for the finite system melting transition.

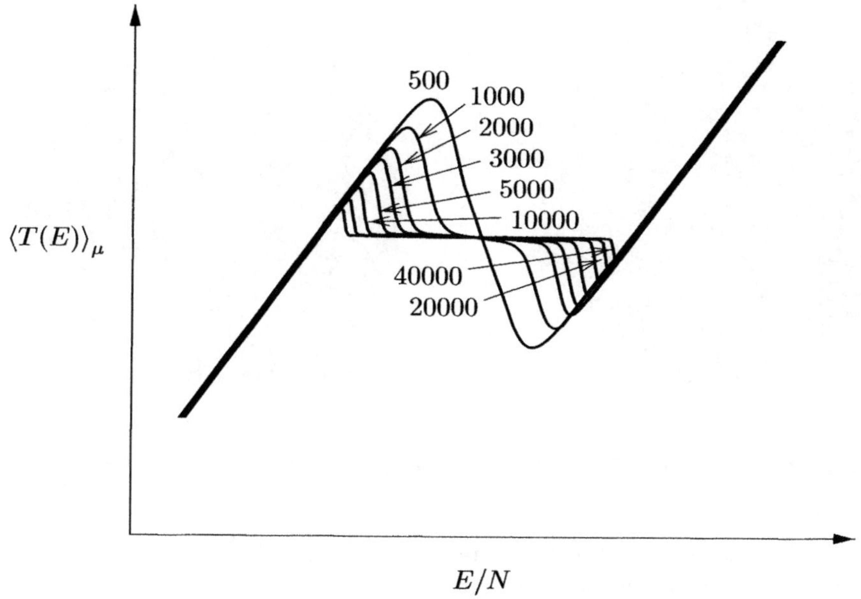

Fig. 22. Size dependence of the microcanonical caloric curve for a model that admits the phase space corresponding to solidlike and liquidlike regions in contact [163].

enable the size dependence of the microcanonical caloric curve to be investigated (Fig. 22). For small systems $\Omega_{\rm mix}(E)$ in phase separated configurations is negligible, but without this term there would be a loop in the bulk limit.

A loop in the isopotential ensemble caloric curve is a necessary (but not sufficient) condition for a loop in $\langle T(E)\rangle_\mu$ [141]. We can see this by writing the canonical probability distribution for the total energy in terms of the probability distributions for the potential and kinetic energy:

$$\begin{aligned} p_{\rm can}(E;T) &= \int_0^E p_{\rm c}(E-E_{\rm K};T)p_{\rm K}(E_{\rm K};T){\rm d}E_{\rm K} \\ &= \int_0^E p_{\rm c}(E-E_{\rm K};T)\frac{E_{\rm K}^{\kappa/2-1}{\rm e}^{-E_{\rm K}/kT}}{\Gamma(\kappa/2)/(kT)^{\kappa/2}}{\rm d}E_{\rm K}. \end{aligned}$$

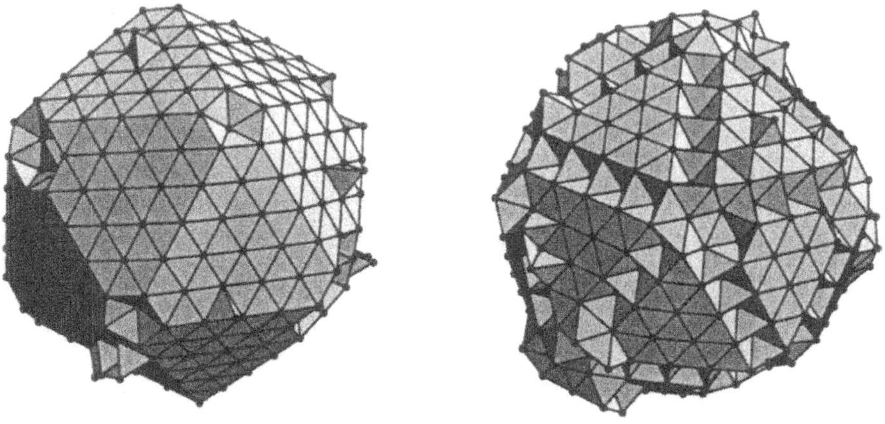

Fig. 23. "Roughened" and reconstructed minima of M_{923} ($\rho = 5$) [183].

This result follows from the the expression for $\Omega_K(E_K)$ derived in equation (6.5), from which we deduce that

$$\begin{aligned} p_K(E_K;T) &= \Omega_K(E_K)e^{-E_K/kT} \Big/ \int \Omega_K(E_K)e^{-E_K/kT} dE_K \\ &= \frac{E_K^{\kappa/2-1}e^{-E_K/kT}}{\Gamma(\kappa/2)} \int \frac{E_K^{\kappa/2-1}e^{-E_K/kT}}{\Gamma(\kappa/2)} e^{-E_K/kT} dE_K \\ &= E_K^{\kappa/2-1}e^{-E_K/kT}/\Gamma(\kappa/2)(kT)^{\kappa/2}. \end{aligned} \tag{7.5}$$

Two groups [174, 175] have used the presence or absence of a loop in the microcanonical caloric curve to classify phase transitions as first or second order in the Ehrenfest sense [176]. Interfacial entropies have also been calculated from the area under the loop following Binder [175, 177, 178]. In these studies the finite size behaviour stems from supercell models of the bulk using periodic boundary conditions. For clusters we expect the loop in the microcanonical caloric curve to first appear and then disappear as a function of size. This behaviour is also likely to be non-monotonic, and hence inferences of bulk properties based upon such considerations must be treated with caution. An alternative approach has been proposed based upon analysis of Lee–Yang zeros [179, 180] of the partition function in the complex temperature plane [181, 182]. However, classification criteria developed for bulk matter cannot produce unambiguous results for finite systems.

Coexistence of different phase-like cluster forms over a range of temperature (canonical ensemble) or total energy (microcanonical ensemble) can be diagnosed using the appropriate restricted thermodynamic potential, as described above [152, 169]. However, short time averages of quantities such as the kinetic energy have also been used to interrogate dynamical coexistence [153, 184–187]. Unfortunately, the choice of a suitable order parameter or dynamical property is somewhat arbitrary, and different choices need not agree with one another. In particular, dynamical criteria based on short time averaging can often reveal coexistence of different phase-like forms without the appearance of inflection points in $S(E)$, or any of the equivalent thermodynamic features described above.

Such ambiguities seem inevitable for finite systems, and lead to some subjectivity in the description of phenomena such as "surface melting" in clusters [183, 188, 189]. For example, short-time-averaged order parameters for LJ_{55} produce multimodal distributions [189], but there are no corresponding features in $\langle T(E)\rangle_\mu$, $S(E)$, etc. [132]. Larger Morse clusters exhibit an anti-Mackay surface reconstruction before they melt, which only produces an inflection in $\langle T(E)\rangle_\mu$ [132]. Surface disordering due to edge- and vertex-migration occurs first, probably analogous to surface roughening (Fig. 23). The temperature at which magic numbers disappear experimentally may correspond to surface reconstruction rather than melting [190].

8 Global optimisation

The global optimisation problem is a subject of intense current interest. Applications of obvious economic importance include travelling salesman-type problems and the design of microprocessor circuitry. In the domain of atoms and molecules, one often wishes to discover the lowest-energy isomer or crystal structure for a system with a given composition. Attempting to predict the three-dimensional structure of a protein solely from its amino acid sequence by computer simulation is a major research effort.

A random conformational search for the global minimum of a typical protein would take an astronomical time. However, it is now generally agreed that the search is not random, but instead the PES is biased towards the global minimum, resulting in efficient relaxation (Sect. 1.1). Hence the amino acid sequences of naturally occurring proteins have presumably evolved to fold rapidly into a unique native structure. In treating any non-trivial global optimisation problem, the principal difficulty arises from the exponentially large number of minima on the PES (Sect. 3.1).

For the cluster of 55 atoms interacting by a Lennard–Jones (LJ) potential (mentioned in Sect. 4.4.4) the number of minima (excluding permutational isomers) is at least 10^{10}, but the global minimum is relatively easy to locate because of the focusing properties of the energy landscape. Most global minima for LJ clusters containing fewer than 100 atoms are based upon icosahedral packing. The exceptions (LJ_{38}, LJ_{75-77} and LJ_{98}) serve as

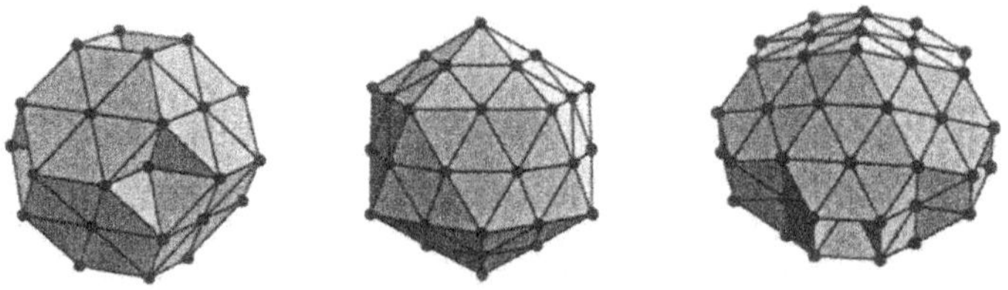

Fig. 24. From left to right: global minima of the LJ potential for 38 atoms (truncated octahedron), 55 atoms (Mackay icosahedron [191]), and 75 atoms (Marks decahedron [192]).

particularly interesting test cases, because the corresponding energy landscapes consist of two families of structures. Two examples, the truncated octahedron for LJ_{38} and the Marks decahedron [192] for LJ_{75}, are shown in Figure 24 [193]. At these sizes, the lowest-energy minimum based upon icosahedral packing acts as a trap, and is widely separated from the true global minimum. Actually, it is quite easy to find the global minima of these clusters by seeding the starting geometry with a core of the right type.

Simulated annealing provided the first generally applicable technique for global optimisation [194]. Here the state of the system is followed by simulation as the temperature is decreased slowly from a high value, in the hope that it will eventually come to rest at the global potential energy minimum. Unfortunately, the free energy global minimum may change at a temperature where energy barriers are too high for the system to escape from a local minimum.

8.1 Basin-hopping global optimisation

There is a simple transformation of the energy landscape that does not change the global minimum, or the relative energies of any local minima [195]:

$$\tilde{E}(\mathbf{X}) = \min\{E(\mathbf{X})\} \tag{8.1}$$

where "min" signifies that an energy minimisation is carried out starting from $\mathbf{X}$. The transformed energy, $\tilde{E}(\mathbf{X})$, at any point, $\mathbf{X}$, becomes the energy of the structure obtained by minimisation. Each local minimum is, therefore, surrounded by a catchment basin [56, 127] of constant energy consisting of all the neighbouring geometries from which that particular minimum is obtained. The catchment basin transformation removes all the transition state regions from the surface and accelerates the dynamics because the system can pass between basins all along their boundary. Atoms can even pass through each other without encountering prohibitive energy barriers.

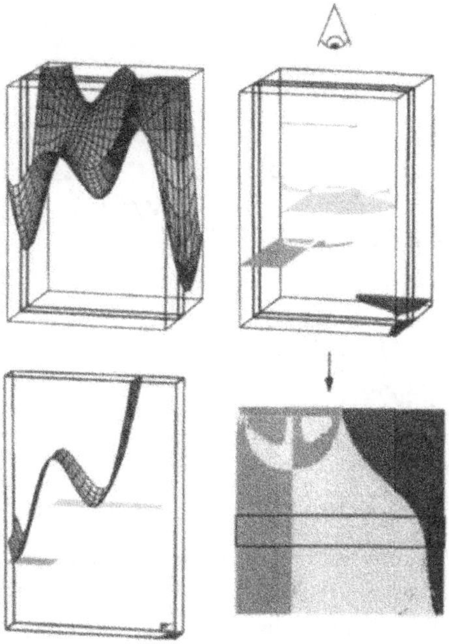

Fig. 25. Illustration of the $\tilde{E}(\mathbf{X})$ energy landscape transformation.

The basin-hopping approach therefore transforms the energy landscape to a discrete set of energy levels corresponding to the energies of local minima (Fig. 25), and must be combined with a search strategy. In the "Monte-Carlo plus energy minimisation" (MCM) procedure steps are proposed by perturbing the current coordinates and carrying out a minimisation from the resulting geometry [23,196,197]. A step is accepted if the energy of the new minimum, $E_{\rm new}$, is lower than the starting point, $E_{\rm old}$. If $E_{\rm new} > E_{\rm old}$ then the step is accepted if $\exp[(E_{\rm old} - E_{\rm new})/kT]$ is greater than a random number drawn from the interval [0,1]. The temperature, T, becomes an adjustable parameter.

The mean number of basin-hopping steps and cpu time required to find the global minimum at a fixed temperature are shown in Figure 26 for LJ_n up to $n = 74$. The results are averages over 100 different random starting points in each case. A simple basin-hopping program in fortran 77 can be downloaded from `http://brian.ch.cam.ac.uk/software.html` and collected results for cluster global minima may be found in the Cambridge Cluster Database [198].

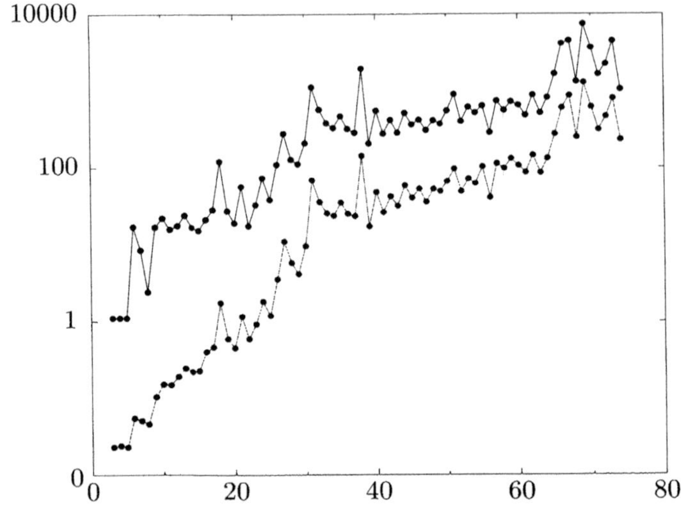

Fig. 26. Statistics for the MCM procedure with LJ_n up to $n = 74$. The cpu time (seconds) in the lower curve is for a 250 MHz Sun Ultra II processor, and the number of basin-hopping steps in the upper plot is dimensionless. The horizontal axis is n, the number of atoms.

I am most grateful to Dr. F. Calvo, Dr. J.P.K. Doye, Dr. M.A. Miller, Dr. A.J. Stone and Prof. T. Taketsugu for their comments on various drafts of these notes.

References

[1] D.J. Wales, J.P.K. Doye, M.A. Miller, P.N. Mortenson and T.R. Walsh, *Adv. Chem. Phys.* **115** (2000) 1.

[2] M. Born and J.R. Oppenheimer, *Ann. Phys.* **84** (1927) 457.

[3] C.B. Anfinsen, in *New perspectives in biology*, edited by M. Sela (Elsevier, Amsterdam, 1964), pp. 42-50.

[4] C. Levinthal, in *Mössbauer spectroscopy in biological systems, proceedings of a meeting held at Allerton House, Monticello, Illinois*, edited by P. DeBrunner, J. Tsibris and E. Munck (Urbana, University of Illinois Press, 1969), p. 22.

[5] C. Levinthal, *J. Chim. Phys.* **65** (1968) 44.

[6] A. Šali, E. Shakhnovich and M. Karplus, *Nature* **369** (1994) 248.

[7] N.D. Socci and J.N. Onuchic, *J. Chem. Phys.* **101** (1994) 1519.

[8] T.R. Sosnick, L. Mayne and S.W. Englander, *Proteins: Struct., Func. Gen.* **24** (1996) 413.

[9] J.D. Bryngelson, J.N. Onuchic, N.D. Socci and P.G. Wolynes, *Proteins: Struct., Func. Gen.* **21** (1995) 167.

[10] R. Zwanzig, A. Szabo and B. Bagchi, *Proc. Natl. Acad. Sci. USA* **89** (1992) 20.

[11] R. Zwanzig, *Proc. Natl. Acad. Sci. USA* **92** (1995) 9801.

[12] P.E. Leopold, M. Montal and J.N. Onuchic, *Proc. Natl. Acad. Sci. USA* **89** (1992) 8271.

[13] H.W. Kroto, J.R. Heath, S.C. O'Brien, R.F. Curl and R.E. Smalley, *Nature* **318** (1985) 162.

[14] H. Frauenfelder, S.G. Sligar and P.G. Wolynes, *Science* **254** (1991) 1598.

[15] P.G. Wolynes, J.N. Onuchic and D. Thirumalai, *Science* **267** (1995) 1619.

[16] N.D. Socci, J.N. Onuchic and P.G. Wolynes, *Proteins: Struct., Func. Gen.* **32** (1998) 136.

[17] N.D. Socci, J.N. Onuchic and P.G. Wolynes, *J. Chem. Phys.* **104** (1996) 5860.

[18] A. Šali, E. Shakhnovich and M. Karplus, *J. Mol. Biol.* **235** (1994) 1614.

[19] H.S. Chan, *Nature* **373** (1995) 664.

[20] A. Gutin, A. Sali, V. Abkevich, M. Karplus and E.I. Shakhnovich, *J. Chem. Phys.* **108** (1998) 6466.

[21] J.P.K. Doye and D.J. Wales, *J. Chem. Phys.* **105** (1996) 8428.

[22] C.B. Anfinsen, *Science* **181** (1973) 223.

[23] D.J. Wales and J.P.K. Doye, *J. Phys. Chem. A* **101** (1997) 5111.

[24] J.P.K. Doye and D.J. Wales, *Phys. Rev. Lett.* **80** (1998) 1357.

[25] J.P.K. Doye, D.J. Wales and M.A. Miller, *J. Chem. Phys.* **109** (1998) 8143.

[26] C.A. Angell, *J. Non-Cryst. Solids* **131-133** (1991) 13.

[27] J.L. Green, K. Ito, K. Xu and C.A. Angell, *J. Phys. Chem. B* **103** (1999) 3991.

[28] W. Kauzmann, *Chem. Rev.* **43** (1948) 219.

[29] G.S. Fulcher, *J. Am. Ceram. Soc.* **8** (1925) 339.

[30] H. Vogel, *Z. Phys.* **22** (1921) 645.

[31] G. Tammann and W.Z. Hesse, *Z. Anorg. Allgem. Chem.* **156** (1926) 245.

[32] J.D. Ferry, L.D. Grandine and E.R. Fitzgerald, *J. Appl. Phys.* **24** (1533) 911.

[33] G. Williams and D.C. Watts, *J. Chem. Soc., Faraday Trans.* **66** (1970) 80.

[34] R. Böhmer, K.L. Ngai, C.A. Angell and D.J. Plazek, *J. Chem. Phys.* **99** (1993) 4201.

[35] F.H. Stillinger, *Science* **267** (1995) 1935.

[36] G.P. Johari and M. Goldstein, *J. Chem. Phys.* **53** (1970) 2372.

[37] G. Adam and J.H. Gibbs, *J. Chem. Phys.* **43** (1965) 139.

[38] C.A. Angell, *Science* **267** (1995) 1924.

[39] P.W. Anderson, B.I. Halperin and C.M. Varma, *Philos. Mag.* **25** (1972) 1.

[40] W.A. Phillips, *J. Low Temp. Phys.* **7** (1972) 351.

[41] R.C. Zeller and R.O. Pohl, *Phys. Rev. B* **4** (1971) 2029.

[42] M. Goldstein, *J. Chem. Phys.* **51** (1969) 3728.

[43] P. Pechukas, *J. Chem. Phys.* **64** (1976) 1516.

[44] P.R. Bunker and P. Jensen, *Molecular Symmetry and Spectroscopy*, 2nd edition, NRC Research Press, Ottawa (1998).

[45] E.B. Wilson, J.C. Decius and P.C. Cross, *Molecular Vibrations* (Dover, New York, 1980).

[46] J.T. Hougen, *J. Chem. Phys.* **37** (1962) 1433.

[47] J.T. Hougen, *J. Chem. Phys.* **39** (1963) 358.

[48] C.J. Tsai and K.D. Jordan, *J. Chem. Phys.* **99** (1993) 6957.

[49] F.H. Stillinger and T.A. Weber, *Phys. Rev. A* **28** (1983) 2408.

[50] F.H. Stillinger and T.A. Weber, *Science* **225** (1984) 983.

[51] F.H. Stillinger, *Phys. Rev. E* **59** (1999) 48.
[52] M. Morse, *Amer. Math. Soc. Colloq. Publ.* **18** (1934).
[53] P.G. Mezey, *Chem. Phys. Lett.* **82** (1981) 100.
[54] P.G. Mezey, *Theoretica Chim. Acta (Berl)* **62** (1982) 133.
[55] P.G. Mezey, *Theoretica Chim. Acta (Berl)* **63** (1983) 9.
[56] P.G. Mezey, *Potential Energy Hypersurfaces* (Elsevier, Amsterdam, 1987).
[57] R.S. Dumont, *J. Chem. Phys.* **97** (1991) 9172.
[58] M.N. Ramquet, G. Dive and D. Dehareng, *J. Chem. Phys.* **112** (2000) 4923.
[59] J.N. Murrell and K.J. Laidler, *J. Chem. Soc., Faraday Trans.* **64** (1968) 371.
[60] K. Fukui, *J. Phys. Chem.* **74** (1970) 4161.
[61] R.G. Pearson, *Acc. Chem. Res.* **4** (1971) 152.
[62] J.W. McIver and R.E. Stanton, *J. Am. Chem. Soc.* **94** (1972) 8618.
[63] R.E. Stanton and J.W. McIver, *J. Am. Chem. Soc.* **97** (1975) 3632.
[64] R.B. Woodward and R. Hoffmann, *Angew. Chem. Int. Ed. Engl.* **8** (1969) 781.
[65] J.D. Goddard and H.F. Schaefer, *J. Chem. Phys.* **70** (1979) 5117.
[66] S.M. Colwell and N.C. Handy, *J. Chem. Phys.* **82** (1985) 1281.
[67] A. Tachibana, I. Okazaki, M. Koizumi, K. Hori and T. Yanabe, *J. Am. Chem. Soc.* **107** (1985) 1190.
[68] P. Valtazanos, S.T. Elbert, S. Xantheas and K. Ruedenberg, *Theor. Chim. Acta* **78** (1991) 287.
[69] T.L. Windus and M.S. Gordon, *Theor. Chim. Acta* **83** (1992) 21.
[70] T. Taketsugu and T. Hirano, *J. Chem. Phys.* **99** (1993) 9806.
[71] R.M. Minyaev and D.J. Wales, *Chem. Phys. Lett.* **218** (1994) 413.
[72] R.M. Minyaev, *Int. J. Quant. Chem.* **49** (1994) 105.
[73] R.E. Leone and P.v.R. Schleyer, *Angew. Chem. Int. Ed. Engl.* **9** (1970) 860.
[74] J.F. Cornwell, *Group theory in physics* (Academic Press, San Diego, 1997).
[75] J.G. Nourse, *J. Am. Chem. Soc.* **102** (1980) 4883.
[76] D.J. Wales, *J. Am. Chem. Soc.* **115** (1993) 11180.
[77] R.G.A. Bone, T.W. Rowlands, N.C. Handy and A.J. Stone, *Mol. Phys.* **72** (1991) 33.
[78] R.G.A. Bone, *Chem. Phys. Lett.* **193** (1992) 557.
[79] D.J. Wales and A.M. Lee, *Chem. Phys. Lett.* **198**, 279 (1992).
[80] B. Friedrich, Z. Herman, R. Zahradnik and Z. Havlas, *Adv. Quant. Chem.* **19** (1988) 257.
[81] A. Banerjee and N.P. Adams, *Int. J. Quant. Chem.* **43** (1992) 855.
[82] D.J. Wales, *J. Chem. Phys.* **113** (2000) 3926.
[83] W.X. Li, T. Keyes, R.L. Murry and J.T. Fourkas, *J. Chem. Phys.* **109** (1998) 9096.
[84] B. Paizs, G. Fogarasi and P. Pulay, *J. Chem. Phys.* **109** (1998) 6571.
[85] O. Farkas and H.B. Schlegel, *J. Chem. Phys.* **109** (1998) 7100.
[86] J. Baker, D. Kinghorn and P. Pulay, *J. Chem. Phys.* **110** (1999) 4986.
[87] R. Lindh, A. Bernhardsson and M. Schutz, *Chem. Phys. Lett.* **303** (1999) 567.
[88] A. Tachibana and K. Fukui, *Theo. Chim. Acta* **49** (1978) 321.
[89] K. Fukui, *Acc. Chem. Res.* **14** (1981) 363.
[90] R.A. Horn and C.R. Johnson, *Matrix Analysis*, (Cambridge University Press, Cambridge, 1985).
[91] P. Valtazanos and K. Ruedenburg, *Theo. Chim. Acta* **69** (1986) 281.

[92] R.M. Minyaev and D.J. Wales, *J. Phys. Chem.* **98** (1994) 7942.

[93] W. Quapp, *J. Chem. Phys.* **114** (2001) 609.

[94] Y. Kumeda and T. Taketsugu, *J. Chem. Phys.* **113** (2000) 477.

[95] D.M. Dennison and J.D. Hardy, *Phys. Rev.* **39** (1932) 938.

[96] H.-S. Sheng, E.F. Barker and D.M. Dennison, *Phys. Rev.* **60** (1941) 786.

[97] M. Kolbuszewski and P.R. Bunker, *J. Chem. Phys.* **105** (1996) 3649.

[98] T.R. Dyke, B.J. Howard and W. Klemperer, *J. Chem. Phys.* **56** (1972) 2442.

[99] G.C. Hancock and D.G. Truhlar, *J. Chem. Phys.* **90** (1998) 3498.

[100] P. Jensen, P.R. Bunker, A. Karpfen, M. Kofranek and H. Lischka, *J. Chem. Phys.* **93** (1990) 6266.

[101] P.R. Bunker, P. Jensen, A. Karpfen, M. Kofranek and H. Lischka, *J. Chem. Phys.* **92** (1990) 7432.

[102] S.C. Althorpe, D.C. Clary and P.R. Bunker, *Chem. Phys. Lett.* **187** (1991) 345.

[103] D.H. Zhang, Q. Wu, J.Z.H. Zhang, M. Vondirke and Z. Bačić, *J. Chem. Phys.* **102** (19965) 2315.

[104] M. Vondirke, Z. Bačić, D.H. Zhang and J.Z.H. Zhang, *J. Chem. Phys.* **102** (1995) 4382.

[105] H.-C. Change and W. Klemperer, *J. Chem. Phys.* **104** (1996) 7830.

[106] W. Klopper, M. Quack and M.A. Suhm, *J. Chem. Phys.* **108** (1998) 10096.

[107] R.C. Cohen and R.J. Saykally, *J. Phys. Chem.* **94** (1990) 7991.

[108] N. Pugliano and R.J. Saykally, *Science* **257** (1992) 1937.

[109] R.J. Saykally and G.A. Blake, *Science* **259** (1993) 1570.

[110] K. Liu, J.D. Cruzan and R.J. Saykally, *Science* **271** (1996) 929.

[111] N. Pugliano and R.J. Saykally, *Science* **257** (1992) 1937.

[112] K. Liu, J.G. Loeser, M.J. Elrod, B.C. Host, J.A. Rzepiela, N. Pugliano and R.J. Saykally, *J. Am. Chem. Soc.* **116** (1994) 3507.

[113] K. Liu, M.J. Elrod, J.G. Loeser, J.D. Cruzan, N. Pugliano, M.G. Brown, J. Rzepiela and R.J. Saykally, *Faraday Discussions* **97** (1994) 35.

[114] K. Liu, M.G. Brown, M.R. Viant, J.D. Cruzan and R.J. Saykally, *Mol. Phys.* **89** (1996) 1373.

[115] M.R. Viant, J.D. Cruzan, D.D. Lucas, M.G. Brown, K. Liu and R.J. Saykally, *J. Phys. Chem. A* **101** (1997) 9032.

[116] M.R. Viant, M.G. Brown, J.D. Cruzan, R.J. Saykally, M. Geleijns and A.V. der Avoird, *J. Chem. Phys.* **110** (1999) 4369.

[117] M.G. Brown, M.R. Viant, R.P. McLaughlin, C.J. Keoshian, E. Michael, J.D. Cruzan, R.J. Saykally and A. van der Avoird, *J. Chem. Phys.* **111** (1999) 7789.

[118] T.R. Walsh and D.J. Wales, *J. Chem. Soc., Faraday Trans.* **92** (1996) 2505.

[119] D. Sabo, Z. Bačić, T. Bürgi and S. Leutwyler, *Chem. Phys. Lett.* **244** (1995) 283.

[120] T. Bürgi, S. Graf, S. Leutwyler and W. Kopper, *J. Chem. Phys.* **103** (1995) 1077.

[121] D.A. McQuarrie, *Statistical Mechanics* (Harper and Row, New York, 1976).

[122] D.J. Wales, *Mol. Phys.* **78** (1993) 151.

[123] D.J. McGinty, *J. Chem. Phys.* **55** (1971) 580.

[124] J.J. Burton, *J. Chem. Phys.* **56** (1972) 3133.

[125] M.R. Hoare, *Adv. Chem. Phys.* **40** (1979) 49.

[126] G. Franke, E.R. Hilf and P. Borrmann, *J. Chem. Phys.* **98** (1993) 3496.

[127] F.H. Stillinger and T.A. Weber, *Phys. Rev. A* **25** (1982) 978.

[128] J.E. Jones and A.E. Ingham, *Proc. R. Soc. A* **107** (1925) 636.

[129] E.M. Pearson, T. Halicioglu and W.A. Tiller, *Phys. Rev. A* **32** (1985) 3030.

[130] I.H. Umirzakov, *Phys. Rev. E* **60** (1999) 7550.

[131] M.P. Allen and D.J. Tildesley, *The Computer Simulation of Liquids* (Clarendon Press, Oxford, 1987).

[132] J.P.K. Doye and D.J. Wales, *J. Chem. Phys.* **102** (1995) 9659.

[133] J. Jellinek and A. Goldberg, *J. Chem. Phys.* **113** (2000) 2570.

[134] K. Huang, *Statistical Mechanics* (Wiley, New York, 1987).

[135] I.R. McDonald and K. Singer, *Discuss. Faraday Soc.* **43** (1967) 40.

[136] A.M. Ferrenberg and R.H. Swendsen, *Phys. Rev. Lett.* **61** (1988) 2635.

[137] A.M. Ferrenberg and R.H. Swendsen, *Phys. Rev. Lett.* **63** (1989) 1195.

[138] C. Bichara, J.P. Gaspard and J.C. Mathieu, *Phys. Lett. A* **119** (1987) 462.

[139] E. Marinari and G. Parisi, *Europhys. Lett.* **19** (1992) 451.

[140] S.F. Chekmarev and I.H. Umirzakov, *Z. Phys. D* **26** (1993) 373.

[141] R.M. Lynden-Bell and D.J. Wales, *J. Chem. Phys.* **101** (1994) 1460.

[142] S.C. Smith, *J. Chem. Phys.* **97** (1992) 2406.

[143] R.S. Dumont, *J. Chem. Phys.* **97** (1991) 9172.

[144] F. Calvo and P. Labastie, *Eur. Phys. J. D* **3** (1998) 229.

[145] I.H. Umirzakov, *Phys. Rev. E* **61** (2000) 7188.

[146] G.F. Roach, *Green's Functions* (Cambridge University Press, Cambridge, 2nd Edn., 1982).

[147] M.A. Miller and D.J. Wales, *J. Chem. Phys.* **107** (1997) 8568.

[148] C. Seko and K. Takasuka, *J. Chem. Phys.* **104** (1996) 8613.

[149] M.A. Miller, *Energy Landscapes and Dynamics of Model Clusters*, Ph.D. Thesis (1999), University of Cambridge.

[150] N. Metropolis, A.W. Rosenbluth, M.N. Rosenbluth, A.H. Teller and E. Teller, *J. Chem. Phys.* **21** (1953) 1087.

[151] J.R. Ray, *Phys. Rev. A* **44** (1991) 4061.

[152] R.M. Lynden-Bell, *Mol. Phys.* **86** (1995) 1353.

[153] R.S. Berry, T.L. Beck, H.L. Davis and J. Jellinek, *Adv. Chem. Phys.* **70B** (1988) 75.

[154] D. Lynden-Bell and R.M. Lynden-Bell, *Mon. Not. Roy. Astron. Soc.* **181** (1977) 405.

[155] J.D. Honeycutt and H.C. Andersen, *J. Chem. Phys.* **91** (1987) 4950.

[156] M. Bixon and J. Jortner, *J. Chem. Phys.* **91** (1989) 1631.

[157] Y. Imry, *Phys. Rev. B* **21** (1980) 2042.

[158] A. Hüller, *Z. Phys. D* **95** (1994) 63.

[159] L. Van Hove, *Physica* **15** (1949) 951.

[160] A.M. Ferrenberg and R.H. Swendsen, *Phys. Rev. Lett.* **61** (1988) 2635.

[161] P. Labastie and R.L. Whetten, *Phys. Rev. Lett.* **65** (1990) 1567.

[162] H.-P. Cheng, X. Li, R.L. Whetten and R.S. Berry, *Phys. Rev. A* **46** (1992) 791.

[163] D.J. Wales and J.P.K. Doye, *J. Chem. Phys.* **103** (1995) 3061.

[164] D.J. Wales and J.P.K. Doye, in *Large Clusters of Atoms and Molecules*, edited by T.P. Martin (Kluwer, Dordrecht, 1996), pp. 241-280.

[165] D. Chandler, *Introduction to Modern Statistical Mechanics* (Oxford University Press, Oxford, 1987).

[166] T.L. Hill, *J. Phys. Chem.* **57** (1953) 324.

[167] T.L. Hill, *J. Chem. Phys.* **23** (1955) 812.

[168] T.L. Hill, *Thermodynamics of Small Systems (Parts I and II)* (Dover, New York, 1994).

[169] J.P.K. Doye and D.J. Wales, *J. Chem. Phys.* **102** (1995) 9673.

[170] D.J. Wales and R.S. Berry, *Phys. Rev. Lett.* **73** (1994) 2875.

[171] D.H.E. Gross, *Phys. Rep.* **279** (1997) 120.

[172] R.S. Berry and D.J. Wales, *Phys. Rev. Lett.* **63** (1989) 1156.

[173] J.P. Rose and R.S. Berry, *J. Chem. Phys.* **98** (1993) 3246.

[174] A. Hüller, *Z. Phys. D* **93** (1994) 401.

[175] D.H.E. Gross, A. Ecker and X.Z. Zhang, *Annalen Physik* **5** (1996) 446.

[176] P. Ehrenfest, in *Collected Scientific Papers*, edited by M.J. Klein (North-Holland, Amsterdam, 1959).

[177] K. Binder, *Phys. Rev. A* **25** (1982) 1699.

[178] J.E. Hunter and W.P. Reinhardt, *J. Chem. Phys.* **103** (1995) 8627.

[179] C.N. Yang and T. Lee, *Phys. Rev.* **87** (1952) 404.

[180] C.N. Yang and T. Lee, *Phys. Rev.* **87** (1952) 410.

[181] N.A. Alves and U.H.E. Hansmann, *Phys. Rev. Lett.* **84** (2000) 1836.

[182] P. Borrmann, O. Mulken and J. Harting, *Phys. Rev. Lett.* **84** (2000) 3511.

[183] J.P.K. Doye and D.J. Wales, *Z. Phys. D* **40** (1997) 466.

[184] J. Jellinek, T.L. Beck and R.S. Berry, *J. Chem. Phys.* **84** (1986) 2783.

[185] T.L. Beck, J. Jellinek and R.S. Berry, *J. Chem. Phys.* **87** (1987) 545.

[186] T.L. Beck and R.S. Berry, *J. Chem. Phys.* **88** (1988) 3910.

[187] D.J. Wales and R.S. Berry, *J. Chem. Phys.* **92** (1990) 4283.

[188] V.V. Nauchitel and A.J. Pertsin, *Mol. Phys.* **40** (1980) 1341.

[189] R.E. Kunz and R.S. Berry, *Phys. Rev. E* **49** (1994) 1895.

[190] M. Schmidt, R. Kusche, W. Kronmüller, B. von Issendorff and H. Haberland, *Phys. Rev. Lett.* **79** (1997) 99.

[191] A.L. Mackay, *Acta Cryst.* **15** (1962) 916.

[192] L.D. Marks, *Phil. Mag. A* **49** (1984) 81.

[193] D.J. Wales, M.A. Miller and T.R. Walsh, *Nature* **394** (1998) 758.

[194] S. Kirkpatrick, C.D. Gelatt and M.P. Vecchi, *Science* **220** (1983) 671.

[195] D.J. Wales and H.A. Scheraga, *Science* **285** (1999) 1368.

[196] Z. Li and H.A. Scheraga, *Proc. Natl. Acad. Sci. USA* **84** (1987) 6611.

[197] Z. Li and H.A. Scheraga, *J. Mol. Struct.* **179** (1988) 333.

[198] D.J. Wales, J.P.K. Doye, A. Dullweber and F.Y. Naumkin, The Cambridge Cluster Database, URL `http://brian.ch.cam.ac.uk/CCD.html`

COURSE 11

CONFINEMENT TECHNIQUE FOR SIMULATING FINITE MANY-BODY SYSTEMS

S.F. CHEKMAREV

Institute of Thermophysics, SB RAN,
630090 Novosibirsk, Russia
and
Novosibirsk State University,
630090 Novosibirsk, Russia

Contents

CONFINEMENT TECHNIQUE FOR SIMULATING FINITE MANY-BODY SYSTEMS

S.F. Chekmarev

Abstract

A novel approach to computer simulation study of a finite many-body system is considered that allows one to gain a detailed information about the system, including its potential energy surface (PES), equilibrium properties and kinetics. The approach is based on a confinement of the molecular dynamics trajectory of the system to the catchment basins on the PES and goes beyond the limitations imposed by the inherent dynamics and statistics of the system that restrict the capabilities of the existing methods. The approach is illustrated by the solution of typical problems for several finite many-body systems (rare-gas and carbon clusters, the molecule of alanin tetrapeptide).

1 Introduction

In these lectures we shall be discussing direct computer simulations of finite many-body systems such as clusters and biomolecules. More specifically, we will work in terms of classical dynamics and statistics, being interested in such things as potential energy surfaces (PESs) of the systems, their equilibrium (thermodynamic) properties and kinetics. We will not touch the electronic structure, optical and magnetic properties and related things that require quantum-mechanical consideration.

Provided that the interatomic potential for the system is known, all information of that sort, which one may want to know about the system, can, in principle, be obtained with the help of the direct computer simulation methods: of the method of molecular dynamics (MD) and/or Monte-Carlo (MC) [1,2]. With time, as more and more information about

This work was supported in part by the Russian Foundation for Basic Research, Grant 99-03-33299, and by the Department of Education of Russian Federation, the Programme "Basic Problems in the Natural Sciences".

interatomic forces comes from the experiment, this way of study of many-body systems becomes more and more promising. However certain problems are encountered on this way.

The first and general problem is that the current computers are not able to compete with nature in performance. To see how large is the gap, consider the frequency of atomic vibrations, which can serve as a characteristic rate for dynamic phenomena in multiatomic systems. For example, for a system which consists of 10^2 atoms interacting through an analytically given semi-empirical potential, one time step of numerical simulation on a current one-node computer typically requires the time of the order of 10^{-2} s. Therefore, one act of vibrations, usually taken to be of the order of 10^2 time steps, requires about 1 s, whereas in reality it is of the order of 10^{-12} s. It follows that the simulation of a microsecond scale process will take 10^6 s, that is about ten days, and the study of this process (collecting reasonable statistics, etc.) will require about one year. At the same time, many systems of interest are much larger in size, and characteristic time scales may expand even to seconds, *e.g.* in the case of biomolecules.

It is evident that such a huge gap between the characteristic rates in nature and in the simulations can hardly be narrowed to a desirable degree by a simple increase of the performance of current computers or by using massively parallel machines, at least in the nearest future. A solution of the problem should rather be sought on the way of devising an appropriate approach to the simulations. Since the proper dynamics of a system is too slow from the computational point of view, such an approach should not adhere this dynamics, however it should allow a reconstruction of the true behaviour of the system from the data obtained in the simulations. This is a goal which the confinement simulations pursue and achieve: going beyond the limitations imposed by the inherent dynamics and statistics, they reduce the computational time to the orders of magnitude and at the same time allow one to obtain a detailed information about the behaviour of a system.

In addition to the previously mentioned general problem there exist two more problems that are specific to many-body systems.

The first one is that a multiatomic system typically possesses a large number of mechanically stable atomic configurations. These configurations are known as *inherent structures* of the system and correspond to local minima on the PES; in the case of clusters and biomolecules they are usually called *isomers* and *conformers*, respectively. For example, for the cluster of 13 atoms interacting through the Lennard-Jones potential (LJ), LJ_{13}, about 1500 geometrically different isomers are known (Sect. 7.1), and for the molecule of alanin tetrapeptide, which consists of 24 atoms, about 1000 conformers (Sect. 7.4). Moreover, the number of the structures grows with

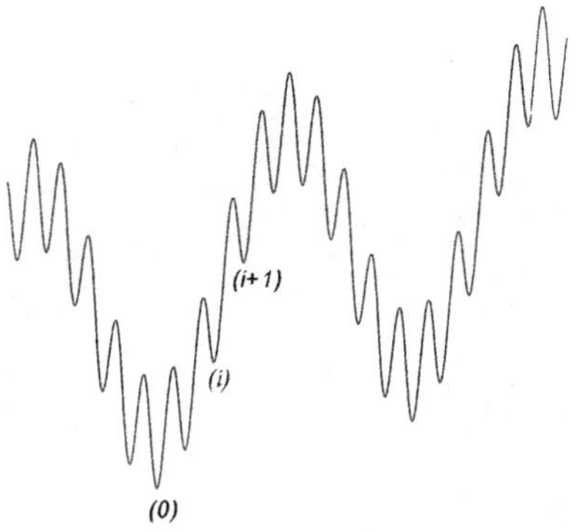

Fig. 1. Schematic picture of a multifunnel potential energy surface.

system size exponentially [3], reaching astronomically large values even for systems of moderate size. For example, for LJ_{55} it is estimated as 10^{21} [4].

A wide variety of inherent structures evidently hinders the simulation study of the system. If, say, we are interested in the properties that characterize the system as a whole (for example, in thermodynamics functions), an unreasonably large time may be required to visit the structures whose contribution is essential. On the contrary, if our goal is specific structures or channels of transitions, another problem arises (which is the inverse side of the previous one): since every particular structure is visited rarely, it is difficult to gain satisfactory statistics for the quantities of interest.

The problem is complicated by the fact that the physical properties of systems of small and intermediate size vary with size strongly and not smoothly [5]. Therefore scaling these properties to finite systems of larger size is difficult.

The other problem is that the PES may be very complex (Fig. 1). Minima may be positioned on very different height and/or be separated by high barriers. Furthermore, they may group together in clusters of minima (so called *superbasins* or *funnels*), which are separated by even higher barriers. In accordance with the Boltzmann distribution, the system predominantly samples the lowest part of the PES. Therefore at reasonable temperatures, or under limited length of the MD (MC) run, it may be trapped in a superbasin.

The latter two problems add complexity to the simulations, making the achievement of ergodic behaviour of the system even more difficult.

The simplest approach which allows one to conduct a comprehensive study of a many-body system, including its PES, equilibrium properties and kinetics, is due to Stillinger and Weber [3]. They suggested to supplement

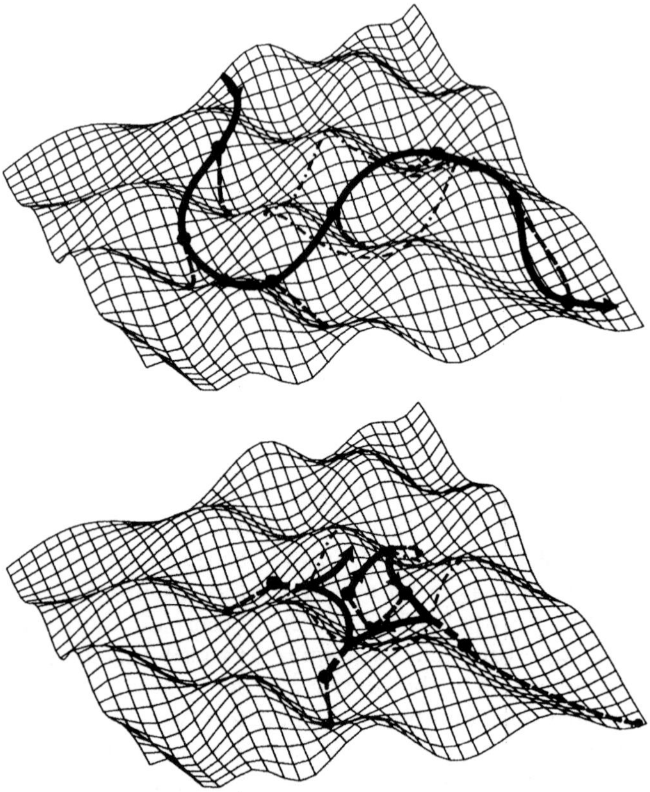
(a)
(b)

Fig. 2. Schematic picture of a 2D potential energy surface. (a) Conventional exploration of the surface, and (b) confinement to the catchment basin. The dash-dot lines show the boundaries of the catchment basin of interest, the thick lines the MD trajectory (in the lower panel, the dashed line segments correspond to the parts of the trajectory outside the basin), and the dashed lines show the quench paths.

the MD (or MC) simulations by a diagnostic quenching of the system at regular intervals. Each quench leads to a certain minimum on the PES, with a variety of quenches leading to the same minimum. The points at which the quenches leading to the same minimum are initiated represent a catchment basin on the PES that surrounds this minimum. The boundaries of the basin are the watershed lines which pass through the saddle points connecting the given minimum with the neighboring ones (Fig. 2). The

PES can thus be separated into the catchment basins representing different inherent structures of the system that are associated with the minima of the basins. Correspondingly, executing the quenching of the system along the MD (or MC) trajectory, the current points of the trajectory can be related to certain catchment basins on the PES, as shown in Figure 2a. In the course of exploration of the PES, the system repeatedly visits different catchment basins. Therefore, one can collect (separately) the phase points for each of the catchment basins. As a result, not only can the PES be surveyed, but the equilibrium properties of particular inherent structures and rates of the transitions between them can also be calculated. In what follows we will refer to this approach to as *conventional* or *SW approach.* The approach is very explicit by the underlying concept and easy to implement, but computationally it is not as efficient as desirable. In fact, it encounters all the problems that were previously mentioned.

For the last two decades, a variety of different methods has been suggested to overcome these problems [6]. The primary concern was to make sampling of the PESs more uniform than in the conventional MD and MC simulations, and correspondingly, to achieve the ergodic behaviour of the system. These methods can roughly be separated into three groups, depending on the underlying concept: *a modification of the PES by introducing an appropriate biasing potential* (umbrella potentials [7–9], multicanonical sampling [10,11], a generalized ensemble approach [12,13], and hyperdynamics [14]), *biasing the temperature* (weighted histogram analysis [15,16] and jump-walking [17] methods, and simulated [18] and parallel [19] tempering), and *a modification of the method of surveying the PES* (the eigenmode method [20], and the activation-relaxation technique [21–23]). In the case of biasing the PES or temperature (the first two groups) the results for the true conditions are recovered from the results of the biased simulations by applying the formulae of statistical mechanics, specifically the Boltzmann law. In contrast to the SW approach, the methods listed are not of multipurpose character; each of them pursues a specific goal: either *the calculation of equilibrium properties* (the weighted histogram analysis method [15,16], umbrella potentials [7–9], simulated [18] and parallel [19] tempering, multicanonical sampling [10,11], generalized ensemble approach [12,13]), or *the study of kinetics* (hyperdynamics [14]), or *surveying a PES* (the jump-walking method [17], the eigenmode method [20], and the activation-relaxation technique [21–23]).

Considerably less attention has been paid to the fact that the system, in accordance with the Boltzmann statistics, repeatedly visits the catchment basins that have been visited. At the same time, if the goal is to survey the PES, such events should evidently be avoided because they bear no new information about the PES. For this purpose the taboo search method could be employed, which has been elaborated in detail for discrete surfaces [24,25] and extended to continuous surfaces [26].

Also note that the problem of surveying a PES is closely related to the global optimization problem (see, *e.g.* [27]). In the latter, the goal is the global minimum of the surface rather than the surface itself. This brings certain specific features in the methods which are used for surveying PESs if (when) they are applied for finding the global minimum [27]. There are also methods which are designed specifically for a search for the global minimum (the simulating annealing [28], and "MC plus energy minimization" procedure [29], and conceptually close to them the basin-hopping method [30], the genetic algorithm [31], and the stochastic tunneling [32]). Some of the concepts that underlie these methods can be useful in application to surveying a PES.

The problem of surveying a PES deserves a more detailed consideration. A knowledge of the PES is of particular importance, since the PES completely defines the behaviour of the system at specific conditions. The understanding of how the PES is built provides a deep insight into dynamics and statistics of the system [1, 33, 34]. Moreover, the knowledge of the PES offers a possibility for writing a master equation that governs system's kinetics [35–40] and thus opens a way for replacing the time consuming simulations by a relatively easy solution of the master equation.

To characterize the PES, one should know, in the first instance, the local minima and the saddles that connect these minima [1]. Of primary importance are the minima, provided that an indication is given which of them are directly connected. With this, finding saddles between the minima becomes a straightforward task, which can be fulfilled using one of the numerous methods suggested for this purpose, *e.g.* the method of slowest slides [41], the TRAVEL algorithm [42], the ridge method [43] or the method of contangency curves [44] (see also Ref. [45], which contains a comprehensive bibliography on these and related methods, as well as a comparison of typical methods in efficiency).

One recently developed approach to surveying a PES invokes successive walks from one minimum to a neighboring saddle, then from this saddle to a next minimum, and so on [20 23]. These walks can be called activation and relaxation phase, respectively [21]. For the activation phase one of the uphill climbing methods [45] can be used, *e.g.* the eigenvector-following method by Cerjan and Miller [46] in its various modifications, as in [20, 23, 47, 48], or a more economical algorithm which does not require evaluation of the full Hessian matrix at each step, as in [21, 22]. For the relaxation phase one of the standard methods, such as the steepest-descent or conjugate gradient method, is employed. If the number of minima is not too large, all, or practically all, minima on the PES can be located by a direct walk from one minimum to another [20,47,48]. For systems of larger size, where the number of minima becomes too large, just a statistical search for the minima is possible. For this purpose a transition between any two nearby minima is taken as an elementary Metropolis step [49], that is the move from one minimum to another is accepted with a probability of $\min[1, \exp(-\Delta E/k_B T)]$, where k_B is the Boltzmann constant, T is the temperature, and ΔE is the change in energy associated with this move, which can be either the energy difference between the minima or the height of the barrier between the minima [21,23].

In these lectures we shall consider a new approach to the simulation study of a finite many-body system. Similar to the SW approach, this approach is of a multipurpose character, allowing one to survey a PES and to study the equilibrium properties and kinetics of a system. At the same time the approach is free from many shortcomings inherent to the SW and other approaches.

2 Key points and advantages of the confinement simulations: General remarks

Before starting a detailed discussion of the approach, let us consider its key points, using for this purpose model PESs depicted in Figures 1, 2.

The basic element of the approach is the confinement of the MD trajectory of the system to the catchment basins on the PES [50–52] (in what follows, instead of "the MD trajectory of the system" and/or "the representative point of the system" we shall say for brevity "the system"; it should not lead to confusion). To initiate simulations, the system is placed into one of the catchment basins, and the MD (MC) run is begun (Fig. 2b). As in the SW method, the system is quenched at regular intervals in order to check if it is still in a given basin or has left it for another, neighboring basin. If left, then depending upon the strategy chosen to survey the PES (see below), the system is either allowed to pass into the basin in which it was found or returned into the given basin for further exploration of this basin (Fig. 2b). To return the system into the basin, a certain procedure of the reversal of the trajectory is employed. In contrast to the conventional dynamics, the system can now be kept in a current basin for an arbitrary long time, and thus all characteristics of the structure corresponding to this basin can be calculated as accurately as desirable. It is essential that though the system does not leave the basin for a time longer than the interval between two subsequent quenches (for brevity, the quenching interval), all neighboring basins can be determined and all attempts to pass into these basins can be recorded. Therefore not only can the thermodynamics (equilibrium) functions for a given basin be calculated but also the rates of transitions into the neighboring basins, and thus the kinetics of the system. Having the option of either allowing the system to pass into a certain basin or not, one can explore just a part of PES that is of interest. Also note that the approach does not require any prior information about the PES landscape: being placed at an arbitrary chosen point on the PES, the system, basin by basin, can explore all the PES, or a desirable part of it.

More can be learned from the consideration of Figure 1, which depicts a two-funnel model PES. Assume for simplicity that all basins are identical except for their minimum energies; in particular, they have the same mean life time, τ, and the same probability, p, for the system to go to the higher neighboring basin (the probability to go in the inverse direction is then $1-p$). Assume also that $p \ll 1$, in which case the conventional dynamics becomes inefficient. According to the above definitions, the mean time for the system to make one step to the higher energy basin is τ/p, and the mean time to go to the neighboring lower energy basins is $\tau/(1-p) \simeq \tau$. Place the system at a bottom of one of the funnels. Then, if the conventional dynamics is used, in which the system freely goes up and down among the basins, the mean time for the system to reach N-th basin from the bottom of the funnel is

$$t_{\rm cnv} = \tau/p^N .$$

In the case of the confinement simulations, we can forbid the system to return to a lower energy basin, since it has been sampled. Therefore, the mean time required to reach the N-th basin is written as

$$t_{\rm cnf} = N\tau/p$$

that is now it increases with the number of basins N linearly and not exponentially, as before. This reveals an essential difference between the conventional and confinement dynamics. In particular, when successively confined to higher and higher positioned basins, the system much easily overcomes the interfunnel barriers. To see how considerable is the gain, set, for example, $N = 3$ and $p = 0.01$, which yields $t_{\rm cnv}/t_{\rm cnf} = p^{1-N}/N = 3333$. Note that for this process the use of constant temperature MD (canonical ensemble) is essential, since in the case of constant energy MD (microcanonical ensemble) the higher the system climbed in potential energy, the lower would be its kinetic energy (temperature) and the slower the rates of transitions. For the constant temperature MD, in contrast, the temperature will be the same in each of the basins, regardless how high the basin is positioned.

This process of basin-by-basin climbing is similar to the process of multistep rising a boat in a channel through the locks. The analogy between these processes becomes even closer if the system is successively confined not to individual basins but to pairs of neighboring basins, in which the upper basin serves as the lower one for the next step of the lockage; then the system can freely pass between two connected basins, exactly as a boat in the lock when the water in different parts of the lock levels off.

3 Methods for generating phase trajectories

To implement the confinement technique, first of all we need a method to generate a phase trajectory of the system, *i.e.* a sequence of points in phase space. In general, both MD and MC methods are applicable for this purpose. A difference is that the MD gives this sequence as a function of time, while the MC simply generates a chain of random points obeying an equilibrium statistical distribution, canonical [49] or microcanonical [19], depending on the conditions under consideration [53]. Correspondingly, the MD method can be used to simulate kinetics of a system directly, whereas the MC requires an additional hypothesis to relate the intervals between the points to physical time. Therefore we shall consider the MD method.

Similar to the MC methods, there are two types of the MD methods that are used to simulate constant energy and constant temperature conditions, respectively.

3.1 Conventional molecular dynamics

The *conventional* MD involves a numerical integration of the coupled Newton equations

$$m\frac{\mathrm{d}^2\mathbf{r}_i}{\mathrm{d}t^2} = -\frac{\partial U}{\partial \mathbf{r}_i} \tag{3.1}$$

where t is the time, U the potential energy of the system, $\mathbf{r}_i$ the coordinates of i-th atom, and m is the atomic mass (for simplicity, we assume all atoms to be identical).

If no restraints are imposed, these equations describe dynamics of a free system. Correspondingly, along a phase trajectory there are conserved the total energy $E = \sum_{i=1}^{n} \mathbf{p}_i{}^2/2m + U$, linear $\mathbf{P} = \sum_{i=1}^{n} \mathbf{p}_i$ and angular $\mathbf{L} = \sum_{i=1}^{n} \mathbf{r}_i \times \mathbf{p}_i$ momenta of the system, where $\mathbf{p}_i = m\mathrm{d}\mathbf{r}_i/\mathrm{d}t$ is the momentum of i-th atom, and n is the number of atoms in the system. Issuing phase trajectories at different initial conditions which satisfy these integrals of motion, *i.e.* starting at different points of phase space that are located at the intersection of the hypersurfaces $E = E_0$, $\mathbf{P} = \mathbf{P}_0$ and $\mathbf{L} = \mathbf{L}_0$, one can form the *microcanonical* ensemble of the systems for specific values of E_0, $\mathbf{P}_0$ and $\mathbf{L}_0$.

This dynamics is also known as *Newtonian* or *constant energy* MD.

3.2 Stochastic molecular dynamics

The simplest version of *stochastic* MD is based on a numerical integration of the coupled Langevin's equations

$$m\frac{\mathrm{d}^2\mathbf{r}_i}{\mathrm{d}t^2} + \alpha\frac{\mathrm{d}\mathbf{r}_i}{\mathrm{d}t} = -\frac{\partial U}{\partial \mathbf{r}_i} + \mathbf{\Phi}_i(t) \qquad (3.2)$$

where α is viscosity, and $\mathbf{\Phi}_i(t)$ are random (Langevin's) forces due to thermal fluctuations in the system at given temperature T. These forces have mean zero and are related to the viscosity by the fluctuation-dissipation theorem:

$$\langle \mathbf{\Phi}_i(t) \rangle = 0, \quad \langle \Phi_i^j(t)\Phi_{i'}^{j'}(t+\tau) \rangle = 2\alpha k_{\mathrm{B}} T \delta_{ii'} \delta_{jj'} \delta(\tau) \qquad (3.3)$$

where k_{B} is the Boltzmann constant, the upper index at Φ stands for a component of the vector $(j, j' = 1, 2, 3)$, the angular brackets denote an ensemble average, and $\delta_{kk'}$ and $\delta(\tau)$ are the Kronecker and Dirac deltas, respectively. This MD simulates constant temperature conditions, with a specific value of the temperature being a result of the balance between Langevin's forces and viscous friction. For a free system none of E, $\mathbf{P}$ and $\mathbf{L}$ is conserved along the phase trajectory. Issuing phase trajectories at different points of phase space, one can form *canonical* ensemble of the systems that corresponds to a given temperature T.

This dynamics is also known as the *Langevin* or *constant temperature* MD.

The Langevin MD requires an additional comment. This dynamics is often used to simulate in an economic way a subsystem which is a part of the system of considerably larger size, with the rest of the system being considered as a heat bath to which the subsystem is coupled. Some examples of this are given by a cluster in vapor surroundings, a biomolecule in solvent, or a cluster on the solid surface. Langevin forces however do not reproduce actual forces coming from the surroundings. Equations (3.2, 3.3) rather imply that the system is embedded into a structureless fluid which equally (though in a random way) affects every atom, regardless either this atom is, say, within the system or at its surface. It can be shown (see, *e.g.* [54]) that at $t \to \infty$ the Langevin MD leads to the canonical (Boltzmann) distribution at the given temperature, and it can thus be used to calculate the equilibrium (thermodynamic) properties of a system. As for kinetics, no well justified estimates are available, except for some reasonable qualitative considerations [54, 55]. Therefore in most cases kinetics calculated on the basis of the Langevin equations can pretend just to a qualitative description of the process.

A more realistic description of the behaviour of a system can be achieved on the basis of the generalized (integral-differential) Langevin equation, which does not assume a white noise for random forces (3.3) and accordingly a constant viscosity α [56]. A problem of application of the generalized Langevin MD to a specific system is that one has to know the kernel responsible for a retarded effect of the frictional forces. This kernel is however specific for a specific system and thus, to be obtained, requires additional information about the system (either from experiment or more accurate dynamic calculations, see, *e.g.* [57]).

Another, recently suggested generalization of the Langevin dynamics makes the heat bath partly structural, – it endows particles of the heat bath with a finite mass [58].

4 Identification of atomic structures

The next basic component of the confinement technique is the procedure of identification of the catchment basin in which the system is currently located.

4.1 Quenching procedure

Each catchment basin can be characterized by its minimum. Therefore we should find a way of how to relate a current point in the MD trajectory to the corresponding minimum. One possibility is to quench the system. In the case of Newtonian MD one can remove the kinetic energy of the system and then allow the system to relax, repeating this procedure until the kinetic energy comes to zero. In the case of the Langevin MD one can simply set and keep temperature zero. In both cases the system should come to the minimum of the basin. A more preferable way, however, is to exclude dynamic effects from the consideration completely and to work solely in terms of the PES.

The simplest and most commonly used method of this sort is the steepest descent method, which involves a numerical integration of the equations

$$\frac{d\mathbf{r}_i}{dt} = -\frac{\partial U}{\partial \mathbf{r}_i} \tag{4.1}$$

with the atomic configuration corresponding to the current point of the MD trajectory being taken for the starting point. These equations describe an overdampted motion of atoms in the system at $T = 0$. They can be obtained from (3.2, 3.3), if, assuming the viscosity to be large, one neglects the inertia of the atoms (the first term in the left-hand side of (3.2)) and set $T = 0$ in (3.3). For this reason, the steepest-descent procedure is often referred to as quenching too.

A certain disadvantage of the steepest-descent method is its slow convergence to the minimum: the closer the system to the minimum, the lower is $\partial U/\partial \mathbf{r}_i$ and the slower approaches the system the minimum. This defect can be remedied in part by choosing the time step for the integration of (4.1) to be proportional to $|\partial U/\partial \mathbf{r}_i|^{-1}$.

A more radical way to avoid slow convergence to the minimum is to use the Newton-Raphson algorithm in the final stage of the search, where $|\partial U/\partial \mathbf{r}_i|$ becomes sufficiently small. This algorithm is based on quadratic approximation to the surface, and in the range where such approximation is valid it enables the attainment of the minimum by a single step

$$\Delta \mathbf{R} = -\mathbf{H}^{-1}\frac{\partial U}{\partial \mathbf{R}}$$

where $\mathbf{R} = (\mathbf{r}_1, \mathbf{r}_2, ..., \mathbf{r}_n)$, $\Delta\mathbf{R} = (\Delta\mathbf{r}_1, \Delta\mathbf{r}_2, ..., \Delta\mathbf{r}_n)$ is the space increment, and $\mathbf{H}$ is the Hessian (*i.e.* the matrix of the second order derivatives of U). Therefore when the system is close to the minimum, and hence the PES is close to parabolic, just a few Newton-Raphson steps are usually required to reach the minimum.

4.2 Characterization of a minimum

As soon as the minimum is found, it should be characterized for the purpose of subsequent comparison with the other minima involved. In the PES terms each of the minima can be characterized by its energy $U_{\min}$ and spectrum of the eigenvalues of the Hessian $H_1, ..., H_{3n}$; the latter define the curvature of the basin at its bottom. For nonlinear (linear) atomic configuration six (five) of the eigenvalues – three for translational and three (two) for rotational degrees of freedom – are equal to zero, and the others, $3n-6$ $(3n-5)$, corresponding to vibrations, are positive.

Alternatively, since every minimum is associated with a certain inherent structure (isomer, conformer), physical characteristics of these structures can be used to distinguish the minima. $U_{\min}$ stands then for the energy of the structure, and instead of the Hessian eigenvalues H_i the normal mode frequencies ω_i are usually used, which are related to the H_i as $\omega_i = \sqrt{H_i}$.

For geometrically different structures, it is usually sufficient to compare the structures only for their minimum energies. To make sure that no occasional coincidence in energy happened, one can compare the spectra of the frequencies as well, even not the spectra themselves but the geometrical means of the frequencies $\overline{\omega} = (\prod_{i=1}^{N}\omega_i)^{1/N}$, where $N = 3n-6$ $(3n-5)$. All other characteristics are rarely used for this purpose, rather they may be of interest from physical point of view (the point group and its order h, the principal momenta of inertia $I_{\alpha i}$ $(\alpha = 1, 2, 3)$, which characterize rotational properties of the structure, etc.).

In certain cases, it may also be necessary to distinguish between permutational structures, *i.e.* the structures that are geometrically identical but differ by the order in which the atoms are labeled. This is achieved by calculating the distance in configuration space $R^2_{\alpha\beta} = \sum_{i=1}^{n} (\mathbf{r}_{\alpha,i} - \mathbf{r}_{\beta,i})^2$ after superposition of the structures by the rotation of one of the structures around its center-of-mass; here α and β label the structures, and $\mathbf{r}_{\alpha(\beta),i}$ stand for atomic coordinates counted from the corresponding center-of-mass.

5 Confinement procedures

A general purpose of the confinement of the system to a catchment basin is to keep the behaviour of the system under control: if the system attempts to leave the basin, either to allow it to do this or to return the system into the basin for further sampling of the basin. In the former case the MD trajectory of the system is simply continued in a usual fashion, but in the latter case a new trajectory should be initiated in the basin. Our goal then is to specify the initial conditions (atomic coordinates and velocities) for this new trajectory so that it would reproduce a "proper" behaviour of the system in the basin. More rigorously, the phase trajectories obtained at the confinement of the system to a basin should be consistent with the trajectories, in the part related to the basin, that are obtained in the conventional simulations, *i.e.* when the system freely explores the PES, entering and leaving basins. Correspondingly, for a correct reproduction of the equilibrium properties and kinetics we need a statistically representative set of such trajectories for each catchment basin involved.

Let the system leave one of the basins and consider how a new trajectory can be initiated in it. We shall discuss two methods to do this. The sequence of operations in both cases is the same: the system is quenched at regular intervals, τ_{quench}, along the MD trajectory, and the current quenched structure is compared with that corresponding to the minimum of the given basin. If the system is found in a neighboring basin, a new trajectory is initiated in the given basin.

5.1 *Reversal of the trajectory at the boundary of the basin. Microcanonical ensemble*

Conceptually most simple method is a reversal of the trajectory at the boundary of the basin (Fig. 3a). In this case, first of all, the boundary has to be accurately located. For this, when the system is found outside the basin, the quenching interval τ_{quench} is reduced by half, and the trajectory is issued again from the last point where, at quenching, the system was found in the given basin. This procedure is repeated until the boundary is fixed with an accuracy of one time step of numerical integration of equations (3.1).

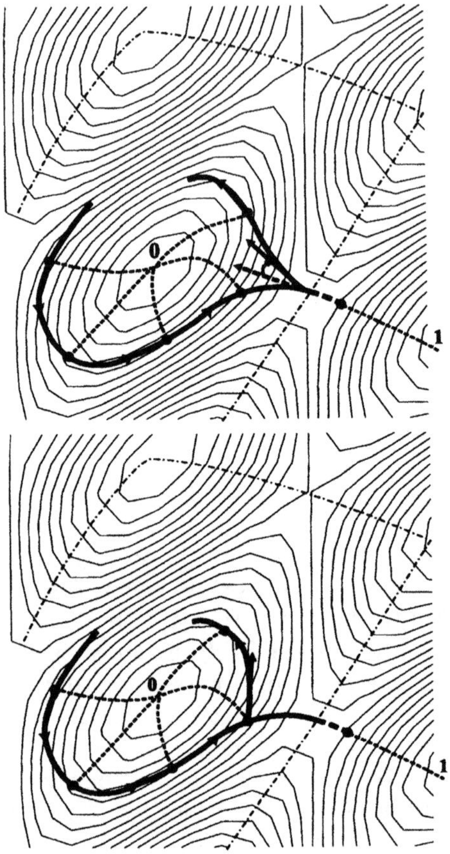

(a)

(b)

Fig. 3. Initiating a new MD trajectory in the catchment basin (topographical view of the 2D potential energy surface). Solid lines show the MD trajectory, the dash-dot lines the boundaries of the catchment basins of interest, and the dashed lines the quench paths. Figures 0 and 1 indicate the minima of the given and neighboring basins, respectively. (a) Reversal of the MD trajectory at the boundary of the basin, and (b) initiating the MD trajectory at the point of the last quenching in the basin. Solid arrows indicate the directions of new trajectories at the points of their initiating, and the dashed arrow in the upper panel shows the direction of the reversed trajectory before scattering of atomic velocities.

The next step is the reversal of the trajectory. Different methods to perform it can be suggested. They may vary from a "pure" reversal, when the velocities of all atoms change to the opposite ones, to a reflection of the trajectory from the boundary of the basin, which may, in turn, be specular, diffusive, or specular-diffusive. Since our concern is to obtain a statistically representative set of trajectory, it is essential that the reversed trajectory would deviate from the forward one, along which the system came to the boundary, as early as possible. The "pure" reversal is the most simple procedure, and, in principle, it furnishes the desirable effect due to the exponential divergency of phase trajectories [59]. Indeed, since the behaviour of the system beside the boundary is essentially nonlinear, one can expect that the reversed trajectory will deviate from the forward one in a short time. However, one can also expect that the variety of phase points generated by such trajectories beside the boundary will be poor as compared to that in the internal part of the basin. In this respect, the reflection of the trajectory looks much more promising, particularly if it is a diffusive reflection. However, to perform it, the orientation of the boundary should be known, the determination of which is a complex and time consuming procedure in the case of multidimensional surfaces.

To resolve this problem, there was taken a compromise between the above variants of the reversal: the velocities of all atoms at the point of the reversal, first, changed their signs, and second, in order to amplify the divergence of phase trajectories, they were subjected to a uniform random scattering. Specifically, the velocities were scattered within a β part of their mean value. The results of the simulations were found just slightly dependent upon a specific value of β (when it varied from 0.2 to 1.0), and $\beta = 0.4$ was used as a regular value. The act of the reversal was followed by a correction of the integrals of motion of the system: the total energy E, linear $\mathbf{P}$ and angular $\mathbf{L}$ momenta of the system after the reversal were fitted to those before the reversal by shifting and rescaling atomic velocities.

To test this procedure, there were conducted confinement and conventional (SW) simulations (*i.e.* in the latter case the system was also diagnostically quenched at regular intervals as in the confinement simulations). For this, there was taken 13-atom cluster with atoms interacting through a pairwise additive Lennard-Jones (LJ) potential

$$U = \sum_{i=1}^{n} \sum_{j<i} 4\epsilon \, [(\sigma/r_{ij})^{12} - (\sigma/r_{ij})^{6}].$$

Here r_{ij} is the distance between i-th and j-th atoms, and ϵ and σ are, respectively, the characteristic energy and length, which are specific for a specific system (for example, for Ar $\epsilon = 119.8$ K and $\sigma = 3.904 \times 10^{-8}$ cm). ϵ and σ are usually taken as the corresponding units of the LJ potential. Together with the atomic mass, m, they constitute so called LJ systems of units, in which, in particular, the time unit is $\sqrt{m\sigma^2/\epsilon}$ (2.15×10^{-12} s for Ar). In what follows, when giving numerical values for LJ clusters, we will use this system of units. The cluster did not execute both overall translation and rotation, specifically there was $\mathbf{P} = \mathbf{L} = \mathbf{R} = 0$, where $\mathbf{R} = \sum_i m\mathbf{r}_i/M$ is the center-of-mass position ($M = nm$ is the total mass of the cluster).

The system of Newtonian equation (3.1) was numerically integrated using the algorithm by Schofield [60], the time step $\tau = 0.0025$ (5.3×10^{-15} s for Ar). Quenching was executed at intervals of 64 time steps, with each quench being continued until the eigenvalues of the Hessian that correspond to the translational and rotational degrees of freedom reached the value of 10^{-6}. To reveal possible dissociation (evaporation) of the cluster, the following criterion was used: the cluster was considered to be in a dissociated state if it divided into fragments (possibly consisting of a single atom) which were spaced from each other by the distance of 2.6 or larger. If dissociation occurred in the course of the conventional simulations, the MD trajectory was terminated, and a new one was started. In the case of the confinement simulations dissociation was automatically prevented because the system was confined to the basins corresponding to bound-state structures.

To compare the confinement and conventional simulations, we will consider two characteristics, which are related to the equilibrium properties and kinetics, respectively (Figs. 4–6).

Let us start with the *equilibrium* properties. In the confinement simulations the system was confined to the catchment basin corresponding to a specific isomer: the atoms are labeled in a certain order, no inversions are expected, but various orientations of the cluster are possible (since the set of orientations is continuous, such basin is a sort of closed valley rather than the true basin). Then at the conditions under consideration ($\mathbf{P} = \mathbf{L} = \mathbf{R} = 0$) the phase volume associated with this isomer can be written as

$$\tilde{G}(E) = \int \mathrm{d}\mathbf{r}^{(3n)}\mathrm{d}\mathbf{p}^{(3n)}\delta(\mathbf{R})\delta(\mathbf{P})\delta(\mathbf{L})\Theta(E - H) \qquad (5.1)$$

where $\mathrm{d}\mathbf{r}^{(3n)}$ and $\mathrm{d}\mathbf{p}^{(3n)}$ are $3n$-dimensional elements of the volume in configuration and momentum space, respectively, δ the delta function, Θ the step function, $H = \sum_i \mathbf{p}_i^2/2m + U$ the cluster Hamiltonian, and the integral with respect to atomic coordinates is taken over the catchment basin under consideration. To pass to the corresponding dimensionless quantity, the total number of states in quasi-classical approximation, the factor $1/(2\pi\hbar)^N$ should be assigned to the integral, where $\hbar$ is the Planck constant, and $N = 3n - 6$ is the number of vibrational degrees of freedom in the cluster (the atomic configurations are supposed to be non-linear).

$\tilde{G}(E)$ completely defines all equilibrium (thermodynamic) properties of the system within the given basin, for example, the entropy is $\tilde{S}(E) = k_\mathrm{B} \ln \tilde{G}(E)$. Therefore a good reproduction of the dependence $\tilde{G} = \tilde{G}(E)$ in

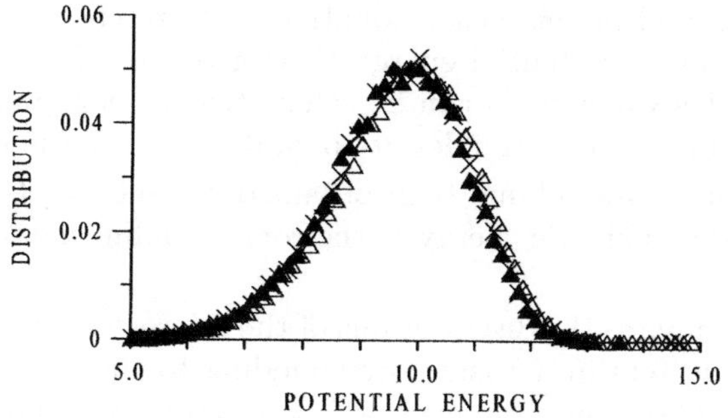

Fig. 4. Distribution of the microstates over the potential energy of the system in a basin corresponding to the ground-state isomer of LJ_{13}; microcanonical ensemble, $E = 15$. Crosses represent the conventional simulations, and the triangles show the results of the confinement simulations: solid triangles are for the reversal of the MD trajectories at the boundary of the basin, and the empty triangles for initiating the trajectory at the point of the last quenching in the basin. The energy is counted from the ground-state isomer energy, $U_{\min} = -44.3268$.

the confinement simulations guarantees a correct calculation of the equilibrium properties. A comparison of this sort will be presented in Section 6.1.

Here we consider another characteristic, which is closely related to $\tilde{G}(E)$ but more illustrative for our current purposes. Integrating over momenta in (5.1) (see Ref. [61] for how it can be done), we obtain

$$\tilde{G}(E) = C \int \mathrm{d}\mathbf{r}^{(3n)} \delta(\mathbf{R}) \frac{(E-U)^{N/2}}{(I_1 I_2 I_3)^{1/2} \Gamma(N/2+1)} \, \Theta(E-U) \tag{5.2}$$

where $C = (2\pi)^{N/2} m^{3N/2}/M^{3/2}$, I_α ($\alpha = 1,2,3$) are the principal momenta of inertia of the current atomic configuration, and Γ is the gamma function. Correspondingly, the density of states $\tilde{\rho}(E) = \mathrm{d}\tilde{G}/\mathrm{d}E$ is

$$\tilde{\rho}(E) = C \int \mathrm{d}\mathbf{r}^{(3n)} \delta(\mathbf{R}) \frac{(E-U)^{N/2-1}}{(I_1 I_2 I_3)^{1/2} \Gamma(N/2)} \, \Theta(E-U). \tag{5.3}$$

Introducing the configuration density of states as

$$\Omega(U') = C \int \frac{\mathrm{d}\mathbf{r}^{(3n)} \; \delta(\mathbf{R}) \; \delta(U-U')}{(I_1 I_2 I_3)^{1/2} \Gamma(N/2)}.$$

Equation (5.3) can be rewritten in the form

$$\tilde{\rho}(E) = \int \mathrm{d}U \; \Omega(U) \; (E-U)^{N/2-1} \, \Theta(E-U). \tag{5.4}$$

Then the integrand in this latter equation will present the distribution of the microstates over potential energy U at given E. The thermodynamic functions (at this value of E) can be calculated as certain moments of this distribution, in particular, the density of states $\rho(E)$ as the moment of zero order (with the phase volume then obtained as $\tilde{G}(E) = \int_0^E \tilde{\rho}(E)\mathrm{d}E$), the mean potential and kinetic energy as the corresponding moments of the first order, etc.

Figure 4 compares the distributions of the microstates over the potential energy of the system in a basin corresponding to the ground-state isomer ($U_{\text{min}} = -44.3268$). The total energy $E = 15$ corresponds to the liquid-like region (see Fig. 10a).

To characterize *kinetic* properties, we shall consider the distribution of life times of the system in a basin. Whereas the distribution of the microstates (Fig. 4) tells us nothing about how these microstates follow each other in time, this characteristic indicates, to a certain degree, either the microstates are met in the proper time order (within a single trajectory) or not. In the conventional simulations the life time was calculated as the interval between two subsequent acts of entering and leaving the basin, and at confinement of the system, as the interval between two subsequent reversals of the trajectory.

Figure 5a compares the distributions of life times in the same catchment basin for LJ_{13} as in Figure 4. In general, the data are in good agreement. A small exception are the smallest times, which correspond to the events when the system does not go far away from the boundary of the basin. For an insight into this disagreement, Figure 5b separates life times into two groups: one group for the events when the system leaves the given basin for the basin from which it came, and another for the events when the system leaves this basin for another basin. It is seen that though the actual dynamics of the system at small times is different for these two groups of events, the confinement technique likely offers some similar dynamics. The latter dynamics seems to be satisfactory for the second group of the events, when at small times the system simply crosses the given basin (such events are characteristic of the regions beside the saddle points of high order). However, for the first group of the events, when the system returns to the basin from which it came, a simple reversal of the trajectory at the boundary of the basin is evidently not able to reproduce the dynamics of the system at small times.

This difference in dynamics however weakly affects the rates of the transitions between the basins, which are the basic quantities in kinetics (see Sect. 7.2). Figure 6 compares the conventional and confinement simulations for the rates of the transitions from the ground-state isomer to its permutational copies and to several low lying excited-state isomers, for which

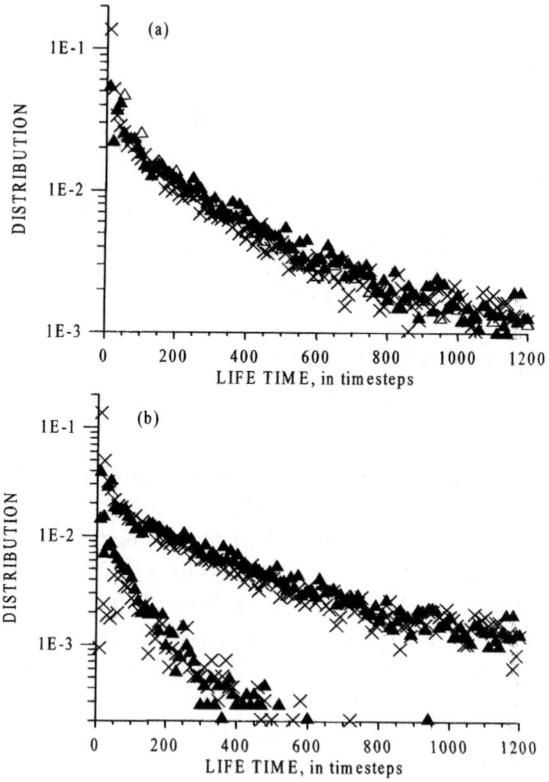

Fig. 5. Distribution of life times of the system in the basin corresponding to the ground-state isomer of LJ_{13}; microcanonical ensemble, $E = 15$. (a) The complete set of the events. (b) All events are separated in two groups: the system leaves the given basin for the basin from which it came (the lower set of the data) and for another basin (the upper set of the data). The notations are the same as in Figure 4.

reasonably good statistics was achieved in the conventional simulations. In both cases the rates were calculated as the mean numbers of the corresponding events per unit time; the only difference was that in the conventional simulations there were counted the actual acts of the transitions, whereas in the confinement simulations the attempts of the system to leave the basin. These rates indicate, to a certain degree, how the acts of the transitions are distributed among different channels.

Note that the distributions of life times in Figure 5, including those obtained in the conventional simulations, deviate from the exponential dependence characteristic of the Poisson decay process (Sect. 7.2, Eq. (7.11)). There are two reasons for this. First, these distributions accumulate the

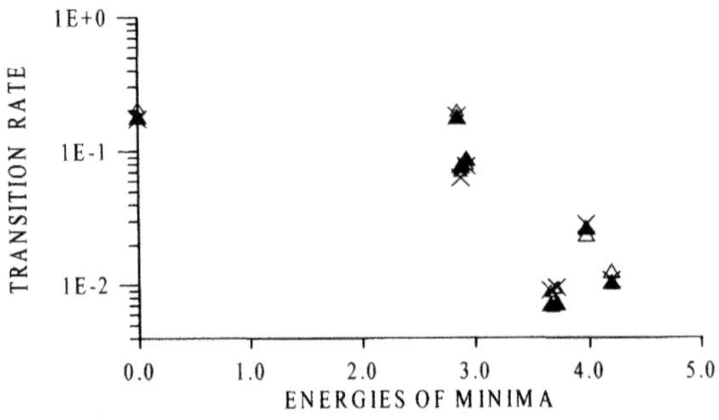

Fig. 6. Rates of the transitions from the ground-state isomer of LJ_{13}; microcanonical ensemble, $E = 15$. The notations are the same as in Figures 4 and 5.

events for many channels, which are characterized by different mean life times (these times are the reciprocal values of the transition rates shown in Fig. 6), and second, the process is not pure stochastic because of a complicated dynamics of the system at small life times [62,63].

Conceptually very explicit, this procedure of initiating new trajectories in catchment basins suffers from certain disadvantages. First, an accurate location of basin's boundary requires many quenches, which are very time consuming. Second, as our experience shows, the reversal of the trajectory may happen to be unsuccessful. At low and moderate energies, until saddles of high order are involved, the system successfully returned to the basin every time when the reversal was executed. However, as the energy increased, not always every act of the reversal furnished the desired result: after a few time steps within the true basin the system attempted to leave it again (which typically happened when the system was in a region where the number of imaginary frequencies was as large as ten or higher). As a result, a new act of the reversal had to be executed in a short time. Furthermore, at even higher energies, when the system had a pronounced tendency towards a decay (evaporation), it happened that the phase trajectory "cycled": the same sequence of the basins (but not of the same phase points!) repeated many times.

5.2 Initiating the trajectory at the point of the last quenching within the basin. Microcanonical and canonical ensembles

From computational point of view, a more promising way is to start a new trajectory immediately at the point where the system was found in the true basin last time. In this case we avoid the time consuming procedure of locating the boundary of the basin, and, presumably, the effect of "cycling"

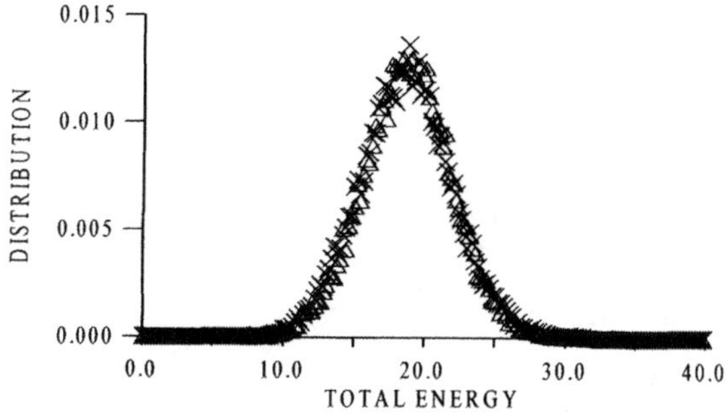

Fig. 7. Distribution of the microstates over the total energy of the system in a basin corresponding to the ground-state isomer of LJ_{13}; canonical ensemble, $T = 0.35$. Crosses represent the conventional simulations, and the triangles show the results of the confinement simulations with the MD trajectories initiated at the points of the last quenching in the basin. The energy is counted from the ground-state isomer energy, $U_{\min} = -44.3268$.

the trajectory. However, instead, new issues arise: if these truncated trajectories are able to reproduce a proper behaviour of the system in the basin, and how atomic velocities should be chosen to start these trajectories. The simplest choice for atomic velocities is to take them as random values obeying the Maxwell distribution, with the kinetic energy of the system being equal to that at the point of the last quenching in the basin. It is understood that in the case of Newtonian dynamics (*microcanonical* ensemble) the atomic velocities at the starting point are corrected so that they satisfy the integrals of motion, exactly as in the preceding section.

This procedure of initiating the trajectory in the case of *microcanonical* ensemble is illustrated by the corresponding data in Figures 4–6. The quenching interval $\tau_{\text{quench}} = 50$ time steps, and its variation did not affect the results noticeably. For the life time (Fig. 5) there is taken the interval between the first and last quenching at which the system was found in the given basin. This time evidently differs from the true life time, presumably by the order of τ_{quench}, and this difference may be seen in Figure 5 at the times comparable with τ_{quench}.

Figures 7–9 compare the results of the conventional and confinement simulations in the case of Langevin's dynamics (*canonical* ensemble). Equations (3.2, 3.3) were integrated using the numerical algorithm by Biswas and Hamann [64]; the time step is equal to 0.01, and the friction coefficient to 3.0 (for Ar, 2.15×10^{-14} s and 1.4×10^{12} s^{-1}, respectively). For testing, as before, a basin for the ground-state isomer of LJ_{13} is taken. The temperature, which is equal to 0.35 in the LJ units, also corresponds to

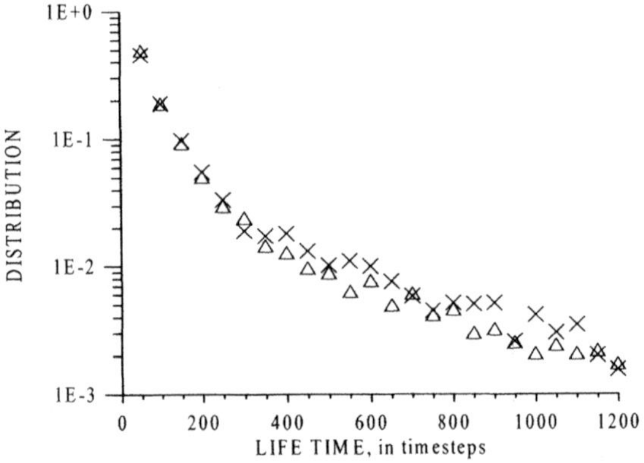

Fig. 8. Distribution of life times in the basin corresponding to the ground-state isomer of LJ_{13}; canonical ensemble, $T = 0.35$. The complete set of the events (as in Fig. 5a). The notations are the same as in Figure 7.

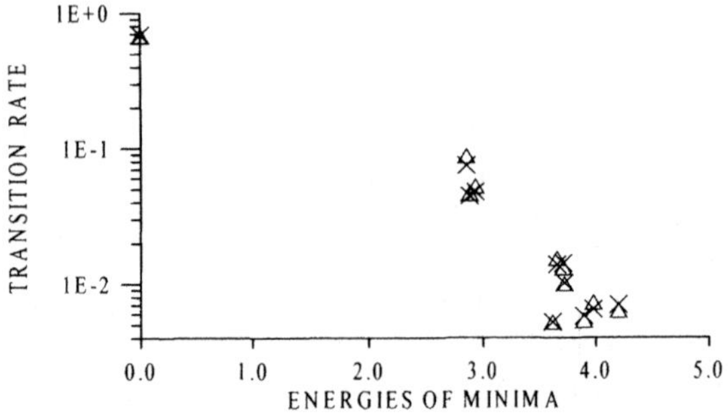

Fig. 9. Rates of the transitions from the ground-state isomer of LJ_{13}; canonical ensemble, $T = 0.35$. The notations are the same as in Figures 7 and 8.

the liquid-like region (note that microcanonical and canonical caloric curves for a system of finite size are characteristically different [65], in particular, compare Figs. 10a and 15). To see how the equilibrium properties are reproduced, the distribution of the microstates over the total energies of the system is considered. According to the relation

$$\tilde{Z}(T) = \int \mathrm{d}E \; \tilde{\rho}(E) \exp(-E/k_{\mathrm{B}}T)$$

where $\tilde{Z}(T)$ is the partition function, the role of this distribution in *canonical* ensemble is similar to the role which the distribution over the potential energy played in *microcanonical* ensemble (compare this equation with (5.4)).

6 Confinement to a selected catchment area. Some applications

For a catchment area we understand here either a catchment basin or a set of connected basins. In all applications *microcanonical* ensembles of the systems are considered, and, correspondingly, Newtonian MD is employed for the simulations. New trajectories in catchment basins are initiated by the reversal of the trajectories at the boundaries of the basins, as explained in Section 5.1.

6.1 Fractional caloric curves and densities of states of the isomers [51, 52]

The caloric curve is a relation between the total energy of the system and its temperature. For microcanonical ensemble, the total energy has a specific value, and the temperature is defined through the mean kinetic energy $\langle E_{\rm kin}\rangle$ according to the relation $\langle E_{\rm kin}\rangle = (N/2)\ k_{\rm B}T$; here N is the number of active degrees of freedom, in particular, if n-atom system does not execute overall translation and rotation and its configuration is nonlinear, $N = 3n - 6$. Instead of the temperature, the mean kinetic energy itself is often considered. Then the caloric curve is a dependence of $\langle E_{\rm kin}\rangle$ on E. In simulations $\langle E_{\rm kin}\rangle$ is calculated as

$$\langle E_{\rm kin}\rangle = \frac{1}{N}\sum_{j=1}^{N}\sum_{i=1}^{n}\mathbf{p}_i^{(j)2}/2m \tag{6.1}$$

where j labels the points taken into account, and N is the total number of the points. If these points are related to an specific isomer, a (fractional) caloric curve for this isomer is obtained, and if to all isomers accessible at given E, then the total caloric curve.

Figure 10a presents caloric curves for LJ_{13}. The cluster executes no overall translation and rotation. Numerical values are given in the LJ units. There are shown the caloric curve for the ground-state isomer, those for several low-lying excited-state isomers, and also the total caloric curve. The total caloric curve was calculated in a usual fashion, allowing the system to explore the PES freely. To calculate caloric curve for a specific isomer, the system was confined to one of the catchment basins corresponding to this isomer. Characteristics of the isomers are given in Table 1 (in this section we use the linear frequency of vibrations ν, which is related to the angular frequency ω as $\nu = \omega/2\pi$). Figure 10a clearly indicates that the caloric curves for excited-state isomers are very similar in appearance, and they are remarkably different from the caloric curve for the ground-state isomer. Moreover, if one counts the total energy for each of the isomers from its minimum energy, these curves follow each other with a good accuracy, even

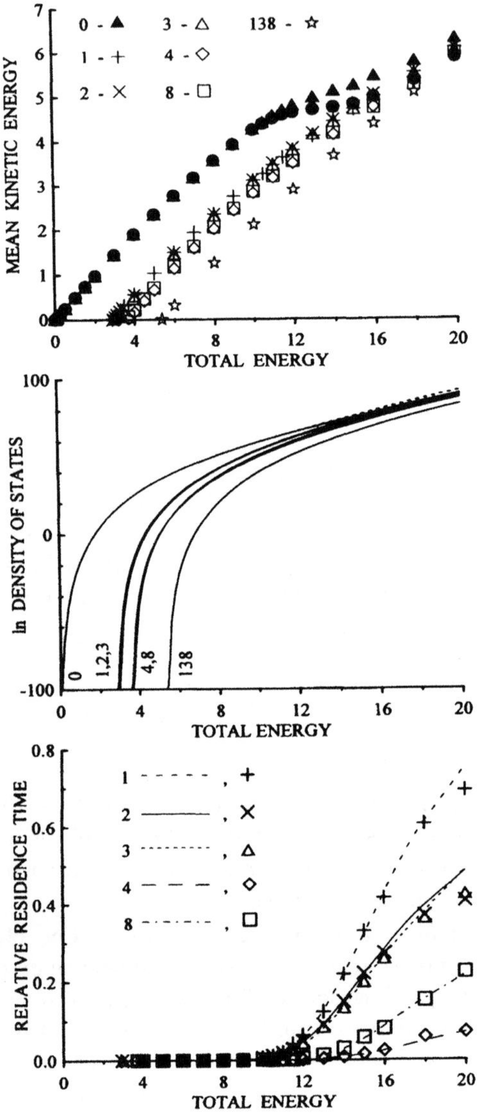

Fig. 10. LJ_{13}, microcanonical ensemble: isomers are numbered according to Table 1, and the total energy is counted from the ground-state isomer energy $U_{\min} = -44.3268$. (a) Caloric curves for individual isomers and the total caloric curve (solid circles). (b) Absolute fractional densities of states and the total density of states (the dashed line). (c) Relative densities of states (lines) in comparison with the relative residence times (symbols).

Table 1. Characteristics of isomers for LJ_{13}. The isomers are numbered according to their order in the isomer energy spectrum, 0 labels the ground-state isomer. $U_{\min,r}$ is the minimum energy of r-th isomer, $\overline{\nu}_r$ the geometrical mean of the normal frequencies, and h_r the point group order.

Isomer	$U_{\min,r}$	$\overline{\nu}_r$	h_r
0	−44.326801	1.750316	120
1	−41.471979	1.636509	2
2	−41.444597	1.635673	2
3	−41.394398	1.635996	2
4	−40.758513	1.638231	4
8	−40.670170	1.632131	1
138	−38.975810	1.553827	6

though the inherent structures of these isomers are quite different (compare, in particular, the point group order h_r, Tab. 1).

These caloric curves make it possible to calculate the absolute densities of states, both for the system as a whole and for specific isomers.

The total phase volume for the system is defined by equation (5.1) except that the integral is taken over all points in configuration space. Following Stillinger and Weber [3], break the integral into the parts corresponding to geometrically different inherent structures (isomers). Each such structure is represented by a set of $2n!/h_r$ geometrically identical copies, where $2n!$ is the order of the complete nuclear permutation and inversion group (CNPI) [66], and h_r is the order of the point group of r-th structure. In turn, to each of the copies there corresponds a certain catchment basin, how it was defined when we wrote equation (5.1) in Section 5.1. Therefore the phase volume associated with r-th isomer can be written as

$$G_r(E) = (2n!/h_r)\tilde{G}(E) \tag{6.2}$$

with $\tilde{G}(E)$ given by equation (5.2), and the total phase volume can be calculated as

$$G(E) = \sum_r G_r(E).$$

Correspondingly, the density of states for r-th isomer and the total density of states are calculated as

$$\rho_r(E) = \mathrm{d}G_r/\mathrm{d}E \quad \text{and} \quad \rho(E) = \sum_r \rho_r(E).$$

As one can find from (5.2)

$$\mathrm{d}\ln\tilde{G}_r/\mathrm{d}E = N/\left(2\langle E_{\mathrm{kin},r}\rangle\right) \tag{6.3}$$

where $\langle E_{\mathrm{kin},r}\rangle$ is the mean kinetic energy for r-th isomer. This equation is simply an analogue of the thermodynamic equation $T\mathrm{d}S = \mathrm{d}E$, with $S = k_\mathrm{B} \ln \tilde{G}_r$ and $T = 2\langle E_{\mathrm{kin},r}\rangle / N k_\mathrm{B}$ being the entropy and temperature, respectively. A similar equation is valid for the whole system.

With the characteristics of the isomers and the caloric curves being known, equation (6.3) can be numerically integrated to give $G_r(E)$ or $G(E)$, depending on which of the caloric curves, fractional or total, is used. To find the arbitrary factor in the integral, the solution can be matched in the lowest energy limit to the harmonic solution

$$\tilde{G}_{\mathrm{h}r}(E) = \frac{8\pi^2 (E - U_{\mathrm{min},r})^N}{\Gamma\,(N+1)\,\overline{\nu}_r^N}$$

where the factor $8\pi^2$ accounts for various orientations of the cluster. When calculating $G(E)$, the harmonic solution for the ground state isomer should be used. To circumvent the difficulties which one meets at attempting a direct matching, the right-hand part of (6.3) was divided into two parts: the harmonic one, which is N/E in accordance with the equipartition principle, and the rest $N\,[1/(2\langle E_{\mathrm{kin},r}\rangle) - 1/E]$, which is an anharmonic correction. The first part is integrated analytically, that provides an automatic matching to the harmonic solution. Figure 10b shows the densities of states so obtained from the caloric curves of Figure 10a.

Calculating the density of states from the caloric curves found by confinement of the system to the catchment basins, we dealt just with local regions of the PES. This offers a unique opportunity to address the fundamentals of statistical mechanics the ergodic hypothesis and the hypothesis of equal *a priori* probabilities. The former states that the time and ensemble averages are equivalent, and the latter that all microstates which are consistent with macroparameters of the systems contribute to the ensemble with the same weight. Following these statements, the system, in particular, should visit the region of the PES corresponding to r-th isomer with the probability

$$P_r\,(E) = \rho_r(E)/\rho\,(E) \tag{6.4}$$

where $\rho_r(E)$ and $\rho(E)$ are the density of states for r-th isomer and the total density of states, respectively. Therefore, we can count the relative residence time for a pair of isomers, r and q, in the conventional simulations and compare it with the corresponding relative density of states $f_{rq}(E) = P_r(E)/P_q(E) = \rho_r(E)/\rho_q(E)$ obtained from the confinement simulations. The residence time is calculated as the number of quenches leading to the corresponding basin (it should not be confused with the life time, which was considered in Sect. 5; for a difference between them see Sect. 7.2). Figure 10c presents such a comparison for the excited-state isomers from the Figure 10b (for a reference isomer, q, the ground-state isomer is taken).

Data for isomer #138 is not presented because this isomer was visited in the conventional simulation so rarely that we were unable to gain satisfactory statistics for it. It is seen that the system in the conventional simulations behaves well in accordance with the previously mentioned statements, at least on interbasin scale.

6.2 *Rates of the transitions between catchment basins. Estimation of the rate of a complex transition by successive confinement [50, 52]*

For a model system in this case we consider 15-atom carbon cluster with atoms interacting through the semi-empirical Brenner potential [67] modified in [68].

Let us study first the interconversion of two specific structures, Figure 11. For these structures we take the linear chain ($U_{\min} = -88.9509$ eV) and the ring ($U_{\min} = -92.1344$ eV), with the latter being the ground-state isomer for this cluster. To emphasize advantages of the confinement simulations (and, partly, to accelerate simulations), let the total energy be rather high, $E = -58.95$ eV, which corresponds to the temperature equal to approximately 8800 K.

At first, we confined the system to the basin for the linear chain and counted all attempts of the system to leave this basin for all other basins that surround the latter. As could be expected, the chain predominantly attempted to dissociate (the relative number of events $p_{\mathrm{d}} = 0.877$), Figure 11a. All other attempts were related to its conversion either into the ring ($p = 0.016$) or into other bound structures (a ring with one or two tails, a double ring, a bowl, etc., $p_{\mathrm{b}} = 0.107$ in total). The absolute value of the rate of the transitions from the linear chain to the ring, corresponding to the given value of p, was found equal to 5.89×10^{10} s^{-1}.

These values of p can serve a good illustrations to the efficiency of the confinement simulations: if the conventional approach were used, more than 98 percent of the acts of the transitions would take the system out of the region of the PES that is of interest.

Then the system was confined to the catchment area consisting of two basins, for the chain and ring, so that it could undergo free interconversion between these structures (Fig. 11b). The transition rate for the chain-to-ring conversion was obtained here as 5.92×10^{10} s^{-1}, which is in excellent agreement with the previously given estimate obtained on the basis of counting the number of attempts to leave the chain structure basin (Fig. 11a). It is pertinent to note that though the ring is lower in energy than the chain, the rate of the backward process, *i.e.* of the ring-to-chain conversion, is considerably higher (1.34×10^{12} s^{-1}). According to the detailed balance, this implies that the population of the chain structure basin (the residence time of the system in it) is higher than that for the ring structure, which,

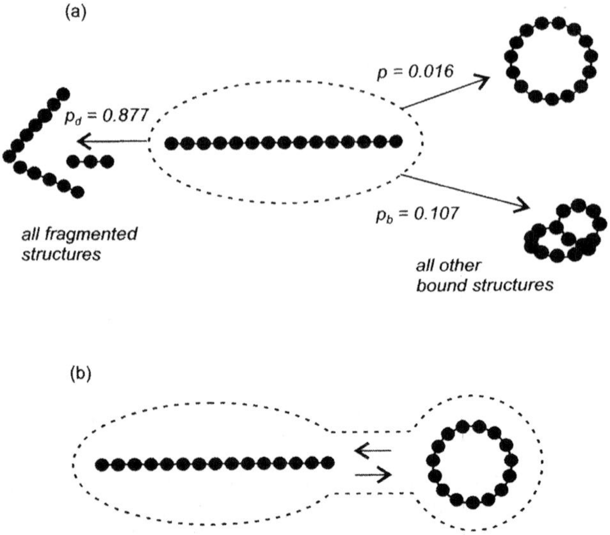

Fig. 11. Schematic picture of investigation of C_{15}: (a) confinement of the system to the catchment basin for the chain structure (the arrow-like and bowl-like structures are just to indicate the classes of dissociated and bound structures, respectively); (b) confinement to the catchment area for the chain and ring structures.

in turn, can be explained by a higher value of the conformation entropy of the chain as compared to the ring.

Each of the above approaches can be used for efficient estimating the rate of a complex transition, when the initial and terminal structures are separated by a set of intermediate structures. Specifically, the system can be successively confined to the catchment areas corresponding to two neighbouring structures (starting from the first pair), or more economically, to the catchment basins corresponding to the initial and all the intermediate structures. We have discussed this procedure for a model PES (Fig. 1) and found it much more efficient than the conventional simulations.

Figure 12 illustrates this procedure of successive confinement for the calculation of the rate of the transition from the linear chain to a triple-ring structure ($U_{\min} = -85.3930$ eV) for a specific channel of the transition. Starting from the linear chain, the system was then successively confined to the basins for the ring and the double-ring ($U_{\min} = -87.4679$ eV) structures. In each of the basins the system was kept until it made at least 50 attempts to pass into the basin to which it was supposed to be confined next. The arrows in Figure 12 show the probability for the system to leave

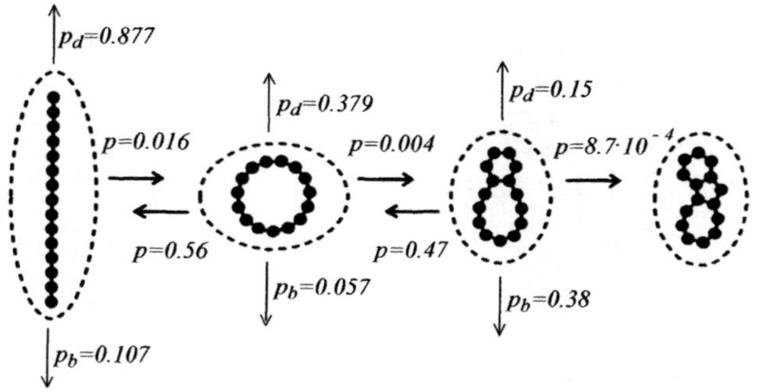

Fig. 12. Schematic picture of estimating the rate of the transition in the sequence of the C_{15} isomers. For details, see the text.

the corresponding structure: either following a channel of interest (p), or dissociating (p_d), or converting into the other bound structures (p_b). The rate of the first stage of the process (the chain-to-ring conversion), which corresponds to $p = 0.016$, is 5.92×10^{10} s^{-1}. Then the rate of the transition from the linear structure to the triple-ring can be estimated by multiplying together the above rate and the corresponding values of p for two subsequent forward transitions. It is obtained as 2.1×10^5 s^{-1}. This process is obviously too slow to study it by conventional simulations. With the time step of numerical integration of Newtonian equations being equal to 5.3×10^{-16} s, even a simple observation of an act of the transition from the linear chain to the triple ring would require the order of 10^{10} time steps. Correspondingly, to calculate the rate of the transition with satisfactory statistics, this value had to be increased by one or two orders of magnitude.

6.3 Creating a subsystem of a complex system. Self-diffusion in the subsystem of permutational isomers [52, 63]

It is common that "real" many-body systems are too complex for theoretical study, and the clusters is the case. The data obtained in numerical simulations (or experiments) represent some averages over a variety of the isomers. Because of this, comparison of analytical predictions with simulation or experimental data turns out to be a real challenge: on the one hand, theoretical consideration of a variety of geometrically different isomers is impracticable, and on the other, introducing some averages, to account for different characteristics of the isomers, may disguise possible defects of the theory.

To this end, it may be useful to consider a subsystem, which, on the one hand, retains many of characteristic properties of the system, and on the

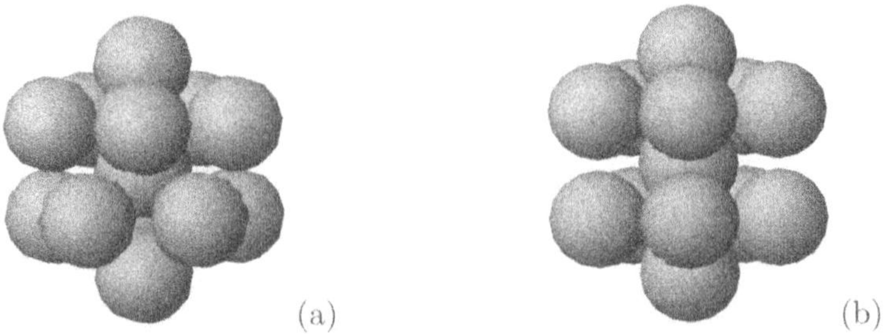

Fig. 13. (a) The ground-state isomer in LJ_{13}, icosahedron, $U_{\min} = -44.3268$. (b) The transition state configuration, dodecahedron, $U = -41.5552$.

other hand, it is much easier for theoretical considerations. Specific isomers that we have considered are, in fact, simple examples of such subsystems. Another, a more sophisticated subsystem is the manifold of permutational isomers, which allows us to study some aspects of self-diffusion. Specifically we shall consider the ground-state isomer of the LJ_{13} (Fig. 13a), whose permutational copies directly transform into each other through the saddles of the first order (transition states), Figure 13b. This transition is the rotation of six surface atoms with respect to the other six surface atoms. The PES of this subsystem consists of identical catchment basins directly connected through identical saddle barriers, and thus the subsystem ideally suits for theoretical considerations.

In the simulations, the MD trajectory of the system was confined to the area of the PES corresponding to the permutational copies of the ground-state isomer, *i.e.* just the transitions among the catchment basins for these copies were allowed. The self-diffusion coefficient was estimated in the conventional way, as

$$D = \frac{1}{6n}\frac{\mathrm{d}}{\mathrm{d}t}\sum_{i=1}^{n}\langle(\mathbf{r}_i(t) - \mathbf{r}_i(0))^2\rangle$$

within the time interval where the mean square displacement $\langle(\mathbf{r}_i(t) - \mathbf{r}_i(0))^2\rangle$ grew linearly with time (the angular brackets denote an ensemble average over a set of initial points $(\mathbf{r}_i(0), ..., \mathbf{r}_n(0))$, in which the MD trajectories started).

Figure 14 compares the self-diffusion coefficient for this subsystem and that for the whole system, *i.e.* for the system freely exploring the PES (the conventional MD). It is seen that their behaviour in both cases is similar, so that such a simple subsystem can be used as a model system to study diffusion in the "real" system, which is incomparably more complex. A detailed the simulation study of this subsystem has been conducted in [63], with the results employed to verify a theoretical model for self-diffusion.

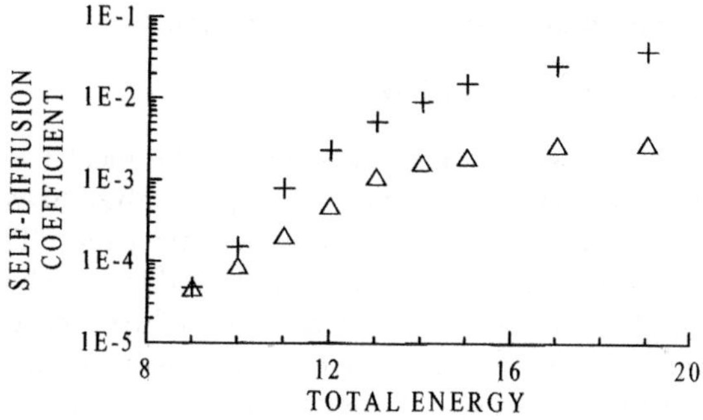

Fig. 14. Self-diffusion coefficients in LJ_{13}: the triangles are for the subsystem of the permutational ground-state isomers, and the crosses are for the system freely exploring the PES. The total energy is counted from the ground-state isomer energy, $U_{\min} = -44.3268$.

The coincidence of the points at the lowest energy in Figure 14 reflects the fact that the transition state involved is the lowest saddle for LJ_{13}. Therefore, at small energies the self-diffusion in the cluster starts as a permutational self-diffusion.

7 Complex study of a system by successive confinement

Originally meant for a detailed study of local regions of a PES, some examples of which were given in the preceding section, the confinement technique was also found to be efficient for a complex study of a system, including its PES, equilibrium properties and kinetics [69].

For this purpose the MD trajectory of a system is successively confined to various catchment basins on the PES. To initiate the simulations, the system is placed into one of the basins, and the MD run is began. As before, the system is quenched at regular intervals in order to check if it is still in a given (current) basin or has left it for another basin. If left, then the system is either allowed to pass into a certain new basin or returned to the given basin. In the first case the trajectory is continued in the usual fashion, and in the second case it starts at the point of the last quenching within the given basin (Sect. 5.2). Since both the transition probabilities and thermodynamic functions can be calculated as averages of the corresponding quantities over the points associated with the given basin, this way of confinement of the system is completely consistent with determining its kinetics (Sect. 7.2) and equilibrium properties (Sect. 7.3).

7.1 Surveying a potential energy surface. Strategies

7.1.1 Strategies to survey a surface

The strategy of how to survey a PES is one of the most essential elements of the successive confinement approach. It prescribes to which basin the system should be confined next and for how long time. These basins are chosen from the basins that have been encountered (as the basins into which the system attempted to pass when it was confined to preceding basins) but have not been sampled yet. The choice of the basins depends on the goal of the study and may involve different criteria. Let us list some possible strategies.

1. If the system is small enough so that a complete sampling of the total PES is possible, we can simply allow the system to pass into each new basin that has not been previously sampled and study it in turn.

2. If the system is large, the goal may be

i) to survey just a part of the PES that is of interest, as for example, in Section 6.3, when the self-diffusion in the subsystem of permutational isomers was studied, or

ii) to obtain a "coarse" survey of the total PES; in this case it seems reasonable to give preference to the basins for which their inherent structures differ most from those for the initial basin and basins which have been already sampled.

3. If one is interested in the dynamics of the system, the preference may be given to basins with lower transition states.

It is essential that in all above cases a prior information about the PES landscape is not required: being placed at any point on the PES, the system, basin by basin, can successively explore all the PES, or every connected part of it that is of interest.

7.1.2 A taboo search strategy. Fermi-like distribution over the minima

Let us consider in more detail the simplest strategy, which allows the system to pass into each new basin encountered that has not been previously sampled [70]. By the underlying concept, it is very close to so called taboo search [24, 25], which has been mentioned in the Introduction. To realize this strategy, different protocols are possible. We will consider the following two.

Protocol I. This corresponds to the direct taboo search in the terminology of [24, 25]. The system is kept in each of the basins until among the basins into which it attempts to pass a new basin appears that has not been sampled. To embody this protocol, a list of the basins is made and updated that indicates the basins that have been already sampled (in fact, this list contains the characteristics of the corresponding quenched structures that are necessary for their identification and comparison, Sect. 4.2).

Protocol II. This corresponds to a partial stochastic sampling [24,25] and assumes that the system is kept in each of the basins until it makes a certain number of attempts, $k_{\rm att}$, to pass into neighbouring basins. In this case two lists of the basins are made and updated: one indicates the basins that have been sampled, as in protocol I, and the other contains information about the basins that have been encountered but not sampled yet. Such information includes the characteristics of the minima of the basins (to recognize the basins) and atomic coordinates and velocities of the system at the points at which the system was found in the basins; these coordinates and velocities are employed to start the MD trajectories when (if) the system is allowed to pass into these basins. The basins to pass into are then taken from this second list of the basins. The simplest choice is to take them in the order in which they were encountered. However, the preference may be given to basins with any desirable property. One example of such choice, the preference to the basins with the lowest minimum energy, will be discussed later in Section 7.4.

If no new minima are found while the system is confined to each of the basins encountered, the system is returned into the first basin of the second list of the basins, and the exploration of the PES is continued until the system visits every basin $k_{\rm ret}$ times (a new trajectory is thus initiated in each of the basins $k_{\rm att}k_{\rm ret}$ times in total). If an isomer whose permutational copy has been already sampled is encountered, in both cases (protocols I and II) the transition into the basin corresponding to this isomer is forbidden (if, of course, the permutational copies are not of specific interest, as it was in Sect. 6.3).

Optimal values of $\tau_{\rm quench}$, $k_{\rm att}$ and $k_{\rm ret}$ may be dependent upon many factors: the nature and size of the system, the type of the dynamics (Newtonian or Langevin's), the conditions at which the simulation is conducted (the value of total energy or temperature), and the goal of the study. Therefore, specification of these parameters, to a large degree, is the matter of trial and comparison. In other words, one should try different values of the parameters and compare obtained results for the convergence. Nevertheless, certain recommendations can be given. For example, for an efficient surveying a PES, $\tau_{\rm quench}$ should evidently be of the order of the life time of the system in a basin. This is for both protocols I and II. As for $k_{\rm att}$ and $k_{\rm ret}$ in protocol II, the results of the search are dependent upon the product of these quantities rather than upon their individual values. Moreover, in a wide range of $k_{\rm att}k_{\rm ret}$, provided that none of the factors is unreasonably small, this dependence is rather weak. In the simulations presented below

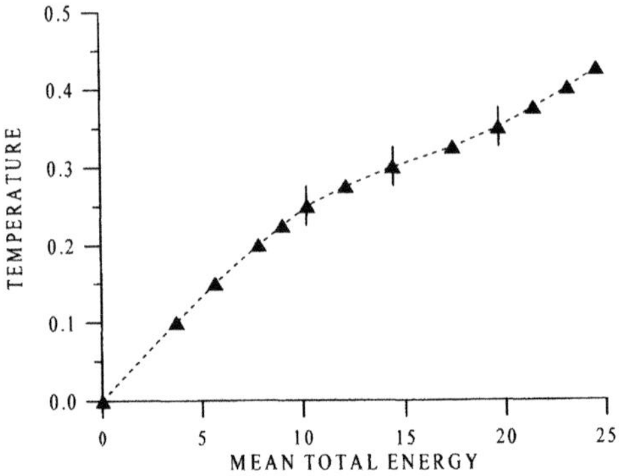

Fig. 15. Caloric curve for LJ_{13}, canonical ensemble.

the following values of these parameters were typically used $\tau_{\text{quench}} = 50$ time steps, $k_{\text{att}} = 20$ and $k_{\text{ret}} = 20$.

To illustrate protocols I and II, consider surveying the total PES for LJ_{13}. The Langevin MD will be used, as described in Section 5.2. In protocol II the basins to pass into are taken from the list of the basins to be sampled in the order in which these basins are encountered. Specifically, the simulations were conducted at three characteristic temperatures which are indicated by bars at the (canonical) caloric curve for LJ_{13} shown in Figure 15. They correspond to the solid-like state region (more definitely, to approximately the upper bound of this region, $T = 0.25$), to the melting region ($T = 0.3$) and to the liquid-like state region ($T = 0.35$). All values in this section, here and below, are given in the LJ units.

Note that in this case the cluster executes the overall translations and rotation. Therefore, for the quantitative comparison of this curve with the microcanonical caloric curve of Figure 10a the mean total energy in Figure 15 should be reduced by the value of $3k_{\text{B}}T$ corresponding to the translational and rotational degrees of freedom.

The total number of geometrically different minima found for LJ_{13} grew every work: Hoare and McInnes [71], employing a growth algorithm to seek for candidate structures, found 988 minima. Later Tsai and Jordan [20], who used a direct search by walking on the PES with the help of the eigenmode method, increased this number up to 1328. Finally, just recently Doye *et al.* [47] and Ball and Berry [48], who also employed a direct search but a different modification of the eigenmode method, found 1467 and 1478 minima, respectively. The search for the total number of minima is complicated by the fact that there is no criterion to decide when the exploration of the PES can be regarded to be complete. Also note

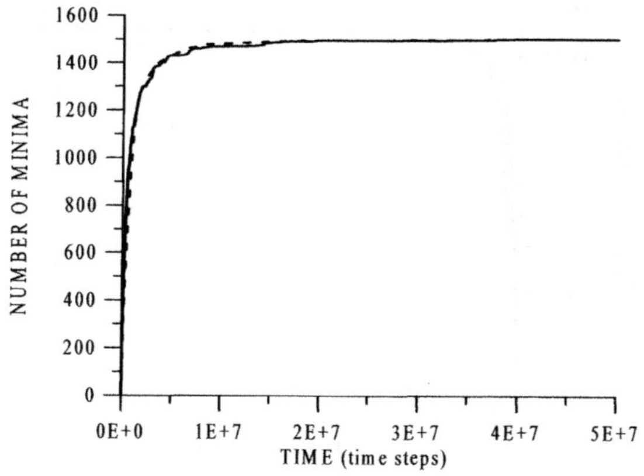

Fig. 16. Number of minima located with time in the confinement simulations: protocol I (solid line) and protocol II (dashed line). LJ_{13}, canonical ensemble, $T = 0.35$.

that the search for specific minimum is a sort of the NP-hard problem (in application to clusters, see, *e.g.* [72]).

Figure 16 presents a typical dependence of the number of minima located with time under protocols I and II. The temperature is for the liquid-like state ($T = 0.35$). In both cases the system was initially placed into the basin corresponding to the ground-state isomer (Fig. 13a). Though k_{ret} (protocol II) did not reach its maximum value ($k_{\mathrm{ret}} = 20$) for each of the basins, both runs were terminated at 5×10^7 time steps, where in the case of protocol II no new minima were found for approximately 1×10^7 last steps. The total number of minima found under protocols I and II is 1504 and 1506, respectively. In both cases it slightly exceeds the maximum number of minima reported in previous works.

As can be seen from the figure, in the initial stage of the search, when the list of sampled minima is almost empty, the number of minima in the search under protocol I grows a bit faster than that under protocol II. However, in the final stage, when that list is almost complete, the situation changes: in the case of protocol I the system has to dwell in a current basin for a long time until a basin which has not been sampled yet is encountered. Moreover, it may happen that the minima which were missed in the initial and intermediate stages of the search are not accessible from the basins in which the system dwells in the final stage. Therefore, protocol II seems to be more preferable for surveying a PES than protocol I.

Figure 17 shows typical results of surveying the PES for the temperatures indicated in Figure 15. Three upper (thick) curves correspond to

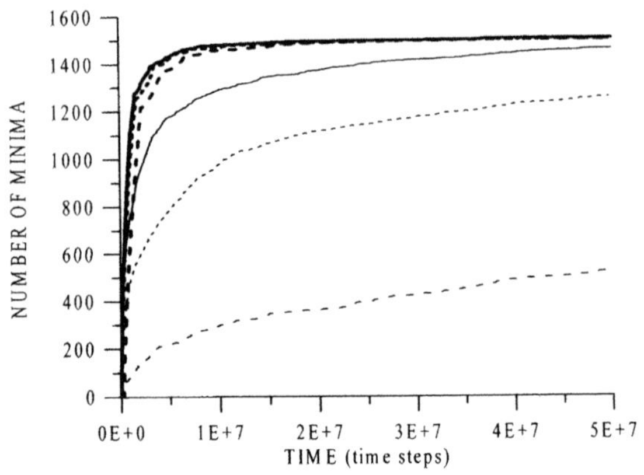

Fig. 17. Number of minima located with time in the confinement simulations under protocol II (thick curves) and in the conventional simulations (thin curves): the long-dash lines correspond to $T = 0.25$, the short-dash lines to $T = 0.3$, and solid lines to $T = 0.35$. LJ_{13}, canonical ensemble.

the successive confinement taboo search (SCTS) under protocol II, and the other (thin) curves to the direct search by the SW method (SWDS); in both cases the quenching interval was the same, 50 time steps. First, as it could be expected, the SCTS is faster than the SWDS. Second, and somewhat surprising, the SCTS results are just slightly dependent upon the temperature. By the time when the last of 1506 minima was located at $T = 0.35$ ($t = 4 \times 10^7$ time steps), there were found 1502 and 1499 minima at $T = 0.3$ and $T = 0.25$, respectively, with total number of minima in both latter cases being 1502. This is particularly surprising in comparison with the SWDS simulations, where the decrease of the temperature from the liquid-like state ($T = 0.35$) to the solid-like one ($T = 0.25$) results in a radical reduction of the number of minima.

The energy spectrum of the minima found in these simulations is shown in Figure 18.

It should be noted that the conventional (SWDS) simulations presented in Figure 17 are not "pure" conventional: if an act of dissociation (evaporation) of the cluster was indicated, the MD trajectory was continued as in the confinement simulations, that is a new trajectory was initiated at the point where the system was found in bound state. Otherwise, at higher temperatures the trajectories have to be terminated in a short time: at approximately 6×10^4 and 3×10^4 time steps for $T = 0.3$ and $T = 0.35$, respectively.

Fig. 18. Energy spectrum of minima for LJ_{13}. The energy is counted from the ground-state isomer energy, $U_{\min} = -44.3628$.

The difference between the SWDS and SCTS simulations is well illustrated by the distributions of the number of quenches over the minima to which these quenches lead. The SWDS distribution, by its nature, is a steady distribution; it depends on the elapsed time only in that the statistics becomes richer with time. In contrast, the SCTS distribution is unsteady, its front propagates with time.

The SWDS distribution originates from the Boltzmann distribution. For the PES divided into catchment basins, as in Section 6.1, the total partition function of the cluster of n atoms is written [3] as

$$Z = \sum_i \exp(-U_{\min,i}/k_\mathrm{B}T) Z_i(T) \tag{7.1}$$

where

$$Z_i = \frac{1}{(2\pi\hbar)^{3n}} \frac{1}{h_i} \int \exp\{-H_i/k_\mathrm{B}T\} \; \mathrm{d}\mathbf{r}^{(3n)} \mathrm{d}\mathbf{p}^{(3n)} \tag{7.2}$$

is the partition function for i-th basin. Here h_i is the order of the point group, and $H_i = \sum_{j=1}^{n} \mathbf{p}_j^2/2m + U_i$ is the Hamiltonian of the cluster associated with i-th basin (the potential energy U_i is counted from the basin minimum energy $U_{\min,i}$). In the given representation, Z and Z_i have a meaning of the total numbers of states and the number of states associated with i-th basin, respectively. According to (7.1, 7.2), the mean number of the quenches which lead to the minimum of i-th basin can be estimated as

$$\langle N_i \rangle = N_\mathrm{tot} \exp(-U_{\min,i}/k_\mathrm{B}T) Z_i/Z \tag{7.3}$$

where N_tot is the total number of quenches, and the angular brackets denote the ensemble average.

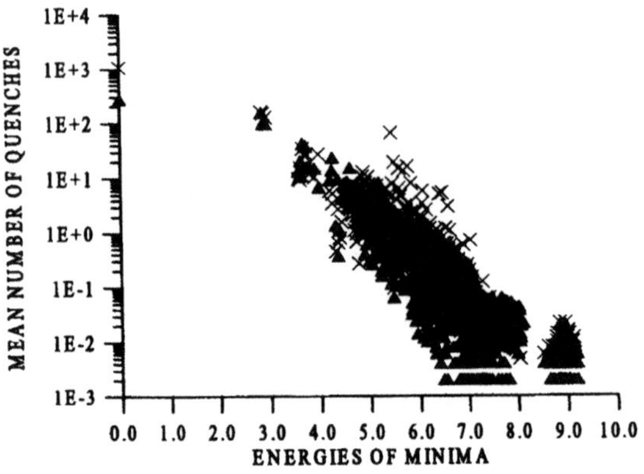

Fig. 19. Distribution of the mean number of quenches over the minima in the conventional simulations. Triangles correspond to the simulation data, and crosses to equation (7.3), with Z_i estimated in the harmonic approximation (7.4). LJ_{13}, canonical ensemble, $T = 0.35$.

Figure 19 compares the SWDS distribution with equation (7.3), in which Z_i are estimated according to the harmonic approximation [73]

$$Z_i(T) = \left(\frac{k_B T}{\hbar\overline{\omega}_i}\right)^{3n-6} \frac{(2k_B T)^{3/2}(\pi I_{1i} I_{2i} I_{3i})^{1/2}}{h_i \hbar^3} \tag{7.4}$$

where $\overline{\omega}_i$ and $I_{\alpha i}$ ($\alpha = 1, 2, 3$) are, respectively, the geometrical mean of normal frequencies and the principal momenta of inertia for the inherent structure corresponding to the minimum of i-th basin; these characteristics, and also h_i, were determined in the course of simulations. The first and second multipliers in the right-hand side of (7.4) correspond to the vibrational and rotational degrees of freedom, respectively; a contribution of the translational degrees of freedom is omitted, since it is the same for all isomers. A discrepancy between the simulation data and the estimate in Figure 19 should be attributed to anharmonic effects, because in the liquid-like region, to which the value of $T = 0.35$ corresponds, the anharmonicity is appreciable.

The SCTS distribution is essentially different from the SWDS distribution: while the intrabasin statistics remains to be the Boltzmann statistics, the interbasin statistics, in the part of visiting the basins that have been visited, is inherently the Fermi statistics. To find mean occupations numbers for the basins in this case, we can consider the collection of the isomers as an ideal gas, with i-th isomer contributing to the partition function as $[Z_i(T)\exp(-U_{\min,i}/k_B T)]^{N_i}$, where N_i takes the values 0 and 1. Then, following the standard procedure (see, *e.g.* [74]), we obtain [70]

$$\langle N_{\mathrm{FL},i} \rangle = \frac{1}{1 + Z_i(T)^{-1} \mathrm{e}^{(U_{\min,i} - \mu)/k_{\mathrm{B}}T}} \tag{7.5}$$

where μ, playing a role of chemical potential, is determined by the condition $\sum_i \langle N_{\mathrm{FL},i} \rangle = \langle N_{\mathrm{tot}} \rangle$. In the case of small values of the mean occupation numbers (the second term in the denominator is much larger relative to unity), which corresponds to the Boltzmann statistics, this equation transforms into (7.3). The appearance of such a distribution in classical simulations may seem somewhat unusual. However, in fact, there is no a great surprise in this, since the quantum character of the Fermi statistics in its conventional applications is due to the Pauli exclusion principle and not to the statistics itself.

Equation (7.5) describes the equilibrium distribution. Therefore it is applicable to the SCTS distributions if the latter are of quasi-equilibrium character. Further, the system under consideration, LJ_{13}, possesses a finite number of the states (minima), therefore equation (7.5) is correct until the front of the distribution reaches the upper bound of the energy spectrum. Fortunately, the quasi-equilibrium character of the SCTS distributions holds well practically for all elapsed times, until the front of the distribution comes into the upper bound of the energy spectrum (see Fig. 21 given below). Also note that though equation (7.5) is appropriate to protocol I rather than to protocol II, at the elapsed times, when the time required to execute k_{att} attempts to leave a current basin is negligible in comparison with the elapsed time, this equation can be applied to protocol II too.

Figure 20 compares the SCTS distribution for $t = 8 \times 10^4$ time steps with equation (7.5), where Z_i are calculated in the harmonic approximation (7.4), as in Figure 19. The distribution does not include the repeat quenches leading to the sampled minima. Correspondingly, μ in (7.5) was determined from the condition that $\langle N_{\mathrm{tot}} \rangle$ is equal to the total number of quenches included into the distribution.

The front of the SCTS distribution proceeds with time as Figure 21 shows. For an illustrative purpose, to avoid an overlapp of the data for different elapsed times, the distributions are given in a smoothed form: each of the points represent an average number of quenches falling into one of 25 boxes into which the energy range was uniformly divided. All curves, except for curve 6, even curve 1, are of quasi-equilibrium pattern, that is the corresponding distributions are described by (7.5), approximately with the same accuracy as in Figure 20. Curve 6 clearly demonstrates the effect of the finite number of the states (basins).

For each current elapsed time t the sum of $\langle N_i \rangle$ over minima, $\langle N(t) \rangle = \sum_i \langle N_i(t) \rangle$, presents the number of minima which probably will be found to this time in an individual run, as in Figure 17.

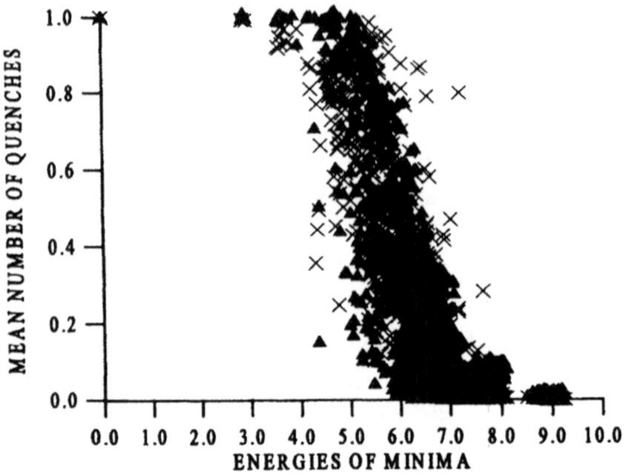

Fig. 20. Distribution of the mean number of quenches over the minima in the confinement simulations (protocol I). Triangles correspond to the simulation data, and crosses to equation (7.5) with Z_i calculated in the harmonic approximation (7.4). LJ_{13}, canonical ensemble, $T = 0.35$.

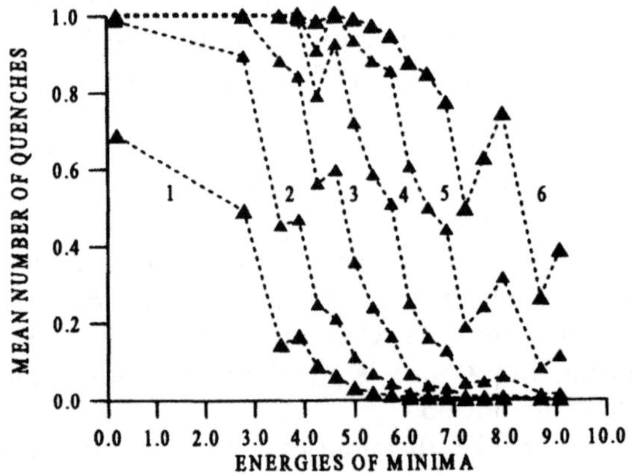

Fig. 21. Evolution of the SCTS distributions (protocol I) with time. Triangles indicate the simulation data, and the dashed lines are to guard the eye. Labels 1,2,3,4,5 and 6 correspond to the elapsed times 1.25×10^3, 5×10^3, 2×10^4, 8×10^4, 3.2×10^5 and 1.28×10^6 time steps, respectively. LJ_{13}, canonical ensemble, $T = 0.35$.

Unfortunately, such a relatively simple picture of surveying the PES, leading to a propagating Fermi-like distribution, is characteristic only of systems whose PESs consist of a single funnel, as for LJ_{13} and some other clusters (*e.g.* LJ_{19} and LJ_{55} [47]). For multifunnel PESs, similar to those shown in Figure 1 and will be shown later in Figures 23 and 25, this picture is destroyed because of the overlap of the energy spectra in different funnels. However, the employed strategy, by itself, turns out to be quite efficient for surveying such PESs (Sect. 7.4).

7.2 Kinetics

To study kinetics, a master equation is usually used. It is a loss-gain equation and describes the time evolution of the probability $P_i(t)$ for finding the system in a state i

$$\frac{\mathrm{d}P_i(t)}{\mathrm{d}t} = \sum_{j=1}^{N} [\tilde{W}_{ij} P_j(t) - \tilde{W}_{ji} P_i(t)] \tag{7.6}$$

where N is the total number of states, and $\tilde{W}_{ij}$ is the probability of the transition from state j to i per unit time. In the matrix notations, (7.6) acquires the form

$$\frac{\mathrm{d}\mathbf{P}}{\mathrm{d}t} = \mathbf{WP} \tag{7.7}$$

where $\mathbf{P}(t) = \{P_i(t)\}$ is the vector of probabilities, and $\mathbf{W} = \{\tilde{W}_{ij} - \delta_{ij} \sum_j \tilde{W}_{ji}\}$ $(i, j = 1, ..., N)$ is the transition probability matrix. The equation is linear and completely determines $\mathbf{P}(t)$ at $t > t'$ by $\mathbf{P}(t')$ (*i.e.* it implies that the process to be considered is a Markov process).

Being interested in basin-to-basin kinetics, we shall consider a catchment basin for a state. Then $P_i(t)$ will present the probability to find the system in basin i, and W_{ij} can be identified with the rate of the transition from basin j to i. W_{ij} depend on the conditions under consideration, in particular, on the total energy of the system and on the temperature for the microcanonical and canonical ensemble, respectively.

One difficulty in application of the master equation to a specific problem is the specification of the transition rates. For this purpose the transition state theory is usually employed in its simplest (harmonic approximation) version. Specifically, according to the Rice-Ramsperger-Kassel-Marcus (RRKM) theory [75], the rate constants for a particular channel of transition (l) from basin i to j are defined by the expressions: for *microcanonical* ensemble

$$k_{ji}^{(l)}(E) = \frac{\prod_{m=1}^{N} \nu_{i,m}}{\prod_{m=1}^{N-1} \nu_{l,m}^{*}} \left(\frac{E - \Delta U_l^{*}}{E} \right)^{N-1} \tag{7.8}$$

and for *canonical* ensemble

$$k_{ji}^{(l)}(T) = \frac{\prod_{m=1}^{N} \nu_{i,m}}{\prod_{m=1}^{N-1} \nu_{l.m}^{*}} \exp(-\Delta U_l^{*}/k_{\mathrm{B}}T). \tag{7.9}$$

Here N is the number of vibrational degrees of freedom ($= 3n - 6$ for non-linear atomic configurations), $\nu_{i,m}$ and $\nu^*_{l,m}$ are, respectively, the frequency of m-th normal mode at the minimum of the basin i and at l-th transition state between basins i and j, and ΔU^*_l is the height of the barrier corresponding to the transition state l, which is counted from the bottom of basin i (generally speaking, the right-hand sides of (7.8) and (7.9) should also be multiplied by the factor h_i/h^*_l to account for a possible difference between the minimum and the transition state in symmetry). The transition rate W_{ji}, entering equation (7.7), is then calculated as the sum of $k^{(l)}_{ji}$ over all channels of the transitions from basin i to j, $W_{ji} = \sum_l k^{(l)}_{ji}$.

In certain cases (7.8) and (7.9) give reasonably good estimates for k_{ji} (see, *e.g.* [39] and the discussion therein). Nevertheless, one can hardly be *a priori* sure that these formulae will work well in every specific case, particular if the basins are shallow or, that is the same, if the temperature is high. In this cases anharmonic effects should be taken in account, which complicates the problem to a large degree.

Confinement simulations allow a direct calculation of the transition rates. While the system is confined to a given basin i, one can calculate the transition probabilities, Q_{ji}, that the system will be found in a certain basin j at the subsequent quenching corresponding to the time interval, τ_{quench}. Specifically, Q_{ji} are calculated as

$$Q_{ji} = n_{ji} / \sum_j n_{ji} \tag{7.10}$$

where n_{ji} is the number of times that the system is found in basin j at the subsequent quenching when at the previous quenching it was in basin i.

It should be noted that j may be equal to i in (7.10). This is presented by diagonal components of the matrix, Q_{ii}, and corresponds to the possibility that the system did not leave basin i, or that it left this basin for a time less than τ_{quench} and then returned to it. Therefore, if the probability of system's re-entering is negligible, Q_{ii} bears information about the average value of the life time of the system in i-th basin, $\tau_{\text{life},i}$. Specifically, if the life times follow the Poisson distribution

$$Q_{ii} = \exp(-\tau_{\text{quench}}/\tau_{\text{life},i}). \tag{7.11}$$

Successively confining the system to various basins on the PES, we can obtain the transition probability matrix $\mathbf{Q} = \{Q_{ij}\}$ at the conditions of the simulation: at constant total energy of the system (microcanonical ensemble) or at constant temperature (canonical ensemble). This matrix is related to the transition rate matrix $\mathbf{W}$ as

$$\mathbf{Q} = \exp(\mathbf{W}\tau_{\text{quench}}). \tag{7.12}$$

Therefore, with the transition probabilities obtained by successive confinement, we can calculate transition rates. For this, for example, taking the quenching interval τ_{quench} to be sufficiently small, the right-hand side of equation (7.12) can be expanded in a power series, and keeping the linear term yields

$$\mathbf{W} = (\mathbf{Q} - \mathbf{I})/\tau_{\text{quench}}. \tag{7.13}$$

However, kinetics can be described directly in the terms of the transition probabilities, if we treat the transitions between the basins as a Markov process in the discrete time domain of quenching intervals. In the framework of this formalism, the transition probability matrix $\mathbf{Q}$ after n successive τ_{quench} steps is

$$\mathbf{Q}(n\tau_{\text{quench}}) = \mathbf{Q}^n(\tau_{\text{quench}}) \tag{7.14}$$

and the time evolution of the vector of probabilities obeys the equation

$$\mathbf{P}(t) = \mathbf{Q}(n\tau_{\text{quench}})\mathbf{P}(0) \tag{7.15}$$

where $t = n\tau_{\text{quench}}$, and $\mathbf{P}(0)$ is the initial vector of probabilities.

7.3 Equilibrium properties

In a stationary state, the vector of probabilities $\mathbf{P}$ $(= \mathbf{P}^{\text{st}})$ satisfies the equation

$$\mathbf{P}^{\text{st}} = \mathbf{Q}\mathbf{P}^{\text{st}} \tag{7.16}$$

and, if the transition probability matrix $\mathbf{Q}$ is irreducible (*i.e.* there exists a path from one basin to any other), this vector corresponds to the equilibrium distribution of residence probabilities, *i.e.* $\mathbf{P}^{\text{eq}} = \mathbf{P}^{\text{st}}$. $\mathbf{P}^{\text{eq}}$ can be calculated in two ways. One way is to solve (7.16) directly, and the other is to evaluate the matrix $\mathbf{Q}(n\tau_{\text{quench}})$ (7.14) at $n \to \infty$ and then to substitute the result in (7.15) in order to obtain $\mathbf{P}^{\text{eq}}$.

From the point of view of statistical mechanics, P_i^{eq} present relative numbers of states associated with certain basins. Specifically, for *microcanonical* ensemble, in accordance with (6.4)

$$P_i^{\text{eq}}(E) = \rho_i(E)/\rho(E) \tag{7.17}$$

and for *canonical* ensemble, according to (7.1, 7.2)

$$P_i^{\text{eq}}(T) = Z_i(T)/Z(T) \tag{7.18}$$

with the factor $\exp(-U_{\min,i}/k_{\text{B}}T)$ being included into $Z_i(T)$. With $P_i^{\text{eq}}(E)$ $[P_i^{\text{eq}}(T)]$ having been calculated in the course of successive confinement, these equations give us $\rho_i(E)$ $[Z_i(T)]$ with an accuracy of a mutual factor depending on E $[T]$. According to the formulae of statistical mechanics,

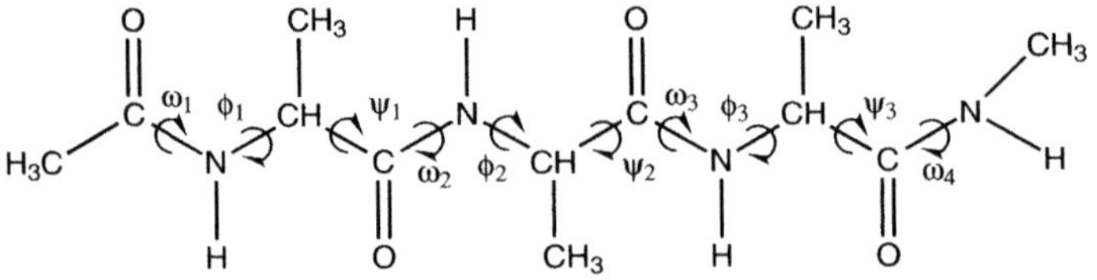

Рис. 1:

Fig. 22. The tetrapeptide $N - acetyl - (Ala)_3 - methylamide$ (TP). The dihedral angles ϕ and ψ (soft torsions) are on each side of the C^α carbons, and ω on each amide bond.

$\rho_i(E)$ determines the absolute value of the mean kinetic energy $\langle E_{\mathrm{kin}} \rangle$ as a function of E (*i.e.* the microcanonical caloric curve for i-th structure), and $Z_i(T)$ the mean total energy $\langle E \rangle$ as a function of T (the canonical caloric curve for this structure). Other characteristics, such as entropy and free energy, can be calculated from $P_i^{\mathrm{eq}}(E)$ and $P_i^{\mathrm{eq}}(T)$ just as relative values with respect to certain reference functions depending on E and T, respectively. To obtain the absolute values of these characteristics, it is sufficient to calculate the absolute value of $\rho_i(E)$ $[Z_i(T)]$ for one of the structures, which makes it possible to find $\rho(E)$ $[Z(T)]$ from (7.17) [(7.18)]. An example of such calculations was given in Section 6.1.

7.4 Study of the alanine tetrapeptide

This method of successive confinement was applied to a complex study of the molecule of alanine tetrapeptide (TP), $N - acetyl - (Ala)_3 - methylamide$, in solvent [69]. This molecule is one of the shortest peptides that can form a full helical turn and thus is of considerable interest as a model for the formation of secondary structures [33, 35, 38]. This molecule is complex enough to have a multifunnel PES, and at the same time it is simple enough to allow a direct simulation that made possible to compare the confinement and conventional simulations. The covalent structure of the TP is shown in Figure 22.

To conduct simulations, there was used the CHARMM program [76] with the ACS implicit representation of the solvent [77] and the polar hydrogen parameters (param19) [78]. The Langevin MD was employed, the friction coefficient being equal to 64 ps^{-1} and the time step to 1 fs. For quenching, the steepest descent method was combined with the adopted-basis set Newton-Raphson minimization [76]. To find heights of saddle barriers between neighboring minima, the TRAVEL algorithm [42] was employed.

To study the system, we applied the simplest strategy of successive confinement that was described in Section 7.1.2, specifically protocol II. Despite its simplicity, this strategy was expected to provide an efficient sampling of PES of a multifunnel pattern that was anticipated for the TP from previous studies [35, 38]. Indeed, turn to Figure 1. Let the system be at the bottom of one of the funnels and allow it to pass just into the basins not been previously sampled. Then, since the number of the basins at the bottom is finite, new basins in this region inevitably come to end, and the system will become explore basins on a higher level of the funnel. Thus, basin by basin, the system will climb the surface until it reaches an interfunnel barrier, then it will overcome this barrier and pass into the corresponding neighboring funnel. At reasonable temperatures (as assumed), the flow of representative points of the system at the upper part of any funnel is directed towards the bottom. Therefore, when the system is confined to a basin at the upper part of the funnel into which it has passed, the basins that lie lower than the current basin will appear with a higher probability than the others. As a result, the system, basin by basin, will descent towards the bottom of this funnel and then, as low-lying basins come to end, it will climb the surface again. In this way, funnel by funnel, the system can explore the total PES.

Surveying the PES was performed first at $T = 500$ K, with the total length of the run of 5×10^7 time steps. The basins were sampled in the order in which they were encountered when the system was confined to preceding basins. For this time, 408 minima and 4800 transitions states were found.

To organize information about the explored fragment of the PES, we use the disconnectivity graph representation introduced in [38], Figure 23. This graph shows which minima are connected by pathways lying below a certain energy threshold. All minima are grouped into funnels depending of the height of the barriers between them: the minima separated by relatively low barriers are considered as connected and thus related to the same funnel, and those separated by high barriers are considered as disconnected and thus related to different funnels. The graph reveals five clearly defined funnels: the central one, which contains the α-helix and β-strand conformers (Fig. 24), and four side funnels, that are separated from it by the barriers of the order of 10 Kcal/mol. The consideration of dihedral angles of the conformations associated with the minima has showed that all conformations in each of the funnels are characterized by a certain set of values of ω_i, and these sets vary from one funnel to another. With the values of $\omega = 180^o$ and $\omega = 0^o$ corresponding, respectively, to the *trans* and *cis* form of a peptide group, we found that the central funnel is associated with all *trans* conformation of the TP, whereas the others represent mixed *trans*/*cis* conformations. The successive confinement thus made it possible to cross the energy barriers which are as high as for the *trans*-to-*cis* transitions in peptides.

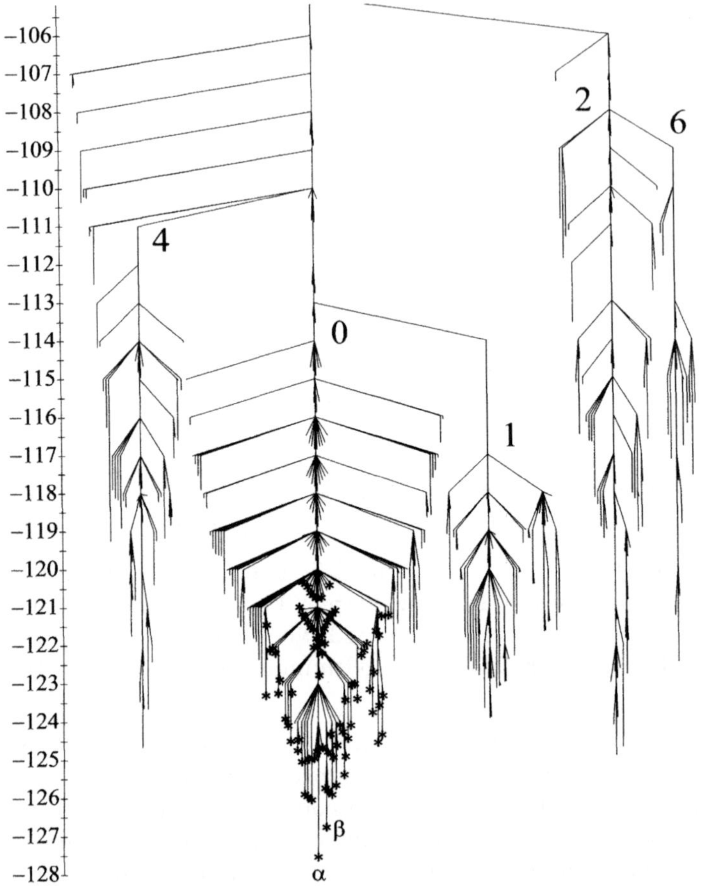

Fig. 23. Disconnectivity graph for the sampled fragment of the PES of the TP. Energy level spacing is 1 kcal/mol. α and β indicate the minima corresponding to the α-helix and β-strand conformers, respectively. Stars indicate the minima that were found at the direct MD exploration of the PES. Figures label the funnels according to the rule given in the text.

To compare the confinement and conventional approaches, the conventional Langevin MD simulations were also conducted. At the same total length of the run (5×10^7 time steps) only 97 minima and 670 transition states were found, all located in the lower part of the central funnel, where the MD trajectory was started. These minima are indicated in Figure 23 by stars.

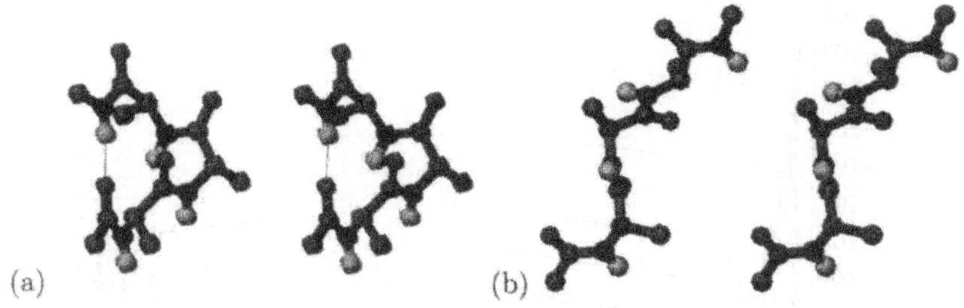

Fig. 24. Stereo view of (a) α-helix ($U_{\min} = -127.5923$ Kcal/mol) and (b) β-strand ($U_{\min} = -126.8182$ Kcal/mol) conformers. Prepared with help of the MOLMOL program [79].

To label funnels in Figure 23 (and in Fig. 25 below), binary digits are used. Namely, with 0 standing for the *trans* form of the peptide group and 1 for the *cis* form, the label of a certain conformation of the molecule is taken as the value of $i_1 i_2 i_3 i_4$, where i_k is equal either to 0 or 1 ($k = 1, 2, 3, 4$), and the count is made from left to right. For example, label 0 means that all groups are in the *trans* form, 2 (= 0100) that the second group is in the *cis* form and the others are in the *trans* form, and 6 (= 0110) that the second and third groups are in the *cis* form and the other two are in the *trans* form.

The next step was to find all 16 funnels corresponding to different 2^4 combinations of the *trans* and *cis* elements. For this purpose a more "course" survey of the PES was employed in the framework of protocol II: the preference was given to the basins with the lowest values of $U_{\min}$. In this case once such a basin occurred to be in a different funnel, the system stopped further exploration of the given funnel and moved to this new funnel. Besides, to accelerate calculations, the temperature was elevated to $T = 800$ K. Figure 25 presents the disconnectivity graph built on the data obtained for 6.4×10^6 time steps.

Kinetics was studied at the normal (room temperature) conditions, $T = 300$ K. For the total length of the run of 3.6×10^7 time steps, 69 minima and 289 transition states were found. In each of the basins the system made 150 attempts to pass into other basins. The conventional simulation, to locate approximately the same number of minima and transition states (53 and 217, respectively), took a time 3 times longer.

Table 2 compares the transition probabilities for the transitions from the β-strand conformer. There are shown channels for which statistics in the case of the conventional simulations could be considered as satisfactory (not less than 100 events). The data are in reasonable agreement. Note that the absolute numbers of events in the confinement simulations are less than

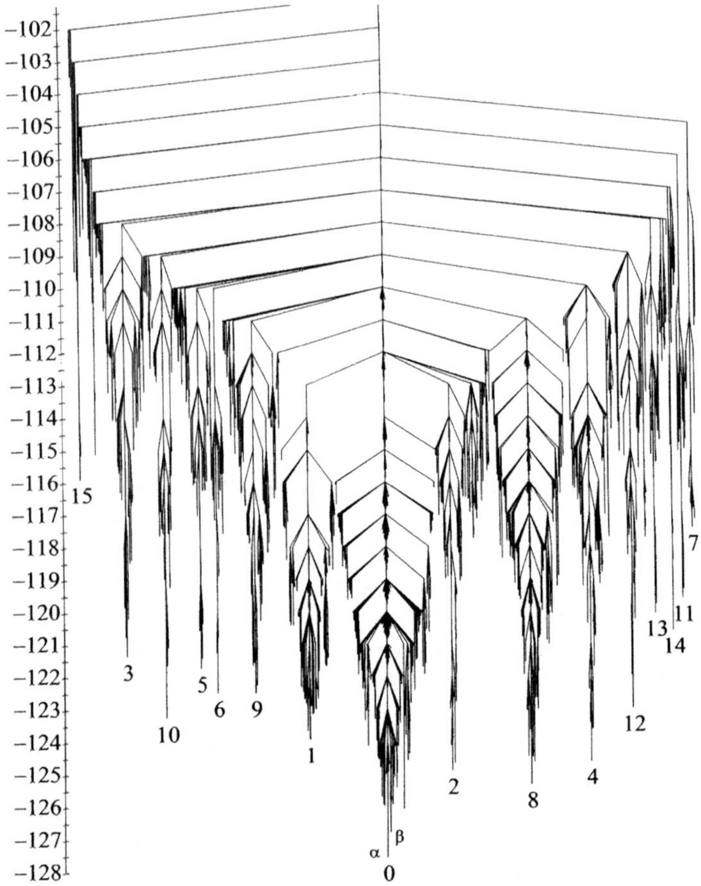

Fig. 25. Disconnectivity graph for the total PES of the TP. The notations are as in Figure 23.

those in the conventional simulations approximately by factor of 50, whereas the length of the run was only 3 times shorter. This is a consequence of more homogeneous distribution of the events over basins in the confinement simulations as well as a reflection of the fact that the β-strand conformer is thermodynamically most stable among the others (see below).

Table 3 compares the equilibrium residence probabilities for a set of low-lying conformers that were calculated directly in the conventional simulations and indirectly, from Eq. (7.16), in the confinement simulations. In general, the data are also in satisfactory agreement, approximately to within the statistical error (remind that in the confinement simulations the

Table 2. Transition probabilities, Q, and absolute numbers of the events, n, to escape from the basin for the β-strand conformer. SW and SC label the results of the conventional and confinement simulations, respectively. $U_{\min}$ indicates the minimum energy (in Kcal/mol) of the conformer into which the system passes from β-strand conformer.

$U_{\min}$	n^{SC}	n^{SW}	Q^{SC}	Q^{SW}
(β) -126.8182	961	44792	.861883	.869849
−125.8585	65	2874	.058296	.055812
−125.8013	73	3016	.065471	.058570
−124.8512	7	302	.006278	.005865
−126.0183	4	223	.003587	.004331
−125.9602	4	197	.003587	.003826

Table 3. Equilibrium residence probabilities: P^{SW} correspond to the conventional MD simulations, and P^{SC} to equation (7.16) with the transition probabilities Q^{SC} found in the confinement simulations (some of them are given in Tab. 2). $U_{\min}$ is the conformer minimum energy, α and β label the α-helix and β-strand conformer, respectively. n^{SW} is the absolute numbers of the acts of the transitions into (from) a given basin that were observed in the conventional simulations. Energies are given in Kcal/mol.

$U_{\min}$	n^{SW}	P^{SC}	P^{SW}
(α) -127.5923	20	.00890	.00654
−125.0197	77	.00143	.00158
−124.8416	16	.00029	.00018
(β) −126.8182	6702	.54508	.54080
−125.8585	3098	.04165	.04052
−125.8013	3432	.05840	.05509
−124.8512	526	.00567	.00605
−126.0183	832	.09673	.08513
−125.9602	285	.11417	.11482
−125.0428	519	.00848	.00663
−125.9658	633	.07392	.08243

number of the acts of the transitions from each of the conformers was equal to 150).

As one can see from Table 3, though the α-helix conformer is lower in energy than the β-strand one, the residence probability (or, that is the same, the relative residence time) for the latter is considerably larger. It is exactly the situation which we had in Section 6.2, when the chain structure for C_{15} was populated higher than the ring structure. Equation (7.18) can be rewritten as

$$P_i^{\rm eq}(T) = Z_i(T)/Z(T) = \exp(-F_i/k_{\rm B}T)/Z$$

where $F_i = F_i(T)$ is the free energy of i-th conformer. Therefore, the residence time for a certain structure is determined by its free energy $F = E - TS$ and not by energy E. The conformation entropy for the β-strand conformer is evidently higher than that for the α-helix, therefore β-strand turns out to be lower in free energy and, as a results, higher populated than the α-helix.

8 Concluding remarks

Going beyond the limitations imposed by the inherent dynamics and statistics, which restrict the capabilities of the existing simulation methods, the confinement technique offers a wide flexibility in the study of a many-body system, including the exploration of its PES and the calculation of the thermodynamic properties and kinetics. Depending on the goal of the study, one can investigate either the total PES or a local part of it that is related to the structures and/or channels of transitions of interest. It is essential that information about the PES, thermodynamic properties and kinetics can be obtained simultaneously, *i.e.* in the course of a single molecular dynamics run.

The approach is still in its infancy. One of the problems, which persists for the systems of large size, concerns the identification of inherent structures. Generally speaking, every method which employs mapping of the MD or MC trajectory onto the minima on the PES encounters this problem. In this work, to identify the structures, we involved quenching. It is the most general approach, being applicable to atomic and molecular systems of different nature, but at the same time, it is the most time consuming element in the simulations. Every quench usually takes the order of 10^3 time steps. Therefore, if we, say, quench the system every 100 time step of numerical integration of the equations of motion, the time is mostly spent for quenching. This is evidently unacceptable in the case of the system consisting of hundreds or thousands of atoms. Therefore, devising and development of time saving methods of identification of the structures appears to be a central problem for a future application of the confinement technique to systems of larger size. One method, suitable for the systems with well defined bond lengths, is the adjacency matrix method [80], in which the bonds between atoms are analyzed (broken or not) rather than the positions of the atoms.

Let us also mention some directions of further development of the confinement simulations.

For an elementary cell to which the MD trajectory was confined, we considered a catchment basin. However, in principle, any suitably defined region of configuration (or phase) space can be used for this purpose. For example, for biomolecules, the configuration (conformation) space can be separated into confinement regions by the values of dihedral angles or by the number of native contacts. The confinement simulations then can be used for a guided construction of so called statistical energy surfaces, which are defined in terms of energy, conformation entropy and a progress variable (with the number of native contacts usually serving for the latter). The knowledge of such surfaces is essential in the context of protein folding problem [33, 34].

The potential energy and interatomic forces in the confinement simulations need not necessarily be given by a means of empirical or semi-empirical potential. They can also be calculated with the use of quantum mechanical treatment of electrons. Correspondingly, the *ab initio* molecular dynamics (AIMD) can be employed. Following Car and Parrinelo [81], various AIMD schemes have been developed that can be applied (and partly have been applied) to simulation studies of cluster dynamics (see, in particular, [82] and the papers cited therein).

Though we considered finite systems, the approach is equally applicable to the study of massive blocks of material, in which case the system is subject to periodic boundary conditions (note that the SW approach was first applied to such systems [83]).

Finally note that in many imaginable scenarios the confinement simulations are consistent with parallel-processing computations. First of all, this concerns the study of kinetic and equilibrium properties, where good statistics is required.

I am grateful to the organizers of the School, Professor C. Guet and Professor P. Hobza, for the invitation to present this material to graduate and post-graduate students. Some results are new, in particular those obtained in recent work with Professor M. Karplus and Dr. S. Krivov [69], whose collaboration is gratefully acknowledged. I would like also to thank S. Krivov for fruitful discussions in the course of preparation of the manuscript.

References

[1] R.S. Berry, *Chem. Rev.* **93** (1993) 2379; *J. Phys. Chem.* **98** (1994) 6910.

[2] D.L. Freeman and J.D. Doll, *Ann. Rev. Phys. Chem.* **47** (1996) 43.

[3] F.H. Stillinger and T.A. Weber, *Phys. Rev. A* **25** (1982) 978.

[4] F.H. Stillinger and D.K. Stillinger, *J. Chem. Phys.* **93** (1990) 6013.

[5] J. Jortner, *Z. Phys. D* **24** (1992) 247.

[6] The references given below do no pretend to covering numerous publications in this field. We cite only a small part of them, either the original paper, where the underlying concept is described, or one of recent papers, which shows the state of the art and contains correspoding references.

[7] G.M. Torrie and J.P. Valleau, *J. Comput. Phys.* **23** (1977) 187.

[8] M. Mezei, *J. Comput. Phys.* **68** (1987) 237.

[9] C. Bartels and M. Karplus, *J. Phys. Chem. B* **102** (1998) 865.

[10] B.A. Berg and T. Neuhaus, *Phys. Rev. Lett.* **68** (1992) 9.

[11] S. Ono, N. Nakajima, J. Higo and H. Nakamura, *Chem. Phys. Lett.* **312** (1999) 247.

[12] M.R. Lemes, C.R. Zacharias and A. Dal Pino, *Phys. Rev. B* **56** (1997) 9279.

[13] U.H.E. Hansmann, F. Eisenmenger and Y. Okamoto, *Chem. Phys. Lett.* **297** (1998) 374.

[14] A.F. Voter, *Phys. Rev. Lett.* **78** (1997) 3908.

[15] A.M. Ferrenberg and R.H. Swendsen, *Phys. Rev. Lett.* **61** (1988) 2635.

[16] P. Labastie and R.L. Whetten, *Phys. Rev. Lett.* **65** (1990) 1567.

[17] D.D. Frantz, D.I. Freeman and J.D. Doll, *J. Chem. Phys.* **93** (1990) 2769.

[18] A.P. Lyubartsev, A.A. Martinovski, S.V. Shevkunov and P.N. Vorontsov-Velyaminov, *J. Chem. Phys.* **96** (1992) 1776.

[19] F. Calvo, J.P. Neirotti, D.L. Freeman and J.D. Doll, *J. Chem. Phys.* **112** (2000) 10350.

[20] C.J. Tsai and K.D. Jordan, *J. Phys. Chem.* **97** (1993) 11227.

[21] G.T. Barkema and N. Mousseau, *Phys. Rev. Lett.* **77** (1996) 4358.

[22] N. Mousseau and G.T. Barkema, *Phys. Rev. E* **57** (1998) 2419.

[23] J.P.K. Doye and D.J. Wales, *Z. Phys. D* **40** (1997) 194.

[24] F. Glover, *ORSA J. Comput.* **1** (1989) 190; *ORSA J. Comp.* **2** (1990) 4.

[25] R. Battiti, in *Modern Heuristic Search Methods*, Chapter 4, edited by Rayward-Smith V.J., Osman I.H., Reeves C.R. and Smith G.D. (Wiley, New York, 1996), pp. 61-83.

[26] D. Cvijović and J. Klinowski, *Science* **267** (1995) 664.

[27] D.J. Wales and H.A. Scheraga, *Science* **285** (1999) 1368.

[28] S. Kirkpatrick, C.D. Gelatt and M.P. Vecchi Jr., *Science* **220** (1983) 671.

[29] Z. Li and H.A. Scheraga, *J. Mol. Struct. (Theochem.)* **179** (1988) 333.

[30] D.J. Wales and J.P.K. Doye, *J. Phys. Chem. A* **101** (1997) 5111.

[31] D.M. Deaven and K.M. Ho, *Phys. Rev. Lett.* **75** (1995) 288.

[32] W. Wenzel and K. Hamacher, *Phys. Rev. Lett.* **82** (1999) 3003.

[33] C.M. Dobson, A. Šali and M. Karplus, *Angew. Chem. Int. Ed.* **37** (1998) 868.

[34] J.N. Onichic, Z. Luthey–Schulten and P.G. Wolynes, *Annu. Rev. Phys. Chem.* **48** (1997) 545.

[35] R. Czerminski and R. Elber, *J. Chem. Phys.* **92** (1990) 5580.

[36] R.S. Berry and R. Breitengraser Kunz, *Phys. Rev. Lett.* **74** (1995) 3951.

[37] K.D. Ball, R.S. Berry, R.E. Kunz, F.Y. Li, A. Proykova and D.J. Wales, *Science* **271** (1996) 963.

[38] O.M. Becker and M. Karplus, *J. Chem. Phys.* **106** (1997) 1495.

[39] K.D. Ball and R.S. Berry, *J. Chem. Phys.* **109** (1998) 8557.

[40] B. Vekhter and R.S. Berry, *J. Chem. Phys.* **110** (1999) 2195.

[41] R.S. Berry, H.L. Davis and T.L. Beck, *Chem. Phys. Lett.* **147** (1988) 13.

[42] S. Fischer and M. Karplus, *Chem. Phys. Lett.* **194** (1992) 252.

[43] I.V. Ionova and E.A. Carter, *J. Chem. Phys.* **98** (1993) 6377.

[44] A. Ulitsky and D. Shalloway, *J. Chem. Phys.* **106** (1997) 10099.

[45] F. Jensen, *J. Chem. Phys.* **102** (1995) 6706.

[46] C.J. Cerjan and W.H. Miller, *J. Chem. Phys.* **75** (1981) 2800.

[47] J.P.K. Doye, M.A. Miller and D.J. Wales, *J. Chem. Phys.* **111** (1999) 8471.

[48] K.D. Ball and R.S. Berry, *J. Chem. Phys.* **111** (1999) 2060.

[49] N. Metropolis, A.W. Rosenbluth, M.N. Rosenbluth, A.H. Teller and E.J. Teller, *J. Chem. Phys.* **21** (1953) 1087.

[50] S.F. Chekmarev and S.V. Krivov, *Chem. Phys. Lett.* **287** (1998) 719.

[51] S.F. Chekmarev and S.V. Krivov, *Phys. Rev. E* **57** (1998) 2445.

[52] S.F. Chekmarev and S.V. Krivov, *Eur. Phys. J. D* **9** (1999) 201.

[53] In fact, the MC methods usually generate sequences of the points in configurations space, *i.e.* of atomic configurations, but this is not crucial, since each of the configurations can easily be complemented by random distribution of atomic velocities.

[54] H. Risken, *The Fokker-Planck Equation* (Springer, Berlin, 1989).

[55] A.F. Voter, *J. Chem. Phys.* **106** (1997) 4665.

[56] R. Kubo, *Rep. Progr. Phys.* **29** (1966) 255.

[57] J. Tully, *Acc. Chem. Res.* **14** (1981) 188.

[58] S.M. Kast and J. Brickmann, *J. Chem. Phys.* **104** (1996) 3732.

[59] V.I. Arnold and A. Avez, *Ergodic Problems of Classical Mechanics* (Addison-Wesley, New York, 1989). In application to clusters see R.J. Hinde and R.S. Berry, *J. Chem. Phys.* **99** (1993) 2942.

[60] P. Schofield, *Comput. Phys. Comm.* **5** (1973) 17.

[61] T. Çaǧin and J.R. Ray, *Phys. Rev. A* **37** (1988) 247.

[62] M.A. Miller and D.J. Wales, *J. Chem. Phys.* **107** (1997) 8568.

[63] S.V. Krivov and S.F. Chekmarev, *Eur. Phys. J. D* **9** (1999) 205.

[64] R. Biswas and D.R. Hamann, *Phys. Rev. B* **34** (1986) 895.

[65] M. Bixon and J. Jortner, *J. Chem. Phys.* **91** (1989) 1631.

[66] P.R. Bunker, *Molecular Symmetry and Spectroscopy* (Academic, New York, 1979).

[67] D.W. Brenner, *Phys. Rev. B* **42** (1990) 9458.

[68] V.A. Schweigert, A.L. Alexandrov, Yu.M. Morokov and V.M. Bedanov, *Chem. Phys. Lett.* **235** (1995) 221.

[69] S.V. Krivov, S.F. Chekmarev and M. Karplus (to be published).

[70] S.F. Chekmarev, *Phys. Rev. E* **64** (2001) 036703 (1-9).

[71] M.R. Hoare and J.A. McInnes, *Adv. Phys.* **32** (1983) 791.

[72] G.W. Greenwood, *Z. Phys. Chemie* **211** (1999) 105.

[73] L.D. Landau and E.M. Lifshitz, *Statistical Physics* (Pergamon, London, 1965).

[74] A. Isihara, *Statistical Physics* (Academic, New York – London, 1971).

[75] W. Forst, *Theory of Unimolecular Reactions* (Academic, London, 1973).

[76] B.R. Brooks, R.E. Bruccoleri, B.D. Olafson, D.J. States, S. Swaminathan and M. Karplus, *J. Comput. Chem.* **4** (1983) 187.

[77] M. Schaefer and M. Karplus, *J. Phys. Chem.* **100** (1996) 1578; M. Schaefer, C. Bartels and M. Karplus, *J. Mol. Biol.* **284** (1998) 835.

[78] E. Neria, S. Fischer and M. Karplus, *J. Chem. Phys.* **105** (1996) 1902.

[79] R. Koradi, M. Billeter and K.J. Wuthrich, *Mol. Graphics* **14** (1996) 51.

[80] S. Sugano, *Microcluster Physics* (Springer, Heidelberg, 1991).

[81] R. Car and M. Parrinello, *Phys. Rev. Lett.* **55** (1985) 2471.

[82] V. Bonačić–Koutecký, J. Jellinek, M. Wiechert and P. Fantucci, *J. Chem. Phys.* **107** (1997) 6321.

[83] F.H. Stillinger and T.A. Weber, *Phys. Rev. A* **28** (1983) 2408.

COURSE 12

MOLECULAR CLUSTERS: POTENTIAL ENERGY AND FREE ENERGY SURFACES. QUANTUM CHEMICAL *AB INITIO* AND COMPUTER SIMULATION STUDIES

P. HOBZA

J. Heyrovský Institute of Physical Chemistry, Academy of Sciences of the Czech Republic and Center for Complex Molecular Systems and Biomolecules, 18223 Prague 8, Czech Republic

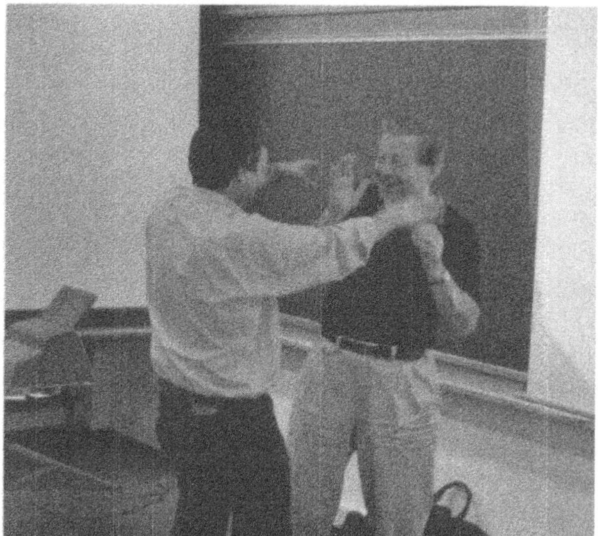

Contents

MOLECULAR CLUSTERS: POTENTIAL ENERGY AND FREE ENERGY SURFACES. QUANTUM CHEMICAL *AB INITIO* AND COMPUTER SIMULATION STUDIES

P. Hobza

1 Introduction

Modern chemistry is based on the understanding of the chemical bond. The chemical bond implies the distribution and delocalization of electrons over the entire molecule resulting in a strong, *i.e.* covalent interaction. Modern theoretical *ab initio* quantum chemistry methods have been extremely successful in describing the electronic structure of isolated molecules to a degree of precision that in some cases comes very close to high-resolution spectroscopic results. From the experimental viewpoint, there is quite a variety of spectroscopic methods to look at a molecule in the gas phase, *i.e.* under conditions where it can be considered unperturbed from collisions or other external forces, for instance, in molecular beams from supersonic jet expansions.

Molecules can interact in two ways – reactively (leading to the formation or breaking of a covalent bond) or nonreactively (leading to the formation of a molecular complex). The former is clearly a covalent interaction; the later is termed non-covalent, physical or van der Waals (vdW) interaction. Such interaction leads to formation of molecular complexes called also vdW molecules. In these systems, the properties of the original subsystems are retained to a considerable degree. VdW molecules are considerably weaker [1] than "classical" molecules and a bonding of 15 kcal/mol and bond length of 2 Å can be very roughly considered as the boundary separating "classical" molecules from vdW molecules.

1.1 *The hierarchy of interactions between elementary particles, atoms and molecules*

Before considering vdW interactions more closely, an attempt will be made to consider them in the context of other types of interactions in the microcosmos. Interactions between atomic nuclei and electrons lead to the formation of atoms. Atoms form molecules. Both these interactions are much stronger than interactions between molecules leading either to formation of a new moelcule or to formation of a molecular cluster. In the former case a covalent bond is formed and broken while in the latter case no covalent bond is broken or formed. Stabilization energies of a molecular cluster are much smaller than other stabilization energies.

1.2 The origin and phenomenological description of vdW interactions

What is the source of the attractive effect [1] between molecules? Consider the overlap of the electron clouds in covalent and non-covalent interactions. If a covalent bond is formed between the subsystems when the electron spheres of the subsystems interact, it is then clear that the attraction must result from this overlap. As overlap is unnecessary in vdW interactions, the reason for attraction must be sought elsewhere. The source of the attraction in non-covalent interactions can lie only in the electrical and magnetic properties of subsystems.

Electric charge in molecules is distributed very unevenly and represents what is termed an electric multipole; this may be dipole, quadrupole, octapole, etc. Interactions between electric multipoles, termed electrostatic interactions (E^{ES}), can be attractive or repulsive and is given by the following equation:

$$E^{\mathrm{ES}} = q_R q_T^{-1} + q_R \vec{\mu}_T \mathbf{r}^{-2} + \cdots + \vec{\mu}_R \vec{\mu}_T \mathbf{r}^{-3} + \cdots + \mathbf{Q}_R \mathbf{Q}_T \mathbf{r}^{-5} + \ldots \qquad (1.1)$$

where q, $\vec{\mu}$ and $\mathbf{Q}$ designate charge, dipole and quadrupole moments, respectively of subsystems R and T ($\vec{\mu}$'s are vectors and $\mathbf{Q}$'s are tensors). The magnitude of the contributions in expansion (1) usually decreases sharply when passing to higher multipoles.

Electric multipoles form an electric field in its surroundings. This field induces an electric multipole in other system in the vicinity of the original multipole. The ability of the system to be polarized is described by important characteristics of the system, called the polarizability. There is an interaction between the original and the induced multipole; this is always attractive (in contrast to interactions between permanent mulptipoles) and is termed induction or polarization interaction. The multipole expansion of the induction energy (E^{I}) is given by equation (2),

$$\begin{aligned} E^{\mathrm{I}} = &-0.5\alpha_R \left[\mathbf{q}_T^2 \mathbf{r}^{-4} + \vec{\mu}_T^2 (3\cos^2\theta_T + 1)\mathbf{r}^{-6} + \ldots\right] \\ &- 0.5\alpha_T \left[\mathbf{q}_R^2 \mathbf{r}^{-4} + \vec{\mu}_R^2 (3\cos^2\theta_R + 1)\mathbf{r}^{-6} + \ldots\right]. \end{aligned} \qquad (1.2)$$

In equation (2) α, q, and $\vec{\mu}$ designate polarizability, charge and dipole moments, respectively, of system R or T and θ_T is the angle formed by the axis of system T and the line connecting the centers of gravity of the two systems. Both above-mentioned components of the interaction energy, electrostatic and induction, can be described in terms of classical physics and both were defined at the beginning of this century. Experience, *e.g.* the possibility of liquefying rare gases, indicates that still another attractive force must exist, acting between uncharged, unpolar systems. London [2] gave a theoretical interpretation to these forces on the basis of quantum mechanics in 1930. As the same oscillator strengths appear in the expression for this force as in the expression for the dispersion of light, the energy corresponding to this interaction is termed the dispersion energy (E^{D}). Similar to induction energy, dispersion energy is always attractive. It is apparent from the multipole expansion of the dispersion energy

$$E^{\mathrm{D}} = -C_6 r^{-6} - C_8 r^{-8} - C_{10} r^{-10} - \ldots \tag{1.3}$$

that the terms of the expansion are proportional to the sixth, eight and tenth power of the reciprocal distance.

It is apparent from equations (1–3) that electrostatic interactions act over the longest range, followed by induction and dispersion interactions. It would thus seem that dispersion interactions are not as important as the first two types; however the contrary is true. Dispersion interactions, *e.g.* between biomacromolecules, represent a very important component of the total interaction energy and are usually more important than the induction energy.

Physical interpretation of the dispersion interactions is neither simple nor unambiguous; the origin of these interactions can be described qualitatively as follows: the oscillation of atomic nuclei and electrons leads to the formation of a time-variable dipole (in general, multipole) even in non-polar, spherically symmetric systems. The value of this dipole depends on the instantaneous positions of the nuclei and electrons, varies in time and equals zero when averaged over time. At the instant when the time- variable multipole is not equal to zero, it acts inductively on the neighboring molecules, and the induced multipole interacts with the time-variable multipole. This is thus interaction between a time-variable multipole and an induced multipole.

So far, attractive forces have been considered; only electrostatic forces are repulsive for certain mutual orientations of the subsystems. If the interacting systems are not to collapse, there must be another force preventing the subsystems from approaching too closely. This force, termed exchange-repulsion, becomes important when the subsystems approach one another closely, to a distance at which the overlap of the electron spheres is non-zero. Interaction between two subsystems leads to an equilibrium between the attractive and repulsive forces, and a minimum appears on the curve describing the dependence of the total interaction energy on the distance between the interacting systems. The minimum depth is related to the stability of the given vdW molecule: complexes held together by dispersion forces have the lowest stabilization energy (1–4 kcal/mol, depending on the size of the molecule); vdW molecules containing hydrogen bonds are characterized by a stabilization energy of 2–8 kcal/mol per hydrogen bond; and, finally, ionic complexes, with a stabilization energy of up to 30 kcal/mol, are most stable.

2 Calculation of interaction energy

It was stated above that non-covalent, vdW interactions differ from covalent interactions. Here the question will be posed: can quantum chemistry describe non-covalent interactions as successfully as covalent interactions? The answer is unambiguously "Yes". However, calculations in this field are one of the most difficult tasks facing quantum chemistry [3].

Just as in other regions of quantum mechanics and in quantum chemistry as a whole, both methods for approximate solution of the Schrödinger equation (the variation and perturbation methods) are important in the study of non-covalent interactions. In the variation method, the interaction energy, ΔE, accompanying the formation of a supersystem by association of subsystems, is given by the difference between the energies of the supersystem (E^{T}) and of the subsystems (the energy of the i-th subsystem is E_i)

$$\Delta E = E^{\mathrm{T}} - \Sigma E_i. \tag{2.1}$$

This expression is very simple. However, the values of E^{T} and ΣE_i often differ by only tenth to thousandths of cal/mol and the value of E^{T} for small and medium molecules equals 10^6 to 10^{10} kcal/mol; thus the calculation is numerically very difficult. A further difficulty lies in the fact that a method with central position in quantum chemistry, the SCF Hartree–Fock (HF) method, does not describe the correlation of the electron motion completely, and thus a certain portion of the total energy, called correlation energy, is neglected. Energies of atoms and molecules calculated by the HF method are subject to an error, termed the correlation error. The correlation energy forms only a small portion (a few percent) of the total energy of the subsystems and of supersystem. However, participation of the correlation energy in the interaction energy is far more important. In some types of complexes (*e.g.* Ar...Ar), it is the only attractive force. Even in stronger complexes involving hydrogen bonds (water dimer, adenine...thymine), it forms an important part of the total interaction energy (10–30%). Thus, neglecting the correlation energy can lead to considerable errors.

At first glance, the perturbation method appears far more suitable for calculation of non-covalent interactions. The interaction energy is calculated directly, rather than as the difference between two large, almost identical numbers. In this approach, ΔE is expressed as the sum of several contributions following from first and second order perturbation calculations, consisting primarily of the coulombic energy (E^{C}), in which the electrostic multipole interaction forms the predominant attractive contribution, the induction energy (E^{I}), the dispersion energy (E^{D}) and the exchange-repulsion energy (E^{ER}). Despite of the problems involved in the variation method (see later), the vast majority of calculation of the interaction energy of various types of complexes is carried out using this method. This is because the variation method is formally simple and straightforward, and because standard quantum chemical programs can be employed in this calculation. The perturbational method is well suited for determining interactions between rigid systems, but its use is limited for systems for which many intramolecular and intermolecular degrees of freedom have to be optimized simultaneously. The perturbational approach, in the form of symmetry-adapted perturbation theory (SAPT) [4], has been used for highly accurate calculations of rigid systems, which provide a benchmark for supermolecular calculations.

There are several advantages of the supermolecular approach which are profitable for studies of large clusters: i) it is applicable to any type of a molecular cluster; ii) it yields a wave function which can be used to derive various properties of the system; iii) highly accurate interaction energies can be obtained, provided that sufficiently large basis sets of atomic orbitals are used and a major portion of the of the correlation energy is covered; and iv) many-body interactions and charge transfer effects are explicitly taken into account. There are however, also drawbacks of this method and two inconsistencies should be mentioned: the size inconsistency and the basis set inconsistency. The size inconsistency does not represent any problem nowadays, since the most efficient methods for calculation of correlation energy are size-consistent (the HF method is size-consistent as well). The basis set inconsistency is a serious obstacle and leads to the basis set superposition error (BSSE). The BSSE is purely mathematical artifact due to the fact that different basis sets are used for energy evaluations of supersystem and the subsystems. The supersystem possesses a larger basis set, which (artificially) increases its stabilization energy, and this effect is referred to as BSSE. Jansen and Ross [5], and Boys and Bernardi [6] independently introduced the function counterpoise (CP) method, which eliminates the BSSE. The principle of the method is simple: subsystems are not treated in their own basis sets but in the basis of the whole complex. The CP procedure is routinely used for evaluation of corrected stabilization energy, which is, much less basis set dependent than uncorrected stabilization energy. The use of CP procedure is, however, also important for determination of the structure of a complex. Currently, complexes are optimized using the standard supermolecular gradient optimization and the BSSE correction is applied only to a subsequent evaluation of interaction energy, *i.e.* after the optimization is finished. This is not correct, since the structure of a complex is optimized on the standard potential energy surface (PES) and not on the theoretically more justified CP-corrected PES. Optimization of the cluster structure with a small or medium basis set without consideration of BSSE can lead to a completely wrong structure. The problem of CP-corrected gradient optimization was solved recently [7,8] and the respective method allows evaluation of the CP-corrected gradient and Hessians of a complex at any *ab initio* level.

The correlation energy is much smaller than the HF energy; nevertheless the role of correlation energy is topical and cannot be neglected. The first treatment really applicable to large systems is the Møller–Plesset (MP) perturbational theory [9]. The correlation energy is determined as a sum of the second, third, fourth, and higher contributions. The most economical and thus widely used is the simplest version of the MP method, the second-order MP theory (MP2). It is applicable to extended complexes (up to hundred of atoms) and gives surprisingly good estimates of the correlation energy. For accurate calculations the use of the coupled cluster theory [10] in the form of CCSD(T) method [11] is recommended. The method covered single and double excitations up to the infinite order, and the same is true for a part of the quadruple and hexatuple excitations; triple excitations are evaluated in a non-iterative way after performing a CCSD calculation. The method is iterative but nonvariational, size consistent, and represents the most robust tool for calculation of the correlation energy of large complexes.

The choice of the atomic orbital (AO) basis set represents a complicated problem, and a basis set as large as possible should be used. The basis set applied should contain polarization and diffuse functions, and to obtain reliable stabilization energies, more sets of these functions are desirable. Among various basis sets, the correlation-consistent basis sets of Dunning [12], starting from the aug-cc-pvDZ (augmented correlation consistent polarized valence double-zeta), can be recommended. This basis set contains diffuse s-, p- and d-functions, which are essential for proper description of interaction energy.

Use of lower-level theoretical methods (density functional theory, semiempirical quantum-chemical methods). The use of popular density functional theory (DFT) [13] in the realm of molecular clusters is very attractive because the method is less computationally demanding than beyond HF methods. Because the DFT energy includes an exchange term and a contribution to the electron correlation energy, it was believed that it is well suited for description of molecular clusters. The current DFT methods are known [14] to work well for H-bonded systems in terms of structure, dipole moments, energetics, and vibrational properties. Its general use for molecular clusters is very insecure because none of the existing functionals describe the London dispersion energy [14]. The DFT also fails for another important class of molecular complexes, namely charge transfer complexes, where these methods predict unrealistically deep energy minima.

Use of currently available semiempirical quantum-chemical methods (such as AM1, PM3 or NDDO) for study of molecular clusters is also very limited since approximations used in these methods do not allow any reasonable description of interaction energy.

3 Vibrational frequencies

Formation of a molecular cluster gives rise to *intermolecular* vibrations. Intermolecular frequencies are usually much smaller than intramolecular ones and are typically around 100 cm^{-1}, frequently even below 50 cm^{-1}. Harmonic frequencies are easily determined even for large non-covalent clusters using Wilson FG analysis [15] and the respective procedure is available in various commercially distributed codes. Vibrational frequencies are, however, not harmonic and "anharmonic" effects should be considered. However, the term "anharmonic vibrations" is somewhat misleading since it implies the harmonic approximation as a starting point. Often, intermolecular vibrations in clusters involve hindered rotations and it makes no sense at all to use the language of harmonic oscillators. Similarly "anharmonic" corrections to the harmonic frequencies can often go into the wrong direction. Given these cautions, here we keep the term "anharmonic" in the sense of non-harmonic. The PES obtained by high-level correlated *ab initio* calculations can be used for evaluating highly accurate anharmonic vibrational frequencies. The agreement between calculated frequencies and their experimental counterparts is usually much more satisfactory when anharmonic effects are fully taken into account. This requirement is especially true for non-covalent clusters with a high degree of non-rigidity.

The standard approach to the vibrational problem for large non-covalent clusters is perturbation theory [16, 17]. If the zero order Hamiltonian (usually harmonic oscillator) is a good approximation to the true vibrational Hamiltonian, perturbation theory represents a very efficient and reliable tool for calculating vibrational frequencies. In the traditional approach, the matrix representation of the molecular Hamiltonian is diagonalized by means of successive contact transformation. This procedure will fail, however, in the case of an accidental resonance. This is often the case if we deal with large systems with many vibrational modes. In this case the terms connecting the resonant levels have to be treated variationally [17,18]. Recently, considerable progress was achieved in developing new procedures based on perturbative treatment for the calculation of anharmonic frequencies [18–20]. The potential energy function is constructed as a low order polynomial (up to the fourth order) expressed in normal coordinates. The force constants are obtained by least-squares fitting of energies, gradients, and hessians calculated at geometries close to the global minimum on the PES. If it is not possible to work with extended basis sets (due to the cluster size) then energies, gradients and hessians should be obtained from the CP-corrected PES. The main advantage of that approach stems from its computational efficiency. The number of hessians required scales linearly with the number of vibrational modes. Thus, the method can even be used for large systems while respecting the full dimensionality of the problem. The applicability of the procedure is, however, less straightforward for non-covalent clusters since they are non-rigid systems. The vibrational dynamics of floppy systems cannot be described in the framework of a single-reference Hamiltonian and therefore, the perturbation series used are necessarily strongly divergent. In such a case the only alternative is a more exact treatment of the large amplitude vibrational modes including all relevant parts of the coordinate space. This requires calculation of the global PES that becomes computationally prohibitive even for systems with few degrees of freedom. However, the number of large amplitude motions is usually a small fraction of the total number of vibrations. Consequently, the large amplitude vibrations can be removed from the perturbative treatment and the Schrödinger equation for the effective large amplitude Hamiltonian is solved variationally.

Literature on higher-dimensional anharmonic vibrational calculations of non-covalent clusters based on *ab initio* correlated calculations includes a variational six-dimensional intermolecular vibrational frequency calculation for the adenine... thymine Watson–Crick base pair [21] and a twelve-dimensional vibrational frequency calculation for the water dimer by perturbation theory [22]. Also a six-dimensional frequency calculation for the water dimer based on various empirical potentials was reported recently by LeForestier *et al.* [23] Other rigorous treatments of vibrations in clusters were reported.

4 Potential energy surface

Generally, the PES of molecular clusters is very rich and contains a large number of energy minima. The global minimum can generally be detected experimentally relatively easily, while elucidation of secondary minima (and saddle points) is much more difficult. These minima can be theoretically ascribed only if the whole PES is known. The number of energy minima increases very rapidly with the cluster size and the number of subsystems. While there is just one energy minimum on the water dimer PES [24–27] there are 11 minima on the uracil dimer PES [28], 284 minima on the benzene... Ar_8 PES [29] and more than 1000 minima on the adenine... thymine... $(H_2O)_2$ PES [30]. The localization of every minima is tedious, if not impossible, by standard methods based on experience and chemical intuition. It is necessary to use some effective search method for determining the geometry of the energy minima: these methods are based on computer experiments, mostly on molecular dynamic (MD) and Monte-Carlo (MC) simulations in combination with quenching technique.

Generally, MD and MC simulations can be performed at any theoretical level yielding energy and forces including nonempirical level. We have witnessed enormous progress over recent years in so called *ab initio* MD, specifically in first-principles MD by the Car–Parrinello method [31, 32]. This approach represents very significant progress particularly because it is parameter-free, includes all many-body terms as well as charge-transfer and can be used even for formation and/or breaking of covalent bonds. The use of the Car–Parinello method for molecular clusters is, however, limited due to the fact that the plane-wave density functional method (as well as any other density functional method) does not cover the London dispersion energy (see later). The use of the method for charged or polar clusters where dispersion energy plays minor role is justified [33, 34]. The vast majority of MD simulations are (and will be in the near future) based on an empirical potential. Such an approach is applicable for molecular clusters but cannot be applied for MD simulations of chemical reactions. The quality of MD simulations depends critically on the performance of the simulation technique but also on the quality of the empirical potential used. This fact is very frequently ignored and it is believed that long MD simulations always yield reliable results. However, the combination of high-level MD simulations with poor potential leads only to poor results. An empirical potential should correctly describe all intermolecular and intramolecular degrees of freedom for any type of molecular cluster. These requirements are extremely strict and usually only a certain empirical potential is applicable to a specific class of molecular clusters. We must bear in mind that more elaborated techniques of quantum chemistry (*e.g.* semiempirical or *ab initio* Hartree–Fock) also fail to simultaneously describe all types of molecular clusters (*e.g.* H-bonded and London dispersion clusters). Present empirical potentials also fail for charge-transfer complexes and in situations where many-body terms are important. Probably the only way to control the quality of an empirical potential is to compare its performance (stabilization energy, structure and geometry, vibrational frequencies) with that of nonempirical correlated *ab initio* calculations. Comparison should be made for selected representatives of the studied class of molecular clusters.

Amar and Berry [35] introduced a very efficient quenching technique for the evaluation of cluster structures; the method was originally developed by Stillinger and Weber [36] for liquids. The principle of the technique is as follow: in the course of MD simulation, after an arbitrary number of steps, the simulation is stopped and energy minimization (quench) is performed. The minimal energy and coordinates of the minimum found are stored and then the MD starts again from the point where it was stopped. After a sufficiently long simulation, a list of nearly all stationary points of the cluster is obtained. The completeness of the list could be verified by starting the quenching at a different temperature and by changing the number of steps.

5 Free energy surface

Populations of various structures of a cluster which are proportional to the change of the Gibbs free energy can be determined by computer simulations for NVT canonical and NVE micro-canonical ensembles (N, V, E and T refer to the number of molecules in a system, its volume, energy and temperature) combined with the quenching method [35, 36]. In the NVT canonical ensemble, the cluster is in temperature equilibrium with the surroundings, and accordingly, the NVT ensemble gives information of the behavior of the cluster when it is interacting with the surrounding. In the NVE microcanonical ensemble all the systems have the same energy and each system is individually isolated. Clearly, the NVE and NVT ensemble results must coincide in the limit of infinitely large systems [37, 38]. However, small molecular clusters are all but infinite size systems. This has important consequences for the probability $P(E) = \Omega(E)\exp(-E/kT)$ (where $\Omega(E)$ is the density of states with energy E) of finding the NVT ensemble at energy E. Whereas small systems approach the one degree of freedom limit $P(E) = \exp(-E/kT)$, large systems with high densities of states possess canonical probabilities which approach a δ-function peaked at the energy of a corresponding microcanonical ensemble [38].

The calculation of populations from quenching is possible in a rather narrow energy interval. The energy should be sufficiently high to allow a high frequency of interconversions among different isomers. The basic advantage of this technique is the fact that it enables an evaluation of relative populations for all structures on the PES, *i.e.* to pass from the PES to the free energy surface (FES). Such calculations require long simulation runs and are thus feasible only for relatively small clusters. Using the free energy perturbation/MD procedure we obtain relative ΔG values for much larger clusters but only for small sections of the PES. These calculations need significantly less simulation time.

The comparison of theoretical and experimental results require some comments. First, the temperature T of the cluster formation in a particular experiment should be known. The stabilization enthalpy at 0 K and the PES characteristics can be only used if T is zero or very low. In many cases, however, temperature is higher and entropy term should be considered. This means that the FES characteristics and not the PES characteristics should be utilized for comparison with experimental results. Care should be further paid to the type of simulations, NVE or NVT. Simulations in the NVE ensemble give properties of a cluster that does not interact with the surroundings while simulations in the NVT ensemble correspond to a situation when the cluster is in thermal equilibrium with the surroundings.

6 Applications

6.1 Benzene... Ar_n clusters

Benzene... Ar_2 [29]. Two structures of the titled dimer were obtained by the gradient optimization of the potential (obtained by fitting to the *ab initio* MP2 PES of the benzene... Ar cluster) in the form of equation (5).

$$U = A/R^{13.305} - B(1 - C/R)/R^6. \tag{6.1}$$

The global minimum ($\Delta E = -2.25$ kcal/mol) corresponds to the structure having both argons on the C_6 axis and localized on the opposite sides of the benzene molecule (structure (1|1). The local minimum ($\Delta E = -2.03$ kcal/mol) corresponds to the structure having both argons on one side of the benzene molecule; abbreviation (2|0) is used for this structure. The MD simulations shows that less stable local minimum is more populated than the global minimum due to an entropy term.

Benzene... Ar_3 *and Benzene...* Ar_5 *clusters* [29]. The global minimum and the first local minimum of the former complex (both having all three argons on one side of the benzene molecule – abbreviation (3|0)) were populated less than the second local minimum having a two-sided structure (2|1). The relative populations of these stationary points are 40, 13, and 46%. The first local two-sided minimum of the latter complex with the (4|1) structure is populated more than that of the global one-sided one having (5|0) structure. Similarly as in the previous cluster three lowest-energy minima are populated much more than the remaining stationary points. For both clusters we obtained a similar picture – the two-sided local minima were populated more than one-sided global minima.

Benzene... Ar_8 [29]. The quenching technique localized 156 stationary points for the cluster at 25.9 K. The clearly dominant population corresponds to minimum (3|2|3) having a bridgelike (pocket) structure. This local minimum is 0.17 kcal/mol less stable than the global minimum with the symmetrical (7|1) structure. Seven argons on one side constitutes the full one-side solvation, forming a "flowerlike" structure which is energetically very favorable. Evidently, the global minimum (7|1) could have only two realizations while the bridgelike structure (3|2|3) could have considerably more. According to the Boltzman relation among entropy (S), Boltzman's constant (k_b), and number W, which is the number of ways in which the systems in the sample can be arranged still having the same total energy,

$$S = k_b \ln W \tag{6.2}$$

the entropy of the (3|2|3) structure is expected to be higher than that of the (7|1) structure. (Let us add that the same argument is valid also for lower benzene... Ar_n clusters with inclusion of the benzene... Ar_2 one.)

No one-sided structure of the complex was observed during simulations. This conclusion agrees with the experimental data in [39]: "in the benzene... Ar_8 spectrum, no feature is observed in the frequency range corresponding to the one-sided isomers".

Concluding remarks. With the exception of the very low temperatures, the global minimum of the benzene... Ar_n clusters was not most polulated. The low populations of the global minima having one-sided structures and high populations of two-sided and bridged structures are explained by the entropy term. The PES and FES of titled clusters differ considerably and benzene... Ar_8 represents a textbook example where the global minimum is populated 5–8 times less than the less stable minimum with a bridgelike structure.

6.2 Aromatic system dimers and oligomers

The benzene-benzene interaction is of key importance as a prototype of interaction prevailing in the aromatic π-systems [40]. Its understanding is crucially important for the interpretation of diverse phenomena such as base pair stacking in DNA, intercalation of drugs into DNA, crystal packing of aromatic molecules, formation of tertiary structure of proteins or porphyrin aggregation. Despite extensive experimental [41–51] and theoretical efforts [52–60] our knowledge of the structural and dynamical properties of the benzene dimer is still limited. This concerns even the structure of the global minimum and the number of other stationary points on the potential energy surface.

The first step toward understanding the dimer structure was achieved by Arunan and Gutowsky [50] who reported well resolved rotational spectra of the dimer. With the pulsed nozzle Fourier transform microwave spectrometer, the authors obtained rotational constants from which the center of mass separation was estimated. The evaluated distance was consistent with the T-shaped dimer structure.

A major step forward has recently come from the investigation of the structure and vibrational dynamics of the benzene dimer using the non-empirical model (NEMO) empirical potential parameterized using CCSD(T) calculations [61]. In this study, gradient and hessian calculations using the NEMO potential revealed the existence of just one energy minimum corresponding to the T-shaped structure. The parallel-displaced structure which has been believed to be an energy minimum (and sometimes even the global minimum) was clearly identified as a transition state structure separating two equivalent T-shaped dimer structures.

Due to a relatively high transition barrier ($\sim$170 cm^{-1}), interconversion tunneling is unimportant in the energy region spanned by the available rotational spectra [50] and is thus neglected. The dimer undergoes a nearly free internal rotation along the axis connecting the benzene centers of mass in the T-shaped equilibrium geometry and a hindered internal rotation (the barrier being $\sim$46 cm^{-1}) along the axis that is perpendicular to the "nearly free" internal rotation axis.

The tunneling splitting observed in the rotational spectrum [50] is probably due to this hindered rotation. An analysis assuming the latter rotation as an independent motion and using purely vibrational tunneling splitting indicates that the genuine value of the hindered rotation barrier is nearly twice higher than its *ab initio* value. Similarly, the difference $\Delta R = 0.25$ Å between the equilibrium ground state value for the distance of the mass centers of the benzene monomers from *ab initio* calculation and experiment is strong evidence that the theoretical potential is much shallower than the correct one. In a complimentary stimulated Raman study, some bands were observed around 3 to 10 cm^{-1} region [49]. These bands can now be assigned to the nearly free rotation and the "energy minimum path" bending motion.

Benzene trimers and tetramers were studied [62] using the same empirical potential used for the benzene dimer. The important advantage of the NEMO potential is that it includes induction energy, allowing the recovery of the many-body energy terms. The induction three-body term is, however, only one of many three-body terms. The others, involving exchange and dispersion might be larger but their evaluation (especially dispersion) is rather difficult. For the benzene trimer, three energy minima were found. The most stable cyclic trimer structure explains the experimentally observed spectral shifts and binding energies [63]. The calculated harmonic frequencies conform to the experimentally found intermolecular vibrations [64]. For the tetramer five structures were obtained. The experimentally found conformer [65,66] does not represent the cyclic lowest-energy structure but rather a cyclic trimer with a monomer attached to one side. This can be understood from entropy considerations. These calculations show that a carefully calibrated empirical potential is well suited to describe larger benzene clusters that are out of reach for accurate *ab initio* quantum chemical calculations.

Many-body terms were shown to be important in the case of water clusters. When the polarity of the subsystems is decreasing, many-body terms are expected to be less important. We have, however, shown [62] that for the cyclic benzene trimer the three-body term contributes about 5% to the overall stabilization. Moreover, the three-body contribution can be attractive as well as repulsive. Three-body terms in benzene oligomers are thus not negligible. Theoretical analysis has further shown [62] that the four-body terms can be omitted. The importance of three body interactions was confirmed in a study of the naphthalene trimer using *ab initio* calculations and an MM3 force field [67]. The lowest energy structure was found to be the C_{3h} cyclic structure which is consistent with experimental results [68].

6.3 Nucleic acid–base pairs

As is well known, genetic information is stored and duplicated in DNA, which can thus be considered one of the most important molecules in our life. The functionality of DNA is predetermined by its unique 3-dimensional, right-handed, double helical structure. The structure of DNA is influenced by various contributions, among them the H-bonding and stacking interactions of nucleic acid (NA) bases play an important role.

The experimental characterization of base pairs is extremely difficult and only very few reliable studies are available. There is still only one published gas phase experiment on the energetics of H-bonded NA base pairs [69] and gas phase experimental data on base stacking are absent; the same is true for the structure of NA base pairs. Presently high level quantum chemical calculations and computer simulations represent the only tools to obtain some reference data on structure, energetics, vibrational frequencies and other properties of NA base pairs. These data are essential for understanding the function and properties of NA base pairs and are also very important for the verification and/or parameterization of empirical potentials for molecular modeling of bio-macromolecules and their interactions.

The competition between base stacking and H-bonding of NA bases was first investigated by *ab initio* methods about ten years ago. More recently, *ab initio* calculations on these clusters including correlation energy have become feasible and have produced more reliable results [70, 71]. These calculations now essentially provide a conclusive picture of the interaction of NA bases in the gas phase and can be summarized as follows:

(i) H-bonded pairs stabilized by electrostatic interactions are more stable than the stacked structures, which are stabilized by dispersion interactions;

(ii) for a consistent description of H-bonding and stacking interactions, the theoretical procedure must include the London dispersion energy, ruling out the use of Hartree-Fock *ab initio* and density functional methods. Also semi-empirical methods of quantum chemistry cannot be used;

(iii) intermolecular vibrational frequencies of H-bonded base pairs are almost constant and do not depend on stabilization energy. For all NA base pairs, the buckle and propeller twist vibrations are the lowest and they are all in the region of 4–30 cm^{-1} in the harmonic approximation;

(iv) in-plane intermolecular harmonic vibrations of the adenine$\cdots$thymine NA base pair are systematically overestimated with respect to anharmonic vibrations by about 50% while the absolute differences among anharmonic and harmonic intermolecular out-of-plane vibrations is lower than 15%. The 50% error in vibrational frequencies is large and care should be paid to the use of harmonic intermolecular vibrational frequencies of NA base pairs. All anharmonic intermolecular frequencies are, however, systematically overestimated in comparison with harmonic ones. This provides an opportunity to use the standard harmonic approach to obtain at least an approximate description of the lowest vibrational frequencies of NA base pairs. It must be remembered that anharmonic treatment of such large clusters is extremely tedious;

(v) the stability order of hydrogen-bonded NA base pairs is not changed when passing from the ΔE to the ΔG description. Though the different NA base pairs vary greatly in stabilization energy, the entropy term is always nearly constant;

(vi) the entropy contribution differs considerably for hydrogen-bonded and stacked structures. This implies that energetically less favorable stacked structures can be favored compared to the H-bonded ones when passing from the ΔE to the ΔG description. To obtain the free energy and hence a reliable thermodynamic characterisation for H-bonded and stacked base pairs, computer experiment simulations are required, mainly molecular dynamics simulations;

(vii) the order of stability of various dimer structures of NA base pairs is different for the PES and FES. This indicates that experimental data should be compared with the FES and not with the PES. However, this would only be correct for experiments carried out at temperatures much higher than 0 K. The necessity to use the FES in contrast to the PES will be particularly important for those NA base pairs for which the difference in stabilization energy between H-bonded and stacked structures is small and hence the entropy contribution becomes most significant. In these cases, even the structure of the global minimum may be different for both surfaces;

(viii) among various empirical potentials tested, the *Amber* potential with the Cornell *et al.* force field [72] was found to best reproduce the *ab initio* H-bonding and stacking stabilization energies and is at present recommended for computer simulations of DNA and RNA. On the basis of the analysis of *ab initio* and empirical potential data, it is evident that future force field generations should include an anisotropy and a polarisation term. The polarisation term would allow the inclusion of many-body effects;

(ix) it is surprising that such a good agreement between theoretical and experimental values concerning structure and function of DNA oligomers is found for standard MD simulations since only pair potentials are applied. This agreement is even more surprising when the highly ionic and polar character of DNA and surrounding water is considered. It is thus quite possible that the agreement obtained is fortuitous and is (as frequently in the field of non-covalent interactions) due to compensation of errors.

The financial support provided by grant No. A4040904 of the Grant Agency of the Academy of Sciences of the Czech Republic is gratefully acknowledged.

References

[1] P. Hobza and R. Zahradník, *Intermolecular Complexes* (Academia, Prague and Elsevier, Amsterdam, 1988).

[2] F. London, *Z. Phys. Chem. B* **11** (1930) 222.

[3] P. Hobza and R. Zahradník, *Chem. Rev.* **88** (1988) 871.

[4] B. Jeziorski and W. Kolos, in *Molecular Interactions*, Vol. 3, part 1, edited by H. Ratajczak, and W.J. Orvill-Thomas (Wiley, New York, 1982); B. Jeziorski, R. Moszczynski and K. Szalewicz, *Chem. Rev.* **94** (1994) 1887.

[5] H.B. Jansen and P. Ross, *Chem. Phys. Lett.* **3** (1969) 140.

[6] S.F. Boys and F. Bernardi, *Mol. Phys.* **19** (1970) 553.

[7] S. Simon, M. Duran and J. Dannenberg, *J. Chem. Phys.* **105** (1996) 11024.

[8] P. Hobza and Z. Havlas, *Theor. Chem. Acc.* **99** (1998) 372.

[9] R.J. Bartlett, *Annu. Rev. Phys. Chem.* **32** (1981) 359.

[10] J. Čížek, *J. Chem. Phys.* **45** (1966) 4256; *Adv. Chem. Phys.* **14** (1969) 35.

[11] K. Raghavachari, G.W. Trucks, M. Head–Gordon and J.A. Pople, *J. Chem. Phys. Lett.* **157** (1989) 479; R.J. Bartlett and J.F. Stanton, in *Reviews in Computational Chemistry*, Vol. 5, Chap. 2, edited by K.B. Lipkowitz and D.B. Boyd (VCH Publishers, New York, 1994).

[12] T.H. Dunning Jr., *J. Chem. Phys.* **90** (1989) 1007; *ibid.* **98** (1993) 1358.

[13] P. Hohenberg and W. Kohn, *Phys. Rev. B* **864** (1964) 136; W. Kohn, and L. Sham, *J. Phys. Rev. A* **140** (1965) 133.

[14] P. Hobza, J. Šponer and T. Reschel, *J. Comput. Chem.* **11** (1995) 1315.

[15] E.B. Wilson, J.D. Decius and P.C. Cross, *Molecular Vibrations* (McGraw-Hill, New York, 1955).

[16] M.R. Aliev and J.K.G. Watson, in *Molecular Spectroscopy: Modern Research*, Vol. III, edited by K. Nahari Rao (Academic Press, New York, 1985).

[17] I.M. Mills, in *Molecular Spectroscopy: Modern Research*, Vol. I, edited by K. Nahari Rao and C.W. Mathews (Academic Press, New York, 1972), p. 115.

[18] P.E. Maslen, N.C. Handy, R.D. Amos and D. Jayatilaka, *J. Chem. Phys.* **97** (1992) 4233.

[19] N.C. Handy, P.E. Masle, R.D. Amos, J.S. Andrews, C.C.W. Murray and G.J. Laming, *Chem. Phys. Lett.* **197** (1992) 506.

[20] O. Bludský, V. Špirko, R. Kobayashi and P. Jorgensen, *Chem Phys. Lett.* **228** (1994) 568.

[21] V. Špirko, J. Šponer and P. Hobza, *J. Chem. Phys.* **106** (1997) 1472.

[22] P. Hobza, O. Bludský and S. Suhai, *Phys. Chem. Chem. Phys.* **1** (1999) 3073.

[23] R.S. Fellers, L.B. Braly, R.J. Saykally and C. LeForestier, *J. Chem. Phys.* **110** (1999) 6306.

[24] M.W. Feyereisen, D. Feller and D.A. Dixon, *J. Phys. Chem.* **100** (1996) 2993.

[25] S.S. Xantheas, *J. Chem. Phys.* **104** (1996) 8821.

[26] A. Halkier, H. Koch, P. Jorgensen, O. Christianses, I.M. Beck Nielsen and T. Helgaker, *Theor. Chem. Acc.* **97** (1997) 150.

[27] P. Hobza, O. Bludský, S. Suhai, *Phys. Chem. Chem. Phys.* **1** (1999) 3073.

[28] M. Kratochvíl, O. Engkvist, J. Šponer, P. Jungwirth and P. Hobza, *J. Phys. Chem. A* **102** (1998) 6921.

[29] J. Vacek, P. Hobza, *J. Phys. Chem.* **99** (1995) 17088.

[30] M. Kratochvíl, J. Vacek, P. Hobza, in preparation.

[31] G. Galli and M. Parrinello, *Ab initio molecular dynamics: Principles and practical implementation*, in *Computer Simulation in Material Science*, edited by M. Meyer and V. Pontikis (Kluwer Acad., 1991).

[32] M. Parrinelo, *Solid State Comm.* **102** (1997) 107.

[33] R.N. Barnett, U. Landman, *Phys. Rev. B* **48** (1993) 2081.

[34] G.J. Martyna and M.E. Tuckerman, *J. Chem. Phys.* **110** (1999) 2810.

[35] F.G. Amar and T.R.S. Berry, *J. Chem. Phys.* **85** (1986) 5943.

[36] F.H. Stillinger and T.A. Weber, *Phys. Rev. A* **25** (1982) 978.

[37] M.P. Allen and D.J. Tildesey, *Computer simulation of liquids* (Oxford University Press, Oxford, 1987).

[38] D. Chandler, *Introduction to modern statistical mechanics* (Oxford University Press, Oxford, 1987).

[39] M. Schmidt, J. Le Calve and M. Mons, *J. Chem. Phys.* **98** (1993) 6102.

[40] Benzene itself represents an important system used frequently for modelling aromatic B-systems playing a vital role in biomolocules. Let us mention, *e.g.* studies of benzene.alkali metal cations giving new insight into the ion channel transport problem (for references see J.C. Ma and D.A. Dougherty, *Chem. Rev.* **97** (1997) 1303; D.A. Dougherty, *Science* **271** (1996) 163.)
[41] K.C. Janda, J.C. Hemminger, J.S. Winn, S.E. Novick, S.J. Harris and W. Klemperer, *J. Chem. Phys.* **63** (1968) 1419.
[42] K.H. Fung, H.L. Selzle and E.W. Schlag, *J. Phys. Chem.* **87** (1983) 5113.
[43] K.O. Börnsen, H.L. Selzle and E.W. Schlag, *J. Chem. Phys.* **85** (1986) 1726.
[44] W. Scherzer, O. Krätzschmar, H.L. Selzle and E.W. Schlag, *Z. Naturforsch. A* **47** (1992) 1248.
[45] O. Krätzschmar, H.L. Selzle and E.W. Schlag, *J. Phys. Chem.* **98** (1994) 3501.
[46] V.A. Venturo and P.M. Felker, *J. Chem. Phys.* **99** (1993) 748.
[47] E.G. Cox, D.W. Cruickshank and J.A. Smith, *Proc. R. Soc. London Ser.* **A** (1958) 247.
[48] A.W. Garrett and T.S. Zwier, *J. Chem. Phys.* **96** (1992) 3402.
[49] B.F. Henson, G.V. Hartland, V.A. Venturo and P.M. Felker, *J. Chem. Phys.* **97** (1992) 2189.
[50] E. Arunan and H.S. Gutowsky, *J. Chem. Phys.* **98** (1993) 4294.
[51] T. Ebata, M. Hamakado, S. Moriyama, Y. Morioka and M. Ito, *Chem. Phys. Lett.* **199** (1992) 33.
[52] G. Karlström, P. Linse, A. Wallqvist and B. Jönssen, *J. Am. Chem. Soc.* **105** (1983) 3777.
[53] P. Hobza, H.L. Selzle and E.W. Schlag, *J. Chem. Phys.* **93** (1990) 5893.
[54] P. Hobza, H.L. Selzle and E.W. Schlag, *J. Phys. Chem.* **97** (1993) 3937.
[55] P. Hobza, H.L. Selzle and E.W. Schlag, *J. Am. Chem. Soc.* **116** (1994) 3500.
[56] P. Hobza, H.L. Selzle and E.W. Schlag, *Chem. Rev.* **94** (1994) 1767.
[57] P. Hobza, H.L. Selzle and E.W. Schlag, *J. Phys. Chem.* **100** (1996) 18790.
[58] K.S. Law, M. Schauer and E.R. Bernstein, *J. Chem. Phys.* **81** (1984) 4871.
[59] M. Schauer and E.R. Bernstein, *J. Chem. Phys.* **82** (1985) 3722.
[60] S. Sun and E.R. Bernstein, *J. Phys. Chem.* **100** (1996) 13348.
[61] V. Špirko, O. Engkvist, P. Soldán, H.L. Selzle, E.W. Schlag and P. Hobza, *J. Chem. Phys.* **111** (1999) 572.
[62] O. Engkvist, P. Hobza, H.L. Selzle and E.W. Schlag, *J. Chem. Phys.* **110** (1999) 5758.
[63] H. Krause, B. Ernstberger and H.J. Neusser, *Chem. Phys. Lett.* **184** (1991) 441.
[64] M.W. Schaffer, P.M. Maxton and P.M. Felker, *Chem. Phys. Lett.* **224** (1994) 544.
[65] B.F. Henson, V.A. Venturo, G.V. Hartland and P.M. Felker, *J. Chem. Phys.* **98** (1993) 8361.
[66] A. De Meijere, F. Huisken, *J. Chem. Phys.* **92** (1990) 5826.
[67] C. Gonzales and E.C. Lim, *J. Phys. Chem. A* **103** (1999) 1437.
[68] P. Benharash, M.J. Gleason and P.M. Felker, *J. Phys. Chem. A* **103** (1999) 1442.
[69] I.K. Yanson, A.B. Teplitsky and L.F. Sukhodub, *Biopolymers* **18** (1979) 1149.
[70] J. Šponer and P. Hobza, in *Encyclopaedia of Computational Chemistry*, edited by P.V.R. Schleyer, N.L. Allinger, T. Clark, J. Gasteiger, P.A. Kollman, H.F. Schaefer III and P.R. Schreiner (John Wiley & Sons, Chichester, 1998), p. 777.
[71] P. Hobza and J. Šponer, *Chem. Rev.* **99** (1999) 3247, and references there.
[72] W.D. Cornell, P. Cieplak, C.I. Bayly, I.R. Gould, K.M. Merz, D.M. Ferguson, D.C. Spellmeyer, T. Fox, J.E. Caldwell and P. Kollman, *J. Am. Chem. Soc.* **117** (1995) 5179.

Seminars by invited speakers

S. Berry: *Phase changes in Small Systems*

V. Bonacic-Koutecky: *Theoretical exploration of ultrafast spectroscopy of small clusters*

J. Belloni: *Agrégats atomiques et photographie*

C. Bréchignac: *Charge instabilities in atomic clusters*

F. Hekking: *Supraconductivity in small systems*

V. Kresin: *Cluster Polarization and Long-Range Forces*

J. Jortner: *Perspective in Cluster Physics*

R. Lavery: *Properties of individual biological molecules*

E. Suraud: *Metallic clusters in Intense Fields*

L. Woeste: *Applications of Atomic Clusters in Photography and Atmospheric physics*

Achevé d'imprimer par Corlet, Imprimeur, S.A.
14110 Condé-sur-Noireau (France)
N° d'Imprimeur : 55765 - Dépôt légal : décembre 2001

Imprimé en U.E.

GPSR Compliance

The European Union's (EU) General Product Safety Regulation (GPSR) is a set of rules that requires consumer products to be safe and our obligations to ensure this.

If you have any concerns about our products, you can contact us on ProductSafety@springernature.com

In case Publisher is established outside the EU, the EU authorized representative is:

Springer Nature Customer Service Center GmbH
Europaplatz 3
69115 Heidelberg, Germany

Batch number: 10755248

Printed by Printforce, the Netherlands